S0-CAH-081

Proceedings of the Fifth International
CONGRESS ON CATALYSIS

Proceedings of the Fifth International Congress on Catalysis,
Miami Beach, Fla., 20-26 August, 1972.

CATALYSIS

edited by

JOE W. HIGHTOWER
Department of Chemical Engineering
Rice University, Houston, Texas, U.S.A.

VOLUME I

1973

NORTH-HOLLAND PUBLISHING COMPANY-AMSTERDAM • LONDON
AMERICAN ELSEVIER COMPANY, INC. - NEW YORK

Library of Congress Catalog Card Number: 62–26910
North-Holland ISBN: 0 7204 0271 9
American Elsevier ISBN: 0 444 10494 1

PUBLISHERS:

NORTH-HOLLAND PUBLISHING COMPANY – AMSTERDAM
NORTH-HOLLAND PUBLISHING COMPANY, LTD – LONDON

SOLE DISTRIBUTORS FOR THE U.S.A. AND CANADA

AMERICAN ELSEVIER PUBLISHING COMPANY, INC.
52 VANDERBILT AVENUE
NEW YORK, N.Y. 10017

PRINTED IN THE NETHERLANDS

PREFACE

The Fifth International Congress on Catalysis was held in the USA at Palm Beach, Florida, August 21-25, 1972. Co-sponsored by the International Union of Pure and Applied Chemistry, the Congress was organized and hosted by The Catalysis Society of North America. *Professor Paul H. Emmett* was Honorary Chairman, while *Dr. Vladimir Haensel* served as Executive Chairman. Chairmen for the Sub-Committies included *J. E. McEvoy* (Advertising and Publicity), *F. G. Ciapetta* and *J. G. Larson* (Financial and Budget), *Ms. G. A. Mills* (Ladies' Hospitality), *W. H. Flank* and *Ms. R. A. Svacha* (Meeting Arrangements and Social Functions), *A. E. Hirschler* and *A. Farkas* (Paper Review), *R. L. Burwell, Jr.* and *W. K. Hall* (Paper Scheduling), *J. W. Hightower* (Publications), *W. K. Leaman* (Registration), *J. G. Larson* (Treasurer), and *H. Heinemann* (Grants).

Although originally scheduled for New York City, the organizers moved the meeting to Miami, Florida, to avoid the congestion and high prices in NYC during summer months. When the President's political party switched its national convention to Miami at the same time, the meeting site was again changed to avoid any possible conflict; all agree that the change to Palm Beach, Florida, was a very wise decision. The elegant facilities of the Breakers Hotel provided a dignified and relaxed atmosphere conducive to maximum enjoyment of the Congress. Many of the approximately 500 delegates from 32 countries took advantage of the lovely Florida weather by sunning themselves or swimming in the calm Atlantic Ocean. Group social events included a get-acquainted cocktail party and a congress banquet. These were supplemented by "hospitality suites" kindly provided by the several catalyst manufacturers.

The theme of the Congress was "The Science of Catalysis." In addition to 107 papers selected from nearly twice that many which were submitted, six invited lectures by outstanding internationally recognized leaders in catalysis rounded out the program. Preprints of all papers were sent to the delegates six weeks before the meeting. Except for the invited lectures, the papers were presented in two parallel sessions with thirty minutes allocated for each paper--15 minutes for presentation and 15 minutes for discussion. Each commentator was asked to submit his oral question or comment in writing immediately after the session. These were then collected and sent to the authors for their written response. Among the papers which generated considerable excitement among the delegates were those on particle size effects, partial oxidation, supported homogeneous catalysts, and mechanism of reactions over pure or mixed oxides. The complete manuscripts, comments, and responses are included in the two volumes of the Proceedings.

The Sixth International Congress will be held in London, England, during the summer of 1976; *Professor C. Kemball* will be the organizing chairman. Officers of the International Congress include *Professor G. K. Boreskov* - President, *Professor C. Kemball* - President Elect, *Professor R. L. Burwell, Jr.* - Vice President, *Dr. D. A. Dowden* - Secretary, and *Professor J. Haber* - Treasurer.

The only sadness associated with the meeting surrounds the untimely death of three men who have been leaders in the International Congress for many years. Pictures and biographical information about each follow.

About 150 people actively participated in organization of this meeting. We are all indebted to them for their splendid efforts. The Editor would particularly like to thank *A. Farkas* for writing the Necrology section and helping edit several of the papers. He also thanks others on the Publications Committee (*R. L. Burwell, Jr., T. R. Hughes, J. V. Kennedy,* and *J. T. Richardson*) for their editorial help. Finally, our deep appreciation is due *Mrs. Ruby Rost* who did all the typing of the discussion and several manuscripts in their camera-ready form.

The personal contribution of $10,000 by *W. Clement Stone,* a noted Chicago Philanthropist, made it possible for a substantial number of students and young instructors to participate in the Congress. Other financial supporters for which we are grateful are included in the lists of sponsors and contributors which follow.

Sponsors

Air Products and Chemicals
Champlin Petroleum Co.
Chemetron Corporation
Cities Service R & D
E. I. du Pont de Nemours
W. R. Grace Co.
Gulf Oil Corporation
Harshaw Chemical Co.
International Nickel Co.
Mobil Oil Corporation
Nalco Chemical Co.
National Science Foundation
Olin Corporation
Phillip Petroleum Co.
Standard Oil Co. of California
Tennessee Eastman Co.
Union Carbide Corp.
Union Oil Co. of California
Universal Oil Products Co.

Contributors

Allied Chemical Corporation
Celanese Corporation
Esso Research & Engineering Co.
GAF Corporation
Hercules, Inc.
PPG Industries
Quaker Oats Co.
Stauffer Chemical Co. Foundation
Shell Oil Company
Standard Oil Co. (Indiana) Foundation
The Standard Oil Co. (Ohio)
Sun Oil Company
Texaco, Inc.

The North-Holland Publishing Company (Amsterdam) has agreed to publish the Proceedings. The Organizing Committee expresses their appreciation to the managing editor, *Drs W. H. Wimmers,* for his preparation and shipment of both the Preprints and the Proceedings.

Houston, Texas
January 10, 1973

Joe W. Hightower
Editor of Proceedings

Jan Hendrick de Boer

1899-1971

Jan Hendrick de Boer was born in Ruinen, Holland, in 1899. He received his Ph.D. in Groningen and joined Philips Gloeilampenfabriek in Eindhoven. His ingenuity in developing technological applications for scientific findings was soon recognized, and he was rewarded by promotion to laboratory head. The war forced Dr. de Boer to move to London in 1940 as director of the Dutch Laboratory of the Ministry of Supply - a position he held until 1945. From 1946 to 1950 he was head of the Unilever Laboratory in England, and then he became associated with Staasminen in Limburg. In addition to these industrial jobs, Dr. de Boer also taught at the Delft Technological University. In 1962 he was appointed president of the Scientific Council for Nuclear Energy. In the period 1964-71 Dr. de Boer was president of the International Congress on Catalysis. Dr. de Boer was one of those rare individuals who could perform brilliantly with equal virtuosity as a scientist, technologist, teacher or administrator, in both physics and chemistry, in any one of several languages, and whose technical knowledge was excelled only by his personal charm. In his research work he covered a dazzling variety of topics, all with superior competence and creativity. De Boer is the author of innumerable papers and patents on the separation of hafnium and zirconium, color centers in solids, photosensitive layers, fluorescence, phosphorescence, photoelectric tubes, rectifiers, electron emission, photographic materials, X-ray screens, sound recording and, of course, adsorption and catalysis. Dr. de Boer received many awards and honorary degrees and was elected to several academies. When he retired in 1969, his former students and coworkers dedicated a book to him entitled "Physical and Chemical Aspects of Adsorbents and Catalysts" - a truly remarkable record of the accomplishments of de Boer and of his school. But de Boer was not to enjoy his retirement for long. On April 26, 1971, he died, and the scientific community of the whole world mourns his passing.

Frank George Ciapetta

1915-1972

Frank George Ciapetta was born in Philadelphia in 1915. After receiving the B.A. and M.A. degrees from Temple University, he joined the Atlantic Refining Company in 1939. In 1946 as a du Pont fellow Frank returned to academic work at the University of Pennsylvania where he was awarded a Ph.D. degree in 1947. Then followed a most successful eight-year research campaign on the applications of catalysis to hydrocarbon chemistry, half of which was spent at Atlantic and half at Socony as a research associate. This work covered virtually all typical refinery reactions, including alyklation, isomerization, cracking and reforming, and culminated in Frank's receiving the coveted Precision Scientific Company Award of the American Chemical Society in Petroleum Chemistry in 1955. The same year Frank joined W. R. Grace & Co. as research director of the Davison Chemical Division where he was promoted to vice president of research in 1968. Frank's enthusiasm for, and success in, matters catalytic continued at Davison and led to some remarkable developments in cracking, auto exhaust and polymerization catalysts. Frank was the author of about seventy papers and patents and participated actively in many scientific and technical organizations. He was chairman of the Petroleum Research Fund Advisory Board and president of the International Congress on Catalysis when he suddenly died on February 7, 1972. Everybody who has known Frank will remember him for his technical accomplishments and his leadership qualities; in his passing we mourn the loss of a true friend.

Alfred Ernest Hirschler

1911-1972

Alfred Ernest Hirschler was born in 1911 in Bluffton, Ohio. After graduating from his hometown college in 1933, he went to Ohio State for his Ph.D. which he recieved in 1937. The same year he joined the Sun Oil Company as a research chemist. In 1940 he was promoted to group leader and became a research associate in 1960. Al's research work concerned solution thermodynamics, oxidation of hydrocarbons, fuel combustion, physical properties of hydrocarbons, separation and purification of hydrocarbons by adsorption, purity determination by freezing point measurements and cracking catalysts. Al devoted a number of years to the study of the acidity of these catalysts with particular emphasis on the correlation between acidity and catalytic activity and on the source and nature of acidity. The work on catalyst acidity became well-known and Al was recognized as an authority. Early in 1972 Al was selected to receive the Philadelphia Catalysis Club Award "for outstanding contributions to the advancement of catalysis." The award was to have been presented in May, but Al died suddenly on February 23, 1972.

Al was an active member of the American Chemical Society, the Philadelphia Catalysis Club and other technical organizations. The chairmanship of the paper review committee of the Fifth International Congress on Catalysis was Al's last professional assignment. As he did with all other tasks, Al accomplished this one too with methodical dispatch and exemplary efficiency. Al was a hard-working, reliable researcher with a sincere and forthright character. He was mild-mannered and modest, but stood firmly for his convictions and had no difficulty in holding his own in a scientific discussion.

Al is sadly missed by his friends, coworkers and colleagues.

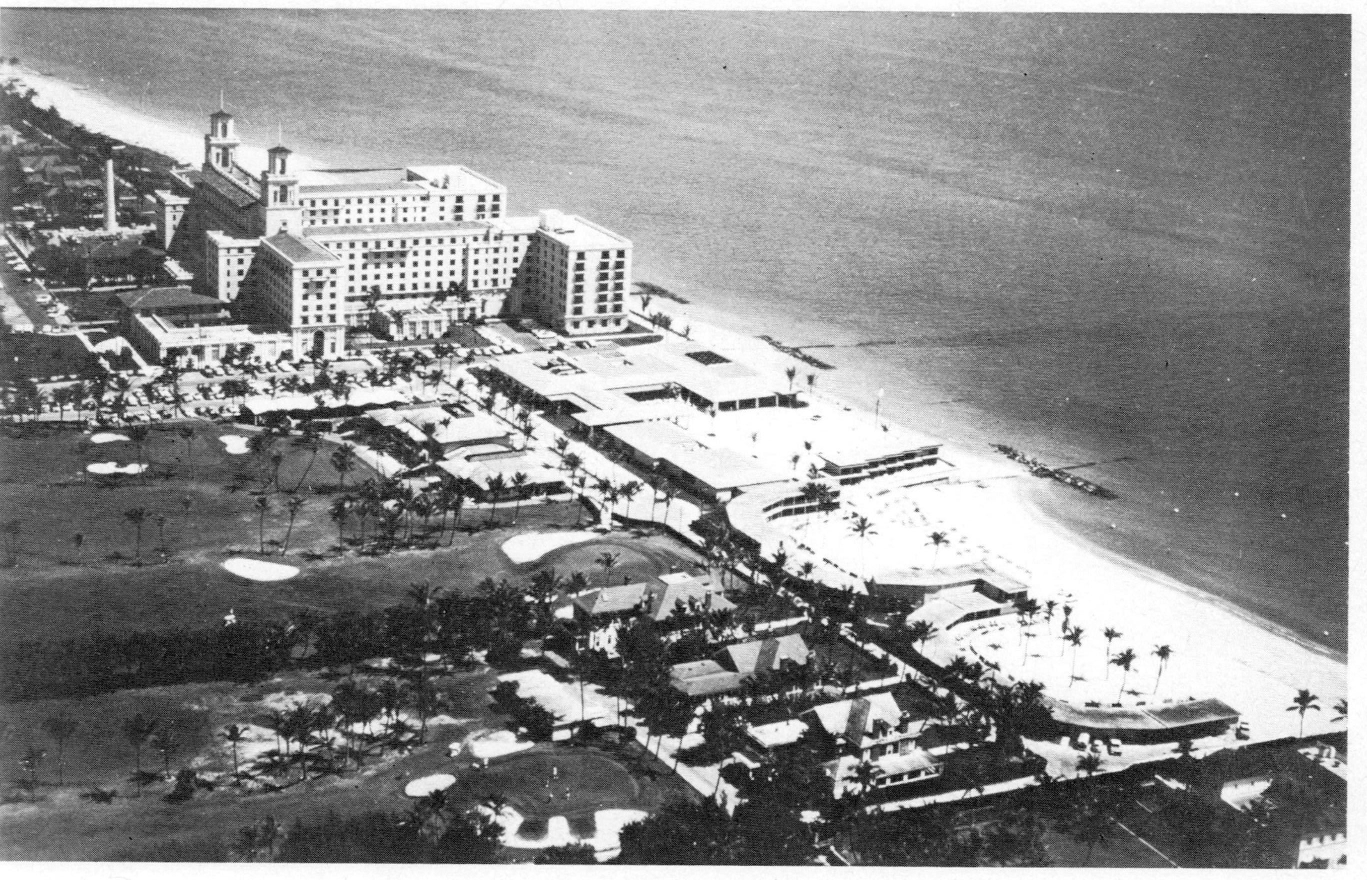

The Breakers Hotel - Palm Beach, Florida

CONTENTS VOLUME I

OPENING REMARKS

DR. V. HAENSEL, *Chairman of Organizing Committee, Fifth International Congress on Catalysis*

Delegates to the Vth International Congress on Catalysis, it is a distinct pleasure to greet you on behalf of the Organizing Committee and the many workers who have given so much of their time and talent to what we hope will be a most successful conclave on catalysis.

During the last 12 months we have suffered substantial grief along the way. We have lost DeBoer, Ciapetta and Hirschler. At this time I would like the delegates to observe a moment of silence in memory of these untiring workers in catalysis and contributors to the success of catalytic congresses.

You are aware of the difficulties in finding a new home for the Vth Congress. As far as I know, this is the first Catalysis Congress which had to move on a moment's notice some 60 miles or 100 kilometers from its original destination, which we so carefully prepared some three years ago. In our world, there is room for science and politics as well as scientists and politicians, however, we did not wish to expose the catalytic scientist to the rougher elements which nowadays inevitably clutter up the arena of a political convention.

So now we are here, and, if some things are not exactly to your liking, please forgive us since we have tried to do our best. If you like them, tell us, we like to have a little praise for our efforts. Similarly, since we have had to select some 110 papers from nearly twice that number originally submitted, we faced some very difficult decisions in accepting some and rejecting others. It was not a choice between poor and excellent, but most of the time it was a choice among very good papers, some of which fitted the theme of the Congress somewhat better than others.

We hope that most of you have had an opportunity to read the preprints and to select the ones in which you would like to participate as a discussor or as a listener. Having done some of this homework myself I have made certain observations that I would like to share with you.

The theme of the Congress is the Science of Catalysis, and there is no doubt that these papers represent by and large some thought provoking contributions. The experimental procedures have become more elegant and thus the interpretations more intriguing. At the same time, one cannot avoid the thought that some of the work digs merely deeper in the territorial imperative, which is quite natural for someone who is the top expert in his own field. And yet it appears to me that these brilliant workers could contribute so much in other perhaps more relevant areas.

How does one go about finding these more relevant areas? This is where the gulf between the industrial and academic sectors is the greatest. The shroud of secrecy overhanging industrial research does a great deal not only to obscure the fine contributions already made in industry, but quite frequently it obscures the existence of the problem itself. It is no wonder that a recent editorial in "Science" pointed out that many professors convey the impression to their students that applied industrial research is a second rate occupation for second rate people.

This Congress, through its communicative powers, can do much to bridge this gulf. It is essential that we do this in the spirit of trying to extend both the understanding and the application of catalysis. You will note in your program that we acknowledge in particular the gift from Mr. W. Clement Stone, who sends his best wishes for the success of the Congress, underscoring this Congress as an opportunity to provide solutions to problems of concern to humanity.

What are the problems of concern to humanity that catalysis may help in solving? So let us carve out the territorial imperative where we believe catalysis can help. The area of energy is a very pertinent one. Here we are concerned with improved methods of conversion of raw materials into products that we need. These raw materials are oil, coal, shale and tar sands. The products we need are gasoline, jet fuels, heating oils and synthetic natural gas.

Catalysis is deeply involved in these conversion systems, and most of you will be amazed to learn how much basic catalytic research is required to help resolve the applied problems. A number of these can be done at the university level if the members of the academic fraternity are willing to work on them. At the same time industry must be willing to share these problems with the academic society.

The same concept applies to another problem, this one connected with the use of the products derived from our raw materials. These products, as well as the raw materials, are burned and in the course of the combustion process we get into all sorts of pollution problems. Here you will be amazed at the amount of basic research work that needs to be done in conjunction with the catalytic processing schemes that have been proposed for these pollution problems. Because of the urgency of the pollution problem much of the basic work remains undone and, here again, industry needs the cooperation of the academia to undertake some very basic aspects of the applied problems.

Thus, the gulf is wide but can be narrowed by cooperation and giving on both sides. Right now, if you don't mind a facetious remark on my part, I believe that the academic and industrial fraternities are separated by one atom only. This is the nitrogen atom. The academic people are decomposing N_2O while the industrial people are convering NO. Now if we can only get rid of that one additional nitrogen atom in the academic fraternity we shall be on the same wave length.

DR. G. ALEX MILLS, President of the Catalysis Society

It is my privilege on behalf of the 1,571 members of the Catalysis Society of North America to express appreciation to all those who have contributed to make this Vth International Congress on Catalysis a success--speakers, discussion leaders, attendees and organizers.

It is to the group or collective activities of those involved in the science of catalysis that I address you briefly. First, in this regard we must applaud Dr. Vladimir Haensel, Chariman of the Organizing Committee and his hard working Committee.

It is the naive belief of some that science can flourish solely by the action of individual scientists. But, while scientific discoveries are indeed accomplished by individuals, it is also true that much more can be accomplished by uniting for the advancement of certain types of important activities, namely those which involve exchange of ideas.

The exchange of information and ideas is the essence required for creativity. New concepts are usually arrived at by discussion and argument, frequently controversial and always exciting. Just inspect the wide range of references in a technical article. Now add to that the benefits of discussion with the author. How I wish we could measure and document the value of our Catalysis Society activities so as to demonstrate their importance and to guide us in functioning even more effectively.

Here in North America we have organized for action in catalytic affairs by forming the Catalysis Society. We do not have a Catalysis Institute as some countries do. Since the Catalysis Society is a unique organization, I hope that a brief review and some comments on our activities will be of interest to those from abroad as well as to members of the Society.

Evidently, we are doing something right because we have become increasingly active for nearly a quarter of a century, having grown in membership since the inception of the Catalysis Club of Philadelphia in 1949 and the formation of the International Congress on Catalysis in 1956. There are now approximately 1,600 members, located in eight clubs or local societies, seven in widely distributed locations in the United States and one in Canada.

No doubt one of the secrets is that we are a loose organization--loose in the sense that we have not been confined by rules and regulations, but have been concerned with the fundamental objective of providing a fellowship dedicated to the advancement of catalysis. In other words, our main concern has been with getting things done.

Another and even greater reason for the success of the Catalysis Society is the enthusiasm of a large number of our members. It is interesting to note that many of the same persons who were members of that original small group who founded the Philadelphia Club in 1949--23 years ago--are still active and are here tonight. I am happy to count myself one of them.

It occurs to me that catalysis is a sort of disease, and, in fact, an incurable disease because here are individuals bitten by the catalysis bug a quarter of a century ago who have not yet recovered. And in fact, even worse, they have in the meantime infected many others who have now caught this incurable disease. May I introduce five of those sort of beneficial "Typhoid Marys" who are here today from the original group.

Dr. Alex Oblad, Dr. Heinz Heinemann, Dr. Roland Hansford,
Dr. Adalbert Farkas, and Dr. Charles Plank.

Now let us look briefly at what the Catalysis Society does. The principal accomplishment has been to increase knowledge and interest in catalysis by providing a means for dissemination of knowledge and interchange of ideas. This has resulted, I believe, in significant additional progress in both the science and application of catalysis. (In monetary terms, catalytic discoveries in the last two decades have resulted in creation of many billions of dollars of wealth as well as making possible technology and products which would not otherwise have been possible). Increased interest has been possible chiefly through the local meetings organized by the 8 individual societies which have provided addresses by experts, discussion and personal contacts. These meetings are usually held monthly in addition to annual one or two day symposia.

The Catalysis Society has also contributed to the advance in Catalysis by its sponsorship of the International Congresses which provide for discussion on a world wide basis--29 nations are represented here. The Society has also held two North American meetings--with the third scheduled for San Francisco in February 1974. At the meeting earlier 1971 in Houston, Texas, Dr. Hightower, Chairman, reported an attendance of 300 people--a remarkable 20 percent of total membership.

The Catalysis Society is also proud of its sponsorship of a society lectureship which has just been named the Ciapetta Lectureship in honor of the late Frank Ciapetta. This Lectureship carries a modest honorarium and a suitably inscribed plaque. We feel the distinguished lecturers have brought us distinction. These have included: Dr. Boer, Dr. Teichner, and Dr. Frank Stone.

Our current lecturer is Dr. George Schuit and we want to publicly acknowledge our debt to him. I am happy to report that Dr. Eischens has recently proposed Dr. Fripiat Lecturer for 1972-1973, and he has accepted this appointment.

The Society has rendered a service in publications as well. Dr. Hightower made available, for example, the papers in English of the Fourth Congress held in Moscow.

It gives me a certain amount of personal pleasure to tell you that we have received a measure of academic recognition, having recently participated as a "learned society" in the inauguration of two University Presidents.

A further activity which is popular with members is the Catalysis Newsletter, issued 2 or 3 times a year, alerting all members to meetings and events. The first and also current editors are present: Jim McEvoy and Maurice Mitchell.

Recently, two substantial awards were arranged by Dr. Heinemann and his committee. The Paul Emmett Award in Fundamental Catalysis, sponsored by the Davison Division of W. R. Grace Company and the Eugene J. Houdry Award in Applied Catalysis sponsored by the Houdry Process Co., Sun Oil Co., Oxy Catalyst Co. and members of the Houdry family. These awards are for $2,000 and $2,500 respectively. We believe that the publicity and financial rewards will contribute to the advancement of catalysis. I would like to recognize the winners of the first awards: Dr. Herman Bloch (Houdry) and Dr. Richard Kokes (Emmett). Nominations for the second award are due September 1, 1972.

It is now necessary to say what we have failed to do. We members of the Catalysis Society are also members of a larger society--the American community. But, in my view we are not adequately relevant and responsive to the needs of this greater society. As a consequence of not being relevant and responsive, we in the United States have seen a loss of support for the science of catalysis by both government and industry. In my opinion there is something wrong with a meeting on catalysis in 1972 which does not have a strong program dealing with the catalysis for prevention of pollution. And there are other areas where catalysis could have profound social benefits. I know that some of my colleagues will object to my viewpoint. But I want to emphasize that I am not advocating a meeting with papers full of empirical industrial testing data. What I am advocating is that we begin to apply science of catalysis to problems such as pollution prevention, problems of concern to our society. I call on each of you to re-examine your selection of scientific research to include this important factor of usefulness to our society. I can only regard this as an opportunity to make catalysis of greater importance.

In closing, I should mention that all the officers and most of the members of the Board of Directors will this year complete their four year appointment. We look to a new team with new ideas and vigor to make the Catalysis Society more relevant, more responsive, and an even greater force in our highly industrialized society. Thank you.

PROFESSOR G. K. BORESKOV, President, International Congress on Catalysis

This is the fifth time that we have assembled an international meeting of scientists working in the area of catalysis. It is parcitularly fitting that this fifth anniversary Congress is being held in the United States which hosted the first Congress in 1956.

In connection with this I would like to remind you that we owe the establishment of the International Congress to the initiative and energy and energy of our American Colleagues. A great deal of effort was required to bring together the scientists active in the area of catalysis and to transform the Congress into the truly international organization that it is today. Much of this credit belongs to Professor deBoer whose success was due both to his high professional standing and to his warm, friendly personality.

There is now no question about the viability and productivity of this international cooperative effort devoted to the study of catalysis. Catalysis has provided the scientific basis for technological progress in a variety of areas and has materially benefited humanity. Its potentialities are by no means exhausted. Catalysis can be a powerful tool for solving the problem of how to protect the human race and nature in general from the harmful effects of technological progress.

I would like to wish success to all the participants in the Fifth International Congress on Catalysis, and I hope that this meeting will promote further international cooperation in this area, will speed up the theoretical understanding of this interesting phenomenon and will also lead to further practical utilization of catalysis for the benefit of all humanity.

PROFESSOR PAUL H. EMMETT, Honorary Chairman, Fifth International Congress on Catalysis

As honorary chairman of the program for the Vth International Catalysis

Congress it has fallen to my lot to say a few words about the future of catalysis. I do so with a full realization that I am not a prophet and that the pathways ahead are confused and not easy to follow. The most one can hope to do is to point out a few of the sign posts that have put in an appearance and call attention to the direction to which they are pointing.[1]

It takes little imagination to see that catalysis is entering an age of automation and computerized operation. Already surface area, pore sizes, pore distribution and in a few cases complete reaction rate measurements are all carried out automatically with little attention from the operator. This will become more and more the common thing rather than the exception.

There seems to be a certainty that the recent upsurge of interest in homogeneous catalysis and particularly in the catalytic activity of metallic complexes is destined to throw a great deal of light on heterogeneous catalysis. Indeed some of the papers that will be given at the present meeting show that progress is being made in actually fixing the inorganic complexes on to a surface in such a manner as to produce a heterogeneous catalyst having the same basic characteristics as the homogeneous catalyst but in addition much greater stability and flexibility. This work also points to a much better understanding of the detailed nature of the metallic complexes that are essential components of our body catalysts, enzymes.

Methods for utilizing controlled beams of x-rays, electrons, positive ions, neutral molecules and photons in surface studies are multiplying so fast that it is difficult to keep abreast of new developments.[2] In the past we have hoped that our catalytic surfaces were of a known degree of cleanliness but we had no way of being sure. Today with surfaces prepared at 10^{-10} mm pressure, analyzed for general conformation by low energy electron diffraction, and assured of purity by Auger spectroscopy, one, for the first time, has an opportunity to associate types of catalytic reactions with the inherent properties of clean surfaces. If promoters or alloys are being used it becomes increasingly possible to differentiate between the composition of the surface of the catalyst and its interior. In short a trip to some of the present laboratories that are focusing these new techniques on the activity of surfaces as catalysts impels one to expect a much more definitive appraisal of the effect of chemical composition and surface topography on catalytic behavior. It is safe to predict that progress, in this approach, will be fast and very revealing.

It augers well for the future when papers begin to appear in which theoretical calculations show relationships between the orbital characteristics of the reactant molecules and those of the catalyst surface. Two such examples come to mind. The first is a paper by Mango and Schackschneider[3] in which they pointed out that the surface orbitals of transition metal oxides should make them excellent catalysts for the disproportionation reactions. A series of these reactions had been described by Banks and Bailey[4] and later by Bradshaw, Howman, and Turner.[5] A second example is to be found in the calculations by Ruch[6] which show that the molecular orbitals of the nitrogen molecules are such as to make molecular chemisorption of nitrogen on iron surfaces reasonable. Furthermore, they showed that the 111 face of the iron would be more capable of adsorbing nitrogen molecules than the 100 or 110 planes in agreement with the claims that 111 face of iron is many times more active for ammonia synthesis than the 100 or 110 planes.

Catalysts have often been referred to as the "heartbeats" of our industrial chemical processes. Probably the biggest application has been in the field of petroleum refining. With the advent of our present antipollution activities catalysis seems destined to gain prominence in at least two new areas. It will be essential in preparing high octane fuels to replace some of the additives that seem likely to be banned from gasoline products of the future. Secondly, catalysis appears to be an essential part of any program to eliminate carbon monoxide, hydrocarbons, and nitric oxides from the automobile exhaust gases. Expansion of its use in space exploration also seems likely. In a word catalysis of the future is likely to increase to the point of being even more vital to our existence than it has been in the past.

In a word, I am optimistic about the future of catalysis because of the many new research tools and approaches that are being introduced, because of the increasing interest in basic work both in homogeneous and heterogeneous catalysis and because of the ever increasing industrial demand for new, different, and better catalysts to help solve the problems that are confronting us. However, I am optimistic most of all because of the fine group of young scientists throughout the world who are tackling with enthusiasm and skill the many problems that we of an older generation have left unsolved.

REFERENCES

1) P. H. Emmett, Transactions of the New York Academy of Sciences Series II, Vol. 31, pages 188-202.
2) G. A. Somorjai, "Principles of Surface Chemistry," Prentice-Hall, New Jersey (1972).
3) F. D. Mango and J. H. Schacktschneider, J. Am. Chem. Soc. 89, 2484 (1967).
4) R. L. Banks and G. C. Bailey, Ind. Eng. Chem. Products Res. Development 3, 176 (1963).
5) C. P. C. Bradshaw, E. J. Howman, and L. Turner, J. Catalysis 7, 269 (1967).
6) R. Brill, E. L. Richter, and E. Ruch, Angew. Chem. 6, 882 (1967).

Invited Lecture A

HYDROGENATION AND RELATED REACTIONS OVER METAL OXIDES

R. J. Kokes
Department of Chemistry, The Johns Hopkins University
Baltimore, Maryland U.S.A. 21218

1. INTRODUCTION

Hydrogenation of olefins and related reactions constitute a major area of catalytic chemistry; hence, it is not surprising that such reactions have been the subject of intensive study. From the very beginning, however, most basic studies of these reactions dealt with metals rather than metal oxides. Initially, this prejudice stemmed from the fact that metals appeared to be much more active than the oxides. More recently, however, it has become apparent that properly activated metal oxides have significant activity at or even below room temperature.[1-3] Furthermore, although the same overall reactions occur over metals and metal oxides, the surface reactions are, in some respects, simpler over the oxides. Thus, it is useful to consider metal oxides as a separate class of heterogeneous hydrogenation catalysts. In the presentation that follows we shall take this approach. First, we shall attempt to summarize the principal differences between metal oxides and metals. Then, we shall discuss some features of hydrogen "activation", hydrogenation, and isomerization, that seem peculiar to oxides. Finally, we shall conclude with some speculation on the nature of the active sites. In this discussion, we have singled out zinc oxide as a prime example of a metal oxide, but we shall try to fit these results into a context that includes other oxides. Such a course is obviously dangerous insofar as each metal oxide has its own unique characteristics. Nevertheless, it does not seem too simplistic to believe that, given the similarities that do exist, some extrapolation is justifiable, and that the resulting overview, speculative to be sure, may suggest possible avenues for future research.

2. REACTIONS OVER METAL OXIDES

2.1 Comparison of Metals and Metal Oxides

Hydrogenation of ethylene over metals is presumed to occur via the following steps:[4]

$$H_2(g) + 2* \rightleftarrows 2H\text{-}* \qquad (1)$$

$$C_2H_4(g) + 2* \rightleftarrows *\text{-}CH_2\text{-}CH_2\text{-}* \qquad (2)$$

$$*\text{-}CH_2\text{-}CH_2\text{-}* + H\text{-}* \rightleftarrows 2* + *\text{-}CH_2\text{-}CH_3 \qquad (3)$$

$$*\text{-}CH_2\text{-}CH_3 + H\text{-}* \rightarrow C_2H_6 + 2* \qquad (4)$$

In the above, the symbol * stands for a bond available from a surface atom. Isomerization of more complex olefins, which often accompanies hydrogenation, is presumed to occur by a variation on step 3, viz.:

$$R\text{-}CH_2\text{-}CH{=}CH\text{-}R' + 2* \rightleftarrows R\text{-}CH_2\text{-}\underset{*}{CH}\text{-}\underset{*}{CH}\text{-}R' \qquad (5a)$$

$$R\text{-}CH_2\text{-}\underset{*}{CH}\text{-}\underset{*}{CH}\text{-}R' + H\text{-}* \rightleftarrows R\text{-}CH_2\text{-}\underset{*}{CH}\text{-}CH_2\text{-}R' + 2* \qquad (5b)$$

$$R\text{-}CH_2\text{-}\underset{*}{CH}\text{-}CH_2\text{-}R' + 2* \rightleftarrows R\text{-}\underset{*}{CH}\text{-}\underset{*}{CH}\text{-}CH_2\text{-}R' + H\text{-}* \qquad (5c)$$

$$R\text{-}\underset{*}{CH}\text{-}\underset{*}{CH}\text{-}CH_2R' \rightleftarrows R\text{-}CH{=}CH\text{-}CH_2\text{-}R' + 2* \qquad (5d)$$

Support for this picture is quite detailed and has been summarized by a number of writers.[4-6] We should like to note specifically, however, the following:

a) Metal hydrogenation catalysts are effective for H_2-D_2 exchange (and parahydrogen conversion) under the same conditions as they effect olefin hydrogenation. This suggests that hydrogen "activation" involves formation of adsorbed hydrogen atoms as suggested by step 1.
b) Addition of deuterium to light ethylene leads to ethane products of the form $C_2H_{6-x}D_x$ (as expected if step 3 is readily reversible). Moreover, efficient isomerization of olefins requires hydrogen as a cocatalyst[4,7] (as expected if the sequence 5a-d is operating). Both these observations suggest that alkyl formation and its reversal play a major role in hydrogenation and related reactions over metals.

Although the above sequence does account for most of the observations on metals, there are some mechanistic inadequacies in this scheme.[8,9] Furthermore, IR studies of supported metals suggest that, in addition to the hydrocarbon species pictured above, a variety of carbonacious residues formed by disproportionation and dimerization are also present on the surface.[10]

It has been suggested that an analog of the sequence 1 through 4 is the principal pathway over oxide hydrogenation catalysts.[2] As with metals, we find that effective oxide catalysts for olefin hydrogenation are also effective for hydrogen-deuterium exchange; this result again suggests that hydrogen "activation" involves formation of adsorbed hydrogen atoms as suggested by step 1. <u>Unlike the metals</u> addition of deuterium to light ethylene yields $C_2H_4D_2$; hence, although alkyl formation may still occur, <u>alkyl reversal is not a significant reaction</u>. This requires a different mechanism for olefin isomerization, and it has been suggested that an allyl (rather than an alkyl) is the intermediate over some oxides.[1,2] There is still another difference between metals and metal oxide catalysts. For at least one oxide (ZnO) adsorbed ethylene can be recovered as such without the disproportionation and polymerization found for metals. Thus, not only are experiments with deuterium simplified by the non-occurrence of alkyl reversal but physical studies of surface species are facilitated by the non-occurrence of residue-forming side reactions.

2.2 Hydrogen Activation on Metal Oxides

The spectrum of hydrogen adsorbed on zinc oxide (Figure 1)[2] is characterized by two strong bands at 3489 cm^{-1} and 1709 cm^{-1}. Eischens, Pliskin and Low,[11] who were the first to report these bands, assigned them

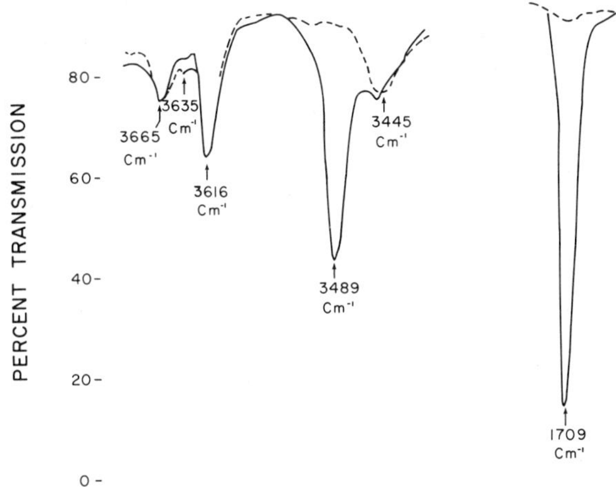

Fig. 1 Spectrum of Hydrogen on Zinc Oxide. The broken line is the spectrum for the degassed catalyst.

to an OH and ZnH species, respectively. A number of subsidiary experiments, including frequency shifts with deuterium, assure that this assignment is essentially correct. Thus, hydrogen adsorption on zinc oxide can can be represented as follows:

$$H_2 + \text{Zn-O} \rightarrow \overset{\text{H}}{\text{Zn}}\text{-}\overset{\text{H}}{\text{O}}$$

where Zn-O represents a metal-oxide pair site on the surface. Further experiments suggest that the hydrogen that gives rise to the bands is limited to 5 to 10% of the surface and is reversible (type I[2]). The irreversible hydrogen chemisorption (type II[2]) does not contribute to the observed IR spectrum. Since only the type I hydrogen exchanges readily with gaseous deuterium[2,12] it is clear that type I rather than type II is the "activated" hydrogen. No such IR bands have been reported for other oxides, but this mode of adsorption has also been suggested for chromia.[1]

Further information on the nature of the active sites is obtained by examination of the product ethanes formed by additions of hydrogen-deuterium mixtures to ethylene. Twigg[13] was the first to use this technique; he showed that the deuterium distribution in the product ethanes formed by a 50:50 H_2-D_2 mixture over nickel was the same for the unequilibrated mixture (about 0% HD) as for the pre-equilibrated mixture (about 50% HD). Since ethane formed from pure deuterium shows the characteristic isotope smear, alkyl reversal may provide the atomic mixing of hydrogen prior to reaction needed to obtain the observed result. No such possibility exists, however, for the addition of deuterium to 2-butyne over palladium which yields essentially pure cis-2-butene-2,3-d_2.[14] From their studies of this reaction with H_2:D_2 mixtures Meyer and Burwell[14] conclude that atomic mixing precedes the addition reaction so that the hydrogen isotopes lose their molecular identity. It has been noted that these results,[15] apparently generally characteristic of metals,[16] can be explained if we assume that, on metals, there are always enough <u>adjacent</u> sites for hydrogen dissociation so that isotopic mixing can occur prior to the addition step.

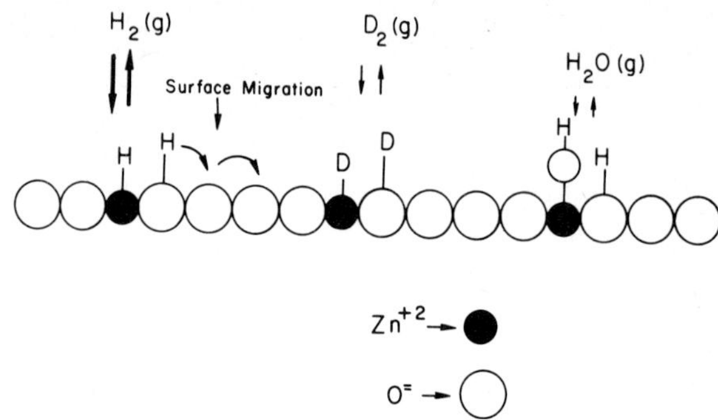

Fig. 2 Schematic Picture of the Active Sites

Over chromia and zinc oxide,[15] reaction of ethylene with H_2:D_2 mixtures yields ethane products that reflect the molecular identity of the reacting gas. Unequilibrated H_2-D_2 mixtures yield predominantly C_2H_6 and $C_2H_4D_2$; reaction with equilibrated H_2:D_2 mixtures yields predominantly C_2H_5D. A similar result has been reported for Co_3O_4 catalysts.[3] Two possible interpretations of these results are:

i) The gaseous hydrogen reacts directly with the adsorbed olefin.
ii) The sites for dissociative adsorption of hydrogen are isolated and non-interacting.

We prefer the second of these alternatives, at least for zinc oxide.

A schematic view of a surface consisting of isolated, non-interacting sites is shown in Figure 2. Clearly, site-to-site migration is required for atomic mixing of hydrogen isotopes. If migration is slow compared to adsorption-desorption, one would expect reaction with ethylene to lead to product ethanes that reflect the <u>molecular</u> composition of the reacting hydrogen mixture. This picture also suggests that the (non-magnetic) para hydrogen conversion, which can occur by adsorption-desorption alone, ought to be much faster than the H_2-D_2 equilibration. Recently, Conner[17] has measured the para hydrogen conversion on zinc oxide pretreated with dry oxygen in order to minimize the number of paramagnetic centers stemming from non-stoichiometry. As expected for the proposed model, he finds the para hydrogen conversion is more than an order of magnitude faster than the hydrogen-deuterium exchange.

Further support for the model pictured in Figure 2 is offered by recent IR studies of isotope effects for the type I hydrogen adsorption.[18] At low temperatures type I adsorption is rapid but it is irreversible. Under these conditions, the <u>kinetic</u> isotope effect can be determined from the relative intensities of adsorbed hydrogen and deuterium bands that appear after competitive adsorption from isotopic mixtures. Near room temperature reversible isotherms can be obtained for type I hydrogen; under these conditions, the <u>thermodynamic</u> isotope effect can be determined from the relative band intensities developed by adsorption from isotopic mixtures. Preliminary calculations of these effects (based on the isolated site model and the positions of the band for adsorbed hydrogen and deuterium) predict rather unusual effects for the adsorption of hydrogen deuteride. One can imagine two modes of adsorption:

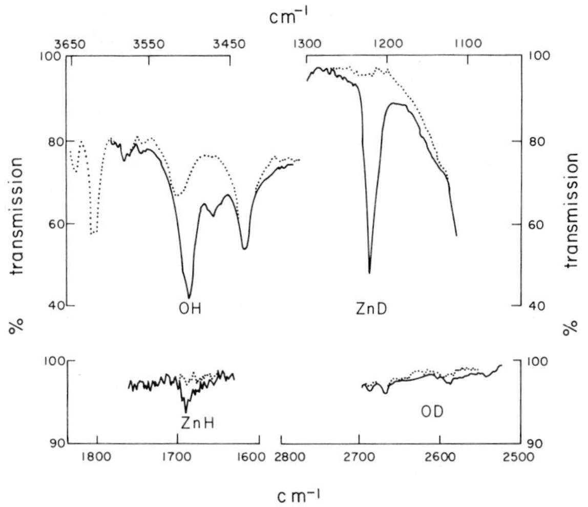

Fig. 3 Spectrum of HD Adsorbed on ZnO at -195°C.
The broken line is the spectrum for the degassed catalyst.

$$HD + Zn\text{-}O \rightarrow \overset{H\ \ D}{Zn\text{-}O}\ (HD)$$

$$\rightarrow \overset{D\ \ H}{Zn\text{-}O}\ (DH)$$

According to the computations, HD should be larger when equilibrium is achieved and DH should be larger when adsorption is kinetically controlled. Figure 3 shows the results obtained for the adsorption of HD at -195°C; not only is the expected kinetic effect (DH preferred) observed but it is quite large. Figure 4 shows the change that occurs after warming to room temperature in the presence of gas. As the sample warms the exchange reaction (which was nil at low temperatures) sets in and, by the time room temperature is reached, the gas in contact with the sample becomes the equilibrated mixture. Nevertheless, quantitative estimates suggest that at room temperature HD is preferred by a factor of roughly three over DH. The changeover from DH to HD on the surface as the sample is warmed starts at about -60°C. Since at these temperatures the adsorption is still essentially irreversible, this suggests (but does not require, see reference 12) that the changeover involves equilibration <u>on the surface</u>, i.e., site-to-site migration. Such facile site-to-site migration would seem to require relatively loose bending vibrations for the adsorbed species. The proposed quantitative interpretation[18] of equilibrium isotope effects does, in fact, require such loose bending vibrations.

It seems quite clear that the dissociative adsorption on zinc oxide which is effective in hydrogen activation occurs on a limited number of metal-oxide pair sites.[2] Poisoning experiments on Co_3O_4[3], Cr_2O_3[1] and Al_2O_3[19] suggest that the number of effective sites for hydrogen activation is limited; hence, it seems reasonable to suppose that, on these oxides,

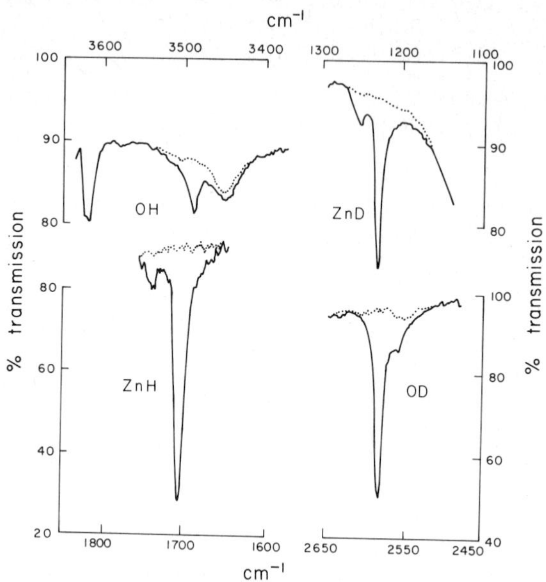

Fig. 4 Spectrum of HD Adsorbed on ZnO at -195°C after Warm-up. The broken line is the spectrum for the degassed catalyst.

also, dissociative adsorption occurs on isolated, non-interacting, metal-oxide, pair sites. This conclusion, however, does not preclude the occurrence of a Rideal-Eley mechanism[6] for the hydrogen-deuterium exchange. In fact, Tamaru and co-workers[12] in an elegant study combining rate measurements with infrared observations, conclude that the exchange reaction on zinc oxide involves interaction of a molecular species with dissociatively adsorbed hydrogen. Such a molecular species might also play a role in the para-hydrogen conversion.

The results of Van Cauwelaert and Hall[19] on alumina impose severe restrictions on the mechanism of hydrogen-deuterium exchange and the related para hydrogen conversion. On the basis of esr and chemical analysis of their catalysts, they concluded that neither paramagnetic impurities nor surface paramagnetism were present, and, hence, the paramagnetic conversion mechanism[20] was <u>not</u> responsible for the para hydrogen conversion. At high temperatures the exchange and conversion occur at comparable rates; at low temperatures the conversion is much faster than exchange. Both reactions are poisoned by chemisorption of carbon dioxide. For alumina pretreated at 500°C, the lethal dose corresponds to a few percent of the surface and is the same for both conversion and exchange. At first glance, it might appear that these results are explicable in terms of surface migration. The common poisoning and common rate would be expected when migration is fast; the common poisoning and faster conversion rate would be expected when migration is slow. The temperature dependence, however, makes this explanation unlikely. Thus, to conform to their conclusions,[19] we must consider the possibility that two different mechanisms operate for conversion and exchange over the same sites. Van Cauwelaert and Hall[19] suggest the following:

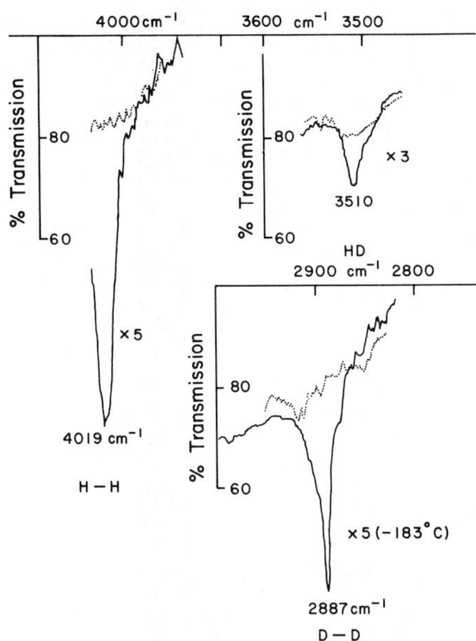

Fig. 5 Spectrum of Molecular H_2, HD and D_2
The broken line is the spectrum for the degassed catalyst.

1) the same sites adsorb hydrogen dissociatively at high temperature (where exchange takes place) but molecularly at low temperatures.

2) the para-hydrogen conversion occurs for the molecular species in the absence of a paramagnetic site.

The first suggestion suffers insofar as it ignores the requirement for site-to-site migration. Strictly on an ad hoc basis, we suggest this difficulty could be obviated if we suppose that the same sites adsorb both molecular and dissociated hydrogen. At sufficiently high temperature exchange and conversion could occur via a Rideal-Eley mechanism involving adsorbed molecular and dissociated species. If the Rideal-Eley exchange step has a much higher activation energy than the (new) conversion mechanism, this would be equally compatible with the observations. Furthermore, such a Rideal-Eley step, involving a weakly adsorbed molecular species, would be consistent with the conclusions of Tamaru et al[12] that the exchange on zinc oxide involves a molecular species.

Figure 5 shows the spectrum of zinc oxide at low temperatures in the presence of gaseous hydrogen, hydrogen deuteride, and deuterium. The hydrogen adsorption, termed type III adsorption[21] is quite weak and the bands disappear after brief evacuation. The isotopic shifts clearly show we are dealing with a molecular species. The band positions, however, are rather strongly shifted from the corresponding positions of the gas-phase Raman bands; for example, the shift for hydrogen is about 142 cm^{-1}. This shift is nearly five times that found for physically adsorbed hydrogen on porous glass.[22] Since, in simple interpretations,[23] solvent shifts are proportional to the interaction potential, the interaction potential for this adsorbed hydrogen would be nearly five times that for hydrogen on porous glass for the same interaction distance. This difference in interaction is also evident in the extinction coefficient; the extinction coefficient for hydrogen on zinc oxide is greater than that reported for porous glass by a factor of 65. Combined IR and adsorption studies[24)] suggest that the type III adsorption occurs on roughly 10% of the surface; hence, this amount is comparable (within a factor of two) to the number of sites that effect dissociative type I adsorption.

Earlier studies[2)] have shown that water selectively poisons the sites for type I chemisorption, presumably by the type of adsorption depicted in Figure 2. If we poison the catalyst with water and expose the catalyst to deuterium, neither type I nor type III adsorption (judged by IR) occurs.[24)] As we reactivate the catalyst in stages by evacuation at successively higher temperatures, the bands for type I and type III adsorption reappear and grow in concert at successive stages of the reactivation. Thus, it appears that the molecular type III and the dissociative type I adsorption can occur simultaneously on the same sites. Further details of the spectrum support this view.[25]

The bands observed for adsorbed hydrogen stem from a transition that involves a change in only the vibrational quantum number, n. Such bands are forbidden in gaseous hydrogen and become allowed due to a dipole moment induced by the bond to the surface. It can be shown[26)] that for a freely rotating hydrogen molecule with such an induced dipole moment, the selection rules are those for the Raman Spectrum,[26)] i.e. $\Delta n = 1$, and $\Delta J = 0, \pm 2$ where J is the rotational quantum number. Then the spectrum contains many bands,[27)] corresponding to changes in the quantum number n and one of the several possible changes in the rotational quantum number, J. Most diatomic molecules in condensed phases, e.g. nitrogen, have a barrier to rotation considerably greater than the separation of rotational levels. In this limit, the rotations become torsional oscillations and only Raman transitions corresponding to changes of one vibrational quantum are allowed; transitions corresponding to $\Delta J = \pm 2$ become combination bands which, to a first approximation are forbidden. For hydrogen, however, the moment of inertia is so small that the separation of rotational levels is unusually large. This means that quite a high barrier is required before it is large compared to the separation of rotational levels; hence, even in the liquid[28)] the observed Raman transitions correspond to those expected for a freely rotating molecule.

We have examined the change that occurs in the rotational energy levels for deuterium as a rotational barrier is imposed. We chose a potential function in which the most stable orientation is along a given axis and the potential has axial symmetry. Figure 6 shows a correlation diagram for this model. The left side is the energy level sequence in the absence of the barrier; the right side shows the limit approached for very high barriers as described by Stern.[29)] We have indicated the components of degenerate levels as separate levels connected by brackets and have indicated symmetric and antisymmetric levels by dashed and solid lines respectively. Intermediate values were computed by the procedures used by White and Lassatre[30)] and King and Bensen[31)]; in fact, in this region, Figure 6 represents a modest extension of their computations. For clarity we have compressed the energy scale as the barrier is increased; the zero, however, was not changed and the arrow (equal to 1 kcal) gives a gauge of the energy scale. The most significant aspect of this diagram is that relatively high barriers must be imposed before the energy levels corresponding to J = 2 approaches those characteristic for a torsional oscillator. Only when the rotation becomes so restricted that it is

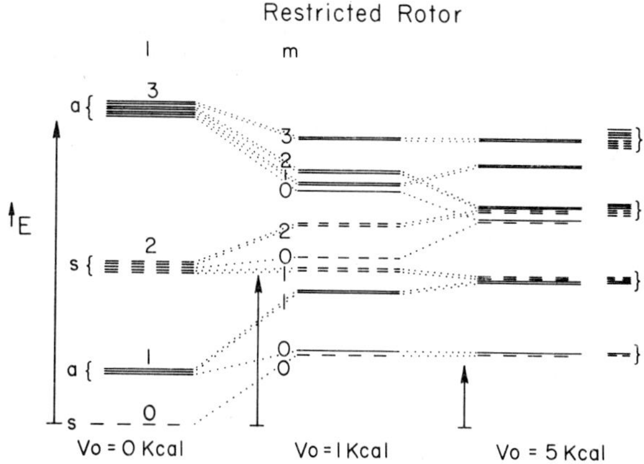

Fig. 6 Energy Levels for Restricted Rotor
See text for details.

effectively a torsional oscillation can we be sure that the observed transitions will be limited to those involving changes in only the vibrational quantum. For barriers well below this limit we would expect a number of well-separated bands due to transitions of vibrational quantum number <u>and</u> rotational quantum number.[27,33] A careful search of the spectrum of adsorbed deuterium reveals no bands other than that corresponding to a change in only the vibrational quantum; hence, we conclude that the first few rotational levels for gaseous hydrogen become effectively torsional oscillations when it is adsorbed and that the barrier to rotation must be quite high. Reasons for this conclusion were stated in terms of a specific oversimplified model for restricted rotation; it seems clear, however, that the same qualitative conclusions will be reached for more realistic models.

Restricted rotation may have significant effects on ortho-para conversion rate. At low temperatures a large part of the ortho-para conversion involves transitions from the lowest to the next to lowest rotational level; these levels correspond to wave functions that are symmetric and antisymmetric with respect to nuclei exchange. As pointed out by Van Cauwelaert and Hall[19] the transition probability includes a factor of the type: $\frac{\langle\psi_0|H'_m|\psi_1\rangle^2}{(E_1-E_0)^2}$ where H'_m represents the perturbation that allows the transition to occur and the ψ's and E's correspond to wave functions and energies for the initial and final states. From Figure 6 we see that the denominator gets quite small for a high rotational barrier; hence, from this effect alone we would expect the transition probability to be enhanced. In the limit of high barriers E_1 and E_0 become degenerate, independent of the form of the barrier. As this limit is approached, conversion mechanisms that are ineffective for freely rotating gaseous hydrogen, e.g. nuclear magnetism, may be enhanced to the point where they become effective for adsorbed hydrogen. In this context, if one judges by the term in the denominator alone, the effect of a given barrier on the transition probability should be much more pronounced for tritium than for hydrogen. In line with this it is observed[33] that the conversion for solid tritium is much higher than one would predict from the rate for solid hydrogen on the basis of Wigner's formulation,[34] which assumes a freely rotating molecule. When this assumption is obviously valid, i.e. for conversion in the gas phase by paramagnetic catalysts, the relative rates for tritium and hydrogen conform to Wigner's theory.[33)]

Until more explicit account is taken of the effect of restricted rotation on other terms in the transition probability, the possibility of such an altered mechanism for para-hydrogen conversion must be viewed as extremely speculative. Nevertheless, the suggestion from the IR data that rotation is severely hindered makes this proposal of Van Cauwelaert and Hall[19)] an attractive one for experimental testing.

2.3 Hydrogenation of Ethylene

At room temperature ethylene adsorbs reversibly on zinc oxide and is recoverable as such by brief evacuation. The initial heat of adsorption is 14 kcal, corresponding to a chemisorptive bond considerably weaker than that found for metals.[35)] Table 1 lists the observed bands for ethylene adsorbed on zinc oxide and compares them to corresponding bands for liquid ethylene and ethylene in a π-complex. Two features are evident for the π-complex:[38)]

1) Bands forbidden in the gas phase become allowed for the π-complex and with two exceptions are near the positions of the fundamentals for gaseous ethylene.

2) The double bond stretching vibration (1621 cm^{-1} for liquid ethylene) shifts to lower frequencies and the symmetric bending vibration (1342 cm^{-1} for liquid ethylene) shifts to higher frequencies.

The similarity of the observed bands of chemisorbed ethylene to those for the π-complex together with other features of the adsorption[2)] suggest that this species be classified as a surface π-complex. Thus, in contrast to ethylene on nickel,[10)] ethylene on zinc oxide retains its olefinic character.

TABLE 1
Observed Fundamental Bands for Chemisorbed Ethylene and Related Systems

$C_2H_4(\ell)$* (cm^{-1})	π-Complex** (cm^{-1})	Adsorbed ethylene† (cm^{-1})
3105 (I)	3094	3140†† (3122)
3075 (R)	3079	3055 (3078)
3008 (R)	3013	2993 (3000)
2980 (I)	2988	2984
1621 (R)	1518	1600
1435 (I)	1426	1438
1342 (R)	1419	1451

*For gaseous ethylene bands labeled I and R are infrared and Raman active, respectively. See Stoicheff (36) for liquid ethylene.

**For the salt $KPtCl_3(C_2H_4)$(37).

†Some of these bands shift as a function of pressure. The numbers listed are for the lowest pressures of ethylene (22 mm); the numbers in parentheses are the positions at the highest ethylene pressures (240 mm). In part these bands are due to physically adsorbed ethylene.

††This band is broad and weak. The position is uncertain to ± 10 cm^{-1}.

Combined IR and kinetic studies[2] provide evidence that the surface π-complex reacts with adsorbed hydrogen in two steps: first, it forms an alkyl radical, and then, the alkyl radical reacts to form ethane. Details of the spectrum assigned to the ethyl intermediate suggest it is bound to the zinc half of the active site. Since formation of ethane occurs by step-wise addition of hydrogen, one atom at a time, it is reasonable to suppose that the reacting hydrogen is dissociated rather than molecular hydrogen. This interpretation, plus the assumption that alkyl reversal does not occur, yields a mechanism consistent with the observed half-order dependence in hydrogen pressure.

The fact that the reactive chemisorbed ethylene forms a loosely held, π-bonded species may account for the lack of alkyl reversal. On metals ethylene apparently forms two σ carbon-metal bonds.[10] Alkyl reversal requires a trade of a σ carbon-hydrogen bond for a σ carbon-metal plus a metal-hydrogen bond. When the energy released in formation of the metal-carbon plus metal-hydrogen bond is enough to compensate for the rupture of the carbon-hydrogen bond, alkyl reversible is energetically favorable. For zinc oxide, however, the π-bonding of ethylene to the surface is much weaker than the σ-bonding of ethylene to metals; initial heats of adsorption on metals are three to four times greater than the measured heat (14 kcal) on zinc oxide. Thus, on zinc oxide and similar oxides, alkyl reversal is likely to be unfavorable. Perhaps, in order to form a bond strong enough to favor alkyl reversal, ethylene must form two σ metal-carbon bonds. On isolated metal-oxide pair sites such bonding would be impossible, ethylene would be present as the more weakly held π-bonded species, and alkyl reversal would not occur.

Recently, Ozaki and co-workers[3] have studied the hydrogenation of ethylene over Co_3O_4 by a combination of traditional mechanistic techniques, but they did not make use of IR techniques. Their results are, in many respects, similar to those found with zinc oxide. Details of their study, however, led them to the conclusion that hydrogenation involves a molecular species. At first glance, this conclusion would suggest that zinc oxide and Co_3O_4 are essentially different kinds of catalyst; but, in fact, the difference may be less profound than it seems. Zinc oxide does adsorb molecular hydrogen on the same sites where dissociative adsorption occurs. Perhaps the same holds true for Co_3O_4 but the adsorption energetics are sufficiently different so that the molecular hydrogen dominates olefin hydrogenation on Co_3O_4 and dissociative hydrogen dominates the reaction on zinc oxide.

2.4 Hydrogenation and Isomerization of Higher Olefins

For the metal oxides that yield dideutero-ethane on the addition of deuterium to light ethylene, alkyl reversal does not occur significantly; hence, an alternative to the alkyl reversal mechanism for olefin isomerization is needed. Burwell et al[1] have suggested that (on chromia) double bond isomerization occurs *via* an allyl intermediate as follows:

$$R\text{-}CH_2\text{-}CH{=}CH\text{-}R' + 2{*} \rightarrow R\text{-}\underset{\textstyle *}{CH}\text{-}CH{=}CH\text{-}R' + H\text{-}{*}$$

$$R\text{-}\underset{\textstyle *}{CH}\text{-}CH{=}CH\text{-}R' \rightarrow R\text{-}CH{=}CH\text{-}\underset{\textstyle *}{CH}\text{-}R'$$

$$R\text{-}CH{=}CH\text{-}\underset{\textstyle *}{CH}\text{-}R' + H\text{-}{*} \rightarrow 2{*} + RCH{=}CH\text{-}CH_2\text{-}R'$$

This seems like a plausible alternative to the alkyl reversal mechanism that may be applicable to other metal oxides. If, however, formation of the alkyl species occurs readily, adsorption of higher olefins with allylic hydrogens should be dramatically different from that for ethylene. This is found to be the case on zinc oxide.[2)]

Propylene adsorbs readily on zinc oxide at room temperature. Part of this adsorbed propylene is readily removed by brief evacuation at room temperature; the more strongly bound propylene is removable only by prolonged evacuation at 125°C. Essentially, all of the adsorbed propylene can be recovered unchanged by this degassing procedure; hence, adsorption is not accompanied by significant irreversible chemical reaction. Infrared studies of this strong and weak adsorption, however, suggest we are dealing with two different chemical species. The bands associated with the weakly bound species, particularly in the C=C region, suggest this is a surface π-complex similar to that found for adsorbed ethylene. The strongly bound species, X, however, has an IR spectrum, rich in structure, which is markedly different from that expected for a simple π-complex. Figure 7 shows the spectrum of this strongly held species in the C-H and O-H region. Its most striking feature is the appearance of a relatively strong OH bond (3593 cm^{-1}), on adsorption which indicates that dissociation occurs. The five bands in the C-H region suggest that, if the process is simple, the hydrocarbon fragment contains five C-H bonds. This, together with the fact that X yields no ZnH band but prevents the formation of Zn-H and O-H in gaseous hydrogen, implies X forms as follows:

$$C_3H_6 + \text{Zn-O} \rightarrow \begin{matrix} C_3H_5 & H \\ | & | \\ \text{Zn} & - \; \text{O} \end{matrix}$$

The hydrocarbon fragment, tentatively identified as C_3H_5, was characterized in more detail by studies of the infrared spectrum of six differently labelled deuterium isomers of propylene. The results are summarized, in abbreviated form, in Table 2; the "Surface hydrogen fragment" was identified as an OH if a band appeared near 3593 cm^{-1} and as an OD if a band appeared near 2653 cm^{-1}. In those cases where the spectrum changes with time, the "Surface hydrogen fragment" refers to the initial spectrum. These data show that the C_3H_5 fragment is formed by removal of an allylic hydrogen.

TABLE 2
Spectrum of Chemisorbed Propylenes

Isomer adsorbed	Surface hydrogen fragment	Spectrum
I CH_3-CH=CH_2	O-H	Stable
II CD_3CD=CD_2	O-D	Stable
III CH_3CD=CH_2	O-H	Stable
IV CH_3-CH=CD_2	O-H	Changes
V CD_3-CH=CH_2	O-D	Changes
VI CD_3-CH=CD_2	O-D	Stable

The bond between the allyl and the zinc might be expected to be similar to that found for allyl complexes of transition metals. Two types of allyl complexes are known[40)] and these can be represented as follows:

M-CH_2-CH=CH_2 (σ)

$$\begin{matrix} & & H & & \\ & & | & & \\ H & & C & & H \\ & \diagdown \; // & & \diagdown\diagdown \; \diagup & \\ & C & M & C & \\ & | & & | & \\ & H & & H & \end{matrix} \quad (\pi)$$

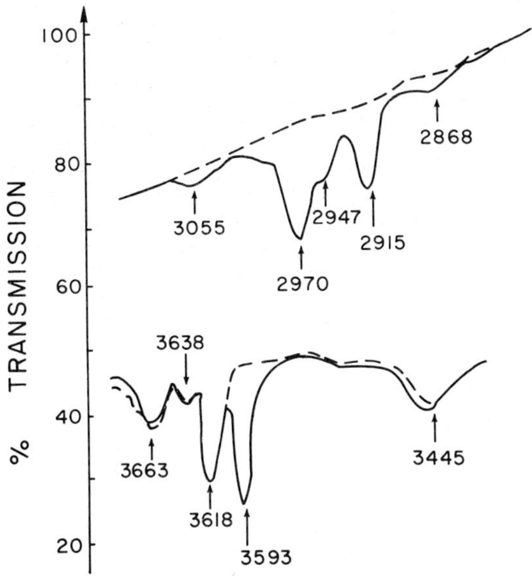

Fig. 7 Spectrum of Chemisorbed Propylene on ZnO in the CH and OH Region
The broken line is the spectrum for the degassed catalyst.

For the σ-complex, the hydrocarbon ligand retains much of its olefinic character; for the π-complex, the C-C bonds in the planar hydrocarbon ligand have only partial double bond character. Which of these more nearly represents the C_3H_5 bound to the surface zinc? An answer is provided by the initial spectrum of compounds IV and V of Table 2: if a σ-allyl complex forms, the hydrocarbon spectrum of IV and V should be markedly different; if a π-allyl complex forms, the hydrocarbon spectrum for IV and V should be identical. The latter occurs, and hence, we conclude that propylene dissociates to form a π-allyl species. Detailed analysis of the spectrum, somewhat speculative, leads to the conclusion that the surface π-allyl is planar, has a C-C-C bond angle of 120°C and C-C bonds with one-half double bond character.[41)]

Combined kinetic and ir studies provide convincing evidence that this surface π-allyl is a reactive intermediate in the "isomerization" of labelled propylenes.[41)] In deuterogenation of propylene, dideutero addition to the π-complex is the preferred reaction,[41,42)] but the π-allyl species is sufficiently reactive that exchange with propylene, and deuterium scrambling in the alkane product, is quite pronounced whereas it is absent for ethylene. Thus, there is physical evidence for the formation of a π-allyl species on zinc oxide and it is a kineiically significant species in hydrogenation and isomerization.*

Studies of the infrared spectrum and isomerization have also been carried out with butenes.[2,43)] The results are more complex for butenes than for propylene inasmuch as five surface species are expected: π-cis-butene-2, π-trans-butenes-2, π-butene-1, syn π-allyl anti π-allyl. All five species have been detected by IR studies and these results, coupled with kinetic studies, provide strong evidence that π-allyls are intermediates in the double bond migration for butenes.[43)] In order to explain the cis-trans isomerization, however, it appears that a σ-allyl, present in such small concentrations that it is not detectable, or a dynamic π-allyl[45] plays a role in the reaction.

At present, the evidence for formation of π and π-allyl species on oxides other than zinc oxide is based on mechanistic inference.[1,44] It would be imprudent to assume that all the results for zinc oxide apply to all other oxides. Yet, the observed kinetic characteristics of these physically characterized species on zinc oxide correspond closely to those kinetic characteristics which were the basis for their postulated existence on other oxides. Certainly, the species on zinc oxide must have many similarities to those on other oxides.

3. ACTIVE SITES ON METAL OXIDES

3.1 Acid-Base Pair Sites

If the active site is a metal-oxide pair, the metal half should be electrophilic and the oxide half should be nucleophilic. Accordingly, one might expect the characteristics of the site to be those of a Lewis acid-base pair. This model has been developed and applied to alumina by several investigators,[46] and Burwell's[1] "coordinatively unsaturated" sites on chromia can be viewed as acid-base pair sites. Such labels are comforting but they are worthwhile only if they have predictive value or systematize the experimental observations. The term "surface acid-base pair" is imprecise in the sense that our framework for assessing acidity and basicity is provided by behavior in solution. Extension of these concepts to adsorption at a fixed point is clearly an overextrapolation; nevertheless, if this site label is to be useful, this is what must be done. In so doing, however, let us not lose sight of the uncertainties involved.

For acid-base pairs, in the sense described above, the following reaction would be possible for any Bronsted acid, HX:

$$\text{H-X} + \text{Zn-O} \rightarrow \text{X-Zn-O-H}$$

If the OH bond is the same for all X and the Zn-X interaction is non-specific, the extent of this reaction might be expected to be determined by factors similar to those that define the acidity of HX in solution. Then strong acids would react readily; weak acids would react not at all. In a list of acids ranked according to pK, we would expect to find a critical value of pK such that acids well below this value dissociate and acids well above this value do not. We have adsorbed on zinc oxide and recorded the spectrum of the series of acids listed (according to pK*) in Table 3. We have taken the view that dissociation has occurred only if an OH band is observed. Hydrogen bonding effects are troublesome with this criterium; nevertheless, we have evidence for dissociation for all species with a pK less than 35 whereas no dissociation occurs for those acids with a pK greater than 35. This criterium[25] can pinpoint the nature of the disso-

*Combined ir and esr studies show that the π-allyl is also a reactive surface species in oxidation reactions over zinc oxide.[43] The π-allyl has been postulated[44] as an intermediate in many catalytic oxidation reactions, but this evidence for such species is largely mechanistic inference.

ciation with impressive precision. For example, we find toluene dissociates; experiments with deuterium labelled toluene reveals the hydrogen is removed only from the methyl group. In line with this benzene does not dissociate. Thus, in both this example and the ethylene-propylene case, we conclude dissociation occurs via allyl hydrogens but not via vinyl hydrogens.

TABLE 3
Acidities of Carbon Acids

Acid	Anion	pK[a]
H_2O	OH^-	16
RCH_2OH	RCH_2O^-	18
$RCOCH_2R$	$RCOCHR^-$	19-20
$H-C\equiv C-H$	$H-C\equiv C^-$	25
NH_3	NH_2^-	34
$CH_3-CH=CH_2$	$CH_2-CH=CH_2^-$	35
C_2H_4	$C_2H_3^-$	37
C_2H_6	$C_2H_5^-$	42

*These values, based, in part, on kinetic acidities, were taken from the compilations in March[47)] and Kosower.[48)] Since entries for the same compound did not agree, the values listed in Kosower were scaled to force agreement for the same compound.

3.2 Polar Character of Adsorbed Species

In terms of the acid-base model one would expect the bound hydrogen to have cationic character and the bound X to have anionic character. Thus, the π-allyl species should have anionic character. The selectivity of olefin isomerization provides a mechanistic test of this assumption. In the homogeneous, base-catalyzed butene-1 isomerization[49)] (which surely involves allyl carbanions) the initial cis-trans ratio is as high as 40. If a basic component is dispersed on silica[50)] this cis selectivity is still evident. By way of contrast over metals[4)] or acidic catalysts[51)] the cis/trans ratios are near unity. Over zinc oxide we obtain initial cis/trans ratios of about 10, which suggest that the π-allyl has carbanion character. This selectivity pattern is found for a number of oxides. The highest initial values reported for such oxides are: Al_2O_3-5,[51)] CaO-8,[52)] MgO-16[52)] and Cr_2O_3-15.[53)] For both chromia (and alumina[51)]) the ratio depends on the severity of degassing. For example, mild (~400°C) degassing of chromia yields a catalyst with an initial cis-trans ratio of 3; rigorous degassing (750°C) raises this ratio to 15; prolonged rigorous degassing (750°C) yields initial ratios of the order of 50.[17)] Thus, on the basis of the cis/trans ratios, it appears that isomerization over well degassed metal oxides probably proceeds via a surface π-allyl with appreciable anionic character. Accordingly, we might expect reactions of hydrocarbons over metal oxides to resemble those that occur in homogeneous bases.

3.3 Allene-Acetylene Isomerization

The base-catalyzed isomerization of methyl acetylene to allene is presumed to occur as follows:

*The writer is aware of the confusion regarding kinetic and thermodynamic acidity, but he has chosen to ignore it in this discussion.

$$B + CH_3\text{-}C{\equiv}CH \rightarrow BH^+ + [CH_2{\overset{\dots}{=}}C{\overset{\dots}{\equiv}}CH]^-$$

$$[CH_2{\overset{\dots}{=}}C{\overset{\dots}{\equiv}}CH^-] + BH^+ \rightarrow B + CH_2{=}C{=}CH_2$$

This reaction pathway is quite similar to that for olefin isomerization via an allyl insofar as it involves the 1,3 shift of a proton aided by a base and the intermediate (in this case the propargyl anion) involves delocalized C-C bonds. If, as suggested, zinc oxide looks like a base to hydrocarbons, it should catalyze this reaction via a propargyl intermediate. A recent study of this reaction utilizing kinetic IR techniques and deuterium tracers[54)] shows that isomerization does occur at room temperature and a propargyl species is the likely intermediate. In this instance, at least, the acid-base model for the active sites has had predictive value.

4. CONCLUSIONS

Metal oxide hydrogenation catalysts are similar in a number of respects. They all appear to have a limited number of sites, presumably metal-oxide pairs, capable of activating hydrogen. On these sites, dissociative adsorption occurs which plays a role in hydrogen-deuterium exchange. On some of these oxides, molecular hydrogen adsorbs on these same sites, and may also play a role in hydrogen-deuterium exchange, parahydrogen conversion and hydrogenation. Olefin isomerization probably occurs via a $\underline{\pi}$-allyl species with some carbanion character. Accordingly, these metal-oxide catalysts show some similarities to homogeneous base catalysts. Despite these similarities, these catalysts do have individual characteristics. Thus, although we might reasonably expect that, if zinc oxide forms a $\underline{\pi}$-allyl, chromia will also form a $\underline{\pi}$-allyl, we must be cautious in assuming that the interaction is such that these $\underline{\pi}$-allyls on different catalysts will undergo the same reactions under the same conditions. In this light it is not surprising to find that addition of deuterium to propylene yields scrambled deuterium in the product propane over zinc oxide but essentially clean dideuteropropane over chromia. In sum, metal-oxide catalysts constitute a class of similar catalysts; they are not, however, identical in their function.

5. ACKNOWLEDGEMENT

Acknowledgement is made to the donors of the Petroleum Research Fund, administered by the American Chemical Society, for support of much of this research. This research was also aided by funds from the National Science Foundation under grant no. GP 34034X.

REFERENCES

1) R. L. Burwell, Jr., G. L. Haller, K. C. Taylor and J. F. Read, Advan. Catal. Relat. Subj. 20 (1969) 1 and references therein.
2) R. J. Kokes and A. L. Dent, Advan. Catal. Relat. Subj. 22 (1972) 1 and references therein.
3) K. Tanaka, H. Nihira and A. Ozaki, J. Phys. Chem. 74 (1970) 4510.
4) G. C. Bond, Catalysis by metals (Academic Press, London, 1962).
5) T. I. Taylor, Catalysis (Reinhold Publishing Corp., New York, N.Y. 1957) Vol. V, pp. 289-316, P. H. Emmett, Ed.
6) D. D. Eley, Catalysis (Reinhold Publishing Corp., New York, N.Y. 1955) Vol. III, pp. 49-77, P. H. Emmett, Ed.
7) J. P. Bartek, Ph.D. Thesis, 1970, The Johns Hopkins University.
8) W. M. Hamilton and R. L. Burwell, Jr., Actes 2nd Congr. Int. Catal., Paris, 1960 1 (1961) 987.
9) J. J. Rooney and G. Webb, J. Catal. 3 (1964) 488.
10) R. P. Eischens and W. A. Pliskin, Advan. Catal. Relat. Subj. 10 (1958) 1.
11) R. P. Eischens, W. A. Pliskin and M. J. D. Low, J. Catal. 1 (1962) 1.
12) S. Naito, H. Shimizu, E. Hagiwara, T. Onishi and K. Tamaru, Trans. Faraday. Soc. 67 (1971) 1519.
13) G. H. Twigg, Discussions Faraday Soc. 8 (1950) 152.
14) E. F. Meyer and R. L. Burwell, Jr., J. Am. Chem. Soc. 85 (1963) 85.
15) W. C. Conner and R. J. Kokes, J. Phys. Chem. 73 (1969) 2436.
16) G. C. Bond and J. Turkevich, Trans. Faraday Soc. 49 (1953) 281.
17) W. C. Conner, unpublished results.
18) R. J. Kokes, A. L. Dent, C. C. Chang and L. T. Dixon, J. Am. Chem. Soc. 94 (1972) 4429.
19) F. H. Van Cauwelaert and W. K. Hall, Trans. Faraday Soc. 66 (1970) 454.
20) B. M. W. Trapnell, Catalysis (Reinhold Publishing Corp., New York, N.Y. 1955) Vol. III, Chapter I, P. H. Emmett, Ed.
21) C. C. Chang and R. J. Kokes, J. Am. Chem. Soc. 93 (1971) 7107.
22) N. Sheppard and D. J. C. Yates, Proc. Roy. Soc. Ser A 238 (1957) 69.
23) P. R. Monson, Jr., H. Chen and G. E. Ewing, J. Mol. Spectrosc. 25 (1968) 501.
24) C. C. Chang, L. T. Dixon and R. J. Kokes, unpublished results.
25) C. C. Chang and R. J. Kokes, unpublished results.
26) E. U. Condon, Phys. Rev. 41 (1932) 759.
27) M. F. Crawford and J. R. Dagg, Phys. Rev. 91 (1953) 1569.
28) G. Herzberg, Molecular spectra and molecular structure. I. Spectra of diatomic molecules (Van Nostrand, New York, N.Y. 1950) p. 533.
29) T. E. Stern, Proc. Roy. Soc. Ser A 130 (1930) 551.
30) D. White and E. N. Lassatre, J. Chem. Phys. 32 (1960) 72.
31) J. King and S. W. Bensen, J. Chem. Phys. 44 (1966) 1007.
32) See, for example, G. E. Ewing and S. Trajmar, J. Chem. Phys. 41 (1964) 814 and J. Chem. Phys. 42 (1965) 4038.
33) E. W. Albers, P. Harteck and R. R. Reeves, J. Am. Chem. Soc. 86 (1964) 204.
34) E. Wigner, Z. Phys. Chem. B 23 (1933) 28.
35) G. C. Bond, Catalysis by metals (Academic Press, London, 1962) p. 77.
36) B. P. Stoicheff, J. Chem. Phys. 21 (1966) 755.
37) J. Pradilla-Sorzana and J. P. Lacher, Jr., J. Mol. Spectrosc. 22 (1967) 180.
38) D. B. Powell and N. Sheppard, Spectrochim. Acta 16 (1958) 69.
39) J. W. Hightower and W. K. Hall, J. Am. Chem. Soc. 89 (1967) 778.
40) W. R. McClellan, H. H. Hoehn, H. N. Cripps, E. L. Muetterties and B. W. Howk, J. Am. Chem. Soc. 83 (1961) 1601.

41) A. L. Dent and R. J. Kokes, J. Am. Chem. Soc. 92 (1970) 1092, 6709, 6718.
42) S. Naito, T. Kondo, M. Ichikawa and K. Tamaru, J. Phys. Chem. 76 (1972) 2184.
43) R. J. Kokes, Catalysis Reviews 6 (1972) 1.
44) H. H. Voge and C. R. Adams, Advan. Catal. Relat. Subj. 17 (1967) 151.
45) G. Wilke, B. Bogdanovic, P. Hardt, P. Hambach, W. Keim, M. Kröner, W. Oberkirch, K. Tanaka, E. Steinrucke, D. Wolter and H. Zimmerman, Angew. Chem. Int. Ed. Engl. 5 (1966) 151.
46) D. J. C. Yates and P. J. Lucchesi, J. Chem. Phys. 35 (1961) 243; J. B. Peri, J. Phys. Chem. 69 (1965) 211, 231; H. Pines and J. Manassen, Advan. Catal. Relat. Subj. 16 (1966) 49.
47) J. March, Advanced organic chemistry: reactions, mechanism and structure (McGraw-Hill Co., New York, N.Y., 1968) p. 219.
48) E. M. Kosower, Physical organic chemistry (John Wiley and Sons, Inc., New York, N.Y., 1968) p. 27.
49) S. Bank, A. Schriesheim and C. A. Rowe, Jr., J. Am. Chem. Soc. 87 (1965) 3224.
50) W. O. Haag and H. Pines, J. Am. Chem. Soc. 82 (1960) 387.
51) W. O. Haag and H. Pines, J. Am. Chem. Soc. 82 (1960) 2488.
52) H. Hattori, N. Yoshii and K. Tanabe, paper 10, Vth International Congress.
53) I. Iwai, Mechanism of molecular migrations (John Wiley and Sons, Inc., New York, N.Y., 1969) Vol. 2, p. 73, B. S. Thyagarajan, Ed.
54) C. C. Chang and R. J. Kokes, J. Catalysis, in press.

Invited Lecture B

COORDINATION AND CHEMISORPTION: A NEW KIND OF PHOTOCHEMICAL EXCITATION?

R. Ugo - Istituto di Chimica Generale - Milan University (Italy)

There are no doubts that chemists have been fascinated in the last few years by one particular aspect of the growing Science of Catalysis, which involves the investigation of the electronic aspects of coordination and chemisorption. The final goal is to give a rational answer to a few questions such as how and why molecules increase or change their reactivity by interaction with the catalyst. Many different attempts have been made to obtain this goal: the easiest and perhaps the most exciting is the so called chemical approach to catalysis.
In the last years, we have had an increasing tendency to relate intermediates formed on the surface of some heterogeneous catalysts to coordination or organometallic complexes by considering chemisorption on metals (mainly transition metals)[1] and on metal oxides (again particularly transition metal oxides)[2] as a localised coordination process from the gas phase to the vacant sites of the solid surface.
Of course, heterogeneous catalysis is not such an easy problem to be fully satisfied by this simple description, however by this approach the large amount of informations available from solution and solid state physicochemical studies on transition metal complexes may be usefully extended and used to explain some aspects of the reactivity of chemisorbed molecules and, consequently, to give a better insight into the microscopic mechanism of catalysis.
This paper will deal with the discussion on the electronic perturbation related to the unusual catalytic reactivity of coordinated or chemisorbed molecules. Such electronic perturbations are often so strong, as we will discuss later, that the usual way of discussing the reactivity of coordinated or chemisorbed molecules in terms of their ground state geometries and electronic structures may be dangerous. In fact, except a few cases, the interaction of different molecules with surfaces or metal complexes cannot be considered as a simple perturbation or a weak polarisation but as a rather complex electronic interaction where exchange terms are often the most important. The actual qualitative theory of bonding between a coordinated molecule and a metal atom or ion is derived from a simple, symmetry based, molecular orbital approach. We can take as example the case of olefins using the Chatt-Dewar picture of the bonding (Fig.1)[3].

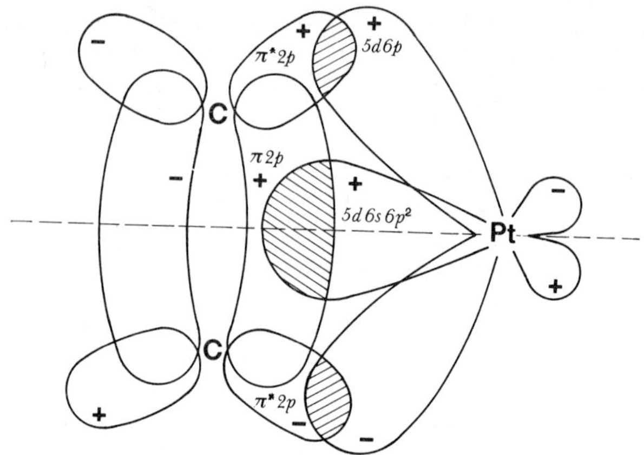

Fig.1 The platinum-olefin bonding

The σ donor pair of the ligand interacts with an empty metal orbital of σ symmetry, giving place to a low energy σ bonding molecular orbital in which the electron donor pair is located. This process corresponds in first approximation to a charge-transfer from the ligand to the metal. However, when the ligand has empty low lying orbitals of π symmetry, an electron back-donation can take place from the metal to the ligand via a π molecular orbital. In conclusion, in metal complexes we have an electronic distribution arising from a series of electron transfer processes as metal to ligand and ligand to metal which however does not necessarily bring to compensation. For instance, very important investigations by ESCA on carbonyl[4] and olefin complexes[5,6] of transition metals in low oxidation state, have indicated the localisation of a negative charge on the ligands.Besides, some authors have indipendently studied the reversible addition of covalent molecules to low oxidation state metal complexes, e.g. the so called oxidative addition to Vaska's compound (Fig.2)

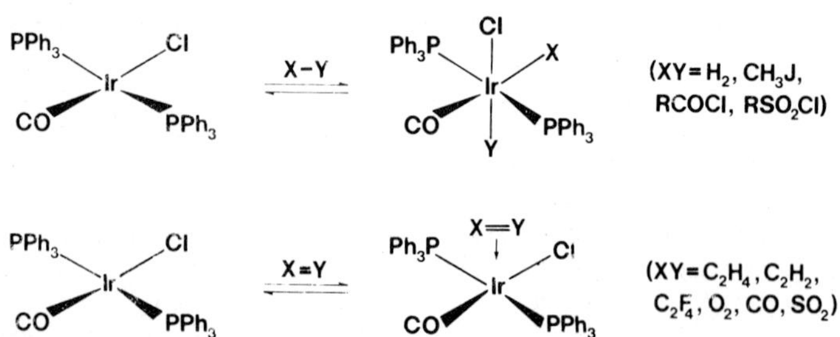

Fig. 2 Some oxidative additions to Vaska's compound

By considering the amphoteric character of the covalent addendum, these authors concluded that this reaction may be considered as an acid-base interaction, where the d metal electronic density behaves as a donor base[1,7]. Of course, such a situation is typical of so called "soft" low oxidation state transition metal complexes, particularly of the second and third row, where π back-bonding is rather extensive. When the bonding occurs with a metal ion in high oxidation state, which may be classified as "hard" centre[8], occupied π-orbitals of sufficiently low energy are not available on the metal and consequently π back-bonding from the metal to the ligand is usually less important. In this latter case the bonding is either only electrostatic or slightly covalent but with a directional σ charge transfer from the ligand to the metal.

Consequently, in a general discussion on homogeneous catalysis[8] two classes of catalysts may be considered: soft catalysts involved in the interaction with soft centres (unsaturated hydrocarbons, carbon monoxide, oxygen etc.) and hard catalysts involved in the interaction with harder centres (hard acids and bases, nucleophilic and electrophilic points of attack on some organic molecules etc.). Of course, such a classification is rather naif, being very difficult to define clearly the border line between the two classes. However for sake of simplicity I would like to extend it to heterogeneous catalysts where as well we have soft surfaces (transition metals and their alloys, some transition metal oxides having metal-type conductance) and hard or less soft surfaces (mainly oxides, acidic and basic catalysts, semiconductors). In the first case chemisorption can be best compared to coordination in low oxidation state metal complexes and in the latter case to coordination in high oxidation state complexes, where the effect of the bonding with the metal ion can be discussed in terms of crystal field theory[9]. These comparisons, which are the basis of the chemical approach to heterogeneous catalysis, are not only formal, as one could suppose at first sight, but they have been continously supported by new chemical and physical evidences. For instance, when we compare the energy of metal-hydrogen bond formation (Table 1), either by adsorption on a clean metal surface or by addition to a metal complex, we find values which are not only similar, but also very little dependent from the d electronic properties of both the metal surface and the metal complex. This point seems to find additional support from the vibrational spectra of the hydrogenated species[11]: in some two dozen of isostructural derivatives of type $H_2IrX(CO)L_2$ each of the two iridium-hydrogen stretching frequencies (ν_{Ir-H_1} at c.a. 2200 cm^{-1} and ν_{Ir-H_2} at c.a. 2100 cm^{-1}) varies only within 40 cm^{-1}. Now ν_{Ir-H}'s for chemisorbed hydrogen on alumina-supported metals have been reported to be in the same spectral range (2120 and 2050 cm^{-1} for iridium, 2105-2050 cm^{-1} for platinum[1,11]).

In conclusion, the strength of the hydrogen-metal bonding is not only invariant (within a few kilocalories) all through a series of metal surfaces and metal complexes, but also shows values completely similar in both cases. Hydrogen is also a covalent molecule little demanding from the electronic point of view, and consequently hydrogen activation by transition metal

Table 1 - Energies of metal-hydrogen bond formation

Metal	E_{M-H} (kcal mole^{-1})	Complex	E_{M-H} (kcal mole^{-1})
Ir, Rh, Ru	c.a. 64.6	$Ir(PPh_3)_2(CO)X$	64
Pt, Pd	c.a. 65.6	$[Co(CN)_5]^{3-}$	57
Co	63.6	$(\pi C_5H_5)_2Mo$	c.a. 58
Fe	68.5	$IrL_2(CO)X$	c.a. 63
Ni	67		

X = Cl, Br, I L = PR_3

surfaces is a "facile reaction" by Boudart's definition[12]. A recent kinetic investigation[13] has indicated that this characteristic seems to be maintained in the case of hydrogen reaction with transition metal complexes. In fact in the case of the reaction:

$$Ir(CO)(Cl)\left[P(pC_6H_4X)_3\right]_2 + Y_2 \longrightarrow IrY_2(CO)Cl\left[P(pC_6H_4X)_3\right]_2$$

$$Y_2 = H_2,\ O_2; \quad X = OCH_3,\ CH_3,\ H,\ F,\ Cl$$

we have obtained for both hydrogen and oxygen (Fig.3) pseudo-Hammett relationships corresponding to the equation

$$\log \underline{K}_{25} = \rho\,\sigma_p + \beta\ .$$

The hydrogen reactivity is not too much affected (ρ = -0.7) by the real electron donor density localised on the iridium atom (measured through Hammett's constants) whilst in the case of oxygen reactivity we have observed a much greater influence (ρ = -2.2) of the π donor properties of the iridium atom.
In conclusion, many close similarities have been disclosed in the coordinative reactivity of metal surfaces and metal complexes when a simple reaction such as hydrogen activation is considered. These similarities would thus strongly support the refined model of metal surfaces proposed by Bond[14], in which localised and "individual" orbitals may be used for bond formation with chemisorbed species as in simple metal complexes. In Bond's approach, the electronic states of transition metal surfaces are considered, on the ground of the crystal lattice perturbation of the d metal orbitals, as formed by largely delocali

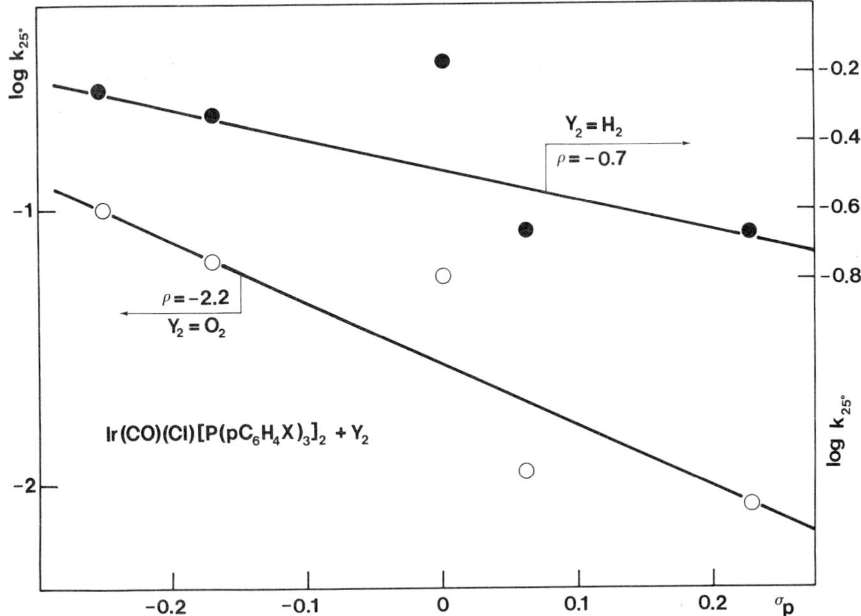

Fig.3 Pseudo-Hammett relationship in oxidative addition reactions.

sed d bands of t_{2g} symmetry together with d localised orbitals of e_g symmetry, which are emergent from metal surfaces (Fig. 4). These latter localised orbitals are generally partially empty and together with p_z orbitals may thus form the vacant sites of chemisorption.

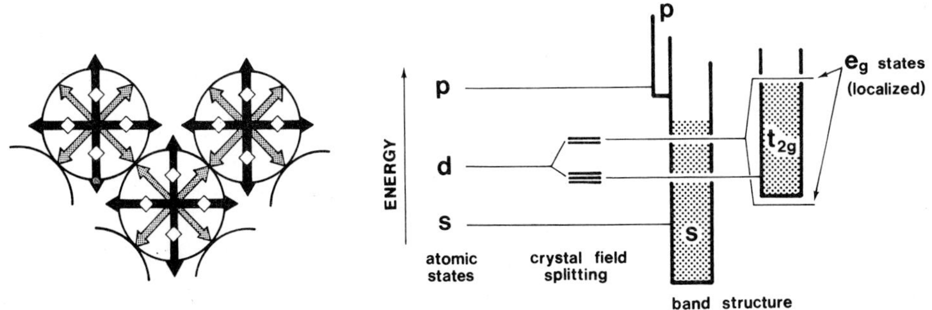

Fig. 4 Diagrammatic emergence of orbitals from (100) surface and electron energy bands and states of a f.c.c. transition metal after Bond (see ref. 1).

As a matter of fact, it was already pointed out few years ago[1] that such electronic description is qualitatively similar, as far as localised orbitals are concerned, to that proposed in the case of square planar tetragonal d^8 complexes or trigonal d^{10} complexes. Of course Bond's description of electronic surface states of transition metals was mainly qualitative and based on the achievements of solid state physical properties, but no direct evidence or detection of these localised orbitals was given. The application of new physical techniques has recently helped to clarify some of these uncertainties and to show that spatial arrangement of some chemisorption complexes resembles the structure of complex ions of transition metals. For instance, Muller and coworkers[15] have used Bond's model for the interpretation of field ionization process, where field ion images were interpreted as projections of regions where the fully occupied orbitals of inert image gas atoms overlap with the localised orbitals of the surface metal atoms.
A more direct evidence of well localised surface orbitals has been obtained from studies on chemisorbed carbon monoxide. Carbon monoxide chemisorption on metal surfaces has been first investigated by infrared spectroscopy in the case of many transition metals[16]. The absorption bands, corresponding to CO stretching, show the presence of both terminal (c.a. 2000 cm^{-1}) and bridging (c.a.1900-1800 cm^{-1}) chemisorbed carbon monoxide, whose energies are completely similar to those of bonded carbon monoxide in simple cluster carbonyl compounds[17]. Such infrared studies thus provided evidence for a close similarity of the bonding of carbon monoxide in surface species and in metal carbonyls, but did not permit the definition of the geometrical arrangements of the adsorbate and adsorbents. Only very recently Mason and coworkers[18] have developed an adeguate Fourier analysis of LEED intensities which was applied to investigate the surface cristallography of carbon monoxide chemisorption on (100) platinum surface. The final structure (obtained with an accuracy of 0.1-0.2 A), reported in Fig.5, is identical with that predicted by Bond (particularly in the case of hydrogen adsorption). This author pointed out in fact that the direction of emergence of the e_g and t_{2g} metal orbitals from the (100) face of a face-centred-cubic metal requires that a bridging ligand must occupy a site of four-fold symmetry better than one of two-fold symmetry.

In Fig. 5 the structure of a transition metal cluster compound (where multicentre metal bonded carbon monoxide ligands are also present) has been reported just to show the very close structural similarities between metal clusters and covered metal surfaces. Such similarities are not only geometrical, but also electronic, as it may be already inferred from the values of carbon monoxide stretching frequencies. Mason and coworkers[18] have

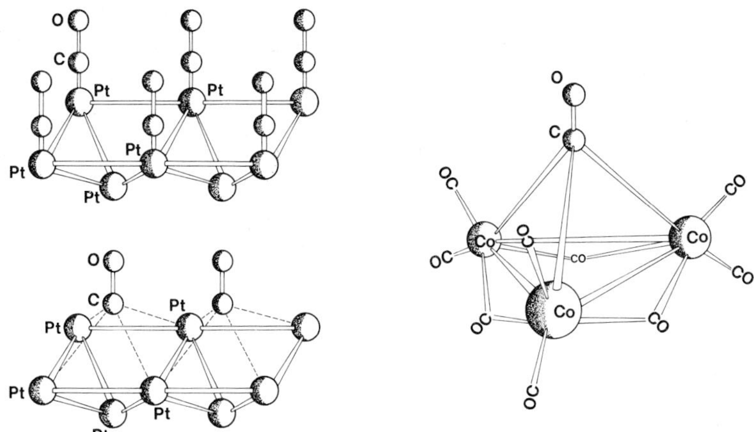

Fig. 5 Surface crystallography of carbon monoxide chemisorption on (100) platinum surface and the structure of $Co_3(CO)_{10}^-$

studied by X-ray induced electron emission spectroscopy the carbon monoxide adsorption on (100) platinum surface. They concluded that this adsorption leads to a net electron transfer from the platinum to the ligands as it was observed in many transition metal carbonyls.

In conclusion, I hope to have substantially demonstrated with these few examples that coordination and chemisorption correspond to similar electronic interactions, where the cooperative electronic properties of the solid play a minor role. This statement is certainly true for metal surfaces, where we have the largest direct evidence, but presumably we can easily extend it to many transition metal oxides, where however we have only some dispersed spectroscopic data of localised electronic surface states and chemisorption bonding[2] and a less simple description of the electronic surface states[19].

The extention to non transition compounds and to semiconductors is definetively more doubious.

In the second part of this paper I will now try to use the informations available from studies on metal complexes to interprete the activation of chemisorbed molecules and some aspects of their reactivity.

The major emphasis will be given to aspects involving soft interactions only, as they cover a large part of catalytic homogeneous and heterogeneous reactions (e.g. catalysis of olefins, aromatic and acetylenic compounds, carbon monoxide etc.).

Now if we consider again the Chatt-Dewar approach (Fig.1) of the bonding of an olefin to a transition metal, we can see that the interaction of the olefin with the transition metal corresponds to a flow of the π electron density of the olefin into its π^* antibonding orbital which is populated by the metal d electron

density via the π back-bonding mechanism. In few words, by the help of the metal we obtain in first approximation the following excitation: $\pi^2 \pi^* \longrightarrow \pi^1 \pi^{*1}$.
Recently X-ray investigations[24] have shown that large distortions of the molecular geometry occur in soft molecules when they are coordinated to a metal complex. The amount of distortion depends on the type of metal and its d electronic density, related to factors such as oxidation state and the nature and number of the ligands of the coordination sphere. These are the same factors which affect π back-donation. In order to accomodate these factors it was proposed[20] that the part of the molecular orbital function, localised on the coordinated soft molecule and which describes the bonding in these complexes, would contain functions representing excited states of both metal and ligand. In particular the charge density is a weighted sum of densities associated with the fragments in their various individual states. The terms in which the ligands are in their first excited triplet states, appear with largest weightings and lead to a ligand charge distribution very similar to that of an isolated ligand molecule in its first excited triplet state. The geometry of a ligand will therefore spontaneously change on coordination, the forces acting on the nuclei being more nearly those of the triplet excited state than those of the ground state. However other excited states, like singlet states, also contribute to the final geometry and in a recent paper[21], where coordination of butadiene to $Fe(CO)_3$ moiety has been considered, the partecipation of all the excited states has been proposed. The geometries of low lying excited states of many soft small molecules are quite well known and have been obtained from hyperfine rotational and vibrational structure of molecular electronic spectra[22]. In other cases, when the resolution of these spectra is too low, the symmetry and geometry of excited

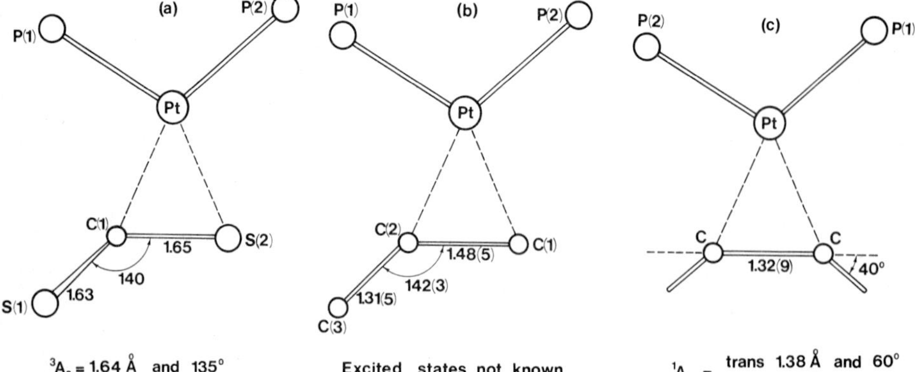

Fig.6(a) Structure of $Pt(PPh_3)_2CS_2$; (b) Structure of $Pt(PPh_3)_2C_3H_4$; (c) Structure of $Pt(PPh_3)_2PhCCPh$.

states may be qualitatively obtained by a simple application of the Walsh rules[23]. I will discuss now a few examples only, although the actual structural evidence is extraordinarily rich[24].
Carbon disulphide in its electronic ground state is linear with a C-S bond length of 1.55 Å; the lowest excited triplet state ($^3A_2$) of CS_2 is bent, the angle being 135° and the mean C-S bond length 1.64 Å. These values are identical, within experimental error, to those found in Pt(0) or Pd(0) addition complexes $(PPh_3)_2MCS_2$ (M = Pt, Pd) (Fig.6a).

The distortion presumably derives in the complexes from the large overlap (and consequently large π back-donation) of the occupied metal d orbitals and the π* antibonding orbitals of carbon disulphide.
Allene $CH_2{=}C{=}CH_2$ is another small molecule similar to CS_2, but the geometries of its excited states are not known in great detail. The Walsh rules state that allene in its excited states must be bent with an unsymmetrical distribution of the π bonds. This is the geometry observed in $(PPh_3)_2Pt(C_3H_4)$ (Fig.6b). Assuming for the lower excited state the geometry and bond lengths obtained from this latter structure, Zocchi and coworkers have recently reinterpreted with satisfactory agreement the molecular electronic spectrum of allene[25]. A few other structures of allene complexes have been recently reported and in all of them the unsaturated hydrocarbon is not linear or symmetric (table 2).

Table 2 - Structural data of allene complexes

Compound	Bending angle	Mean C-C length
$Rh_2(CO)_2(acac)_2(C_3H_4)$	144.5	1.39 ± 0.02
$Rh(acac)\{C_3H_2(CH_3)_2\}_2$	152.6 ; 153.3	1.35 ± 0.05
$Rh(acac)\{C_3(CH_3)_4\}_2$	147.2 ; 148.9	1.35 ± 0.03
$[Cl_2Pt(C_3(CH_3)_4)_2]2CCl_4$	151	1.36 ± 0.01
$(PPh_3)_2PtC_3H_4$	142	1.40 ± 0.08

The distortion values fluctuate from a bending angle of 142°±3 and a mean C-C length of 1.40 ± 0.08 Å (in the case of the Pt(0) complex) to 155°-150° and about 1.35 ± 0.01 Å respectively (in the case of Rh(I) and Pt(II) complexes). The differences of the C-C distances between the free and the coordinated double bond of allene is lower in Rh(I) and Pt(II)(1.30-1.36 Å and 1.37-1.41 Å respectively) than in the Pt(0) complex (1.31Å and 1.48 Å), meaning that the contribution of an excited state to the overall electron density of the coordinated allene is lower in Rh(I) and Pt(II) than in Pt(0) complexes. This is in accord with the π donor properties of the metal atom, which usually increase by decreasing the formal oxidation state of the metal.

An analogous effect has been observed in complexes of acetylenic compounds, where the acetylenic moiety is not longer planar (in Fig.6c the structure of $(PPh_3)_2PtPh_2C_2$ has been reported as an example). The molecule is bent in a cis manner, the C≡C-R bond angle averaging, in zerovalent and low oxidation state complexes (table 3), 140° and the C≡C bond length 1.30-1.33 Å.

Table 3 - Structural data of π-acetylenic (RC≡CR) complexes.

Metal system	R	Bending angle RC≡C	C≡C lenght
—	H	180	1.202
Pt(II)Cl_2(p-toluidine)	$C(CH_3)_3$	163	1.24
Ni(O) $(CH_3)_3NC_2$	Ph	149	1.28
W(O) (CO)	Ph	140	1.30
Ir(I) X(CO) $(PPh_3)_2$	CN	140	1.29
Pt(O)$(PPh_3)_2$	Ph	140	1.36±0.03

The first excited state of C_2H_2 (^1A_u) has a trans bent structure (120° and 1.38 Å being respectively the molecular parameters) whilst the corresponding cis bent excited state (which in free acetilene is slightly destabilised by a non-bonded repulsion interaction) has an angle of 142° and a C≡C bond length of 1.38 Å. These values are very similar to those observed in low oxidation state metal complexes. When the metal has a higher oxidation state (Table 3) both the distortion from the linearity and the increase of the C≡C bond length are lower, as a consequence of the decreased π back-bonding properties of the metal.

We know many structures of π-olefin complexes; in all the structures reported we can observe a certain distortion from the planarity together with an increase of the C=C bond length sometimes up to 1.5 Å [26]. Although this increase can be equally attributed to σ donation to the metal or to π back-donation from the metal, the distortion from planarity may be related only to some population of π* antibonding orbital of the olefin.

It is rather difficult to draw a direct and easy correlation between the geometries of the excited states of the ethylene and those of the olefinic moieties in metal complexes. Some spectroscopic evidence[27] indicates excited states of the ethylene with not only a longer C=C distance (up to 1.69 Å against 1.34 Å) but also with a geometry in which the molecule distorts into a pyramidal configuration for the atoms surrounding each carbon atom.

In the lowest singlet excited state of ethylene (^1B_u), with one electron occupying the antibonding π^* orbital, the most stable configuration is qualitatively defined as one in which the planes, corresponding to the CH_2 groups, are rotated around the C-C bond, at right angles to one another. This is in keeping with the theoretical expectations, since the destabilizing influence of an electron into the π^* antibonding orbital may correspond to the original p_π carbon orbitals with a zero overlap as when the two CH_2 groups are at right angle to one another. This is not the essential stereochemistry of olefin complexes, in which the C=C bond length increases but the olefinic carbon atoms become nearly tetrahedral. However the different geometry of the deformation, which also corresponds to a decreased p_π overlap, may be attributed to some steric hindrance due to the presence of the metal moiety.

Quite clearer is the effect of coordination on the geometry of a delocalised olefinic system, such as butadiene or quasi-butadiene[24]. We have two classes with a not always well defined border line (Fig. 7).

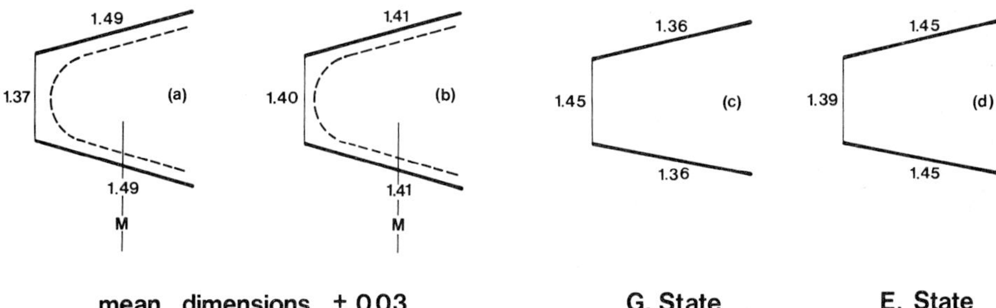

Fig. 7 Geometries of coordinated butadiene fragment and of ground and lower excited state of butadiene.

In class (a) the butadiene fragment is trans to poor π acceptor ligands (such as π-cyclopentadienyl, olefins), in class (b) it is trans to strong π acceptors such as carbon monoxide. The ligands in trans position obviously compete with the butadiene fragment for the d electron density of the metal available for back-donation. In conclusion, we have a strong deviation (Fig.7) from the ground state geometry of the conjugate dienes only when π back-bonding to the butadiene moiety takes place. The final geometry of class (a) complexes is not too far from that of the first excited state of butadiene calculated with a simple Hückel molecular orbital approach (Fig. 7). Therefore, we can say that in class (b) the final geometry roughly corresponds to a mixing of the ground and lowest excited state of free butadiene.

Some authors[28] have criticised the use of excited state structures to explain the type of bonding, particularly in the case of olefinic ligands.

However the discussion of the bonding of molecular oxygen to low oxidation state metal complexes[24], where ionic structures have been proposed as well, gives credit to the use of excited state functions.

Many structures of oxygen adducts of transition metal complexes have been so far reported: in Fig. 8 it is given the O-O bond length of a series of isosteric Ir(I) and Rh(I) complexes, where the metal atom has different π donor properties. The increase of the O-O bond length is parallel to the increase of the basicity of the metal. Besides the structure in which the O-O bond length is 1.65 Å very clearly indicates the shortcomings in discussions of the coordinated oxygen molecule in terms of O_2^{n-} (O_2^- is 1.28 Å, O_2^{2-} is 1.49 Å). If the bonding of oxygen is similar to that of π-olefin complexes, we could imagine that its structure would approximately correspond to the excitation $\pi^4\pi^{*2} \rightarrow \pi^3\pi^{*3}$ in the spin-paired oxygen molecule. The precise amount of electron density transferred to antibonding orbitals of oxygen via the intermediacy of the metal depends on the π donor properties of the complex related to the type of metal and to the remaining ligands of the coordination sphere.

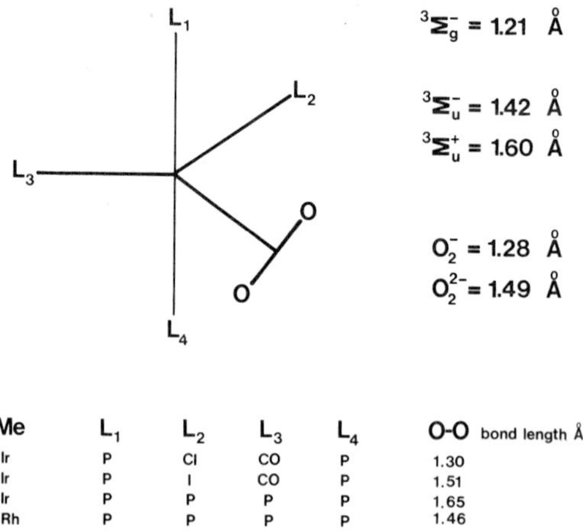

Me	L_1	L_2	L_3	L_4	O-O bond length Å
Ir	P	Cl	CO	P	1.30
Ir	P	I	CO	P	1.51
Ir	P	P	P	P	1.65
Rh	P	P	P	P	1.46

Fig. 8 O-O bond length of a series of isosteric Rh(I) and Ir(I) complexes with different coordination sphere.

The two identified triplet excited states arising from the $\pi^3\pi^{*3}$ configuration are $^3\Sigma_u^+$ and $^3\Sigma_u^-$. The 0-0 distance in the $^3\Sigma_u^+$ state is 1.42 Å, but is about 1.60 Å in the $^3\Sigma_u^-$ state, that is not too far from the value of 1.65 Å. In conclusion, we have a large structural evidence that soft molecules coordinated to soft metal centres (low oxidation state complexes, metal surfaces etc.) may assume electronic structures and geometries very similar to those of their excited states. This is not the case when harder centres of coordination are involved; in fact the structures of molecules coordinated to high oxidation state complexes do not show distortions from their ground state geometries, unless when the polarising effect of the electric field of the hard coordination centre is very high (as with the proton, for instance). In these latter cases, the polarisation of the coordinated molecule may correspond to a small mixing of very low lying excited states which gives place to this distortion.

As a consequence of the previously discussed similarities between coordination and chemisorption we can thus reasonably propose that when a molecule is chemisorbed or coordinated to a metal, some aspects of its reactivity may be related to the reactivity of its electronically excited states. We have indeed many examples of catalytic reactions which agree with this assumption.

Irreversible poisoning in mild conditions of metal surfaces and of homogeneous catalysts by sulphur compounds is a well known experimental fact; however, the mechanism of the irreversible poisoning is still uncertain. We have actually some evidence, obtained from investigations on the interaction of mole

Photochemical reaction	Homogeneous reaction	Heterogeneous reaction
$H_2S \xrightarrow[h\nu]{\lambda = 2000-2550\ \text{Å}} HS + H$	$Pt(PPh_3)_2 + H_2S \rightleftharpoons Pt(PPh_3)_2(H_2S)$ $\rightleftharpoons (PPh_3)_2Pt(H)(SH) + O_2 \rightarrow$ $\rightarrow [(PPh_3)_2PtS]_2 + H_2O$	$H_2S \rightarrow$ M–M $\rightleftharpoons$ H, SH on M–M $\xrightarrow{O_2}$ S bridging M–M
$CS_2 \xrightarrow[h\nu]{\lambda = 2100-2400\ \text{Å}} CS + S$	$Rh(PPh_3)_2Cl + CS_2 \rightleftharpoons Rh(PPh_3)_2(CS_2)Cl$ $\xrightarrow{\text{sulphur acceptor}} Rh(PPh_3)_2(CS)Cl$	M–M $\xrightarrow{CS_2}$ S≡C=S on M–M $\xrightarrow{-CS}$ S bridging M–M

cules such as CS_2 [29], H_2S [30] with low oxidation state complexes, that the thermal irreversibility of the poisoning is probably related to a reversible desulphuration (CS_2 case) or dehydrogenation (H_2S case) reaction (induced by the electronic excitation due to chemisorption or coordination), followed by an irreversible reaction of the electronically excited sulphur species.

Hydrogen activation (for instance the reaction of hydrogen-deuterium exchange or of hydrogenation using either hydrogen or water as hydrogen source) may be also explained with this approach:

Photochemical reaction	Homogeneous reaction	Heterogeneous reaction
$H_2O \xrightarrow[h\nu]{\lambda = 1356-2420\ \text{Å}} HO + H$	$Pt(PEt_3)_3 + H_2O \rightleftharpoons [PtH(PEt_3)_3]OH$	$M{-}M \xrightarrow{H_2O} M(H){-}M(OH)$
$H_2 \xrightarrow[h\nu]{\lambda \simeq 1000-1500\ \text{Å}} 2H$	$Pt(PEt_3)_3 + H_2 \rightleftharpoons (PEt_3)_3PtH_2$	$M{-}M \xrightarrow{H_2} M(H){-}M(H)$
$R{-}CH_2{-}CH{=}CH_2 \xrightarrow[h\nu]{\lambda \simeq 1500\ \text{Å}} R{-}CH{-}CH{-}CH_2 + H$	$R{-}CH_2{-}CH{=}CH_2 + MCl_2 \xrightarrow{-HCl} [(R)CH{-}CH{-}CH \rightarrow M(Cl)]_2$, M = Pt, Pd	π-allyl surface species (e.g. olefin isomerization)
$C_6H_6 \xrightarrow{h\nu} C_6H_5 + H$	$(\pi C_5H_5)_2W + C_6H_6 \rightleftharpoons (\pi C_5H_5)_2W(C_6H_5)H$	Hydrogen-Deuterium exchange of aromatics

Activation of molecular hydrogen is a more facile reaction than water activation in the case of both homogeneous metal complexes and heterogeneous reactions where milder conditions are required when hydrogen is used. This probably reflects the higher tendency of hydrogen to reach excited states and to react photochemically when compared with water.
Another kind of hydrogen activation involves hydrogen abstraction from π delocalised organic molecules (olefins, aromatic hydrocarbons). This is an important step of many catalytic reactions such hydrogen-deuterium exchange in π unsaturated hydrocarbons, isomerisation, dimerisation and oxidation of olefins and so on.

Although the photochemical reactivity of these π unsaturated molecules is not certainly simple and completely known[31], we can draw also in this case a quite satisfactory parallelism between photochemical reactions and formation of surface or coordination species.
We have investigated[1] a little more in detail the hydrogen abstraction from the allyl position of an olefin. This abstraction has been proposed for quite a long time as an important step of many catalytic reactions of olefins (selective oxidation, isomerisation, dimerisation etc.). This step has been recently investigated by Böneman[32], who demonstrated by n.m.r. spectroscopy the existence of a thermal equilibrium (Fig.9).

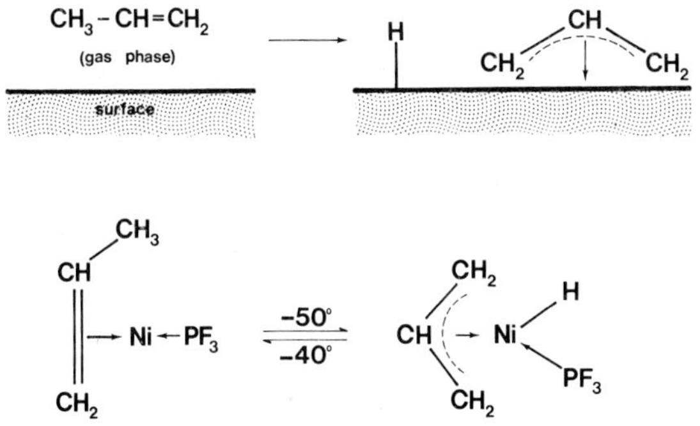

$CH_3-CH=CH_2$ (gas phase) → surface: H + CH_2–CH–CH_2

CH_3–CH=CH_2 → Ni ← PF_3 ⇌ (−50°, −40°) CH_2–CH–CH_2 → Ni(H)(PF_3)

Fig. 9 The hydrogen abstraction from the allyl position of an olefin.

We have thus tried to evaluate with a semiquantitative approach the aptitude of metals to form surface π -allyl species. The energy changes corresponding to the formation of π allyl surface species from olefins are reported in Fig. 10. The first term gives the dissociation energy of the carbon-hydrogen bond of the methyl group, followed by the rehybridization of the carbon atom; the second term gives the change of all the π energy on going from propylene to allyl radical.

On a metal surface we have besides the formation of the metal-hydrogen bond. The states A and B of Fig. 10 reporte respectively the initial state, that is the olefin and the metal not interacting at the minimum distance, and the final state corresponding to the surface allyl species. The reaction coordinate represents the movement of the hydrogen atom and of the olefin from situation A to situation B. The curve b, which represents the formation of the metal-hydrogen bond, can be assumed to be constant for metals of group VIII. Indeed the energy of M-H bond formation is always around 65 ± 2 kcal/mole (Table 1). In Fig. 10 a_1 refers to free propylene without the metal, a_2 to propylene interacting with a metal surface with partial occupation of its π^* antibonding orbital.

	$-\Delta E_{\pi}$ (β units)	X_I(eV)
propylene	0.25	–
propylene exc.	1.31	–
Co ads. propylene	1.30	6.67
Pd ads. propylene	1.11	10.91
Pt ads. propylene	1.10	12.01
Ni ads. propylene	0.68	10.51

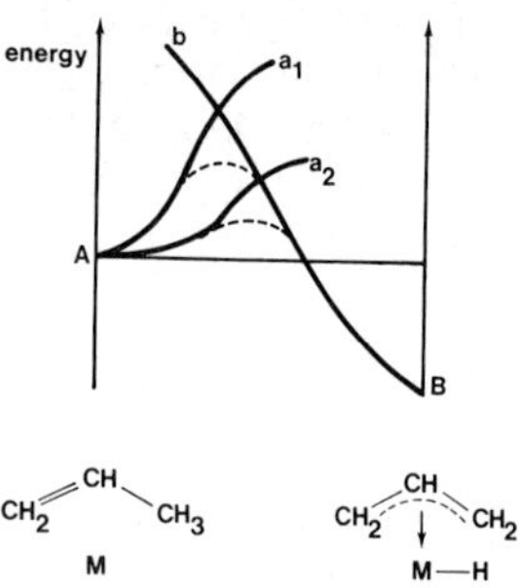

$$\Delta E = D_{C-H} + \Delta E_{\pi}$$

Fig. 10 Energy changes in the reaction path of formation of allyl species.

An approximate calculation of ΔE_{π} can be performed with an extended Hückel method on the π allyl metal complex, taking into account σ and π interactions. The results of this simple calculation are reported in Fig. 10 (where the values of ΔE_{π} are given in $\beta = H_{cc}$ units) together with values of the ionisation potentials of surface states which have been obtained with large approximations[1]. Despite that, our calculations

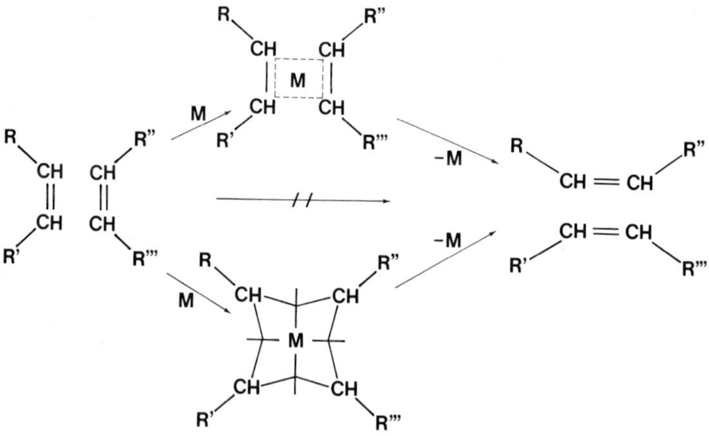

Fig. 11 The disproportionation of olefins

confirm that the interaction with the metal contributes to make easier the formation of π allyl radicals. It is noteworthy also that the strongest effect occurs with cobalt, which is one of the best isomerisation catalysts.

Finally, I would like to mention an example in which is evident a direct connection between photochemistry and catalisys. The olefin disproportionation reaction (Fig.11) is thermally forbidden but photochemically allowed on the basis of the Woodward-Hoffman rule.
However this reaction takes place smoothly in the presence of both homogeneous and heterogeneous transition metal catalysts. Two different interpretations have been proposed in order to explain the role of the transition metal (Fig.11)

Mango[33] has developed a symmetry based approach in which a quasi-cyclobutane intermediate is formed, whilst Pettit[34] has proposed the formation of carbene surface species. Although it appears that this latter interpretation has more experimental evidences, in my opinion one general point rises to attention: in both approaches the transition metal presents, via π back-bonding, a template of atomic orbitals in which the electron pairs of the reacting olefins can interchange and flow into the required regions of space which, without the metal orbitals, cannot be populated. In few words, the presence of a transition metal corresponds to a photochemical excitation of the π electrons of the olefins.
I will thus conclude this paper by saying that many other examples could be added to extend the similarities between some aspects of catalysis and photochemistry[35].
Of course, such analogies could be only accidental or formal, however I believe that they may be the basis of some useful work-hypotheses for the future developments of the Science of Catalysis.

AKNOWLEDGMENTS.
I wish to thank Italian C.N.R. for financial support and Prof. R. Mason for results given before publication.

REFERENCES

1. S. Carrà and R. Ugo, Inorg.Chim. Acta Rev. , 1, 49 (1967) and references therein
2. S. Carrà, R.Ugo and L. Zanderighi, Inorg.Chim.Acta Rev., 3, 55 (1969) and references therein
3. J. Chatt, J. Chem. Soc., 2939 (1953)
4. I.H. Hillier and V.R. Saunders, Mol. Phys., 22, 1025(1971)
5. C.D. Cook, K.Y.Wan, V.Gelius, K.Hamrin, G.Johannson, E. Olsson, K. Siegbahn, C. Bordling and K. Sigbahn, J. Am. Chem. Soc., 93, 1904 (1971)
6. R. Mason, D.M.P. Mingos, G. Rucci and J.A. Connor, Proc. Roy.Soc.(in press)
7. L. Vaska, Acc. Chem. Res., 1, 335 (1968)
8. R.Ugo, Chimica e Industria (Milan), 51, 1319 (1969)

9. D.A. Dowden, Catalysis Rev., 5, 1 (1971)
10. G.C.Bond, "Catalysis by metals" Academic Press (1962)
11. L. Vaska and M.F. Werneke, Trans. New York Acad.Sci., 33, 70 (1971)
12. M. Boudart, A. Aldag, J.E. Bensen, N.A. Dougharty and C.G. Harkings, J. Catalysis, 6, 92 (1966)
13. R. Ugo, A. Pasini, A. Fusi and S. Cenini, J. Am.Chem.Soc., (in press)
14. G.C. Bond, Disc. Far. Soc., 200 (1966) and references therein
15. Z. Knor and E.W. Müller, Surface Science, 10, 21 (1968)
16. See for instance M.L. Hair, "Infrared spectroscopy in surface Chemistry" M. Dekker (1967)
17. P. Chini, Inorg. Chim. Acta Rev., 2, 31 (1968)
18. T.A. Clarke, R. Mason and M. Tescari, Proc. Roy.Soc. (in press)
19. See for instance P. Mark, Catalysis Rev., 1, 165 (1967)
20. R. Mc Weeny, R. Mason and A.D.C. Towl, Disc.Far. Soc., 47, 20 (1969)
21. P.G. Perkins, I.C. Robertson and J.M. Scott, Theoret.Chim. Acta 22, 299 (1971)
22. G. Herzberg "Electronic spectra and electronic structures of polyatomic molecules", Van Nostrand (1966)
23. A.D. Walsh, J.Chem. Soc., 2260 (1953)
24. R. Mason, Nature, 217, 543 (1968) and references therein
25. A. Albinati, F. Maraschini and M. Zocchi, to be published
26. See for instance L. Manojlović-Muir, K.W. Muir and J.A. Ibers, Disc. Far. Soc., 47, 84 (1969)
27. R.N. Dixon "Spectroscopy and structure", p.160, Methuen (1965)
28. J.N. Murrel, Disc. Far. Soc., 47, pag.59 (1969)
29. M.C. Baird and G.Wilkinson, J.Chem.Soc.(A), 865 (1967); Chem. Comm., 267 (1966)
30. R. Ugo, G. La Monica, S.Cenini, A. Segre and F. Conti, J.Chem. Soc. (A), 522 (1971)
31. J.G. Calvert and J.N.Pitts Jr., "Photochemistry",J. Wiley (1966)
32. H. Bonemann, Angew.Chem. Int. Ed., 9, 736 (1970)
33. F.D. Mango and J.H. Schachtschneider, J. Am.Chem.Soc., 93, 1123 (1971)
34. G.S. Lewandos and R. Pettit, Tetrahedron Lett., 11, 789 (1971)
35. D.N. Shigorin and V.G.Plotnikov, Russian J. Phys. Chem., 40, 160 (1966)

Invited Lecture C

REFINED DEUTERIUM TRACER METHODS IN THE MECHANISTIC STUDIES OF HETEROGENEOUS CATALYSIS

KOZO HIROTA
Department of Chemistry Faculty of Science, Osaka University, Toyonaka, Osaka, 560 Japan*

ABSTRACT: Recent technical progress has made possible, in principle, quantitative location of the deuterated positions in the products of a catalytic reaction, when deuterium is included in one of the reactants. Accordingly, from the analytical data, selection of the most plausible scheme of the reaction among several ones becomes easy. The present talk attempts to show microwave spectrometry to be a very suitable technique for this purpose, adopting hydrogenation schemes of α-olefins and α-acetylenes as examples.

1. INTRODUCTION

First of all, I would like to say that it is my great pleasure and honor to talk about the recent progress of the isotopic tracer method on this occasion. However, there is a Japanese saying that it is ridiculous to preach the doctrine of Buddhism to Saint Buddha. I am afraid the same thing may occur here, preaching the usefulness of the isotopic tracer method to the many experts in this field, particularly Professor Emmett, our honorary chairman.

It is known already that kinetic information is not so powerful by itself in elucidating the reaction scheme or in determining all the processes of which a catalytic reaction consists. Recently, alternative methods based on direct identification of the intermediate were proposed, but they were not successful in many heterogeneous catalysis, due to their too low concentration and too short lifetime, i.e., observable surface species during the reaction were not always the true intermediates. Therefore, the classical technique to analyze all the final products as quantitatively as possible still remains effective in deducing the reaction scheme.

More advanced techniques involving deuterium or other isotopic tracers, were utilized in this field of study as soon as deuterium became obtainable. The deuterium tracer method based on this technique was especially effective in identifying the observable intermediates and providing convincing information about the reaction scheme. Studies of Frakas and Rideal, of Taylor and Morikawa, and of Horiuti and Polanyi in

---

*Present address: Department of Engineering Chemistry, Chiba Institute of Technology, Narashino-City, Chiba Pref. 275.

the early thirties may be cited as the classical examples of the deuterium tracer method.[1] In that decade, deuterium of low concentration (ca. 3%) was practically the unique isotope, and could be used as a pure tracer. Therefore, though their research opened a new way to study the mechanistic problems at that time, its range of application was too confined to determine the complete scheme of the complex heterogeneous catalysis and to show the effectiveness of refined deuterium tracer methods in ruling out some of the schemes hitherto proposed.

Since deuterium of high concentrations became available, their range of application was extended gradually. Development of analytical devices accelerated this tendency. These advances led to new methods to elucidate mechanisms of heterogeneous catalysis, and they are now used widely; for instance, methods of kinetic isotopic effect in the reaction rate and of deuterium distribution in the products. These two methods are found useful to estimate the total scheme of several reactions, so that they were adopted in many papers and will be adopted more and more.

A good example of the kinetic isotope effect is offered by Adams and Jennings[2] on the oxidation of propylene to acrolein over bismuth molybdate. By comparing the rate of various deuteropropylenes, they concluded that the hydrogen atom dissociating from the propylene molecule was from the methyl group of propylene. However, my talk will be restricted to the deuterium distribution method due to the limited time given me. The method can be characterized by the procedure to determine the deuterium content of all isotopic isomers in the products and to trace the behavior of hydrogen during progress of the reactions in cases when one of the reactant molecules includes deuterium.

After the pioneering work of Turkevich[3] in 1950, this method has been widely used by Kemball,[4] Burwell,[5] and others. However, final reaction schemes still cannot be reached in many reactions, because of "experimental degeneracy;" i.e., different theories can give the same deuterium distribution, making selection of the true scheme impossible. To differentiate such a degeneracy, simultaneous use of another tracer, e.g., C-13 may be effective. An alternative attack may be found in quantitative location of the deuterated positions in the products. This refined method, though difficult to accomplish in the past, became applicable by recent technical advances, and more detailed isotopic information on each species is now obtainable, which allows precise discussion of the mechanistic problem without help of kinetic and other methods. If the method is applied in combination with other isotopic tracers, its range of application may become even wider.

The possibility will be shown actually by the research done mostly in our laboratory, Osaka University. I would like to beg your pardon for taking the liberty to indulge my laziness, not covering the literature in the world.

2. THE DEUTERIUM DISTRIBUTION METHOD

In this paper, the deuterium distribution method will be divided into two classes, classical and refined, depending on whether the method is relied solely on mass spectrometry (MS) only or not.

2.1 Classical method

As an example, the classical method is explained first and its range of application is to be pointed out on the catalytic reactions of hydrocarbons with deuterium. In the case of cyclopentane over group VIII metals, following exchange reactions occur:

$$C_5H_{10} + D_2 \rightarrow C_5H_{10-x}D_x + xHD, \ x=0,1,---,10 \qquad (2.1)$$

The initial distribution of $C_5H_{10-x}D_x$ molecules vs. x gave interesting conclusions, according to Kemball[6]) and Burwell,[5]) as Fig. 1 shows: i.e., a sharp maximum appears at cyclopentane-d_5 over both Pd-film and Pd-Al_2O_3 as well as a minimum between d_5 and d_{10} over a wide temperature range. This break at d_5 in the distribution is explained by repetition of the process between di- and mono-adsorbed states shown in Equation (2.2), where X denote H or D. Therefore, this indicates that all five hydrogen atoms in one side of the ring are exchangeable during a single surface residence.

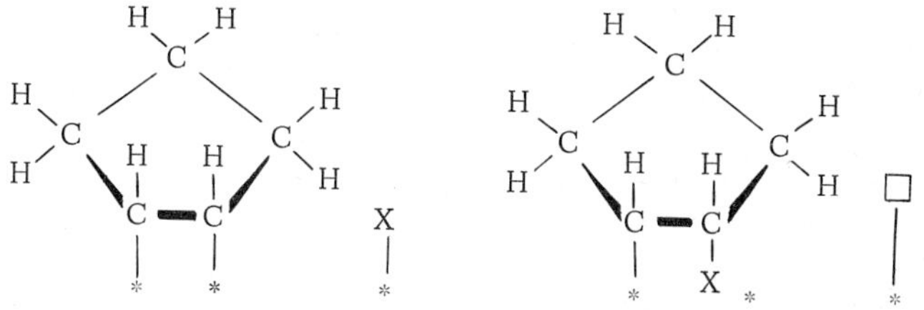

$$\text{[di-adsorbed cyclopentane } C_5H_8(*)_2 + X{-}* \text{]} \rightleftharpoons \text{[mono-adsorbed } C_5H_8X(*) + \square{-}* \text{]} \qquad (2.2)$$

Five hydrogen atoms can be substituted in this way on the same side of the molecule as the original point of attachment. This scheme is widely called the "α-β process." Similar results were given over other group VIII metals.

Amounts of the cyclopentanes containing more deuterium atoms than five are too large to be accounted for by desorption

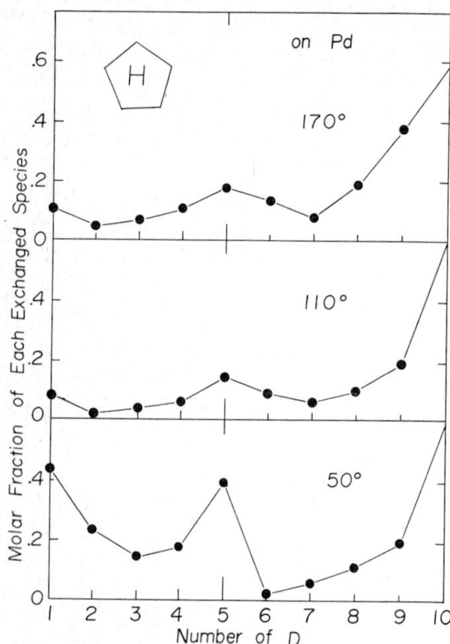

Fig. 1 Initial isotopic exchange patterns of cyclopentane with deuterium over Pd-α-Al_2O_3

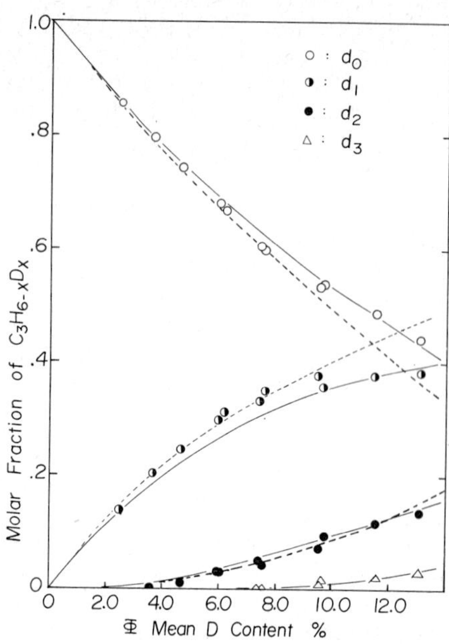

Fig. 2 Fine distribution of deuteropropylenes vs. Φ obtained by reaction of propylene with deuterium over Cu at 25°C

and readsorption of molecules which have suffered exchange at one side. As an explanation, a "turn-over" process of the adsorbed molecule was incorporated into the reaction scheme. This was supported by the effect of increasing temperature which makes the break at d_5 less distinct, because the "turn-over" may have a higher activation energy than desorption. Concerning the details of the turn-over, two intermediates, (i) and (ii), were proposed, but superiority between them could not be decided by the deuterium distribution alone.[7)]

(i) 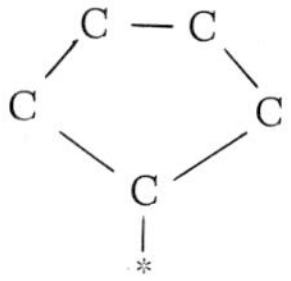 Kemball: cyclic structure C—C / C C / C with attached *

(ii) 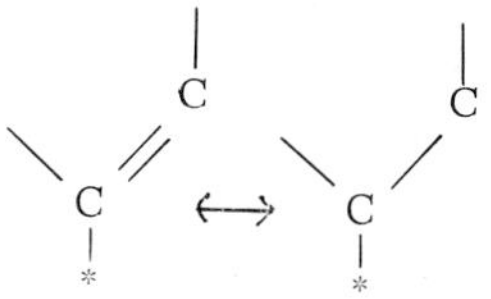 Burwell: C=C(−*) ⟷ C−C(−*) allyl-type structures

Kemball

Burwell

Another good example is the exchange reaction of 3,3-dimethylhexane (A) and 2,2,3-trimethylbutane (B) with deuterium over nickel-kieselguhr catalyst.[8] In both cases, the maximum number may correspond to

(A) 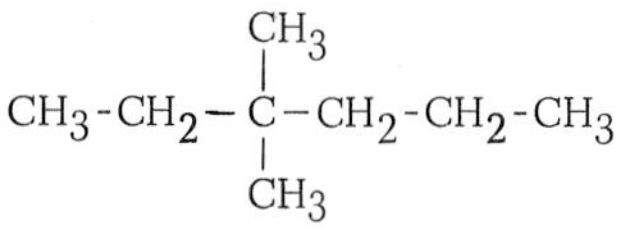 $CH_3-CH_2-C(CH_3)_2-CH_2-CH_2-CH_3$ (B) 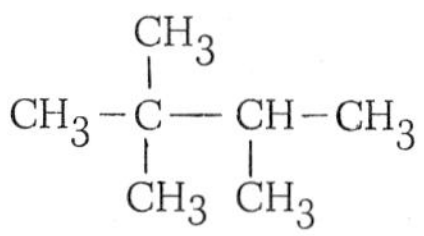 $CH_3-C(CH_3)_2-CH(CH_3)-CH_3$

replacement of hydrogen atoms in the propyl group (A) and the isopropyl group of (B). Such explanation is very plausible, considering the presence of a quaternary carbon atom which can prevent the propagation of the α-β process, ruling out the possibility of the "α-γ process."

In the above discussions, structural knowledge of the molecule is necessary to estimate the activity of hydrogen atoms. There is a different approach from the above, to consider the number of hydrogen atoms similar in activity. The method is based on the assumption of the random distribution of deuterium atoms in the product molecules, first proposed by Wagner et al.[9]

Now, if there are N equivalent hydrogen atoms in the molecule, numbers of each molecule containing n deuterium atoms are given by Equation (2.3) at the equilibrium state, if the isotope effect be neglected.

$$d_n = \frac{N!}{n!(N-n)!} x^n (1 - x)^{N-n}, \quad n \leq N \qquad (2.3)$$

where x is the mean atomic fraction of deuterium in the molecules.

Though Equation (2.3) may be applicable to the system in equilibrium, it has also been applied successfully even to reacting systems. Therefore, Equation (2.3) is called the "random distribution law" and is often used to determine the presence of different hydrogens in a molecule, different active sites on a catalyst and so on. In spite of these successes, use of the law presents some uncertainties: firstly, estimation of the equivalent hydrogen atoms, and secondly, the law is not effective enough to discriminate two possible schemes.

For example, the deuterium distribution of propylene obtained after its reaction with deuterium over copper will be considered.[10] The summarized result is shown by Fig. 2, where full lines correspond to those calculated by Equation (2.3) assuming n=6, while dotted lines to those assuming n=2. Clearly the latter lines coincide with the experiment at low conversion, in spite of small possibility from the structural reason. Actually, as will be shown, methinic hydrogen atoms of propylene are the most exchangeable at the initial stages, so that the exchange is not limited to two atoms. Accordingly, better agreement of the dotted lines with the observed is to be regarded accidental.

In short, even if the classical deuterium distribution method used was useful in many reactions, more refinement in technique would be necessary, especially if the molecules in the products have hydrogen atoms of different bonding. To solve the problem, location of deuterium atoms in each isomer is necessary. A new method devised for this purpose is called to be "hyperfine distribution" method hereafter, while the classical one to be "fine distribution" method, so as to avoid confusion during the discussion.

2.2 New refined method

There are many techniques to locate the deuterated position in a molecule. Among them, the first but complex technique is the usual organic analysis, i.e., deuterium location can be determined by some chemical procedure appropriate to each molecule. The technique, though standard in researches using radioactive tracers, has been used rarely in those using deuterium tracers. A good example belonging to this category is given by Sachtler and de Boer[11] in the catalytic formation of acrolein from propylene over bismuth molybdena, using propylene labelled with ^{14}C ($H_2{}^{14}C{=}CHCH_3$, $H_2C{=}{}^{14}CHCH_3$ and $H_2C{=}CH{-}{}^{14}CH_3$). They analyzed the acrolein produced by photolyzing the acrolein to yield CO and C_2H_4 and by determining the radioactivity contained in both gases. They concluded the surface intermediate formed on the catalyst to be an allyl radical, which indicates dissociation of a hydrogen atom at the methyl group, as mentioned already.

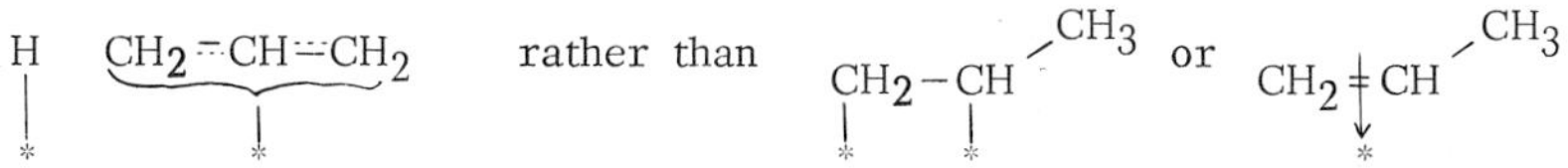

$$\underset{*}{\overset{\mathrm{H}}{|}} \quad \underbrace{\mathrm{CH_2 \overset{...}{-} CH \overset{...}{-} CH_2}}_{*} \quad \text{rather than} \quad \underset{*}{\mathrm{CH_2}} - \underset{*}{\mathrm{CH}}\diagup^{\mathrm{CH_3}} \quad \text{or} \quad \mathrm{CH_2} \underset{*}{\downarrow} \mathrm{CH}\diagup^{\mathrm{CH_3}}$$

It should be more general to adopt some physicochemical technique in quantitative assignment of the deuterated positions in the products. In addition to the mass spectrometry used to determine the fine distribution, several spectral techniques are available recently; e.g., infrared (IR), NMR and microwave spectrometry.

(a) IR technique is widely used now and its effectiveness is known well to give information about surface intermediates, as Eischens[12)] first showed. Indication of formate species on metal catalysts offered a powerful evidence to determine the dehydrogenation scheme of formic acid.[13)] Another example, determination of the deuterated positions of aromatic compounds, may be cited in the study of their reaction with deuterium oxide.[14)] Generally speaking, IR technique is only effective for qualitative rather than for quantitative purposes. Fig. 3 shows the IR spectra of the propylene after reaction of deuterium oxide with propylene over Pd and Ni. Though deuterium distributions of both samples are similar, as Table 1 indicates,[15)] a clear difference is shown in their spectra at the region of C-D stretching bands. Quantitative confirmation of this discrepancy, however, had to resort to some other technique.

Another weak point of this technique may be accidental overlapping of key bands as well as their broadness which makes even its qualitative application often restricted. Thus, IR is not a pertinent technique to the present purpose.

(b) Mass spectrometry utilized previously to locate the deuterated position is still an effective method, especially when other techniques cannot be applied. However, it has several demerits.

First, care must be taken to avoid deuterium exchange of some functional group (OH, NH_2, etc.) in the ion source of mass spectrometers. In such cases, before determination of the deuterium content in the group, the easily exchangeable deuterium is substituted by protium and then the sample is subjected to the MS analysis. By use of such procedure, deuterium distribution in the skeletal part of several alcohols was investigated by Patterson and Burwell[16)] recently. An unexpected result was obtained in the exchange reaction of deuterium with 2-butanol over a copper catalyst; i.e., OH group could not be exchanged as easily as α-hydrogen, contrary

TABLE 1
Deuterium distribution in propylene after the exchange with D_2O

	Ni-a	Ni-b	Pd	$Ni\text{-}Al_2O_3$
C_3H_6 mmole	4.6	4.6	4.6	6.0
D_2O mmole	10	10	10	19
Cat. gr	1.65	1.65	0.50	5.0
Temp. °C	25	17	25	19
Time hr	25	25	150	359
d_0	51.6	58.4	51.2	63.6
d_1	36.8	33.3	30.6	29.8
d_2	10.1	7.3	10.7	5.9
d_3	1.6	1.0	1.6	—
d_4	—	—	—	—
$\Phi = 1/6 \sum nd_n$	11.6	8.5	10.4	7.4

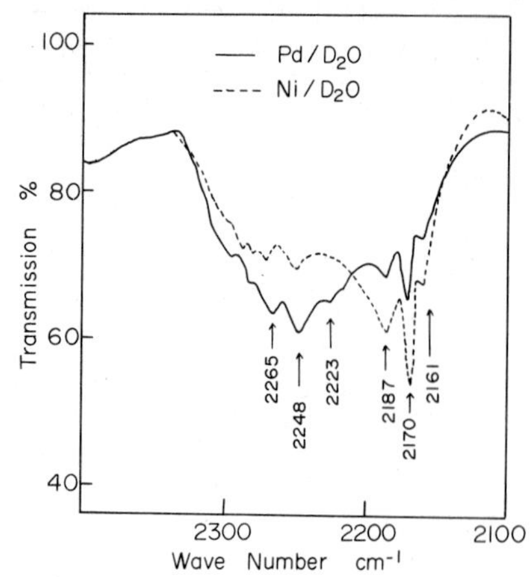

Fig. 3 IR spectra of produced propylene over Ni-a and Pd of Table 1

to the cases of primary and secondary alcohols; the result is quite different from that over Pd and group VIII metals, which can exchange hydrogen in the OH group easier. Thereby, an assumption about the effect of isotopic substitution by the MS is not only necessary, but it may also effect the fragmentation scheme. As reported by Field and Franklin[17] the ratio of scission probability of an individual C-H bond to that of a C-D bond in CH_3D reaches as much as 1.8. An isotopic effect of this order is often reported on other compounds. Another point to be mentioned may be the recent situation that reliability of MS has decreased in quantitative analysis in spite of the rapid progress of apparatus in resolving power and sensitivity.

(c) NMR technique may be applicable even to large molecules, and give a very quantitative result. Several researchers applied this technique in addition to MS to the mechanistic studies; e.g., to show unsymmetrical addition of deuterium across the double bond,[18a] and geometrical isomerization of labelled pentadiene[18b] over group VIII metals. This technique was also found effective to determine the nature of exchangeable hydrogen atoms of benzene derivatives.[19] Recently, the technique was applied to the catalytic deuterogenation of alkylcyclohexanone over group VIII metals.[20a]

Thereby, a new NMR shift reagent, $Eu(dpm)_3$ or $Pr(dpm)_3$, was used, so that stereoselectivity in the produced cyclohexanol could be determined quantitatively. This interesting result will be presented at this conference.[20b)]

Other points to be taken care in its application can be consulted to the textbooks easily.

(d) Microwave spectrometry was first applied, being suggested by Professor Morino, to elucidation of the discrepancy between Table 1 and Fig. 3,[15,21)] and then to the noncatalytic rearrangement study of cyclobutyl cation.[22)] Although examples are still limited to a few compounds now, these will increase in future due to (i) the high sensitivity of the technique, (ii) its high reliability, and (iii) the possibility of locating deuterium atoms precisely. Therefore, this technique may be the most powerful and is the most promising in the mechanistic study with the deuterium distribution method. Before entering into details of its application, some of the results are mentioned briefly.

In the exchange reaction of deuterium oxide with propylene, four kinds of propylene-d_1 could be determined quantitatively. The results shown in Table 2[23)] indicate clear differences of catalytic activity among several metals which is not possible with the classical deuterium distribution in Table 2; e.g., methynic hydrogen is the most active over Ni, nearly all the hydrogen atoms have the same activity per atom over Pd, while trans-1-d_1 is more readily produced than is cis-1-d_1 over Pt. This finding will be explained later together with that of deuterogenation of propylene. Furthermore, the last column shows that alumina-supported nickel behaves similarly in activity to palladium, though silica-supported nickel is similar to pure nickel (not shown in the table).

TABLE 2

Deuterium distribution of propylene-d_1 produced by the exchange reaction:

$$C_3H_6 + xD_2O \rightarrow C_3H_{6-x}D_x + xHDO \quad (ca.\ 25°C)$$

Catalyst	Pd	Ni	Pt	Rh	$Ni\text{-}Al_2O_3$[15b)]
$\Phi = 1/6 \sum nd_n$	10.4	8.5	14.7	ca. 6	7.4
$CH_2DCH{=}CH_2$	59.2±2.2	22.2±2.7	19.5±2.1	49.0±0.6	45.3±2.5
t-$CH_3CH{=}CHD$	13.1±1.3	9.5±2.1	47.4±4.8	19.6±0.9	12.5±1.5
c-$CH_3CH{=}CHD$	13.7±1.5	9.9±2.2	5.8±1.4	15.5±0.9	11.3±1.7
$CH_3CD{=}CH_2$	14.1±1.3	58.1±5.1	27.3±4.5	15.9±0.8	30.1±2.8

$$\underset{\textrm{3-d}_1}{\overset{\textrm{CH}_2\textrm{D}}{\underset{\textrm{H}}{\,}}\textrm{C=C}\overset{\textrm{H}}{\underset{\textrm{H}}{\,}}}\qquad \underset{\textrm{trans-1-d}_1}{\overset{\textrm{CH}_3}{\underset{\textrm{H}}{\,}}\textrm{C=C}\overset{\textrm{H}}{\underset{\textrm{D}}{\,}}}\qquad \underset{\textrm{cis-1-d}_1}{\overset{\textrm{CH}_3}{\underset{\textrm{H}}{\,}}\textrm{C=C}\overset{\textrm{D}}{\underset{\textrm{H}}{\,}}}\qquad \underset{\textrm{2-d}_1}{\overset{\textrm{CH}_3}{\underset{\textrm{D}}{\,}}\textrm{C=C}\overset{\textrm{H}}{\underset{\textrm{H}}{\,}}}$$

Another important application is to self-exchange reactions where identification is impossible by MS alone. Propylene-3-d_1 was isomerized over Ni, Ni-Al_2O_3, Pd, Pt and Rh at 30°C or 40°C,[24] and the relative amounts of products formed initially followed the order trans-$CH_3CH=CHD$ > cis-$CH_3CH=CHD$ > $CH_3CD=CH_2$. Since d_0 and d_3 species were also found, the reaction must proceed intermolecularly, dissociating a hydrogen atom from the methyl group as the first step.

This refined technique can be extended widely to the study of other reactions, but it is limited to compounds which have a finite dipole moments. Nevertheless, due to its high sensitivity, even the mocrowave spectra of the compounds having a very small dipole moments are measurable. Some of them are shown in Table 3, so that they can be used.

TABLE 3
Dipole moments of several hydrocarbons whose microwave spectra were determined

Hydrocarbons	Dipole Moment
Ethane-d_3	0.01078 ± 0.0009 debye
Propylene	0.364 ± 0.003
Propane	0.083 ± 0.001
Allen-d_2	0.0034

(This table was offered by Dr. Shuji Saito, Sagami Chemical Research Centre)

Microwave spectra of 1-butene and cis-2-butene offer another example. Tamaru *et al.*[25] discussed the mechanism of butene isomerization over o-toluene sulfonic acid-silica, and by use of kinetic data they concluded that the carbonium ion mechanism makes the major contribution to the isomerization of n-butene rather than a concerted mechanism. Recently a similar method of research was reported by Yasumori *et al.*[26] on the homogeneous catalysis by Pt(II)-Sn(II) chloride complex.

Microwave spectra of propanes were also used in the mechanistic study of homogeneous hydrogenation of propylene in benzene by Ueda. Details will be explained later.

3. APPLICATION OF THE DEUTERIUM DISTRIBUTION METHOD

As examples of application of the method to discussion of reaction schemes, hydrogenation of α-olefins and acetylenes over metal catalysts are very appropriate because they have been investigated in detail; however they still have several uncertain points in their final schemes.

3.1 Catalytic hydrogenation of α-olefins

Main processes of hydrogenation of α-olefins over metal catalysts may be expressed by Equation (3.1), where * denotes a chemisorption site on the catalyst:

$$\left.\begin{array}{l} RCH{=}CH_2 \xrightleftharpoons{1} \underset{*}{\overset{|}{RCH}}-\underset{*}{\overset{|}{CH_2}} \\ H_2 \xrightleftharpoons{2} \begin{cases} \underset{*}{\overset{|}{H}} \\ \text{- - - - - - - -} \end{cases} \end{array}\right\} \xrightleftharpoons{3} \left.\begin{array}{l} \underset{*}{\overset{|}{RC_2H_4}} \\ \underset{*}{\overset{|}{H}} \end{array}\right\} \xrightarrow{4} RC_2H_5 \qquad (3.1)$$

The above scheme might be called the standard one, because several points have to be modified and extended,[27] after the proposal of Horiuti and Polanyi. The essential point of this scheme concerns with inclusion of the half-hydrogenated species RC_2H_4 as a surface intermediate. Among these processes, process 4 is slow compared to other ones, in spite of its large free energy decrease, so that process -4 can be neglected, while process -3 cannot.

Most persuasive evidence was first given by the use of deuterium distribution on the catalytic hydrogenation to ethylene by Turkevich et al.[3] In this case, not only various deuterated ethanes $C_2H_{6-x}D_x$ (x=0,1,---,6) but also various deuterated ethylenes $C_2H_{4-y}D_y$ (y=0,1,---,4) were produced, making C_2H_6 the major initial product, not $C_2H_4D_2$, contrary to the simple scheme of direct addition of a hydrogen molecule. This result, though it seemed curious at first, is very reasonable according to the scheme, Equation (3.1), where the reactant H_2 is replaced with D_2. Because of the strong adsorbability of α-olefins, D-* is rapidly replaced with H-*, as Equation (3.2) shows, by process -3', which decreases the concentration of the chemisorbed RC_2H_3D species at the same time:

$$\underset{*}{\underset{|}{RCH}}-CH_2D \;\left(\text{or } RCHD-\underset{*}{\underset{|}{CH_2}}\right) + \underset{*}{\underset{|}{\square}}\,\underset{*}{\underset{|}{\square}} \begin{array}{l} \xrightarrow{-3'} \underset{*}{\underset{|}{RCH}}-\underset{*}{\underset{|}{CHD}} + \underset{*}{\underset{|}{H}} \\ \xrightarrow{-3''} \underset{*}{\underset{|}{RCH}}-\underset{*}{\underset{|}{CH_2}} + \underset{*}{\underset{|}{D}} \end{array} \qquad (3.2)$$

Since in spite of process -3' the chemisorbed RC_2H_3 remains much larger in concentration than the chemisorbed RC_2H_2D, it reacts with H-*, producing RC_2H_4-* and then RC_2H_5 predominantly.

In spite of the elegant results, there are still two questions which cannot be answered unambiguously.

(a) Points at issue: The first point at issue concerns especially ethylene hydrogenation over nickel; i.e., the possibility whether hydrogen or deuterium can be adsorbed independently of ethylene or not. Twigg proposed Equation (3.3) instead of process 3 in the standard scheme, where a wavy line denotes the physically sorbed state.

$$\underset{*}{\underset{|}{RCH}}-\underset{*}{\underset{|}{CH_2}} + \underset{*}{\underset{\wr}{D_2}} \begin{array}{l} \xrightarrow{5} \underset{*}{\underset{|}{RCH}}-CH_2D + \underset{*}{\underset{|}{D}}\,\underset{*}{\underset{|}{\square}} \\ \xrightarrow{5'} RCHD-\underset{*}{\underset{|}{CH_2}} + \underset{*}{\underset{|}{D}}\,\underset{*}{\underset{|}{\square}} \end{array} \qquad (3.3)$$

In this modified scheme, process 5 and 5' are indispensable instead of process 2 for deuterium to be adsorbed directly on nickel surface. If so, an additional process is necessary for Equation (3.3) to conform with the fact that C_2H_6 is initially the major product. Processes -6 and -5 as the reverse process of Equation (3.3) ought to be possible, as well, viz.

$$\underset{*}{\underset{|}{RCH}}-CH_2D + \underset{*}{\underset{|}{D}}\,\underset{*}{\underset{|}{\square}} \begin{array}{l} \xrightarrow{-6} \underset{*}{\underset{|}{RCH}}-\underset{*}{\underset{|}{CHD}} + \underset{*}{\underset{\wr}{HD}} \\ \xrightarrow{-5} \underset{*}{\underset{|}{RCH}}-\underset{*}{\underset{|}{CH_2}} + \underset{*}{\underset{\wr}{D_2}} \end{array} \qquad (3.4)$$

According to process -6, only HD is producible. In the case of higher α-olefins, similar processes have to be assumed, of course.

Possibility of this modified scheme can be ruled out by the analysis of the ratio of H_2/HD during the reaction.

Because if the mechanism is true, H-* cannot be larger in concentration than D-*, so that $H_2/HD << 1$, especially at the initial stages, while if process -2 of Equation (3.1) or the recombination process is rapid, $H_2/HD > 1$. According to isotopic measurements done in the reaction of propylene with deuterium,[28] H_2/HD is always larger than two up to 20% conversion (Fig. 4). Equation (3.3) may not, therefore, be a major process to produce the adsorbed H and D atoms.

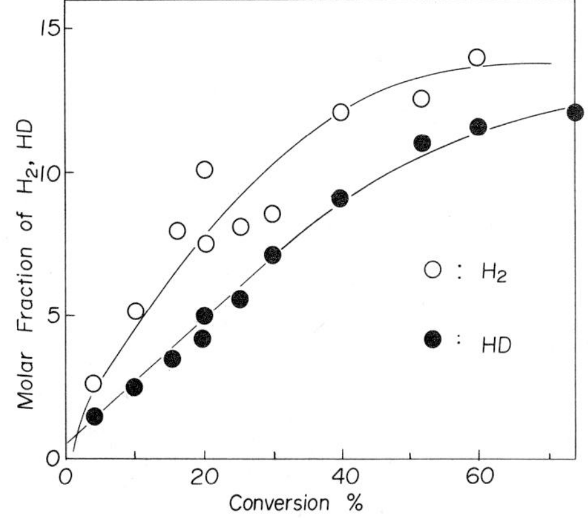

C_3H_6: D_2 = 1:1
Catalyst: Reduced Copper
Temp: 25°C

Fig. 4 Molar fraction of H_2 and HD in the gaseous phase during the reaction of propylene with D_2

It may be said that this conclusion was obtained by studying the hfd of a product, hydrogen. This method is already used widely.

Rigorous investigation on the scheme Equation (3.1) may be possible in cases when an α-olefin other than ethylene is selected, because ethylene is too simple to eliminate the "experimental degeneracy" occurrable by the discussion of deuterium distribution only. Therefore, propylene is suitable for the mechanistic study on hydrogenation of olefins and will be adopted hereafter to resolve the points at issue.

In order to explain the major product $RCH=CH_2$ during the initial stages of the reaction, assuming the modified scheme, an isotopic redistribution process between the species I is proposed in addition to Equations (3.3) and (3.4). They are shown by Equations (3.5a) and (3.5b), where the latter is a nonobservable exchange reaction:

$$\underset{\substack{\text{I}\\\text{(normal propyl species)}}}{CH_3\text{-}\underset{|\atop *}{CHD}\text{-}CH_2} + CH_3\text{-}\underset{|\atop *}{CH}\text{-}\underset{|\atop *}{CH_2} \xrightarrow{a} CH_3\text{-}\underset{|\atop *}{CD}\text{-}\underset{|\atop *}{CH_2} + \underset{\text{III}}{CH_3\text{-}CH_2\text{-}\underset{|\atop *}{CH_2}} \quad (3.5a)$$

$$\xrightarrow{b} CH_3\text{-}CHD\text{-}\underset{|\atop *}{CH_2} + CH_3\text{-}\underset{|\atop *}{CH}\text{-}\underset{|\atop *}{CH_2} \quad (3.5b)$$

$$\underset{\substack{\text{II}\\\text{(isopropyl species)}}}{CH_3\text{-}\underset{|\atop *}{CH}\text{-}CH_2D}$$

Similar relations may exist concerning isomeric half-hydrogenated species II instead of the species I. Now, a special assumption is necessary for the nondeuterated species III to be produced more easily than for both species I and II. If not, preponderant formation of propane-d_0 at the initial stage similarly observed as in the case of ethylene cannot be explained, even though H-* thereby necessary to produce propane has to be formed via Equation (3.2) or some other process.

Besides, explanation of the "hfd" which differs according to catalysts, may become too complicated. This means that plausibility of Equations (3.5a) and (3.5b) would decrease, especially when there is some explanation which does not require both processes as well as Equation (3.3) at all.

(b) Characteristics of Each Group VIII Metal: As shown already by the comparison of Table 1 and 2, hfd can give a clear difference of characteristics of each catalyst. Another example of the same conclusion is given, according to our studies, by the deuterium distribution of propylene obtained by its catalytic reaction of propylene with deuterium over Cu, Ni, Pd, Pt and Ru when the degree of deuteration Φ is similar. Here it was difficult to derive some clear-cut distinctions in their activity from the fine distributions, because qualitative difference between them is too small. However, if the deuterated positions of propylene were determined their different characteristics would be clarified by the hfd of propylene-d_1. Ueda _et al._[23] obtained the result in the propylene deuterogenation shown in the lower lines in Table 4 by use of the microwave spectroscopy.

Hfd of each group VIII metal in Table 4 is very similar respectively to that in Table 2, giving support to the assumption that process 4 in Equation (3.1) is slow. Their results

TABLE 4

Hfd of propylene-d_1 produced in the reaction of propylene with D_2 on Group VIII metals D_2/C_3H_8=0.5 mmole; C_3H_6=1-2 mmole

	Ni	Cu	Pd	Rh	Pt
Catalyst gr	5.0	10	0.50	0.1	0.50
Temp. °C	20	25	20	30	20
Time	6.9min	11hr	10.2min	45min	20min
Conversion %	80	70	80	30	80
Φ	11.6	11.5	7.7	ca. 7	11.5
$CH_2DCH=CH_2$	36.2±2.2	26.4±0.3	54.0±2.9	34.8±0.3	21.1±1.7
t-$CH_3CH=CHD$	11.4±1.8	7.8±0.5	16.4±1.5	21.6±1.1	29.1±3.3
c-$CH_3CH=CHD$	11.7±1.8	7.1±0.2	13.4±2.1	20.2±0.8	13.1±2.5
$CH_3CD=CH_3$	40.7±2.9	59.0±2.4	16.1±2.4	23.5±1.6	26.7±3.6

can be classified into three kinds: ---1) Ni, Cu; 2) Pd, Rh; 3) pt. Class 1) indicates the preponderance of normal propyl species I on the surface rather than isopropyl species II, Class 2) indicates nearly equal exchangeability of H atoms of propylene, while Class 3) shows different concentrations of trans- and cis-1-d_1 species. The last stereospecific activity can be explained easily if the adsorbed species over Pt is a dissociative species IV, but not its isomeric species V, probably due to the steric hindrance in the adsorbed layer. This chemisorbed state is supported by the finding that trans-d_1 species is produced more than cis-d_1 species.

$$\begin{array}{c} CH_3 \\ | \\ H-C \\ \quad \diagdown\!\!\diagdown \quad \diagup H \\ \qquad C \\ \qquad | \\ \qquad * \end{array} \quad IV \qquad \qquad \begin{array}{c} H \\ | \\ CH_3-C \\ \quad \diagdown\!\!\diagdown \quad \diagup H \\ \qquad C \\ \qquad | \\ \qquad * \end{array} \quad V$$

Preponderance of CH_3CDCH_2 in Class 1) indicates the formation of n-propyl species to be larger than that of isopropyl intermediate, shown by Equation (3.6), occurs via α-path.

$$\begin{array}{c} CH_3 \\ \diagdown \\ CH-CH_2D \\ | \\ * \end{array} \begin{array}{l} \xrightarrow{\alpha} \begin{array}{ccc} H & + & CH_2-CH\diagup CH_2D \\ | & & | \quad\; | \\ * & & * \quad\; * \end{array} \\ \xrightarrow{\beta} \begin{array}{c} CH_3\diagdown CH-CHD + H \\ \quad\; | \quad\;\; | \qquad\;\; | \\ \quad\; * \quad\;\; * \qquad\;\; * \end{array} \end{array} \longrightarrow \begin{array}{c} H \;\square\; \square \\ | \quad | \quad | \\ * \quad * \quad * \end{array} + C_3H_5D \qquad (3.6)$$

It would be noteworthy that this process is a prototype of isomerization observable from 1-butene to 2-butene as already mentioned.[21b)]

Recently,[29)] α-alumina was found to have a different catalytic activity in the above hydrogenation reaction from group VIII metals, according to Tamaru et al., i.e., the produced monodeuteropropylene at 30°C did not contain $CH_3CD{=}CH_2$ and isotopic mixing proceeds between methyl and methylidene groups. The latter finding can be explained by the isomerization process shown in Equation (3.6), without producing isopropyl species II at all.

Another interesting result is the hfd of propanes as well as propylene obtained with a $RhCl(PPh_3)_3$ catalyst at 25°C by Ueda,[30)] using equimolar mixture of C_3H_6 and D_2. Between 40-60% conversion, C_3H_6 occupied more than 96% of propylenes, while hfd of propane is as follows:

$CH_2DCHDCH_3$	$CH_2DCH_2CH_3$	CH_3CHDCH_3	$CH_3CH_2CH_3$
75%	5.5%	7%	12%

i.e., two D atoms apparently add to the double bond. This finding indicates that the homogeneous metal-complex catalysis is quite different in scheme from the heterogeneous one. Another example to use hfd of propane is to be reported Ueda herself.

(c) Kinetic treatment of the deuterium distribution: To determine the reaction scheme, kinetic treatment of the deuterium distributions would be useful, if possible. However, such an investigation has not been published.

The attempt of this kind was first proposed by Keii[31)] and Kemball.[32)] Both researchers predicted the deuterium distributions of ethylene and ethane in the deuterogenation reaction of ethylene, based on the standard reaction scheme, determining the relative specific rates of each process. Kemball's theory, easier in application, was applied to several hydrogenation reactions of olefins and of acetylene,[33] but Twigg's modified scheme could not be checked with this treatment.

This situation may be overcome by a more general method, which can treat the time-course of hfd in the product molecule. The theory proposed by Yasuda[34] is one such refined theoretical method. After this theory could predict the time-course of the fine distribution obtained by the deuterogenation of propylene shown by Fig. 2 with success, it was applied to the kinetics of the isomerization of trans-deuteroethylene (T) on palladium reported by Flanagan and Rabinovitch.[35]

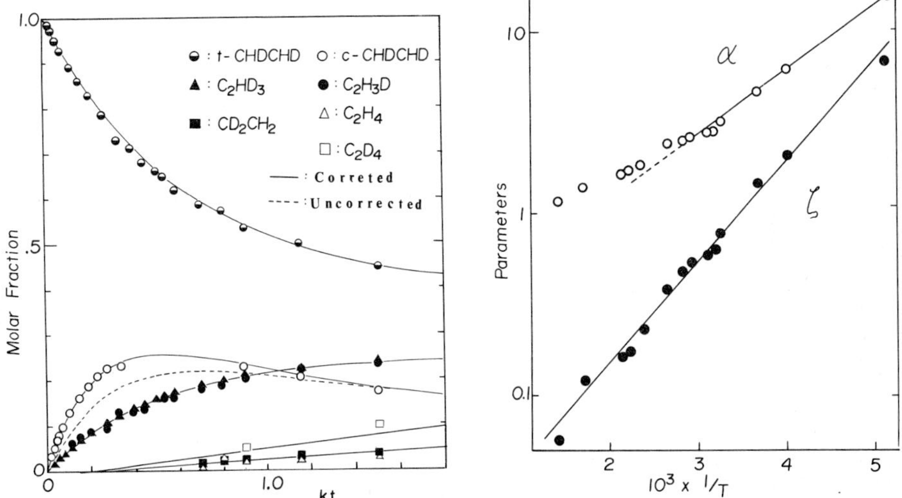

Fig. 5 Experimental values vs. calculated in the isomerization of trans-dideuteroethylene
A. Time-course of the reaction, B. Parameters vs. Temp.

Fig. 5 indicates that a distinct discrepancy exists between the observed (open circles) and calculated (dotted line) values of cis-d_2-species, though the other species do not give such a discrepancy. Therefore, they introduced an isotope effect defined by α=(prob. of C-H rupture)/(prob. of C-D rupture) and could explain the result. The result can also be explained[36] by assuming a competitive "direct isomerization" process as well as the process via the half-hydrogenated state; i.e., a competitive process 7 must be added to processes 3 and -3 in Equation (3.7) by the following reason.

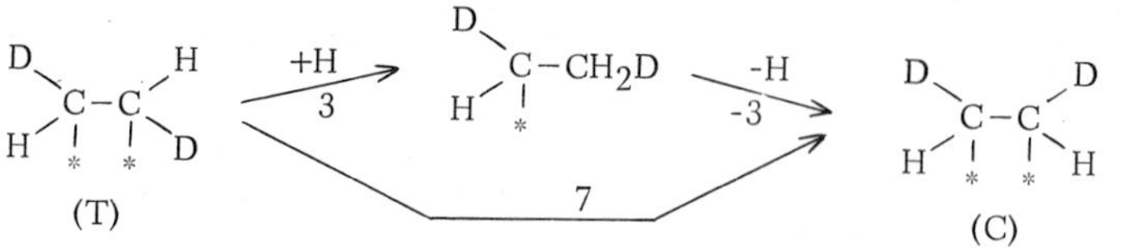

$$\underset{(T)}{\overset{D\ \ \ \ \ H}{\underset{H\ \ *\ \ *\ \ D}{C-C}}} \xrightarrow[3]{+H} \underset{*}{\overset{D}{\underset{H}{C}}}-CH_2D \xrightarrow[-3]{-H} \underset{(C)}{\overset{D\ \ \ \ \ D}{\underset{H\ \ *\ \ *\ \ H}{C-C}}} \quad (T)\xrightarrow{7}(C) \tag{3.7}$$

By adopting a suitable ratio ζ of the rate via processes 3 and -3 to that via process 7, the dotted line of cis-d_2-species shifts to the full line, which is recalculated under the new assumption. Thus, the discrepancy disappears. It is very interesting that Hansen[37)] recently proposed the same process to be possible on tungsten.

Another point to be mentioned in Yasuda's treatment lies in the conclusion that the assumption of a kinetic isotope effect is unnecessary to explain the experimental results, as opposed to the attempts to explain the data based on the effect.[32b)]

The third point to be suggested is the possibility that process 7 via a radical intermediate (R) which makes the isomerization possible. Its plausibility will be mentioned later.

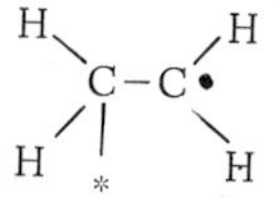

$$\underset{(R)}{\overset{H\ \ \ \ \ H}{\underset{H\ \ *\ \ \ \ H}{C-C\bullet}}}$$

Such investigations to analyze the time-course of deuterium distributions are still few, but these will undoubtedly increase as computors are made available to analyze the data, and this would contribute much to the mechanistic problem.

3.2 Catalytic hydrogenation of α-acetylenes

The scheme of deuterogenation of α-acetylenes over metal catalysts is generally thought to be similar to that of α-olefins, Equation (3.1).[38)] However, some modification must be introduced into it, because, since stereoisomers are possible in the produced olefins, a special problem of selectivity arises in the present reaction. This situation is convenient for the refined deuterium tracer method to be applied.

(a) Selectivity: Repeatedly, it has been reported in the hydrogenation of acetylenes that more olefins than paraffins were produced. As an example, our result on the reaction of methylacetylene with deuterium over six group VIII metals at 25°C is shown by Fig. 6.[39)]

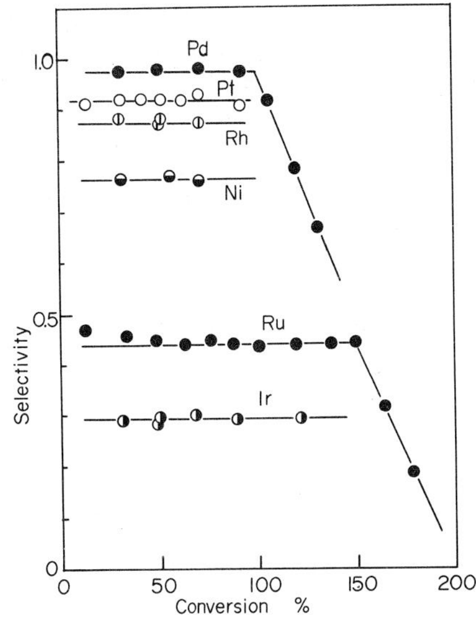

$$S = \frac{\text{propylene}}{\text{propylene+propane}}$$

Fig. 6 Selectivity in the catalytic deuterogenation of $CH_3C \equiv CH$

The selectivity S is especially high on palladium, and also high on Pt and Rh, but smaller on Ru and Ir. It is noteworthy that S is nearly unity until ca. 100% conversion on Pd and constant at some value more than 100% on other metals.

According to the fine distribution on Pd, propylenes containing more than three deuterium atoms are not observable at 90% conversion, and hardly at 110% conversion, indicating the possibility that addition of hydrogen or deuterium occurs only at the triple bond but does not at the double bond, as long as acetylene remains in the gas phase. The same tendency is indicated on other metals as shown by Table 5. This characteristic was already explained by the poisonous effect of acetylene on the palladium surface, i.e., the produced propylene is not only desorbed into the gas phase, but also its readsorption becomes difficult. Nevertheless, more confirmative experiments must be carried out in order to be able to define on the complete mechanism. It will be given by the hfd of the deuteropropylenes, shown in Table 6, by which mechanism of the stereoselectivity on the hydrogenation can be studied.

Table 5
Fine Distribution of Methylacetylene and Propylene (25°C)

	Conv. (%)	Methylacetylene(%) d_0	d_1	d_2	Propylene (%) d_0	d_1	d_2	d_3	d_4	d_5	d_6	Φ/100
Pd	50	98.1	1.9	—	1.1	11.1	79.2	8.6	—	—	—	1.95
Pt	50	96.9	3.1	—	1.1	14.0	75.0	9.3	0.6	—	—	1.94
Rh	50	62.9	37.1	—	3.4	20.5	47.9	24.5	1.9	1.0	0.8	2.07
Ir	50	98.8	1.2	—	2.7	13.5	56.7	20.2	2.7	2.3	1.9	2.21
Ru	50	85.1	14.9	—	1.8	11.9	45.2	33.1	3.3	2.6	2.1	2.40
Ni	70	98.6	1.4	—	1.8	18.8	49.6	29.1	0.7	—	—	2.08

Table 6
Hyperfine Distribution of Propylene-d_1 and -d_2

(A)

	Conv. (%)	Propylene-d_1 (%) c-1-d_1	t-1-d_1	2-d_1	3-d_1
Pd	50	5.5±0.3	49.7±6.2	44.8±4.7	—
Pt	50	8.5±0.4	42.7±4.8	48.8±4.5	—
Rh	70	17.4±0.5	37.0±2.0	42.8±2.7	2.8±0.5
Ir	50	7.0±0.3	46.2±2.7	45.5±4.1	1.4±0.4
Ru	70	16.2±0.9	42.8±6.0	37.7±3.6	3.3±0.3
Ni	70	16.7±0.9	37.1±3.0	45.3±6.7	0.9±0.3

(B)

	Conv. (%)	Propylene-d_2 (%) 1,1-d_2	t-1,2-d_2	c-1,2-d_2	1,3-d_2	2,3-d_2	3,3-d_2
Pd	50	0.9±0.1	10.6±0.7	88.5±0.6	—	—	—
Pt	50	3.4±0.2	14.8±1.3	81.8±1.0	—	—	—
Rh	70	20.8±1.2	21.2±3.0	53.3±1.8	1.4±0.3	1.5±0.2	1.8±0.3
Ir	50	7.1±0.3	8.5±0.4	82.7±0.8	0.4±0.1	0.9±0.2	0.5±0.1
Ru	70	19.8±1.3	17.4±1.6	58.2±1.7	1.3±0.6	2.7±0.5	0.6±0.1
Ni	70	15.9±2.4	25.5±1.9	57.7±1.9	—	0.9±0.2	—

Table 5[39] indicates also another noteworthy point that methylacetylene-d_1 is different over different catalysts, i.e., it is interesting that among the metals having smaller selectivities, Ru and Rh produce much more $CH_3C{\equiv}CD$ than Ir. This offers another example which indicates a difference of catalytic activity of each metal.

(b) Stereoselectivity: In the case of substituted acetylenes, the product olefins have several stereoisomers.

For instance, dimethylacetylene is hydrogenated to cis-2-butene over Pd-Al_2O_3 with practically 100% selectivity, producing no trans-2-butene, according to Burwell.[40)] Detailed results on the hfd of the products from methylacetylene over several catalysts indicate also the same correspondence with regard to the added position of deuterium.[39b)] Here, palladium is shown the most stereoselective in producing the cis-1,2-d_2 species, as expected (cf. Table 6 (B)). Iridium and platinum give about the same selectivity, while Ru, Ni and Rh show a smaller selectivity. Thus, the stereoselectivity sequence obtained now is not the same as in the usual selectivity.

$$Pd > Pt > Ir >> Ru > Ni > Rh$$

Secondly, according to the hfd of propylene-d_2 deuterated positions are limited mainly to three, and cis-1,2-d_2 is the predominent species. The methyl group is hardly exchangeable, especially over Pd, Pt and Ni. This finding cannot be drawn conclusively from the fine distribution alone, even though the maximum amount of deuterium of each metal is three. It is noteworthy that this is in sharp contrast with the reaction of propylene with deuterium shown in Tables 2 and 4, where during the reaction deuterium can substitute for protium in all the CH bonds in propylene, especially over palladium. This difference must originate not only from the strong adsorptive power of methylacetylene on the Pd surface, forming an associative state, but also by the steric hindrance for the methyl group to come near the surface. The reason why the propylene remains inert without deuterium exchange may be ascribed to the above but not to its short life on the metal.

These findings indicate that hfd can give us a detailed knowledge which the fine distribution cannot do.

(c) Reaction scheme: From the above findings, hydrogenation may proceed over Pd by the processes shown in Equation (3.8), where X denotes H or D.

$$CH_3\text{-}C{\equiv}CH \xrightarrow{1} \underset{*\ \ \ *}{CH_3\text{-}C{=}C\text{-}H} \xrightarrow[+X]{2} \underset{\text{II}}{CH_3\text{-}C(*){=}C(H)\text{-}X} \xrightarrow[+X]{3} CH_3\text{-}C(X){=}C(H)\text{-}X \tag{3.8}$$

$$CH_3\text{-}C{\equiv}CH \overset{1'}{\rightleftharpoons} \underset{*}{CH_3\text{-}C{\equiv}C} + \underset{*}{H}$$

$$\underset{*\ \ \ *}{CH_3\text{-}C{=}C\text{-}H} \xrightarrow[+X]{2'} \underset{\text{II'}}{X\text{-}C(CH_3){=}C(*)\text{-}H} \xrightarrow[+X]{3'} X\text{-}C(CH_3){=}C(X)\text{-}H$$

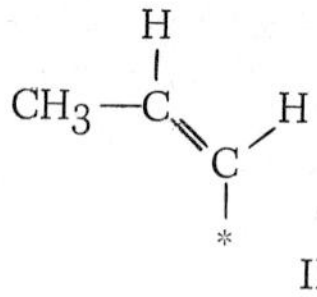

By this scheme, preponderant production of the cis-1,2-d_2 species can be explained. Both diadsorbed and monoadsorbed species can be assumed planar, because the adsorbed state II" probably cannot be attained due to the large steric hindrance between adsorbed species. The attacking hydrogen to II or II" from the C-* direction would produce only the cis-1,2-d_2 species.

CH_3—C(H)=C(H)—*   II"

This explanation is interesting from the findings that cis- and trans-species of dideuteroethylene are not produced in equal amounts by the reaction of acetylene and deuterium, while cis-2-butene is produced with a good stereoselectivity by the hydrogenation of 2-butyne.[40] It means that there is some steric effect in the reaction of acetylene, while there is much in that of 2-butyne having methyl groups. Our methylacetylene gives an intermediate behavior.

The above scheme alone, however, cannot be said satisfactory, though sufficient to explain the experiment. The trans-1,2-d_2 species, the next largest in amount to the cis-d_2 species, is not accounted for by the scheme, since there is little change for acetylene to be readsorbed due to the strongly adsorbed acetylenic species.

Another process which can predict the production of trans-species without desorption, and, if possible, of all the propylenic isomers might be proposed, assuming radical intermediates IV and IV' shown in Equation (3.9), i.e., processes 4, 4', 5 and 5' must be added to step 2 in Equation (3.8) as competitive steps, making the production of the isomeric species, V', possible from IV and IV'.

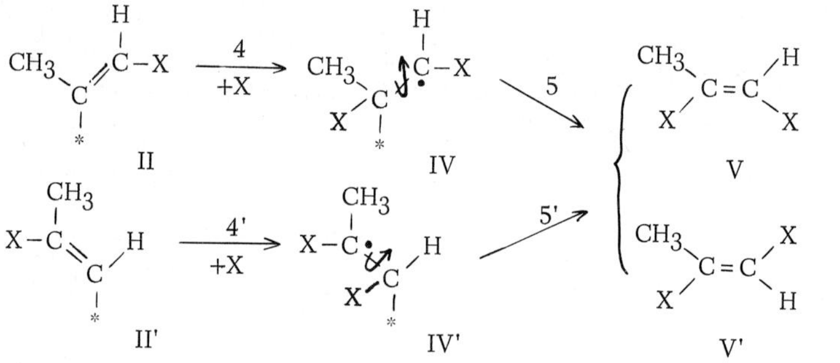

$$\mathrm{CH_3{-}C(*){=}CH{-}X\ (II)\ \xrightarrow[+X]{4}\ CH_3{-}CX(*){-}\dot{C}H{-}X\ (IV)\ \xrightarrow{5}\ \left\{\begin{array}{l}\mathrm{CH_3(X)C{=}C(H)X\ (V)}\\ \mathrm{CH_3(X)C{=}C(X)H\ (V')}\end{array}\right.}$$

$$\mathrm{X{-}C(CH_3){=}CH(*)\ (II')\ \xrightarrow[+X]{4'}\ X{-}\dot{C}(CH_3){-}CHX(*)\ (IV')\ \xrightarrow{5'}\ V,\ V'} \qquad (3.9)$$

In equation (3.9), IV and IV' are sigma radicals, so that they can rotate freely around the C-C bond. This extended scheme can explain the production not only of trans-1,2-d_2, but also of cis-1-d_1, depending on whether X is H or D. However, for the scheme to be accepted plausible, it might be desirable to derive the relative probabilities of producing all the eight isomers of deuteropropylenes quantitatively. If it is succeeded, reliability of the scheme would increase very much and investigation of other one would be felt unnecessary.

To evaluate the relative probability of producing each propylene isomer, ratios of the isotopic adsorbed species, [H]/[D] and $[CH_3C{\equiv}CD]/[CH_3C{\equiv}CH]$ are defined by x and y, respectively, and also the ratio of specific rate of process (9) to specific rate of process (8) by α.

Detailed formulation of the probability can be derived by easy arithmetic; the results are summarized in Table 7. In Table 8, their calculated and observed values over palladium, platinum and nickel are compared assuming the three parameters (x, y and α) which can be obtained from Tables 5 and 6. The agreement between them is fully satisfactory, especially over Pd and Pt. However, if the observed ratios of gaseous methylacetylenes are adopted for y, no good agreement can be obtained. This indicates that rate of the desorption process of methylacetylene is too small for the concentration of $CH_3C{\equiv}CD$ in the gas to reach equilibrium with that in the adsorbed phase, so that the former process remains smaller at least on Pd and Pt. Contrarily, the equilibrium is estimated to obtain faster in the cases of Rh and Ru, judging from the fact that $CH_3C \equiv CD$ was observed in a large amount.

Consequently, the hydrogenation scheme of methylacetylene seems to be accounted for by combination of Equations (3.8) and Equation (3.9) with α unique for each catalyst, especially over Ni, Pd and Pt catalyst. But the scheme may become more complicated to some extent over Ir, Ru and Rh, considering the presence of highly deuterated propylenes up to d_6-species. Actually, similar attempts of calculation based on the data over Ir, Ru and Rh, failed to obtain a good agreement between theory and experiment. It is noteworthy hereby that Ir shows the same character, though the metal produces $CH_3C{\equiv}CD$ in an amount as small as Pd. This suggests that other processes must be considered in addition to the total scheme in the cases of Ir as well as Ru and Rh.

(d) Remark on the result obtained by the analysis of hfd: A radical intermediate was also proposed for the hydrogenation of methylacetylene as well as to explain kinetics of isomerization of trans-ethylene-d_2. Of course, life of the

Table 7

Relative Probability of Producing Each Propylene Isomers a)

Species	From $CH_3-\overset{*}{C}\equiv CH$	From $CH_3-\overset{*}{C}\equiv CD$	Total probability
d_3		$y\cdot(1+\alpha)$	$y\cdot(1+\alpha)$
c-1,2-d_2	$1+\alpha/2$	$xy\cdot\alpha/2$	$1+(1+xy)\cdot\alpha/2$
t-1,2-d_2	$\alpha/2$	$xy\cdot(1+\alpha/2)$	$xy+(1+xy)\cdot\alpha/2$
1,1-d_2		$xy\cdot(1+\alpha)$	$x\cdot[y\cdot(1+\alpha)]$
2-d_1	$x\cdot(1+\alpha)$		$x\cdot(1+\alpha)$
t-1-d_1	$x\cdot(1+\alpha/2)$	$x^2y\cdot\alpha/2$	$x\cdot[1+(1+xy)\cdot\alpha/2]$
c-1-d_1	$x\cdot\alpha/2$	$x^2y\cdot(1+\alpha/2)$	$x\cdot[xy+(1+xy)\cdot\alpha/2]$
d_0	$x^2\cdot(1+\alpha)$		$x\cdot[x\cdot(1+\alpha)]$

a) $[\overset{*}{H}]/[\overset{*}{D}] = x$, $[CH_3-\overset{*}{C}\equiv CD]/[CH_3-\overset{*}{C}\equiv CH] = y$, $\frac{\text{Rate of proc.(3.9)}}{\text{Rate of proc.(3.8)}} = \alpha$

Table 8

Comparison of Calculated Relative Abundance of Deutero-propylene with the Observed (25°C) a)

Catalyst / Species	Pd obs.	Pd calc.	Pt obs.	Pt calc.	Ni obs.	Ni calc.	Rh obs.	Rh calc.
d_3	0.13	0.13	0.16	0.17	1.02	1.02	1.01	1.35
c-1,2-d_2	1.00	1.00	1.00	1.00	1.00	1.00	1.00	(1.00)
t-1,2-d_2	0.13	0.13	0.20	0.20	0.44	0.44	0.40	(0.40)
1,1-d_2	0.01	0.01	0.03	0.02	0.28	0.26	0.39	(0.39)
2-d_1	0.08	0.09	0.11	0.12	0.30	0.30	0.33	0.29
t-1-d_1	0.08	0.08	0.10	0.10	0.24	0.25	0.29	0.29
c-1-d_1	0.01	0.01	0.02	0.02	0.11	0.11	0.13	(0.12)
d_0	0.01	0.01	0.02	0.01	0.06	0.07	0.12	0.08
x		0.08		0.10		0.25		0.29
y		0.12		0.14		0.86		1.34
α		0.28		0.46		0.67		0.02

a) Numerals in parentheses are used to determine x, y and α.

radical intermediates may be too short to be measurable, but be sufficiently long to assume a role of usual reaction intermediates. This conclusion is interesting because a radical intermediate is often observed on metal oxide catalysts by ESR, and a different radical species is also proposed on acetylene hydrogenation by Bond and Wells.[41]

Second point to be mentioned is the confirmation that

the selectivity of Ni is similar to Pd and Pt, although Ir shows much different behavior.

4. CONCLUSION

This refined deuterium tracer technique using microwave spectrometry is not limited to only a few applicable reactions, considering that isotopic distributions of C-13, N-15 and O-18 can be used simultaneously. Accordingly, the refined technique of the deuterium distribution method can contribute very much to determination of the total reaction scheme, especially of complex heterogeneous reactions, in the future.

ACKNOWLEDGEMENT

The author wishes his sincere thanks to Emeritus Professor Yonezo Morino, University of Tokyo, and Professor Eizi Hirota, Kyushu University, for their kind suggestion and help on the application of microwave spectrometry, and also to a number of his coworkers in Osaka University cited in references.

REFERENCES

1) cf. T. I. Taylor, in "Catalysis," edited by P. H. Emmett, Reinhold Publ. Co., New York, Vol. V (1957) p. 257.
2) C. R. Adams and T. J. Jennings, J. Catalysis 3 (1964), 550.
3) J. Turkevich et al., Discuss. Faraday Soc. 8 (1950) 352.
4) C. Kemball, Proc. Roy. Soc. A207 (1951) 539.
5) R. L. Burwell, Jr. and W. S. Briggs, J. Amer. Chem. Soc., 74 (1952) 5096.
6) C. Kemball and J. R. Anderson, Proc. Roy. Soc. A226 (1954) 472.
7) C. Kemball, Catal. Rev. 5 (1971) 33.
8) H. C. Rowlinson, R. L. Burwell and R. H. Tuxworth, J. Phys. Chem. 59 (1955) 225.
9) C. D. Wagner, J. N. Wilson, J. W. Otvos and D. P. Stevenson, J. Chem. Phys. 20 (1952) 338, 1331.
10) K. Hirota, N. Yoshida, S. Teratani and T. Kitayama, J. Catalysis 13 (1969) 306.
11) W. M. H. Sachtler and N. H. de Boer, Proc. 3rd Intern. Congr. Catalysis 1 (1965) 252.
12) R. E. Eischens et al., J. Phys. Chem. 58 (1954) 1059, etc.
13) K. Hirota, K. Kuwata and Y. Nakai, Bull. Chem. Soc. Japan 31 (1958) 863, etc.

14) cf. R. L. Burwell, Jr., et al., Adv. in Catalysis 20 (1969) 1; K. Hirota and T. Ueda, J. Phys. Chem. 72 (1968) 1976.
15) (a) K. Hirota and Y. Hironaka, J. Catalysis 4 (1965) 602; (b) Y. Hironaka and K. Hirota, Bull. Chem. Soc. Japan 38 (1965) 1558.
16) W. R. Patterson and R. L. Burwell, Jr., J. Amer. Chem. Soc. 93 (1971) 833.
17) F. H. Field and J. L. Franklin, "Electron Impact Phenomena," Academic Press, New York (1957).
18) (a) G. V. Smith and J. R. Swoap, J. Org. Chem. 31 (1966) 3904; G. V. Smith, J. Catalysis 5 (1966) 152; (b) P. B. Wells and G. R. Wilson, Discuss. Faraday Soc. 41 (1966) 237.
19) R. R. Frazer and R. N. Renaud, J. Amer. Chem. Soc. 88 (1966) 435; K. Hirota and T. Ueda, Proc. Intern. Congr. Catalysis, 3rd. Amsterdam 2 (1965) 1238.
20) (a) Y. Takagi, S. Teratani and J. Uzawa, Chem. Comm. (1972) 280; (b) Y. Takagi, S. Teratani and K. Tanaka, Preprint No. 52.
21) (a) K. Hirota, Y. Hironaka and E. Hirota, Tetrahedron Lett. (1964) 1654; (b) K. Hirota and Y. Hironaka, Bull. Chem. Soc. Japan 39 (1966) 2638.
22) H. Kim and W. D. Gwinn, Tetrahedron Lett. (1964) 2535.
23) T. Ueda, J. Hara, K. Hirota, S. Teratani and N. Yoshida, Zeits. physik. Chem., N. F. 64 (1969) 64.
24) T. Ueda and K. Hirota, J. Phys. Chem. 74 (1970) 4216.
25) Y. Sakurai, Y. Kaneda, S. Kondo, E. Hirota, T. Onishi and K. Tamaru, Trans. Faraday Soc. 67 (1971) 3275.
26) K. Hirabayashi, S. Saito and I. Yasumori, J. Chem. Soc. Faraday I (1972), in press.
27) G. C. Bond, "Catalysis by Metal," Academic Press, London (1962) Chap. XI.
28) N. Yoshida, D. Imai and K. Hirota, Nippon Kagaku Zasshi 91 (1970) 1026.
29) (a) Y. Sakurai, T. Onishi and K. Tamaru, Bull. Chem. Soc. Japan 45 (1972) 980; (b) Trans. Faraday Soc. 67 (1971) 3094.
30) (a) T. Ueda, Preprint No. 29; (b) private communication.
31) T. Keii, J. Res. Inst. Catalysis, Hokkaido Univ. 3 (1953) 36; J. Chem. Phys. 22 (1954) 144; Proc. 4th Intern. Congr. Catalysis, Moscow (1968).
32) C. Kemball, J. Chem. Soc. (1956) 735; (b) C. Kemball and P. B. Wells, ibid. A (1968) 444.
33) G. C. Bond, G. Webb and P. B. Wells, J. Catalysis 12 (1968) 157.

34) Y. Yasuda and K. Hirota, Zeits. physik, Chem., N. F. 71 (1970) 170, 195.
35) T. B. Flanagan and B. S. Rabinovitch, J. Phys. Chem. 60 (1956) 724, 730.
36) Y. Yasuda and K. Hirota, J. Catalysis, in press.
37) R. S. Hansen *et al.*, J. Phys. Chem. 74 (1970) 3446, 3298.
38) G. C. Bond, *loc. cit.* Chap. XII.
39) (a) K. Hirota, N. Yoshida, S. Teratani and S. Saito, J. Catalysis 15 (1969) 425; (b) N. Yoshida, Doctoral Thesis, Osaka Univ. (1971).
40) W. M. Hamilton and R. L. Burwell, Jr., Actes de II Congr. Intern. Catalysis (Paris) (1960) p. 987.
41) G. C. Bond and D. P. Wells, J. Catalysis 6 (1966) 397.

ADDENDA

1°) The possibility that H-* is much larger than D-*, as shown by Fig. 4, is suggested already by Twigg (Disc. Faraday Soc. 8 (1950) 90).
2°) In future, the use of a combined gas chromatograph-mass spectrometer may be promising to analyze time-course of deuterium distribution (R. S. Dowie, C. Kemball, J. C. Kempling and D. A. Whan, Proc. Roy. Soc. A327 (1972) 491).

Invited Lecture D

THE USE OF HEAT-FLOW MICROCALORIMETRY IN HETEROGENEOUS CATALYSIS RESEARCH

P. C. GRAVELLE
Institut de Recherches sur la Catalyse (C.N.R.S.), Villeurbanne, France

1. INTRODUCTION

In contrast with the general expansion of the scientific litterature, the annual production of articles describing thermochemical measurements in adsorption and heterogeneous catalysis has remained, for the last decade, at an almost constant and low level : less than one hundred articles per year deal with heats of adsorption and less than a dozen refer specifically to heats of chemisorption.

Yet, the need for accurate thermochemical data in this field is evident and has been emphasized in all textbooks on adsorption and heterogeneous catalysis. The recent development of gas-chromatographic techniques for measuring isosteric heats of reversible adsorption is, indeed, a proof that this need is not purely academic but results from very practical problems. However, the number of experimentalists who resort to adsorption calorimetry is still very limited, though the direct calorimetric method is often the best method to determine thermochemical data. We believe that the apparently surprising discredit in which adsorption calorimetry has been confined for too many years is caused by very commonplace reasons. The multiplicity of the adsorption calorimeters which have been described, the discussions among specialists on the experimental details of the construction and testing of these calorimeters and, above all, the unavailability of commercial instruments which can be readily adapted to the study of gas-solid interactions have, in this field as in some others, "contributed in a negative way to the value of calorimetry in the eyes of the scientific and technological community" and probably explain why "many scientists with an interest in energetics make careful detours around the field" [1].

Now, heat-flow microcalorimeters of several types are commercially available and we have shown recently that they can easily be adapted to the study of gas-solid interactions, provided that some simple preliminary experiments and calibration tests are carefully made [2]. A survey of the different uses of this calorimetric technique in the fields of adsorption and heterogeneous catalysis is therefore timely. Before presenting several applications of heat-flow microcalorimetry in these fields, we shall, however, briefly describe the instrument and outline the related theory in order to place more easily the experimental results in a proper perspective.

2. HEAT-FLOW MICROCALORIMETRY

According to the generally accepted definition, the term "heat-flow calorimeter" is used in connection with calorimeters in which i) a path is clearly defined mechanically for the flux of heat (or a constant fraction of it) between the inner calorimetric vessel, in which the thermal phenomenon under study is produced, and the outer medium, called heat sink in this case, and ii) the intensity of the heat flux may be measured along this path as a function of time [3].

In most heat-flow calorimeters operating at low or moderate temperatures, heat flows by conduction along a heat conductor connecting the inner vessel to the heat sink. The quantity of heat flowing along the heat conductor is evaluated from the temperature difference between the ends of the conductor. Thermocouples or semiconducting thermoelements have been used for this purpose and several arrangements have been proposed. The heat-flow calorimeters which can easily be adapted to adsorption studies have been recently reviewed [2].

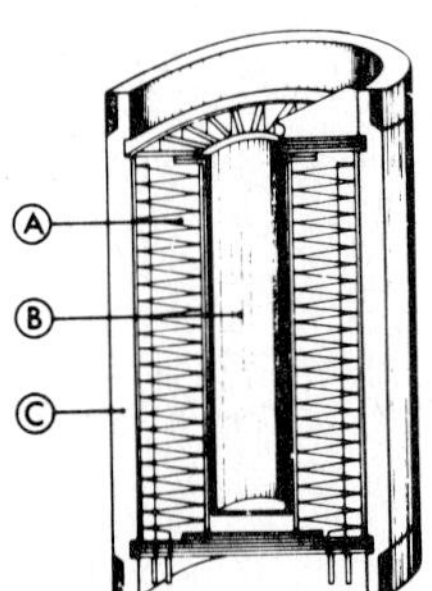

Fig. 1. TIAN-CALVET microcalorimetric element [4]. Batteries of thermocouples (A) ; Inner calorimetric vessel (B) ; metal cylinder wedged in the heat sink (C).

In the TIAN-CALVET microcalorimeters [4] which have been used for most examples presented in the following Sections, heat flows along the wires of a large number of differential thermocouples (500 in the standard model), associated in a series of batteries ("thermopile", A in fig. 1). Each battery is affixed normal to the outer wall of the calorimetric vessel (B) in such a way that the active junctions of the couples completely surround the inner vessel. The reference junctions are in thermal contact with the inner wall of a metal cylinder (C) which is wedged into a cavity of the same shape in a thermostated metal block (the heat sink). In order to increase the stability of the thermopile e.m.f., in the absence of production or absorption of heat in the calorimeter vessel, and therefore the sensitivity of the calorimeter, the thermocouples of two "calorimetric elements" (fig. 1), placed in the same metal block, are connected differentially. In our experiments, fluctuations of the total e.m.f. correspond usually to apparent changes of thermal power of the order of 10^{-5} μW/s for periods extending over days [5]. The stability of the record base-line is thus extremely good and slow or very slow thermal phenomena can be easily studied with TIAN-CALVET microcalorimeters.

Provided that the quality of all material used in the construction of the microcalorimetric element allows it and that a suitable thermostat is available, the heat sink of a heat-flow calorimeter may be maintained at temperatures in a wide range. TIAN-CALVET microcalorimeters of the standard

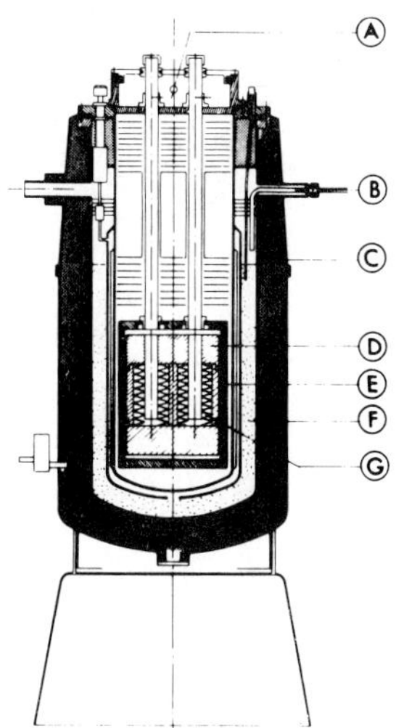

Fig. 2. Vertical cross-section of a low-temperature TIAN-CALVET microcalorimeter. Evacuated space (A) ; liquid-nitrogen inlet (B) ; insulation (C) ; heat-sink (D) ; thermostat (E) ; liquid nitrogen (F) ; microcalorimetric elements (G). By courtesy of SETARAM Co, Lyon, France.

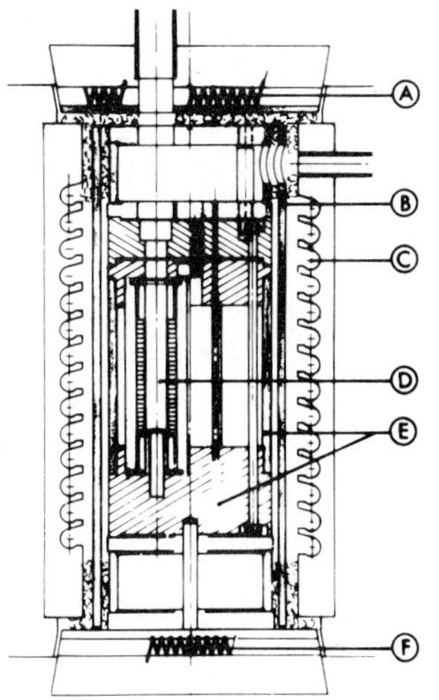

Fig. 3. Vertical cross-section of the heat sink and thermostat of a high-temperature TIAN-CALVET microcalorimeter. Top (A) and bottom (F) electrical heaters ; metal canister (B) ; electrical heater (C) ; microcalorimetric element (D) ; heat sink (E). By courtesy of SETARAM Co, Lyon, France.

model may be used between room-temperature and 200°C. The temperature range has been still more extended with models which are now available : the low-temperature calorimeter, presented in fig. 2, may be used from 67°K to 473°K and the new high-temperature calorimeter, presented in fig. 3, may be used from room-temperature up to 1273°K. The whole temperature range in which gas-solid interactions are usually investigated is therefore accessible to calorimetric measurements and a single TIAN-CALVET microcalorimeter may be used to tackle a variety of problems in adsorption and heterogeneous catalysis.

Now, if it is supposed that i) the temperature of the heat sink is constant and uniform and ii) the temperature of the inner vessel is uniform the heat balance between the inner vessel and its surroundings, when heat is liberated or absorbed in the calorimeter cell, may be written in a simple way.

Part of the heat, dQ, produced or absorbed at any time, dt, increases or decreases the temperature of the calorimetric vessel from $T_o + \theta$ to $T_o + \theta \pm d\theta$. If μ is the heat capacity of the inner vessel and of its contents, the fraction of the heat which remains in the calorimetric vessel amounts to $\mu d\theta$. The other fraction of the heat produced or absorbed at any time flows from or to the inner vessel. The heat flux, during time dt, is proportional to the temperature difference, θ, between the inner vessel and its surroundings. Finally

$$dQ = \mu \, d\theta + p \, \theta \, dt \quad \text{or} \quad \frac{dQ}{dt} = p \, \theta + \mu \frac{d\theta}{dt} \tag{1}$$

This basic equation of heat-flow calorimeters is called the TIAN equation after the name of the french scientist who described the first heat-flow calorimeter in 1924 [6].

Since the e.m.f. of the thermopile, which is recorded continuously by a potentiometric device, is proportional to the temperature difference between the inner vessel and the heat sink, the ordinate, Δ, at time t, of the calorimetric curve (the thermogram) is also proportional to θ :

$$\Delta = g \, \theta$$

Eq. 1 may be therefore written :

$$\frac{dQ}{dt} = \frac{p}{g} \Delta + \frac{\mu}{g} \frac{d\Delta}{dt} \tag{2}$$

The quantity of heat evolved or absorbed by the phenomenon under study is obtained by integrating TIAN equation. If the integration limits are chosen in such a way that they include the whole thermal phenomenon, the total heat is directly proportional to the area under the thermogram,

$$Q = \frac{p}{g} \int_{t}^{t'} \Delta \, dt \tag{3}$$

the value of the second term in eq. 2 being zero for both integration limits. The heat capacity of the calorimeter cell and of its contents, μ, does not appear in eq. 3. Changes of the heat capacity of the calorimetric cell influence the shape of the thermogram but not its area. When a heat-flow calorimeter with a suitable sensitivity is used, it is thus possible to study thermal phenomena occuring in or on poor heat-conducting substances. This property of heat-flow calorimeters is especially useful in heterogeneous catalysis research since many industrial catalysts present a large heat capacity.

The ratio, p/g, which defines the sensitivity of the instrument may change depending upon the calorimeter itself, the amplification and recording line, the temperature of the thermostat, etc... In the case of TIAN-CALVET microcalorimeters, it varies usually from $\sim$ 1 µW to $\sim$ 30 µW per a 1 mm-deflection on the chart paper. Because of the large sensitivity of these calorimeters, it is possible to use small samples of adsorbent (30 to 200 mg) and, thereby, to limitate interparticular diffusion.

TIAN equation indicates, moreover, that the heat power dissipated or absorbed, at any instant, in the calorimetric cell is simply related to the ordinate of the recorded thermogram. Thermokinetics of the processes taking place in the calorimeter cell may be, therefore, directly deduced from the calorimetric data, at least when the conditions underlying TIAN equation are reasonably well satisfied, i.e., when the evolution of the thermal process is slow [4].

TIAN equation cannot, however, be used to determine the kinetics of fast processes and the complete determination, by means of FOURIER equation, of the heat transfer in a real calorimeter appears to be, in the general case, hopelessly complicated. But, all transfers of heat and electricity being governed by the linear ONSAGER relations, the whole calorimetric system behaves, in normal operation, as a single linear system with localized constants and thus, in the domain of LAPLACE transforms, any heat input, f(t) (the phenomenon under study) is related to the recorded thermogram, g(t), by the fundamental equation of linear systems (supposed initially at rest) :

$$G(p) = H(p) \cdot F(p) \qquad (4)$$

where G(p) and F(p) are the LAPLACE transforms of g(t) and f(t). H(p), the transfer function of the system, is the transform of the calorimeter response to a DIRAC unit impulse [7]. The experimental determination of the calorimeter response to a unit impulse, or, better, to its integral, a unit step which can be generated by JOULE heating, completely characterizes the calorimeter and therefore allows, at least theoretically, to deduce the kinetics of any thermal phenomenon from the experimental thermogram. Methods which allow to solve eq. 4 by means of the state-functions theory and/or the time-domain matrix methods have been recently described [8]. The use of computers is required and calorimetric data must be digitized before processing.

The range of linearity of a given calorimeter must however be carefully determined. This may be achieved by verifying the basic laws of linear systems. The unit-step response, $u_i(t)/P_i$, which can be calculated from the

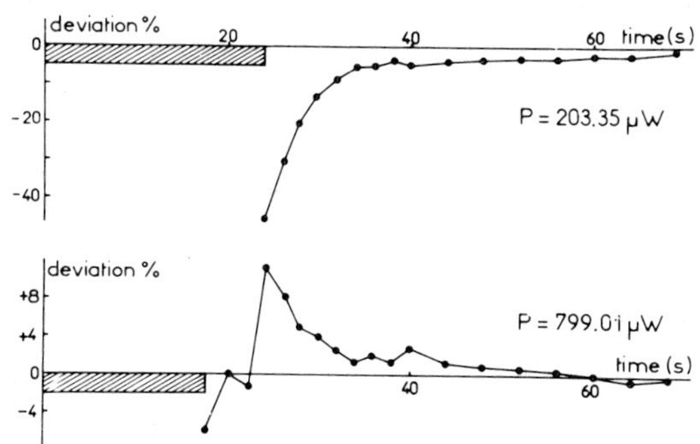

Fig. 4. Experimental verification of the linearity of a TIAN-CALVET microcalorimeter heated at 200°C, for two values of the thermal power. Reprinted from [9], with permission.

calorimeter response, $u_i(t)$, to a JOULE heating generating a thermal power P_i, should, for instance, remain constant for all thermal powers within the range of linearity and identical, at any instant, to the mean unit-step response. The deviations, from the mean, of two particular unit-step responses recorded with a TIAN-CALVET microcalorimeter are, as an example, presented on fig. 4 [9]. Forty seconds after the beginning of the JOULE heating, when the amplitude of the unit-step response is still less than 3 % of its final value, the deviation from linearity is less than 4 %. The larger deviations observed before 40 seconds (fig. 4) are caused by the errors in the determination of small outputs, the amplitude of which is close to that of the noise in the recording line. It has been shown moreover [9] that the time gap (shaded zones in fig. 4) which separates the initial instant of the experiment when the JOULE heating is switched on and the apparent beginning of the record is, in any case, the length of time which is needed for the signal to emerge from the noise. This is another proof of the calorimeter linearity [9].

The heat pulses or steps which are needed to determine the transfer function of a calorimeter are usually produced by JOULE heatings. However, heat paths within the calorimetric vessel may be different during JOULE heatings and the actual experiments. In order to assess the validity of the transfer function, it is necessary to perform preliminary experiments, similar to the experiment which is now described [10].

A dose of nitrous oxide was introduced in the constant-volume line connected to a TIAN-CALVET microcalorimeter heated at 200°C. A sample of nickel oxide, placed in the part of the line immersed into the calorimeter, was used as a catalyst for the decomposition of the reactant into nitrogen and oxygen. During the course of the reaction, a stopcock placed along the line was opened and closed at regular intervals (respectively 1 and 2 minutes) so that all or a part only of the gas remaining in the volumetric

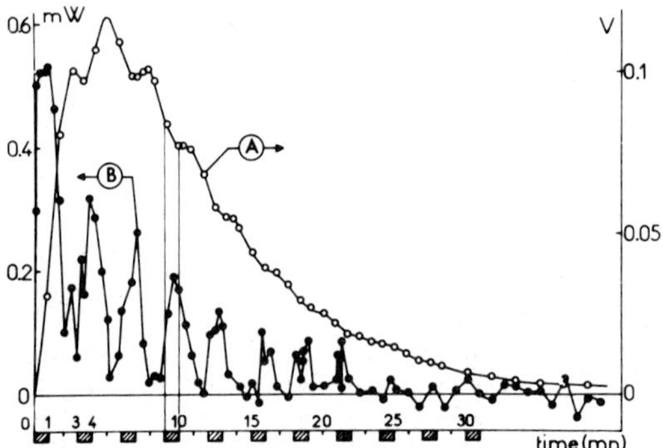

Fig. 5. Experimental (A) and corrected (B) thermograms for the catalytic decomposition of nitrous oxide at 200°C in a reactor whose volume is periodically changed (see the text) [10].

line was able to react at the catalyst surface. The shaded zones along the time axis in fig. 5 indicate the 1 mn-periods during which the stopcock was opened and thence all molecules in the line had a free access to the catalyst. The thermogram, recorded during this experiment (fig. 5, A) shows that, at least at the beginning of the experiment, the experimental procedure perturbs the catalytic process. As the reaction proceeds, however, the fluctuations of the thermal flux apparently become negligible. The corrected thermogram, deduced from the experimental data by application of eq. 4 and use of the transfer function determined by electrical calibration, is also presented on fig. 5 (B). The corrected thermogram shows, in contrast with the experimental calorimetric curve, that, in all cases, the successive openings and closings of the stopcock produce intense increases or decreases of the heat flux evolved by the reaction. Moreover, all thermal fluctuations detected on the corrected thermogram are perfectly synchronized with the details of the experimental procedure. This is, for instance, particularly shown for the 4th opening of the stopcock (guide lines in fig. 5), although the experimental thermogram only indicates a decrease of the heat flux for the same part of the experiment.

By means of the methods which have been outlined, it is therefore possible to determine, with accuracy, the thermokinetics of fast thermal processes occuring in the inner cell of a heat-flow calorimeter. These methods may, of course, find many applications in the fields of adsorption and heterogeneous catalysis.

Summing up the preceding discussion, the advantages of heat-flow calorimeters in heterogeneous catalysis include :

i) the ability to measure small heat outputs of short or long duration.

ii) the possibility to study poor heat-conducting solids with a large heat capacity.

iii) a very high sensitivity, if needed, and, generally, a good degree of accuracy.

iv) a very extended temperature range.

Finally v) heat-flow calorimeters are very nearly isothermal in operation and they can provide thermokinetic data.

3. SURFACE CHARACTERIZATION

Physical adsorption is commonly used to determine the surface structure of adsorbents and to measure their surface area. Gravimetry and volumetry are the usual techniques for such studies. Several hypotheses concerning the energetics of the adsorption process are however made in the course of the analysis of the data. Adsorption calorimetry can therefore provide more detailed informations about the surface structure than gravimetry or volumetry. Heat-flow calorimeters have the qualities required for these investigations.

The variations, at 25°C, of the integral heat of adsorption (and desorption) of water vapor on two samples of alumina, in the complete range of relative pressures, are, as an example, presented on fig. 6 [11]. The precision

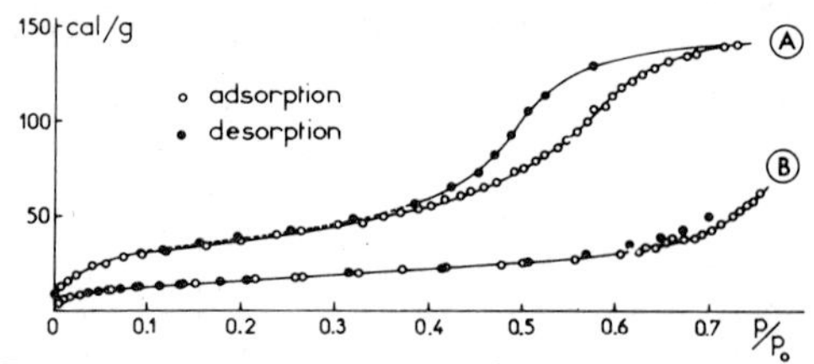

Fig. 6. Integral heats of adsorption and desorption of water vapor on "eta" (curves A) and "theta" (curves B) alumina at 25°C, plotted as a function of the relative equilibrium pressure. Reprinted from [11], with permission.

in the determination of both adsorption and desorption branches of the isotherms compete favorably with that of any volumetric or gravimetric techniques, especially if it is considered that, in order to avoid diffusional limitations, small samples were used (the surface area of the samples was, in both cases, close to 20 m^2, the specific surface area of "eta" and "theta" alumina being respectively 220 and 70 m^2/g). Moreover, the accuracy of the calorimetric "isotherms" is particularly good since the stability of the base line and the high sensitivity of TIAN-CALVET calorimeters provide a very sensitive criterion for the attainment of equilibrium : it is thus possible to detect, at any relative pressure, the adsorption of $\sim 10^{-2}$ µmole per hour of a gas being adsorbed with the production of 10 kcal/mole.

The curves presented on fig. 6 are particularly characteristic of the adsorption process and of the adsorbent surface since the integral heat, i. e. the total heat given off by the system when the relative pressure of the adsorbate is increased from zero to any particular value, is related to the adsorbed amount and to the average heat of adsorption of the adsorbed species. Curve B indicates, thus, that "theta" alumina has a very low porosity, if any. On the contrary, the large hysteresis loop of De Boer's type B or E, which is obtained in the case of "eta" alumina, reveals the presence of pores. Very similar results have been obtained by adsorption of nitrogen at 77°K on similar samples [12].

Heat-flow calorimeters present an additional important advantage for physical adsorption studies. Multilayer adsorption usually produces significant changes of the heat capacity of the sample [13]. When adiabatic or pseudo-adiabatic calorimeters are used for such studies, it is therefore necessary to recalibrate the instrument after the adsorption of each new increment of adsorbate. This increases considerably the complexity of the experimental procedure. As shown in the preceding Section, the operation of a heat-flow calorimeter does not require such frequent calibration experiments, the calibration ratio being independant of the heat capacity of the calorimetric cell and of its contents.

The calorimetric study of the successive adsorptions of separate incre-

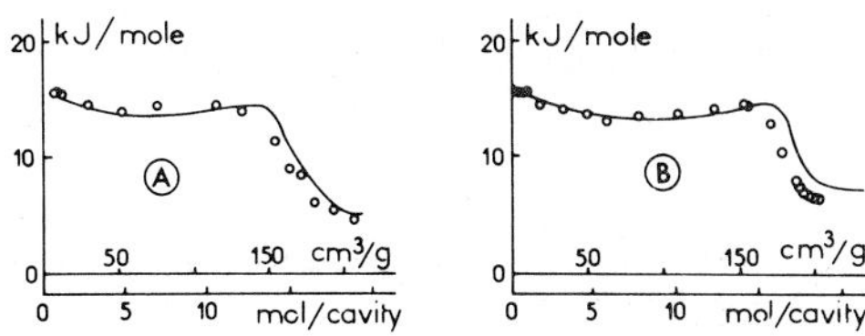

Fig. 7. Differential heats of adsorption of oxygen on zeolite 5A (curve A) or 13X (curve B) at 77°K, determined by the continuous (line) or incremental (open circles) methods. Reprinted from [15], with permission.

ments of adsorbate, which is often undertaken to determine the differential heats of adsorption or, as in fig. 6, the variations of the integral heat as a function of the relative equilibrium pressure of the adsorbate, is time-consuming, especially when a sensitive calorimeter is used. The stability of the record base-line, obtained with heat-flow calorimeters, however, allows the use of faster procedures. A flow of adsorbate may be, for instance, continuously introduced onto the sample of adsorbent, placed in the calorimeter cell. If the adsorption process is fast enough, as it usually is in physical adsorption, and if the flow rate is small enough, the system may be, at any instant, close to equilibrium. The recorded thermogram is, then, a direct measure of the differential heat of adsorption as a function of coverage [14]. Fig. 7 shows that in the case of the adsorption of, e.g., oxygen on two different zeolites (5A and 13X) at 77°K, both continuous and incremental methods yield very similar results [15]. The continuous curves in fig. 7 were obtained in a few hours ; it took several days to determine the same curves by the adsorption of separate increments of adsorbate (circles in fig. 7).

The chief advantage of the continuous method for studying physical adsorption is that it gives a more detailed representation of the process than the incremental one. This can be illustrated by the thermogram presented on fig. 8 which refers to the adsorption of argon on Graphon at 77°K [16]. After the initial increase of the heat due to the thermal inertia of the calorimeter, the differential heat increases regularly to a maximum which corresponds to the completion of the first layer in agreement with BEEBE's calori-

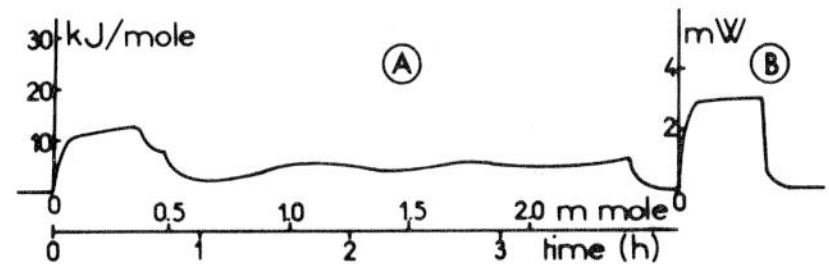

Fig. 8. Thermograms recorded during the continuous adsorption of Argon on Graphon at 77°K (A) and during a calibration test (B). Reprinted from [16], with permission.

metric results for the same system [17]. Completion of the second and third layers are also clearly accompanied by maxima of the differential heat. In all cases, the maxima are explained by lateral interaction of adsorbate molecules on the very uniform surface [16,17,18]. Using an incremental procedure, BEEBE et al. [17,18] also obtained some indications of such maxima but the evidence was not very convincing being based on a very small number of experimental points. The accident which appears on the curve in fig. 8 after the first maximum corresponding to the completion of the first layer has been tentatively interpreted as a change in the physical state of the adsorbed layer (from mobile to localized adsorption) [19]. Such a detail which may be very significant would probably either not be determined or be disregarded with the incremental method.

For a correct application of the continuous method, however, equilibrium must be nearly achieved at any instant. In some particular cases, this condition cannot be fulfilled. Heat-flow calorimetry has indicated, for instance, that in the case of the adsorption of argon or nitrogen, at 77°K, on a porous silica gel, heat is still evolved after the stabilization of the equilibrium pressure of the adsorbate [20]. This has been considered as an indication of surface diffusion of adsorbed molecules from low-potential to high-potential adsorption sites. In such cases, the continuous method must be used with extreme caution and the incremental method probably yields more reliable results.

Heat-flow calorimetry may be also used to characterize the chemical affinity of a catalyst with respect to a gaseous reactant and to determine the subsequent reactivity of the adsorbed species. The calorimetric oxygen-hydrogen titration method for determining the exposed metal area on oxide-supported platinum catalysts, which is presented in one of the papers submitted to this Congress [21], is, for instance, a typical application of heat-flow calorimetry in this field.

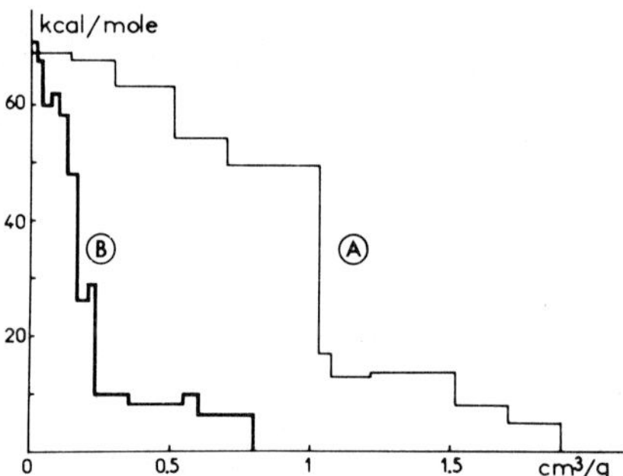

Fig. 9. Differential heats of adsorption of oxygen, at 30°C, on samples of nickel oxide prepared by the dehydration of nickel hydroxide at 240°C either under vacuum (A) or in the presence of water vapor (B) [23].

Several examples related to the adsorption of oxygen on nickel-oxide catalysts will be presented to illustrate this point. Gravimetric and volumetric measurements have indicated that, at 30°C, the affinity towards oxygen of samples of nickel oxide prepared by dehydration of nickel hydroxide at 240°C either in a vacuum or in the presence of water vapor (4.6 torr) is very different [22,23]. The vacuum-prepared sample has a large affinity for oxygen and adsorbs it readily at pressures as low as 10^{-4} torr. The hydrated sample, however, has a much smaller affinity for oxygen. This is indeed confirmed by heat-flow calorimetry as shown in fig. 9 [23]. But the calorimetric results indicate also that the heat of the oxygen adsorption on the small number of reactive sites which subsist on the hydrated solid is very similar to that determined in the case of the vacuum-prepared sample. Moreover, the subsequent interaction of carbon monoxide, at 30°C, with the pre-adsorbed species yields, in both cases, very similar heats (80 to 85 kcal/mole) [23]. In both cases, carbon dioxide is formed. The hydroxylation of the nickel-oxide surface considerably decreases the number of sites which are very reactive with respect to oxygen but it does not influence the reactivity of the small number of remaining free sites. This is an unexpected result which could not, probably, be determined as easily with other experimental techniques as with heat-flow microcalorimetry.

The high sensitivity which characterizes many heat-flow microcalorimeters is, in many cases, required if the "energy spectrum" of the catalyst surface, with respect to a given reactant, is to be determined accurately. Such precise determinations are often indeed of a paramount importance since apparently small changes in the catalyst preparation or pretreatment may induce important modifications of its energy spectrum and, thereby, of its catalytic activity. Such an example is presented in fig. 10 where a 50-degree increase in the preparation temperature of nickel oxide catalysts (from 200 to 250°C) is shown to considerably modify the energy spectrum of these samples with respect to oxygen at 30°C [24]. This modification explains as was demonstrated earlier, their different catalytic activities in the room-temperature combustion of carbon monoxide [24].

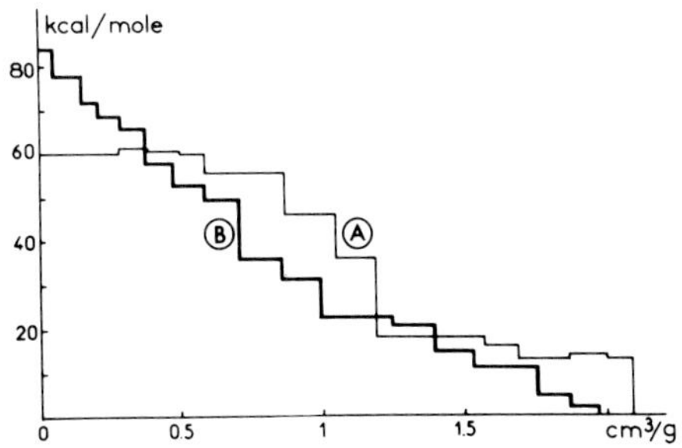

Fig. 10. Differential heats of adsorption of oxygen, at 30°C, on samples of nickel oxide prepared by the dehydration of nickel hydroxide, in vacuo, at 200°C (A) and 250°C (B) [24].

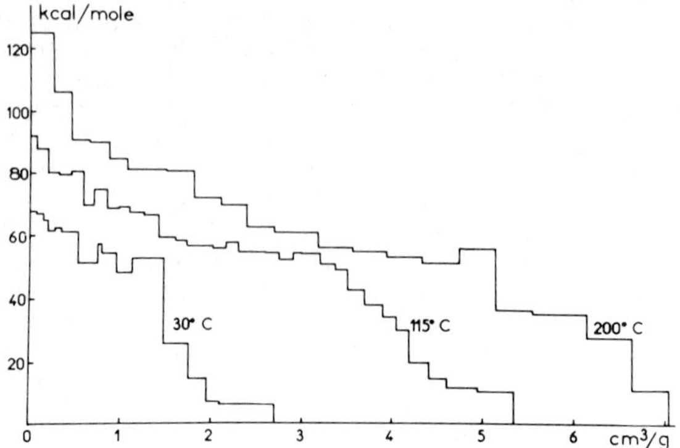

Fig. 11. Differential heats of adsorption of oxygen on nickel oxide, measured at 30°C, 115°C and 200°C [25].

Moreover, the energy spectrum must be determined at the temperature of interest : temperature changes often induce a modification of the surface structure of a catalyst and, thereby, of its affinity towards a reactant. Fig. 11 shows, for instance, that the number of sites, reactive with respect to oxygen, which are present on the surface of a nearly stoichiometric nickel oxide, increases when the temperature increases and that their average reactivity, as measured by the heat of adsorption, increases simultaneously [25].

A progressive decrease of the adsorption rate for the successive increments is frequently observed in chemisorption studies. Thermogram A in

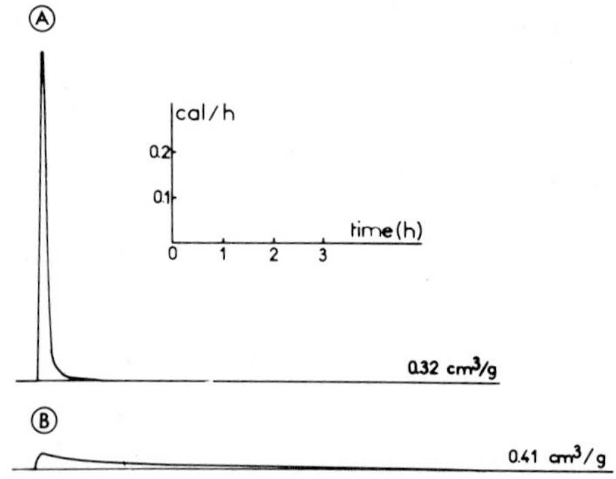

Fig. 12. Thermograms recorded during the adsorption of doses of oxygen on nickel oxide at 200°C [26].

fig. 12 was recorded during the adsorption of one of the first increments of oxygen on nickel oxide at 200°C. Thermogram B was recorded at the end of the same experiment [26]. The ability of heat-flow microcalorimeters to study either fast or slow thermal processes is therefore extremely useful in many chemisorption experiments.

The preceding examples have shown that, in chemisorption, the energy spectrum of the surface sites is quite sensitive to any modification of the surface structure of the adsorbent. It appears therefore that, in the general case, one should not rely upon "generally accepted" values for the heat of adsorption of a given adsorbate on a given adsorbent. When needed, heats of adsorption should be actually measured, the temperature, the catalyst pretreatment and all experimental conditions being identical to those of, for instance, the catalytic reaction. Versatile calorimeters are therefore required : heat-flow microcalorimeters are very convenient for this purpose.

4. REACTION MECHANISMS

The adsorption of a reactant is the first step of the mechanism of a catalytic reaction. This step is followed by surface interactions between adsorbed species or between a gaseous reactant and adsorbed species. In many cases, these interactions may be detected by the successive adsorptions of the reactants in different sequences. Heat-flow calorimetry can be used with profit for such studies [2,27]. The differential heats of the successive adsorptions, compared for different surface coverages with thermodynamic data in thermochemical cycles, frequently yield useful informations on reaction mechanisms. These informations must however be ascertained by studies of the catalytic reaction itself. Heat-flow calorimeters can also be used for these studies. As an example, the volumetric and calorimetric results of the successive and alternate adsorptions of carbon monoxide and oxygen, at 300°C, on a sample of cerium oxide, $CeO_{1.985}$, are reported in table 1 [28]. A heat-flow calorimeter of the EYRAUD type was used for this study [2,29].

TABLE 1

Successive and alternate adsorptions of carbon monoxide and oxygen on ceria at 300°C [28].

V_{CO} (cm^3/g)	V_{O_2} (cm^3/g)	V_{CO}/V_{O_2}	Q_{CO} kcal/mole	Q_{O_2} kcal/mole	$Q_{CO} + \frac{1}{2} Q_{O_2}$ kcal/mole
0.319	0.153	0.479	51	70	86
0.318	0.158	0.496	42	69	77
0.314	0.150	0.476	44	58	73
0.318	0.159	0.500	44	58	73
0.310	0.169	0.545	44	50	69

The total volume of carbon monoxide, adsorbed in successive portions, was limited, during all CO-adsorptions of the sequences $CO-O_2$, to very small values (0.3 cm^3/g, table 1, column 1), compared to the adsorption capacity of the sample ($\sim$ 9 cm3/g) [28]. During all oxygen adsorptions, the introduction of successive doses of the gas was repeated until the heat evolved was equal to 5 kcal/mole, which was considered as the lower limit for the heat corresponding to the strong adsorption of this adsorbate. The corresponding volumes of adsorbed oxygen are given in table 1 (column 2). In columns 4 and 5, the average heats of adsorption of, respectively, carbon monoxide and oxygen are reported.

Under these experimental conditions, the oxidation of carbon monoxide to carbon dioxide takes place. Carbon dioxide is indeed detected in the cold trap placed near the sample. The fact that the adsorption of oxygen at 300°C on the freshly prepared sample is minute (0.06 cm^3/g) but occurs to a larger extent on a sample containing preadsorbed carbon monoxide (table 1) is an additional evidence of the catalytic combustion [28]. Moreover, the reactants are consumed, during all successive adsorptions, in a proportion which is, within experimental error, in agreement with the normal stoichiometry for the reaction (table 1, column 3) :

$$CO_{(g)} + \frac{1}{2} O_{2\ (g)} \longrightarrow CO_{2\ (g)} \qquad (5)$$

The decrease of the heats of adsorption which is observed for both reactants when the number of adsorption sequences increases and which is especially pronounced in the case of the adsorption of oxygen (table 1) indicates that the surface interactions, which initially take place on reactive surface sites, move on to less reactive sites as successive adsorptions are performed on the same sample. This result suggests a progressive inhibition of the reactive sites by the reaction product, carbon dioxide.

The comparison of the heats measured during the carbon monoxide and oxygen adsorptions in each sequence with thermodynamic data confirms this hypothesis. The sum of the heat released by CO-adsorption and half the heat released by O_2-adsorption, interactions which, together, yield carbon dioxide should be equal to $\sim$ 68 kcal/mole, the heat of formation of gaseous carbon dioxide in the homogeneous phase. Calculated values for $Q_{CO} + 1/2\ Q_{O_2}$, given in table 1 (column 6) exceed, for all sequences but the last, 68 kcal/mole. A surface interaction occurs which does not yield gaseous carbon dioxide and, yet, consumes both reactants in the same proportion as reaction 5; it must yield adsorbed carbon dioxide :

$$CO_{(g)} + \frac{1}{2} O_{2\ (g)} \longrightarrow CO_{2\ (ads)} \qquad (6)$$

A calorimetric study of the adsorption of carbon dioxide on ceria at 300°C has shown, however, that reaction 6 should evolve $\sim$ 110 kcal/mole when all generated carbon dioxide (0.3 cm^3/g) remains adsorbed on the most reactive surface sites. The lower values in table 1 indicate that inhibition by adsorbed carbon dioxide proceeds but progressively, a fraction of generated carbon dioxide being desorbed during all adsorption sequences. Inhibition is completed after the 4th adsorption sequence.

A calorimetric study of the combustion of carbon monoxide on ceria at 300°C has confirmed the inhibition of reactive surface sites by the reaction product [28]. When small doses of the stoichiometric reaction mixture successively react on the catalyst, the heat of reaction initially rises to 80 kcal/mole and decreases to 68 kcal/mole when ~ 3 cm^3/g of mixture have reacted. Simultaneously, the high initial catalytic activity decreases to its steady state value.

Heat-flow microcalorimeters can also be used to study the structural modifications of the catalyst surface which frequently occur in the course of the catalytic reaction, especially when the catalyst is initially in a divided and unstable state, and produce the decrease of its activity. An example of the catalyst "break-in" phenomenon is given by the catalytic combustion of carbon monoxide in excess oxygen at ~ 200°C on samples of nickel oxide prepared at 200 or 250°C by the vacuum dehydration of the hydroxide[26]. Continuous use of a 5 mg-sample of the catalyst at 208°C (p_{CO} = 44.7 torr) during 44 hours causes, for instance, a tenfold decrease of its activity (from 147 cm^3 CO_2/h initially to 14 cm^3 CO_2/h). Thermal sintering due to the reaction exothermicity does not explain completely the deactivation since a presintering of the catalyst during 20 hours, at 308°C, temperature which is not attained in the catalytic reactor, only produces a twofold decrease of the activity (73 cm^3 CO_2/g compared to 147 cm^3 CO_2/g). Irreversible interactions between the reactants and the catalyst probably contribute to the deactivation.

The calorimetric study of the catalytic reaction confirms the existence of such catalyst modifications [26]. On fig. 13, are reported the experimental heats recorded, at 200°C, during the reaction of successive doses of CO-O_2 mixtures, containing an excess of oxygen, at the surface of divided (curve A) and sintered (curve B) nickel oxide. Heats exceeding 45 kcal/mole of

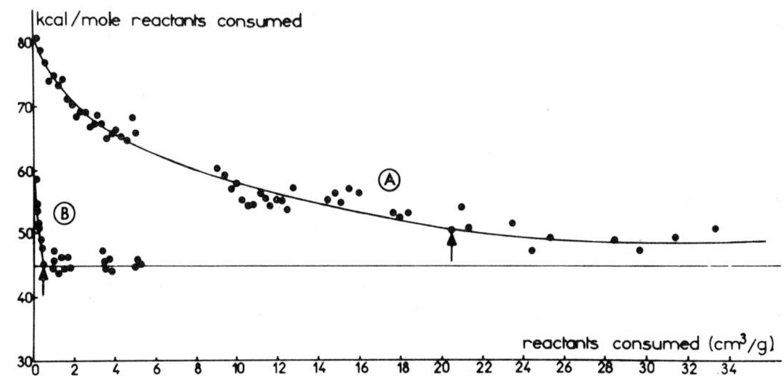

Fig. 13. Differential heats for the catalytic combustion, at 200°C, of carbon monoxide in excess oxygen (volumetric composition of the mixtures : 50 % CO, 50 % O_2, curve A ; 57 % CO and 43 % O_2, curve B) at the surface of a divided nickel oxide, prepared by the vacuum dehydration of nickel oxide at 200°C (curve A) [26] and at the surface of a sintered nickel oxide, prepared in air at 600°C from basic nickel carbonate and outgassed in vacuo at 500°C (curve B).

mixture, i.e., 68 kcal/mole CO, measured at the beginning of both experiments are related, in each case, to the adsorption of oxygen, introduced in excess : after the reaction of all doses on the left of the arrows (fig. 13), the final pressure is very low ($\sim 10^{-4}$ torr). When the catalyst surface contains a maximum amount of oxygen irreversibly adsorbed under these conditions, excess oxygen in all doses (right side of the arrows in fig. 13) remains in the gas phase, both reactants being then consumed in a normal proportion. All carbon dioxide formed is, within experimental error, desorbed to the gas phase and condensed in a cold trap. No inhibition by carbon dioxide occurs.

The difference between the normal heat for the reaction (45 kcal/mole of mixture) and the experimental heats for the doses on the left of the arrows (fig. 13) allows to calculate, mixtures composition being known, the differential heats of adsorption of oxygen in excess. In the case of the reaction on sintered nickel oxide, they vary from 40 to 117 kcal/mole O_2, in correct agreement with the heats measured in a direct oxygen adsorption. In the case of the reaction on divided nickel oxide, the calculated heats (80 to 180 kcal/mole O_2), however, considerably exceed those given by a direct experiment (fig. 11). During the reaction of the reaction mixture, at the beginning of the experiment with divided nickel oxide, interactions occur which are not simply related to the combustion of carbon monoxide and the adsorption of oxygen and which produce the evolution of extra heat. Moreover, the heats (average value : 49 kcal/mole) recorded when the reaction is performed on the divided catalyst containing a maximum amount of adsorbed oxygen (doses on the right of the arrow on curve A, fig. 13) still exceed 45 kcal/mole. Since both reactants are then consumed in a normal proportion and since carbon dioxide is desorbed to the gas phase, the heat produced by the catalytic process, itself, <u>must</u> be equal to the theoretical value. The excess heat which is experimentally recorded must be given off by the catalyst which thereby becomes, energetically, more stable and, catalytically, less active. The modification of the catalyst internal energy is not observed in the case of the sintered solid : the experimental heat for all doses on the right side of the arrow (fig. 13, curve B) are, within experimental error, equal to the theoretical heat.

A calorimetric study of the successive adsorptions of the reactants on the divided nickel oxide has given some informations on the surface interactions which cause the catalyst stabilization and deactivation. The differential heats for the interaction, at 200°C, of carbon monoxide with an oxygen-precovered sample of divided nickel oxide are given in fig. 14 (differential heats of adsorption of oxygen, at 200°C, on the same sample have been presented in fig. 11). The differential heats are, initially, high and almost constant ($\sim$ 55 kcal/mole) ; they decrease, however, when a total volume of carbon monoxide close to 6.5 cm^3/g has reacted with the solid, to values similar to those measured when a freshly prepared sample is reduced by carbon monoxide, with the formation of metal cristallites [30]. The thermograms, the integration of which has given the differential heats for doses A to J in fig. 14, are presented in fig. 15. They indicate that the surface reactivity

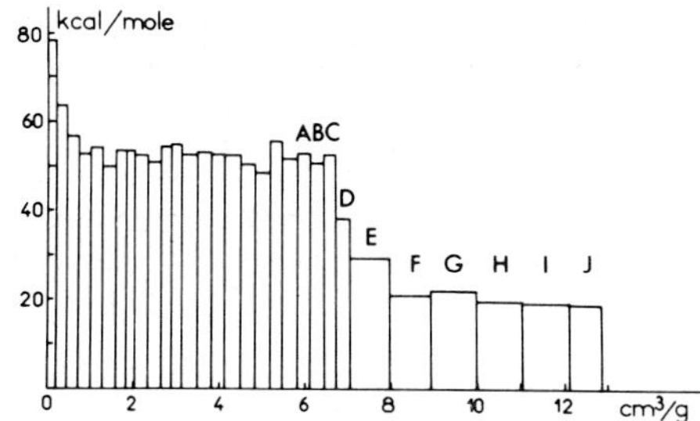

Fig. 14. Differential heats for the interaction, at 200°C, of carbon monoxide with a sample of divided nickel oxide, containing preadsorbed oxygen ($\sim$ 7 cm^3 O_2/g). Reprinted,with permission, from [27].

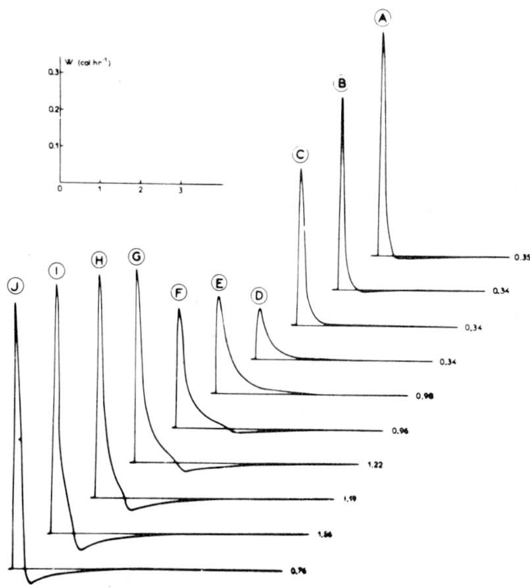

Fig. 15. Thermograms recorded during the interaction of doses A to J (fig. 14) of carbon monoxide with nickel oxide, containing preadsorbed oxygen. Reprinted from [27] with permission.

towards carbon monoxide is, first, high (thermograms A to C in fig. 15 corresponding to doses A to C in fig. 14) but that the decrease of the differential heats (doses D to J in fig. 14) is correlated with a very evident modification and evolution of the thermogram profile (thermograms D to J in fig. 15). It has been already shown that these thermograms are characteristic for the reduction process [30]. Heat-flow calorimetry provides thus a very sensitive method for determining the amount of carbon monoxide which interacts with adsorbed oxygen, without reducing the oxide surface.

The volume of carbon monoxide ($\sim$ 6.5 cm^3 CO/g) which has reacted with adsorbed oxygen in the course of the experiment reported on fig. 14, is not very different from the volume of irreversibly preadsorbed oxygen ($\sim$ 7 cm^3 O_2/g, fig. 11). The stoichiometry of the interaction suggests, therefore, the formation of a surface carbonate. However, all molecules of carbon monoxide in all successive doses are, within experimental error, oxidized to carbon dioxide which is condensed in a cold trap. A possible explanation of the apparent contradiction of the results could be that a fraction, close to 50 %, of adsorbed oxygen, the fraction chemisorbed with the production of the highest heats, behaves as surface lattice oxygen ions. The interaction of carbon monoxide, at 200°C, with samples containing limited amounts of strongly adsorbed oxygen was therefore studied [31]. In all experiments, the same stoichiometric ratio CO/O_2, equal to 1, was determined. It was concluded that the ratio cannot be apparent ; surface carbonate species are indeed formed but they must spontaneously decompose to yield carbon dioxide and surface oxygen :

$$CO_{(g)} + 2\ O_{(ads)} \longrightarrow CO_{3(ads)} \longrightarrow CO_{2(g)} + O_{(s)} \qquad (7)$$

From the calorimetric data, it has been possible to show that $O_{(ads)}$ and $O_{(s)}$ are not equivalent : the surface interactions, summarized in eq. 7,

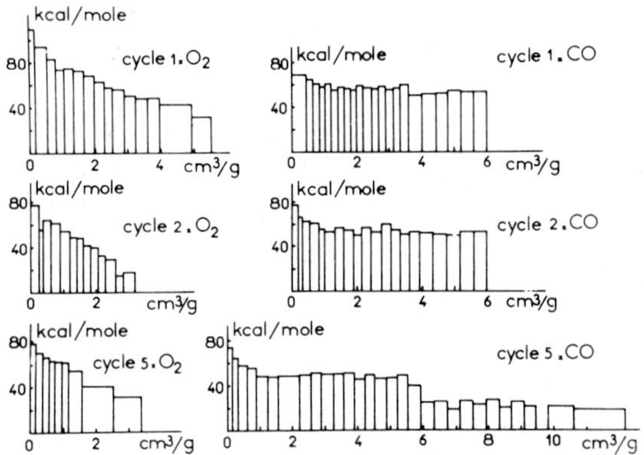

Fig. 16. Differential heats for the successive and alternate adsorptions of oxygen and carbon monoxide on divided nickel oxide at 200°C [26]. Cycles 3 and 4 are not presented for clarity.

produce the incorporation of adsorbed oxygen in the surface lattice, with the release of $\sim$ 12.5 kcal/atom [26].

The incorporation of excess oxygen in the surface lattice, which is a consequence of the interactions between reactants at the surface of the divided solid, is, however, limited to small amounts of oxygen, and, probably, cannot account for the observed deactivation of the catalyst in the course of its use. In order to eventually detect other surface modifications, alternate and successive adsorptions of oxygen and carbon monoxide were studied calorimetrically at 200°C. The results for some sequences (1, 2 and 5) are reported in fig. 16. The adsorption of carbon monoxide was, in all sequences but the last, limited to $\sim$ 6 cm^3 CO/g in order to avoid the reduction of the oxide surface (surface reduction does not occur during the catalytic reaction of a mixture containing an excess of oxygen). Doses of oxygen were introduced until no heat effect could be detected.

The differential heats for the adsorption of carbon monoxide ($\sim$ 6 cm^3 CO/g) are, for all sequences, identical to those measured during the preceding experiment (fig. 14). Carbonate species are, therefore, probably formed. They must spontaneously decompose, as before, since a stoichiometric amount of carbon dioxide is detected in the cold trap after each adsorption of carbon monoxide. The formation of the carbonate species requires, however, the participation of labile surface anions since oxygen from the gas phase is now consumed (cycles 2 and 5 in fig. 16) in a proportion corresponding to the normal stoichiometry for the combustion of carbon monoxide :

$$CO_{(g)} + O_{(ads)} + O_{(labile)} \longrightarrow CO_{3(ads)} \longrightarrow CO_{2(g)} + O_{(s)} \qquad (8)$$

The participation of lattice ions in the reaction mechanism (eq. 8) is indeed confirmed by the calorimetric data. The sum of the average heats for the successive adsorptions of the reactants (Q_{CO} + 1/2 Q_{O_2}) exceeds, during all sequences, the theoretical value ($\sim$ 68 kcal/mole CO) : the excess heat (10 to 16 kcal/mole CO) must be given off by the catalyst. The stabilization of labile lattice anions is, thus, likely to occur since it agrees with all experimental evidence.

A final test of the accuracy of the reaction mechanisms (eq. 7 and 8), deduced from the successive adsorptions of the reactants, was to calculate, from the calorimetric results for the catalytic reaction at 200°C on the divided nickel oxide (curve A in fig. 13), the differential heats of adsorption of oxygen in excess, by assuming that interactions 7 and 8 take place. The calculated heats are compared to the experimental heats of adsorption of oxygen in fig. 17. Their excellent agreement is an additional proof of the self-consistency of the proposed mechanisms. Heat-flow calorimetry can be thus utilized not only to determine probable reaction mechanisms but, also, to study the probable surface interactions (the incorporation, in the lattice, of adsorbed and labile oxygen ions, in the preceding example) which may cause structural modifications of the catalyst surface and, thereby, its deactivation, in the course of the catalytic reaction.

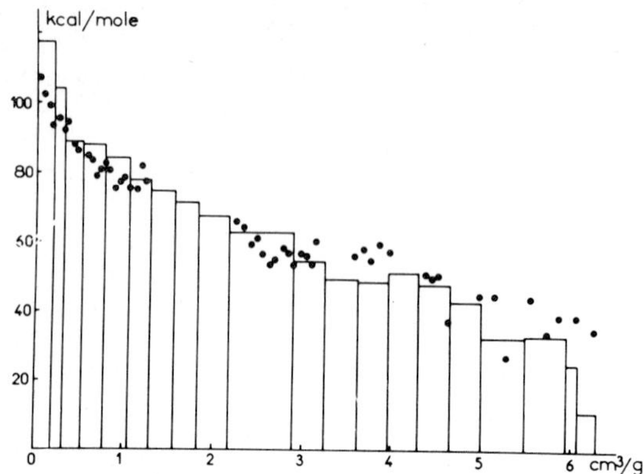

Fig. 17. Differential heats of adsorption of oxygen on divided nickel oxide, as calculated from the results in fig. 13, by means of eq. 7 and 8 (see the text) (filled circles) and as measured experimentally (stepped curve) [26].

5. ADSORPTION AND REACTION KINETICS

The possibility offered by heat-flow calorimeters to measure, at any instant, the heat flux produced by the thermal process under study is one of the most important advantages of calorimeters of this type,as already indicated in Section 2. A correction of the thermograms, for the thermal lag in the calorimeter, is, of course, necessary when a complete analysis of thermokinetics is required but the direct examination of recorded thermograms frequently reveals significant informations about the process taking place in the calorimeter. It has been shown, for instance, that the thermograms D to J in fig. 15 and the evolution of their profile are characteristic of the processes (germination of metal cristallites and metal-oxide interfacial reaction) occuring during the reduction of a metal oxide [30].

The thermograms corresponding to adsorption or catalytic processes often reveal, thus, secondary phenomena which may influence the main process under investigation but would otherwise remain undetected. Curves A and B (reaction yield versus time) in fig. 18 show that the catalytic activity of a sample of titanium oxide, after a treatment in a reducing atmosphere (H_2) at 500°C during 15 hours, decreases but slightly when two stoichiometric mixtures of carbon monoxide and oxygen ($p_{CO} = \sim 0.6$ torr) successively react on its surface at 500°C. The thermograms (fig. 18), recorded at 500°C during the same experiments, however indicate that when the reaction mixture first contacts the fresh catalyst a transient but intense heat effect (A) appears

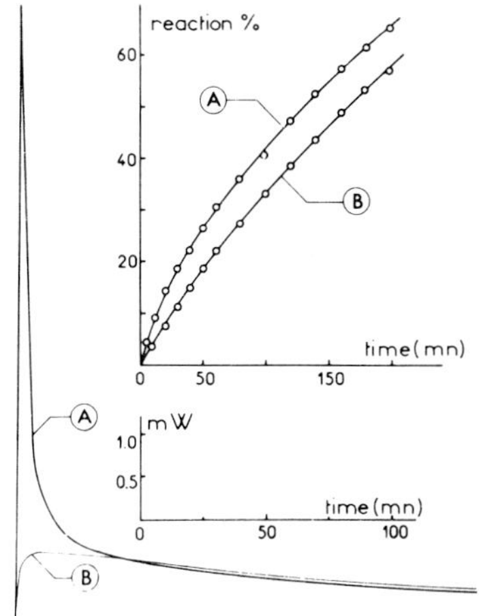

Fig. 18. Reaction yield as a function of time and thermograms for carbon-monoxide oxidation at 500°C on titanium oxide (anatase). Tests A and B were performed successively on a sample pretreated in a reducing atmosphere (H_2) at 500°C during 15 hours.

which does not exist during the 2nd reaction test (B).

By using small doses of reaction mixture, it has been possible to determine the differential heats corresponding to the transient phenomenon ($\sim$ 150 kcal/mole) and to relate these high heats to the irreversible interaction of oxygen from the mixture with surface sites created during the reducing treatment [2,32]. The volume of oxygen involved in this surface interaction, which produces a modification of the surface structure and, thereby, of the surface properties of the catalyst, is however small ($\sim$ 0.5 cm^3/g) and this surface interaction is not detected in a usual reactor.

In the case of relatively slow thermal processes, the thermograms can be, simply but correctly, analyzed by means of TIAN equation 1 (error $<$ 10%). Fig. 19 represents, for instance, the percentage of heat evolved, as a function of time, during several interactions taking place, at 30°C, on the surface of a gallium-doped nickel oxide [33]. These interactions are the steps of two possible reaction mechanisms for the room-temperature combustion of carbon monoxide on this catalyst :

<u>Mechanism 1</u> :

$$\frac{1}{2}\, O_{2(g)} + Ni^{2+} \longrightarrow O^-_{(ads)} + Ni^{3+} \qquad (9)$$

$$O^-_{(ads)} + CO_{(g)} + Ni^{3+} \longrightarrow CO_{2(g)} + Ni^{2+} \qquad (10)$$

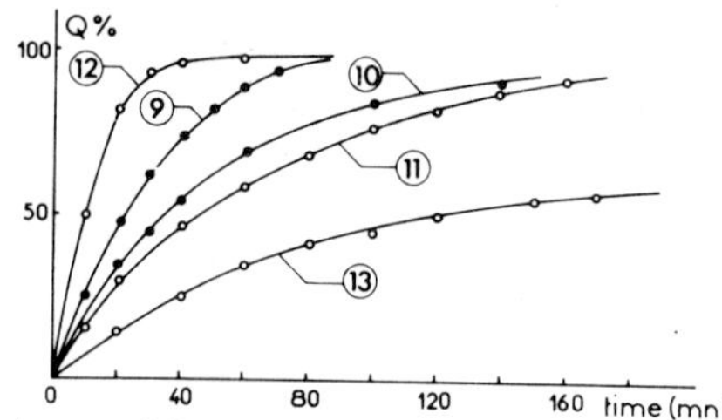

Fig. 19. Percentage of heat evolved as a function of time, at 30°C, for several interactions at the surface of gallium-doped nickel oxide. Reaction mechanism 1 (filled circles) ; Reaction mechanism 2 (open circles). Reprinted from 33) with permission.

Mechanism 2 :

$$CO_{(g)} \longrightarrow CO_{(ads)} \qquad (11)$$

$$CO_{(ads)} + O_{2(g)} + Ni^{2+} \longrightarrow CO_{3(ads)}^{-} + Ni^{3+} \qquad (12)$$

$$CO_{3(ads)}^{-} + CO_{(g)} + Ni^{3+} \longrightarrow 2\ CO_{2(g)} + Ni^{2+} \qquad (13)$$

It is clear, from the results in fig. 19, that the interaction of carbon monoxide with preadsorbed oxygen is the slowest step of mechanism 1. Interaction between $CO_{3(ads)}^{-}$ species and carbon monoxide is the slowest step of mechanism 2. Moreover, interaction 13 is slower than the adsorption of carbon monoxide (eq. 11). The rate-determining step of mechanism 2 probably involves adsorbed carbon monoxide. This interpretation of the calorimetric data is in agreement with the conclusions of the kinetic study of the same reaction at the surface of a sample of pure nickel oxide on which mechanism 2 actually governs the catalytic reaction 34). On the gallium-doped nickel oxide, however, mechanism 1 which allows faster reaction rates probably prevails, this change of reaction mechanism being the cause of the larger activity of gallium-doped nickel oxide compared to the sample of pure nickel oxide 24).

The thermograms, the analysis of which has given the results presented in fig. 19, were recorded during the adsorption or interaction of very small amounts of reactants for a given surface coverage of the catalyst. Interaction rates in fig. 19, therefore, characterize the reactivity of a small number of sites on which interactions 9 to 13 actually take place, as shown by thermochemical cycles. The same method may be used for different surface coverages and may give informations not only about rate-limiting steps but also about the distribution of active sites on the catalyst surface.

Because of their high sensitivity, heat-flow microcalorimeters can be used to study the thermokinetics of processes which are less easily investigated by more conventional techniques. The decomposition of nitrous oxide on

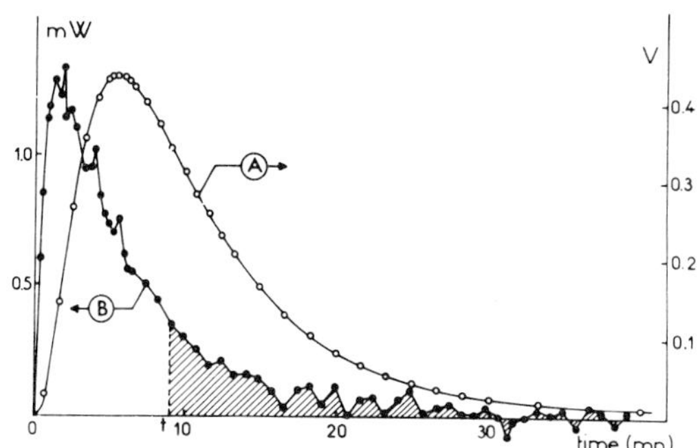

Fig. 20. Experimental (A) and corrected (B) thermograms for the catalytic decomposition of nitrous oxide (4×10^{-2} cm^3) at the surface of a sample (77 mg) of nickel oxide at 200°C [10].

oxide surfaces with the formation of gaseous nitrogen and <u>adsorbed</u> oxygen is an example of such processes since it is limited by the small adsorption capacity of many oxide surfaces for oxygen. It is generally assumed that this is a monomolecular process :

$$N_2O_{(g)} \longrightarrow N_{2(g)} + O_{(ads)} \qquad (14)$$

and this reaction is sometimes used to generate adsorbed oxygen atoms or atomic species. Thermogram A in fig. 20 was recorded during the decomposition of a small volume of nitrous oxide (4×10^{-2} cm^3) at the surface of nickel oxide at 200°C [10]. All generated oxygen remained on the oxide surface. The corrected thermogram B in fig. 20 was deduced from the experimental one by means of the techniques outlined in Section 2. In the case of a simple monomolecular process, the reaction rate is evidently at any instant proportional to the quantity of reactant remaining in the system. The heat flux at any instant, i.e., the ordinate of thermogram B in fig. 20, is a measure of the reaction rate. The total heat evolved, after time t, is evidently produced by the reaction of all reactant remaining in the system at time t. If it is supposed that all sites involved in the process present a uniform reactivity towards oxygen and a uniform catalytic activity (assumptions which are reasonably valid when a small fraction of available surface sites is involved), a plot of the ordinate of thermogram B as a function of the area under thermogram B from time t to the end of the process (shaded area in fig. 20) should therefore yield a straight line. Fig. 21 shows that this is indeed the case. Reaction 14 is a monomolecular process. The first-order rate constant calculated from the slope of the straight line is 2.20 $sec^{-1}.m^2$.

A particularly important application of kinetic studies by means of

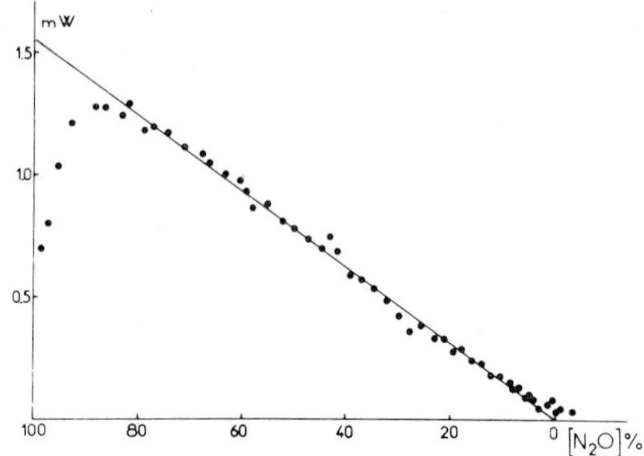

Fig. 21. Representation of the calorimetric data in fig. 20 (thermogram B) according to first-order kinetics [10].

sensitive heat-flow microcalorimeters, in the field of heterogeneous catalysis, is the determination of the activity spectrum of a catalyst. This may be achieved by repeating experiments and calculations similar to those reported in figs. 20 and 21 either for progressively larger doses of reaction mixture or on a catalyst which is treated, before the calorimetric experiments, by progressively larger amounts of a catalyst inhibitor.

6. CONCLUSIONS

By means of examples taken from recent studies, it has been attempted to show that the applications of heat-flow calorimetry in the fields of adsorption and heterogeneous catalysis are very diversified. Heat-flow calorimeters have been used to characterize catalyst surfaces, to study the adsorption spectra of reactants, to determine probable reaction mechanisms and, even, to evaluate the activity spectra of catalytic surfaces. Many examples have referred to simple adsorption or catalytic processes but heat-flow calorimeters can probably be used, with success, in the case of more complex systems. Because of their sensitivity, their reliability, their extended temperature range of operation, their increasing availability, it is therefore believed that heat-flow microcalorimeters can contribute, in a significant manner, to future progress in the Science of Catalysis.

The author's thanks are due to Mrs. M. GRAVELLE, to Pr. J.L. PETIT, to Drs. C. BRIE, G. EL SHOBAKY, G. MARTY, G. REY and J.P. REYMOND, for their active collaboration, to Pr. B. CLAUDEL and Dr. M. BREYSSE, to Pr. G. VENTURELLO and Dr. G. DELLA GATTA, to Dr. J. ROUQUEROL, for communicating still unpublished informations and to Pr. S.J. TEICHNER for a critical reading of the manuscript.

REFERENCES

1) S. Sunner, Summer School on Calorimetry, Villeurbanne, France, Sept. 1971 (not published).
2) P.C. Gravelle, Advan. in Catalysis, 22 (1972), 191.
3) H.A. Skinner, J.M. Sturtevant and S. Sunner, in"Experimental Thermochemistry" Vol 2 (H.A. Skinner Ed.), Interscience, John Wiley, New-York, (1962), p. 163 ; H.A. Skinner in "Biochemical Microcalorimetry" (H.D. Brown Ed.), Academic Press, New York, (1969), p. 7 ; F.M. Camia, J. Chabert, P.C. Gravelle, M. Laffitte, J.L. Macqueron, J.L. Petit and P. Tachoire, Rev. Gen. Therm., 7 (1968), 895.
4) E. Calvet and H. Prat, "Microcalorimétrie, applications physicochimiques et biologiques", Masson, Paris (1956) ; E. Calvet, H. Prat and H.A. Skinner, "Recent Progress in Microcalorimetry" Pergamon, New York (1963).
5) P.C. Gravelle, Rev. Gen. Therm., 8 (1969), 873.
6) A. Tian, C.R. Acad. Sci. Paris, 178 (1924), 705.
7) A. Papoulis, "The Fourier Integral and its Applications" McGraw-Hill, New York (1962).
8) C. Brie, M. Guivarch and J.L. Petit, Proc. 1st Intern. Conf. Calorimetry and Thermodynamics, Varsaw, 1969, Pol. Sci. Pub. , (1971) p. 73 ; C. Brie, Ph. D. Thesis, N° 44, Université Claude Bernard, Lyon, 1971.
9) C. Brie, J.L. Petit and P.C. Gravelle, C.R. Acad. Sci. Paris, 273B (1971), 1.
10) P.C. Gravelle, C. Brie and J.L. Petit, 97ème Congr. Nat. Soc. Savantes, Nantes, 1972 (to be published).
11) G. Della Gatta, B. Fubini and G. Venturello, Colloq. Int. Cent. Nat. Rech. Sci. Paris, 201 (in the press).
12) G.C. Bye, J.G. Robinson and K.S.W. Sing, J. Appl. Chem., 17 (1967), 138.
13) G.L. Kington and P.S. Smith, J. Sci. Instr., 41 (1964), 145.
14) J. Rouquerol, Colloq. Int. Cent. Nat. Rech. Sci., Paris, 201 (in the press).
15) F. Rouquerol, S. Partyka and J. Rouquerol, Colloq. Int. Cent. Nat. Rech. Sci., Paris, 201 (in the press).
16) J. Rouquerol, Proc. Int. Symp. on Surf. Area Determination, Butterworths London, (1970), 309.
17) R.A. Beebe and D.M. Young, J. Phys. Chem., 58 (1954), 93.
18) R.A. Beebe, B. Millard and J. Cynarski, J. Am. Chem. Soc., 75 (1953), 839.
19) J. Rouquerol and F. Rouquerol, Cah. Therm., Soc. Fr. Therm. Ed. , 1B (1971), 86.
20) W. Muller and D. Schuller, Colloq. Int. Cent. Nat. Rech. Sci., Paris, 201 (in the press).
21) J.M. Basset, A. Théolier, M. Primet and M. Prettre, this Congress.
22) S.J. Teichner, R.P. Marcellini and P. Rué, Advan. in Catalysis, 9 (1957), 458.
23) G. Rey, Ph. D. Thesis, N° 73, Université Claude Bernard, Lyon (1972).

24) G. El Shobaky, P.C. Gravelle and S.J. Teichner, J. Catalysis, 14 (1969), 4.
25) P.C. Gravelle, G. Marty and S.J. Teichner, Proc. 1st Intern. Conf. on Calorimetry and Thermodynamics, Varsaw, 1969, Pol. Sci. Pub. , 1971, p. 365.
26) G. Marty, PH. D. Thesis, N° 695, Université de Lyon (1970).
27) P.C. Gravelle and S.J. Teichner, Advan. in Catalysis, 20 (1969), 167.
28) M. Breysse, M. Guénin, B. Claudel and J. Véron, personal communication (to be published in J. Catalysis).
29) M. Richard, R. Isaac and L. Eyraud, Mesures, 310 (1963), 379.
30) P.C. Gravelle, G. Marty and S.J. Teichner, Bull. Soc. Chim., (1969), 1525.
31) G. Marty, P.C. Gravelle and S.J. Teichner, Cah. Therm. (Soc. Fr. Therm., Ed.) 1B (1971), 59.
32) J.P. Reymond, thesis N° 20, Université Claude Bernard, Lyon (1971).
33) P.C. Gravelle, G. El Shobaky and S.J. Teichner, J. Chim. Phys., 66 (1969), 1760.
34) J. Coué, P.C. Gravelle, R.E. Ranc, P. Rué and S.J. Teichner, Proc. 3rd Intern. Congr. Catalysis, Amsterdam, 1964, North-Holland Pub., Amsterdam, (1965), p. 478.

Invited Lecture E

SURFACE REACTIONS AND FIELD INDUCED SURFACE REACTIONS INVESTIGATED BY FIELD ION MASS SPECTROMETRY

JOCHEN BLOCK
Fritz-Haber-Institut of the Max-Planck-Gesellschaft
Berlin-Dahlem, Germany

ABSTRACT: Recent developments in field ion mass spectrometry are reviewed. The mechanism of ion formation cannot only be described by a pure wave mechanical electron transition (tunneling). Due to molecular surface interactions, reactions such as proton transfer, charge transfer and heterolytic bond cleavage are responsible for the ion formation as well. These reactions are induced or enhanced by the artificial electrical surface fields of > 10^7 V/cm. For surface reactions in heterogeneous catalysis surface species are analyzed with a sensitivity for single particles and a geometrical resolution within the atomic scale. Advantages and disadvantages in studying these processes are described for different ion separating and detecting techniques, which have been used in combination with different metal and oxide emitters.

Investigations on field induced surface reactions concern proton transfer during the interaction of hydrocarbons, hydrazine, acetone and butanol. Reactions, which form surface carbonium ions are distinguished by an extraordinary field- and temperature-dependence of field ion currents. Recent results of the mechanism of carbonium ion formation on zeolites are supported by new results of their IR spectra.

The application of field pulses in connection with a time of flight analysis of desorbed ions permits kinetic studies of surface reactions and allows one to measure the field distortion of these processes. The surface selective chemisorption of hydrocarbons on tungsten and platinum has been examined. For n-butane the carbon skeleton is rearranged even at 300°K on different cleaned Pt surfaces. Depending on the hydrogen pressure, higher or lower molecular weight residues are formed.

1. INTRODUCTION

One of the major objectives in catalysis research is concerned with the elucidation of the reaction mechanism of surface reactions. The detailed knowledge of the structure of intermediates in catalytic reactions and of the mechanism for activating certain molecules to undergo reactions selectively, is important for understanding and improving catalysts.

In recent years, new methods have been developed for

characterizing catalyst surfaces, chemisorption structures and the molecular behavior at solid surfaces. During the last decade a deeper insight has been gained for various surface systems by using LEED, Auger-Spectroscopy, ESCA, etc. but these have been predominantly applied to investigate the surface interaction of very simple gas molecules. Thus, these new results so far have only limited importance for real catalytic reactions. The uncertainty in interpreting experimental data of newly developed techniques in surface research still creates restrictions on what can be studied.

However, in industrial reactions it is necessary to use extremely complicated reacting schemes even though the fundamental knowledge in the science of catalysis is still not available. The challenge of the present work is to narrow this gap between platonic surface research on simple systems and the study of real catalysis.

Field ion mass spectrometry (FI-MSP) is based on E. W. Müller's[1)] invention of the field ion microscope (FIM) and was developed by Inghram and Gomer[2)] and Beckey.[3)] There are recent review articles on the principles of field ionization[4-6)] and on the mass spectrometry of field ions.[7-9)] In this survey on surface reactions and field induced surface reactions, we shall therefore restrict ourselves to a concise description of the pertinent essentials which are required for an understanding of the following.

2. CAPABILITY AND APPLICABILITY OF FI-MSP

In principle, this technique will identify a single surface atom or molecule which can be correlated with an individual surface site on an atomic scale. In principle, FI-MSP can resolve kinetic mechanisms of surface reactions in the μ-sec time scale and those of monomolecular decay processes of field ions with a time resolution close to the molecular vibrational modes of about 10^{-12}s. In principle, it is also possible to identify high molecular weight compounds like polymers and biomolecules without decomposition. However, in none of these cases is the use of FI-MSP a routine method. The intricacy of the field ionization phenomenon demands that special arguments and skillful experiments be found to permit correlating the experimental data with the actual mechanism of a reaction. This general statement holds for all the areas, where FI-MSP has found application so far : metallurgy,[9-11)] gas phase analysis,[7)] evaporation of solids[12-13)] and structure determination of nonvolatile biomolecules.[14)]

3. TECHNICAL DEVELOPMENTS

In field ionization, a neutral gas molecule approaches a metal surface and is at the same time exposed to an extremely high electrostatic field strength ($F \lesssim 5 \cdot 10^8$ V/cm). An electron then is transferred from the molecule to the metal surface by a wave mechanical tunneling phenomenon (which needs approximately 10^{-16}s for s-electrons). A positive ion thus formed is immediately accelerated by the external field and can be used in a FIM for imaging the surface or focused into a mass spectrometer for mass analysis.

Originally magnetic sector instruments were used for mass discrimination of field ions. For surface reactions, they have the disadvantage that the initial surface structures after gas adsorption are difficult to determine, since only one mass can be focused at any time. Thus processes in the important microsecond or millisecond time scale range can not be revealed. On the other hand, the mass resolution is favorable in this instrument and, in addition, an energy analysis of field ions can be performed because of the laws of ion focusing in magnetic sector fields. In this way we can discriminate between ions desorbed from a metal or from an oxide layer of a surface.

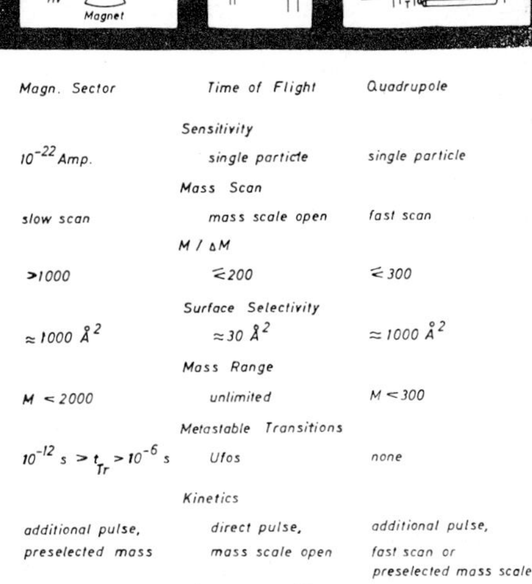

Magn. Sector	Time of Flight	Quadrupole
	Sensitivity	
10^{-22} Amp.	single particle	single particle
	Mass Scan	
slow scan	mass scale open	fast scan
	$M / \Delta M$	
>1000	$\lesssim 200$	$\lesssim 300$
	Surface Selectivity	
≈ 1000 Å^2	≈ 30 Å^2	≈ 1000 Å^2
	Mass Range	
$M < 2000$	unlimited	$M < 300$
	Metastable Transitions	
10^{-12} s $> t_{Tr} > 10^{-6}$ s	Ufos	none
	Kinetics	
additional pulse, preselected mass	direct pulse, mass scale open	additional pulse, fast scan or preselected mass scale

Fig. 1 Comparison of mass spectroscopic techniques

The time-of-flight (TOF) mass spectrometer has its principle advantage for kinetic studies. Repetitive field pulses can be applied for short times. Since species in the entire mass range can be analyzed at once, any ion can be observed from the very beginning of a chemisorption process of chemical surface reaction. However, the mass resolution is inferior and does not permit measurements of the energy distribution of field ions.

Recently, quadrupole mass filters (QMF) were used for field ion studies.[15,16] They are advantageous because they are compact and easy to use in UHV, have fast response and sufficient sensitivity. Since the mass scale is independent of the ion energies, they give an accurate analysis of surface species.

In early developments of field ion microscopy, refractory metals with high mechanical strength like tungsten were preferred for the field ionization of imaging gases such as helium or hydrogen. In FI-MSP the chemical reactivity is a much more important consideration because field corrosion - field enhanced oxidation for instance - usually determines the stability of an emitter system. Various metals like Pt, Ni, Fe, Ag and semiconductors like Si and Ge have been used for field ion studies. Needles of organic material were formed to fieldionize molecules[17)] and even oxide emitters like ZnO[18,19)] were prepared from single crystals. In each case, however, exceptional precautions were required as long as surface interactions with pure solids were to be investigated. One of the most important questions in FI-MSP is to determine whether field ions are created directly at a clean bulk emitter material or in conjunction with chemisorbed layers and surface contaminants. There are several methods that provide evidence to answer this important question.

4. IONIZATION AT SURFACES

We pose the question of how a neutral molecule is ionized by the influence of an external field at a surface. The two main competetive effects, which have to be considered, are the extremely high external electric field and the interaction of gas molecules with surfaces. Ions investigated by FI-MSP may, in general, be created by one of four different modes (Fig. 2): Firstly, field ionization can occur as a gas molecule approaches the solid surface without any chemical interaction. Secondly, molecular ions can be produced by proton transfer reactions (or an ion abstraction) in connection with an intermolecular interaction with a chemisorbed layer. Thirdly, charge transfer complexes can be formed due to the molecular interaction in high fields and at surfaces and finally by the field induced heterolytic bond cleavage at surface sites, stimulated by the external electric field. In many cases the different mechanisms of ion formation may be involved simultaneously. Mere FI and classic surface ionization (Langmuir-Kingdon) are the boundaries of a large domain of chemical bond formation and ion desorption.

4.1 Field ionization

A He-atom may be field ionized simply by a deformation

of the Coulomb potential within the atom. The electron tunnels out of the molecule into the metal. As the field strength at the emitter surface is increased, ions may be formed at considerable distances of more than 10 Å in front of the solid surfaces, where chemical interactions are negligible. If this field ionization process applies, the ionization probability D simply depends on the ionization potential I_p of the particular atom or molecule and the work fundtion Φ of the surface.

$$D \approx \exp\{-A(I_p-\Phi)^{1/2} \cdot I_p^{1/2} \cdot F^{-1}\}\, f(F, I_p) \qquad (1)$$

(A = const.; f(F,I) = corrections that can be neglected to a first approximation). The minimum field strength F_0 for field ionization, which is an important magnitude, is then linear function of $(I_p-\Phi)\, I_p^{1/2}$. This relationship has been found experimentally fulfilled only for pure rare gases on cleaned metal tips.[20] Even under these conditions, a field stabilized adsorption layer of highly polarized rare gas atoms seems to be present, which is stabilized by the interacting field energy ($E_F = \mu F \cdot 1/2 \cdot \alpha\, F^2 + \ldots$) where μ is the surface dipole and α the polarizability of the surface complex. Evidence for this field stabilized adsorption of rare gas atoms has been obtained from field evaporation studies with the "atom probe"[9] where predominantly metal-rare gas complexes have been found.

In gas mixtures, for instance He/H_2, the field ionization of He is promoted,[21.22] i.e. ionization is achieved at lowerfield strength in gas mixtures. Further evidence for intermolecular interactions, which are involved in the mechanism of ion formation, comes from measurements of onset fields with adsorption layers of various gases.[23] In contrast with the expectations of equation (1), the change in work function resulting from adsorption does not explain the observed shift to higher or lower onset fields of field ionization. In the same sense, field ions of aromatic hydrocarbons within the homologeous series of benzene (napthene ... perylene) could be produced at decreasingly low external fields (< 0.1 V/Å). Thus, intermolecular interactions at surfaces are influencing the ionization process.[24] Electron tunneling of an isolated gas atom or molecule without chemical interaction at surfaces is an exceptional case.

4.2 Proton transfer

In electrochemistry, one of the most common mechanisms of ion formation is the heterolytic dissociation of polar bonds. In FI-MSP proton transfer reactions also predominate under certain conditions. Field ion mass spectra often also yield molecular ions which are known from solutions in superacids.[25]

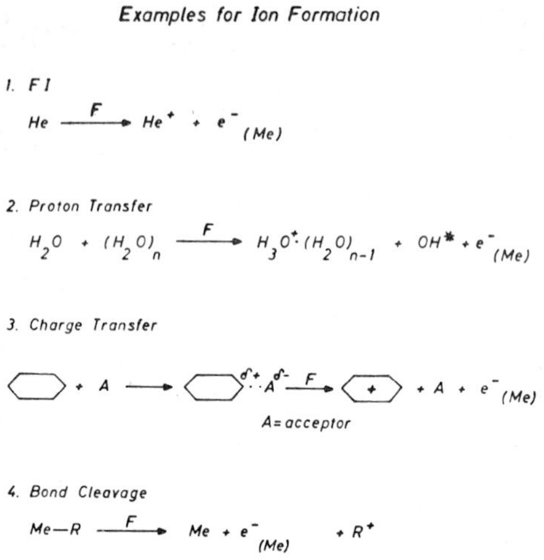

Examples for Ion Formation

1. FI

$He \xrightarrow{F} He^{+} + e^{-}_{(Me)}$

2. Proton Transfer

$H_2O + (H_2O)_n \xrightarrow{F} H_3O^{+} \cdot (H_2O)_{n-1} + OH^{*} + e^{-}_{(Me)}$

3. Charge Transfer

$\bigcirc + A \longrightarrow \bigcirc^{\delta+} \cdots A^{\delta-} \xrightarrow{F} \oplus + A + e^{-}_{(Me)}$

A = acceptor

4. Bond Cleavage

$Me{-}R \xrightarrow{F} Me + e^{-}_{(Me)} + R^{+}$

Fig. 2 Modes of ion formation

Adsorbed multilayers of polar gas molecules may be formed on the emitter surface even at gas pressures of $<10^{-4}$ torr[26,27] because of the high gas compression in the electric field. Under these conditions proton transfer and proton attachment on molecules occur according to Onsager's theory of electrolytic dissociation or because of interactions with chemisorbed molecules or ions. For several cases the energetics of these processes are known and detailed mechanisms of ion formation have been developed.

4.3 Charge transfer complexes

The field enhanced charge transfer between molecular structures at the emitter surface is another mechanism involved in field ion production. Several examples are known where the presence of electron acceptors considerably reduces the minimum ionization field strength. Naphthalene, for instance, is ionized as a parent molecular ion at 50% reduced field strength if the well known electron acceptor chloranil is admixed.[24] In contrast to proton transfer reactions where the p + 1 molecular ion predominates, parent molecular ions (p) are found after charge transfer. In certain cases, electron donor-acceptor complexes could be identified as stable molecular ions.[27]

4.4 Bond cleavage

The field induced heterolytic bond cleavage at surfaces is one of the important processes for catalysis research. Field desorption of surface complexes is used to identify chemisorption structures. At elevated fields, field evaporation of metal atoms can be observed; chemisorbed molecules are frequently evaporated together with these metal atoms. Field evaporating surface atoms or molecules may be singly or multiply charged. Brandon[28] and Tsong and Müller[5] tried to develop a theory of field evaporation. The variety of observation however can still not be explained theoretically. During the field evaporation of selenium, for instance, molecular ions with as many as 33 Se-atoms and with single, double and four-fold charge could be observed.[12,29)]

5. CHEMISTRY IN EXTREMELY HIGH ELECTROSTATIC FIELDS

We may ask how a molecule behaves in the space in front of the emitter, where a field strength F of the order of 10^7 to 10^8 V/cm is considered under these conditions. If it is assumed that the field is homogeneous for the dimensions of a molecule, a new variable ΔF has to be introduced in order to describe the behavior of matter. In analogy to high temperature or high pressure chemistry, the new area "high field chemistry" has to be investigated in order to understand field ion mass spectroscopy.

For neutral molecules, a large field strength causes considerable changes in molecular properties like polarity, electron density distribution, bond length, vibrational excitation, molecular orientation, reactivity and thermodynamics. At present, only a few examples of the change in molecular properties can be cited. Eigen[30)] showed that the thermodynamic equilibrium of a chemical reaction is shifted, if it has an electrical momentum defined by the difference in polarity of reactants and products. This field effect has been found in FI-MSP for the association of formic acid.[31)] The reaction 2 HCOOH $\rightleftarrows (HCOOH)_2$ has an electrical momentum, since the dipole moment of the monomeric form μ_{mom} = 1.35. Debye is compensated in the planar form of the dimer (μ_{dim} = 0). Consequently, with increasing field strength, the direction with increasing momentum (monomeric HCOOH) is favored.

Extremely high electric fields are not unusual in normal chemistry at interfaces. Thus Sheppard and Yates,[32)] Monod[33)] and Rabo[34)] demonstrated that fields of the same magnitude are interfering in various systems. Our field ion experiments will be related to those systems later on.

6. SURFACE REACTIONS OBSERVED UNDER STATIONARY ELECTRIC FIELDS

An unperturbed catalytic reaction can only be observed if the external electrostatic field does not change the reaction conditions in surface chemistry. This will only apply in exceptional cases. However, field ion mass spectra depend on the chemical nature of the emitter material and display specific surface interactions. There are mainly three reasons why catalytic surface reactions are perturbed: the change of the energy hyperface of the reaction due to polarization, contamination of the surface by neutral radicals formed during fast ($< 10^{-12}$s) fragmentation of field ions, and field induced chemisorption of polar or ionic molecules on surface sites which are catalytically active.

In addition, there is a kinetic argument to be considered. The residence time of a molecule on the emitter surface may be extremely small and insufficient to permit the catalytic reaction step. The turnover number,[35] i.e. number of product molecules formed per surface site per sec., may be in the order of 10^{-5} to 10^{+2} for usual reaction regimes in heterogeneous catalysis. The chance to find the field ion as a reaction product within an investigated area of 1000 surface atoms depends on the turnover number and is reasonably high only for a very fast reaction. If we consider that during field ionization a total ion current of 10^{-11} amps. may be desorbed from the whole emitter surface ($2\pi r^2$, $r \approx 1000$ Å) of $6 \cdot 10^{-10}$ cm^2, each surface site produces 10^3 to 10^4 field ions but only 10^2 catalytic reaction products in a very fast reaction.

On the other hand, surface mobility of adsorbed species will allow reactants and products to diffuse from the emitter shank area, where only low electrical fields are present, into the ionizing zone. On multicomponent emitter systems surface diffusion and spillover has been observed during the interaction of hydrocarbons.[36] Thereby, it has to be considered, that field directed surface mobility represents a major part of the supply into the ionizing area of the emitter.

If surface diffusion is excluded, only very fast surface processes such as a chemisorption with low activation energy can be measured directly under permanently applied electric fields, i.e. normal reactions in heterogeneous catalysis are frequently much too slow for this rapid analyzing method.

In spite of these restrictions, there are many examples where products of a surface reaction can be observed under steady field conditions. The dissociative chemisorption of cyclohexane on platinum forming the C_6H_{11} surface compound[37] could be investigated, as well as the dehydrogenation of substituted cyclohexanes.[38] Benzene, in contrast, is ionized as $C_6H_6^+$ and only at specific surface areas $C_6H_4^+$ could be obtained, indicating a two site chemisorption complex. Hydrocarbons, like n-heptane form $C_7H_{16}^+$ and $C_7H_{15}^+$ ions on Pt, from a physically adsorbed and chemisorbed state respectively. At the presence of an oxide layer, like As_2O_3, the dissociative chemisorption is much more predominent and accordingly the $C_7H_{15}^+$ ion.[13]

Using a sensitive sampling technique capable of

identifying minor ion intensities with relative concentrations of 10^{-6}, we have tried to investigate directly the deuterium exchange with neopentane on palladium:[39]

$$(CH_3)_4 \cdot C + 6\ D_2 \rightleftarrows (CD_3)_4 \cdot C + 6\ H_2$$

It could be shown that this reaction follows a single-step exchange mechanism, i.e. all of the intermediates $(CH_3)_3$-$CH_2D \cdot C$, $(CH_3)_2\ (CH_2D)_2 \cdot C$, etc. are in equilibrium with the gas phase. Evidence for this mechanism could however only be found by an indirect technique, using FI-MSP as a tool for analysis of products desorbing thermally from a usual PD catalyst surface.

Proton transfer reactions are very fast and their products familiar in FI-MSP. The field ionization of methane yields the CH_4^+ parent ion and some other ions of minor intensity. If water molecules are admixed the CH_5^+ ion is obtained and may be exclusively formed at higher water doses (besides H_3O^+, $H_3O^+\ (H_2O)_n$ and H_2O^+).[40-42] The energy of proton attachment onto CH_4 is 5.3 eV (for H_2O 7.2 eV) and the reaction cross section with H_2O is high. It is known from homogeneous molecule-ion reactions that these cross sections very much depend on polarity, dipole interaction, etc. Accordingly in a mixture of CH_4 and H_2 no CH_5^+ could be observed,[43] although the ion desorption is promoted.

Obviously, proton transfer often occurs between tightly chemisorbed surface structures and gas molecules. All the ions observed during the field ionization of acetone could be explained by Röllgen and Beckey[44] in this way. Surface radical sites will tightly bind molecules even in an ionic form. For instance acetone may be chemisorbed and react in the following way (Fig. 3a). The field induced chemisortpion of an ion is followed by a proton transfer reaction, which yields the ion with the mass number M 59. The radical M 57 can be field ionized as such or react with another acetonyl radical for dimerization (Fig. 3b) and polymerization in order to form needles of organic material. In a similar way reactions forming the ions CH_3-C-CH_2^+, $CH \cdot C \cdot (OH) = CH_2^+$, CH_3 - C (O-CH_2) = CH_2^+ and $CH_3)_2 \cdot$ CO C $(CH_3)_2^{++}$ could be correlated with molecular data and used as an explanation for all of the mass signals observed in FI-MSP. There is, however, still some doubt whether ionic structures can be stabilized at metal surfaces or whether they are neutralized by backdonation from higher electronic states near the Fermi level.

Another example of field induced proton transfer reactions was observed during investigations on the heterogeneous decomposition of hydrazine.[45] On a platinum surface $N_2H_5^+$,

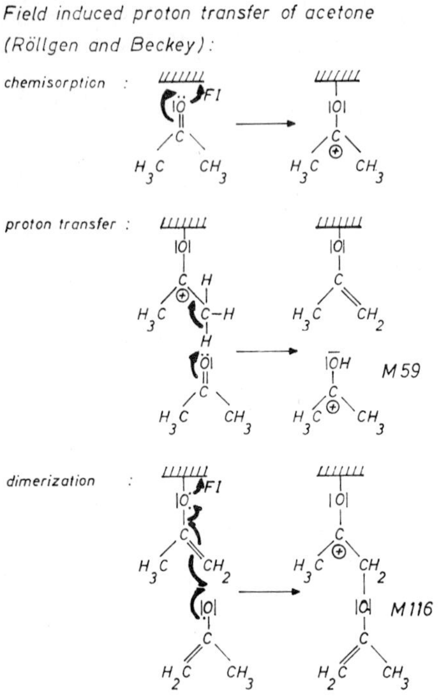

Field induced proton transfer of acetone (Röllgen and Beckey):

Fig. 3a Field induced chemisorption of acetone with subsequent proton transfer

b Dimerization as initial step for polymerization

$N_2H_4^+$ and $N_2H_3^+$ ions are the most frequent ones. The formation of these ions can be understood by the following mechanism. A hydrazine molecule can be chemisorbed to the metal by the free electron pair (Fig. 4). Proton transfer with a second molecule yields M 31 and M 33. Gaseous or physically adsorbed molecules are found simultaneously. The temperature dependence of these ion intensities is shown in Fig. 4. On heating the Pt emitter to 175°C, the intensity of the $N_2H_5^+$ ions decreases by more than two orders of magnitude. This can be explained by the exothermic - and field enforced - heat of chemisorption of N_2H_4. The same argument explains the temperature dependence of the $N_2H_4^+$ signal, which first decreases in intensity, but increases when bimolecular reactions are less frequent. The N_2H_3 - radical is probably more strongly bound to the surface. Thus the field desorption has an activation barrier. At higher temperatures (≈ 200°C), where products of a catalytic reaction are expected NH_4^+ and NH_3^+ are found. There is no evidence of NH_2^-, NH- or N-surface species in accordance[46] with the mechanism of the catalytic hydrazine decomposition which showed no isotope exchange in the reaction $^{14}N^{14}NH_4 + {}^{15}N^{15}NH_4 \rightleftarrows {}^{14}N^{14}N + 4H_2$. In contrast, under reaction conditions the molecular ion N_2H_7 can be observed (Fig. 4b).

7. SURFACE REACTIONS INVOLVING CARBONIUM IONS

Carbonium ions have been formulated as intermediates of

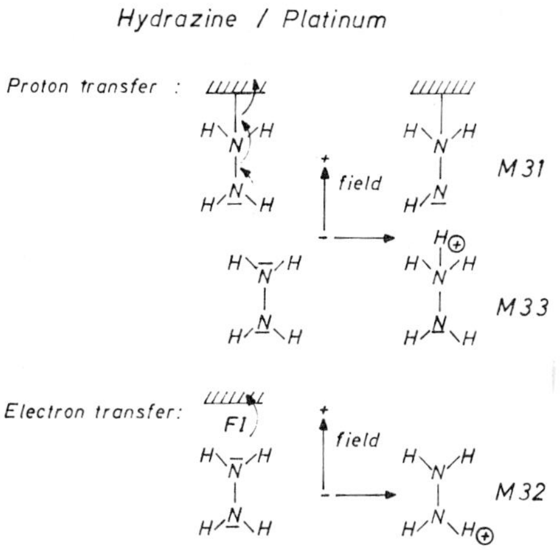

Hydrazine / Platinum

Proton transfer :

field

M31

M33

Electron transfer:

FI

field

M32

Fig. 4a Field reactions of hydrazine

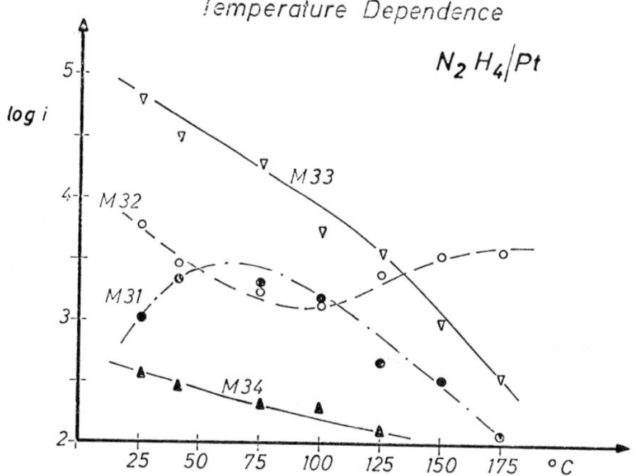

Fig. 4b Temperature dependence of ion intensities

various catalytic reactions like isomerization, hydrogenation and dehydration. Schwab's findings[48] on donator acceptor reactions also involve ionic or highly polarized surface species. The formation of ionic intermediates of a surface reaction which is not induced by external electric fields poses an interesting question for investigations in the field ion mass spectrometer.

Suppose a chemical reaction $RX + cat. \rightarrow R^+ + cat. X^{(-)}$

forms a carbonium ion R^+ without having an external field applied. The possibility of desorbing this ion from the surface and detecting it in the mass spectrometer will be different from a field ionization process. If the bond energy of the carbonium ion to the catalyst surface is small or if a high mobility of R^+ at the surface can be assumed, we may expect to be able to desorb these particles easily and have an exceptionally selective method to detect these ionic surface structures.

Experiments were first performed with stable carbonium ions of triphenylmethyl ($(Ph_3)\cdot C$) compounds.[49] The results disclose principal differences for the ion desorption from metal and oxide surfaces. For these experiments zeolite grains were fused to a Pt emitter or Pt Wollaston wire (Fig. 5). By this method a platinum surface was compared with a surface of a Ca-Y-zeolite. Zeolites were chosen on account of a misleading idea. Rabo[34] has evaluated extremely high electrostatic fields on these surfaces and we intended to compare these surface fields with those of field ionization. A molecular beam of Ph_3C compounds was directed to the emitter surface and ionized while the electric field was applied. At the pt-surface molecular ions $(Ph_3)CX^+$ (X = Cl, Br, Oh, H) are formed by field ionization only if a minimum external field strength of > 10^7 V/cm (ΔU > 4 kV for this particular emitter) is expected (Fig. 6). Approximately the same field strength was necessary to ionize hydrocarbons like cyclohexane. Besides the parent molecular ion, fragments and metastable transitions were found on platinum clearly indicating the usual field ionization mechanism. In accordance with this mechanism, the intensity of field ions decreased with increasing emitter temperature.

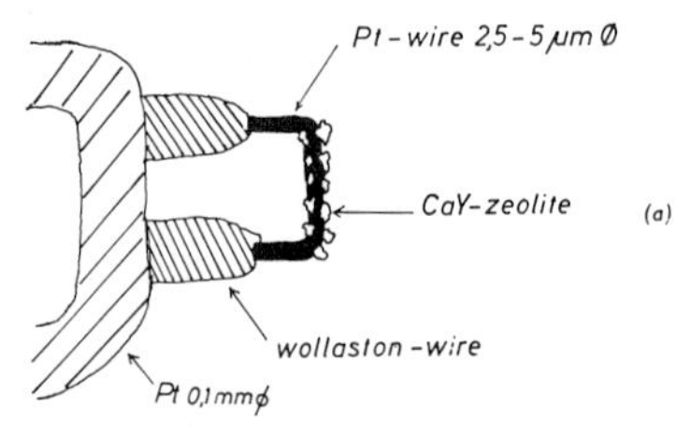

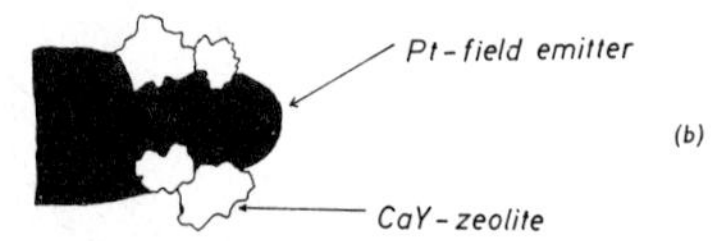

Fig. 5 Field emitter systems with zeolite and platinum surfaces

At the zeolite surface no parent molecular ions were observed and only carbonium ions Ph_3C^+ could be found. These ions were desorbed without necessarily having the external field applied. However, a minimum reaction temperature was required. With increasing field strength ion intensities

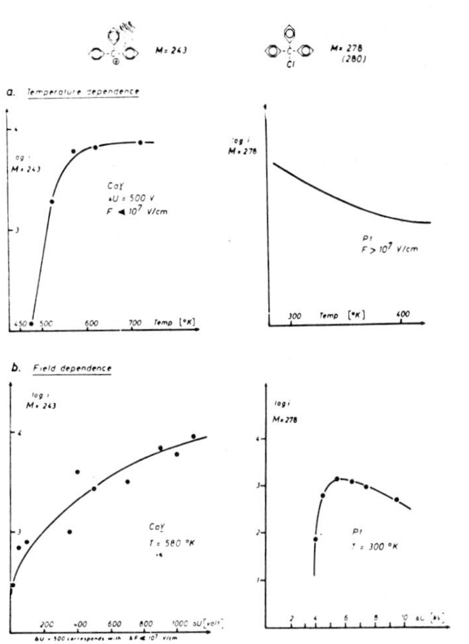

Fig. 6 Ion emission of Ph_3CCL from Pt-surfaces and CaY-zeolite surfaces.

a The temperature dependence at the zeolite and the Pt-surface.

b The field dependence at the two surfaces (the decrease in intensity at elevated voltages on Pt is due to a defocusing ion optic)

increased according to a $\sqrt{F}$-law which proves that a Schottky barrier exists for the ion desorption (Fig. 7). The ease of ion formation increases in tne sequence X = OH, Cl, Br. No ions were formed with X=H, COOH. The ion formation furthermore increased by simultaneously adsorbing electron acceptors ($(CN)_2C-C(CN)_2$) or by intramolecular substitution $[(CH_3)_2N \cdot C_6H_4]_2 \cdot C_6H_5 \cdot C \cdot X$ which polarized the C - X bond. Using exactly the same samples, Karge[50] investigated the formation of carbonium ions by IR-spectroscopy. As the result (Fig. 8a) the ionic species $(Ph_3)\ C^+$ was identified by the characteristic absorption line $\gamma_5(C_6H_5\text{-}C)$ at 1357 cm^{-1}. This line is absent if C_6H_5CH is adsorbed by the CaY-zeolite (Fig. 8b). There is further indication that carbonium ion formation with $(C_6H_5)_3C \cdot Cl$ occurs even at 100°C, (i.e. temperatures which are much lower than the desorption temperatures), that Lewis acid centers are involved in the mechanism of ion formation and that these ions will also be produced in the interior channels of the zeolite lattice.

The thermal, field stimulated desorption of molecular ion has already been observed before. In a series of publications Zandberg[51] and Ionov[52] have reported about ion emission from oxide surfaces. These oxide surfaces were not characterized in detail and it was attempted to find an explanation in terms of the

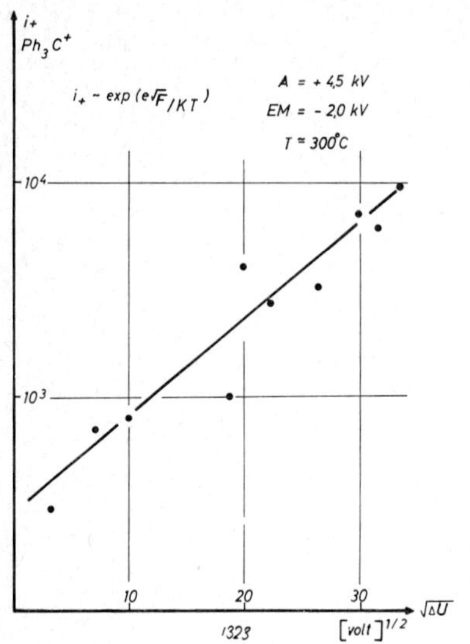

Fig. 7 Schottky barrier for Ph_3C^+ desorption. The intensity of the carbonium ion depends linearity on the square root of the applied field strength

Saha-Langmuir equation.

In order to investigate carbonium ions while catalytic reaction proceeds, the dehydration of tert-butanol on a Pt/zeolite surface was investigated.[53] The ion emission starts at a relatively low electric field (ΔU=1.2 kV between emitter and first electrode, compared with $\Delta U >$ 2.5 kV for benzene). Field ion products are however screened by the extreme instability of the tert-butanol molecular ions, which decompose instantaneously ($<10^{-12}$s) according to

$$CH_3-\overset{CH_3^+}{\underset{OH}{C}}-CH_3 \rightarrow CH_3-\overset{CH_3}{\underset{OH}{C}}{}^{\oplus} + CH_3$$

M 74 M 59

Therefore M 59 is the principal ion, found during field ionization of this molecule at a metal surface.[54] This is also true for our emitter system. Increasing the temperature, however, increases the ion intensities M 57, M 58 and M 56 (Fig. 9). These molecules

$$CH_3-\overset{CH_3}{\underset{OH}{C}}-\underset{\overline{cat.}}{CH_2} + H \xrightarrow{-\,OH} CH_3-\overset{CH_3}{\underset{\overline{cat.}}{C}}-CH_2$$

M 73 M 56

are expected as intermediates of a catalytic reaction and could give M 57 (or as indicated above) M 56. The ion M 58 would be expected in connection with a proton transfer. On the other hand, a temperature dependent Frank-Condon transition could be responsible for these signals. Final conclusions can be drawn only after investigations with field pulses.

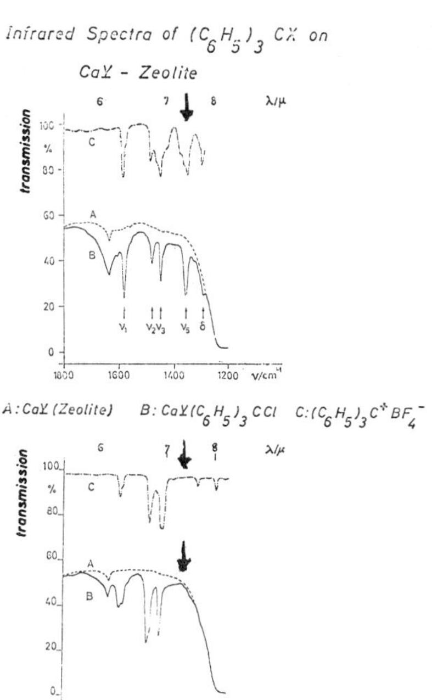

Fig. 8 Infrared spectra of different Ph_3CX compounds in comparison with adsorbed molecules on a Cay-zeolite. These spectra indicate the carbonium ion formation of compounds with X=Cl at the zeolite surface.[50)]

8. FIELD PULSE INVESTIGATIONS

The field disturbance of surface processes can be widely suppressed, if field pulses instead of stationary electric fields are applied[2,37,55)] For a certain reaction time τ_R(> 100 ns) surface processes may proceed at zero or small electric fields (given by the potential U_r) . Products formed during this time are desorbed by a field pulse with fast rise time. The amplitude of this pulse ($U_D \lessapprox 30$ kV), the pulse duration (τ > ns), and the repetition time frequency < 70 kc can be changed within convenient limits.

By changing the reaction time τ_R, kinetic studies can be performed [56)] in the milli- and micro-second time scale and by changing the reaction field (U_r) the field dependence of chemical reactions in extremely high fields ($10^6..10^7$ V/cm) can be studied. The desorption field (U_d) has to be chosen sufficiently high in order to field desorb, field evaporate or field ionize reactants or reaction products. This surface analysis is performed selectively on an area of > 30 $Å^2$ of a single crystal face which can be identified by FEM or FIM.

From FEM studies it is known[57)] that hydrocarbons display characteristic face selective surface interactions with metals. FEM images are like fingerprints, typical for each hydrocarbon. On tungsten, for instance, each of the n-alkanes

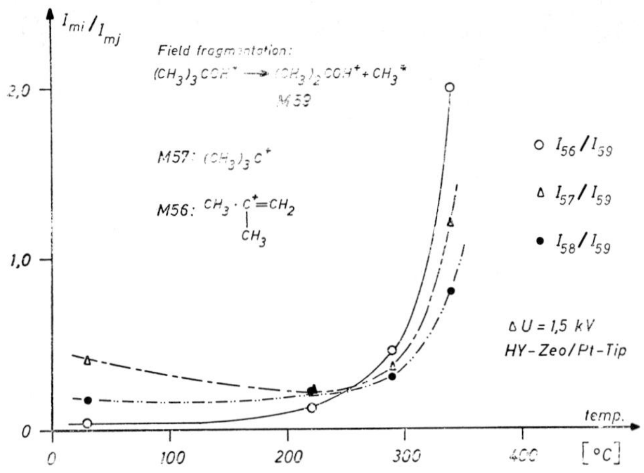

Fig. 9 Decomposition of test-butanol on CaY-zeolite, relative intensities of fragments M 56, M 57 and M 58 in dependence of the catalyst temperature.

with C_1 to C_3, C_2H_4 and hydrogen can be identified by the special appearance of the FEM pattern. As Fig. 10 demonstrates, particular surface areas of a tungsten single crystal change the emission current for each of the adsorbed gases in a characteristic manner. It has to be suspected that this is due to a surface selective adsorption.

By using TOF-analysis we have studied the interaction of hydrocarbons on different single crystal planes of platinum and tungsten emitters[58)] in order to compare reaction products in different areas of the surface.

First, the TOF mass spectrum of n-butane/hydrogen (1:5) on a Pt-(111) surface is described in Fig. 11. The mass signals of the molecular ion (M≈58) and the fragment C_2H_5 (M 29) are shown in the upper right. The region of the parent mass is shown with higher resolution on the upper left. It is clearly seen, that masses 59, 58 and 57 are formed. These mass spectra are obtained by using an ion counting and sampling technique with computer data evaluation, calculated mass scale and intensity distribution which are directly machine plotted. The lower abscissa indicates the time of flight (μ-sec.) the upper the mass scale. Experimental conditions of experiments are given on the right hand side of the diagram. Ion intensities are related to Langmuirs of gas adsorption.

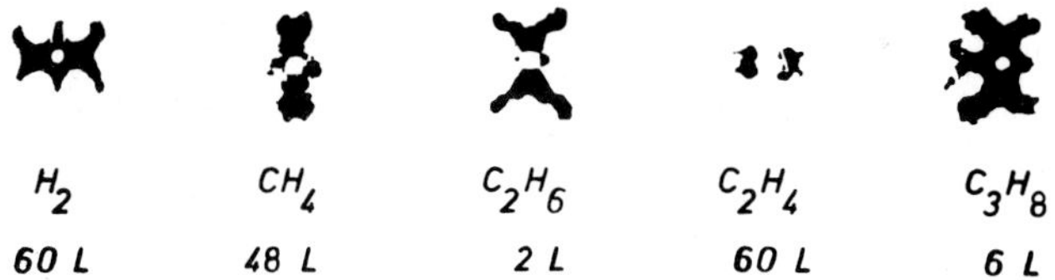

Fig. 10 FEM images of hydrocarbons chemisorbed on tungsten at 298°K. Specific differences of these images are related to a surface specifity of chemisorption (L=Langmuirs).

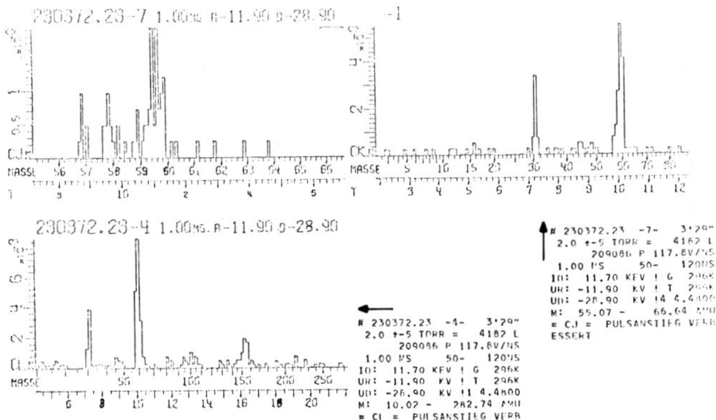

Fig. 11 TOF-mass spectra, n-butane chemisorption on the (111) surface of platinum (computer plotted): parent ion and fragments (upper right), parent ion mass scale in higher resolution (upper left), higher mass region (lower spectrum). Data for experiments are given (←, ↑ : data and no of spectrum, time of measurement in top line, next, pressure and adsorption in L), number of pulses applied and pulse rise time, 10 ion energy, G, T, gas and tip temperature and M = mass scale plotted.

The lower left hand mass spectrum indicates that higher masses are formed during the interaction of n-butane with the

Pt (111)-surface. This is a general observation made during these investigations. Without the presence of hydrogen, C-C bonds of hydrocarbons are split and high molecular weight hydrocarbons are formed on a previously cleaned Pt-surface even at 300°K. This deposit is formed preferentially on high index surface planes, as indicated by FEM images. Obviously, these surface compounds are mobile on low index planes and migrate to sites, where they are tightly bound. It is impossible to remove this deposit by thermal evaporation. The molecular weight of the compounds formed depends on the hydrogen pressure: with excess of hydrogen products with lower molecular weight predominate.

Since surface compounds formed under these conditions are in general very tightly bound to the surface, field desorption needs extremely high field amplitudes and metal-hydrocarbon complexes are often observed. This is clearly indicated in the mass spectra of Fig. 12 where high values of U_D had been applied. The mass scale which has low resolution gives on the left hand mass signals in the M 58 region and

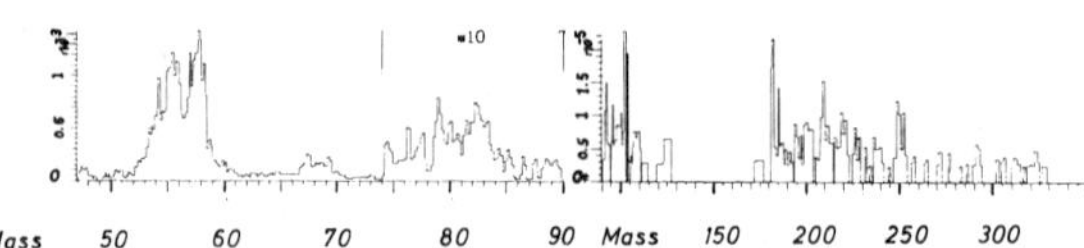

Fig. 12 TOF mass spectra of n-butane/Pt, showing hydrocarbon polymerization and desorption of Pt-hydrocarbon complexes at higher field desorption amplitudes.

below the M 72 (C_5) and M 86 (C_6) regions. An other region of the mass scale right hand side shows hydrocarbons with C_7 and intense mass signals in the region of platinum and platinum compounds (> M 190).

On a contaminated surface, ion desorption proceeds much easier (at lower fields) than on a clean one. Experimentally, at the beginning of an experiment the intensity of ion emission increases during the first 100 to 500 s at constant reaction conditions ($p \approx 10^{-5}$ torr, T = 300°K). Obviously a deposit is formed at the Pt surface and only after a partial coverage has been achieved, the ion emission can be observed at convenient fields. This may be demonstrated in one example in Fig. 13. Ion intensities of n-butane are shown as a function of the desorption amplitude U_D. Starting the experiment with U_D = 18.9 kV ($\approx 2 \cdot 10^8$ V/cm) none or very small ion intensities are observed. With proceeding reaction time, increasing ion intensities are desorbed, leading to a hysteresis in the experimental dependence on U_D as indicated by arrows in Fig. 13 at the beginning of the experiment. Although the desorption field strength diminished after taking subsequent experimental points (several minutes each) ion currents grow to a maximum at 16.9 kV. After this time of incubation, ion intensities depend reversibly on U_D as to be expected from the usual field dependence of ionization probabilities. The ion M 59 is formed by proton transfer. The ions M 57 and M 29 may be fragments of the parent ion or products of a surface reaction. The latter process seems to be more probable since the fragmentation pattern of butane field ions at elevated fields is different and includes fragments of M 15 and M 43 which cannot be found in our experiments. In general, field impulse measurements of the interaction of n-butane with different Pt surfaces reveal a dissociative chemisorption, involving the C-C bond cleavage at the metal surface. In accordance with observations of other authors a variety of hydrocarbons is formed. The mean molecular weight depends on the hydrogen concentration which is actually present.

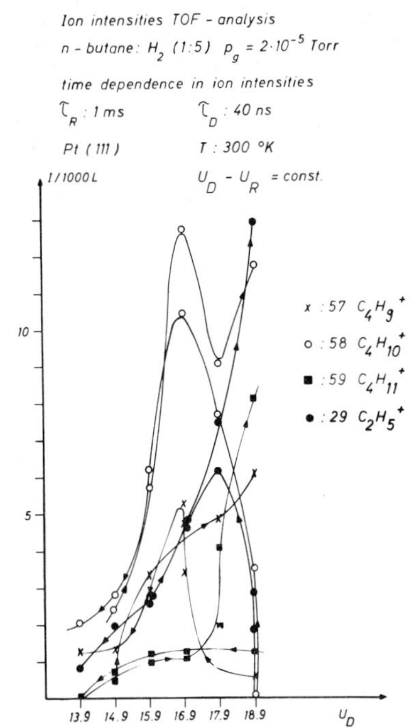

Fig. 13 Incubation times in ion detection due to surface formation n-butane on (111) Pt.

9. ACKNOWLEDGEMENTS

The described unpublished results of this review have been obtained in cooperation with: G. Abend, K. Bätjer, W. A. Czanderna, O. Frank, A. Martin, W. A. Schmidt, P. Solymosi, H. Thimm and M. S. Zei. The Deutsche Forschungsgemeinschatt and the Senat of Berlin (ERP-Fonds) generously supported this research project.

REFERENCES

1) E. W. Müller, Z. Physik 131, 136 (1951).
2) M. Inghram and R. Gomer, Z. Naturforsch. 10 a 863 (1955).
3) H. D. Beckey, Z. Naturforsch. 14a, 712 (1959).
4) G. Ehrlich, Advances in Catalysis and Rel. Subj. 14, 255 (1963) Academic Press.
5) E. W. Müller and T. T. Tsong "Field Ion Microscopy" Elsevier 1969.
6) K. M. Bowkett and D. A. Smith, "Field-Ion Microscopy," North Holland publ. 1970.
7) H. D. Beckey, "Field Ionization Mass Spectrometry," Akademie Verlag, Berlin 1971.
8) J. Block, Advances in Mass Spectrometry, Vol. 4, 791, Inst. of Petrol. 1968, E. Kendrick editor.
9) E. W. Müller, Naturwissensch. 57, 222 (1970).
10) P. J. Turner and M. J. Southon, in Dynamic Mass Spectrometry, Vol. 1, 147 (1970) Heyden & Son. D. Price & J. E. Williams editors.
11) S. S. Brenner and J. T. McKinney, Surface Sci. 20, 411 (1970).
12) H. Saure and J. Block, Advances in Mass Spectrometry Vol. 5, 404, Inst. of Petrol 1971, A. Quayle editor.
13) K. A. Becker, J. Block and H. Saure, Berichte der Bunsenges, 75, 406 (1971).
14) H. D. Beckey, G. Hoffmann, K. H. Maurer and H. U. Winkler, Advances in Mass Spectr. Vol. 5, 626, Inst. of Petroleum 1971, A. Quale editor.
15) T. Utsumi and O. Nishikawa, J. Vac. Soc. and Techn. 9, 477 (1972).
16) A. Martin and J. Block, MeBtechnik in press.
17) H. D. Beckey, H. Heinsing, H. Hey and H. G. Metzinger, Advances in Mass Spectr. Vol. 4, 817, Inst. of Petr. 1968, E. Kendrick editor.
18) J. Marien, R. Leysen and H. van Hove, phys. stat. sol. (a) 5, 121 (1971).
19) K. P. Frohmader, Solid State Commun 7, 1543 (1969).
20) E. W. Müller, Advances in Electronics and Electron Phys. 13, 83 (1960).
21) E. W. Müller, S. Nakamura, O. Nishikawa and S. B. McLane, J. Appl. Phys. 36, 2496 (1965).
22) L. W. Swanson, D. E. Reed and A. E. Bell, 14th Field Emission Symp., Washington, D.C. 1967.

23) W. A. Schmidt, O. Frank and J. Block, Berichte der Bunsengesellschaft 75, 1240 (1971).
24) J. Block, Zeitschr. f. Physik. Chem. N. F. 64, 199 (1969).
25) G. A. Olah, Chem. Eng. News 45, 76 (1967).
26) See reference 7, page 190.
27) B. E. Job and W. R. Patterson, "Proc. Conf. on Newer Phys. Methods in Struct. Chem.," p. 129, United Trade Press, London, 1967.
28) D. G. Brandon, Brit. J. Appl. Phys. 14, 474 (1963).
29) H. Saure and J. Block, Intern. Journ. of Mass Spectr. and Ion Phys. 7, 157 (1971).
30) K. Bergmann, M. Eigen and L. De Maeyer, Ber. Bunsenges. 67, 819 (1963).
31) J. Block and P. L. Moentack, Z. Naturf. 22a, 711 (1967).
32) N. Sheppard and D. J. C. Yates, Proc. Roy. Soc. A238, 69 (1956).
33) P. Monod et al., J. Phys. Chem. Sol. 27, 727 (1966).
34) J. A. Rabo, C. L. Angell, P. H. Kasai and V. Schomaker, Disc. Farad. Soc. 41, 328 (1966).
35) Schlachter and M. Boudart, J. Catal. 24, 482 (1972).
36) J. Block, H. Thimm and M. S. Zei, Symposium Mobility of Surface Molecules, Louvain Sept. 1972.
37) J. Block, Z. phys. Chem. N. F. 39, 169 (1963).
38) J. Block, IV Intern. Congr. on Catalysis, Reprints of Papers for the Catalysis Society, compiled by J. W. Hightower, Vol. 4, 1601 (1969).
39) P. Hindennach and J. Block, Ber. Bunseng. 75, 993 (1971).
40) A. E. Bell, L. W. Swanson and D. Reed, Surf. Sci. 17, 418 (1969).
41) J. Block and P. L. Moentack, Techn. Rep. 36/62 Union Carb. Europ. Res. Assoc.
42) H. D. Beckey, Z. Naturf. 17a, 1103 (1962).
43) K. Bätjer, Dissertation F.U. Berlin in preparation.
44) F. W. Röllgen and H. D. Beckey, Surface Sci. 11, 69 (1970).
45) J. Block, Z. phys. Chem. N.P., in print.
46) J. Block and G. Schulz-Ekloff, J. Catalysis, in preparation.
47) W. A. Schmidt, Angew. Chem. 80, 151 (1968).
48) G.-M. Schwab, Mém. Soc. Rog. Sci. Liège, tom 1, 31 (1971).
49) J. Block and M. S. Zei, Surface Sci. 27, 419 (1971).
50) H. Karge, private communication.
51) E. J. Zandberg et al., Theor. and Exper. Chem. (UdSSR) 7, 363 (1971).
52) N. I. Ionov, Soviet Physics. Techn. Phys. 14, 542 (1969).
53) M. S. Zei, unpublished.
54) H. D. Beckey and P. Schulze, Z. Naturf. 21a 214 (1966).

55) F. W. Röllgen, Dissertation Bonn 1970.
56) J. Block, H. Thimm and K. Zühlke, J. Vac. Soc. and Technol. 7, 63 (1970).
57) S. Hellwig and J. Block, Surf. Sci. 29, 523 (1972).
58) H. Thimm, G. Abend, unpublished.

Invited Lecture F

KINETICS OF HETEROGENEOUS CATALYSIS

M. I. TEMKIN
Karpov Institute of Physical Chemistry, Moscow USSR

ABSTRACT: The theory of the reaction kinetics of heterogeneous catalysis is an important branch of the science of catalysis. This branch has more practical application than any other in the present state of the theory of catalysis. Therefore one is justified in examining some general questions in the theory of the kinetics of heterogeneous catalysis. The present discussion will deal with the dependence of the rate of reaction on the concentration of reactants in the vicinity of the surface of catalyst and on the temperature of the surface. This dependence is determined directly if the reaction takes place in the kinetic region.

1. COMPLEX REACTIONS IN IDEAL ADSORBED LAYERS

The simplest mathematical expression of the regularities of adsorption and reactions on the surfaces of solids is based on the model introduced by Langmuir (1,2). It may be called the model of the ideal adsorbed layer analogously to the terms "ideal gas," and "ideal solution" (Langmuir himself used the term "simple adsorption"). From Langmuir's model it follows that the law of mass action applies to the elementary reaction rates on the surface, including the processes of adsorption and desorption, and also to the equilibria in which adsorbed particles participate; just as the law of mass action applies to reactions in ideal volume systems. The difference is only that the mathematical expression of the mass action law for the surface reaction contains the surface concentrations of reacting species along with the concentrations of free surface sites, if the latter are required for the reaction. The mathematical expression may in addition contain volume concentrations. In the interpretation of the kinetics it is not necessary to distinguish between the adsorption and desorption stages and the surface chemical interaction stages.

The law of mass action deals with elementary reactions. A heterogeneous catalytic reaction is always a set of elementary reactions whose number may be large. The complex reaction may proceed simultaneously in several directions. How may one describe such a reaction? A simple and complete answer to this question exists in the case of stationary reactions, and it has wide application.

In order to discuss the question of the rates of stationary complex reactions, let us use as a concrete example the reaction of the hydrogenation of nitric oxide[3] which was studied recently in our laboratory. An industrial method for the preparation of hydroxylamine sulfate was worked out in F.G.R. on the basis of this reaction. An aqueous solution of sulfuric acid is saturated with nitric oxide and hydrogen. The catalyst used is platinum supported on activated carbon. In addition to hydroxilamine there are formed ammonia and nitrous oxide as by-products. The kinetics of this reaction turned out ot be rather singular. On increasing the partial pressure of nitric oxide over the solution, the rate of formation of hydroxylamine, as well as ammonia, proceeds through a maximum and then falls to practically zero. The rate of formation of nitrous oxide on the other hand increases continuously.

The observed dependences were explained on the basis of the following reaction mechanism (see Table 1). Please do not pay any attention for the present to the column and line outside the frame (identified as $N^{(4)}$). The letter Z in the scheme designates the site on the surface, i.e. stage 1 is the adsorption of hydrogen in the molecular form, etc. The columns $N^{(1)}$, $N^{(2)}$, $N^{(3)}$ contain the stoichiometric numbers.[4] If the chemical equations of the stage are multiplied by the number in one of the columns and added up, then all intermediates --Z, ZH_2, ZH_2NO, etc. cancel out and we obtain one of the overall reaction equations; using the column $N^{(1)}$, we obtain the equation of line $N^{(1)}$, etc.

Three routes given by the vector-columns $N^{(1)}$, $N^{(2)}$, and $N^{(3)}$ form a basis of routes. This means, first that they are linearly independent and secondly that any other route of the reaction is a linear combination of them. For example the stoichiometric numbers of the route $N^{(4)}$ are equal to one-half of the sums of the numbers of routes $N^{(1)}$ and $N^{(2)}$. Consequently, the following vector equations holds true

$$N^{(4)} = \frac{1}{2}\,(N^{(1)} + N^{(2)}).$$

In addition to this linear combination of basic routes one may form a number of others.

In the given example the overall equations of the basic routes are linearly independent; in the general case this is not necessary.[5]

The rate of an elementary reaction is the number of reaction acts which take place in unit time per unit reaction space, i.e. per unit volume if the reaction is homogeneous or per unit of surface area if the reaction is heterogeneous. The rate of a state is the difference between the rates of the forward and reverse elementary reactions which together compose the stage. Let us call the number of stage runs the

TABLE 1
Mechanism of the hydrogenation of nitric oxide

	$N^{(1)}$	$N^{(2)}$	$N^{(3)}$	$N^{(4)}$
1) $Z + H_2 \rightleftarrows ZH_2$	3	5	2	4
2) $ZH_2 + NO \rightarrow ZH_2NO$	2	2	2	2
3) $ZH_2NO + ZH_2 \rightarrow Z + ZH + NH_2OH$	2	0	0	1
4) $2ZH \rightleftarrows ZH_2 + Z$	1	-1	1	0
5) $ZH_2NO + ZH_2 \rightarrow Z + ZOH + NH_3$	0	2	0	1
6) $ZOH + ZH \rightarrow 2Z + H_2O$	0	2	0	1
7) $2ZH_2NO \rightarrow 2ZH + H_2O + N_2O$	0	0	1	0

$N^{(1)}$; $2NO + 3H_2 = 2NH_2OH$
$N^{(2)}$; $2NO + 5H_2 = 2NH_3 + 2H_2O$
$N^{(3)}$; $2NO + H_2 = N_2O + H_2O$
$N^{(4)}$; $2NO + 4H_2 = NH_2OH + NH_3 + H_2O$

difference between the number of acts of the forward and reverse elementary reactions; then the rate of a stage is the number of its runs taking place in unit time per unit reaction space.

In order to define the rate of a complex reaction let us introduce the term "run along route." The run along a route is made up of the stage runs each taken in number equal to the stoichiometric number of the stage for the given route. The result of this corresponds to the overall chemical equation of the route.

In the stationary course of a reaction, the amounts of intermediates do not change. This is possible only if the elementary acts of the reaction taking place in the system group themselves in the route runs: each particle of intermediate formed must be expended. Runs by all kinds of different routes, not necessarily basic routes, make up the reaction. Let two runs along route $N^{(4)}$ have taken place. According to the vector equation given above, one may distribute the comprising state runs between the routes $N^{(1)}$ and $N^{(2)}$. This amounts to one run along route $N^{(1)}$ and one run along route $N^{(2)}$. In the same way all the remaining runs which do not proceed by basic routes may be substituted by runs along basic routes. As a result, practically all state runs for a sufficiently long period of time can be distributed among the basic routes. This justifies one to introduce the concept of "rate of reaction along a basic route,"[6] which is equal by definition to the number of runs along the basic route in unit time per unit reaction space, provided that all elementary reaction acts are grouped in runs along basic routes.

Let us designate the stoichiometric number of stage s by the route p as $\nu_s^{(p)}$. One run along the basic $N^{(p)}$, corresponds to $\nu_s^{(p)}$ runs of stage s. From this there follows the stage steady-state conditions (6, 5):

$$\sum_{p=1}^{P} \nu_s^{(p)} r^{(p)} = r_s - r_{-s}; \quad s = 1,2...S.$$

Here $r^{(p)}$ - is the rate along basic route $N^{(p)}$, r_s and r_{-s} are the rates of the forward and reverse elementary reactions which together comprise stage s, P is the number of basic routes, S - the number of stages. Expressing r_s and r_{-s} according to the mass action law one obtains equations which express the relationship between rates along routes and concentrations. The number of equations is equal to the number of stages S. Besides these we may have balance equations showing the relationship between the concentrations of intermediates. In the reaction under consideration the balance equation is:

$$\theta_Z + \theta_{ZH_2} + \theta_{ZH_2NO} + \theta_{ZH} + \theta_{ZOH} = 1$$

where θ_Z is the fraction of free sites on the surface, θ_{ZH_2} - the fraction of the surface covered by H_2 molecules, etc. The overall number of equations is always equal to the number of unknowns, which are the rates along basic routes and concentrations of intermediates.

By using the balance equations, one may decrease the number of unknown concentrations of intermediates by the number of these equations. The intermediates whose concentrations after this are not excluded were called independent intermediates by Horiuti. Their concentrations are determined together with reaction rates along the basic routes by the stage steady-state conditions.

In deriving the kinetic equations one often uses additional simplifying assumptions. One assumes that some stages are in equilibrium and that the degrees of surface covering by some of the intermediates are negligibly small.

The method of deriving kinetic equations of a complex stationary reaction in an ideal system which uses the law of mass action and the stage steady-state conditions is quite general. It may be applied to catalytic reactions as well as to reactions of any other type, for example to stationary chain reactions. There are available methods which simplify the mathematical analysis of complex cases.[6)]

2. THE ATTAINMENT OF STATIONARY REACTION RATES

A knowledge of the kinetics of an industrial catalytic process in stationary conditions may not be sufficient. It may be necessary to know how quickly the stationary regime is established on changing the conditions of the reaction. Let us suppose that a reaction, for the sake of simplicity a single route reaction, proceeds in a stationary way for a long time under unchanging conditions; these conditions then suddenly are changed and assume new values which are kept constant. The reaction rate begins to change and assymptotically attains a new value which corresponds to the stationary reaction under new conditions. The following method similar to that used by Semenov in chain theory,[7] may be used in many cases for mathematical description of such transitory phenomena. Let us divide the independent intermediates of the given reaction in two classes: in the first class are those intermediates whose concentrations are not in equilibrium in the transitory conditions and in the second whose concentrations are in quasi-equilibrium. A species of the first class will be called a long-lived intermediate. Let us add to the basic routes of the stationary reaction, routes whose overall chemical equations describe the formation of long-lived intermediates. After this, the reaction may be considered as quasi-stationary, only with an increased number of routes. One may apply to it the stage steady-state conditions. We calculate the rate of formation of the long-lived intermediates. Then knowing their initial concentrations (from the solution of the equations for the stationary reaction), we obtain by integration the concentration of the long-lived intermediates as a function of time and hence the rates along the routes.

In a number of reactions we have one long-lived intermediate. For example, in the synthesis of ammonia, adsorbed nitrogen is such an intermediate, the concentrations of the adsorbed intermediates formed in the hydrogenation of adsorbed nitrogen being in quasi-equilibrium with the product of the reaction. The synthesis of ammonia is a one route reaction. Consequently in order to describe the transitory regime in the synthesis of ammonia we must examine the reaction with two basic routes - one of them leads to the formation of ammonia, the other - to the formation of adsorbed nitrogen.

In the simplest cases, the observed rate of reaction r will change with time according to the following law:

$$r = r_\infty - (r_\infty - r_0)e^{-\frac{t}{\tau}}$$

where t designates time, r_∞ and r_0 designate the value of r at $t = \infty$ and $t = 0$, τ is a constant, the relaxation time. On the basis of the reaction mechanism hypothesis and using the above method one can derive an expression for τ, which will contain the rate constants of the elementary reactions and the concentrations of the reaction participants. An experimental determining of the reaction rate relaxation time and its dependence on the concentrations of the reaction participants may serve for checking the reaction mechanism hypothesis, independent of the check which yields a determination of the stationary reaction rate. The relaxation time always satisfies the inequality $\tau < L/r_\infty$, where L is the number of adsorption sites per unit area of the catalyst surface. Calculations for different reactions in ordinary conditions yield the expected value of the order of magnitude of τ, - about one second. The experimental determination of such values is difficult. One of the possible ways of determining the relaxation time is to slow up the reaction, carrying it out at lower pressures and temperatures.

In addition to the rapid relaxation processes in the adsorbed surface layer considerably slower relaxation processes may take place caused by changing of the phase composition of the catalyst, or by "deep adsorption," i.e. penetration of the gases in the sub-surface layers of the solid. For example, silver which is a catalyst for the oxidation of ethylene to ethylene oxide is capable of deep adsorption of a large quantity of oxygen.[8,9] Experiments, carried out in our laboratory by Kurilenko,[10] showed that when in a mixture of ethylene and oxygen reacting on silver a part of oxygen is exchanged for nitrogen, the rate of reaction diminishes slowly. At 218°C there was required about 20 hours for establishing the new stationary reaction rate.

The influence of the medium on the catalyst is discussed extensively by Boreskov.[11]

3. REACTIONS IN REAL ADSORBED LAYERS

Physical chemistry would be much simpler and a more exact science, if the laws of ideal systems were always obeyed. In actuality the behavior of real systems differs from that of ideal ones sometimes very significantly. In the case of gases, gaseous mixtures, and liquid solutions, the method developed by Lewis, which is based on the notions of fugacity, activity, and activity coefficient is used by physical chemists for description of the properties of the real systems. Developments in the field of real adsorbed layers on the surfaces of solids proceeded by a different course. In this case the concept of non-uniform surface proved to be serviceable.

H. S. Taylor used the model of a non-uniform surface widely for qualitatively explaining a series of phenomena. In the most general form this model, however, is too indefinite for use in quantitative applications. A decisive step in removing this indefiniteness consists in assuming the existence of a relationship between the thermodynamic and kinetic characteristics of the surface sites. A prototype for the mathematical expression of this relationship is the relationship between the reaction rate constant and the equilibrium constant established by Brönsted for the reactions of acid catalysis. The theoretical basis of this and analogous relationships for homogeneous reactions was developed by Evans and Polanyi.[12)] For heterogeneous catalysis on non-uniform surfaces the above relationship was postulated in connection with the theory of the kinetics of ammonia synthesis.[13)] It was then applied to other heterogeneous reactions and to electrode processes on non-uniform surfaces.[14)] At the present time it is based on a large quantity of experimental data in the field of the kinetics of heterogeneous catalysis and electrode processes.

It may be formulated thus: the rate constants of an elementary reaction k_s on various sites of the non-uniform surface and the corresponding equilibrium constants K_s are related in the following way:

$$k_s = g\,K_s^a; \quad 0 \leq a \leq 1 ,$$

where g and a are constants. The quantity a is called the transfer coefficient. Therefore the rule expressed by the given equation and analogous rules for homogeneous reactions, may be called "rules of transfer" or more fully "rules of transfer of equilibrium changes on the reaction rate." The coefficient a is usually written α when adsorbed particles are formed; for the reverse direction it is written as β. It is easy to see that $\alpha + \beta = 1$. It is common that $\alpha = \beta = 1/2$. The rate of reaction is the sum of the contributions of different surface sites, i.e., it is expressed as an integral under whose sign there is placed the unknown distribution function, which characterizes the non-uniform surface. If the kinetics of reaction is known from the experiment and the character of the non-uniformity is sought then one must solve an integral equation. Fortunately in simple reaction mechanisms, the problem reduces to an integral equation which has a very elegant, exact solution. This equation is

$$f(x) = \int_{-\infty}^{\infty} \frac{g(\xi)\,d\xi}{1+e^{\xi-x}}$$

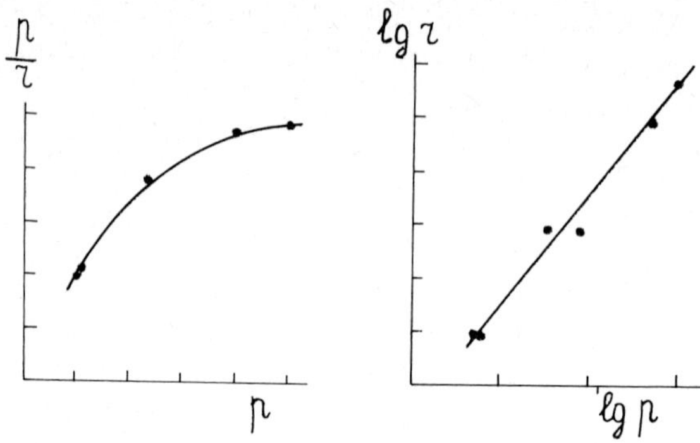

nitrogen pressure which would correspond to the equilibrium with the actual pressures of ammonia and hydrogen in the reaction mixture, P_{NH_3} and P_{H_2}:

$$P_{N_2} = \frac{1}{K} \frac{P_{NH_3}}{P^3_{H_2}}$$

where K is the equilibrium constant. With the help of this notation one may describe the kinetics of the synthesis of ammonia in the following way:

$$r=r_+-r_-;\ r_+=k_\pm \frac{P_{N_2}}{p^m_{N_2}};\quad r_-=k_\pm\, p^{1-m}_{N_2};\quad 0 \le m \le 1.$$

Here $k_\pm$ and m are constants, P_{N_2} is the actual nitrogen pressure. Let us assume that the nitrogen fugacity on the catalyst surface equals P_{N_2}. this means that the adsorbed nitrogen is in equilibrium with gaseous H_2 and NH_3. Then r_+ should correspond to the rate of nitrogen adsorption, and r_- - to the rate of nitrogen desorption. If the surface were uniform, the rates of nitrogen adsorption and desorption would obey the equations

$$r_+ = k_\pm \frac{P_{N_2}}{P_{N_2}+b},$$

and

$$r_{-} = k_{\pm} \frac{P_{N_2}}{P_{N_2}+b} .$$

Instead of this, fractional orders of P_{N_2} are present in the equations. The kinetic equation for the synthesis of ammonia was confirmed by measurements in the pressure range from 1/4 to 500 atm. If the fractional power relationship were an approximation of the hyperbolic one it certainly could not be obeyed at changing the pressure 2000 fold. The problem of the character of the surface non-uniformity, leading to the kinetics of ammonia synthesis, may be reduced to the integral equation mentioned above. Its solution shows[16] that:

$$\phi(\xi) = Ae^{\gamma\xi}$$

where $\phi(\xi)$ is the distribution function: the number of sites with the value ln b from ξ to $\xi + d\xi$ is equal to $\phi(\xi)d\xi$, A and γ being constants,

$$\gamma = \alpha - m.$$

If $\gamma = 0$, then $\phi(\xi) = A$. This even distribution corresponds to the logarithmic adsorption isotherm. In the original derivation of the kinetic equation for the synthesis of ammonia [13,14] such even distribution was assumed, in which case the power in the kinetic equation m is equal to the transfer coefficient α. The exponential distribution with $0 < \gamma < 1$ corresponds to the Freundlich adsorption isotherm, with the gas pressure to the power γ. Thus the kinetic equation for the synthesis of ammonia is not necessarily related to the logarithmic isotherm. It may be related to the Freundlich isotherm, and the possibility of an exponential non-uniformity with $-I < \gamma < 0$ is also not excluded.[17]

Research on the synthesis of ammonia using promoted iron catalysts often yields the value m = 0.5, as in the first publications:[13] recently this value was again confirmed;[18] other values, however were also obtained (see for example (19)). It is difficult to imagine a change in the transfer coefficient α, it is natural to think, that the difference in iron catalysts may be due to the character of the non-uniformity, i.e., the value of γ.

In order to determine α, one needs to know not only m but γ, i.e., it is necessary to establish the form of adsorption isotherm of nitrogen on the catalyst. This was done for the catalyst with m = 0.5, by studying the equilibrium of adsorbed nitrogen with gaseous hydrogen and ammonia.[20] Determining the covering of the surface at different values of P_{N_2}, one obtains the adsorption isotherm. In this way it was found, that the surface was evenly non-uniform, i.e., $\gamma = 0$, and consequently $\alpha = 0.5$.

The kinetics of ammonia synthesis on cobalt corresponds m = 0.2; on nickel m = 0.3.[21] It was natural to suppose, that on those metals $\alpha = 0.5$; then the nitrogen adsorption isotherm should follow the Freundlich's equation. This supposition was confirmed experimentally, in the investigation of the nitrogen adsorption isotherm on cobalt and nickel, carried out in our laboratory by N. M. Morozov and L. I. Lukyanova by the above method. The results of the experiments which remained unpublished do lead to the Freundlich isotherm with the power $\gamma = 0.3$ in the case of cobalt and $\gamma = 0.2$ in the case of nickel so that $\alpha = m + \gamma = 0.5$ in all three metals of the iron group. Therefore, the original derivation of the kinetics of ammonia synthesis[13,14] supposing $\gamma = 0$, is actually true for that catalyst which was investigated in the first works, but for other catalysts a more general approach is required.[17]

The above discussion shows that on catalysts used in ammonia synthesis we meet with "biographical" non-uniformity according to the terminology of S. Z. Roginsky, as it was assumed in the early researches. Recent research in our laboratory using isotope exchange between nitrogen on the surface of the ammonia synthesis catalyst with ammonia in the gaseous phase at equilibrium with hydrogen confirmed the existence of biographical non-uniformity.[22]

In other cases there may predominate the so-called "induced non-uniformity," i.e., the effect of the mutual influence of adsorbed particles. Thus in the adsorption of hydrogen on platinum and palladium the change in the strength of the adsorption bond with the surface coverage corresponds to the theoretically calculated one, on the basis of the surface electronic gas model, if one assumes that each hydrogen atom donates 1 electron to the metal surface layer, and if one uses the known values of the effective mass of the electron in these metals.[23]

In the investigations of adsorption equilibrium there were obtained many times both logarithmic and fractional power adsorption isotherms. Also there were observed many times adsorption kinetics corresponding to those isotherms, described by the known equations of Zeldovich-Roginsky and Bangham respectively. There were often found fractional reaction orders.

Therefore, we may assume that the exponential law of

surface non-uniformity including the often occurring case of even non-uniformity in conjunction with the rule of transfer gives a sufficient empirical basis for the interpretation of the reaction kinetics on non-uniform surfaces.

Among the more important industrial catalytic reactions where such an approach proved fruitful, in addition to the synthesis of ammonia one can name the contact oxidation of sulfur dioxide.[24]

REFERENCES

1) I. Langmuir, J. Amer. Chem. Soc., 40, 1361, 1918.
2) I. Langmuir, Trans. Faraday Soc., 17, 621 (1921-1922).
3) N. N. Savodnik, N. V. Kul'kova, D. M. Dokholov, V. L. Lopatin, M. I. Temkin, Kinetika i kataliz (in print).
4) J. Horiuti, J. Res. Inst. Catalysis, Hokkaido Univ., 5, 1 (1957).
5) M. I. Temkin, in the Coll. "Mechanism and Kinetics of Complex Catalytic Reactions," Pub. "Nauka," Moscow, 1970, p. 57.
6) M. I. Temkin, Doklady Acad. Nauk USSR, 152, 156 (1963).
7) N. N. Semenov, Zhur. Fiz. Khim., 17, 187 (1943). N. N. Semenov, "Concerning Several Problems of Chemical Kinetics and Reactivity," (Russ.) 2 Ed. Acad. Sci. USSR, Moscow 1958, p. 460.
8) N. V. Kul'kova, M. I. Temkin, Dokl. Acad. Nauk USSR, 105, 1021 (1955).
9) R. P. Kayumov, N. V. Kul'kova, M. I. Temkin, Kinetika i Kataliz (in print).
10) A. I. Kurilenko, Thesis, Karpov Institute of Physical Chemistry, Moscow, 1960, see also A. I. Kurilenko, N. V. Kul'kova, N. A. Rybakova, M. I. Temkin, Zhurn. Fiz. Khim. 32, 797 (1958).
11) G. K. Boreskov, Kinetika i Kataliz, 13, 543 (1972).
12) M. G. Evans, M. Polanyi, Nature, 137, 530 (1936), Trans. Faraday Soc., 32, 1333 (1936).
13) M. I. Temkin, in the review of V. Zarinsky and N.Tunitsky, Uspekhi Khimii 7, 1092 (1938). M. Temkin, V. Pyzhev, Zhur. Fiz. Khimii 13, 841 (1939); Acta Physicochim. URSS, 12, 327 (1940).
14) M. I. Temkin, Zhur. Fiz. Khimii 15, 296 (1941).
15) E.C. Titchmarsh, Introduction to the Theory of Fourier Integrals, Oxford, 1937. M. Temkin, V. Levich, Zhur. Fiz. Khimii 20, 1441 (1946).
16) A. A. Khomenko, L. O. Apelbaum, M. I. Temkin, Kinetika i Kataliz, 7, 671 (1966).

17) M. I. Temkin, in the Coll. "Problems of Kinetics and Catalysis," VI, Pub. Acad. Sci. USSR, Moscow-Leningrad, 1949, p. 54.
18) E. N. Shapatina, V. L. Kuchaev, M. I. Temkin, Kinetika i Kataliz, 12, 1476 (1971).
19) A. Nielsen, J. Kjaer, B. Hansen, J. Catalysis, 3, 68 (1964).
20) A. E. Romanushkina, S. L. Kiperman, M. I. Temkin, Zhur. Fiz. Khim., 27, 1181 (1953).
21) N. M. Morozov, E. N. Shapatina, L. I. Lukyanova, M. I. Temkin, Kinetica i Kataliz, 7, 688 (1966).
22) E. G. Boreskova, V. L. Kuchaev, B. E. Pen'kovoi, M. I. Temkin, Kinetika i Kataliz, 13, 358 (1972).
23) M. I. Temkin, Kinetika i Kataliz, 13, 555 (1972).
24) G. K. Boreskov, "Catalysis in the Production of Sulfuric Acid" (Russ), Goskhimizdat, Moscow-Leningrad, 1954.

Paper Number 1

THE INFLUENCE OF THE MATRIX ON THE CATALYTIC ACTIVITY OF TRANSITION METAL IONS IN OXIDE SOLID SOLUTION

A. CIMINO, F. PEPE and M. SCHIAVELLO
Centro di Studio sulla "Struttura ed Attività Catalitica di Sistemi di Ossidi," Istituto Chimico, Università di Roma, Roma, Italy

ABSTRACT: The catalytic N_2O decomposition on the solid solution systems CoO-ZnO, CuO-MgO, CuO-ZnO and the spinel $CoAl_2O_4$, has been investigated. The results obtained, together with previous data on the solid solutions CoO-MgO, NiO-MgO, and NiO-ZnO, allow a comparison of the role played by the matrices in the activity. It is shown that a given transition metal ion is more active in MgO than in ZnO solid solution. In order to distinguish whether the effect is due solely to different coordination or also to a direct intervention of the surface, a study of the spinel-structure activity has been performed. It has been established that: i) the coordination symmetry affects the activity, and ii) the role of the matrix surface in reaction steps can be important.

1. INTRODUCTION

Investigation of the catalytic activity of solid solutions of transition metal ions (t.m.i.) in inert oxides offers, in principle, a means of investigating the influence of interionic interactions. This type of investigation has been carried out in our laboratory for a simple reaction, namely the N_2O decomposition. It has been found possible to show that the intrinsic activity of each t.m.i. is higher in dilute solid solutions than in concentrated ones. The use of solid solutions, however, allows one, in principle, to study two additional effects, by an appropriate choice of the matrix: a) the influence of the coordination symmetry of the active ion (imposed by the matrix); b) intervention of the matrix surface in reacttion steps, such as diffusion and desorption of products. This paper presents the results of an investigation of the activity of systems containing Co^{2+} ions in ZnO and in $CoAl_2O_4$, Cu^{2+} ions in MgO and in ZnO, which, together with previous data on Ni^{2+} in MgO[1,2], ZnO[3], and $MgAl_2O_4$[4] and Co^{2+} in MgO[5] allow a comparison between the different t.m.i. to be made. The two effects a) and b) can then be evaluated.

2. EXPERIMENTAL

2.1 Catalytic activity. The N_2O (initial pressure about 60 Torr) was decomposed up to a maximum of 1%. The progress of the reaction was followed by measuring the amount of non-condensed gas produced[6]. Calculation of the velocity constant was based on 1st order kinetic law: $-dP_{N_2O}/dt = kP_{N_2O}$ which was found to be obeyed at the small decomposition percentages investigated here. All data are expressed as absolute velocity constants in cm min^{-1}. The amount of catalyst used was about 0.2 g. Catalysts were initially conditioned by outgassing in vacuum ($P = 10^{-5}$ torr) at 480°C for 4 hrs; outgassing was also carried out between runs, for a period of 30 min., using the same initial conditioning temperature. A higher conditioning temperature was chosen in some cases, after determination of the activity with 480°C outgassing.

2.2 Catalysts. The catalysts and their main features are listed in Table 1, which also reports some of the relevant catalysts previously studied. Preparations were performed by impregnating a given amount of the matrix (MgO, ZnO) with a comparable volume of a nitrate solution containing the t.m.i. in the required concentration. The soaked mass was dried at 110°C, heated at 600°C, then ground and fired at the desired temperature for the appropriate time, specified for each systems. The $CoAl_2O_4$ specimen was prepared by impregnating γ-Al_2O_3 with a comparable volume of reagent grade $Co(NO_3)_2$. After treatment at 600°C, the ground specimen was divided into two portions, and both were pelletted and fired for 70 hr at 1200°C. One portion, designated as SACo(1200), was quenched from 1200°, the other, SACo(800), was submitted to an annealing at 800°C for 40 hrs and then quenched from 800°. The specimens used for the catalytic work were previously ground. All the catalysts were characterized using X-ray methods, magnetic susceptibilities, reflectance spectroscopy and chemical analysis. Electron spin resonance techniques were also applied to the CuO-MgO system[7]. All catalysts listed in Table 1 were found to be true solid solutions. Details of the systems CuO-MgO and CuO-ZnO and CoO-ZnO and $CoAl_2O_4$ will be given elsewhere.

3. EXPERIMENTAL RESULTS

3.1 Pure copper oxides. Because of its electronic configuration (d^9), the Cu^{2+} ion has a high Crystal Field Stabilization Energy, E_c, for the square planar configuration ($= -12.28\ D_q$) and a lower E_c for the octahedral ($=-6.0\ D_q$) and tetrahedral ($= -1.78\ D_q$) ones. The influence of the E_c value, together with the inherent instability of Cu^{2+} at the high temperatures necessary for solid state reactions, renders the preparation of solid solutions of Cu^{2+} ions more difficult than for other transition metal ions[8]. It was possible, however, to prepare specimens CuO-MgO and CuO-ZnO which were true solid solutions according to X-rays (phase and lattice parameters determina-

TABLE 1
Catalysts and their main features

Catalyst	Symbol	T.m.i. content (atom%)	Surface area* (m^2g^{-1})	E_a** ($kcal \cdot mole^{-1}$)
CuO-MgO 0.1	MCu 0.1	0.12	38	25
CuO-MgO 0.5	MCu 0.5	0.49	48	22
CuO-MgO 1	MCu 1	1.12	18	22
CuO-ZnO 0.1	ZCu 0.1	0.2	0.3	25,25(650),25(750)
CuO-ZnO 3	ZCu 3	3.1	0.3	25,25(650)
CuO	CuO	-	6.6	25
Cu_2O	Cu_2O (650)	-	0.9	23(650)
Cu_2O	Cu_2O (700)	-	0.9	23
CoO-ZnO 0.1	ZCo 0.1	0.1	0.1	32(600),28(700)
CoO-ZnO 1	ZCo 1	1.1	0.2	35(600),31(700)
CoO-ZnO 5	ZCo 5	5.3	0.1	42(600),36(700)
$CoAl_2O_4$	SACo(800)	-	0.8	29
$CoAl_2O_4$	SACo(1200)	-	0.9	20
CoO-MgO 1	MCo 1	1.22	14	22.5
NiO-MgO 1	MN 1	1.0	14	19
NiO-ZnO 0.5	ZN 0.5	0.56	0.5	40
ZnO	ZO(1000)	-	0.6	33(650)
ZnO	ZO(1200)	-	0.2	35(600),35(700)

* B.E.T. method, krypton adsorption
** In parentheses, outgassing temperatures when higher than 480°C

tions) and optical reflectance spectroscopy (absence of CuO or Cu_2O characteristic absorption [8] in the region of 1000 nm). The reasons outlined above rationalize the tendency to segregation of copper (II) oxide, and/or to reduction to copper (I). Segregation was found after catalysis in ZnO based catalysts, in accordance with the expected order of stability. It is therefore important to investigate the catalytic activity of the pure copper oxides, and to check the influence of various heat and outgassing treatments on the pure oxides.

Earlier studies[9,10] have given clear indication that Cu_2O is more active than CuO. The state diagram for Cu-Cu_2O-CuO system[11] shows that Cu_2O is formed under vacuum at temperatures above 500°C. Below these temperature, surface reduction (as opposed to bulk) can occur, with formation of surface Cu_2O, or, at least, of Cu^{1+} ions and oxygen vacancies. The variation of the catalytic activity of CuO with outgassing temperatures has been explored, and is shown in fig. 1. CuO (ex carbonate, in air, 300°C for 24 hr) outgassed at 300° for 4 hr showed very low activity (not reported in the graph). Outgassing at 480° for 4 hr gave rise to higher activity (line e), but a larger increase of activity was noticed when the oxide was prepared by outgassing at 650° for 30 min (line f). Allowance has been made for the variation of surface area, which shrank from 6.6 to 0.9 m^2g^{-1}. Treatment with N_2O at 650°, followed by outgassing at 480°, led to a drop in activity, whilst outgassing at 650° restored the initial level.

The importance of the initial outgassing temperature is shown by an additional experiment, whereby a CuO specimen was subjected to an outgassing ($P = 10^{-5}$ Torr) at 700°C for 7 hr (surface area 0.9 m^2g^{-1}), but later submitted to a 480°C outgassing for 30 min, between N_2O decomposition experiments. The catalytic activity (line g) was found to be maintained at a high level.

There is no doubt that Cu^{2+} ions are far less active than Cu^{1+} ions. It is also questionable whether the low (but measurable) activity of CuO is indeed attributable to Cu^{2+}.The constancy of the apparent activation energy, E_a, suggests that in all cases Cu^{1+} ions are involved, only their number (hence the preexponential factor) being changed.

3.2 Cu^{2+} in MgO. Three specimens, MCu 0.1, MCu 0.5, MCu 1, were prepared at 1200°C in air, for 5 hr. The catalytic results are described in fig. 1; the reference activity of the matrix MgO[6] is also reported (line a). The main features of the system are: a) the catalytic activity rises with Cu^{2+} content; b) E_a decreases with increasing Cu^{2+} content.

In order to clarify the dependence of the activity upon Cu^{2+} ions concentration, and the nature and symmetry of the active sites in the catalysts, E.S.R. experiments have been performed[7]. The E.S.R. spectra of the same MCu (untreated) catalysts used in catalysis have shown the presence of two signals attributable to Cu^{2+} ions in MgO. The first signal ($g = 2.19$) can be attributed to isolated ions in octahedral symmetry, hence to bulk Cu^{2+} ions, and it is largely insensitive to treatments like outgassing, oxidizing atmosphere etc. The second signal ($g_{\parallel} = 2.352$, $g_{\perp} = 2.063$) which can be attributed to Cu^{2+} ions in sites whose symmetry has been lowered by an axial component is believed to be due to Cu^{2+} ions close to the surface, and it varies strongly on treatment. The outgassing treatment used in catalysis, in particular, destroys part of this axial signal. It is believed that the axial (surface) Cu^{2+} ions can partly be reduced to the diamagnetic Cu^{1+} ions on the surface: low temperature oxidation, in fact, restores this signal to a large extent.

The total amount of axial Cu^{2+} signal destroyed by the outgassing (hence the amount of Cu^{1+} created) increases with Cu^{2+} ion content.

The activity of all CuO-MgO specimens can therefore be explained in terms of Cu^{1+} species the amount of which increases with increasing initial Cu^{2+} content. These results are in line with earlier studies on CuO, Cu_2O[9,10] and with the present work.

3.3 Cu^{2+} in ZnO. Two specimens, ZCu 0.1 and ZCu 3 prepared at 1000°C for 140 hr and quenched from 1000°, in addition to the matrix ZnO, (Specpure, Johnson and Matthey) tested after treatment in air at 1000°C, for 140 hr, were investigated,and the results are reported in fig. 2. The catalytic activity is only slightly dependent on composition, but it is markedly increased by the increasing outgassing temperature. Moreover, the E_a values are constant, and equal to those found for pure copper oxides. The ZCu 3 specimens exhibited a variation of lattice parameter and of the reflectance spectra after catalysis, which showed that segregation of copper oxide had occurred. The activity of the specimen can thus be attributed to segregated copper oxide, whose reduction to Cu_2O, as for the pure oxide, leads to an active phase. The small copper content of ZCu 0.1 does not allow a positive demonstration of the segregation process. However, the similar behaviour of ZCu 3 strongly suggests that the same phenomenon may have occurred for the low content specimen.

In conclusion, whilst the CuO-ZnO system has not led to measurable levels of activity of solid solution, it can be stated that the activity of Cu^{2+} ions in tetrahedral symmetry must be very low, and comparable (if not less active) to that of the matrix itself. Indeed, an activity higher than that of the matrix can be found only when segregated copper oxides is formed, i.e. copper ions not in tetrahedral symmetry.

3.4 Co^{2+} in ZnO. Three catalysts have been examined, namely ZCo 0.1, ZCo 1 and ZCo 5, and the catalytic results are shown in fig. 3. The specimens were prepared at 1200°C, for 5 hr, and quenched from this temperature. The matrix ZnO (ex-carbonate, 1200° 5 hr) has also been tested. No appreciable activity was found below 500°C with the 480° outgassing temperature. The main features of the system are: a) the addition of Co^{2+} ions to ZnO enhances the activity of the catalysts; b) the increase in activity reaches a maximum for t.m.i. content of about 1% (ZCo 1); c) high outgassing temperatures increase the activity and lower the activation energy for all the concentrations of t.m.i. studied. Such an effect is absent for the pure matrix ZnO: in fact ZnO shows a substantially constant activity and a constant activation energy, when outgassed at 600° and 700°C; d) a decrease of activity per cobalt ion is found when the t.m.i. concentration exceeds 1%. The activation energy decrease with decreasing cobalt content confirms the trend found on the MCo system[5].

3.5 Co^{2+} in $CoAl_2O_4$. Specimens SACo(800) and SACo(1200) have been studied. It is known that cobalt spinel is almost completely normal. Studies of cationic distribution, however, show that on passing from a lower to a higher temperature the percentage of tetrahedral cobalt is decreased[12].

N_2O decomposition experiments have shown (fig. 4) that:
a) the 1200° quenched sample, SACo(1200), is more active, by a factor of ten, than the 800° quenched sample SACo(800);
b) the temperature range explored for an appreciable activity is higher than that explored on CoO-MgO samples but definitely lower than that necessary for ZCo samples. It can be inferred that the increasing octahedral cobalt content favours catalytic decomposition.

4. DISCUSSION

Earlier work on the decomposition of N_2O had shown that a scale of activity can be drawn for pure oxides[10]. We have found that ions which are active in the pure oxides are also active when atomically dispersed in MgO. Furthermore, the activity of each t.m.i. is larger in a diluted than in a concentrated solid solution. This fact, which will not be discussed here in detail, shows that the matrix does not prevent the fast occurrence of the N_2O decomposition steps. The steps can be schematized as follows:

$$N_2O(g) + t \longrightarrow N_{2(g)} + O\text{-}t \qquad (1)$$

$$O\text{-}t + m_1 \longrightarrow t + O\text{-}m_1 \qquad (2)$$

$$O\text{-}m_1 + m_2 \longrightarrow m_1 + O\text{-}m_2 \qquad (3)$$

$$O\text{-}m_i + O\text{-}m_j \longrightarrow m_i + m_j + O_{2(g)} \qquad (4)$$

where t is a t.m.i., m is a lepton belonging to the matrix (cation, anion or vacancy). No specification of the charge is made.

The reaction steps (2), (3) and (4) represent oxygen migration and desorption processes. They must be taken into account when dealing with solid solutions, since direct desorption from neighbouring O-t centres is less likely here than in pure oxides. The rate of the catalytic process is therefore dependent on how fast all the above steps can proceed. Step (1) (rupture of N_2O bond and formation of O-t bond) depends on the nature of the t.m.i. and of its available orbitals. Steps (2) will again depend on the t.m.i. (rupture of t-O bond), but it will also depend on the ease with which the O atom can be accepted by a neighbouring site, hence on the nature of the matrix. Also steps (3) and (4) will obviously

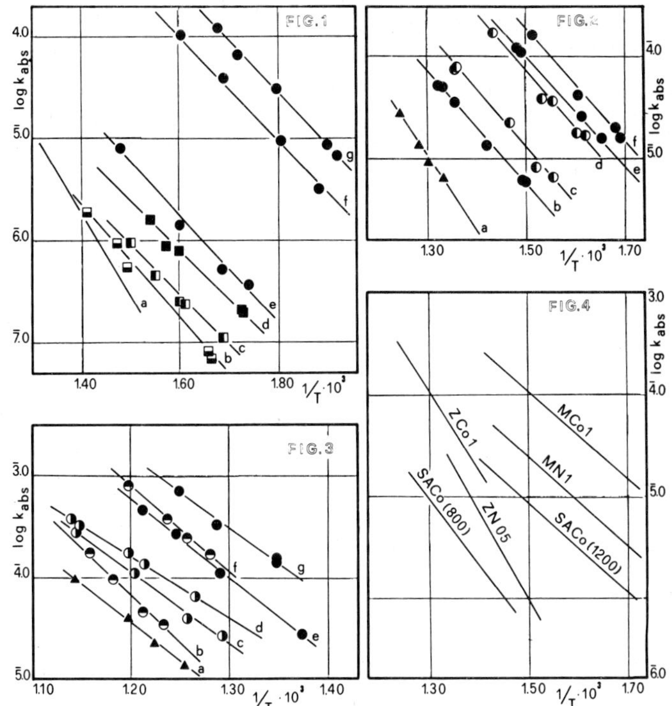

Arrhenius plots for pure oxides and for solid solutions. Symbols are illustrated in Table 1. Outgassing temperatures specified after symbol, between bars / /.

Fig. 1. a: MgO/480/; b: MCu 0.1/480/, c: MCu 0.5/480/; d: MCu 1/480/; e: CuO/480/; f: Cu_2O(650)/650/; g: Cu_2O(700)/480/.

Fig. 2. a: ZO(1000)/650/; b: ZCu 0.1/480/; c: ZCu 3/480/; d: ZCu 3/650/; e: ZCu 0.1/650/; f: ZCu 0.1/750/.

Fig. 3. a: ZO(1200)/600/; b: ZCo 5/600/; c: ZCo 0.1/600/; d: ZCo 0.1/700/; e: ZCo 1/600/; f: ZCo 5/700/; g: ZCo 1/700/.

Fig. 4. SACo(800)/480/; ZN 0.5/480/; ZCo 1/600/; SACo(1200)/480/; MN1/480/; MCo 1/480/.

depend on the matrix.

The experimental data can now be examined, with the help of fig. 4, which reports the Arrhenius plots of activities of nickel solid solutions[1,2,3)] and of cobalt solid solutions[5)]. The activity is markedly reduced on passing from MgO-based to ZnO-based solid solutions. A decrease of activity can also be inferred for copper-containing catalysts since the initial level of ZCu catalysts (before extensive segregation) is lower than the level of MCu catalysts. The segregation of copper oxide in the former catalysts prevents a quantitative comparison to be made. Having shown that the matrix has a marked effect on the activity one can inquire whether both points a) and b) (see Introduction) can be responsible for the effect.

The presence of effect a) is clearly shown by the spinel matrices. Raising the temperature of quenching increases the catalytic activity for $CoAl_2O_4$. The higher the quenching temperature, the larger is the population of cobalt ions in octahedral sites: One might suspect that some change of the spinel surface (such as its defect state) occurs with different temperature treatments, and that the higher catalytic activity is linked to that effect. However, for the nickel spinel[4)] the reverse influence of temperature was true: specimens quenched from higher temperatures were less active. The lower the temperature the higher is the population of nickel ions in octahedral sites. Therefore a consistent picture can be drawn:a high octahedral occupancy by the active ions leads to a high catalytic activity. It is no wonder then that the ZnO based catalysts (ions in tetrahedral sites) are less active.

One can inquire why a cobalt or a nickel ion in tetrahedral symmetry is less active than in octahedral symmetry. A simple way to look at the problems is to relate the activity to metal-oxygen bond strength[3)], and to state that the shorter distances observed in tetrahedral arrangements reflect a difficulty in the cleavage of the M-O bond. Other approaches such as the E_C (Crystal Field Stabilization Energy) values can be taken into account[13,14,15)]. It is difficult, however, to evaluate the relative importance of E_C with respect to total energy, and, secondly, to define a mechanism for the oxygen addition-desorption process. It is noteworthy that the observed order (Co>Ni) can be predicted for octahedral arrangement[5,13)], and that the same order would be predicted for ions in tetrahedral symmetry if displacement mechanism (type S_N2) is assumed, as in a tetrahedral ⟶ square pyramid transformation[13)].

The presence of an effect b) can be deduced on an indirect basis only. It has been pointed out before that oxygen migration must occur to achieve oxygen desorption. Thanks to the extensive work of Winter[16)] and of Boreskov[17)] it is possible to have information on the behaviour of several oxides with regard to oxygen exchange and oxygen equilibration. In particular, Winter demonstrated two points, relevant to the present discussion: i) the ability of the whole oxide surface to undergo exchange; ii) a parallelism between the N_2O decomposition and the oxygen exchange. The ZnO surface, if oxygenated, is less able than the MgO surface to accomplish the oxygen exchange reaction. It should be remarked that during N_2O decomposition atomic oxygen is produced on the surface and a small percentage of N_2O decomposition can give coverages comparable to maximum oxygen coverages obtainable from molecular oxygen. The ZnO surface can thus be considered in a state where surface oxygen is less mobile. An explanation has been given[3], by considering that atomic oxygen can move by formation of a peroxide on the surface, a process more likely on MgO than on ZnO

$$O \text{ (from } N_2O) + O^{2-} \longrightarrow O_2^{2-}$$
$$O_2^{2-} + O^{2-} \longrightarrow O^{2-} + O_2^{2-}$$

A superoxide O_2^- mode could equally well be invoked, and it is known[18,19] that O_2^- can be formed. However, irrespective of the explanation, it can be seen that if the mobility of oxygen is reduced, the ZnO-based catalysts will show a reduced ability for oxygen migration. The reduced diffusion acts in the same direction as point a) above.

A short comment can be made on the value of E_a, which is markedly increased in passing from MgO-based solid solutions to ZnO-based ones. The increase of E_a can reflect the increase in the bond strength. It may also be brought about by a reduced mobility of oxygen, which can tend to shift the oxygen desorption mechanism from that illustrated by equations (3) and (4) to one involving anions more tightly bound and close to t, eq. (1).

In conclusion, when dealing with catalytically active ions dispersed in a matrix, the catalytic activity cannot be considered a property of the ionic species, but the resultant of a favourable electronic configuration (thereby implying both nature and site symmetry) and of a favourable participation of matrix surface in steps successive to the primary step brought about by the active ion. In this context it is interesting to recall the work by Rudham et al.[20], who have shown that t.m.i. in zeolites tend to behave differently for N_2O decomposition from how they behave in MgO. A further point is the possibility of relating the influence of a matrix on the behaviour of supported non-noble t.m.i. to effects of the types here discussed. An example has been illustrated for nickel on alumina[21].

AKNOWLEDGEMENT

The Authors wish to thank Drs. S. De Rossi and G. Ferraris and Mr. C. Angeletti for assistance to the experimental part.

REFERENCES

1) A. Cimino, R. Bosco, V. Indovina and M. Schiavello, J. Cat.,

5 (1966), 271
2) A. Cimino, V. Indovina, F. Pepe and M. Schiavello, J. Cat., 14 (1969), 49
3) M. Schiavello, A. Cimino and J. M. Criado, Gazz. Chim. Ital. 101 (1971), 47
4) A. Cimino and M. Schiavello, J. Cat., 20 (1971), 202
5) A. Cimino and F. Pepe, J. Cat., in press
6) A. Cimino and V. Indovina, J. Cat., 17 (1970), 54
7) Unpublished results from this laboratory
8) F. H. Chapple and F. S. Stone, Proc. Brit. Ceramic Soc., (1964), 45
9) R. M. Dell, F. S. Stone and P. F. Tiley, Trans. Faraday Soc. 49 (1953), 201
10) F. S. Stone, Advan. Catal Rel.Subj., 13 (1962), 1
11) J. Bloem, Philips Res. Rep. 13 (1958), 167
12) S. Greenwald, S. J. Pickart and F. H. Grannis, J. Chem. Phys. 22 (1954), 1597
13) D. A. Dowden and D. Wells, Actes 2me Congr. Int. Catal., 1960 Paris, 2 (1961), 1499
14) J. Haber and F. S. Stone, Trans. Faraday Soc., 60 (1964), 192
15) K. Klier, J. Catal. 8 (1967), 14
16) E. S.R. Winter, J. Chem. Soc. A, (1968), 2889
17) G. K. Boreskov, Disc. Faraday Soc., 41 (1966), 263
18) A. J. Tench and P. Holroyd, Chem. Commun. (1968), 471
19) A. J. Tench and T. Lawson, Chem. Phys. Lett. 7 (1970), 459
20) R. Rudham and M. K. Sanders, papers presented at a Symposium on "Chemisorption and Catalysis", Institute of Petroleum, London, 1970
21) M. Schiavello, M. Lo Jacono and A. Cimino; J. Phys. Chem., 75 (1971), 1051

DISCUSSION

J. J. F. SCHOLTEN

You did a number of experiments with pure zinc oxide. Did you find any influence of the pretreatment conditions (temperature and time of pumping) on the catalytic activity? It might be imagined that the number of interstitial zinc atoms in the surface, which depends on the pretreatment conditions, influences the catalytic behavior (see paper No. 22).

M. SCHIAVELLO

For the decomposition of N_2O we studied three samples of ZnO of different sources, submitted to different firing temperature and to different conditioning. Two ZnO were from $ZnCO_3$ Carlo Erba (reagent grade) fired at 1200°C for 5 hrs. The other was from Johnson-Matthey "specpure" and was fired at 1000°C for several hours. The results for one ZnO ex carbonate were already published[1] and they showed that the catalytic activity and the apparent activation energy (Ea) were not changed by outgassing the specimen at 480°C or 650°C. Two portions of the other ZnO ex carbonate were studied: a) on a first portion it was found an increase in activity (factor 1.4) by increasing the outgassing temperature from 600° to 700°C without variation in Ea; b) on the second portion increasing the outgassing temperature from 480°

to 700°C resulted in an increase in activity (factor 1.3), with constant Ea. A further increase of the outgassing temperature from 700° to 750°C resulted in a decrease of activity of about a factor 3, without a change in Ea. Finally the ZnO J.M. was studied outgassing at 480°, 600°, 700° and 750°: for each increase of the outgassing temperature there was an increase in activity while the Ea varied within 3 Kcal/mole. In conclusion all these data show that there can be an influence of the outgassing temperature on the catalytic activity. The influence, however, is not always a positive one. The phenomenon requires further investigation in order to assess the possible causes.

1) M. Schiavello, A. Cimino and J. M. Criado, Gazz. Chim. Ital., 101, 47 (1971).

G. M. SCHWAB

1) Do your ESR results concerning two signals, out of which the one can be assigned to a Cu^{2+} species outside the lattice, coincide with the very similar conclusions Voitlaender and Lumbeck achieved on CuO in alumina during a redox cycle?

2) Did you by any chance make experiments with illumination of your catalysts? This would be interesting because UV light is known to loosen the Zn-O bond.

M. SCHIAVELLO

1) The ESR study of Lumbeck and Voitlander[1] is concerned with CuO on δ-Al_2O_3. Their specimens were fired at 600°C. In spite of the different matrices and of the different firing temperatures used, the results obtained by us on our CuO-MgO specimens agree in a general way with those of Lumbeck and Voitlander. Their spectra are more complex than ours, but in conclusion two different types of Cu^{2+} were found both in δ-Al_2O_3 and MgO, one in the bulk, the other outside the lattice.

2) We did not make experiments with illumination.

1) H. Lumbeck and J. Voitlander, J. Cat. 13, 117 (1969).

G. M. SCHWAB

You found an increase of specific activity of Cu^{2+} in a matrix of ZnO by increasing the outgassing temperature from 480-650°C. This must refer to the activity per ion, for the total activity goes down because of the loss of surface. Now, we found many years ago that pure CuO loses activity by heating above 600°C, but that this loss can be avoided by addition of comparable amounts of ZnO, thus acting as a stabilizer or a structural promoter. My question is whether this effect is still observable in the case of your very dilute solutions of CuO in ZnO.

M. SCHIAVELLO

We have found in our CuO-ZnO specimens an increase of activity without change in the apparent activation energy (Ea) by increasing the outgassing temperature from 480° to 650°C. There is no loss in surface area in the CuO-ZnO specimens in this range of temperature. It may be useful to recall that they were prepred at 1000°C for 140 hrs and that their surface area is low (0.3 m^2g^{-1}). The increase of activity is attributed to segregation of Cu^{1+} on the surface for the reason given in the paper.

The situation is different from the pure oxides for which there is a decrease of surface area from 6.6 to 0.9 m^2g^{-1} in the temperature range 480-650°C. The variation of surface area is taken into account in the value of k_{abs} (see text - pp. 1-4). Your finding that pure CuO loses activity above 600°C can be attributed to the shrinkage of surface area. This shrinkage can be limited by the presence of ZnO.

C. NACCACHE

Why is the Cu^{2+} signal at 2.19 attributed to isolated ions in octahedral symmetry? Generally isolated copper ions in octahedral symmetry, according to the Jahn Teller effect, give axial ESR spectra with two g values.

M. SCHIAVELLO

The ESR spectra of Cu^{2+} ions in octahedral sites in MgO have been discussed by several authors.[1,2] The symmetrical line at g = 2.190 arises from a dynamic Jahn-Teller effect. Only when the spectrum is recorded at very low temperature (around liquid helium temperature) an anisotropic signal is found. Moreover when Cu^{2+} ions are located on and close to the surface the J-T distrotion is "frozen out" in a fixed direction also at room temperature, and a signal with two g values is found. Details on the CuO-MgO system studied by ESR will soon appear.

1) J. W. Orton, P. Auzin, J. H. E. Griffiths and J. E. Wertz, Proc. Phys. Soc. 78, 554 (1951).
2) R. E. Coffman, J. Chem. Phys. 48, 609 (1968).

Paper Number 2

THE DECOMPOSITION OF ISOPROPANOL OVER CORUNDUM-PHASE SOLID SOLUTIONS OF Cr_2O_3-Al_2O_3

F. PEPE* and F. S. STONE**
School of Chemistry, University of Bristol, England

ABSTRACT: Isopropanol decomposition has been studied at 300-425°C over α-Al_2O_3, α-$Cr_xAl_{2-x}O_3$ and α-Cr_2O_3. The catalysts have been prepared at 1350°C to ensure complete isomorphism. Both pulse and continuous flow techniques have been used.

Activities, selectivities and activation energies for formation of propylene (E_{PRO}) and acetone (E_{ACE}) are reported. α-Al_2O_3, predominantly dehydrogenating, becomes a dehydrating catalyst when only 0.1% Cr is dispersed in solid solution. Beyond 1% Cr the dehydration activity declines. This pattern is matched by a minimum in E_{PRO} at 0.1% Cr in the range $0 < x < 0.2$. At $x = 0.2$ the catalysis is mainly dehydrogenation with $E_{ACE} < E_{PRO}$, but at $x = 2.0$ (α-Cr_2O_3) the favoured reaction on well-outgassed oxide reverts to dehydration.

The results establish that the activity of α-Cr_2O_3-Al_2O_3 is not a monotonic function of solute concentration. Promoter action and the generation of acid centres in α-Al_2O_3 are discussed.

1. INTRODUCTION

A number of recent studies have drawn attention to the significance of employing oxide solid solutions in fundamental catalytic work.[1-4] Using a diamagnetic oxide as matrix, the properties of transition metal ions in chemisorption and catalysis can be studied with control over the electronic interactions between them and, for dilute solutions, there is the additional possibility of generating active centres around isolated solute ions.

Studies of this kind have hitherto dealt with catalytic activities for single reactions, e.g. NH_3 decomposition,[1] N_2O decomposition,[5,6] C_2H_4 hydrogenation,[7] p-H_2 conversion and H_2-D_2 exchange.[3,4] It was considered that valuable additional information would be obtained by studying the influence of transition metal ion concentration on the selectivity of a solid solution catalyst towards two reactions proceeding

* Present address: Istituto di Chimica Generale, Università di Roma, Rome, Italy.

** Present address: School of Chemistry and Chemical Engineering, University of Bath, Bath BA2 7AY, England.

simultaneously. Alcohol decomposition is an attractive reaction for such experiments, and alumina in the corundum phase (α-Al_2O_3) is a suitable matrix. Corundum forms solid solutions with a number of transition metal sesquioxides and in alcohol decomposition pure α-Al_2O_3 gives a convenient balance of dehydration and dehydrogenation to serve as a reference point for changes in selectivity caused by the incorporation of transition metal ions. In the present paper we report on the influence on the activity and selectivity of alumina in the catalytic decomposition of isopropanol when increasing quantities of Cr^{3+} ions are introduced in the homogeneous solid solution α-$Cr_xAl_{2-x}O_3$.

Dehydrogenation of isopropanol leads to acetone (ACE) and dehydration to propylene (PRO): if p is the partial pressure of each product, the selectivity S can be defined as $S = p_{ACE}/(p_{ACE} + p_{PRO})$. Besides studying the changes in S as a function of catalyst composition, it is important to know whether S changes during the course of establishing the steady state with a given catalyst. We have examined this aspect by carrying out both pulse experiments and continuous flow experiments. The latter are preferable for evaluating kinetic parameters, but the former provide information on the variation of S with adsorption strength. This may be achieved by examining product formation when a number of reactant molecules equivalent to 1 monolayer is pulsed and products are collected for (a) a short time, viz. partial recovery experiments (PRE) and (b) a long time, viz. total recovery experiments (TRE). Recognising that not all sites on the surface are of equal adsorption strength, a comparison between PRE and TRE reveals whether there are selectivity differences for conditions of weaker and stronger adsorption respectively.

2. EXPERIMENTAL

α-$Cr_xAl_{2-x}O_3$ solid solutions (designated AC) were obtained by impregnating high-purity boehmite with CrO_3 and firing at 1350°C; [8] α-Al_2O_3 and α-Cr_2O_3 were also prepared at 1350°C. The following samples were used: α-Al_2O_3 (designated A), α-Cr_2O_3 (designated C), and solid solutions AC 0.1, AC 1, AC 7 and AC 10, where the numeral following AC denotes the nominal number of chromium atoms per 100 aluminium atoms.[8] Thus AC 1 is α-$Cr_{0.02}Al_{1.98}O_3$. The surface areas (BET, Kr, 77°K) of all catalysts were $1.9 \pm 0.1\ m^2g^{-1}$, except α-Cr_2O_3 at $1.3\ m^2g^{-1}$.

The flow system was of conventional design. Isopropanol (B.D.H. Analar, purified by distillation) was dosed into a helium stream either as a pulse (2.2×10^{18} molecules) or continuously (partial pressure 32 torr). The reactor was a 10 mm silica tube containing the powdered catalyst (0.55 g in pulse experiments, 1.1 g in continuous flow experiments) mounted on a sintered glass disc sealed across the tube. Analysis for isopropanol, acetone and propylene* was carried out by gas chromatography using a 1 metre column of 20% Carbowax 20M on Chromosorb at 100°C and a katharometer as detector. A metal valve arrangement beyond the reactor enabled a small-volume section to be used in pulse experiments as a cooled trap in which acetone, propylene and unchanged isopropanol could be condensed for a fixed time and then flash-evaporated into the chromatograph. In steady state flow experiments a similar volume served as a sampling device.

* In the temperature range used in the present work, isopropyl ether was not formed as a dehydration product.

Isopropanol and acetone are sufficiently strongly adsorbed on α-Al_2O_3, α-Cr_2O_3 and AC that overlap of peaks occurs in pulse experiments if GLC analysis is attempted without the use of a trap. In this work the trap beyond the reactor was cooled at -195°C and for partial recovery experiments (PRE) a standard trapping time (measured from time of injection of the pulse) of 2.5 min was adopted: this was the minimum time consistent with sweeping the dead space and obtaining sufficient product for accurate analysis. For total recovery experiments (TRE), a standard trapping time of 60 min was adopted. This was more than adequate at the higher reaction temperatures, but for strictly total recovery at the lowest temperature used, i.e. for the sum of the collected isopropanol, acetone and propylene to equal the injected isopropanol, a time of many hours would have been needed. A 60-minute trapping represented a convenient compromise, since this was long enough to ensure at least 90% recovery in all cases and did not significantly distort the TRE pattern. The helium flow rate was 40 $cm^3 min^{-1}$ in pulse experiments and the catalyst pretreatment was an initial outgas at 10^{-5} torr and 450°C overnight. Between pulses the catalyst was flushed at 450°C in helium for 30 min.

The procedure in continuous flow experiments was to outgas the catalyst and, after cooling to the first reaction temperature, to admit the isopropanol/helium stream. When a steady state had been attained, the conversion and selectivity were noted, and the temperature was then changed. The flow rate F was 80 $cm^3 min^{-1}$, which lay in the region of low conversion where the plot of product formation vs 1/F was linear, and the respective activation energies for production of acetone and propylene were determined conventionally from plots of log p_{ACE} (or log p_{PRO}) vs 1/T. Two series of experiments were conducted, Series I in which the catalysts were outgassed initially at 450°C overnight and Series II in which they were outgassed at 850°C for 1 hour.

The maximum temperature used in the catalytic studies was 425°C. A check showed that there was no decomposition of isopropanol in the absence of catalyst below 430°C.

3. RESULTS

3.1 Pulse experiments

The selectivities in isopropanol decomposition over α-Al_2O_3 and three solid solutions (AC 0.1, AC1 and AC 7) are summarised in table 1. The activities for α-Al_2O_3 and AC 1 are illustrated in fig.1, where results are expressed in terms of α, the percentage of the injected isopropanol which has reacted in 2.5 min (PRE) or in 60 min (TRE) to give acetone or propylene respectively. α-Al_2O_3 at 300-350°C is seen to be principally a dehydrogenating catalyst, but the introduction of very small amounts of chromium leads to the development of a predominant dehydrating capacity. The induced dehydration activity is so high that with AC 0.1 it can be observed at 170-200°C to the complete exclusion of dehydrogenation. As the chromium content is increased, the dehydration activity weakens and the selectivity S increases. S is independent of whether partial recovery or total recovery is studied (table 1). Thus the strongly adsorbed species of isopropanol decompose to give the same fractions of acetone and propylene as the weakly adsorbed species.

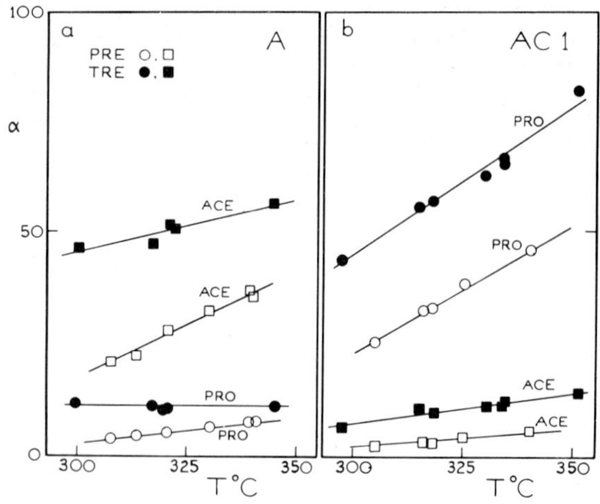

Fig.1 Percentage conversion (α) to acetone (ACE) and propylene (PRO) versus temperature for pulse experiments on (a) α-Al_2O_3 and (b) AC 1. Open symbols refer to partial recovery experiments (PRE) and filled symbols to total recovery experiments (TRE).

TABLE 1

Selectivities in isopropanol decomposition over α-Al_2O_3 and α-Cr_2O_3-Al_2O_3

Catalyst	Expt	Temp.(°C)	S	Remarks
A	PRE	308-340	0.84	Mainly dehydrogenation
	TRE	300-345	0.83	
AC 0.1	PRE	163-205	0.00	Wholly dehydration
	TRE	172-202	0.00	
AC 1	PRE	277-340	0.10	Predominantly dehydration
	TRE	297-352	0.15	
AC 7	PRE	291-346	0.48	Equivalence of dehydrogenation and dehydration
	TRE	302-350	0.49	

The ratios between corresponding conversions in TRE and PRE at a given temperature tend to unity as the temperature is increased (table 2). A rise in temperature not only increases the rate of decomposition, but also favours desorption, thereby lessening the distinction between strongly-held and weakly-held adsorbate. However, over the range between 310° and 340°C where by interpolation we have data on three catalysts in both TRE and PRE, it is significant to note certain differences. Thus in dehydration, the ratio α(TRE)/ α(PRE) is very large on AC 7, less on α-Al_2O_3 and least on AC 1.

TABLE 2
Values of the ratio α(TRE)/α(PRE) as a function of temperature

T(°C)	Dehydration			Dehydrogenation		
	A	AC1	AC7	A	AC1	AC7
310	2.8	1.9	4.0	2.1	3.2	4.0
320	2.2	1.7	3.3	1.9	2.7	3.6
330	1.7	1.5	2.8	1.7	2.4	3.4
340	1.4	1.4	2.5	1.5	2.0	3.0

This confirms that AC 1 affords a relatively weak chemisorption for isopropanol in a mode which leads readily to dehydration and desorption of propylene.* In dehydrogenation, however, AC 1 behaves similarly to the other solid solutions. The most active catalyst, in this case α-Al_2O_3, is again the one which shows the lowest α(TRE)/α(PRE) values.

3.2 Steady-state flow experiments

The selectivity variations as a function of chromium content noted in the pulse experiments were further examined in continuous flow experiments. AC 10 and α-Cr_2O_3, not investigated in pulse experiments, were included in these studies.

The time taken to reach a steady state was 6-12 hours at temperatures of 300-310°C, but only about 2 hours at 350°C. In Series I, where conversions were higher than in Series II, the temperature range most expedient for study was 300°-370°C, whilst in Series II it was 340°-420°C. Experiments were conducted by selecting temperatures in an arbitrary sequence; the results were reproducible, and independent of the sequence taken.

Results for Series I and II are presented together in table 3. In fig.2 we show the results for Series II in the form of Arrhenius plots. Solid lines refer to dehydration and dashed lines to dehydrogenation: experimental points are omitted from the lower group of curves for the sake of clarity. The data for α-Cr_2O_3 in fig.2 have been corrected to

* Although not shown in the table, the (TRE)/ (PRE) values for AC 0.1 in dehydration are 1.0 over the whole range from 170° to 200°C.

TABLE 3
Continuous flow experiments: percentage dehydrogenation and dehydration

	SERIES I (450°C outgas)			SERIES II (850°C outgas)		
Catalyst	Temp (°C)	Acetone (%)	Propylene (%)	Temp (°C)	Acetone (%)	Propylene (%)
A	319	2.4	0.12	346	0.18	0.17
	325	3.2	0.21	351	0.25	0.30
	337	6.4	0.37	370	0.59	0.74
	341	7.6	0.52	388	1.2	1.6
	347	8.5	0.80	392	1.3	1.7
	354	13.2	1.2	396	1.5	1.6
	361	14.4	1.7	410	2.7	3.5
	365	15.8	2.0	412	2.8	2.9
AC 0.1	311	0.40	0.26	350	0.52	13.6
	324	0.83	0.60	372	0.80	24.4
	333	1.6	0.90	390	1.4	32.1
	347	3.4	1.8			
	360	3.7	3.2			
	370	12.0	4.7			
AC 1	304	0.40	0.15	367	1.3	5.0
	324	1.3	0.52	370	2.4	5.1
	333	2.0	0.80	375	1.4	6.8
	344	3.6	1.2	391	2.4	10.8
	364	9.0	3.0	395	3.3	11.6
				398	2.5	14.8
				418	4.3	22.8
AC 10	329	2.2	0.12	345	0.44	0.12
	344	4.5	0.37	366	0.94	0.37
	352	6.7	0.55	381	1.6	0.74
	364	10.3	1.1	394	1.9	0.92
				418	7.0	3.5
C	324	1.7	1.6	351	0.30	3.0
	333	2.8	2.5	377	0.71	5.2
	344	4.2	4.1	390	1.1	6.45
	353	6.3	5.3	418	2.9	10.1

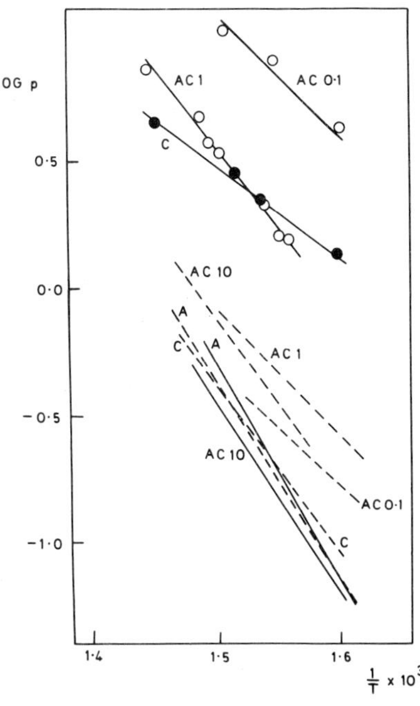

Fig.2
Plots of log p vs. 1/T, where p is the partial pressure (in torr) of acetone (dashed lines) or of propylene (solid lines) in the effluent from the reactor in steady-state flow experiments.

allow for its lower specific surface area of 1.3 m^2g^{-1}, and the activities of all the catalysts refer to the same surface area (2 m^2). The activation energies for dehydrogenation and dehydration in Series I and II are summarised in table 4, and the respective variations of selectivity with chromium content at the median temperature of 350°C are shown in fig.3.

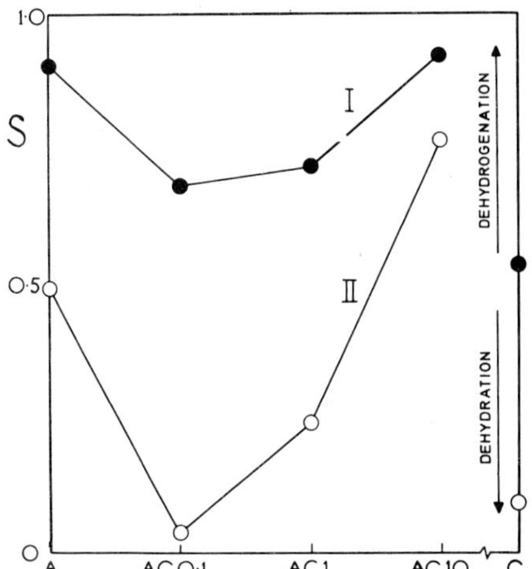

Fig.3 Selectivity S vs. chromium content for Series I (filled circles) and Series II (open circles).

The pattern of an induced activity and selectivity for dehydration observed in the pulse experiments when traces of chromium ions are introduced into α-Al_2O_3 is seen to be fully upheld by the steady-state flow experiments (fig.3). As in the pulse experiments, the selectivity towards dehydration decays when the chromium concentration exceeds a few per cent. The high dehydration activity found for AC 0.1 and AC 1 in pulse experiments is observable in the continuous flow experiments at temperatures of 350-400°C provided the catalyst has been outgassed at a high temperature (fig.2).

TABLE 4

Activation energies for dehydrogenation and dehydration (kcal $mole^{-1}$) derived from conversions in steady-state flow experiments.

Catalyst		α-Al_2O_3	AC 0.1	AC 1	AC 10	α-Cr_2O_3
Series I	Dehydrogenation	36	43	37	34	32
	Dehydration	49	37	35	45	29
Series II	Dehydrogenation	35	22	23	29	29
	Dehydration	39	22	28	35	16

4. DISCUSSION

The selectivity towards dehydrogenation shown by α-Al_2O_3 is in agreement with the results of previous workers who have studied alcohol decomposition,[9,10)] although Rubinshtein and Dulov[11)] in an investigation of isopropanol decomposition over several different preparations of α-Al_2O_3 did not observe $S > 0.5$. The selectivity in our work is reduced to 0.5 if the catalyst is outgassed at 850°C. Table 3 shows that this is not due to an increase in dehydration activity, but to a genuine loss in dehydrogenation activity. The trend of decreasing dehydrogenation with increasing outgassing temperature is a general one for all the catalysts. The effect is most probably due to a reduction in the number of dehydrogenation centres rather than to an increased difficulty for reaction, for table 4 shows clearly that the activation energies for acetone production all decrease on passing from Series I to Series II. A feature implicit in this change in dehydrogenation activity is that the centres active in dehydrogenation are different from those active in dehydration; a similar conclusion has been drawn from studies on zinc oxide by Krylov and Fokina.[12)]

For the catalysts with high selectivity towards dehydrogenation (α-Al_2O_3, AC 10) the activation energy for dehydrogenation is consistently less than that for dehydration. However, this situation is seen to reverse when dehydration activity is high (AC 0.1, α-Cr_2O_3). This implies that the number of active centres is not necessarily a dominant criterion of activity: the energetics of the transformation are also important.

The most striking feature in the results is the appearance of a pronounced catalytic activity for dehydration when very small quantities of chromium ions are introduced into the α-Al_2O_3 matrix. This is established in the pulse experiments and confirmed by the continuous flow experiments. The solid solution with the highest dehydration activity is AC 0.1, but the effect is strong also at AC 1. As the chromium concentration is increased further, the selectivity returns to a value similar to that shown by α-Al_2O_3. The activation energy for dehydration passes through a minimum between α-Al_2O_3 and AC 10.

At AC 0.1 only one in 1000 of the cations in the solid is a chromium ion. Magnetic susceptibility studies on the AC solid solutions show that for AC 0.1 the magnetic moment of the chromium is 3.88 Bohr magnetons, the value expected for Cr^{3+}(d^3, spin-only), and in e.p.r. only the ruby (δ) spectrum is present.[8] If there was aggregation of chromium on the surface (or in the bulk) the magnetic moment would be reduced and the e.p.r. spectrum would show the β spectrum characteristic of pair-wise interaction. We conclude, therefore, that the surface chromium ions in AC 0.1 are present as <u>isolated</u> ions, i.e. they are surrounded by Al^{3+} and O^{2-}, and there are no adjacent Cr^{3+} ions.

We must now ask ourselves why the presence of these centres in such small numbers confers such a high dehydration activity on the solid. α-Al_2O_3 is made up of linked AlO_6 octahedra, and in the surfaces formed by cleavage the oxygen co-ordination around Al^{3+} will frequently be reduced from six to five. This will also apply to well-outgassed α-Al_2O_3. When Cr^{3+} ions occasionally replace Al^{3+} ions, the special situation arises of a 5-coordinate surface Cr^{3+}. Now with Cr^{3+} (d^3) there is a driving force favouring 6-coordination due to crystal field (or ligand field) stabilization. On high temperature treatment the surface may adapt to a state of lowest free energy by oxygen transferring from an adjacent Al ion so as to produce a 6-coordinate Cr^{3+} and leave a 4-coordinate Al^{3+}.[13] The anomalous low-coordinate Al^{3+} adjacent to the isolated Cr^{3+} ions will act as a Lewis acid. Such centres are believed to participate in alcohol dehydration either by a carbonium ion mechanism or a concerted mechanism involving also basic sites.[14-16] An alternative hypothesis is that the mere presence of isolated Cr in the structure leads to electron-deficiency (and hence Lewis acidity) at the adjacent Al atoms because of an increased covalency in the Cr-O bond compared with the Al-O bond. This mechanism would not require an actual change in oxigen co-ordination. Either of these models qualitatively accounts for the generation of dehydration activity in AC 0.1 and AC 1. We should not be particularly concerned that the number of centres is small. The activation energy is low for these centres and there is ample precedent from the cracking catalyst field to remind us that carbonium ion reactions can be vigorously sustained on a small number of acidic centres. The decline in dehydration activity beyond AC 1 suggests that there is an optimum situation for dehydration when the active centres are isolated.

The range from AC 1 to AC 10 is the region of chromium concentration where Cr^{3+} ions become adjacent to other Cr^{3+} ions, and electronic interaction occurs. Magnetic studies reveal this very clearly.[8] Pulse experiments show an increase in dehydrogenation at concentrations above AC 1 and at AC 10 the activation energy for dehydrogenation is lower than that for dehydration (table 4). There is an analogy here with the catalytic activity in H_2-D_2 exchange, which shows a marked increase in rate constant between AC 1 and AC 10.[4] Hydrogen adsorption at 400°C also peaks in this range of concentration.[17] The delocalisation of charge which is revealed in the magnetic studies evidently facilitates the reactions involving adsorbed hydrogen, and we attribute the development of dehydrogenation activity between AC 1 and AC 10 to this effect.

α-Cr_2O_3 has a dehydrogenation activity comparable with α-Al_2O_3, but a much higher dehydration activity. The activation energy for propylene formation on α-Cr_2O_3 was the lowest of all the catalysts studied. Dosing water vapour to the isopropanol/helium stream led to the release of acetone, so we envisage that water is able to displace adsorbed acetone and poison dehydrogenation centres. Dehydrogenation of hydrocarbons is a characteristic reaction with chromia, but this property is suppressed in the present case by the competition and inhibition caused by the water-producing reaction.

The results of this solid solution study show that incorporation of chromium ions into the α-Al_2O_3 matrix leads to catalytic properties which are particular to certain concentrations. The behaviour is not monotonic with increasing chromium content. Oxide solid solutions containing transition metal ions have catalytic properties *per se* which we regard as being defined by induced surface structures (active centres), by the incidence of a controlled amount of electronic coupling between transition ions, and by the effect on the electron distribution in the metal-oxygen bonding in the solvent oxide.

ACKNOWLEDGMENTS

We are indebted to Dr.J.C.Vickerman for preparation and characterisation of the solid solutions and for helpful discussions. We also acknowledge support of this work by the NATO Research Grants Programme which made possible the stay of one of us (F.P.) in the University of Bristol.

REFERENCES

1) E.G.Vrieland and P.W.Selwood, J.Catal., 3 (1964), 539.
2) A.Cimino, M.Schiavello and F.S.Stone, Disc.Faraday Soc., 41 (1966), 350.
3) P.W.Selwood, J.Am.Chem.Soc., 88 (1966) 2676.
4) F.S.Stone and J.C.Vickerman, Z.Naturforsch., 24a (1969) 1415.
5) A.Cimino, R.Bosco, V.Indovina and M.Schiavello, J.Catal., 5 (1966) 271.
6) A.Cimino, V.Indovina, F.Pepe and M.Schiavello, Proc.4th Int.Congr. Catalysis (Moscow 1968), 1 (1970) 168 (paper 12).
7) H.Schaefer and E.Buchler, Z.Naturforsch., 22a (1967) 2117.
8) F.S.Stone and J.C.Vickerman, Trans.Faraday Soc., 67 (1971) 316.
9) G.-M.Schwab and E.Schwab-Agallidis, J.Am.Chem.Soc., 61 (1949) 1806.
10) O.V.Krylov, 'Catalysis by Non-metals', Academic Press (1970) p.118.
11) A.M.Rubinshtein and A.A.Dulov, Zhur.Neorg.Khim., 4 (1959) 1498.
12) O.V.Krylov and E.A.Fokina, Kinetika i Kataliz, 1 (1960) 542.
13) F.S.Stone, Chimia, 23 (1969) 490.
14) V.A.Dzisko, M.S.Borisova, N.S.Kotsarenko and E.V.Kusnetsova, Kinetika i Kataliz, 3 (1962) 728.
15) O.V.Krylov, Zhur.Fiz.Khim., 39 (1965) 2656.
16) H.Pines and J.Manassen, Adv.Catal., 16 (1966) 49.
17) F.S.Stone and J.C.Vickerman, unpublished results.

DISCUSSION

Z. G. SZABÓ

I believe that the idea of employing solid soultions of analogous compounds, as in the case of this paper by Pepe and Stone, can be of quite general validity and can thus contribute to the elucidation of catalytic effects. I also agree fully with the statement that the number of active centers is not necessarily a dominant criterion for activity; the energetics of the transformation are also important. Even today it is not superfluous to emphasize it.

The discussion of selectivity given by the authors is fairly attractive, but my opinion is that if the solid is fired in air a chemical change of Cr(III)-oxide into Cr(VI)-oxide can also be responsible for the increased dehydration. Such an oxidation has been observed by several authors and also by us in chromite spinel formation. A small amount on the surface is already enough to display high dehydration activity. At higher Cr(III)-oxide concentration the Cr(III)-Cr(VI) interaction can make the effect of Cr(VI) disappear. This opinion is supported also by the view that an increased covalency of the oxygen ions facilitates dehydration, as emphasized in our paper presented at the 3rd International Congress to which McCaffery and co-workers refer in Paper 3.

I should also like to refer to the fact that dehydration activity is always observed at lower temperatures than dehydrogenation. The dehydrogenation activity can be measured on samples with poor dehydrating activity. The necessarily increased temperature of the dehydrogenation not only increases the velocity of dehydration, but it also increases the covalency of the oxide. In this way the dehydrogenation centers cannot be supplied with sufficient alcohol molecules to display their effectiveness. We cannot therefore exclude the possibility that samples AC 0.1 and AC 1 might also dehydrogenate as the others do, if the dehydration action were not taking away all the alcohol.

F. S. STONE

The preparation temperature of our catalysts is well above that at which Cr(VI) is stable. I feel confident that at 1350°C in air the chromium in Cr_2O_3-Al_2O_3 is present as Cr(III). There is, of course, a possibility that some $Cr(VI)O_4$ centers might develop on the surface of α-$Cr_xAl_{2-x}O_3$ in the course of a subsequent catalytic reaction at 250-350°C, especially in those surface sites where the oxygen ligancy is down to four, but oxidizing conditions would be necessary. Now we do not have such conditions. Moreover, it should be noted that it is at very low concentrations of chromium that we observe the highest dehydration activity. Covalent $Cr(VI)O_4$ centers are least likely to occur at low concentrations: at 1% Cr and below the matrix is much less covalent than at high Cr contents, and also the chances of finding surface chromium ions (as opposed to aluminum ions) with less than the average coordination of five oxygen ligands are slender. Thus, whilst I do not question the activity of chromates in dehydration, I am not inclined to link our dehydration activity with chromate formation.

With regard to the dehydrogenation, Fig. 2 shows that AC 0.1 and AC 1 do indeed give acetone in similar amounts to the other catalysts at 350-400°C.

B. DELMON

First, concerning Professor Szabó's remarks, I completely agree with Professor Stone that, in the preparation conditions he used, there is certainly no substantial amount of the high valence chromium ions in the α-phases.[1]

My question to Professor Stone is that I just wonder what is actually the increase of the dehydration activity between 0.1% and 7% chromia in alumina.

The conspicuously low dehydrogenating activity in this composition range (which is the range where the solid solutions exhibit the ruby color) is consistent with our own results on the dehydrogenation of isobutane at higher temperatures on the whole series of Al_2O_3-Cr_2O_3 solid solutions.[1] So I agree with this point. But I wonder whether one could not explain the apparent increase of the dehydrating activity by simply assuming that there is less inhibition of the dehydration reaction by the products of the dehydrogenation reaction, notably acetone, because of the much lower rate of dehydrogenation reaction. As this point is very important for Professor Stone's interpretation, I would like to ask him about the arguments he has for excluding such a hypothesis.

1) C. Marcilly and B. Delmon, J. Catalysis, 24, 336 (1972).

F. S. STONE

The variations in dehydration activity are most clearly seen in Table 3.

In reply to the second point, evidence against the inhibition hypothesis is that when water vapour is injected into the reactant isopropanol stream it produces a release of acetone and hydrogen from the catalyst. This implies that acetone and hydrogen are adsorbed relatively weakly, and so we do not regard these products as inhibiting for the dehydration reaction.

H. PINES

You imply that the introduction of small amounts of chromia to alumina is responsible for the formation of Lewis acid sites to give dehydration via a carbonium ion mechanism. There is enough evidence in the literature that the removal of water from 1- and 2-alcohols occurs via a trans-elimination reaction or E_2 mechanism. What evidence have you that the dehydration of isopropyl alcohol to propylene occurs via a carbonium ion mechanism?

F. S. STONE

We have no evidence of our own that the dehydration occurs via a carbonium ion mechanism in preference to other mechanisms. You are right to point out that we have rather stressed this particular mechanism and that, in spite of its popularity, this may no longer be justified.

H. NOLLER

I too would like to draw your attention to the fact that not only a carbonium ion mechanism but also a concerted mechanism has been observed for dehydration. For instance, a concerted mechanism was found over γ-Al_2O_3. However, if your explanation is correct, I would consider it possible that over AC 0.1 you might have a carbonium ion (or E_1) mechanism, since the acidity (or electron pair acceptor strength) of the actual site should be very high.

R. L. BURWELL, JR.

Water resulting from dehydration would presumably adsorb at the coordinately unsaturated sites. Is there any possibility that the effect of 1% Cr^{3+} might, in part, result from changes in the nature of this adsorption? In particular, what would be the effect of exposing the catalyst to extra water?

F. S. STONE

I agree that it is logical to expect that water produced in the reaction will be adsorbed at the coordinately unsaturated sites, and the heat of adsorption is presumably different on Cr^{3+} (or Cr^{3+}-O^{2-}) than on Al^{3+} (or Al^{3+}-O^{2-}). An explanation of the high dehydration activity for isolated Cr along these lines does not readily follow, but it is certainly a possibility.

One effect of exposing the catalyst to extra water has been mentioned in my reply to Professor Delmon. The activity and selectivity of 450°-outgassed oxide in comparison to 850°-outgassed oxide (Table 3 and Fig. 3) might be said to illustrate a second effect of water on the catalyst, in this case as residual OH.

M. M. BHASIN

Have you made any measurements on the amount and strength of the acid-base or ion-pair sites on these solids of 2 m^2/g surface? These aluminas may very well have a substantial number of these sites which play an important role in governing the selectivity of the reaction.

F. S. STONE

As yet we have no information on the amount or strength of the acid-base sites. Their determination is a rather difficult assignment with solids of such low surface area. The sites of greatest interest may only have a density comparable with the chromium content (1 per 10^2-10^3 surface cations). Do you have a method to recommend?

M. M. BHASIN

In reply to Professor Stone's question, measurements are possible using some of the more sensitive methods employing various acids and bases.

Paper Number 3

THE INFLUENCE OF THE MOBILITY OF OXYGEN IN TRANSITION METAL OXIDES ON THEIR SELECTIVITIES IN ISOPROPYL ALCOHOL DECOMPOSITION

E. F. McCAFFREY, D. G. KLISSURSKI* and R. A. ROSS
Department of Chemistry, Lakehead University, Thunder Bay, Ontario, Canada

*On leave from Academy of Sciences of Bulgaria, Sofia, Bulgaria

Abstract: The decomposition of isopropyl alcohol has been studied on niobium and molybdenum oxides and on oxides of the first transition metal series and the exchange kinetics of surface oxygen with gaseous oxygen-18 determined for seven of these oxides. Apparent activation energies, rates of isopropyl alcohol dehydration and dehydrogenation and rates of isotopic exchange of oxygen have been established. The complex character of isopropyl alcohol decomposition has been shown by the detection of mesityl oxide, 4-methyl-2-pentanone and 4-methyl-1,3-pentadiene as secondary products in dehydrogenation catalysis.

Relationships have been proposed between the selectivities in isopropyl alcohol decomposition and (i) the mobility of oxygen as measured by isotopic exchange, (ii) the ionicity of the metal-oxygen bond in the oxides and (iii) selectivities in the oxidation of methyl alcohol. These correlations may be of general value in predicting the catalytic behaviour of the oxides in other redox reactions.

1. INTRODUCTION

Recent studies of the catalytic activity and selectivity of manganese oxides in the decomposition of isopropyl alcohol have shown that dehydrogenation and dehydration reactions occur predominantly while at temperatures above 250°C, condensation products are present to a small extent[1,2]. It has been demonstrated also that the dehydrogenation activity of a series of non-stoichiometric manganese oxides can be correlated with the binding energy of the last gram atom of oxygen in the oxide[2].

Earlier, a relationship was established[3] between the selectivities of transition metal oxides in the catalytic oxidation of methyl alcohol and their activation energies in the isotopic exchange of gaseous oxygen. Further, a comparison of catalyst selectivities in (i) isopropyl alcohol decomposition[4] and (ii) methyl alcohol oxidation[3,5] reveals that those oxides which catalyse the dehydrogenation of isopropyl alcohol also catalyse the complete oxidation of methyl alcohol, while catalysts with selectivities which favour dehydration of isopropyl alcohol are active in the selective oxidation of methyl alcohol to formaldehyde.

In an attempt to establish whether these various concepts may be linked and even shown to have general implications, studies of the decomposition of isopropyl alcohol have been carried out on oxides of the first transition metal series and on niobium and molybdenum oxides.

The activities of seven of these catalysts have been measured also in the oxygen-18 exchange reaction.

2. EXPERIMENTAL

2.1 *Apparatus*

The decomposition of isopropyl alcohol was measured with the all-glass flow system described earlier[1]. The organic reaction products were analyzed at 80°C by a Beckman GC5 gas chromatograph with a flame ionization detector using stainless steel columns, 72 x 0.125 in. o.d., packed with "Carbowax 1540" on "Teflon 6" with nitrogen as carrier gas.

The isotopic exchange of O^{18} was measured in a 'static' circulation system[6]. Samples were first evacuated to 10^{-4}mm. for 4 h. at 50°C above the highest temperature to be used in the rate measurements. The O^{18} content was 10 to 20% of the total oxygen, 10 mm., admitted to the reactor. Samples were withdrawn into evacuated ampoules at 15 min. interval and analyzed with a Hitachi-Perkin Elmer RMU-7 mass spectrometer.

Surface areas were determined by low temperature krypton adsorption (BET; 77°K; A_m = 21.5 Å[2]).

2.2 *Materials*

a) Metal oxides. Details of the preparation of the manganese oxides have been given[1,2].

Scandium (III) oxide (99.90%), vanadium (V) oxide (99.90%), zinc (II) oxide (99.99%) and niobium (V) oxide (99.99%) were obtained from Alfa Inorganics Inc. Titanium (IV) oxide (as 100% anatase) was supplied by British Drug Houses. All of these oxides were heated to 600°C in air for 10 h. prior to their use in the experiments.

Chromium (III) oxide (99.90%) was prepared by calcining the hydroxide at 600°C in air for 10 h. The hydroxide was precipitated from a chromium nitrate solution at pH 10, washed free of nitrate ions with distilled water and dried at 100°C before calcination[5].

Iron (II)(III) oxide was prepared by reducing iron (III) oxide (99.99%-- Alfa Inorganics Inc.) in flowing hydrogen (60ml.min^{-1}) at 600°C for 12 h.[7].

Cobalt (II)(III) oxide (99.90%) was prepared by heating cobalt (II) nitrate at 600°C in air for 10 h.[5] and cobalt (II) oxide (99.90%) obtained by calcination of the carbonate in a stream of nitrogen for 10 h. at 500°C[8].

'Green' nickel oxide (99.90%) was obtained by calcination of nickel (II) nitrate at 600°C for 10 h. in air and copper (I) oxide (99.90%) prepared[9] in a glove box, dried under nitrogen and heated at 600°C in flowing nitrogen for 10 h. X-ray powder diffraction analysis confirmed the copper (I) oxide structure.

Molybdenum (VI) oxide was precipitated by nitric acid from a solution of ammonium molybdate at pH = 1, dried at 100°C and calcined at 630°C for 10 h.[5].

Isopropyl alcohol was "Spectroquality" and its ultraviolet absorption characteristics, Unicam SP800A, were the same as those on its accompanying certificate. Nitrogen (Canadian Liquid Air Ltd.) was 99.99% pure and oxygen-18, 90% in O_2-16, supplied by Isomet Corporation.

3. RESULTS

3.1 *Rates and apparent activation energies of decomposition of isopropyl alcohol*

The particle size of the 1 g. of oxide catalysts used in the isopropyl alcohol experiments was kept below 0.05 mm. and the absence of internal and external diffusion demonstrated by the usual tests[1,10].

All samples were brought to the highest temperature of the experiments in an atmosphere of isopropyl alcohol (300 ml.min^{-1} (NTP); partial pressure, 9.37 mm.) and kept under these conditions for 12 h. to establish the steady activity values used in rate calculations. Generally dehydrogenation activity decreased in this time while dehydration activity increased.

At the end of each catalytic experiment the reaction temperature was raised to the highest value used in the course of the runs and it was noted that no change had occurred in the reaction rate values. Before removal from the reactor, catalysts were cooled to room temperature in nitrogen at 60 ml.min^{-1} (NTP) to prevent spontaneous combustion. They were then examined by X-ray powder diffraction and their surface areas determined for the specific reaction rate calculations.

With the exception of vanadium (V) oxide, iron (II)(III) oxide, nickel (II) oxide, cobalt (II)(III) oxide, copper (I) oxide and molybdenum (VI) oxide, all other oxides were stable under the conditions of the experiment, Table 1. Vanadium (V) oxide was converted to vanadium (III) oxide at 150°C but at 50°C it retained its pale-orange color and analysis by X-ray diffraction confirmed that no change had occurred in the vanadium (V) structure. However, the catalytic activity was too low to allow for rate calculations at temperatures below 50°C. In runs at 74 to 104°C, Table 1, partial reduction of the vanadium (V) oxide occurred as indicated by the color change from pale-orange to brown-green but X-ray powder diffraction showed only the vanadium (V) oxide structure. A similar reduction has been observed in the oxidation of methyl alcohol at 300-350°C[5]. After reaction, the color of cobalt (II)(III) oxide changed from black to grey-green, a color indicative of cobalt (II) oxide and X-ray powder diffraction revealed the presence of a mixture of cobalt (II)(III) oxide and cobalt (II) oxide. When nickel (II) oxide was removed from the reactor specks of a black powder were present. The presence of nickel metal was suspected. At 100°C copper (I) oxide was found to be catalytically inactive. The temperature was then raised to 140°C when the oxide was reduced to metallic copper. Molybdenum (VI) oxide was reduced to a blue powder under these conditions, Table 1.

Both dehydration and dehydrogenation reactions occurred on all samples, except on manganese (III) oxide which showed dehydrogenation activity only. In Table 1, the reaction rates for both dehydrogenation (r_1) and dehydration (r_2) are compared at 200°C along with catalyst selectivities for dehydrogenation and the apparent activation energies.

The Arrhenius plots for the dehydrogenation reactions on manganese (II) oxide, manganese (II)(III) oxide and cobalt (II)(III) oxide had two distinct temperature regions, giving two different values for the apparent activation energies as shown in Table 1.

Table 1

Kinetic parameters for the decomposition of isopropyl alcohol on metal oxide catalysts. (Flow rate, 300 ml.min^{-1} (NTP); partial pressure of isopropyl alcohol, 9.37 mm; r_1, rate of dehydrogenation; r_2, rate of dehydration.)

Catalyst	Temperature range (°C)	$r_1 \times 10^8$ (mole.m^{-2}.sec^{-1}) (200°C)	$r_2 \times 10^8$ (mole.m^{-2}.sec^{-1}) (200°C)	$E_1 \pm 1$ (kcal. mole^{-1})	$E_2 \pm 1$ (kcal. mole^{-1})	Surface area (m^2.g^{-1})	Selectivity $\frac{r_1}{r_1 + r_2}$
Sc_2O_3	200-250	0.0174	0.0045	12	57	9.7	0.79
TiO_2 (anatase)	200-300	0.0133	0.1820	14	42	4.5	0.07
$V_2O_{(5-x)}$	74-104	-	-	22	42	3.6	0.11
Cr_2O_3	150-200	0.9300	0.8300	38	57	10.6	0.53
MnO	150-210	0.0900	0.0030	9	50	5.9	0.96
	210-300			29			
Mn_3O_4	150-180			8	56	16.0	
	180-210	0.1800	0.0050	26			0.97
Mn_2O_3	35- 85	-	-	15	-	14.0	1.00
CoO	170-250	0.7700	0.0120	22	30	8.4	0.98
'Co_3O_4'	145-170			11	31	2.2	
	170-200	2.1200	0.0170	25			0.99
'NiO'	160-250	1.7800	0.0230	17	16	7.2	0.99
ZnO	200-250	0.2370	0.0070	35	38	2.3	0.97
$MoO_{(3-x)}$	130-230	7.6200	257	19	50	0.4	0.03
Nb_2O_5	180-250	-	1.6800	-	46	2.5	-

3.2 *Reaction products*

The primary products from the decomposition of isopropyl alcohol were acetone and propylene. Secondary products detected may be subdivided into: (a) those formed solely by dehydration of the alcohol, i.e. isopropyl ether and (b) those formed by simultaneous dehydration and dehydrogenation, i.e. 4-methyl-1,3-pentadiene, 4 methyl-2-pentanone and mesityl oxide.

Traces of hexadienes, which may result from isomerisation of 4-methyl -1,3-pentadiene, were observed in the products of catalysis from chromium (III) oxide, nickel (II) oxide and zinc (II) oxide. Isopropyl ether up to a maximum amount of 4% v/v of the reaction products was detected from catalysts with a selectivity, S, less than 0.5, Table 1, except in the case of niobium (V) oxide which did not yield any detectable amounts of isopropyl ether. Less than 0.1% v/v of isopropyl ether was also detected in the reaction products from manganese (II) oxide, manganese (II)(III) oxide and nickel (II) oxide. Approximately 0.5% v/v each of 4-methyl-1,3-pentadiene and mesityl oxide were detected in the reaction products on all predominantly dehydrogenating catalysts with the exception of zinc (II) oxide. The formation of 4-methyl-1,3-pentadiene from isopropyl alcohol may be assumed to parallel the formation of butadiene in catalytic reactions involving ethyl alcohol and was detected with oxides which are known to catalyse that reaction[11].

4-methyl-2-pentanone was detected only in the products of catalysis from nickel (II) oxide. This condensation reaction is known to occur on supported nickel metal catalysts[12]. Nine heavier fractions than mesityl oxide, totalling 0.5% v/v, were detected in the products from the reaction on nickel (II) oxide at 200°C.

3.3 *Oxygen-exchange studies*

In the oxygen-exchange experiments 5 g. oxide, particle size 0.5-1.0 mm., was used for each run. Fresh samples of nickel (II) oxide, manganese (III) oxides, chromium (III) oxide and cobalt (II)(III) oxide were examined each time while samples of niobium (V) oxide, molybdenum (VI) oxide and vanadium (V) oxide were regenerated by outgassing to 10^{-4} mm. at 20°C above the working temperature of the subsequent experiments. Since the oxide samples were outgassed at a temperature higher than that at which the exchange measurements were made the reaction may be represented[13]:

$$^{18}O_2 + {}^{16}O_s \rightleftharpoons {}^{16}O{}^{18}O + {}^{18}O_s. \qquad (1)$$

The apparent activation energies for the selected oxides were calculated by the method described[6]. The values obtained were 21, 16, 14, 21, 61, 61 and 57 kcal.mole^{-1} for Cr_2O_3 (300-400°C), Mn_2O_3 (210-300°C), Co_3O_4 (150-250°C), NiO (250-350°C), V_2O_5 (400-500°C), Nb_2O_5 (600-700°C) and MoO_3 (600-700°C) respectively.

4. DISCUSSION

The activiation energy values determined for the isotopic exchange of oxygen on the seven selected oxides were in fair agreement with previous data[13,14,15] and hence it was felt unnecessary to extend the measurements to cover all oxides used in the decomposition experiments. Fig. 1 shows (i) the values of the activation energies for the isotopic exchange determined in these studies and (ii) the selectivities of the oxides in the catalytic dehydration of isopropyl alcohol to propylene. High selectivity towards dehydration is apparent for those oxides which exchange surface oxygen with a high activation energy.

Fig. 2 shows (i) the logarithm of the rates of the isotopic exchange and (ii) the selectivities for acetone formation in the decomposition for seven of the oxides. It may be deduced that those oxides which exchange their surface oxygen at the fastest rates are the most selective towards dehydrogenation. Figs. 1 and 2 are, in a sense, complementary.

Plots of the binding energies of the last g. atom of oxygen in the oxides[16] and the logarithm of the rates of dehydrogenation for all of the oxides showed some scatter but a definite trend was indicated in that dehydrogenation activity was most favoured by oxides with the lowest binding energies. This relationship was similar to that established earlier for a series of non-stoichiometric oxides of manganese[2]. All of these data indicate that selectivities and rates of dehydrogenation of isopropyl alcohol are influenced by the surface mobility of oxygen in the oxides as this has been interpreted in the various ways.

For this oxide series, Fig. 3 shows that selectivity changes in the formation of propylene in the catalytic decomposition of isopropyl alcohol parallel the selectivity changes for the catalytic oxidation of methyl alcohol to formaldehyde[3,5]. Since the activities of oxides in many oxidation reactions have been correlated[17] with oxygen mobility, it is clear that the present results for the catalytic decomposition of isopropyl alcohol could also be used to this end.

The polarising power of an ion, ρ, has been used previously[18] in calculations concerning the ionicity of the metal/oxygen bond in metal oxides when the modulus ($\rho_{cation} - \rho_{anion}$) was regarded as giving a measure of the 'degree of covalency' of the bond. Various attempts[19] have been made to quantify ρ, the simplest with respect to the cation[20,21] is to use the charge/radius ratio. This latter empirical formula can be refined[19], but for the present purpose it does give a useful indication of the importance of polarisation trends in the series of metal oxides in influencing their catalytic selectivity. In Fig. 4, the selectivities of the metal oxides in isopropyl alcohol dehydration are plotted against the charge/radius ratio[19]. The graph obtained supports the hypothesis advanced earlier[21] that the greater the degree of covalency of the oxide, i.e. the higher the value of the ratio, then the more likely dehydration is to be favoured.

On the other hand, an oxide with a low value of the ratio is more ionic in nature and will favour dehydrogenation. These observations may be extended to explain the initial step in the mechanisms of both reactions by using the earlier concept of (i) hydrogen and (ii) hydroxyl coupling[21] to oxygen for (i) dehydrogenation and (ii) dehydration.

Finally, it may be noted that the published values of the apparent activation energies for the dehydrogenation of cyclohexane[7,22] on these oxides are very similar to those for dehydrogenation in Table 1. This similarity may provide further information on the mechanism of dehydrogenation of isopropyl alcohol in that two C-H bonds may be ruptured in the molecule with the subsequent isomerisation of the enolate to acetone

REFERENCES

1) D. G. Klissurski, E. F. McCaffrey and R. A. Ross, Can. J. Chem. 49 (1971) 3778.
2) E. F. McCaffrey, D. G. Klissurski and R. A. Ross, J. Catal. in press.
3) D. G. Klissurski, Intern. Congr. Catalysis, 4th, Paper 36, Moscow, 1968.
4) O. V. Krylov, Catalysis by non-metals, Academic Press, London and New York, 1970.
5) G. K. Boreskov, B. I. Popov, V. N. Bibin and E. S. Kozishnikova, Kinetics and Catalysis, 9 (1968) 657 Eng.
6) V. V. Popovskii and G. K. Boreskov, Kinetics and Catalysis,1 (1960) 530 Eng.
7) P. C. Richardson and D. R. Rossington, J. Catal. 14 (1969) 175.
8) G. M. Dixon, D. Nicholls and H. Steiner, Proc. Intern. Congr. Catalysis, 3rd, Amsterdam 1964, 1 (1965) 815.
9) W. Biltz, Laboratory methods of inorganic chemistry, 2nd Ed., Wiley, New York, 1928.
10) J. M. Thomas and W. J. Thomas, Introduction to the principles of heterogeneous catalysis, Academic Press, London and New York, 1967.
11) S. K. Bhattachryya and N. D. Gangnly, J. Appl. Chem. 12 (1962) 97.
12) V. N. Ipatieff, G. S. Monroe, L. E. Fisher and E. E. Meisinger, Ind. Eng. Chem. 41 (1949) 1802.
13) E. R. S. Winter, J. Chem. Soc. (1968) 2889.
14) G. K. Boreskov, Adv. in Cat. 15 (1964) 285.
15) J. Novakova, Cat. Revs. 4 (1970) 77.
16) K. Kleir, J. Catal. 8 (1967) 14.
17) G. K. Boreskov, V. A. Sazonov and V. V. Popovskii, Dokl. Akad, Nauk. SSSR 176 (1967) 768 Eng.
18) Yu. M. Golutivin, Heats of formation and type of the chemical bond in inorganic crystals, Moscow, 1962, Russ.
19) N. N. Greenwood, Ionic crystals, lattice defects and nonstoichiometry, Butterworths, London, 1968.
20) F. A. Cotton and G. Wilkinson, Advanced inorganic chemistry, 2nd Ed., Interscience Publishers, New York, London and Sydney, 1966.
21) I. Batta, S. Borcsok, F. Solymosi and Z. G. Szabō, Proc. Intern. Congr. Catalysis, 3rd, Amsterdam 1964, 2 (1965) 1340.
22) A. A. Tolstopyatova, V. A. Naumov and A. A. Balandin, Russ. J. Phys. Chem. 38 (1964) 879 Eng.

LIST OF FIGURES

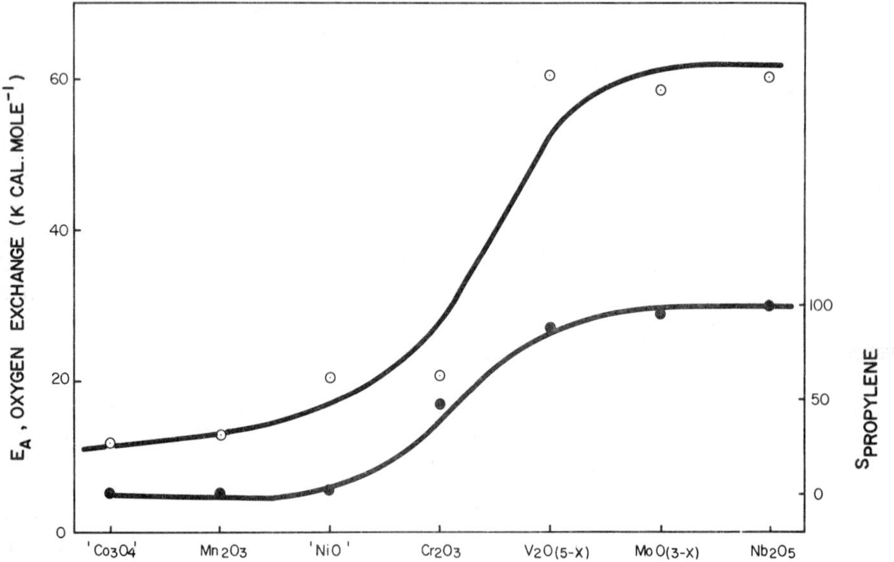

Fig. 1. Relationship between the activation energy of isotopic exchange of oxygen on metal oxides and their selectivities in the catalytic dehydration of isopropyl alcohol.

● Selectivity for propylene formation, Table 1.

⊙ E_A, Oxygen exchange (present values).

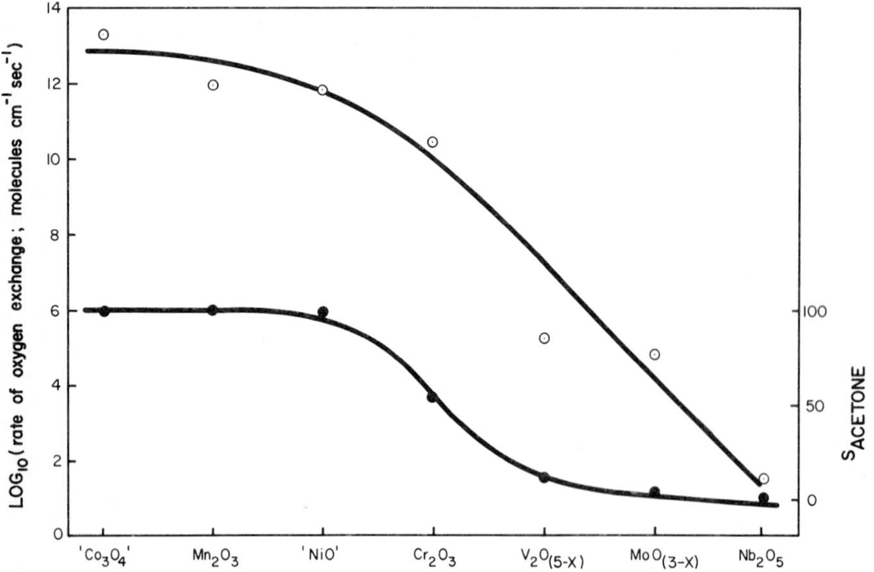

Fig. 2. Relationship between Log_{10} (rate of oxygen exchange; molecules $cm^{-2}sec^{-1}$) on metal oxides at 300°C and their selectivities for the formation of acetone in the catalytic decomposition of isopropyl alcohol.

⊙ Rate of oxygen exchange (present values).

● Selectivity for acetone formation, Table 1.

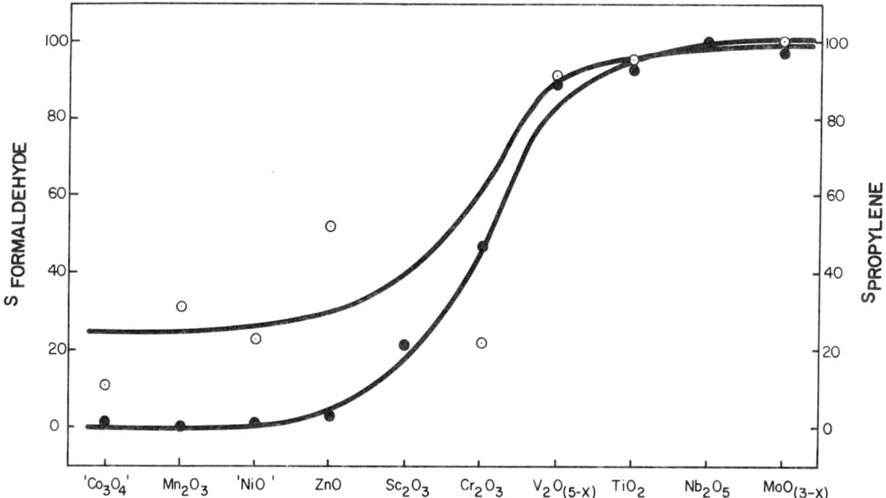

Fig. 3. Relationship between the selectivities of metal oxides for the formation of propylene in the catalytic decomposition of isopropyl alcohol and their selectivities for the formation of formaldehyde in the catalytic oxidation of methyl alcohol.

● Selectivity for propylene formation, Table 1.

⊙ Selectivity for formaldehyde formation (3,5).

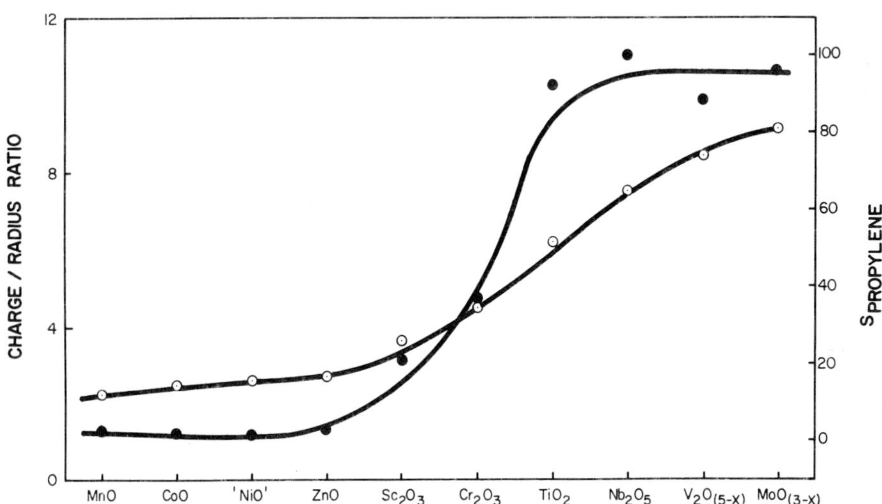

Fig. 4. Relationship between charge/radius ratio of metal oxides and their selectivities for the catalytic dehydration of isopropyl alcohol to propylene.

⊙ Charge/radius ratio.

● Selectivity for propylene formation, Table 1.

DISCUSSION

Z. G. SZABÓ

I was very glad to hear your discussion on the correlation between dehydrating activity and some other properties of oxide catalysts, a view suggested by us some 8 years ago concerning the bond characters. But since that time, I have had the feeling that it should be differentiated depending on whether the increase of covalency is connected with an electron shell deformation or the cations in such oxides are stronger Lewis acids or are both formulations of the same language.

E. F. McCAFFREY

If the charge/radius ratio increases then it is possible that both the Lewis acid strength and the covalency of the metal oxygen bond increase at the same time. However, it seems unlikely that dehydration catalysis would be completely confined to either Lewis or Brønsted acid sites. We note that increased covalency should cause an increase in the strength of any Brønsted acids formed on the surface as intermediates.

H. PINES

I wonder how a study of the kind you have presented can be made without evaluating in detail the mechanism by which the transformation of isopropyl alcohol to the various products has occurred. Even in a presumably simple mechanism such as conversion of isopropyl alcohol to propylenes various mechanisms can be involved.

Was there any study in depth made to insure that the reactions occurred by these mechanisms on the various catalysts mentioned in your paper?

E. F. McCAFFREY

In our work we were studying a decomposition which involved both dehydration and dehydrogenation. We did not emphasize the evaluation of step-by-step reaction mechanisms. However in our work on manganese (II) oxide reported elsewhere[1)] the experimental details were established and possible mechanisms examined. However, we have not been able, as yet, to establish theoretical reaction sequences which when treated by steady-state analysis lead to an expression analogous to that found by experiment.

The general themes expressed in our paper would indeed carry additional weight had it been within our current main purpose to produce detailed mechanistic information. Such work is being pursued in our Department using isotopic methods.

1) D. G. Klissurski, E. F. McCaffrey and R. A. Ross, Can. J. Chem. 49 (1971) 3778.

H. NOLLER

As to the factors determining the dehydration selectivity, I would give more emphasis to the electron pair acceptor strength than to the polarizing power of the cations and the bond character so the attention would be drawn to the polarizing effect of the catalyst upon the reactant. Considering the dehydration on an elimination reaction, the activity for dehydration should increase with the electron pair acceptor strength of cations and the electron pair donator strength of the anion. For the selectivity, the same factors should be important, at least to some extent.

E. F. McCAFFREY

It seems that electron-pair acceptor strength should increase with the charge/radius ratio and therefore with polarizability. Hence, it seems valid to suggest that both electron-pair acceptor strength and polarizability contribute to dehydration activity.

M. M. BHASIN

Have you made any measurements on the amount and strength of the acid-base or ion-pair sites on the various catalysts studied?

E. M. McCAFFREY

No. In the past, catalyst selectivity in the decomposition of alcohols has been related mainly to surface acidity. The results presented here do not reject this influence, but underscore the importance of both the oxidation/reduction properties of the surface and surface acidity.

G. MUNUERA

On Table 1 of your paper activation energies for dehydration and dehydrogenation reactions on the same catalyst are sometimes very different. This actually means that Arrhenius plots may cross, thus affecting the selectivity pattern when selectivities are measured at different temperatures. My question is whether your selectivity pattern remains unchanged at different temperatures within your experimental range (35-300°C)? I wonder whether selectivity is a good choice to compare series of catalysts. I would like to mention in this respect that we have recently studied[1] formic acid decomposition on 3d metal oxides and found that selectivity patterns change with the temperature selected.

1) J. M. Trillo, G. Munuera and J. M. Criado, Cat. Reviews (in press).

E. F. McCAFFREY

This is an excellent point and we are in agreement that such patterns are related to temperature. In our work we found that the selectivities determined at 150 and 300°C were related to temperature but only in a minor way with respect to oxide neighbors in the selectivity series. The general trend was still the same at both temperatures.

H. van BEKKUM

You noticed the formation of several C_6 species.

(i) Are such "dimeric" products formed by consecutive or parallel reactions, i.e., are they also formed from acetone?

(ii) Does an Aldol-type mechanism have to be considered in which an enolate anion is formed on the surface and is attacked by acetone from the gas-phase (Rideal)? I would like to remark then that to obtain 4-methyl-2-pentanone besides a condensation and a dehydration step, also a hydrogenation step is required.

E. F. McCAFFREY

(i) Mesityl oxide was detected when acetone was passed, over freshly prepared manganese (II) oxide.[1) This suggests that a consecutive mechanism occurs. However, this rate of mesityl oxide formation was less than that observed when it was formed in the decomposition of isopropyl alcohol at the same temperature.

(ii) We could only postulate that the formation of mesityl oxide and 4-methyl-2-pentanone occurs through an aldol-type mechanism in which an adsorbed enolate anion is attacked by acetone or isopropyl alcohol. However, it seems that the attack could come from either adsorbed or gaseous species. With regard to the final remark, both the direct reaction of acetone with isopropyl alcohol[2) and the hydrogenation of mesityl alcohol[3) have been suggested as possible mechanisms for the formation of 4-methyl-2-pentanone.

1) D. G. Klissurski, E. F. McCaffrey and R. A. Ross, Can. J. Chem. 49 (1971) 3778.
2) B. H. Davis and P. B. Venuto, J. Catal. (1969) 100.
3) V. N. Ipatieff and V. Haensel, J. Org. Chem. 7 (1942) 189.

Paper Number 4

CATALYTIC CONVERSION OF ACETIC ACID ON METAL OXIDES

T. IMANAKA, T. TANEMOTO, and S. TERANISHI
Department of Chemical Engineering, Faculty of Engineering Science, Osaka University, Osaka, Japan

ABSTRACT: Adsorbed states of acetic acid on metal oxide catalysts have been investigated by means of infrared absorption method, and the reaction mechanism of catalytic conversion of acetic acid into acetone has been studied using deuterium as an isotopic tracer.

The surface species adsorbed on such metal oxides as MgO, CaO, MnO, and ZnO are found to be acetic acid and acetate ion. Besides, ketene appears during the course of reaction. Acetic acid is adsorbed as the protonated one in the cases of ZnO and MnO.

When CD_3COOD is added to the reaction system involving ketene, only CH_2DCOCD_3 is formed on MgO and CaO, while CD_3COCD_3 is also formed on ZnO, where the ratio of CH_2DCOCD_3 : CD_3COCD_3 is 3 : 2.

On the basis of these facts, it is concluded that the conversion of acetic acid on such basic oxides as MgO and CaO proceeds completely through the reaction intermediate involving ketene produced by dehydration reaction. However, on such amphoteric catalysts as ZnO and MnO, the conversion proceeds through the two intermediates which are ascribable to the protonated acetic acid and the ketene.

1. INTRODUCTION

It is well known that metal carbonates and metal oxides are effective catalysts for the conversion of acetic acid into acetone. (ketonization). In general, ketone is formed by dehydration of corresponding carboxylic acid on metal oxides or thermal decomposition of its metal carboxylate. Accordingly, metal acetates on the surface of catalysts can be proposed as active intermediates. However, the ketonization owing to the dehydration process of acetic acid is different from the thermal decomposition of metal acetates. Hitherto, only a few studies have been reported with respect to the mechanism of this reaction by means of several spectroscopic methods. The present paper reports the relation between the adsorbed states of acetic acid and the reaction mechanism, which proceeds on the divalent metal oxides as the catalysts.

2. EXPERIMENTAL

2.1 Materials

Acetic acid used in this work was purified by vacuum distillation after dehydrating with anhydrous magnesium sulphate. Deutero-acetic acid-d_4, CD_3COOD, and diketene were prepared by similar purification. Metal oxides (MgO, CaO, ZnO) were used without further purification, but manganese oxide, MnO, was prepared by vacuum heating its carbonate, $MnCO_3$, at 500°C for 48 hrs.

2.2 Apparatus and procedure

Measurement of acetone formation rate was carried out through the use of a circular vacuum system at 50°C. Mixtures of gaseous acetic acid and argon were used in the measurements. Volume of the vacuum system was about 370 cm^3. Surface area of the catalysts was measured by the BET method with nitrogen. All the catalysts were precalcined at 400∿500°C for about one hour and were cooled down to room temperature. Then acetic acid was introduced into the reaction system. The infrared absorption cell was essentially the same as that described in a previous publication.[1)] For the infrared spectra measurement sample powders were mounted on a potassium bromide plate. After evacuation at 350∿400°C, gaseous acetic acid was introduced into the cell. After the reaction, the reaction mixture was subjected to the mass spectroscopic analysis.

3. RESULTS AND DISCUSSION

3.1 Adsorbed states of acetic acid

Fig. 1 shows the typical infrared absorption spectra of the adsorbed acetic acid on ZnO. The spectra indicate the absorption bands at 1760 (weak), 1725 (strong), 1555 (very strong), 1440 (medium), 1410 (m.), 1340 (w.), and 1275 (m.). In addition, two broad bands also appeared near 2950 (w.) and 2550 (m.) cm^{-1} (curve B). On pumping at 170∿200°C for 2 hrs., the absorption bands at 1760, 1725, 1410, 1340, 1275, 2950, and 2550 cm^{-1} dissapear, then the strong band at 1555 cm^{-1} splits into two bands at 1595 and 1555 cm^{-1}, and a sharp band appears at 1440 cm^{-1} (curve C).

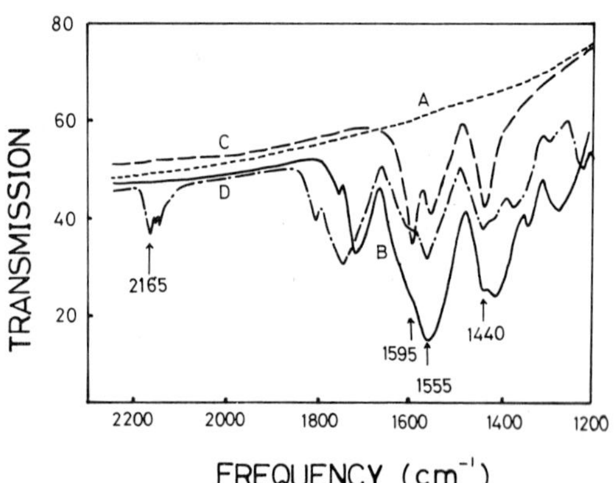

Fig. 1 Infrared spectra of adsorbed acetic acid on ZnO

A: Evacuated at 350°C for 2 hrs. and cooled to room temperature

B: Exposed to acetic acid vapor (9.7 mm) at room temperature for 3 hrs. and evacuated for 2 hrs.

C: Evacuated at 170∿200°C for 2 hrs. and recorded at 200°C

D: Exposed to acetic acid (11.5 mm) at 420°C for 2.5 hrs. and recorded at 434°C

Gaseous acetic acid gives the bands at 1800∿1780 cm^{-1}, which corresponds to the C=O stretching mode of vibration of acetic acid monomer, and a band at 1730 cm^{-1}, due to that of the dimer. Besides, two bands ascribable to the resonance of C-O stretching mode with the OH deformation mode were observed at 1425 and 1295 cm^{-1}, (curve A in Fig. 2), as well as the broad bands due to the OH stretching frequencies of the dimer near 2950 and 2550 cm^{-1}. Since in physical adsorption the perturbing effects of the surface forces on the adsorbate molecule are comparable to those of the surrounding molecules in the liquid state, the frequency shifts would be expected to be of a similar order of magnitude to that observed in changing from the gaseous state to the liquid state. Therefore, the absorption bands with frequency shifts from 1730, 1425, and 1285 cm^{-1}, respectively, to 1725, 1410, and 1275 cm^{-1} can be ascribed to the physically adsorbed acetic acid, and a weak band near 1760 cm^{-1} to that of the monomer and/or that shifted from 1780 to 1760 cm^{-1} to the dimer.

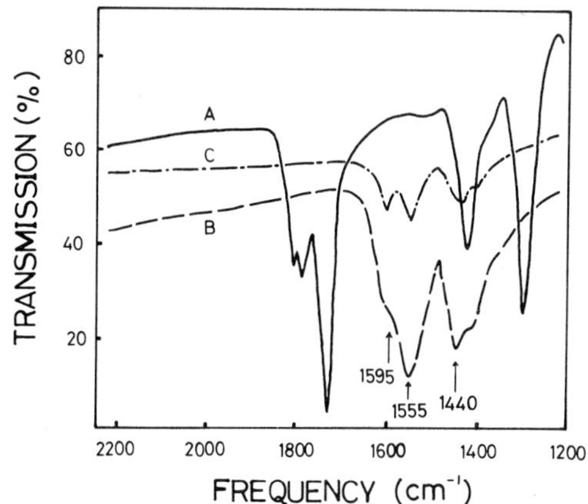

Fig. 2 Infrated spectra of gaseous acetic acid and $Zn(CH_3COO)_2 \cdot 2H_2O$

A: Gaseous acetic acid

B: $Zn(CH_3COO)_2 \cdot 2H_2O$: Evacuated at room temperature for one hr.

C: $Zn(CH_3COO)_2 \cdot 2H_2O$: Evacuated at 200°C for one hr.

The spectra of $Zn(CH_3COO)_2 \cdot 2H_2O$ which is mounted on a potassium bromide plate at room temperature is shown by curve B in Fig. 2. The bands of acetate ions were observed at 1555, 1440, and 1340 cm^{-1}. This finding shows that acetic acid on ZnO is adsorbed as the acetate ion as the result of dissociation of the acid.

At room temperature, it is always possible to observe a weak band at about 1595 cm^{-1} as a shoulder to the acetate band at 1555 cm^{-1}. It was previously reported[2)] that the absorption band at 1595 cm^{-1} is due to the protonated acetic acid, $CH_3COOH_2^+$. This species, therefore, may be the surface intermediate in the dehydration reaction of acetic acid on acidic catalysts. In our previous paper[3)] it was reported that the protonated formic acid is the intermediate for the dehydration reaction of formic acid on V_2O_5 or M_0O_3. Curve C in Fig.2 indicates the spectrum of $Zn(CH_3COO)_2 \cdot 2H_2O$ on heating under vacuum at 200°C. The absorption band appears at 1595 cm^{-1} in this case. Similar behavior is also observed on ZnO and MnO. From these findings, these adsorbed species can be assigned to be the protonated acetic acid.

As shown in Fig. 1 carried out at 434°C, three absorption bands at 2165, 2155, and 2140 cm^{-1} are observed by admitting acetic acid on various metal oxides (curve D). Similar absorption bands are also observed by heating diketene in the gas phase, but no band of diketene is observable at the same temperature. Accordingly, these bands may be ascribable to the P, Q, and R branches of the C=O stretching frequencies of ketene, which is produced by dehydration of acetic acid.

On the basis of these findings, the intermediate species on all the metal oxides are the adsorbed acetic acids and acetate ions and also gaseous ketene. However, in the cases of ZnO and MnO the protonated acetic acid appears also as an intermediate adsorbed species.

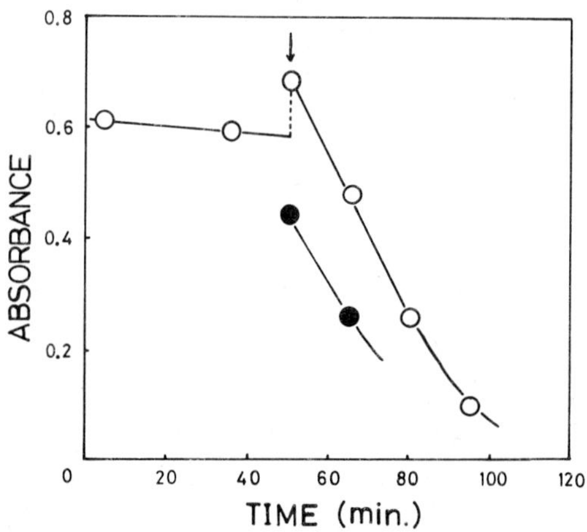

Fig. 3 Effect of addition of acetic acid to the reaction system involving ketene and CaO
O: Band intensity of ketene (2135 cm^{-1}). ●: Band intensity of acetic acid (1780 cm^{-1})
Reaction temperature: 372∿375°C
Total pressure after admitting of acetic acid: 16.4 mm

3.2 Ketonization of acetic acid

In order to study the effect of addition of acetic acid to the reaction system involving ketene formed by heating of diketene, $(CH_2CO)_2 \longrightarrow 2\,CH_2CO$, the infrared absorption method is applied to the system. The result on CaO is shown in Fig. 3, where the band intensity in both characteristic bands at 2135 cm^{-1} due to ketene and at 1780 cm^{-1} due to acetic acid is plotted against time.

Addition of acetic acid to ketene shown by the arrow in the figure leads to the formation of acetone as the product. This finding that ketene increases rapidly by the addition of acetic acid is due to the ketene produced by dehydration of acetic acid. This shows that the dehydration rate of acetic acid is faster than that of the reaction rate of acetic acid with ketene. Similar tendency is also observable on MgO.

In order to elucidate the role of ketene in the reaction, the catalytic dehydration of acetic acid -d_4(CD_3COOD) with diketene was investigated on ZnO, MgO, and CaO. The results determined by mass spectrometry are presented in Table 1. It was assumed that no exchange reaction between hydrogen and deuterium in the reactants occurs in this case. This assumption can be confirmed by the formation of the significant amount of CD_3COCD_3 on ZnO as is shown in Table 1.

Table 1
Ratio of CD_3COCD_3 to CH_2DCOCD_3

Catalysts	MgO[a)]	MgO	CaO	ZnO
Reaction temperature (°C)	397	290	430	380
Initial partial pressure (mm)				
CD_3COOD	5.1	18.0	18.6	15.2
$(CH_2CO)_2$	9.7	11.3	14.7	11.0
Reaction time (min.)	50	60	10	60
Ratio of acetone formed (%)				
CD_3COCD_3	——	——	5	42
CH_2DCOCD_3	100	100	95	58

a) Reaction in the IR cell

The ratio, $R = CH_2DCOCD_3/CD_3COCD_3$, is calculated by comparing the peak intensity of CH_2DCOCD_3 (m/e=62) with that of CD_3COCD_3 (m/e= 64).

In the deuterated acetone produced, its main product is CH_2DCOCD_3 on MgO and CaO, but the ratio, R, was about 0.6 on ZnO.

This result shows that the acetone formed on MgO and CaO is produced by the process through the reaction intermediate involving ketene, but the 60% and the remaining 40% acetone formed on ZnO are produced by the same process and some other process, respectively.

Kuriacose and Swaminathan[5] previously investigated the competition reaction, which involves dehydration and dehydrogenation of acetic acid on Cr_2O_3 by means of gaschromatography, using the mixture of alcohol and acetic acid. From the necessity of two active sites corresponding to both dehydration and dehydrogenation, they proposed that such surface intermediates acylcarbonium ion, CH_3C^+O, and acetate ion were produced on Cr_2O_3. Their confirmation by use of spectroscopic method, however, was not attempted. Therefore, some other species which they did not proposed may exist in such reactions.

3.3 Mechanism

We propose the following mechanism of the ketonization, Eq. (1), based on the intermediates which exhibit the infrared absorption bands.

$$\begin{array}{l} CH_3COOH - \left\{ \begin{array}{l} CH_3COO^-(ad) \\ H^+(ad) \end{array} \right. \\ \quad \backslash -H_2O \\ (CH_2CO)_2 - CH_2CO \end{array} \quad H^+(ad) + CH_2CO \longrightarrow CH_3C^+O ; \quad CH_3C^+O + CH_3COO^-(ad) \longrightarrow \left\{ \begin{array}{l} CH_3COCH_3 \\ CO_2 \end{array} \right. \qquad (1)$$

Since it is generally recognized that the ketene produced by diketene or by dehydration of acetic acid reacts easily with proton, the reaction of ketene and proton formed by dissociation of acetic acid leads to formation of the acylcarbonium ion, CH_3C^+O. The methyl group of acetate ion reacts subsequently with the acylcarbonium ion to form acetone and carbon dioxide .

In the cases of ZnO and MnO, taking into account the appearnce of the absorption band of protonated acetic acid, $CH_3COOH_2{}^+$, following mechanism, Eq. (2), is partially proposed. Adsorbed acetic acid reacts

$$CH_3COOH - \left\{ \begin{array}{l} CH_3COO^-(ad) \\ H^+(ad) \\ CH_3COOH(ad) \end{array} \right. \quad H^+(ad) + CH_3COOH(ad) \longrightarrow CH_3COOH_2{}^+(ad) \longrightarrow \left\{ \begin{array}{l} CH_3C^+O(ad) \\ H_2O \end{array} \right. ; \quad CH_3C^+O(ad) + CH_3COO^-(ad) \longrightarrow \left\{ \begin{array}{l} CH_3COCH_3 \\ CO_2 \end{array} \right. \qquad (2)$$

with the proton formed by dissociation of acetic acid. The protonated acetic acid results in the formation of the acylcarbonium ion. Then the acylcarbonium ion reacts with the adsorbed acetate ion to produce acetone as the product.

Initial reaction rate on metal oxides is examined under a vapor pressure (23 mm) of acetic acid. Fig. 4 shows the Arrhenius plot for the ketonization of acetic acid on four metal oxides at various temperatures in the region of 440∿280°C.

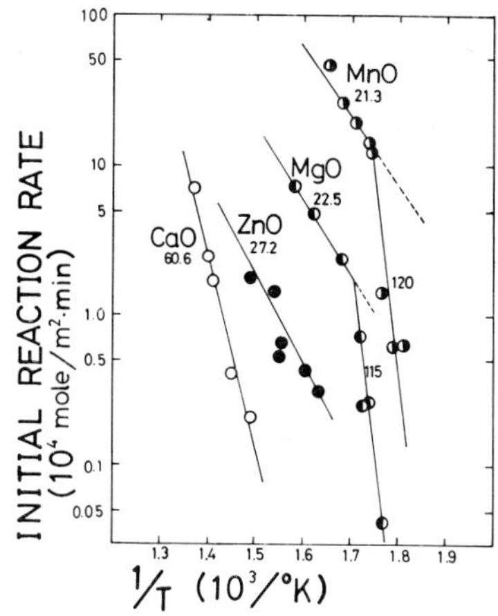

Fig. 4 Arrhenius plot for ketonization of acetic acid on metal oxides
Initial pressure of acetic acid: 23 mm

In the cases of MgO, the curves consist of two lines of different slope. The large activation energies (115 and 120 kcal/mole, respectively) were obtained by the lower temperatures. This can be ascribed to the decrease in catalytic activity by forming surface metal carbonates under reaction temperature below 300°C. Metal carbonates are formed on the surface of the catalysts, because this reaction produces acetone and carbon dioxide. Consequently, it is suggested that the remarkable decrease of the reaction rate based on the decrease of active sites leads to the increase in the apparent activation energy.

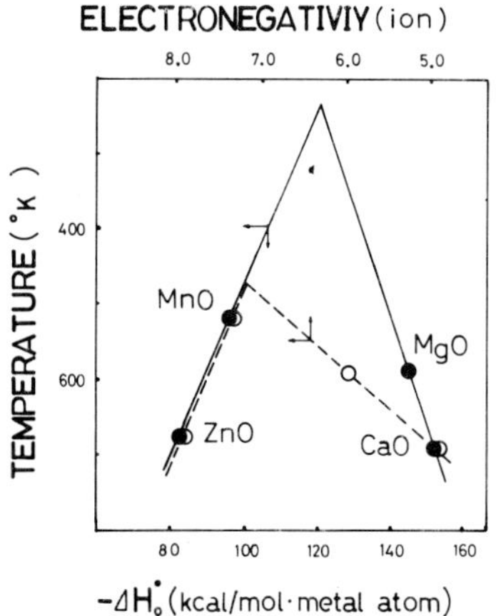

Fig. 5 Catalytic activity versus heat of formation or electronegativity of metal ion
numerals: ΔE kcal/mole

In the cases of MgO, CaO, and ZnO, the reaction order is found to be zero with respect to the pressure of acetic acid. The result shows that the rate determining step of the acetone formation is the surface process and the saturated adsorption of reactant occurs on the surface of the catalysts. A relationship between the absolute temperature for the catalytic activity, which was expressed by the initial rate of 2×10^{-4} mole/min m^2, and heat of formation of the metal oxide calalysts is shown in Fig. 5. The temperature for MnO was obtained by the extraporation of Arrhenius plot in the high temperature region. Each numerical value in Fig. 5 represents the apparent activation energy, ΔE. In Fig. 5, metal oxides are separated into two groups by the difference of the heat of formation. The two groups may be due to the difference between basic oxides (MgO and CaO) and amphoteric oxides (ZnO and MnO). The heat of formation for basic oxides are larger than those for the amphoteric oxides. It appears possible that the catalytic activities on metal oxides become the volcano type with respect to the heat of formation for metal oxides and the electronegativities of metal ions. Consequently, it is presumed that the acetic acid on ZnO and MnO is adsorbed as the protonated acetic acid, because ZnO and MnO are relatively stronger acidic solids in comparison with CaO and MgO.

Hence, it can be concluded that the ketonization of acetic acid on the basic oxides, MgO and CaO, proceeds through the reaction intermediate involving ketene produced by the dehydration reaction, but the ketonization on the amphoteric oxides, ZnO and MnO, proceeds through the two intermediates which are ascribable to the protonated acetic acid and ketene.

REFERENCES

1) M. Adachi, K. Kishi, T. Imanaka, and S. Teranishi, Bull. Chem. Soc. Japan, 40, 1290 (1967).
2) S. Hoshino, H. Hosoya, and S. Nagakura, Can. J. Chem., 44, 1961 (1966).
3) M. Adachi, T. Imanaka, and S. Teranishi, Nippon Kagaku Zasshi, 91, 400 (1970).
4) P. G. Blake and G. E. Jackson, J. Chem. Soc., B 10, 1153 (1968).
5) J. C. Kuriacose and R. Swaminathan, J. Cat., 16, 357 (1970).

DISCUSSION

K. KOCHLOEFL

1) What type of active sites do you propose can operate in the ketonization of acetic acid on ZnO and MgO?

2) Have you observed D-H exchange in the OD group of the unreacted CD_3COOD on ZnO, MgO?

T. IMANAKA

1) The active sites on both ZnO and MgO are considered to be $Zn(CH_3COO)_2$ and $Mg(CH_3COO)_2$; however, another active site on ZnO appears to be the surface hydroxyl group on ZnO.

2) The D-H exchange in the OD group of the unreacted CD_3COOD was not observed on both ZnO and MgO.

J. M. TRILLO

We have recently studied the thermal decomposition of 3d metal acetates prepared as thin layers on the corresponding 3d metal oxides. The results have been compared with those obtained in the ketonization of acetic acid on 3d metal oxides in order to elucidate the possibility of an acetate intermediate. The activity of catalysts has been measured in a flow reactor.

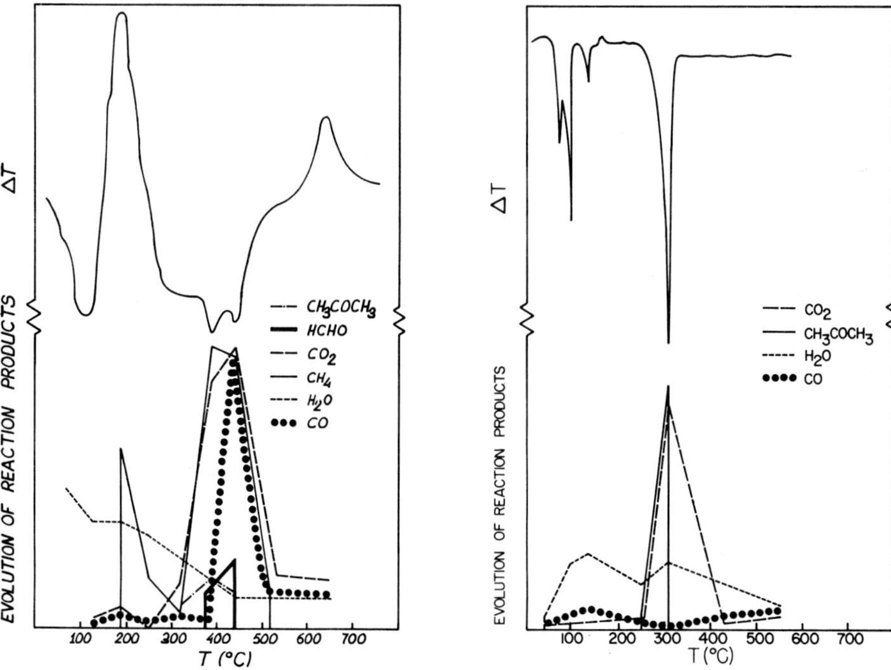

Fig. 1 Fig. 2

Differential thermal analysis of bulk chromium (Fig. 1) and manganese (Fig. 2) acetate (upper curves) and gas phase chromatography analysis of evolved products (lower curves).

The values obtained for the activation energy and frequency factor were 41 kcal·mol^{-1} and 1 x 10^{29} molec. cm^{-2}s^{-1} respectively; both maganese acetate decomposition and acetic acid ketonization on manganese oxide are the same.

Chromium acetate could not be prepared as a thin layer on the oxide, but in order to check the results of Kuriacose et al., which you mentioned in your paper, the thermal decomposition of bulk chromium acetate has been examined. Figure 1 shows its differential thermal analysis trace and the decomposition products observed. The complex decomposition mechanism of the salt does not allow, then, comparison with acetic acid ketonization on chromia. Figure 2 shows the results for manganese acetate.

Therefore, the conclusion drawn in your paper is only valid under certain experimental conditions.

T. IMANAKA

The ketonization of acetic acid occurs under certain experimental conditions, however I think that the distribution of products formed during the ketonization may be due to the acidic and basic properties of the different catalysts.

J. F. FRAISSARD

Our NMR spectra of acids or alcohols chemisorbed on many oxides show a relation between the width and the chemical shift of the lines of adsorbed phase protons and the selectivity of decomposition reactions. Evolutions of these spectra with regard to that of isolated molecule are due not only to changes of screening constants and dipolar interactions of nuclei, but also to their speed. From this study it seems that (1) chemisorption occurs often on the oxide catalysts in a complex form, as for example a dimer form, and (2) the selectivity is not only due to gas-solid interactions but also to the interactions in the molecular complex. For example, the ratio of the frequency of protonic exchange along C=O:::H-O adsorbed dimer acid bond to the number of diffusional jumps per unit time of molecules seems to play an important role in the reaction selectivity.

The key question is: do you have some information about the lifetime and molecular interactions in the adsorbed complex?

T. IMANAKA

Though the physically adsorbed acetic acid was observed at room temperature, the adsorbed state of acetic acid could not exactly be determined whether monomer or dimer. Accordingly, the frequency of protonic exchange along C=O:::H-O adsorbed dimer could not be observed.

The information about the life time and other molecular interactions were not obtained from infrared absorption.

Paper Number 5

THE MECHANISM OF THE WATER-GAS-SHIFT REACTION OVER IRON OXIDE CATALYST

S. OKI,* J. HAPPEL,** M. HNATOW,** and Y. KANEKO*
*Department of Chemistry, Utsunomiya University, Utsunomiya, Japan; **Department of Chemical Engineering, New York University, New York

ABSTRACT: Oxygen-18 was used as a tracer to study the water gas shift reaction over an iron oxide catalyst at temperatures ranging from 400°C to 450°C. The results obtained using oxygen-18 together with data previously reported using deuterium and carbon-14 as tracers were used to evaluate sequence of probable mechanistic steps and to establish which steps are rate-controlling.

Step I as well as step V of the following sequences of steps were determined to be rate controlling.

$$
\begin{array}{l}
CO \xrightarrow{\text{I}} CO(a) \\
H_2O \xrightarrow{\text{II}} \begin{cases} O(a) \\ 2H(a) \end{cases} \\
CO(a) + O(a) \xrightarrow{\text{III}} CO_2(a) \xrightarrow{\text{IV}} CO_2 \\
2H(a) \xrightarrow{\text{V}} H_2
\end{array}
$$

or

$$
\begin{array}{l}
CO \xrightarrow{\text{I}} CO(a) \\
H_2O \xrightarrow{\text{II}} \begin{cases} OH(a) \\ H(a) \end{cases} \\
CO(a) + OH(a) \xrightarrow{\text{IIIa}} HCOO(a) \xrightarrow{\text{IIIb}} \begin{cases} CO_2(a) \xrightarrow{\text{IV}} CO_2 \\ H(a) \end{cases} \\
H(a) + H(a) \xrightarrow{\text{V}} H_2
\end{array}
$$

A methodology is presented which shows the relevance of using more than one tracer to study mechanisms of reaction.

I. INTRODUCTION

It has generally been accepted that the observation of stoichiometric number provided a powerful method of determining the mechanism of chemical reaction,[4),8)] by the use of more than one appropriate isotopic tracer as discussed by Oki and Kaneko[1-3),4)].

The purpose of the present paper is to establish the mechanism of the catalyzed water gas shift reaction

$$CO + H_2O = CO_2 + H_2 \qquad (1)$$

in the presence of iron oxide catalyst by determining the stoichiometric number of isotopic exchange path by means of oxygen-18 in conjunction with the previous results obtained by using deuterium[1)] and carbon-14[2)] as tracer. The present study is also based on the stoichiometric number concept as applied to reactions where more than one rate-controlling step may exist. This concept was originated by Horiuti[5)] and discussed by Happel[6)] for the case where a single rate-controlling step and for more complicated cases, using the concept of reaction paths introduced by Happel[7)].

Principles of the determination of $\nu(r)$, experimental procedures and results are described in what follows.

2. PRINCIPLES OF $\nu(r)$-EVALUATION

According to previous results,[1-3] we assume the reaction path is sequence I described in Table 1.

Consider that mixture of CO, H_2O, CO_2 and H_2 is circulated over iron oxide catalyst and both the partial pressure of CO etc., p^{CO} etc., and the transfer rate of oxygen-18 from H_2O to CO or CO_2 are followed until the equilibrium of the water gas shift reaction is attained.

The apparent stoichiometric number is determined by observation according to the general equation (2)[5] as follows

$$\nu(r) = \frac{-\Delta G}{RT \ln (V_+/V_-)} \tag{2}$$

where $-\Delta G$ is the chemical affinity of overall reaction (1) and V_+ or V_- is the forward or backward unidirectional rate of the exchange path of oxygen atoms. The affinity $-\Delta G$ of reaction (1) is expressed in terms of p^{CO} etc. and equilibrium constant of reaction (1), K_p, as

$$-\Delta G = RT \ln \left[\frac{p^{CO}\, p^{H_2O}}{p^{CO_2}\, p^{H_2}} \cdot K_p \right] \tag{3}$$

The transfer rate of oxygen-18 in CO, H_2O and CO_2 are expressed as

$$t^{CO} = -\frac{d(n^{CO}z^{CO})}{dt} = z^{CO}\, v_{+I} - z^{CO(a)}\, v_{-I} \qquad \text{(i)}$$

$$= z^{CO(a)}\, v_{+III} - z^{CO_2(a)}\, v_{-III} \qquad \text{(ii)}$$

$$(4)$$

$$t^{H_2O} = -\frac{d(n^{H_2O}z^{H_2O})}{dt} = z^{H_2O}\, v_{+II} - z^{O(a)}\, v_{-II} \qquad \text{(iii)}$$

$$= z^{O(a)}\, v_{+III} - z^{CO_2(a)}\, v_{-III} \qquad \text{(iv)}$$

$$t^{CO_2} = \frac{d(n^{CO_2}z^{CO_2})}{dt} = z^{CO_2(a)}\, v_{+IV} - z^{CO_2}\, v_{-IV} \qquad \text{(v)}$$

where t^i is the transfer rate of oxygen-18 in i, n^i or z^i the number of moles or atomic fraction of oxygen-18 in i and v_{+s} or v_{-s} the forward or backward unidirectional rate of step s.

Solving for $z^{O(a)}$, $z^{CO(a)}$ and $z^{CO_2(a)}$ from eq. (4,i), (4,iii) and (4,v), we have

$$z^{CO(a)} = \frac{1}{v_{-I}} \left(z^{CO} v_{+I} - t^{CO}\right) \qquad \text{(i)}$$

$$z^{O(a)} = \frac{1}{v_{-II}} \left(z^{H_2O} v_{+II} - t^{H_2O}\right) \qquad \text{(ii)} \qquad (5)$$

$$z^{CO_2(a)} = \frac{1}{v_{+IV}} \left(z^{CO_2} v_{-IV} + t^{CO_2}\right) \qquad \text{(iii)}$$

Substituting $z^{CO(a)}$ etc. into eq. (4,ii) or (4,iv), we have

$$\left[\frac{V_+}{V_-}\right]^{I,III,IV} = \frac{1}{t^{CO}-vz^{CO}}\left\{\left(t^{CO}-vz^{CO_2}\right)+\left(\frac{v_{+IV}}{v_{-IV}}-1\right)\left(t^{CO}-t^{CO_2}\right)\right\} \quad (6)$$

or

$$\left[\frac{V_+}{V_-}\right]^{II,III,IV} = \frac{1}{t^{H_2O}-vz^{H_2O}}\left\{\left(t^{H_2O}-vz^{CO_2}\right)+\left(\frac{v_{+IV}}{v_{-IV}}-1\right)\left(t^{H_2O}-t^{CO_2}\right)\right\} \quad (7)$$

As described above, since we have two equations, eq. (6) and (7), with three unknown, we need to obtain another equation or additional set of data. However, it is clear that the last term in both equation (6) and (7) could be zero if either the rate of step IV, chemisorption and desorption of CO_2, is very fast where $v_{+IV} = v_{-IV}$ or if we are able to adjust the experimental condition of experimental run such that $t^{CO} = t^{H_2O} = t^{CO_2}$.

In that case we have, from eq. (6) or (7)

$$\left[\frac{V_+}{V_-}\right]^{I,III,IV} = \frac{t^{CO}-vz^{CO_2}}{t^{CO}-vz^{CO}} \quad (6')$$

or

$$\left[\frac{V_+}{V_-}\right]^{II,III,IV} = \frac{t^{H_2O}-vz^{CO_2}}{t^{H_2O}-vz^{H_2O}} \quad (7')$$

On the other hand, if the last term in eq. (6) and (7) do not vanish we can solve for the three unknown values by other methods which are not discussed in the present paper.

Substituting $-\Delta G$ and V_+/V_- respectively from eq. (3) and (6) or (7) into eq. (2), we can obtain the apparent stoichiometric number of path I, III and IV or II, III and IV.

In the case when we assume the other reaction path, i.e. sequence II as shown in Table 1, we also obtain equations quite similar to eq. (6) and (7) or eq. (6') and (7') according to the same method of derivation. It is interesting that this mechanism is based on occurence of the reaction with formic ion as an intermediate because the chemical species of water gas shift reaction are produced by the decomposition of formic acid.

3. EXPERIMENTAL

3.1 Materials

Carbon monoxide was prepared by dehydration of formic acid and purified by passing it through liquid nitrogen trap. Carbon dioxide was formed by decomposition of sodium bicarbonate and purified by vacuum distillation. Hydrogen was passed through silica gel, liquid nitrogen trap, a palladium thimble kept at 380°C and liquid nitrogen trap.

Heavy water traced by oxygen-18, containing 10.9 atom percent ^{18}O was the only tracer used in all experiments, which was supplied by the Research and Development Ltd., Rehovoth, Israel and was used without further purification.

Iron oxide catalyst was kindly presented by Mitsubishi Kasei Co. Ltd. The amount of catalyst used was 0.5 gr in the state of Fe_2O_3 before being reduced to Fe_3O_4 in the working state, the particle size being 8 to 10 mesh. Before starting the measurements, the catalyst was treated several times with a mixture of CO_2 and H_2 followed by CO and H_2O traced by oxygen-18 for about one hundred hours at 500°C. After carrying out one run, the catalyst was evacuated at reaction temperature and used for next run.

3.2 Apparatus and experimental procedure[1-3]

The apparatus was made entirely of pyrex glass in order to minimize oxygen exchange between glass and materials, otherwise the reaction system and experimental procedure were the same as described previously. The reacted gas was drawn into sampling bulbs successively as the water gas shift reaction progressed toward equilibrium. The sampled gas was analyzed by a Hitachi mass spectrometer, RMS-3B type, by fixing the electron accelerating voltage at 80 volt.

4. RESULTS

According to the experimental procedure described above, the results obtained are summarized in tables 2 to 5. Figure 1 shows the time course of partial pressure of CO, P^{CO}, and atomic fraction of oxygen-18 in CO and CO_2, Z^{CO} and Z^{CO_2}, for run 6 by way of example, where the atomic fraction of oxygen-18 in H_2O, Z^{H_2O}, is quite equal to Z^{CO_2} except for the very initial stage of reaction. As shown in tables and Fig. 1 for run 6 as an example, the transfer rate of path II, III and IV, where oxygen exchange occurs between H_2O and CO_2, is very fast (i.e. this path is essentially in partial equilibrium), where $Z^{H_2O} = Z^{CO_2}$ as noted above. This result suggests that a free energy cascade does not exist in the oxygen transfer path II, III and IV. In other words, the rate ratios of unidirectional forward to backward of steps II, III and IV are all unity. Therefore the apparent stoichiometric number for the transfer path II, III and IV is definitely augmented in order of magnitude as shown by ∞ in what follows. On the other hand, the $\nu(r)$ obtained of transfer path I, III and IV is about 2. This suggests that the free energy cascade must be in step I, insofar as there exists a free energy cascade in transfer path I, III and IV.

If this free energy cascade is the only rate-determining step, the observable value of stoichiometric number must be unity but not 2 by assuming the sequence of steps as shown in the table 1. However, the present results are 2 for $\nu(r)^{I,III,IV}$. Also it has previously been reported that the $\nu(r)$ was 2 as determined by using either deuterium[1] or carbon-14[2] as tracer.

5. DISCUSSION

Recently, Glavachek et al[9] have reported that the most suitable relationship for describing the mechanism is that proposed by Temkin and his co-workers[10]. This mechanism corresponds to sequence III in table 1 with the first two steps rate-controlling. For this mechanism, the transfer rate of path I, III and IV or path II, III, and IV should respectively be appropriate to the rate of overall reaction, so that the observable value of $\nu(r)$ should be unity. Therefore our results conflict with Temkin's mechanism but it is interesting that his mechanism

TABLE 1
Possible mechanisms for reaction (1) and observable stoichiometric number

Sequences	Elementary step		Stoichiometric number			
			ν	0	^{14}C	^{18}O
I	$CO \rightarrow CO(a)$	(I)	1	∞	1	1,∞
	$H_2O \rightarrow 2H(a)+O(a)$	(II)	1	1	∞	1,∞
	$CO(a)+O(a) \rightarrow CO_2(a)$	(III)	1	∞	1	1
	$2H(a) \rightarrow H_2$	(V)	1	1	∞	∞
	$CO_2(a) \rightarrow CO_2$	(IV)	1	∞	1	1
II	$CO \rightarrow CO(a)$	(I)	1	∞	1	1,∞
	$H_2O \rightarrow OH(a) + H(a)$	(II)	1	1	∞	1,∞
	$CO(a)+OH(a) \rightarrow COOH(a)$	(IIIa)	1	1,∞	1	1
	$COOH(a) \rightarrow CO_2(a)+H(a)$	(IIIb)	1	1,∞	1	1
	$CO_2(a) \rightarrow CO_2$	(IV)	1	∞	1	1
	$2H(a) \rightarrow H_2$	(V)	1	1	∞	∞
III	$H_2O \rightarrow 2H(a)+O(a)$	(I)	1	1	∞	1,∞
	$CO+O(a) \rightarrow CO_2$	(II)	1	∞	1	1,∞
	$2H(a) \rightarrow H_2$	(III)	1	1	∞	∞
IV	$H_2O \rightarrow H_2O(a)$	(I)	1	1	∞	1,∞
	$CO \rightarrow CO(a)$	(II)	1	∞	1	1,∞
	$CO(a)+H_2O(a) \rightarrow CO_2+H_2$	(III)	1	1	1	1
V	$CO \rightarrow CO(a)$	(I)	1	∞	1	1,∞
	$H_2O+CO(a) \rightarrow CO_2+H_2$	(II)	1	1	1	1

TABLE 2

Observed data, run 1 at 400°C

$(p^{H_2O})_o$ = 13.89 mm Hg $(p^{H_2})_o$ = 14.29 mm Hg
$(p^{CO})_o$ = 13.67 mm Hg $(p^{CO_2})_o$ = 15.28 mm Hg

Time min	p^{CO} mm Hg	z^{CO}	z^{CO_2}	$\left[\frac{V_+}{V_-}\right]^{I,III,IV}$	$\nu(r)^{I,III,IV}$
0	13.67	0	0	-	-
5	13.57	0.0011	0.0136	1.4688	(5.88)
15	12.65	0.0018	0.0310	2.0952	2.71
30	11.90	0.0046	0.0326	2.0502	2.49
50	10.99	0.0060	0.0306	1.9226	2.34
70	10.04	0.0074	0.0300	1.8476	2.03
100	9.12	0.0094	0.0302	1.7801	1.67
150	7.80	0.0167	0.0288	1.4538	1.39

TABLE 3

Observed data, run 2 at 440°C

$(p^{H_2O})_o$ = 14.83 mm Hg $(p^{H_2})_o$ = 14.15 mm Hg
$(p^{CO})_o$ = 13.19 mm Hg $(p^{CO_2})_o$ = 15.22 mm Hg

Time min	p^{CO} mm Hg	z^{CO}	z^{CO_2}	$\left[\frac{V_+}{V_-}\right]^{I,III,IV}$	$\nu(r)^{I,III,IV}$
0	13.19	0	0	-	-
5	12.61	0.0008	0.0301	2.0729	(2.66)
15	11.44	0.0039	0.0340	2.1022	2.17
30	10.15	0.0066	0.0316	1.9155	1.90
50	8.75	0.0118	0.0313	1.7141	1.50
90	7.59	0.0151	0.0301	1.4353	1.20
140	7.02	0.0213	0.0298	1.2036	1.30

TABLE 4

Observed data, run 5 at 450°C

$(p^{H_2O})_o$ = 22.91 mm Hg $(p^{CO_2})_o$ = 9.22 mm Hg
$(p^{CO})_o$ = 22.81 mm Hg $(p^{H_2})_o$ = 4.76 mm Hg

Time min	p^{CO} mm Hg	z^{CO}	z^{CO_2}	$\left[\frac{V_+}{V_-}\right]^{I,III,IV}$	$\nu(r)^{I,III,IV}$
0	22.81	0	0	-	-
10	19.83	0.0030	0.0505	3.5641	2.80
20	17.61	0.0057	0.0490	3.3373	2.27
30	15.84	0.0088	0.0466	3.0229	2.18
40	14.24	0.0085	0.0446	2.7519	1.95
50	13.01	0.0100	0.0429	2.5628	1.74
60	11.69	0.0106	0.0417	2.2732	1.55
70	10.99	0.0135	0.0406	2.0054	1.54
90	9.79	0.0185	0.0400	1.6560	1.41
120	8.69	0.0292	0.0388	1.2290	1.73

TABLE 5

Observed data, run 6 at 450°C

$(p^{H_2O})_o$ = 19.64 mm Hg $(p^{CO_2})_o$ = 20.03 mm Hg
$(p^{CO})_o$ = 19.93 mm Hg $(p^{H_2})_o$ = 11.90 mm Hg

Time min	p^{CO} mm Hg	z^{CO}	z^{CO_2}	$\left[\frac{V_+}{V_-}\right]^{I,III,IV}$	$\nu(r)^{I,III,IV}$
0	19.93	0	0	-	-
7	19.22	0.0022	0.0314	3.0087	2.24
15	18.12	0.0038	0.0336	2.6986	2.23
25	16.92	0.0049	0.0335	2.4492	2.17
40	15.53	0.0055	0.0327	2.1527	2.12
55	16.62	0.0064	0.0325	1.9213	2.16
75	13.12	0.0099	0.0318	1.6422	2.16
100	12.00	0.0119	0.0309	1.4573	2.05
130	11.09	0.0177	0.0305	1.2977	2.04
160	10.41	0.0217	0.0301	1.1565	2.40
180	10.10	0.0235	0.0294	1.1082	2.54

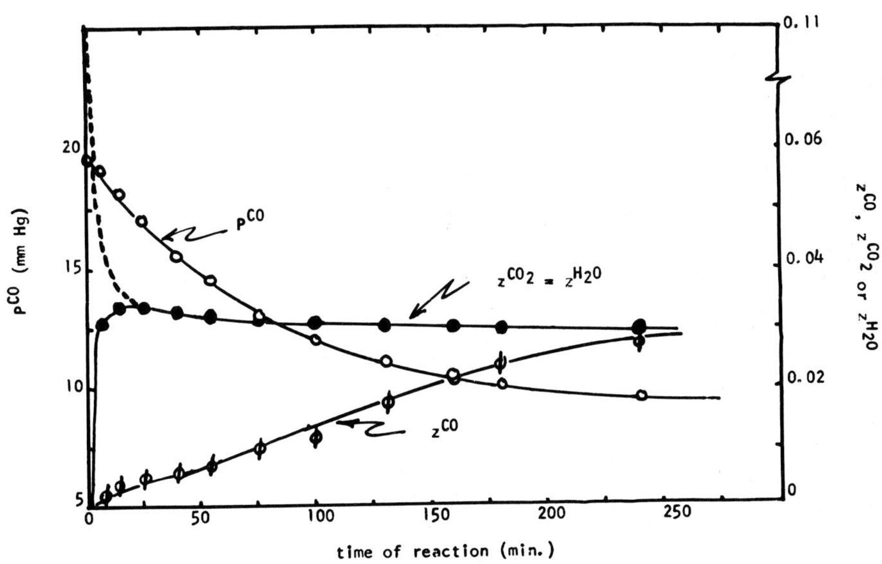

Fig. 1 p^{CO}, z^{CO} and z^{CO_2} vs. time of reaction, run 6 at 450°C. (a dotted line shows z^{H_2O} at initial stage of reaction)

is based on the assumption of more than one rate controlling step[10].

The chemical affinity of a reaction for a single route[11] consisting S steps is given as[5]

$$\Delta G = \sum^{S} \nu(s)\Delta g_s \qquad (s = 1 \cdots\cdots, S) \tag{7}$$

where $\nu(s)$ and Δg_s is the stoichiometric number and Gibbs free energy change of step s respectively and Δg_s is now given in terms of the forward and backward unidirectional rate as

$$-\Delta g_s = RT \ln \left(\frac{V_{+s}}{V_{-s}}\right) \tag{8}$$

Inserting eq. (8) into (7), we obtain

$$-\Delta G = RT \ln \prod^{S} \left(\frac{V_{+s}}{V_{-s}}\right) \tag{7'}$$

Also, we have the rate ratio of transfer path as

$$\left[\frac{V_+}{V_-}\right]^{I,III,IV} = \frac{V_{+I}\ V_{+III}\ V_{+IV}}{V_{-I}\ V_{-III}\ V_{-IV}} \tag{9}$$

or

$$\left[\frac{V_+}{V_-}\right]^{II,III,IV} = \frac{V_{+II}\ V_{+III}\ V_{+IV}}{V_{-II}\ V_{-III}\ V_{-IV}} \tag{10}$$

If we use $[V_+/V_-]^{I,III,IV}$ or $[V_+/V_-]^{II,III,IV}$ instead of V_+/V_- in eq. (2), we have

$$\nu(r)^{I,III,IV} = \frac{RT \ln \prod^{S}\left(\frac{V_{+s}}{V_{-s}}\right)}{RT \ln \left[\frac{V_{+I}\ V_{+III}\ V_{+IV}}{V_{-I}\ V_{-III}\ V_{-IV}}\right]} \tag{11}$$

or

$$\nu(r)^{II,III,IV} = \frac{RT \ln \prod^{S}\left(\frac{V_{+s}}{V_{-s}}\right)}{RT \ln \left[\frac{V_{+II}\ V_{+III}\ V_{+IV}}{V_{-II}\ V_{-III}\ V_{-IV}}\right]} \tag{12}$$

According to the results obtained that the path II, III, and IV is in partial equilibrium, eq. (11) and (12) may be rewritten as

$$\nu(r)^{I,III,IV} = \frac{RT\,\ell n\left(\frac{v_{+I}}{v_{-I}}\frac{v_{+V}}{v_{-V}}\right)}{RT\,\ell n\left(\frac{v_{+I}}{v_{-I}}\right)} = \frac{\Delta g_I + \Delta g_V}{\Delta g_I} \qquad (13)$$

or

$$\nu(r)^{II,III,IV} = \frac{RT\,\ell n\left(\frac{v_{+I}}{v_{-I}}\frac{v_{+V}}{v_{-V}}\right)}{RT\,\ell n\ 1} = \frac{\Delta g_I + \Delta g_V}{0} \qquad (14)$$

Quite similarly, on the other hand, we have another equation for the hydrogen exchange path II and V between H_2O and H_2 as

$$\nu(r)^{II,V} = \frac{RT\,\ell n\left(\frac{v_{+I}}{v_{-I}}\frac{v_{+V}}{v_{-V}}\right)}{RT\,\ell n\left(\frac{v_{+V}}{v_{-V}}\right)} = \frac{\Delta g_I + \Delta g_V}{\Delta g_V} \qquad (15)$$

As published previously[1], the $\nu(r)$ - values obtained in accordance with eq. (15) was about 2 using deuterium as tracer. Therefore, from eq. (13) and (15), we obtain $\Delta g_I = \Delta g_V$ since our results show that the $\nu(r)$-values obtained are 2 as observed by means of either deuterium or oxygen-18 as tracer. This conclusion is also consistent with the fact that the $\nu(r)$ was 2 without correction by using carbon-14[2] as tracer.

From these discussions, we see that the experimental results do not fit in with the sequence III, IV and V in Table 1. If sequence I and II is operative, the free energy cascade must be steps I and V in either case. Moreover, the observation of apparent stoichiometric numbers by the use of more than one isotopic tracer provided a powerful tool of determining the mechanism of this heterogeneous catalysis reaction.

REFERENCES

1) S. Oki and Y. Kaneko, J. Res. Inst. Catalysis 13 (1965), 55; Shokubai 6 (1964), 356; Bull. General Educ. Utsunomiya Univ. 2 (1969), 33.
2) S. Oki and Y. Kaneko, J. Res. Inst. Cat. 13 (1965), 169; 15 (1968), 185.
3) S. Oki, Y. Kaneko, Y. Arai and M. Shimada, Shokubai 11 (1969), 184P.
4) S. Oki and Y. Kaneko, Shokubai 9 (1967), 16; J. Res. Inst. Cat. 18 (1970), 93.
5) J. Horiuti, J. Res. Inst. Cat. 1 (1948), 8.
6) J. Happel, J. Chem. Eng. Progr. Symp. Ser. 63 (72) (1967), 31.
7) J. Happel, J. Res. Inst. Cat. 16 (1968), 305; J. Cat. 20 (1971), 132.
8) Cf. Shokubai Kogakukoza, 1 Chijin-Shokan, Tokyo (1967).
9) V. Glavachek, M. Marek and M. Korzhinkova, Kinetika i Kataliz 9 (1968), 1107.
10) G.G. Shchibrya, N.M. Morozov and M.I. Temkin, Kinetika i Kataliz, 6 (1965), 1057.
11) J. Horiuti and T. Nakamura, "Advances in Catalysis" 17 (1968), 1.

DISCUSSION

J. J. F. SCHOLTEN

On page 5 of your paper you introduce the formic acid mechanism for the water-gas-shift reaction:

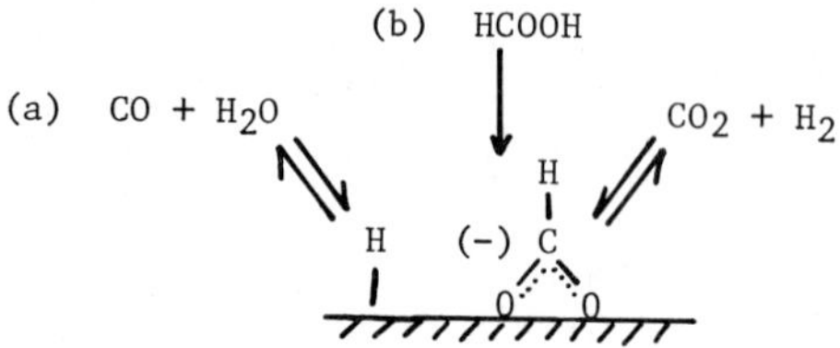

This mechanism, suggested earlier by Professor Mars, corresponds to a stoichiometric number ν=1 for the shift reaction (a).

It is of interest to measure ν for the formic-acid decomposition (b). The value of ν might be 1 or a half (1/2). The last number is valid for: $2\ HCOO^- \rightarrow 2\ CO_2 + H_2$ being the rate determining step. Perhaps ν = 1 and ν = 1/2 are both possible.

J. HAPPEL

Work with formic acid would be of interest. However, even if one obtained a stoichiometric number of one with formic acid decomposition it might indicate that other chemisorption steps such as adsorption of HCOOH are rate controlling. The stoichiometric number of 1/2 would only be obtained if the surface reaction were the rate controlling step.

L. GUCZI

In your preprint you presented the experimental data concerning the calculation of stoichiometric number under different experimental conditions. It seems that as the time is increasing, ν is decreasing. Does the change of the stoichiometric number bear any meaning for the rate-determining step?

S. OKI

According to our experimental results, the apparent stoichiometric number for CO-CO_2 path changes with time.

This result means that step V or I is more important at the stage far from equilibrium or near equilibrium, even if both steps I and V are free energy cascade.

The change of these values may depend on the partial pressure change of components, especially CO and H_2.

P. N. ROSS, JR.

At small reaction times immediately after introduction of the fresh reaction mixture of CO and $H_2{}^{18}O$, do you observe the production of $C^{16}O_2$? The appearance of $C^{16}O_2$ would be indicative of the participation of oxygen of the iron-oxide lattice.

S. OKI

In our experimental condition, we could not observe the production of $C^{16}O_2$ at the initial stage since we used very small amounts of catalyst compared with that of mixture.

However, in the case where we used non tagged mixture with oxygen-18 tagged iron oxide catalyst for extra work, $C^{16}O^{18}O$ was formed at the initial stage in agreement with your suggestion.

This indicates the participation of oxygen of the iron oxide catalyst for the overall reaction.

Paper Number 6

ACTIVE PHASES IN CHROMIA DEHYDROGENATION CATALYSTS. THE ROLE OF ALKALI METAL AND MAGNESIUM ADDITIVES

J. MASSON and B. DELMON*
Laboratoire de Chimie Générale,
Université de Grenoble, Grenoble, France

ABSTRACT: Surface area, catalytic properties (in the dehydrogenation of isobutane), X-ray diagrams and E.P.R. spectra of Al_2O_3 (92.5% mol.) - Cr_2O_3 (7.5% mol.) oxides, pure or promoted with Li_2O, Na_2O, K_2O, Rb_2O, Cs_2O or MgO, were investigated. Li_2O, Na_2O and MgO (class I) have little effect. K_2O, Rb_2O, and Cs_2O (class II) increase the intrinsic activity (i.e. per unit area) and decrease the stability. Specific surface areas of the samples also increase. Class II samples contain no α phase and do not exibit the β_N line in E.P.R. The results suggest that the effective additives (class II) do not act directly through their basicity. Dehydrogenation and fouling are probably parallel reactions on identical Cr^{3+} active sites. The effective additives increase the Cr^{3+} active site content and inhibit recrystallization to α phase.

1. INTRODUCTION

The action of true Cr_2O_3-Al_2O_3 solid solutions in the dehydrogenation of isobutane has been discussed in a recent paper[1)]. It was shown that the intrinsic activity, defined as the molar conversion to isobutene per unit surface area of catalyst, is maximum for Cr_2O_3 rich α-solid solutions. Concerning Al_2O_3 rich samples, α-phases are nearly inactive, whereas γ-phases exhibit relatively high activities. Stability to deactivation decreases from Al_2O_3 rich to Cr_2O_3 rich solid solutions. The mechanism of dehydrogenation and deactivation was discussed in terms of the chromium ions of various charges present in the solid solutions. It was concluded that Cr^{2+} ions were probably responsible for a high initial activity, which is very rapidly inhibited. The useful dehydrogenation centers are Cr^{3+} ions. The difference in activity between the α and γ phases in the Al_2O_3 rich domain was explained by the difference in covalent character of the two lattices.

In addition to chromia and alumina, dehydrogenation catalysts usually contain some alkali promoters. The presence of the latter may have some influence on the structure and texture of the catalyst. An alkali promoter may also have some influence on the mechanism of the catalysis (e.g. dehydrogenation reaction or deactivation processes) because of its basicity. Therefore, it was desirable to investigate the role of these promoters.

The present communication concerns a system containing 92.5% Al_2O_3 and 7.5% Cr_2O_3. According to the previous work and our own measurements, the best catalytic properties are

---

(*) Professor, University of Louvain, Heverlee, Belgium.

obtained for compositions near these proportions. The results presented here concern the influence of Li_2O, Na_2O, K_2O, Rb_2O, Cs_2O and MgO.

Numerous publications have been devoted to the role of promoters in Cr_2O_3-Al_2O_3 dehydrogenation or dehydrocyclization reactions. It is generally observed that the catalytic performances are improved by the addition of K_2O, Rb_2O and Cs_2O [2) 3) 4) 5) 6) 7) 8)]. Some authors mention an increase in activity in the presence of Li_2O or Na_2O [6) 9)], while others, on the contrary, observe a deactivation.[5) 10)].

Mention has already been made of various experimental methods for the study of the catalysts : namely, conductivity[6)], magnetic susceptibility [5) 6)], E.P.R. [3)], radiocrystallographic [4) 11) 12) 13)] and acidity [2) 5)] measurements, and the titration of the oxidation state by chemical methods [3) 5) 6)].

In general, alkaline additives promote the formation of transition structures, like γ or Θ alumina structures [4) 11) 12) 13)]. But the samples are poorly crystallized. C.P. Poole and D.S. Mac Yver [14) 15)] showed that E.P.R. is a more valuable tool for the identification of certain species. Four different E.P.R. lines are mentioned in literature, namely :

- . a line, near 1,600 gauss, which is caused by isolated Cr^{3+} ions (δ line)
- . a line, near 3,400 gauss, with a width of 50 gauss, which has its origin in the Cr^{5+} ions (γ line)
- . two broad (i.e. 300-500 gauss) lines, near 3,500 gauss, which have their origin in Cr^{3+} ions in strong interaction. One of them (β_W) has been attributed to a Al_2O_3-Cr_2O_3 amorphous phase, the other (β_N) to an α solid solution[14)].

In the present study, the samples were subjected to X-ray, E.P.R., surface area and activity measurements.

2. EXPERIMENTAL

The mixed oxide samples containing Al_2O_3, Cr_2O_3 and the oxide of the promoter Me were obtained by quick pyrolysis in air of oxalato-complexes of formula $Me_z(NH_4)_{3-z}\ Al_xCr_y(C_2O_4)_3$, nH_2O [16)]. This method has been shown to give the best homogeneous oxides [17)]. One series of samples was prepared at 900°C (2.30 hrs), the other at 1,050°C (2 hrs). The proportion of additive, Z, is expressed in moles divided by the number of moles of Cr_2O_3 + Al_2O_3. Surface areas were measured by B.E.T., at liquid nitrogen temperature (table 1).

E.P.R. measurements were made at room temperature, with a VARIAN 4501 spectrometer (wave length 3 cm, modulation 100 kHz.). The paramagnetic ion content, I, relative to 1 g of

TABLE 1
Specific surface area of the samples

Additives (Z = 2.5%)		None	Li_2O	Na_2O	K_2O	Rb_2O	Cs_2O	MgO
Surface area ($m^2.g^{-1}$)	900°C series	36	38	33	47	52	46	18.5
	1,050°C series	38	30	32	48	52	47	34

catalyst, is expressed in arbitrary units. The line width, ΔH, is expressed in gauss.

Catalytic activity was measured in a flow apparatus, according to a procedure which has been described previously[1]. Before each measurement, the catalyst sample was heated in air, for 6 hrs, at 770°C, cooled in nitrogen for 1 hr to 588 ± 2°C, which was the reaction temperature. The feed was pure isobutane. The effluents were analysed by gas chromatography. Isobutene is the main hydrocarbon product. Very little isobutane is transformed to propene (1% max.),propane (traces), ethane (traces) or methane (1.5% max.).

The intrinsic activity A is expressed as the isobutene yield per unit surface area of catalyst, at constant feed rate. The stability of the catalyst, i.e. its resistance to poisoning by carbonaceous products formed during the dehydrogenation run, is expressed by the time $t_{1/3}$ necessary for the activity to lose one third of its initial value A^{*}.

3. RESULTS

3.1 Catalytic properties

The variations of the activity with time, for the various samples, is suggested by fig. 1, relative to the series of catalysts prepared at 900°C. Initial intrinsic activities and stabilities for both series are given in table 2. The intrinsic activity is considerably increased by the addition of K_2O, Rb_2O and Cs_2O. The other additives, namely Li_2O, Na_2O and MgO, have only a negligible effect. Stability is always decreased. This effect is especially notable with K_2O, Rb_2O and Cs_2O.

The influence of the additive content was investigated in the case of K_2O. The activity-time curves for the 900°C series are presented in fig. 2. The summerized results for both series are given in table 3. A maximum in the activating effect is observed for values of Z near 2.5 - 3 %.

3.2 X-ray diffraction

All the samples were subjected to radio-crystallographic study. They are all poorly crystallized.

In the samples prepared at 900°C, only a poorly crystallized phase of γ-Al_2O_3 type can be detected.

TABLE 2
Initial intrinsic activities and stabilities

Additive (Z = 2.5%)	900°C series		1,050°C series	
	Initial intrinsic activity (m^{-2})	Stability $t_{1/3}$ (min)	Initial intrinsic activity (m^{-2})	Stability $t_{1/3}$ (min)
None	0.22	80	0.21	75
Li_2O	0.24	70	0.26	50
Na_2O	0.28	60	0.26	40
K_2O	0.68	40	0.54	37
Rb_2O	0.65	38	0.50	50
Cs_2O	0.89	26	0.73	34
MgO	0.44	90	0.26	80

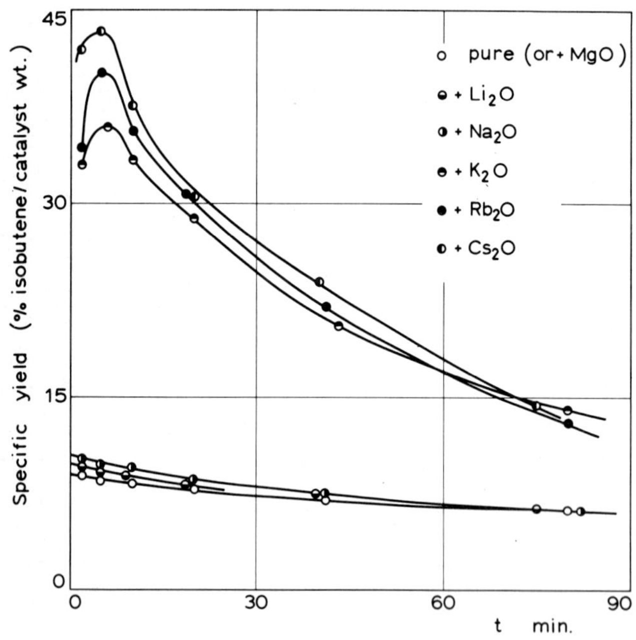

Fig. 1 Specific yield (i.e. isobutene yield per gram of catalyst) (900°C series)

TABLE 3
Influence of the K_2O content on the initial intrinsic activity

K_2O content Z %		0	2	2.5	3	3.5
Initial intrinsic activity	900°C series	0.22	0.33	0.68	0.65	0.42
	1,050°C series	0.21	0.53	0.54	0.64	0.50

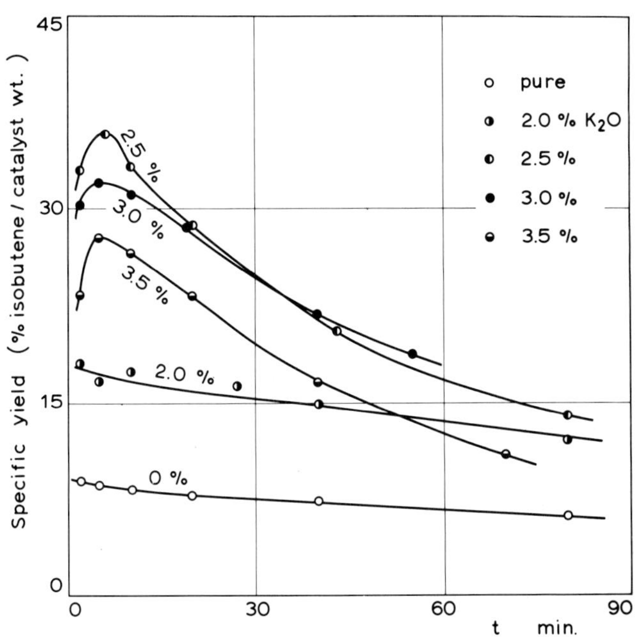

Fig. 2 Specific yield for various K_2O contents (900°C series)

Two distinct phases are present in the series prepared at 1,050°C, namely a Θ-Al_2O_3 and an α-Al_2O_3 type phase. The Θ phase is present in all samples. The α phase is undetectable in the samples doped with K_2O, Rb_2O and Cs_2O, and present in all other samples, including pure Cr_2O_3-Al_2O_3. It thus appears that K_2O, Rb_2O and Cs_2O somehow inhibit the formation of the α phase.

3.3 E.P.R.

The δ E.P.R. line is present in the spectra of all samples, whatever their composition and preparation temperature. Its intensity is practically independant of the nature of the additives.

The γ line is present in all samples prepared at 900°C. Its intensity is quite variable from one sample to another. Since this line is caused by Cr^{5+} ions, (which were shown to play no role in the dehydrogenation processes, but to be only responsible for some initial oxidation of the hydrocarbon[1]), no more attention was given to it. The γ line is hardly detectable in the series prepared at 1,050°C.

All the samples prepared at 900°C are characterized by the presence of a β_W and the absence of any β_N line. The intensity of the β_W line does not vary significantly with the composition of the samples.

The characteristics of the β_N and β_W lines observed in the 1,050°C series are indicated in table 4. The samples clearly fall into two distinct categories. The Li_2O, Na_2O or MgO doped, as well as the pure samples, exhibit a strong β_N line and a very faint, or sometimes undetectable, β_W line. Conversely, for the samples with K_2O, Rb_2O or Cs_2O, the β_W line is very intense, whereas the β_N line is undetectable.

TABLE 4
E.P.R. data for the 1,050°C series : β_W and β_N lines

Additive (Z = 2.5%)	β_W		β_N	
	I (arbitr. units)	ΔH gauss	I (arbitr. units)	ΔH gauss
None	Non detect.	Non detect.	15	500
Li_2O	Non detect.	Non detect.	21	460
Na_2O	Non detect.	Non detect.	15	400
K_2O	29	950	Faint	∿ 400
Rb_2O	36	1,070	Non detect.	Non detect.
Sr_2O	28	1,000	Non detect.	Non detect.
MgO	Non detect.	Non detect.	22	300

4. DISCUSSION

First, it must be stressed that the catalytic behaviour of our samples is quite similar to that of usual, alumina supported, catalysts. Results[1] previously obtained with some industrial catalysts are presented in table 5, and can be compared with those in table 2. Our samples are pure active phases whereas, in industrial catalysts, the active phase is supported and inert material probably contribute to a high amount to the total surface area. As could be expected, intrinsic activities are accordingly higher for our catalysts. Stabilities have quite comparable values. The strong similarity in composition and properties enables us to compare our results with those obtained by various authors on catalysts which had been prepared differently, and to suggest that our conclusions are not restricted to our samples.

TABLE 5
Initial intrinsic activities and stabilities of three industrial catalysts

Catalysts	Initial intrinsic activities (m^{-2})	Stability $t_{1/3}$ (min)
GIRDLER G 41	0.017	44
HARSHAW Cr 1404 P	0.081	> 85
PROCATALYSE ACR 3	0.079	36

The deactivation of chromia-amumina catalysts during a dehydrogenation run has sometimes been attributed to surface acidity (see, for example, ref.[18]), i.e. this acidity would cause the polymerization of the olefin in its transient adsorbed state and, hence, the formation of carbonaceous products progressively covering the surface. A very simple hypothesis concerning the effect of alkaline additives was that they neutralized this surface acidity, with the net result of an overall increase in activity. Our findings suggest that this factor is not relevant. In fact, sodium and potassium, which possess similar basicities and which form very similar salts with the aluminate or the chromite ions, behave quite differently with respect to activity. Moreover, it seems that the samples with effective promoters, i.e. K_2O, Rb_2O and Cs_2O, deactivate more rapidly than those of low activity.

Our results suggest that the determining factor is in relation to the structural properties of the Al_2O_3-Cr_2O_3 oxide. The more important results of our work are summerized in fig.3, where the intensity of the various effects is represented diagrammatically. This figure indicates a clear parallelism, i.e. the addition of potassium, rubidium and cesium induces the following effects simultaneously : a) increase of specific surface area; b) increase of intrinsic activity; c) disappearance of the E.P.R. β_N line; d) disappearance of the α phase.

It is known that progressively higher recrystallization temperatures induce the formation of the alumina phases in the following order : $\gamma \rightarrow \theta \rightarrow \alpha$. Our results give the same conclusion concerning Al_2O_3 phases. The presence of K_2O, Rb_2O or Cs_2O seems to inhibit the successive recrystallizations. The higher specific surface areas of the samples with these additives also suggest that the structural rearrangements are hindered[17]. It

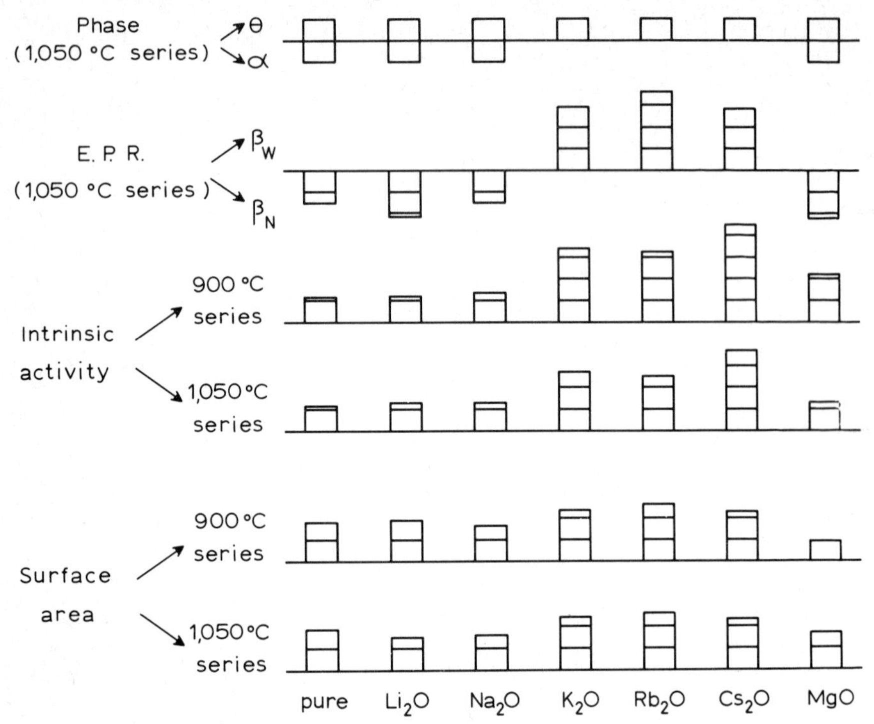

Fig. 3 Correlation between the various properties of the pure and doped Al_2O_3-Cr_2O_3 oxides.

is striking that these additives simultaneously increase the available surface area and the activity per unit area. A common explanation to these effects must be sought in slight changes in the surface state of the oxide, which modify its reactivity in recrystallization and catalysis.

Incidentally, it should be mentioned that our results support those of C.P. Poole and D.S. Mac Yver [15] concerning the relation between the β_N line and the α phases. It is more difficult to attribute the β_W line to some specific crystallized or amorphous phase. We observed it, not only when γ or Θ phases could be detected, but also in completely amorphous phases (when calcined below 900°C).

The mechanism of the various reactions involved in dehydrogenation over chromia-alumina catalysts can be discussed in the light of the above results. The reactions caused by Cr^{5+} or Cr^{2+} will not be reexamined, because the present results do not change any previous conclusion [1]. The important facts concern the reactions catalyzed by Cr^{3+}. It was concluded previously [1] that Cr^{3+}, probably when in a proper matrix environment, is the normal dehydrogenating agent. Moreover, it was strongly suspected that the "normal" slow deactivation is caused by this same ion, possibly in pairs. The present results support this view. Activity results indicate that the surface state of the oxides containing K_2O, Rb_2O or Cs_2O favors the formation of a greater number of active Cr^{3+} ions. As a consequence, more deactivation sites are expected to be present. This explains the more rapid deactivation observed in the active samples. In conformity with the opinion of M.M. Shendrik, G.K. Boreskov and L.V. Kirilyuk [19) 20)] it follows that the formation of carbonaceous products is a reaction parallel to dehydrogenation.

ACKNOWLEDGEMENTS

We are greatly endebted to Prof. Dr. J.M. Bonnier, Prof. Dr. Ph. Traynard and Dr. J. Nechtschein for stimulating discussions. Physical measurements were made possible by the kind permission of Dr. P. Servoz-Gavin and help of Dr. M. Caillet and Mrs. M. Archiprêtre. The financial support of the I.F.P. is gratefully acknowledged.

REFERENCES

1) C. Marcilly, B. Delmon, J. Catal., in press.
2) J.M. Bridges, G.T. Rymer, D.S. Mac Yver, J. Phys. Chem., 66 (1962), 871.
3) A.F. Gremillon, W.R. Knox, J. Catal., 1 (1962), 216.
4) O.D. Sterligov, A.D. Belen'kaya, Izvest. Akad. Nauk S.S.S.R., (1962, N°5), 800.
5) I.V. Nicolescu, M. Gruia, A. Papia, V. Dumitrescu, Rev. Roumaine Chim., 10 (1965), 523.
6) I.V. Nicolescu, M. Spinzi, M. Gruia, A. Papia, V. Dumitrescu, M.G. Vieweg, S. Nowak, H.G. Konnecke, Rev. Roumaine Chim., 11 (1966), 363.
7) B.A. Dadashev, I.G. Sultanova, A.A. Sarydzhanov, S.A. Kasimova, N.R. Abdurakhmanova, Azerb. Khim. Zh., 3 (1969), 17.
8) B.A. Dadashev, A.A. Sarydzhanov, U.S. Gadzhi Kasumov, Azerb. Neft. Khoz., 1 (1970), 36.
9) O.D. Sterligov, N.A. Eliseev, A.P. Belen'kaya, Kin. i Kat., 8 (1967), 141.
10) S. Carra, L. Forni, C. Vintani, J. Catal., 9 (1967), 154.
11) A.M. Rubinstein, N.A. Pribitkova, C.A. Afanasiev, A.A. Slinkin, Kin. i Kat., 1 (1960), 129.
12) A.M. Rubinstein, N.A. Pribitkova, V.M. Akimov, L.D. Kretalova, A.L. Klyachko-Gurvich, Izvest. Akad. Nauk S.S.S.R., (1961, N°9), 1552.
13) D. Cismaru, D. Manolescu, Rev. Roumaine Chim., 15 (1970), 291.
14) C.P. Poole, W.L. Kehl, D.S. Mac Yver, J. Catal., 1 (1962), 407.
15) C.P. Poole, D.S. Mac Yver, Adv. in catalysis, 17 (1967), 223.

16) J. Pâris, R. Pâris, Bull. Soc. Chim. Fr., (1965), 1138.
17) C. Marcilly, B. Delmon, Bull. Soc. Chim. Fr., (1970), 446.
18) R.H. Griffith, J.D.F. Marsh, Contact Catalysis, Oxford Univ. Press, London, 1957, p. 176.
19) M.M. Shendrik, G.K. Boreskov, L.V. Kirilyuk, Kin. i Kat., 6 (1965), 313.
20) M.M. Shendrik, G.K. Boreskov, L.V. Kirilyuk, Kin. i Kat., 8 (1967), 79.

DISCUSSION

J. HAPPEL

Do the authors have any information as to whether the materials in Class I which have little effect are related to the size of the ionic radii of Li, Na or Mg which are small compared with those of K, Rb and Cs? Thus, would one expect that alkaline earth elements like strontium and barium would be effective in view of their larger ionic radii?

		Ionic Radius Å
Class I	Li	0.60
	Na	0.95
	Mg	0.65
Class II	K	1.33
	Rb	1.48
	Cs	1.69
	Sr	1.13
	Ba	1.35

B. DELMON

Parts of our current investigations are intended to ascertain the role of the size of the ionic radii. Our information, at the present time, is insufficient for giving a conclusive answer. The expectation that some alkline earth elements would be effective in view of their larger ionic radii is substantiated in the case of barium, but not strontium.

H. PINES

The use of isobutane as a model compound will not permit one to differentiate the dramatic effect that traces of alkali metals can have on the course of a catalytic reaction. We have the following differences by adding small amounts of Na^+ and K^+ to Cr_2O_3-Al_2O_3

(O) + C=C(C)–C $\xleftarrow{\text{without}}$ (O)–C(C)(C)–C $\xrightarrow{K^+}$ (O)–C=C(C)–C

(O)(CH_3)–CH_3 $\leftarrow$ (O)(CH_3)(CH_3) $\xrightarrow{K^+}$ (O)–CH_3 + CH_4

Models different from isobutane studied in the paper could probably throw more light on the difference between the various chromia-alumina catalysts.

B. DELMON

We thank Professor H. Pines very much for his remarks. The model reactions he mentions are acid catalyzed reactions and, therefore, are necessarily sensitive to the addition of any alkali metal to the catalyst. The purpose of our paper was mainly to explain why the effect of, e.g., sodium or potassium on dehydrogenation catalysts is different. One conclusion is that deactivation by coking does not seem to be related to the acidity of the catalyst. Another conclusion is that the effectiveness of additives in dehydrogenation catalysts does not depend on some neturalization of the surface acidity of the Cr_2O_3 - Al_2O_3 solid solutions, but on structural changes these additives promote or inhibit in these solid solutions.

J. RABO

In the doping of an oxide by a cation of substantially lower valence, in most cases the guest cation floods the surface of the half oxide crystal. Do you have information on how much surface chromium ion was left after the doping process?

Suggestion: the doping cations which aided activity (K, Rb, Cs) are large and of low valence, and consequently their mobility on the surface of the chromium oxide is more facile than that of the smaller cations.

B. DELMON

We agree with Dr. J. Rabo that the presence of alkali metal ions might change the surface composition of the catalyst and this, either by changing the relative proportion in the bulk and at the surface or by some covering of the surface by alkali metal ions. We have no information concerning these changes.

However, we would like to remark that the dehydrogenation processes are obviously dependent on the presence of chromium ions. The intrinsic activity of the catalyst either remains unchanged (Li, Na, Mg), or increases (K, Rb, Cs) after doping. All explanations must therefore take into account the fact that some chromium ions necessarily remain at the surface or close to the surface. But this remark does not preclude that the total number of surface chromium ions (i.e., active and inactive) could decrease upon doping, provided the number of active ions remains the same or increases. The active ions might be the ions in a certain type of site (e.g., tetrahedral).

M. M. BHASIN

Have you done quantitative ESR work to show that surface Cr^{+4} and Cr^{+6} are not present on the catalyst? These Cr species can be stabilized in the presence of alkali and alkaline earth metal ions and can contribute to the dehydrogenation reaction.

B. DELMON

Neither Cr^{4+} nor Cr^{6+} are directly detected by ESR.

Our former results show that neither Cr^{6+} nor Cr^{5+} are the active species in dehydrogenation. The normal dehydrogenating agent is Cr^{3+}.[1] It has, however, been supposed that Cr^{3+} pairs might actually be Cr^{2+} - Cr^{4+} pairs.[2] In this sense, Cr^{4+} might be considered as an active species.

The suggestion of Dr. M. M. Bhasin that the promoter action might be explained by the stabilizing action of alkali or alkaline earth metals on Cr^{4+} or Cr^{6+} ions, is therefore to be understood a little differently. Concerning Cr^{4+}, one could assume a stabilization of the Cr^{2+} - Cr^{4+} form. Concerning all the higher valency ions, including Cr^{4+}, one could alternatively hypothesize that they are necessary precursors of the active Cr^{3+} ions, i.e., that the active Cr^{3+} ions may only be obtained by reduction of these higher valency ions.

1) C. Marcilly, B. Delmon, J. Catal., 24 (1972) 336.
2) F. S. Stone, Chimi, 23 (1969) 490.

J. R. OWEN

If intrinsic activity is plotted versus stability, the data for both the 900°C and 1050°C series of catalysts (except for a couple of points) fall on a smooth curve defining a regular decrease in stability with increase in activity. This appears to indicate that catalyst deactivation (coke formation) is a series rather than a parallel reaction with dehydrogenation--i.e., the formation of more olefins simply leads to the formation of more coke. It should also be pointed out, for possible correlation purposes, that the authors' class II metals have a lower work function, a considerably lower ionization energy, and a considerably higher ionic radius than their class I metals.

B. DELMON

We agree with Dr. J. R. Owen that there is a regular decrease in stability with increase in activity. This correlation, by itself, neither proves nor disproves that coking and dehydrogenation are parallel or series reactions. Such a conclusion would actually necessitate some assumption as to, either, the relative action of the promoter on the parallel reactions, or the influence of olefin surface concentration on the polymerization rate.

Coking has been thought to be similar, in its mechanism, to other polymerizations on solid catalysts. In this respect, it was considered as a series reaction, with a mechanism completely distinct from dehydrogenation, namely catalyzed by surface acidity. Our results only suggest that surface acidity is not the determining factor. The simplest remaining hypothesis to explain the parallel variation of dehydrogenation and coking rates is that these reactions are twin reactions. But we agree with Dr. J. R. Owen that more sophisticated hypotheses could also be considered.

Concerning the other suggestions by Dr. J. R. Owen, the influence of the ionic radius size has already been discussed in our answer to Dr. J. Happel. At the present time, we cannot make any comment on the possible influence of the ionization energy of the additives.

Paper Number 7

ON THE EFFECT OF REGENERATION AND CARRIER ON THE DISPERSION OF Cr_2O_3 IN Al_2O_3-Cr_2O_3-K_2O DEHYDROCYCLIZATION CATALYSTS

H. BREMER, J. MUCHE and M. WILDE[1)]
Section of Process Chemistry, Technische Hochschule für Chemie "Carl Schorlemmer" Leuna-Merseburg, Merseburg, German Democratic Republic

ABSTRACT: The dispersion of Cr_2O_3 on various Al_2O_3-Cr_2O_3-K_2O catalysts for dehydrocyclization (DHG) made by impregnation is compared by means of optical reflectance spectroscopy, EPR-measurements and oxygen chemisortpion. High surface area alumina carriers are necessary for high dispersion of Cr_2O_3.

Studies of the aging and regeneration of these catalysts (independent of the "coke" deposition and burning) showed that the regeneration of aged catalysts by oxidation and subsequent reduction under mild conditions results in a rearrangement of Cr_2O_3 that causes enlargement of the Cr_2O_3 surface. The aging in flowing hydrogen at relatively low temperatures leads to an enlargement of the Cr_2O_3 clusters which results in reduced catalytic activity in the DHC of n-hexane. This process is reversible; the aging can be cancelled by regeneration. At temperatures above 650°C, rapid irreversible aging begins due to the formation of solid Cr_2O_3-Al_2O_3 solutions. This diminishes the oxidizability of chromia and thus decreases its capability of being regenerated.

1. INTRODUCTION

Due to the chemical industry's increasing demand for aromatics, the technical application of the catalytic DHC of paraffins is receiving an ever increasing interest. During the last two decades, numerous papers on the DHC of Al_2O_3-Cr_2O_3 catalysts (e.g.[2)]) and their physical-chemical properties ([3)]) have been published. The dehydrogenation of the paraffin to olefin[4,5)] was found to be the rate controlling step of the DHC. This reaction takes place on the chromia, and Cr^{2+} [6-9)] or Cr^{3+} ions[10,11)] are assumed to be the active centers. Therefore the size of the Cr_2O_3 surface should be of decisive importance for a high catalytic DHC or dehydrogenation activity of Al_2O_3-Cr_2O_3 catalysts. While investigating the influence of various carriers and regeneration methods on the properties of Al_2O_3-Cr_2O_3-K_2O catalysts, we consequently gave special attention to the characterization of the samples as regards the dispersion of the chromia contained in them.

In order to investigate the aging and regeneration independent of "coke" deposition and burning, we simulated the aging by calcining in oxygen followed by reduction in hydrogen flow under relatively mild conditions. In this paper, the terms "aging" and "regeneration" refer exclusively to the above mentioned processes.

2. EXPERIMENTAL

2.1 Preparation of the catalysts

2.1.1 Preparation of the catalysts on different Al_2O_3 carriers

These catalysts contained 16.6% Cr_2O_3 and 3.7% K_2O. They were made by impregnating various aluminas with CrO_3 and $K_2Cr_2O_7$; pretreatment included 24 hour drying at 110°C, 5 hour glowing in air flow at 600°C and 10 hour glowing at 600°C in hydrogen flow.

2.1.2 Preparation of aged and regenerated samples

To study the aging and regeneration, we simulated the aging of catalyst 6450 from VEK Leuna-Werke "Walter Ulbricht" by 5 hour glowing in hydrogen flow at temperatures between 300°C and 900°C. This catalyst contained 20% Cr_2O_3 and 4% K_2O on the commercial γ-Al_2O_3 carrier Leuna-5784. The aged samples are called L 300 to L 900 according to the aging temperature.

We simulated the regeneration of the aged catalyst samples by oxidation for 5 hours at 570°C in oxygen flow, followed by 1 hour reduction in hydrogen flow at 500°C. These samples are referred to as L 600 R to L 900 R.

2.2 Optical reflectance spectroscopy

The measurements were performed by means of a VSU 2 with reflection attachment O/R (VEB "Carl Zeiss" Jena) with the specific alumina carriers serving as white standards for each catalyst. The samples were used as powders (grain size < 0.125 mm), and the measurements were carried out in air.

2.3 EPR

The EPR signals of the β-phase[12] were taken (in the x-band) with an ER-9 from VEB "Carl Zeiss" Jena. Two series of preparations were made. The "evacuated samples" were post-reduced with hydrogen at 500°C and evacuated at this temperature for 1 hour at 10^{-3} torr. The samples were then stored in sealed Rasotherm tubes. The "water-treated" preparations were initially treated as described above. After evacuation at 500°C, they were exposed to saturated steam for one hour at room temperature before the tubes were sealed.

The relative spin intensities were determined according to

$$I_{rel} \sim \Delta H_{max}^2 h$$

ΔH_{max} peak-to-peak width of the first derivative of the signal

h peak-to-peak height of the first derivative of the signal

2.4 Oxygen chemisorption

Oxygen chemisorption was used to determine the Cr_2O_3 surface area as described by Bridges, Mac Iver and Tobin.[13]

The reference sample was pure Cr_2O_3 whose surface area was determined from a nitrogen isotherm (S_{BET} = 37 m^2/g).

2.5 Oxidizability

The average degree of oxidation was determined iodometrically after 5 hour glowing in oxygen flow at 570°C. The iodometric determination according to Weller and Voltz[14] was modified by us in that all procedures were carried out in an atmosphere of purest nitrogen.

2.6 Catalytic activity in the DHC of n-hexane

The catalytic activity was tested in a pulse microreactor (catalyst weight 1 g, 10µl n-hexane per pulse, purest nitrogen as carrier gas). The benzene yields (with reference to the liquid product) were determined at 570°C. It had previously been found that the sequence of the activities of various catalysts was the same in the pulse micro system as in the flow system.

3. RESULTS AND DISCUSSION

3.1 Effect of the carrier on the dispersion of Cr_2O_3

Table 1 shows a number of properties of the catalysts produced on the basis of various kinds of carriers.

Our EPR-measurements were carried out at 26°C, which is below the Néel temperature of α-Cr_2O_3 (approximately 40°C). Due to the dependence of Néel temperature on particle size,[15] the relative spin intensity of the β-phase resonance was expected to be smaller in the presence of larger Cr_2O_3 particles --because of their antiferromagnetic behavior--than in the case of very small particles.[16] Actually, the 3 samples with

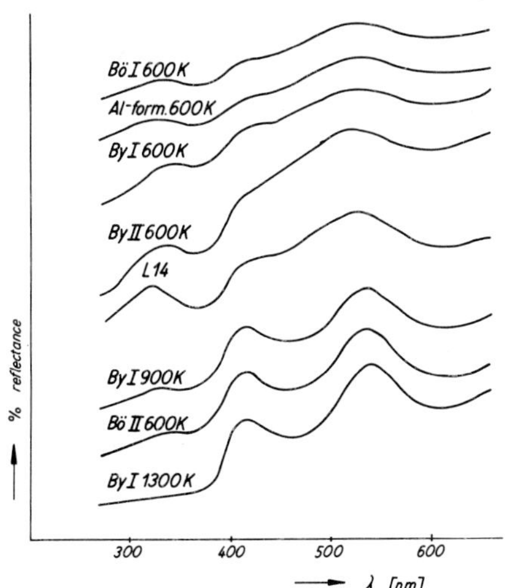

Fig. 1 Reflectance spectra of the catalysts produced on the basis of various Al_2O_3 carriers

a clearly lower relative spin intensity value (By I 1300 K, Bo II 600 K, By I 900 K) were the only ones showing distinct α-Cr_2O_3 X-ray interferences. As can be seen in Fig. 1, these 3 catalysts show far better structured reflectance spectra than the others. This also suggests a more distinct crystallinity of Cr_2O_3 in these samples.

TABLE 1
Physical-chemical properties of the catalysts with various Al_2O_3 carriers

Catalyst	Carrier	S_{BET} (carr.) m^2/g	S_{BET} (catal.) m^2/g	$S_{Cr_2O_3}$ m^2/g	I_{rel}	% benzene in liquid product
L14 (Leuna)	γ-Al_2O_3	222	183	21.0	0.89	29.6
BoI 600 K	γ-Al_2O_3	221	148	12.6	0.82	38.7
BoII600 K	γ-Al_2O_3	82	35	3.0	0.49	7.4
ByI 600 K	η-Al_2O_3	229	155	16.8	0.78	48.8
ByII600 K	η-Al_2O_3	227	149	18.0	0.81	72.5
ByI 900 K	θ-Al_2O_3	103	69	9.0	0.56	5.0
ByI1300 K	α-Al_2O_3		10	0	0.10	0
Al-form 600 K	amorph. Al_2O_3	390	105	4.7	0.86	8.9

The low dispersion of the Cr_2O_3 in the catalysts By I 1300 K, Bo II 600 K and By I 900 K is due to the small surface area of the carrier. Of the remaining 5 catalysts of this series, the samples By II 600 K, By I 600 K, Bo I 600 K and L 14 are characterized by the largest Cr_2O_3 surfaces and the highest catalytic activities. Also, the catalyst Al-form 600 K still contains disperse Cr_2O_3 whose acessibility is limited due to its deposition in ink bottle pores and the blocking of the entrance to these pores.[1)]

It is thus apparent that large surface area carriers are of great importance for a high dispersion of Cr_2O_3 on Al_2O_3. Nevertheless, differences in catalytic activity do occur even if the Cr_2O_3 surface areas are almost equal (see Table 1), although this point will not be discussed here.

3.2 Aging and regeneration

The importance of dispersion of the chromia for the catalytic activity in DHC is also reflected by the investigations of aged and regenerated samples (series L and LR).

Fig. 2 illustrates by the example of the catalysts L 750 and L750 R, that regenerated samples have more poorly structured reflection spectra than the corresponding aged samples and thus contain more highly dispersed Cr_2O_3.

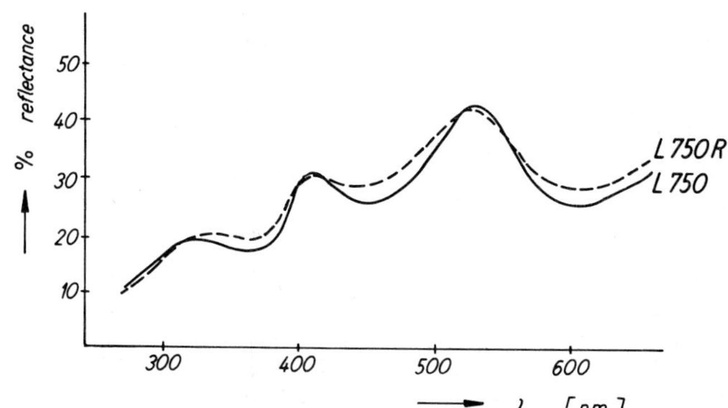

Fig. 2 Reflectance spectra of the catalysts L 750 and L 750 R

In order to compare all samples of the series L and LR at once, we introduce the ratio of the Kubelka-Munk function $F(R_\infty)$ of the absorption maximum at 600 nm and $F(R_\infty)$ of the minimum at 530 nm as a measure of the structural characteristics of the spectra. The long-wave part of the spectrum was chosen because it should not be disturbed by charge transfer transitions. The results are shown in Fig. 3. According to this figure the dispersion of Cr_2O_3 in the aged samples (L-series) should decrease at pretreatment temperatures above 550°C, and the catalysts of the LR-series should contain more disperse Cr_2O_3 than the aged samples. As the increasing ratios for the catalysts of the LR series show, the effectiveness of regeneration (recovery of higher dispersion) drops with increasing pretreatment temperature.

The conclusions drawn from comparison of the reflectance spectra were confirmed by the EPR-investigations. Shvets and Kazanskii[10] having compared the β-phase line width of water-treated samples and evacuated ones found a lower width for water-treated samples due to additional indirect exchange interactions. The extent of this effect should depend on the size of the chromia surface. Table 2 lists the β-phase line widths for the catalysts L 600, L 900, L 600 R and L 900 R.

Greater differences in the line width were measured with the samples of the LR-series than with the catalysts of the L-series; regenerated samples should therefore have a larger Cr_2O_3 surface than aged ones. Both with aged and regenerated samples, the difference in the line width is reduced when the aging temperature is increased.

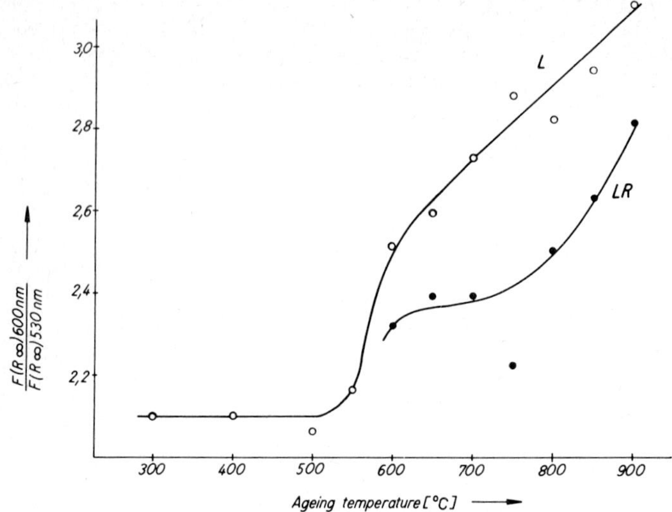

Fig. 3 $\frac{F(R_\infty)\ 600\ nm}{F(R_\infty)\ 530\ nm}$ of the catalysts of the L and LR series

TABLE 2
β-phase line widths (EPR) of evacuated and of water-treated samples

Catalyst	ΔH(ev.) Gauss	ΔH (H_2O) Gauss	ΔH(ev.) - ΔH (H_2O) Gauss
L 600	940	820	120
L 900	1000	960	40
L 600 R	980	800	180
L 900 R	960	880	80

A comparison of the Cr_2O_3 surface areas of the aged (L) and the regenerated samples (LR) shows greater values in the case of the regenerated ones (Fig. 4), stressing again that the dispersion of Cr_2O_3 is increased due to regeneration. This effect is also shown by the BET surfaces.

Regeneration of aged samples by oxidation and subsequent reduction under mild conditions evidently results in a rearrangement of the chromia, by which a larger chromia surface is obtained. As Fig. 5 shows, this fact is also reflected by the results of the catayltic studies.

To interpret the different dependences of the catalytic

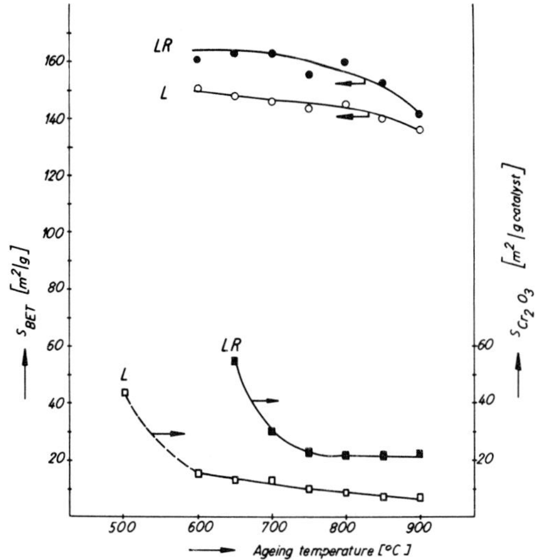

Fig. 4 The surface areas according to BET and the Cr_2O_3 surface areas of the catalysts of the L and LR series

activity on the aging temperature in the L- and the LR-series, the process of aging must be explained in somewhat greater detail.

Above 500°C, and to an even larger extent above 550°C, reversible aging, which is due to the enlargement of the Cr_2O_3 clusters, begins and results in a direct reduction in catalytic activity; however, these effects can be removed by regeneration. The process of reversible aging is essentially complete at 750°C.

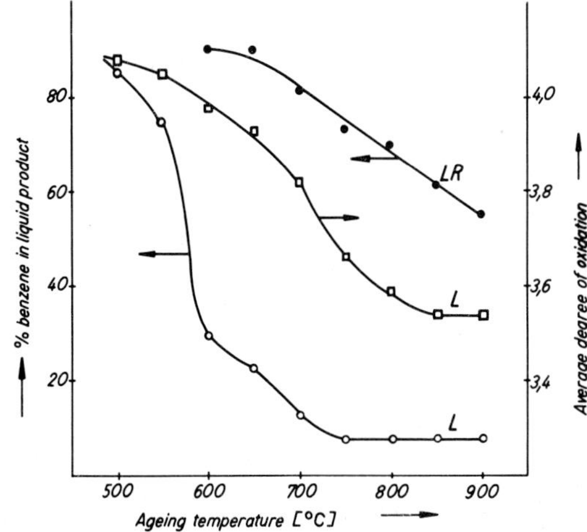

Fig. 5 The catalytic activity in the DHC of n-hexane and the average degree of oxidation of chromium measured on the catalysts of the L and LR series

Above 650°C an irreversible aging process has a large effect on the capability of the catalysts to be regenerated. We explain this process by the formation of solid Cr_2O_3-Al_2O_3 solutions. This is suggested by the widening of the β-phase lines due to the Al^{3+} ions penetrating into the Cr_2O_3 clusters, which reduces the interactions causing the exchange narrowing. The decrease in the oxidizability of the chromia (Fig. 5), beginning at temperatures in excess of 650°C, also confirms this concept.

The irreversible aging has hardly any effect on the catalytic activity of the aged samples because the surface area of the Cr_2O_3 clusters is not further reduced. However, due to the impairment of the oxidizability, the irreversible aging results in decreased regeneration capacity because the regeneration requires the oxidative decomposition of large Cr_2O_3 clusters.

REFERENCES

1) J. Muche and M. Wilde, Thesis, Merseburg, 1971.
2) S. Nowak and H.-G. Vieweg, Chem. Tech. (Leipzig) 22 (1970), 94.
3) C. P. Poole, Jr., and D. S. Mac Iver, Advan. Catal. 17 (1967), 223.
4) G. M. Panchenkov, K. A. Venkatachalam, and Yu. M. Zhorov, Neftekhimiya 4 (1964), 30.
5) Yu. M. Zhorov, K. A. Venkatachalam, and G. M. Panchenkov, Neftekhimiya 6 (1966), 831.
6) L. L. van Reijen, W. M. H. Sachtler, P. Cossee, and D. M. Brouwer, Proc. Intern. Congr. Catalysis, 3rd, Amsterdam, 1964 2, 829.
7) K. I. Slovetskaya, T. R. Brueva, and A. M. Rubinshtein, Kinetika i Kataliz 7 (1966), 566.
8) K. I. Slovetskaya and A. M. Rubinshtein, Kinetika i Kataliz 7 (1966), 342.
9) K. I. Slovetskaya, K. M. Gitis, R. V. Dmitriev, and A. M. Rubinstein, Izv. Akad. Nauk SSSR, Ser. Khim. 1969, 196.
10) V. A. Shvets and V. B. Kazanskii, Kinetika i Kataliz 7 (1966), 712.
11) P. Boutry, R. Montarnal, and C. Blejean, Bull. Soc. Chim. Fr. 1967, 3690.
12) D. M. O'Reilly and D. S. MacIver, J. Phys. Chem. 66 (1962), 276.
13) J. M. Bridges, D. S. Mac Iver, and H. H. Tobin, Actes intern. congr. catalyse, 2e, Paris, 1960 2, 2161.
14) S. W. Weller and S. E. Voltz, J. Am. Chem. Soc. 76 (1954), 4695.
15) C. P. Poole, Jr., and J. F. Itzel, Jr., J. Chem. Phs. 51 (1964), 287.
16) Yu. N. Molin, V. M. Chibrikin, V. A. Shabalkin, and V. F. Shuvalov, Zavodsk. Lab. 32 (1966), 933.

DISCUSSION

B. DELMON

The results of H. Bremer, J. Muche and M. Wilde concern dehydrocyclization. Nevertheless, I am bothered by discrepancies with our own results on dehydrogenation, because the experimental conditions in both researches are similar and because the dehydrogenation is obviously a very important step in dehydrocyclization.

We postulate that solid solutions with transition alumina structures are the true dehydrogenating phases, while Professor H. Bremer attributes the irreversible aging to the formation of solid solutions!

Maybe I could try an explanation. Professor H. Bremer's samples have a high chromia content (i.e., more than 16%). I hypothesize that only part of the chromia is in strong interaction with the γ or η alumina carrier, where it possibly forms solid solutions with high dehydrogenating activities. But there remains much uncombined chromia, which retains the α structure. In my opinion, this α phase has very low catalytic activity. This hypothesis seems to be substantiated by Table 1 in the printed communication and by the indications on the same page, where it is stated that the samples where the α-Cr_2O_3 phase is detected, namely Bö II 600 K, By I 900 K and By I 1300 K have remarkably low catalytic activities. I emphasize that this is not entirely due to the low specific surface area, as proven by the comparison between samples Bö I 600 K (which is active) and By 900 K (which is inactive).

In my opinion, only the free chromia spontaneously recrystallizes upon aging. Our results show that solid solutions of γ or η alumina type are extremely stable, even at 770°C.[1] In addition, there might be the possibility of excess free chromia nucleating the segregation of pure chromia out of mixed oxides.

In addition to these remarks, I would like to add that the aging process adopted by Professor H. Bremer is too severe to be really representative of the aging in industry-like conditions.

Another remark is that the amount of potassium (in proportion to the total possible quantity of γ or η solid solutions) in Professor H. Bremer's samples probably exceeds the optimum value and does not prevent the recrystallization to α-phases as efficiently as it would have done at the level of, say, 2%.

1) C. Marcilly, B. Delmon, J. Catal. 24, 336 (1972).

H. BREMER

With respect to the problem of the specific catalytic activities, we will state our ideas in the answer to the comment of Dr. Owen.

In reply to the question for the industry-like conditions of the aging process adopted by us, we mention that, by investigating the aging process, we received results analogous to those published here with samples heated at only 600°C for varying length of time.

The chemical composition of the catalysts is identical with some commercial ones. The choice of the K_2O content is a problem of optimization:

In addition to the influence on activity and aging one must consider the influence of the K_2O on regenerability and duration of the formation period.

S. K. BHATTACHARYYA

The catalytic activity of chromium oxide and chromia - chromia catalysts depend considerably on the methods of preparation and their pretreatment. We have done considerable work on the thermal characteristics of the oxides by thermal analysis as well as by X-ray diffraction studies. The Cr_2O_3 gels prepared from the nitrates show an endothermic peak by thermal analysis technique ranging up to about 300°C and an exothermic peak at 395°C. This exothermic peak is due to crystallization which does not appear in absence of air. The catalytic action is found to be maximum at a temperature of about 300°C when the dehydration is complete and the Cr_2O_3 gets the maximum surface area if this pretreatment of the oxide is done at about 300°C. The X-ray diffraction study shows that Cr_2O_3 remains amorphous up to this temperature.

In the case of Cr_2O_3-Al_2O_3 system, the presence of Al_2O_3 either retards or inhibits the crystallization of Cr_2O_3. We have found that there are two compositions 29% Cr_2O_3 and 70% Cr_2O_3 where the catalytic activity for some dehydrogenation and dehydration reaction is the maximum. One composition involves structural change of Cr_2O_3 (δ-Cr_2O_3) and this other corresponds to maximum hydrogen chemisorption. The author has not mentioned the role of K_2O in the activity of the Cr_2O_3-Al_2O_3-K_2O catalyst and also in the change of structure of Cr_2O_3-Al_2O_3 system.

H. BREMER

We have not varied the K_2O content and not especially investigated its influence. However, the following detection is interesting: Some of the catalysts had an extremely low oxidizability. From these catalysts we could extract by means of water at 100°C essentially less K_2O than from the others. Evidently, the accessibility of the K_2O is an important prerequisite to the oxidizability and, consequently, to the regenerability of the catalysts.

J. R. OWEN

This paper is of particular interest to me because of the work I reported 25 years ago (J.A.C.S. 69, 2559 (1947)) on the relation between BET surface area(S) and activity(A) of a CrAl catalyst for butane dehydrogenation. In this connection a crossplot of the data in Figures 4 and 5 shows: (1) In the L series there is no increase in A with S (BET of Cr_2O_3) until S reaches a critical value (>145 m^2/g and >10 m^2/g, respectively) and then there is a rapid linear increase in A with S. (2) In the LR series there is a smooth parabolic relation between S_{BET} and A (as I found), and there is a large increase in A with very little increase in $S_{Cr_2O_3}$ followed by a small increase in A with a large increase in $S_{Cr_2O_3}$ and with almost no increase in S_{BET}. The point for $S_{Cr_2O_3}$ of L600R is not shown and would be of interest. I would appreciate the author's comments on these points and also on the considerable difference in activity of the apparently almost identical ByI 600 K and ByII 600 K catalysts.

I too found that the effectiveness of regeneration dropped with increasing treating temperature, probably because of solid solution formation. I also found that active catalysts had a titratable higher-valent chromium content of about 15 percent (as Cr^{+6}) following regeneration.

H. BREMER

The centers, active in DHC and dehydrogenation, contain chromium ions. The first prerequisite to high catalytic activity is therefore a high $S_{Cr_2O_3}$. This prerequisite is realized by using carriers of high surface area and by appropriate pretreatment (reduction at low temperatures, regeneration of aged catalysts). In our opinion the higher activity of the Cr_2O_3-rich α-solution in comparison with that of the Al_2O_3-rich γ-phase in the work of Professor Delmon can only be explained by a higher $S_{Cr_2O_3}$ (unfortunately, values of $S_{Cr_2O_3}$ are not given there). We consider the correlation of activity data with S_{BET} not to be very useful in the case of supported catalysts. If the Cr_2O_3 is distributed more finely on the surface, then an increase in $S_{Cr_2O_3}$ is connected with a decrease in $S_{Al_2O_3}$. Therefore $S_{BET} = S_{Cr_2O_3} + S_{Al_2O_3}$ does not give sure information on the dispersion of chromia, e.g., $S_{Cr_2O_3}$ of the L 650 R in comparison with that of the L 750 R is 32 m^2/g cat. higher, but S_{BET} is only 8 m^2/g cat. higher.

$S_{Cr_2O_3}$ is in the first approximation the activity-determining criterion; nevertheless, there are differences in the specific activities, i.e. the activities related to $S_{Cr_2O_3}$. In our opinion a strong interaction of chromium ions with γ- or η- Al_2O_3 is responsible for a high specific activity. This may be a solid solution, which case is technically less interesting for catalysts made by impregnation, because a lower $S_{Cr_2O_3}$ results from sintering. But a strong interaction of chromium ions with the carrier is also reached by high dispersion of Cr_2O_3 on the surface of the carrier. We want to discuss this conception in connection with an additional diagram.

In the range I the activity and $S_{Cr_2O_3}$ decrease by increasing of aging, down to catalyst L 750. In the case of the catalysts L 750 to L 900 the decreasing $S_{Cr_2O_3}$ in its influence on catalytic activity is compensated by a growing formation of solid solution. In the range II we also note a symbate behavior of $S_{Cr_2O_3}$ and activity. The catalysts with the greatest portion of finely distributed Cr_2O_3 (L 650 R, L 500, L 700 R) are the most active ones. The samples L 650 R and L 600 R are in principle identical, because L 600 as well as L 650 are completely regenerable. The relatively high activity of the catalysts L 750 R to L 900 R can be interpreted as follows: With advancing aging, that part of chromia increases, which is inaccessible to regeneration caused by the formation of a solid solution with alumina. The lower quantity of Cr_2O_3, rearranged on the surface by regeneration, is then characterized by a particularly strong interaction with the carrier surface. This interpretation explains, too, the discontinuity of the curve between L 600 and L 900 R: Both catalysts have the same $S_{Cr_2O_3}$. But on the sample L 900 R there exist smaller particles, since, in this case, a considerable part of the chromia has disappeared from the surface by diffusion into the interior of the carrier. Consequently, these particles show a stronger interaction with the carrier surface.

The cause for the differences in activity between By I 600 K and By II 600 K is to be searched in the different preparations of the carriers: The bayerite I is made by hydrolysis of aluminum iso-propylate, the bayerite II by reaction of aluminum with water.

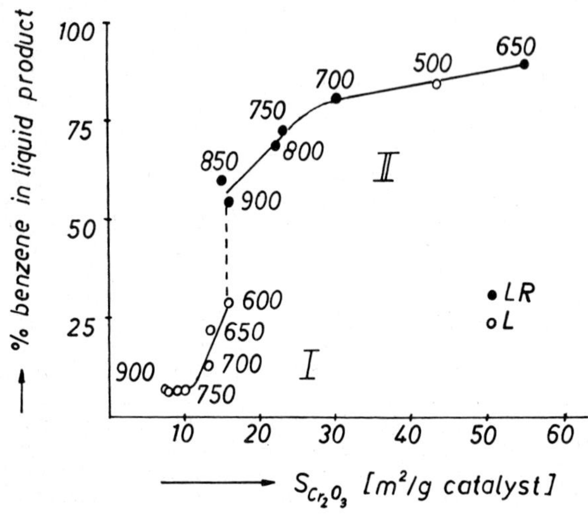

Fig. 1 Correlation between the catalytic activity in the DHC of n-hexane and $S_{Cr_2O_3}$

Paper Number 8

REACTIONS OF THE CATALYTIC SITE IN THE PHILLIPS ETHYLENE POLYMERIZATION SYSTEM

HANS L. KRAUSS
Department of Chemistry, Freie Universität Berlin
Berlin (West), German Federal Republic

ABSTRACT: The coordinatively unsaturated Cr(II) surface compound - the active site of the Phillips polyethylene process - undergoes different types of reactions, such as reoxidation, coordination and π-complex formation. Some of these reactions are described here in more detail: Cl_2 leads to a Cr(III) surface compound with Cr:Cl = 1:1; at higher temperature CrO_2Cl_2 and a 1:2 complex are formed. CO and N_2 form relatively unstable complexes with one and two ligand molecules respectively. Ethylene is linked to the metal most likely by a π-bond prior to polymerization and it uses the same sites for coordination as does carbon monoxide.

"Low polymer synthesis" allowed the use of conventional chemical analytical methods; it was shown, that hydrolytic cleavage of the Cr-carbon bond leads to α-olefins, but splitting by O_2 yields n-aldehydes. Further experiments led to the removal of the Cr(II) together with its alkyl ligands from the surface. Mass spectroscopic data of the complex suggest a Cr-carbon σ-bond.

1. INTRODUCTION

The Phillips process for heterogenous ethylene polymerization[1)] is a typical example of a class of reactions which use a compound of a transition metal supported on an inert oxide as catalyst. The doping is usually performed by impregnating the oxide (silica, alumina, etc.) with an aqueous solution of a suitable compound of a transition metal in high oxidation state like chromate(VI) or molybdate(VI)[2]. An important step in catalyst preparation is the subsequent heat treatment in an oxidizing atmosphere during which the metal compound is chemically linked to the support, forming there a surface compound[3].

Although these steps seem to be quite simple, the composition of the catalyst, and particularly its mode of action, offer many unsolved and challenging questions: in the most thoroughly investigated chromium/silica system for ethylene polymerization the catalytic center was regarded for a long time as chromium(V)[4] in various stereochemical environments[5]; only recently it has been shown[6,7] that the active species actually contains Chromium in the +2 oxidation state formed either in the first reaction step with the monomeric olefin or in a separate reduction step. The formal equation

$$\begin{matrix}-O & & O \\ & \searrow Cr \nearrow & \\ -O & \swarrow \quad \searrow & O\end{matrix} \xrightarrow{\text{red.}} \begin{matrix}-O & \\ & \searrow Cr \\ -O & \nearrow\end{matrix}$$

suggests that the resulting product may exhibit a highly unsaturated coordination sphere. After removal of the oxygen ligands, it should show almost the behaviour and the reactivity of a "free surface". Indeed the reduced surface compound seeks stabilization by adding further ligands of different kinds which may be olefins as in the (likely) first step of polymerization[7], or other donor and acceptor molecules. Most remarcable is the reaction with oxygen, which occurs quite vigorously with orange luminescence[7]: the equilibrium pressure of oxygen was measured to be 10^{-27}atm[8].

Some of these reactions will be dealt with in this paper in more detail and lead to a better understanding of the behaviour of the active surface compound. Finally, experiments will be described which involve the removal of the metal center together with a grown hydrocarbon chain from the surface, thus allowing the use of standard chemical methods of investigation.

2. EXPERIMENTAL

2.1 Preparation and analysis of the "free" Cr(II) catalyst.

The following supports were mostly employed: Kieselgel "Merck" 7733; Aluminosilicate "Davison" with 13 % Al_2O_3, 77 % SiO_2; Aluminium oxide "Merck" 1077 (henceforth designated as Silica, Al-Silica and Alumina respectively). For impregnation an aqueous solution of CrO_3 in appropriate concentration was used.

The support was suspended in a slight excess of solution,filtered and then dried at 10 torr at 110°C. Condensation of liquid water on any particle of the product was rigorously avoided. The next step, activation in a dry stream of oxygen at 500°C, was carried out in a fluidized bed reactor. After 30 min., activation was usually complete; the gas was replaced by argon, and then at 350°C carbon monoxid was blown through for another 30 min. The product was then heated again under argon to 500°C for 10 min. to remove all CO (test by IR, see below), and the cooled under argon to room temperature. Unless otherwise stated, the concentration was chosen so that the final product had a 1 % chromium content. (Al-Silica had to be heated in vacuo to 550°C for 16 hrs. prior to use, to remove NH_4Cl and other impurities.)

Analysis: Cr(VI) : iodometrically after dissolving the sample in NaOH, endpoint determined potentiometrically. Cr(total): as Cr(VI), but oxidizing with H_2O_2 at $p_H = 9$ prior to acidification. Cr(II): suspension in 2n H_2SO_4 under purified argon, titration with $Cr_2O_7^{2-}$ or MnO_4^-, endpoint determined potentiometrically.

All analytical data were calculated on the basis of the weight of "standard support": airdry support, heated at 500°C for 30 min. at 10 torr.

2.2 Stoichiometry of reactions of the free Cr(II) catalyst.

An important problem inherent in reaction studies was the competition of chemisorption on chromium and physical adsorption on the support. Correction for physical adsorption was made by blank experiments in wich the support itself was given the same treatment as the final catalyst; the soaking with the acidic solution of CrO_3 was replaced by treatment with dilute HNO_3 of the corresponding p_H. This correction includes the assumption that the surface chromium does not occupy just those sites on the support surface which are preferred for physisorption. This assumption is supported by experiments which showed that the support alone exhibited the same degree of physical adsorption of nitrogen as did the catalyst in its oxidized form.

Adsorption was determined gravimetrically in cases in which definite and stable products were easily obtained, such as with O_2 and Cl_2.

With N_2 and CO, equilibria occur over a wide range of temperatures. Frontal gas chromatography which measures the volumes of physically adsorbed and chemisorbed gas at given T, p was suitable[9)]. We used a modified Perkin Elmer F 7 Gas Chromatograph. All gases were purified thoroughly; the He carrier gas was freed from N_2 by treatment with Ca metal at 410°C.

2.3 Visible and IR spectra of the catalysts.

Reflection spectra were obtained with samples under different atmospheres in a special cuvette[10)] at temperatures from -100 to + 400°C using a Beckman DK2 reflex photometer.

The infrared spectra were taken from silica pellets compressed at 100 atm. or more simple by using Vycor porous glass[11)] as the support. Again a special cuvette was used which allowed treatment with different gases at temperatures up to 500°C. A Beckman IR 7 was employed.

2.4 Preparation and analysis of solutions containing the chromium-hydrocarbon complex.

A sample of standard Cr(II)/Silica catalyst was evacuated, heated to 100°C and then allowed to react with ethylene at 100 torr for 60 sec. After evacuation, the product was suspended in HCl-saturated, O_2-free methanol. A deep blue solution was formed, from which the chromium compound was extracted into cyclohexane or CCl_4 where it forms a neutral, relatively stable species. Up to 2o % of the originally employed chromium can be extracted into the nonpolar solvent. By evapoaration, a dark blue oil can be isolated.

Cleavage of the Cr-C bond was achieved either by heating with protonic acids or by treatment with oxygen. The organic products were purified by thin layer chromatography (see below) .

2.5. Thin layer chromatography, mass spectroscopy.

These methods were used to separate, concentrate and identify reaction products in small quantities, especially from "low polymer synthesis" experiments.

The "Desaga" apparatus was used for TLC, with Kieselgel G Merck layers. The separated zones were eluted into KBr, which was then used directly for IR. In some cases it was useful to condense the substances from the KBr to a cooled BaF_2 window[12]. Mass spectroscopy was carried out using a Varian CH4 Atlas instrument; again the impregnated KBr powder was directly suitable as sample.

3. RESULTS AND INTERPRETATIONS

3.1. The "free" Cr(II) catalyst

The process of impregnation and activation in the representative case of Silica + CrO_3(aq) forms a mixed anhydride of polysilicic and chromic acid, the latter preferentially as bichromate(VI): Analysis established that - with excess of Cr(VI) during impregnation - one OH group was lost from the support for each Cr(VI) incorporated; the visible spectrum exhibited absorption by bichromate(VI). Depending upon the type of silica used, an upper limit of 70 - 85 % of the OH groups of the support were involved in anhydrid formation.

"Overdoping" with chromium leads to a corresponding amount of Cr_2O_3, which is not able to stabilize Cr(VI) on its surface.

The oxidation number of surface chromium in the activated product reaches 6.0 if there was no intermediate condensation of water during heating steps (chromium content 1 % ca.). With Al-Silica and Alumina this value is slightly lower.

The reduction (carried out with CO to avoid H_2O as reaction product) involves Cr(VI) only and gives an amount of surface Cr(II) proportional to the original surface Cr(VI), with a yield of 98 - 100 %. The final oxidation number is 2.0 - 2.1 , somewhat higher with Al-Silica and Alumina supports expectedly.

The visible spectra of the reduced catalyst containing 0.025 % Cr in argon or in vacuo show a broad maximum at 12,900 $\pm$ 100 cm^{-1}, which correlates with the blue-green colour of the catalyst.

3.2. The reaction with the electron acceptors O_2 and Cl_2

The well known reaction of the surface Cr(II) and oxygen[13] regenerates Cr(VI) almost completely. The visible reflection spectrum corresponds to that of bichromate(VI).

At room temperature chlorine forms a green 1:1 compound of Cr(III). This product is still somewhat sensitive to oxygen. At elevated temperatures (500°C) a brown product is formed which has a Cr:Cl ratio of 1:2 and which does not react further with oxygen[14]. Blank experiments showed that there is no Si-Cl formation under our conditions. CrO_2Cl_2 is formed as a by product but none remains adsorbed.

3.3. The reaction with the n-electron donors N_2 and CO.

At room temperature N_2 forms a blue 1:1 complex in equilibrium. The heat of formation was determined to be ca. 1.5 kcal/mole at 25°C. At lower temperatures a complex with two N_2 molecules is formed, and this is the stable end product at - 100°C. The reaction is reversible.

At 20°C with 1 atm. of nitrogen the absorption maximum (visible reflex spectrum) of surface Cr(II) is shifted to 13,150 ± 100 cm^{-1}. No IR absorption of the N_2 ligand was detected.

CO shows a similar behaviour, forming a 1:1 complex as the stable produkt at - 65°C. At room temperature it has an absorption maximum at 19,000 cm^{-1} (blue-violet), but near - 60°C the maximum shifts reversibly to 20,410 cm^{-1} (brown). For the blue-violet compound the IR spectra exhibit peaks at 2,183.5 / 2,186.5 and 2,192.5 cm^{-1}; at deeper temperatures an additional peak at 2,090.5 cm^{-1} appears. (The first mentioned band at about 2,185 cm^{-1} is the best, quick test for traces of complexed CO in the catalyst.)

This behaviour suggests the existence of two - sterically different - Cr(II) species in the surface which exhibit a different complex formation with CO: without π-back donation (2,185 cm^{-1}) and with π-back donation (2,090 cm^{-1}).

3.4. The reaction with π-electron donors: olefin adducts.

Addition of C_2H_4 to the free Cr(II) surface compound normally causes instantaneous polymerization; but by adding only a limited amount of ethylene at low pressure, an addition complex is formed. The $CH_{\nu 9}$ vibration in this C_2H_4 adduct is at 3,008 rather than at 3,106 cm^{-1},as in the free ethylene. A similar behaviour is found in chemisorbed cyclohexene where the corresponding band is shifted from 3,065 cm^{-1} to 3,021 cm^{-1}. This is to be expected for π-bond formation[15].

While complexation of CO competes with polymerisation,[7] it is remarkable that, after "low polymer synthesis", addition of CO gives the typical IR bands at 2,185 cm^{-1} in the same intensity as prior to polymerisation, while the π-bonded CO with its absorption at 2,090 cm^{-1} does not reappear. This leads to the conclusion that a) CO and C_2H_4 use the same sites at the Cr center for bounding, and b) an alkyl ligand does not hinder formation of this complex.

3.5. The blue solution.

The desire to use "standard" chemical rather than solid state methods to investigate the nature of the Cr-carbon bond led us to experiments with systems where this bond is still intact whereas the metal-surface bond has been split.

Treatment with a proton-acidic acid in a polar solvent like CH_3OH/HCl or CH_3OH/NH_4OH removes the chromium center from the surface after "low polymer synthesis". Almost no cleavage of the Cr-alkyl bond occured at room temperature or below provided an appropriate catalyst system was used and water was strictly excluded. Solutions are best obtained if the Cr carries a carbon chain of 16 to 32 carbon atoms; the complex is neutral in cyclohexane and surprisingly stable.

The Cr(II) in solution no longer has free coordination sites; the ligands are very strongly bonded (otherwise no cleavage of the support-Cr-bond would occur !), and the solutions showed no polymerization activity.

As a first test, the mass spectrum of the complex was run. It showed different fragments containing chromium, all of the common formula $[Cr-(CH_2CH_2)_n]^+$ with $n = 0....16$.

The cleavage of Cr-alkyl in the solutions can be carried out either by hydrolysis at elevated temperature or by oxygen. The first reaction yields α-olefins, the latter the corresponding n-aldehydes formed by a secondary reaction. This was established by mass spectroscopy after separation and purification by thin layer chromatography.

4. DISCUSSION

The formation of the catalyst seems now to be rather well understood. First step is the linkage of Cr(VI), mostly as bichromate, to the support, forming a H_2O sensitive, thermally stable mixed anhydride. The second step, the reduction, leads to coordinatively unsaturated Cr(II) surface compounds. These centers react uniformly with strong electron acceptors like O_2 and Cl_2, forming Cr(VI) and Cr(III) compounds; but with the weak electron donor CO the Cr(II) centers form two different types of addition products - without or with a π- back bonding effect.

The existence of a weak 1:2 complex with nitrogen shows there to be at least two coordination sites available for donor molecules.

Olefines react with formation of a π-complex prior to polymerization, as shown by IR data. As the presence of CO stops the polymerization under certain conditions[7], both molecules seem to use the same site(s) for complex formation. This is confirmed by our IR data: If the active species of the polymerization process is characterized by I, it is reasonable to expect II to be formed by vacuum and CO treatment as was found in the IR experiments.

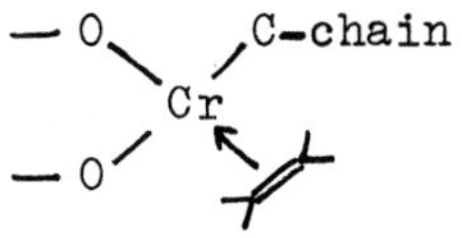

—O C–chain O C–chain
Cr Cr
—O O CO

I 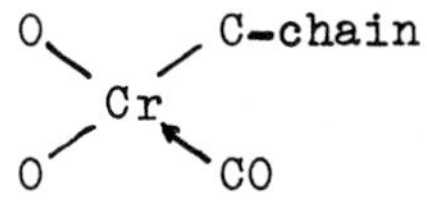 II

While the IR spectra indicate that the monomer is π-complexed, the mass spectra of low polymer-Cr species suggest on the other hand the presence of a Cr-C(chain) σ-bond: The fact that stable fragments $[Cr(C_2H_4)_n]^+$ are found shows a considerable stability of the Cr-C bond, wich is not very plausible for a π-to-cation bond.

All these results support the mechanism for polymerization as proposed by J.P.Hogan[7], with a σ-π change on two sites of the chromium.

DISCUSSION

D. D. ELEY

Second page, reference 7 by Hogan does not in fact give any view as to the oxidation state of Cr in the active site. In two papers with C. H. Rochester and M. S. Scurrell to appear in Proc. Roy. Soc. London, we have studied the polymerization of ethylene on a chromia on silica gel catalyst (Phillips process), the catalyst being supplied by B. P. Chemicals International Ltd. We followed the polymerization by decrease in pressure and by the infra-red technique. We have supported the contention that the active site is Cr^{5+} on the evidence cited in our papers. The suggestion that it involves two chromium atoms in different oxidation, as Cr^{3+} and Cr^{6+} (Miesserov), is not discounted. In a third paper (in press, J. Catalysis) we obtained results on the effects of CO which agree with those of Krauss (page 8). We agree that CO and C_2H_4 occupy the same sites at the Cr center, and that alkyl ligand does not hinder formation of the CO-Cr complex. We failed to see an I.R. band for C_2H_4 held by π-bonding on Cr sites.

The bright yellow color of the solid originally (Cr^{6+}) on reduction by ethylene at 300°C changed to yellow-green at maximum activity, and with further reduction the activity fell and the solid catalyst became green (Cr^{3+}).

H. L. KRAUSS

Several species of chromium present in the catalyst can presumably form the active site: chromium in the oxidation states VI, V, III and II. Chromium (VI) can be ruled out, being not stable with ethylene (redox reaction). Chromium (V) can be ruled out as well; according to our (e.s.r.) experiments the amount of Cr (V) does not exceed ~1% of the total chromium after contact with ethylene, whilst the Cr-atoms participating in the polymerization are approaching 20% (blue, hexane soluble complex!). Both Cr(VI) and Cr(V) could still play a role as precursors of the reduced species; as far as we know there is no difference in behavior towards the reduction either by ethylene itself or by other reducing agents.

The decision concerning Cr(III) and Cr(II) is more difficult, since no direct (wet) chemical analysis of the oxidation number is possible after the contact with ethylene (interfering of polymer organic products). On the other hand the treatment with non-polymerizing organic reductants as well as the reduction with CO lead to products, the activity of which is proportional to their chromium (II) content.[6] This leads us to the conclusion that the active site is Cr(II) only. -Cr (III) formed as a green byproduct in some of the redox reactions is obviously not active in the polymerization reaction.

N. D. PARKYNS

It seems somewhat surprising to me that a stable complex of Cr^{II} and nitrogen should be formed which apparently has no IR absorption band. CO adsorbs on metals to give strong IR bands of variable frequency which is accounted for by variations in the degree of d-π* backbonding. In the present results, adsorption of CO gives bands at around 2100 cm^{-1} and, as the author rightly remarks, this is indicative of σ bonding only with no backbonding. As N_2 is a poorer ligand than CO in general, it is difficult to envisage σ-bonding only operating in the case of the Cr^{II}-N_2 complex and I would be interested to see whether the author could speculate about the type of bonding involved; in any case, an IR-active band should result.

H. L. KRAUSS

There is no doubt that a N_2-complex of the surface chromium (II) is formed (evidence by equilibrium pressure measurements and visible reflection spectroscopy). We agree that an IR-band of the N_2-ligand should be found but in spite of our repeated attempts so far there was no indication of such an absorption. As a possible explanation perhaps we should remember the fact that there is only a low concentration of the nitrogen complex at room temperature/1 atm N_2 (approx. 15% of 1% surface chromium (III)). So we hope to get a reasonable IR-absorption of the N_2-unit at low temperatures where the complex is stable.

Y. I. YERMAKOV

I should like to note that the problem of the oxidation state of chromium which is necessary to obtain active centers may be solved using data about the active center concentration in catalyst. The number of propagation centers may be calculated from the propagation rate constant. Chromium-oxide catalysts in the oxidized form (after activation in the dry air) under suitable conditions have activity as high as 1.5 x 10^4 g of polyethylene/g Cr·hr·atm, with K_p = 10^6ℓ/mole·hr which gives ~15% of chromium included in active center. As these catalysts by analytical methods have no chromium in other oxidation states than Cr^{+6}, it is possible to conclude that the precursor of active centers (or "active component") in catalyst contain chromium in oxidation state 6. Catalysts activated by carbon monoxide reduction have an activity as high as 10^5 g of polyethylene/g Cr·hr·atm; with K_p = 2.5·10^6ℓ/mole·hour. This means that the portion of chromium in active centers reaches ~35% of total supported chromium. The catalysts reduced at 400° have no chromium in oxidation state higher than 3. So the general scheme of the active center formation in chromium oxide catalysts may be written as follows:

active component (Cr^{+6} surface compound)

$\xrightarrow[\text{or by monomer}]{\text{reduction by CO}}$ surface compound of Cr^{+2} (or Cr^{+3})

$\xrightarrow[\text{by monomer}]{\text{alkylation}}$ propagation center (in which the active chromium-carbon bond is present).

H. L. KRAUSS

As I mentioned, it is possible to remove some of the surface chromium after a limited reaction with ethylene (polymerization, formation of a short alkyl chain). A deep blue solution of a chromium complex in cyclohexane can be obtained, with the chromium-to-chain bond still intact. These solutions contain approximately 15 - 20% of that chromium which was originally present in the solid catalyst. The solubility in the hydrocarbon solvent being obviously due to the hydrocarbon "panhandle" ligand attached to the metal, this shows the incorporation of at least 15 - 20% of the used chromium in the polymerization process--in good agreement with the numbers given by Yermakov (~15 - 35%) and by Hogan (~8%).[7] The given scheme of the active center formation in Yermakov's comment is identical with our proposal, except for the oxidation number of the reduced chromium which shall be no other than +2 in our concept, prior to the propagation step.

Paper Number 9

THE CATALYTIC ACTIVITY OF RARE EARTH OXIDES

Kh. M. MINACHEV
N. D. Zelinsky Institute of Organic Chemistry
Academy of Sciences, Moscow, USSR

ABSTRACT: This paper deals with the catalytic properties of the rare earth oxides in the dehydrogenation of cyclohexane, the dehydrocyclization of heptane, the cracking of butane, the hydrogenation and isomerization of olefins, isotopic exchange between deuterium and hydrocarbons, the dehydrogenation, dehydration and ketonization of alcohols, and the oxidation of hydrogen and propylene. The oxides act as highly active catalysts for the hydrogenation and double bond migration of olefins and for conversions of alcohols but they exhibit only moderate activity in the other reactions. The dependence of the catalytic activity of an oxide upon the atomic number of the rare earth is interpreted in terms of its electronic structure.

1. INTRODUCTION

Interest in the correlation of catalytic activity with the electronic structure and position in the Periodic Table of the chemical elements of catalysts led us to study the catalytic activity of the oxides of the fourteen rare earth metals. Since the 4f-subshell is filled by the end of the series, the effect of f-electrons can be ascertained by comparing the catalytic properties of the various rare earths. Further, in recent years, the rare earth oxides have become readily available and the study of their catalytic activity has acquired practical importance.

The present paper reports the catalytic behavior of the rare earth oxides in the conversion of hydrocarbons and of oxygen-containing compounds, the isotopic exchange of oxygen, and the oxidation of hydrogen and propylene. Some of the data have been already published, others are reported for the first time.

2. EXPERIMENTAL

2.1 Catalyst preparation

The rare earth oxides[1-5] were obtained by decomposing hydroxides prepared from nitrates by precipitation with ammonia. The purity of the starting compounds was as high as 99.5%. The main admixtures were other rare-earths. The total content of Fe, Cu, and Ca did not exceed 0.01%. Oxides obtained by direct decomposition of the hydroxides at 800°C in the reaction vessel *in vacuo* were used for ethylene hydrogenation. In other cases, hydroxides were dehydrated at 600-650°C for 4-6 hrs. For butane cracking, ethylene hydrogenation, hydrogen oxidation, butane isomerization, and oxygen isotopic exchange the catalysts were pretreated *in vacuo* (10^{-5} mm Hg) at temperatures of 650, 800, 500, 800 and 500°C, respectively. In other cases, the catalysts were activated by heating in nitrogen at 500°C and regenerated in air at 560°C. The specific surfaces of the oxides, determined from BET nitrogen adsorption, varied in a range of 10 to 50 m^2/g.

2.2 Measurement of catalytic activity

The experiments of butane cracking (400-550°C) ethylene hydrogenation (-120 to +20°C, $H_2/C_2H_4 = 1.2$), oxidation of hydrogen (150-350°C, $H_2/O_2 = 2$), oxygen isotopic exchange with molecular oxygen (200-420°C), and butane isomerization (-30° to + 20°C) were carried out by using either a static or a static circulatory method at pressures of 2, 10, 3, 40, and 50 mm Hg, respectively. In other cases the catalytic activity was determined in a flow reactor at atmospheric pressure. In the experiments reported in 3.1 the oxide activity was determined from the yields of the reaction products. The specific rate constant calculated from the first order rate equation was used to represent the activity of the catalyst in cyclohexane dehydrogenation (520-590°C), heptane dehydrocyclization (530-560°C), butane cracking, and hydrogen oxidation. In the ketonization of n-butyl alcohol (385-430°C), the catalytic activity was determined by using the specific rate constant from a kinetic equation of order 0.3 in starting alcohol. For ethylene hydrogenation and oxygen isotopic exchange, initial rates for various catalysts were compared. For propylene oxidation (360-440°C), the catalytic activity was determined from the yield of carbon dioxide.

It has been shown experimentally that the reactions proceed in the kinetic regime.

3. RESULTS AND INTERPRETATION

3.1 On catalytic activity of the rare earch oxides

Data which qualitatively characterize the catalytic activity of rare-earth oxides are summarized in Table 1. The oxides catalyze the exchange of hydrocarbons with deuterium, hydrogenation and double bond migration of olefins, dehydrogenation of paraffins and olefinic hydrocarbons; dehydrogenation, dehyration and ketonization of alcohols.

The reactivity of hydrocarbons for exchange with deuterium[6)] decreased in the sequence arenes, alkenes, alkanes. The distribution of the deuterated products[6)] accords with a mechanism in which only one hydrogen atom is exchanged for one period of residence on the surface and indicates that the principal surface intermediate is a mono-adsorbed hydrocarbon.

The oxides proved to be highly active catalysts for the hydrogenation of ethylene[4,7)] but only after pretreatment *in vacuo* above 600 to 700°C. Lanthanum and praseodymium oxides catalyze the hydrogenation even at temperatures as low as -120°C. The rare earth oxides show low activity for dehydrogenation of hydrocarbons;[1,8)] dehydrogenation of cyclohexane

TABLE 1
Catalytic reactions on rare earth oxides

Reactant	Catalyst	Temp. °C	Contact Time sec.	Conversion %
Isotopic exchange between D_2 and hydrocarbons (6:1)				
cyclopentane	Gd_2O_3	460	3	4.4
cyclohexane	Gd_2O_3	460	3	1.4
hexane	Gd_2O_3	460	3	16.1
cyclohexene	Gd_2O_3	440	3	47.8
1-methylcyclohexene	Gd_2O_3	440	3	33.8
1-methylcyclopentene	Gd_2O_3	440	3	22.8
benzene	Gd_2O_3	380	3	61.0
Hydrogenation of ethylene, H_2/C_2H_4 = 1.2				
ethylene	Nd_2O_3*	- 25	60	50
ethylene	Pr_2O_3*	-120	7200	50
Dehydrogenation				
cyclohexane	Nd_2O_3	527	65	5.5
cyclohexene	Nd_2O_3	530	16	20
1-methylcyclo-hexene	Nd_2O_3	530	16	56
cyclohexanol	Nd_2O_3	425	20	28
4-heptanol	Nd_2O_3	400	1.8	29
Double bond migration				
1-butene	Dy_2O_3*	- 30	1800	82
1-methylcyclohexene	Nd_2O_3	340	16	6.5
1-methylcyclopentene	Nd_2O_3	395	16	4.6
allylcyclopentane	Er_2O_3	400	65	94.3
1-ethylcyclopentene	Er_2O_3	400	65	20.7
Dehydration				
cyclopentanol	Nd_2O_3	350	6.5	34
1-methyl-2-cyclohexanol	Nd_2O_3	350	6.5	35.4
1-methyl-4-cyclohexanol	Nd_2O_3	350	6.5	32.4
Ketonization				
1-heptanol	Nd_2O_3	430	15	26
1-butanol	Nd_2O_3	430	15	27
butyraldehyde	Nd_2O_3	415	2.5	61

*Weight of catalyst, 4.5g.

proceeds appreciably only above about 500°C. The dehydrogenation of cycloolefins proceeds more rapidly under these conditions. The reactivity of methylcyclohexane exceeds that of cyclohexane both in dehydrogenation and in isotopic exchange with deuterium. The dehydrogenation of alcohols occurs at a lower temperature than that of hydrocarbons.

The yields of the products of double bond migration in olefins depend upon catalyst pretreatment. Neodymium and erbium oxides pretreated at 500°C catalyze double bond migra-

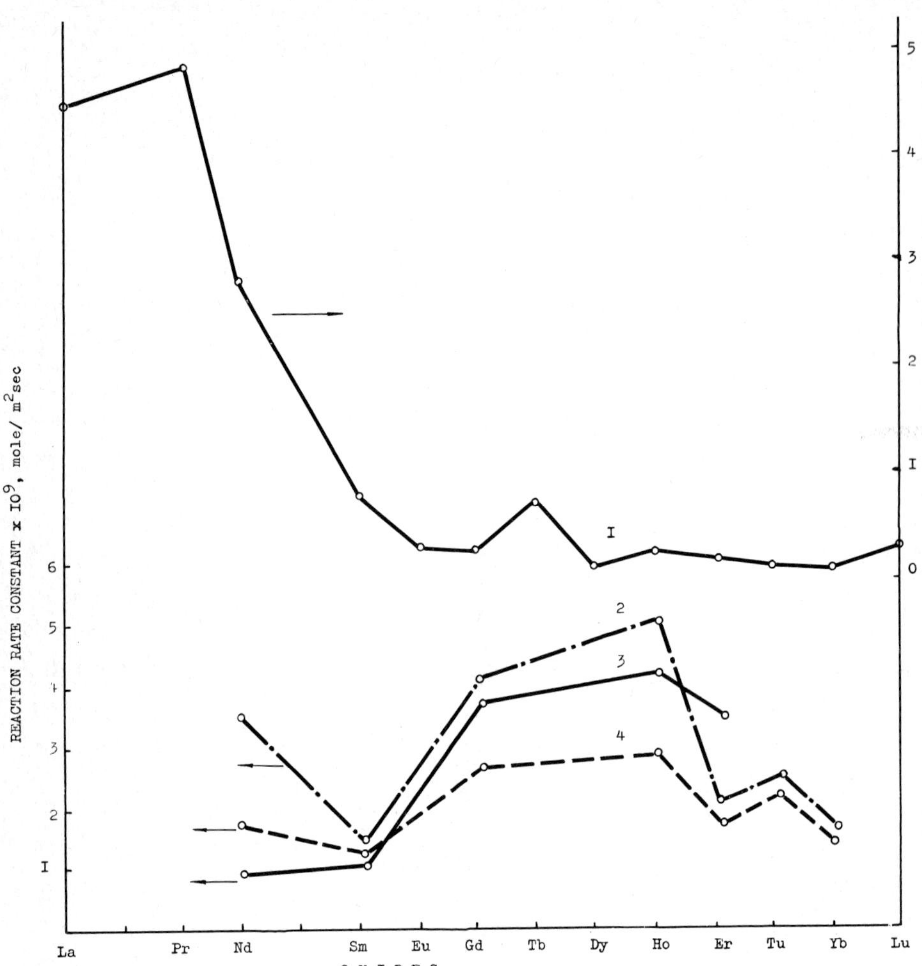

Fig. 1 Variations in catalytic activity of the rare earth oxides in hydrocarbon conversion; 1, ethylene hydrogenation, -78°C; 2, butane cracking, 500°C; 3, heptane dehydrocyclization, 530°C; 4, cyclohexane dehydrogenation, 530°C.

tion in cycloolefins only at 340-400°C. Evacuation at above 700°C increases the catalytic activity of the oxides. For example, dysprosium oxide pretreated in vacuo at 800°C catalyzes double bond migrationin butene at -30°C. Neodymium and lanthanum oxides behave similarly. No structural isomerization has been observed under these conditions.

Dehydration of alcohols on neodymium oxide was unaccompanied by double bond isomerization of the resulting olefins. Thus, 1-methyl-4-hexene was formed from 1-methyl-4-cyclohexanol.

This low activity for double bond migration can be ascribed to the adsorption of the water formed during dehydration. Yields of olefins were nearly independent of the structure of the alcohol.

The yields of ketones from n-butyl and n-heptyl alcohols were about the same. Ketonization of butyraldehyde proceeded at lower temperatures than that of alcohols and resulted in larger yields.

3.2 Variation of catalytic activity among the rare earth oxides.

3.2.1 Hydrocarbon conversions

Data obtained on the specific catalytic activity of rare earth oxides in hydrogenation of ethylene,[4,7] cracking of butane,[2] dehydrogenation of cyclohexane[1] and dehydrocyclization of heptane[1] are summarized in Fig. 1. Our results on the dehydrocyclization of heptane are in good agreement with those of Komarevsky.[10]

The data on ethylene hydrogenation are for sesquioxides. The sesquioxides of praseodymium and terbium were obtained by reduction of the non-stoichiomeric oxides of these elements. The non-reduced oxides of Pr_6O_{11} Tb_4O_7 and cerium dioxide showed no catalytic activity. The last oxide was inactive even after hydrogen treatment at 800°C, and we failed to obtain its sesquioxide under these conditions.[11] The activity of the sesquioxides did not change upon treatment with hydrogen at 700°C.

One can see from Fig. 1 that catalytic activity for the hydrogenation of ethylene by the oxides of the cerium subgroup (from Pr to Gd) varies considerably, whereas the activity of the oxides of the ytterbium subgroup varies much less. The shape of the curve for catalytic activity as a function of the atomic number of the rare earth was about the same in the temperature range, -78 to 20°C. The most active oxides were those of lanthanum and praseodymium. In the oxides from praseodymium to gadolinium, activity decreased monotonically. The variation in activity of the oxides of the second subgroup was insignificant and was within the accuracy of our measurements. The activation energy of ethylene hydrogenation depend but little on the rare earth and was about 5-7 kcal/mole.

The catalytic activity of the oxides in dehydrogenation of cyclohexane, cracking of butane and dehydrocyclization of heptane varied similarly. This indicates that the rate limiting processes of the reactions are analogous. The activation energies for dehydrogenation (40 to 60 kcal/mole) and cracking of butane (15 to 36 kcal/mole) also varied in a parallel fashion and correlated with the effective magnetic moments of the tripositive ions of the rare earths. We suggest that the rate limiting step of the three reactions is that of dehydrogenation of hydrocarbon to olefin. A mechanism involving consecutive steps for the formation of benzene from cyclohexane on holmium oxide has been established by use of ^{14}C. The formation of heptane as an intermediate in dehydrocyclization on chromium oxide has been reported.[13]

3.2.2 Ketonization of n-butyl alcohol

Rates of catalytic ketonization of n-butyl alcohol[5] are

presented in Fig. 2. Praseodymium oxide showed the highest activity, ytterbium oxide, the lowest. Variation in activity of the oxides of the cerium subgroup is larger than that of the ytterbium one. The catalytic activity decreased with increasing atomic number from praseodymium to ytterbium. Praseodymium oxide of an initial composition Pr_6O_{11} is likely to undergo reduction[14] under the conditions of the reaction to Pr_2O_3. The activation energy for ketonization varies negligibly among the rare earth oxides (32 to 38 kcal/mole).

The following mechanism for ketonization of butyl alcohol on the rare earth oxides has been suggested on the basis of kinetic studies of the ketonization of butyl alcohol, butyraldehyde and 4-heptanol and also of hydrogen transfer between butyl alcohol and dipropyl ketone.[5]

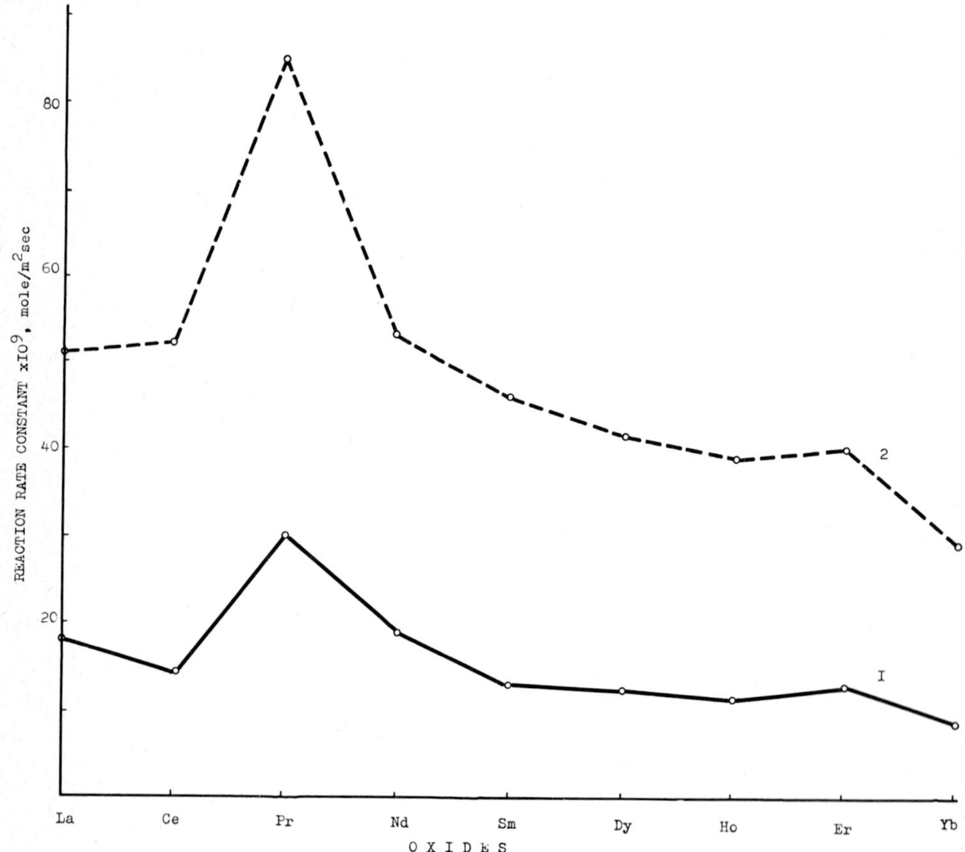

Fig. 2 Specific catalytic activity of rare earth oxides for ketonization of n-butyl alcohol at 400°C (curve 1) and 430°C (curve 2).

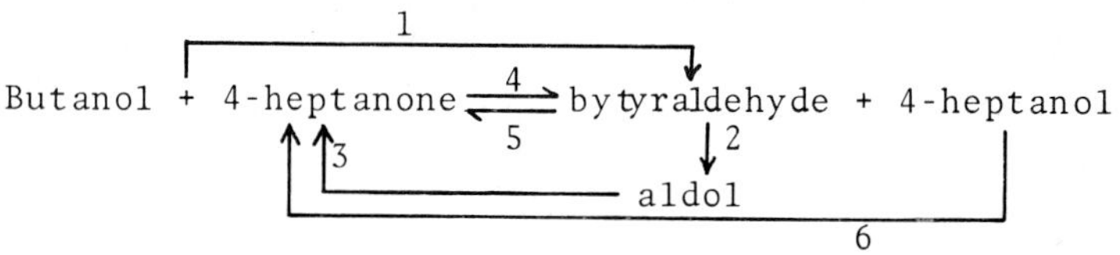

According to this mechanism, ketone is formed mainly by the aldol mechanism, steps 1-3, but steps 4-6 also contribute. On praseodymium and neodymium oxides, the yields of 4-heptanol formed by transfer of hydrogen between butyl alcohol and dipropyl ketone attained about 7-10%.

Since the ketonization of butyraldehyde and the dehydrogenation of 4-heptanol proceed to give larger yields and at lower temperatures than ketonization of butyl alcohol (Table 1), we suggest that the rate limiting process is dehydrogenation of butyl alcohol to butyraldehyde. It follows from the above scheme that variations in the catalytic activity in rare earth oxides for ketonization should parallel that for dehydrogenation of alcohols.

3.2.3 Oxidation of hydrogen and propylene and isotopic exchange between molecular oxygen and lattice oxide ions.

Comparison of the catalytic activity of the rare earth oxides in the hydrogen oxidation[15] with oxygen mobility[3] of the oxides determined from the initial rate of isotopic exchange is presented in Fig. 3. In the cerium subgroup praseodymium shows the highest activity and in the ytterbium subgroup, terbium.

The activation energy for oxidation of hydrogen and for isotopic exchange of oxide ion with molecular oxygen was the same for each of the oxides studied (Sm_2O_3, Nd_2O_3, Gd_2O_3, Pr_6O_{11}, La_2O_3), and varied in this series from 12 kcal/mole to 26 kcal/mole. Thus, one may conclude that the catalytic activity of the oxides is directly connected with the mobility of their oxygen. This suggests[16] that the rate limiting process of hydrogen oxidation is interaction of surface hydrogen in the molecular form with surface oxygen.

Praseodymium and terbium oxides show anomalously high oxygen mobility, probably because of their defect structure. They are intermediates between sesquioxides and dioxides and occur as solid solutions with structures of the fluorite type. There are a large number of defects in the lattice.

For propylene oxidation[17] no such correlation could be obtained between catalytic activity and oxygen mobility. This reaction also exhibits two peaks of catalytic activity in the series of oxides. Here, cerium and terbium oxides showed the maximum activity, whereas praseodymium and terbium oxides showed the highest oxygen mobility (Fig. 3). Despite this apparent discrepancy, we suggest that the binding energy of oxygen with the surface may also determine catalytic activity for the oxidation of propylene. The mobility of the oxygen in the oxides in this case may be different from that in oxygen isotopic exchange and hydrogen oxidation due to the presence of propylene and its oxidation products in the mixture. Besides, carbonates are formed as a result of interaction of the oxides with carbon dioxide. Their presence in a catalyst may also affect the binding energy of oxygen with the catalyst.

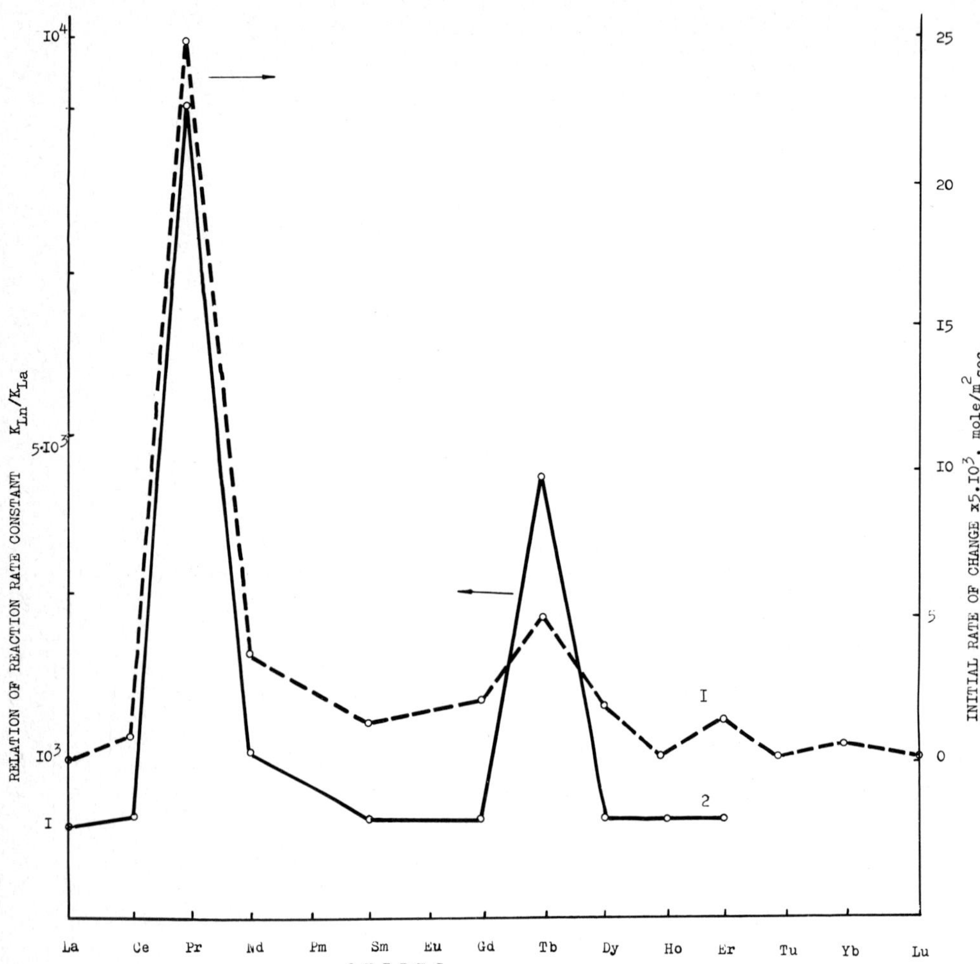

Fig. 3 Initial rate of isotopic exchange of the oxygen of oxides with molecular oxygen at 370°C (1), and their relative activity in hydrogen oxidation at 340°C(2). The activity of lanthanum oxide is taken to be unity.

4. DISCUSSION

Our data show that rare earth oxides act as catalysts in hydrogenation of ethylene, double bond migration in butenes and cycloolefins, dehydrogenation of cyclohexane, dehydrocyclization of heptane, cracking of butane, ketonization, dehydrogenation and dehydration of alcohols, oxygen and propylene oxidation, and exchange of hydrocarbons with deuterium. Structural isomerization of olefins, paraffins and naphthenes does not proceed on rare earth oxides in the temperature range, 400-550°C.

The mobility of oxygen in lanthanum and lutecium oxides has been found to be about the same as that in iron and nickel monoxide.[19] The rate of oxygen exchange between praseodymium oxide and the gas phase[18] is about the same as that of cobalt monoxide and manganese dioxide which are considered to be the most active catalysts for oxidation. Moreover, the rare earth oxides show moderate activity in propylene oxidation.

The catalytic activity of the rare earth oxides in ethylene hydrogenation and double bond migration depends to a great extent on the pretreatment of the oxides. When heated at 500°C, they show low activity. When activated in vacuum at 800°C, they catalyze ethylene hydrogenation at -78°C or lower and double bond migration in butenes at -30°. Only Cr_2O_3 and Co_3O_4 show similar activity among the transition elements of the fourth period.[19,20]

The comparatively low activity of the rare earth oxides in dehydrogenation and dehydrocyclization of hydrocarbons probably results from the following factors: (i) the temperature of the catalyst pretreatment might have been insufficiently high (500°C), (ii) the reactions were carried out at elevated temperatures where partial reduction of the surface oxides could take place, (iii) the carbon dioxide resulting from this reduction could convert the oxides into carbonates of low activity.

The activities of rare-earth oxides in alcohol dehydrogenation can be compared with those of the oxides of Cr, Mn, Fe, Mo, I Sn. In alcohol ketonization, rare-earth oxides are inferior to iron oxide. Their activity is about the same as that of chromium oxide and higher than that of zinc oxide.

In cyclohexane dehydrogenation, butane cracking, heptane dehydrocyclization, and ethylene hydrogenation, the activities of the rare earth oxides vary within one order of magnitude; in isotopic exchange of the oxygen of the oxides with molecular oxygen, and in the oxidation of hydrogen, they vary within 2-3 orders.[18] In this respect, the rare earth oxides differ from those of transition elements of the third and fourth periods where the catalytic activity in the hydrocarbon reactions varies within 2-3 orders, in the reaction of isotopic exchange, within 8 orders.

The dependence of catalytic activity of the oxides on the atomic number of the rare earth depends upon the reaction catalyzed, but the general character of this dependence fits the assumed division of the oxides into cerium and ytterbium subgroups. As a rule, the variations in the activity of the oxides of the cerium subgroup are larger than those of the ytterium subgroup. The electronic structure of the oxides in these subgroups is different. The elements of the first subgroup from Ce to Gd contain f-electrons with parallel spin, whereas, in the subsequent seven elements, the 4f-subshell increasingly contains pairs of electrons with antiparallel spin. Thus, the appearance of the first seven electrons in the 4f-subshell affects the catalytic activity to a greater extent than the subsequent filling of the subshell. In this connection it is interesting to note that the tendency toward anomalous valency in the first subgroup is greater than in the second. Compounds of quadrivalent cerium and praseodymium and of divalent samarium, neodymium, and europium are known. In the ytterbium subgroup, the quadrivalent compounds are characteristic only of terbium, and divalent ytterbium and thulium are less stable than corresponding divalent states of europium and samarium in the cerium subgroup. For butane cracking and cyclohexane dehydrogenation, the activation energies depend upon the effective magnetic moments of trivalent ions of the rare earth. Such dependence has been observed for isopropyl alcohol dehydrogenation,[21] homomolecular oxygen exchange, and carbon monoxide oxidation.[22] Since the values of the effective magnetic moment are connected with the number of f-electrons, the correlation indicates that the activation energies of the reactions depend on the electron structure of 4f-subshells of the rare earths.

REFERENCES

1) Kh. M. Minachev, M. A. Markov, O. K. Shchukina, Neftekhimiya 1, 489, 610 (1961).
2) Kh. M. Minachev, Yu. S. Khodakov, Kinetika i Kataliz, 6, 89 (1965).
3) Kh. M. Minachev, G. V. Antoshin, Dokl. Akad. Nauk, SSSR, 161, 122 (1965).
4) Kh. M. Minachev, Yu. S. Khodakov, V. S. Nakshunov, Neftekhimiya, 11, 824 (1971).
5) Kh. M. Minachev, G. A. Loginov, M. A. Markov, Neftekhimiya 9, 412, (1969); 10, 393 (1970).
6) Kh. M. Minachev, E. G. Vakk, R. V. Dmitriev, Izv. Akad. Nauk SSSR, Otd. Khim. Nauk, 1962, 1086; 1965, 618.
7) Yu. S. Khodakov, V. S. Nakhshunov, Kh. M. Minachev, Kinetika i Kataliz, 12, 535 (1971).
8) Kh. M. Minachev, M. A. Markov, "Probl. Kinetiki i Kataliza," Akad. Nauk SSSR, 11, 223 (1966).
9) Kh. M. Minachev, M. A. Markov, G. A. Loginov, Neftekhimiya 1, 356 (1961); 3, 181 (1963).
10) V. J. Komarewsky, Ind. Eng. Chem., 49, 264 (1957).
11) G. Brauer, H. Gradinger, Z. anorg. allg Chem., 277, 89 (1954).
12) Yu. I. Derbentzev, M. A. Markov, G. V. Isaguliants, Kh. M. Minachev, A. A. Balandin, Dokl. Akad. Nauk SSSR, 155, 128 (1964).
13) M. I. Rosengart, E. S. Mortikov, B. A. Kazanskii, Dokl. Akad. Nauk SSSR, 166, 619 (1966).
14) E. Cremer, Z. phys. chem., A144, 231 (1929).
15) G. V. Antoshin, Kh. M. Minachev, R. V. Dmitriev, Izv. Akad. Nauk SSSR, Ser. Khim., 1967, 1864.
16) G. V. Antoshin, Kh. M. Minachev, M. E. Lokhuaru, J. Catalysis, 22, 1 (1971).
17) Kh. M. Minachev, D. A. Kondratiev, G. V. Antoshin, Kinetica i Kataliz, 8, 131 (1967).
18) O. V. Krylov, "Catalysis with Nonmetals," Khimiya, Leningrad, 1967.
19) D. L. Harrison, D. Nicholls, H. Steiner, J. Catalysis, 7, 359 (1967).

20) Y. Kubokawa, T. Adachi, T. Tomino, T. Ozawa, Proceedings of the Fourth Internationsl Congress on Catalysis, Moscow, 1968, paper 39.
21) A. A. Balandin, "Current Status of the Multiplet Theory in Heterogeneous Catalysis," Nauka, Moscow, 1968.
22) E. S. Artamonov, L. A. Sazonov, Kinetika i Kataliz, 12, 961 (1971).

DISCUSSION

A. OZAKI

Since the catalysts were activated at higher temperatures than the reaction temperature, there might be poisoning effect of water formed by oxidation of H_2. If you had poisoning, what is the activity you plotted?

W. N. DELGASS

We have been studying dehydrogenation of cyclohexane over some rare earth catalysts in a pulsed microcatalytic reactor and have looked at the high temperature activation phenomenon reported in this paper. We find an enhancement in activity of Eu_2O_3 after pretreatment in flowing He at 800°C for 4 hours as compared to similar pretreatment at 550°C. Dehydrogenation is significant only at 500°C, however, so the enhancement in activity is minor compared to the remarkable effect reported by Professor Minachev for hydrogenation of ethylene.

I would like to ask two questions: 1) in considering correlations of rare earth activity with electronic properties, what is the effect of the radial extent of the f orbitals, and 2) what is the evidence for formation of surface carbonates on rare earth oxides?

P. TETENYI

There is a statement in the paper: "the catalytic activity of the oxides in dehydrogenation of cyclohexane, cracking of butane, and dehydrocyclization of heptane varied similarly." I would like to ask: was such a parallelism also observed in respect of exchange of hydrocarbons with deuterium?

R. J. H. VOORHOEVE

In your paper, you show convincingly that Praeseodymium oxide has considerably higher activity than the other Rare Earth oxides. In our work (R. J. H. Voorhoeve, J. P. Remeika, P. E. Freeland, B. T. Matthias, Science, July 28, 1972) on mixed oxides of cobalt or manganese with the Rare Earths we also noted the special characteristics of Pr. You link this in your paper with the well known existence of Pr^{4+}. Do you have any suggestions why this existence of Pr^{4+} would lead to higher activity for, e.g., oxidation of carbon monoxide?

D. D. ELEY

The work of Ashmead, Eley and Rudham (D. R. Ashmead, D. D. Eley, and R. Rudham, Trans. Faraday Soc. 59 (1963) 207; D. R. Ashmead, D. D. Eley, and R. Rudham, J. Catalysis 3 (1964) 280) provides some additional data on the $H_2 + D_2$ reaction on four R.E. oxides activated by pumping to 10^{-6} torr for 150 hours at 673K. (More rigorous treatment, e.g. heating 16 hours in hydrogen at 1028K did not appear to have much effect on the role of parahydrogen conversion at 77K). From the formulae of Table 4 (D. R. Ashmead, D. D. Eley, and R. Rudham, J. Catalysis 3 (1964) 280), one may calculate for K_m, the absolute rate in molecules cm^{-2} sec^{-1} at 4 torr and 500K, the following approximate values: Nd_2O_3 9.6 x 10^{12}; Gd_2O_3 2.4 x 10^{13}; Dy_2O_3 1.6 x 10^{14}; Er_2O_3 3.7 x 10^{13}. It was pointed out that the reaction appears to occur only on a fraction of the total surface, 0.001 for Nd_2O_3 (D. R. Ashmead, D. D. Eley, and R. Rudham, Trans. Faraday Soc. 59 (1963) 207), that is, we have a well marked peak of activity over the short series at Dy_2O_3. It is also perhaps interesting that our apparent activation energies of 5.16 to 6.41 kcal mole are similar to those reported by Minachev for ethylene hydrogenation.

(No written responses to these comments received from authors)

DISCUSSION

(COMMENT BY A. OZAKI)

Kh. M. MINACHEV

The hydrogen oxidation experiments were carried out in a static system with a stoichiometric mixture of reactants; the initial pressure was 3 torr and the catalyst weight was 1 g.

The catalytic activity was estimated from a first order reaction rate constant evaluated at 10% conversion.

The effect of water on the catalyst activity is extremely important and this fact demands special investigation.

(COMMENT BY W. N. DELGASS)

Kh. M. MINACHEV

The results cited by Dr. Delgass are very interesting, although it is difficult to comment on them without knowing the experimental details. In our investigation the hydrocarbon conversion was accompanied by carbon deposition on the catalysts, and so their regeneration by air at temperatures below 700°C brought about the conversion of the rare earth oxides into carbonates which showed low catalytic activity. Carbonate formation was also observed on storage of oxides under air. It was difficult to avoid the partial formation of carbonates in most of our experiments performed at high temperatures. The experiments on ethylene hydrogenation and double bond migration in butenes were carried out at temperatures which excluded carbonate formation.

Our data showed that the structure of the 4f-subshell did not affect the catalytic activity directly but manifested its influence indirectly via some other properties, e.g. basicity or changeable valency. At the same time the possibility of more direct participation of the 4f-orbitals in catalysis could not be excluded.

(COMMENT BY DR. TETENYI)

Kh. M. MINACHEV

We have no evidence for regularity of the change of the catalytic activity in the reaction of hydrogen exchange in hydrocarbons within a whole series of the rare earth oxides. The following sequences of activity were observed for the H-D-exchange of cyclohexane:

440°C $Gd_2O_3 > CeO_2 > Nd_2O_3$

480°C $Gd_2O_3 > Nd_2O_3 > CeO_2$

The activity of gadolinium oxide was two or three times as high as that of other oxides.

(COMMENT BY R. J. H. VOORHOEVE)

Kh. M. MINACHEV

Praesodymium oxide really showed maximum catalytic activity for the hydrogen oxidation, and we explained it by well-marked defectivity of the oxide which included Pr^{3+} and Pr^{4+}. At the same time catalytic activity of praeseodymium oxide for the propylene oxidation was lower than the activity of cerium dioxide. We suggested that it is caused by the partial formation of praseodymium carbonates. Cerium dioxide did not contain carbonates at these conditions. It should be noted that our results were obtained with the simple catalysts, and so it is difficult to use them in the interpretation of the data obtained for a multicomponent system.

(COM MENT BY D. D. ELEY)

Kh. M. MINACHEV

I agree with your opinion that approximate equality of the activation energies in H-D exchange and ethylene hydrogenation could point out the similarity in the chemical nature of active sites for these two reactions. Our data showed that the rate of H-D exchange on Dy_2O_3 at -78°C was 1.5 times as high as the rate of ethylene hydrogenation. However, the difference in the preparation and the activation of the oxides could influence on the number and the strength of these sites.

Paper Number 10

CATALYTIC ACTIVITIES AND SELECTIVITIES OF CALCIUM AND MAGNESIUM OXIDES FOR ISOMERIZATION OF 1-BUTENE

HIDESHI HATTORI, NAOJI YOSHII and KOZO TANABE
Department of Chemistry, Faculty of Science,
Hokkaido University, Sapporo, Japan.

ABSTRACT: The catalytic activities of evacuated calcium and magnesium oxides for the isomerization of 1-butene are much higher than that of silica-alumina. Their selectivities (ratio of cis- to trans-2-butene from 1-butene) were very high (ca.8 in CaO and ca.16 in MgO). The activities and selectivities changed remarkably with change in evacuation temperature.

On the basis of measurements of basic properties, experiments of dehydration and decarboxylation of catalysts, poisoning experiments with NH_3, CO_2, H_2O etc., and infrared studies on hydroxyl and carboxyl groups, the following conclusions about the nature of active sites were drawn. 1) The active sites for the isomerization of 1-butene to cis-2-butene, which are created by dehydration, are mainly basic. 2) The active sites for the isomerizations of 1-butene to trans-2-butene, which are created by decarboxylation, are mainly acidic.

1. INTRODUCTION

The isomerization of butenes which provides a useful tool for investigating properties of catalysts has been studied extensively over solid acid catalysts,[1] but not much over solid base catalysts. Calcium and magnesium oxides which were recently realized to be solid bases[2-8] are highly active and selective for the isomerization reaction[9,10] and it was pointed out that the catalytic action seems to be entirely different from that of solid acid catalysts.[11,12] However, the nature of active sites on the alkaline earth oxides is not known yet.

In the present paper, a characterization of the active sites has been attempted on the basis of kinetic results on the isomerization of 1-butene combined with measurements of basic properties, the degrees of dehydration and decarboxylation of the catalysts, poisoning experiments, and infrared studies on hydroxyl and carboxyl groups of catalysts and on the interaction of deuterated calcium oxide with 1-butene.

2. EXPERIMENTAL

2.1 Catalysts and materials

Calcium and magnesium oxides were prepared by calcining their hydroxides at various temperatures ranging from 200 to 900°C in air for 3 hrs or in vacuo for 2 hrs. The hydroxides were guaranteed reagents of Wako Junyaku Co., which contained a small amount of carbon dioxide. Silica-alumina containing 12 wt.% of alumina was prepared from aluminum isopropoxide and ethylorthosilicate. The 1-butene of pure grade (99.0%) was distilled through molecular sieves 4A kept at temperature of dry ice-acetone. Heavy water used was a Merck's reagent of 99.75% purity.

2.2 Amounts of dehydrated water and decarboxylated carbon dioxide

The above hydroxides were evacuated at room temperature in a vacuum apparatus. After a trap attached to the apparatus was cooled with liquid nitrogen, the hydroxides were heated at desired temperatures for 2 hrs. The amount of gas evolved when liquid nitrogen was replaced by dry ice-acetone or when the trap was kept at room temperature was taken as the amount of decarboxylated carbon dioxide or dehydrated water respectively.

2.3 Basic properties

Basicities at various basic strengths were measured by titrating the sample suspended in benzene with benzene solution of 0.1N benzoic acid, using various indicators. The indicators used are bromothymol blue (pKa=7.1), 2,4,6-trinitroaniline (12.2), 2,4-dinitroaniline (15.0), 4-chloro-2-nitroaniline (17.2), 4-nitroaniline (18.4) and 4-chloroaniline (26.5).

2.4 Kinetic measurements and poisoning experiments

The isomerization of 1-butene was carried out at 30°C or 200°C in a closed circulation apparatus. Metal hydroxides evacuated (10^{-5}mmHg) in situ at various temperatures for 2 hrs were used as catalysts. When metal oxides calcined in air were used as catalysts, they were transferred into a reaction tube and evacuated at 150°C for 1 hr before the reaction. In most cases, the pressure of butenes was approximately 200 mmHg and the amounts of catalysts were 17-25 mg. The volume of the reaction system was 470 ml when 1-butene was reacted over calcium oxide and 1350 ml in other cases. Reaction products were analyzed by a gas chromatography, using a column of dimethyl formamide on alumina.

In poisoning experiments, 4-11 mmHg each of water, carbon dioxide, ammonia and n-butylamine was introduced on calcium oxide evacuated at 500 and 700°C and the reaction was carried out in the same way as above. In the case of magnesium oxide catalyst, several mmHg of ammonia were adsorbed on the catalyst at room temperature for 1 hr. After the catalyst was evacuated at 30-200°C for 1 hr, the reaction was carried out as described above.

2.5 Infrared spectra measurements

For spectra measurements of hydroxyl and carboxyl groups, a metal hydroxide was pressed into a disk and evacuated at various temperatures for 2 hrs in a vacuum in situ cell of quartz having sodium chloride windows. After cooling to room temperature, spectra were measured by using a Hitachi 215 infrared spectrophotometer. For a study on the interaction of 1-butene with catalyst surface, the spectra of deuterated calcium oxide before and after the contact of 1-butene (51 mmHg) were observed. The deuteration of hydroxyl groups was made by repeating three times the procedure of introducing heavy water (10 mmHg) onto calcium oxide kept at 400°C and of evacuating for 5 min. After the deuterated catalyst was evacuated at 500°C for 2 hrs, 1-butene was introduced at room temperature and then evacuated at the same temperature.

3. RESULTS AND INTERPRETATION

3.1 Basic properties of catalysts

The basicities (amounts of base) at various basic strengths of calcium and magnesium hydroxides calcined at various temperatures in air are shown in figs. 1 and 2. As preheating

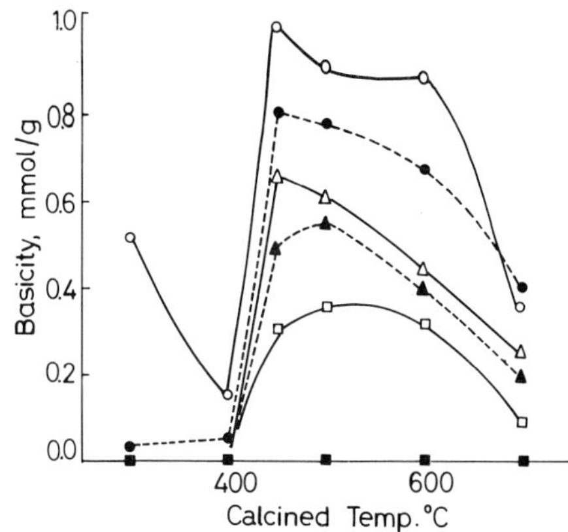

Fig. 1 Basicities at various basic strengths of CaO calcined at various temperatures in air.
○ pKa=7.1, ● pKa=12.2, △pKa=15.0, ▲pKa=17.2
□ pKa=18.4, ■ pKa=26.5

temperature is raised, the basicities at basic strengths of pKa=7.1-18.4 increase rapidly and attain maximum values and then decrease. The basicity maxima appeared at 450-500°C for calcium oxide and at 600°C for magnesium oxide. It was shown previously that the basicity change is not due to the change of surface area in the case of calcium oxide.[2] Comparison of fig. 1 with fig. 2 shows that the basicities at any basic strengths of calcium oxide are higher than those of magnesium oxide. Figs. 1 and 2 further indicate that the basic strength of both calcium and magnesium oxides is not as strong as pKa=26.5.

3.2 Activity and selectivity of calcium oxide

Calcium oxide prepared by calcining its hydroxide *in vacuo* showed pronounced catalytic activity and selectivity for the isomerization of 1-butene. For example, 63% of 1-butene isomerized to cis- and trans-2-butene at 30°C in 20 min when only 17 mg of calcium oxide evacuated at 600°C was used. No other products were found. The activity of the catalyst was about one hundred times higher than that of silica-alumina. However, no reaction occurred even at 200°C in 120 min with 140 mg of calcium oxide prepared by calcining the hydroxide in air for 3 hrs at 350-900°C. The catalytic activity *k* (cf. ref. 9) increases sharply with rise of evacuation temperature as shown by open circles in fig. 3. The ratio of cis- to trans-2-butene at zero conversion, obtained by extrapolation, is highest (ca.8) at an evacuation temperature of 400°C and decreases to less than 1

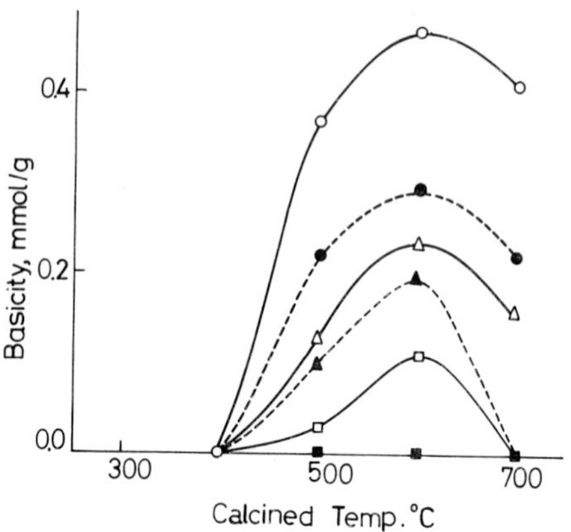

Fig. 2 Basicities at various basic strengths of MgO calcined at various temperatures in air. The same symbols are used as in fig. 1.

above 600°C.

In order to interpret the changes in observed activity and selectivity, the degrees of dehydration and decarboxylation of

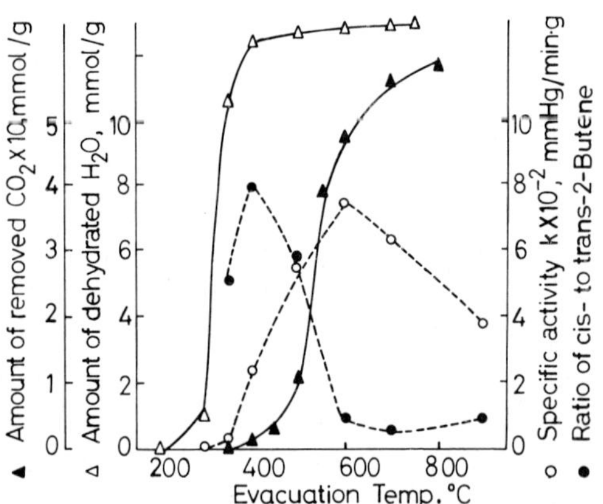

Fig. 3 Degrees of dehydration and decarboxylation from $Ca(OH)_2$ at various evacuation temperatures and activity and selectivity of the catalysts at 30°C.

calcium oxide were examined. As shown in fig. 3, water begins to be dehydrated sharply by evacuation above 300°C and the dehydration is almost complete at 400°C, while carbon dioxide begin to be removed sharply above around 450°C. It is seen in fig. 3 that the evacuation temperature at which dehydration begins almost coincides with the temperature at which activity begins to appear, whereas the evacuation temperature at which decarboxylation begins coincides well with the temperature at which selectivity begins to change, i.e., the temperature at which trans-2-butene begins to predominantly form.

3.3 Activity and selectivity of magnesium oxide

The activity, selectivity and the degrees of dehydration and decarboxylation vs. evacuation temperature are shown in fig. 4. Magnesium oxide becomes active by evacuation at 400-500°C. In the isomerization, cis-2-butene forms predominantly with the catalyst evacuated below 500°C, while a predominant formation of trans-2-butene is observed when evacuated above 500°C. Again, similarly as in the case of calcium oxide, no reaction took place even at 200°C when magnesium oxide was calcined in air. On the other hand, both the dehydration and decarboxylation begin by evacuation at 200°C and the latter is complete at around 700°C (fig. 4).

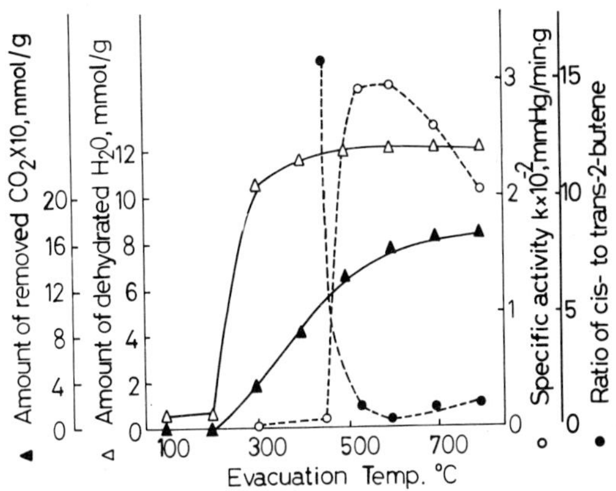

Fig. 4 Degrees of dehydration and decarboxylation from $Mg(OH)_2$ at various evacuation temperature and activity and selectivity of the catalysts at 30°C.

3.4 Poisoning of catalysts with several reagents

The results of poisoning of calcium oxide with ammonia, water, carbon dioxide and n-butylamine are shown in Table 1. Both water and carbon dioxide completely poisoned the active sites for the isomerizations of 1-butene to both cis- and trans-2-butene. The poisoning with ammonia greatly decreased the activities for the isomerization of 1-butene to trans-2-butene of calcium oxide evacuated at 500 and 700°C, also considerably decreased the activity for the isomerization of 1-butene to cis-2-butene of the catalyst evacuated at 500°C. However, the activities of the catalyst at 700°C evacuation for the shift of 1-butene to cis-2-butene were increased by poisoning with ammonia. The poisoning with n-butylamine decreased greatly the shift of 1-butene to trans-2-butene, but only slightly the shift of 1-butene to cis-2-butene. The observed activity increase by poisoning might be attributed to the increase of basic sites by ammonia adsorption.

In the case of magnesium oxide, ammonia greatly retards the activity for the reaction of 1-butene to trans-2-butene,

Table 1

Poisoning of CaO with NH_3, H_2O, CO_2 etc.

Evacuation temp. °C	Poisoning Reagents	Pressure mmHg	Activity of poisoned CaO*** / Activity of CaO without poisoning: 1- to cis-**	1- to trans-
500	NH_3	6.0	0.56	0.38
	H_2O	6.8	0.00	0.00
	CO_2	7.1	0.00	0.00
700	NH_3	4.0-11.0	1.46	0.38
	H_2O	2.2	0.00	0.00
	CO_2	4.7	0.00	0.00
	BA*	3.8	0.77	0.04

*n-butylamine
**1-:1-butene, cis-:cis-2-butene, trans-:trans-2-butene
***reaction temperature:30°C

but only slightly the activity for the reaction of 1-butene to cis-2-butene (Table 2).

Table 2

Poisoning of MgO with NH_3

Evacuation temp. °C	Evacuation temp. of NH_3 °C	Activity of poisoned MgO* / Activity of MgO without poisoning: 1- to cis-	1- to trans-
700	30	1.15	0.25
	110	0.62	0.19
	200	0.86	0.42

*reaction temperature:30°C

3.5 Infrared study on calcium and magnesium oxides

The change in the structure of carbon dioxide included in calcium hydroxide with the change of evacuation temperature was studied by observing the infrared spectra. The spectra are shown in Fig. 5, where the absorption bands at 1560, 1460, 1420 and 1310 cm^{-1} observed when evacuated at 500°C are considered as the bands of carbonates which are formed by splitting of its ν_3 symmetric vibration. The bands at 1460 and 1420 cm^{-1} are assigned to carbonate ion-like monodentate carbonate.[13] These bands almost disappear when evacuated at 600°C.

The observation of spectra in the absorption range of OD and OH showed that the OD band of deuterated calcium oxide did not change by contact with 1-butene and no OH group was newly formed. Therefore, the OD groups do not seem to interact with 1-butene.

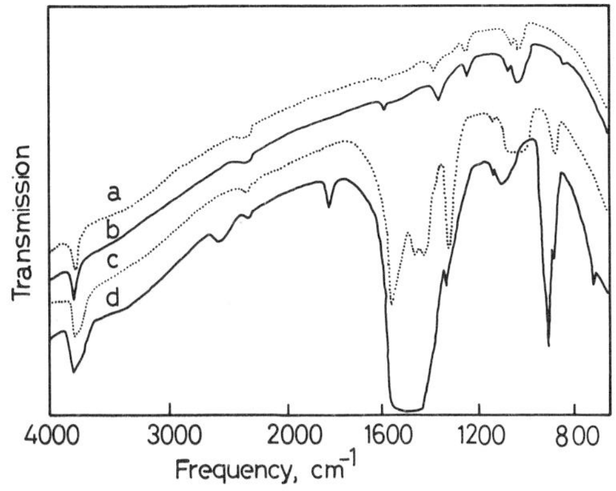

Fig. 5 Infrared spectra of CaO evacuated at various temperatures. a:evacuated at 700°C, b:600°C, c:500°C, d:400°C.

The infrared spectra of magnesium oxide evacuated at various temperatures are shown in fig. 6. The absorption bands at 1520, 1470 and 1330 cm^{-1} were observed when the catalyst was evacuated at 400°C, but the 1520 cm^{-1} band disappeared and the 1330 cm^{-1} band shifted to 1340 cm^{-1} by evacuation at 500°C. The 1330 cm^{-1} band is a superposed one of two bands; one pairing

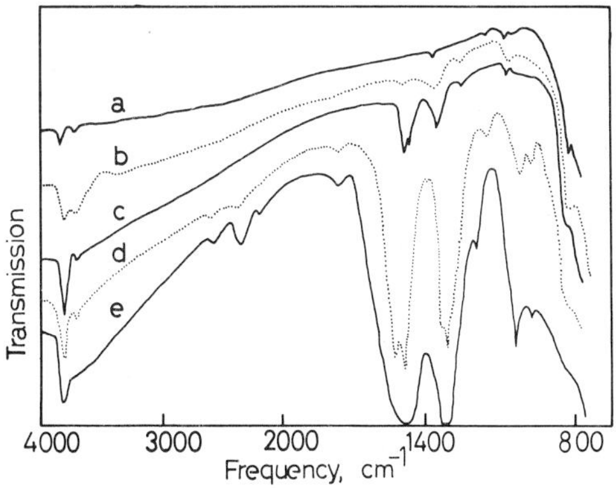

Fig. 6 Infrared spectra of MgO evacuated at various temperatures. a:evacuated at 700°C, b:600°C, c:500°C, d:400°C, e:300°C.

with the 1520 cm^{-1} band and the other pairing with the 1470 cm^{-1} band. The fact that the bands at 1520 and 1330 cm^{-1} disappeared and the bands at 1470 and 1340 cm^{-1} remained indicates that the carbonate showing the 1520 and 1330 cm^{-1} bands is removed by evacuation at lower temperatures.

4. DISCUSSION

Calcium and magnesium oxides showed remarkable activities and selectivities for the isomerization of 1-butene when calcined _in vacuo_, but they did not show any activities when calcined in air. This is because the active sites are completely covered with a small amounts of water and carbon dioxide when the catalysts were calcined in air (see Table 1). Basic properties could be measured on the surfaces of both oxides calcined in air by means of benzoic acid titration, because carbon dioxide or water adsorbed on basic sites was replaced by benzoic acid molecules, though they were not replaced by 1-butene.

There are two types of basic sites on calcium oxide:[2)] one is strongly basic O^{2-} ions which occupy a large fraction of the basic sites and the other is weakly basic OH groups which occupy a small fraction of basic sites.[2)] However, the OH groups are not active for the isomerization of 1-butene according to infrared studies described in 3.5. Since basic O^{2-} ions appear by dehydration and dehydration correlates with the activity for the formation of cis-2-butene from 1-butene (see fig. 3), the basic O^{2-} sites on calcium oxide are considered to be the main active sites for the isomerization. In the case of magnesium oxide, both dehydration and decarboxylation begin at 200°C, but the activity begins to appear by 400°C evacuation (fig. 4). Therefore, the main active sites for cis-2-butene formation seem to be the basic O^{2-} sites which are formed by removing the carbonate having infrared absorption bands at 1520 and 1330 cm^{-1} and not by dehydration (cf.3.5). Though the conditions of catalyst treatment are different, rough correlations observed between basicity increases (figs. 1 and 2) and activity increases for cis-2-butene formation (figs. 3 and 4) also indicate that the active sites are mainly basic. High ratios of cis- to trans-2-butene (ca.8 in CaO and ca.16 in MgO) observed with the catalysts evacuated at 400-500°C further support the above conclusion, because it is reported that preferential formation of cis-2-butene is observed usually on basic catalysts.[14,15)] However, as the active sites for cis-2-butene formation are partially poisoned by ammonia (see the value at 500°C evacuation in Table 1), acidic sites also may play a role in the activity. Both selective poisoning with ammonia or butyl amine and regulation of dehydration temperature offer means of control of selectivity of the isomerization.

The catalytic selectivities of the oxides evacuated above 500°C are different from those evacuated below 500°C. The former catalysts were found to be active for the formation of trans-2-butene from 1-butene. Since the activities for trans-2-butene formation correlate with decarboxylation of the catalysts (cf.3.2, 3.3, 3.5) and are decreased greatly by poisoning with ammonia, the active sites are considered to be mainly acidic, though the participation of basic sites for trans-2-butene formation cannot be excluded. It is likely that the acidic sites are Ca^{++} or Mg^{++} exposed by decarboxylation. In the case of magnesium oxide, the active sites for trans-2-butene formation are created by removing one type of carbonate having infrared absorption bands at 1470 and 1340 cm^{-1}.

Though the nature of active sites on calcium and magnesium oxides seems to have been considerably elucidated, the reaction mechanism for the isomerization of butene over these catalysts is not yet clear. However, the observed high selectivities and acid-base properties of the catalysts strongly suggest that the isomerization over the catalysts evacuated at 400-500°C proceeds via the anionic π-allyl intermediate which was proposed by Kokes[16)] for the same reaction over zinc oxide catalyst, since the latter catalyst also gives high selectivity and acid-base pair sites are considered to be active for the isomerization.

REFERENCES

1) For example, see a review article of J. W. Hightower and W. K. Hall, Chem. Eng. Prog. Symposium Series, 63 (1959), 122.
2) T. Iizuka, H. Hattori, Y. Ohno, J. Sohma and K. Tanabe, J. Catal. 22 (1971), 130.
3) O. V. Krylov, Z. A. Markova, I. I. Tretiakov and E. A. Fokina, Kinet. Katal. 6 (1965), 128.
4) S. Malinowski, S. Szczepanska and J. Sloczynski, J. Catal. 7 (1967), 67.
5) S. Malinowski, S. Szczepanska, A. Bielanski and J. Sloczynski, J. Catal. 4 (1965), 324.
6) G. Kortüm, Angew. Chem. 70 (1958) 651.
7) H. Zeitlin, R. Frei and M. McCarter, J. Amer. Chem. Soc. 87 (1965), 1857.
8) H. E. Zaugg and A. D. Schaffer, J. Amer. Chem. Soc. 87 (1965), 1857.
9) K. Tanabe, N. Yoshii and H. Hattori, Chem. Comm. (1971), 464.
10) Y. Schächter and H. Pines, J. Catal. 11 (1968) 147.
11) A. Clark and J. N. Finch, 4th International Congress on Catalysis, Moscow, (1968), Preprint of Paper No. 75.
12) N. F. Foster and R. J. Cvetanovic, J. Amer. Chem. Soc. 82 (1960), 4274.
13) M. L. Hair "Infrared Spectroscopy in Surface Chemistry," Marcel Dekker, Inc., New York (1967) Chapt. 5.
14) W. O. Haag and H. Pines, J. Amer. Chem. Soc. 82 (1960), 387, 2488.
15) S. Bank, A. Schriesheim and C. A. Rowe, Jr., J. Amer. Chem. Soc. 87 (1965), 3224.
16) L. Dent and R. J. Kokes, J. Amer. Chem. Soc. 92 (1970), 1092, 6709, 6718; R. J. Kokes, to be published in the Herman Pines volume of Intra-Science Chemistry Reports (1971).

DISCUSSION

Z. G. SZABÓ

Some time ago we were also engaged in measuring the surface basicity and found e.g., on MgO the number of centers of strength $pk_a \geq 9.3$ as $2 \times 10^{20}/g$, while Professor Tanabe and his associates got for the strengths $pk_a \geq 12.2$ as $1.8 \times 10^{20}/g$.

They also assume that the cis-2-butene are formed on basic sites, while the trans-product forms on acidic sites. My coworker, Mr. Jover, and I wonder if a more simple and probable explanation would not be this: on basic sites at first only cis-2-butene is formed, and the trans species is formed by isomerization on acidic sites. Perhaps the following scheme could be written:

$$\text{1-Bu} \xrightarrow{k_1} \text{cis-2-Bu} \xrightarrow{k_2} \text{t-2-Bu.}$$

A basic poison can cause the enhancement of cis-2-Bu if k_2 decreases more than k_1. This scheme can also easily explain the course of the cis-trans-ratio in Figure 3.

H. HATTORI

The selectivities shown in Figs. 3 and 4 were obtained by the extrapolation of the ratio of cis-/trans- to zero conversion. If trans-2-butene were formed from cis-2-butene by the successive reaction as you mentioned, the ratio of cis-/trans- would be infinity at zero conversion. Moreover, the rate of the formation of trans-2-butene from 1-butene is faster than that of trans-2-butene from cis-2-butene over the catalyst pretreated at 500°C. This means trans-2-butene can be formed directly from 1-butene.

C. DANIEL

If you suggest a π-allyl intermediate, do you expect to form 1,3-butadiene (or benzene from propylene) on CaO and MgO? In similar work on a Sr catalyst we observed that 1-butene gives butadiene, and some Japanese work reports benzene from bismuth arsenate. Is it a single acid site that gives the allyl intermediate?

H. HATTORI

In our experiments, we could not detect butadienes nor hydrogen. It may be due to the low reaction temperature, 30°C. It may be possible to form butadienes at higher temperature.

We believe that no single acid sites but acid-base pair sites give the allyl intermediates.

H. NOLLER

I would like to draw your attention to steric effects. You have cis-preference when the surface is covered with CO_2, and trans-preference when the surface is not covered.

We have a corresponding result with the dehydrochloridination of 2-butyl chloride over salt catalysts like $NiSO_4 \cdot x\ H_2O$, $CoSO_4 \cdot x\ H_2O$, $ZnSO_4 \cdot x\ H_2O$. When the water content of the catalyst is high, cis-2-butene is preferred; when the water is removed, trans-2-butene is favored. We believe that this is due to steric effects. If there is anything on the surface, the less bulky product is favored, i.e., cis-2-butene. When the surface is uncovered, the thermodynamically more stable trans-2-olefin can be formed.

H. HATTORI

It is difficult to rule out the possibility of the steric effect as you suggested. If the steric effect is the main factor in determining the product ratio cis-/trans-, it would be expected in the poisoning experiments that the ratio depends mainly on the amount of poison and is independent of their properties. We have some results that the ratio depends upon the property of poisoning reagents, and these results will be soon presented elsewhere.

In connection with your comment, we would also like to make a comment on your results that the selectivity of the products from 2-butyl chloride depends on the water content of the catalyst. The acidic properties of your catalysts are quite dependent on the dehydration temperature. For example, $NiSO_4 \cdot XH_2O$ has maximum Bronsted acidity after the dehydration at 250°C and maximum Lewis acidity at 350°C.[1] There is a possibility that the nature of acid sites on the surface controlled the selectivity in your case.

1) Hattori, H., Miyashita, S. and Tanabe, K., Bull, Chem. Soc., Japan, 44, 893 (1971).

O. V. KRYLOV

In connection with the paper of Hattori et al., I would like to note that some results can be explained by existence of unusual basic and acid active sites such as oxygen anions and magnesium cations situated close to vacancies (F- and V-sites). Existence of such sites was proved by Boudart in his recent paper reported at the last congress on reactivity of solids. The same facts were observed also in our laboratory. My opinion is that fine infrared structure observed by Hattori, et al. can be explained in such a way. In this case the number of active sites must be small.

H. HATTORI

We appreciate your comment on the paper. We are trying to determine the nature and the number of active sites on the surface more precisely.

In coincidence with your prediction, a very small number of the active sites can be expected judging from the poisoning experiments.

Paper Number 11

THE CATALYTIC ACTIVITY OF SUPPORTED ZINC AND IRON OXIDES

R. UMA, R. VENKATACHALAM and J. C. KURIACOSE
Department of Chemistry, Indian Institute of Technology
Madras, India

ABSTRACT: $ZnO-TiO_2$, ZnO-NgO, $ZnO-Al_2O_3$ and $Fe_2O_3-Al_2O_3$ catalysts have been studied for the decomposition of 2-propanol. ZnO prepared from the nitrate is catalytically inactive and its X-ray diffraction pattern reveals a defect structure. Similarly prepared ZnO catalysts show the normal X-ray pattern for ZnO and an increased dehydrogenation activity. The activity in $ZnO-TiO_2$ and ZnO-MgO catalysts is due to the ZnO/support interface while that of $ZnO-Al_2O_3$ catalysts to the interface between ZnO and $ZnAl_2O_4$.

Fe_2O_3 loses its inherent dehydrogenation activity when supported on Al_2O_3 and exhibits an increased dehydration activity. This results from an interaction between Fe_2O_3 and Al_2O_3, which facilitates the transformation of α-Fe_2O_3 to γ-Fe_2O_3 does not affect the structural identity of Fe_2O_3.

1. INTRODUCTION

Catalysts containing two or more metal oxides often have activities and selectivities superior to those of the pure oxides. In most cases the origin of higher activity of these solids is not well understood. We selected ZnO supported on MgO, on Al_2O_3 and TiO_2 and Fe_2O_3 on Al_2O_3 for a study of their activity in the catalytic decomposition of 2-propanol for these reasons.

MgO, Al_2O_3 and TiO_2 differ in their structures and cationic charges. They have different types of activity towards the decomposition of 2-propanol but when mixed with ZnO all of them exhibit enhanced dehydrogenation activity. On the other hand, Fe_2O_3 which is a predominantly dehydrogenating catalyst, becomes a dehydration catalyst on mixing with Al_2O_3. These systems provide examples of different types of supports enhancing the same kind of activity, and the same support enhancing two different types of activity. Hence ZnO supported separately on MgO, Al_2O_3 and TiO_2 and Fe_2O_3 on Al_2O_3 have been chosen for a study of their behavior in the catalytic decomposition of 2-propanol with a view to understand the origin of the modification in activity and selectivity of the mixed oxides.

2. EXPERIMENTAL

Weighed quantities of ZnO (May and Baker) are dissolved in A.R. HNO_3, and the supporting oxides are impregnated with this solution. After standing overnight, the slurry is decomposed at 400°C for six hours. The $Fe_2O_3-Al_2O_3$ catalysts are prepared by the coprecipitation of the hydroxides starting from $Fe(NO_3)_3 \cdot 9H_2O$ ("Baker Analyzed" Reagent) and $Al(NO_3)_3 \cdot 9H_2O$ (BDH Analar) and subsequent dehydration at 120°C followed by baking at 500°C.[1)] The composition of these catalysts is given in terms of percentage weight of ZnO or Fe_2O_3 as the case may be.

The catalytic activity of the mixed oxides is determined by decomposing 2-propanol over known weights of the catalyst, in a flow type reactor.[2] The products are analyzed gas chromatographically. The amount of ZnO in the catalysts is estimated by titration against a standard solution of $K_4Fe(CN)_6$ and iron by titration against standard $K_2Cr_2O_7$ solution. Surface areas of the catalysts are calculated from the nitrogen adsorption isotherms using the B.E.T. equation. Electrical resistance is measured by the two probe technique using a Keithley Electrometer. The solid phases present in the catalysts are identified from the X-ray diffractograms or from the powder photographs. CuK_α radiation is used for the ZnO catalysts and Fe radiation (without filter) is used for the catalysts containing iron oxide.

3. RESULTS

Pure ZnO prepared by thermal decomposition of the nitrate is yellow and is inactive for the decomposition of 2-propanol. X-ray diffraction lines of this sample (referred to as inactive ZnO) reveal an unusual sequence for their intensities, quite different from those given by a May and Baker sample. The latter (referred to as active ZnO) dehydrogenates 2-propanol and shows the same sequence of intensities as that reported in literature.[3] X-ray diffraction lines of ZnO in the mixed catalysts are arranged in decreasing order of intensities in Table 1. X-ray diffractograms of ZnO-MgO and ZnO-TiO_2 catalysts show that they exist as mere mechanical mixtures of the components. The 27 per cent ZnO-Al_2O_3 catalyst shows an X-ray pattern corresponding to $ZnAl_2O_4$ and ZnO. The nature of Al_2O_3 present cannot be inferred from the diffractograms. The diffraction lines of ZnO in ZnO-MgO and ZnO-Al_2O_3 catalysts have intensities which correspond to those of active ZnO. In the 10 per cent ZnO-TiO_2 catalyst the lines of ZnO are not sharp enough for a comparison of their intensities.

The variation in surface area with composition for the different ZnO catalysts is shown in Fig. 1.

TABLE 1
Sequence of intensities of X-ray diffraction lines of ZnO in different catalysts

Catalyst	d values for the lines of ZnO in the order of decreasing intensity Å		
	I	II	III
ZnO (reported)	2.476	2.816	2.602
ZnO (May and Baker)	2.463	2.841	2.601
ZnO from the nitrate	2.841	2.463	1.353
14 per cent ZnO-MgO impregnated	2.477	2.818	2.601
27 per cent ZnO-Al_2O_3 impregnated	2.463	2.798	2.601
64 per cent ZnO-TiO_2 impregnated	2.465	2.838	2.605

In the case of $ZnO-TiO_2$ and $ZnO-MgO$ the apparent activation energy for the dehydrogenation reaction is independent of composition. For $ZnO-Al_2O_3$ which has both dehydrogenation and dehydration activities the total conversion is too high for the calculated energy of activation to be meaningful. Fig 2. shows the variation in dehydrogenation activity per gram of the catalyst (specific activity) with composition. For all three series of catalysts the specific activity rises, reaches a maximum and then falls as ZnO content increases.

X-ray evidence shows that in freshly prepared $Fe_2O_3-Al_2O_3$ catalysts, Fe_2O_3 is present as haematite ($\alpha-Fe_2O_3$). This series is designated as $\alpha-Fe_2O_3-Al_2O_3$. The variation in the electrical resistance with temperature of typical catalysts is shown in Fig. 3. The break characteristic of pure Fe_2O_3 is seen only in the case of 85 per cent $Fe_2O_3-Al_2O_3$ catalyst. Catalysts rich in Fe_2O_3 show feeble dehydrogenation activity. The aged catalysts in the oxidized form (referred to as $\gamma-Fe_2O_3-Al_2O_3$) contain $\gamma-Fe_2O_3$. X-ray lines of Al_2O_3 are not detected in any of these catalysts, though in catalysts calcined at 800°C the presence of $\alpha-Al_2O_3$ can be identified. It is likely that Fe_2O_3 has an inhibiting effect on the crystallization of Al_2O_3 since pure Al_2O_3 prepared under similar conditions has the γ-structure. Whether it is the fresh or the aged sample, 2-propanol reduces Fe_2O_3 to the magnetic stage (Fe_3O_4) during its decomposition. In $\alpha-Fe_2O_3-Al_2O_3$ catalysts the reduction is only partial and in $\gamma-Fe_2O_3-Al_2O_3$ catalysts reduction to magnetite is complete. The changes in the activity of a 66 per cent $Fe_2O_3-Al_2O_3$ catalyst brought about by treatment with different acidic and basic reagents are given in Table 2.

TABLE 2
Effect of pretreatment on the activity of a ferric oxide-alumina (66% Fe_2O_3) catalyst

No.	Description of the treatment of catalyst	Relative Activity	
		De-hydration	De-hydrogenation
1	--	1.00	1.00
2	Treatment with acetic acid and regeneration	0.72	2.86
3	Treatment with propionic acid and regeneration	0.44	4.45
4	Treatment of the acid treated catalyst with pyridine and regeneration	0.44	4.63
5	Four alternate treatments with phenol and 2-propanol mixture in equal proportions and regeneration	1.00	0.96
6	Treating the above catalyst with 50 mole per cent acetic acid and 2-propanol mixture and regeneration	0.65	2.15

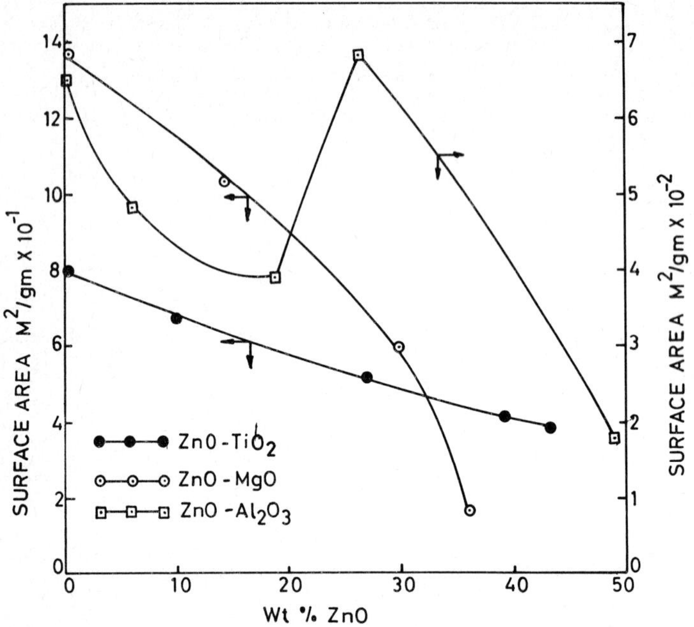

FIG.1. VARIATION OF SURFACE AREA WITH COMPOSITION

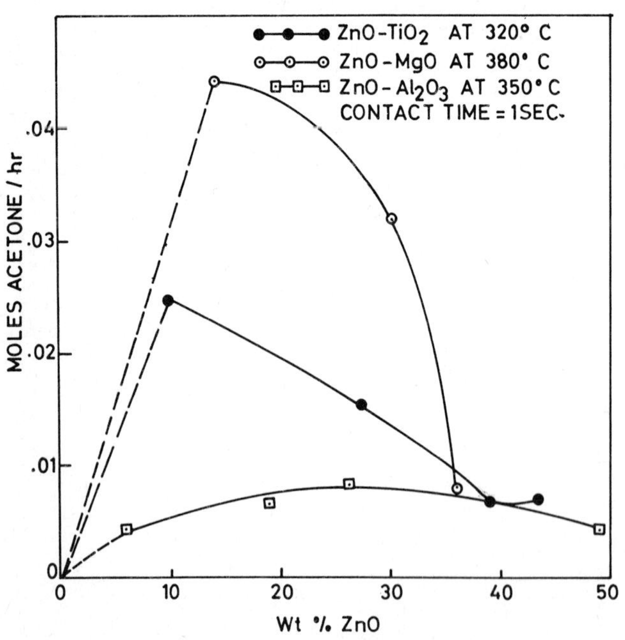

FIG.2. VARIATION OF DEHYDROGENATION ACTIVITY PER GRAM OF CATALYST WITH COMPOSITION.

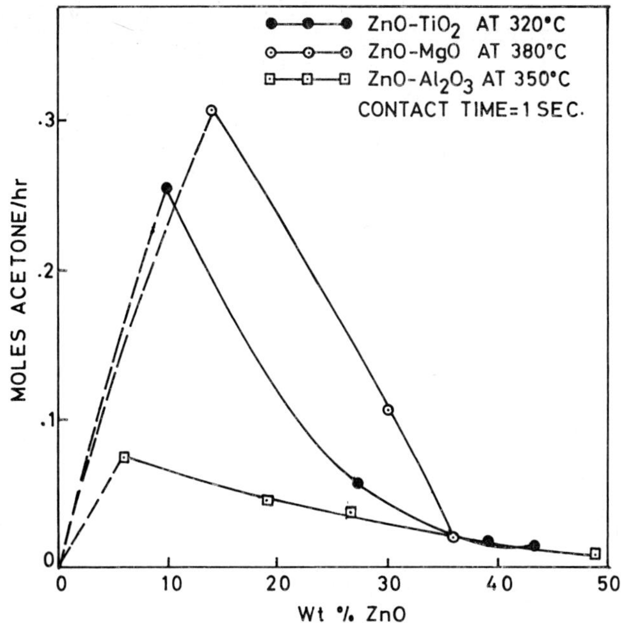

FIG.3. VARIATION OF DEHYDROGENATION ACTIVITY PER GRAM OF ZnO WITH COMPOSITION.

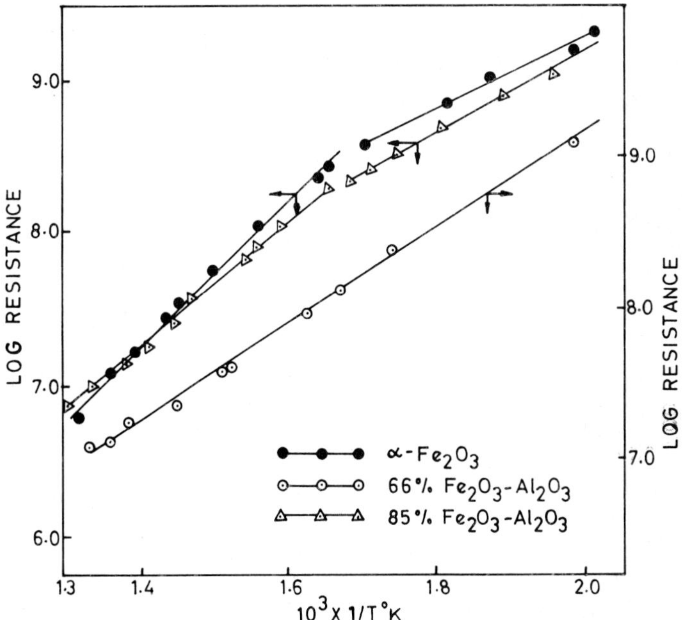

FIG.4. VARIATION OF THE ELECTRICAL RESISTANCE WITH TEMPERATURE.

4. DISCUSSION

The unusual sequence of intensities of X-ray lines shown by the inactive sample of ZnO indicates an alteration of the atomic positions in the lattice which can result in an unfavorable geometry for the dehydrogenation reaction. There is a gradation in the catalytic properties of the supporting oxides: Al_2O_3 is a purely dehydrating catalyst, MgO is purely dehydrogenating and TiO_2 is predominantly dehydrating with a slight dehydrogenation activity also. All the mixed oxide catalysts, however, have dehydrogenation activities superior to those of the component oxides. ZnO and TiO_2 are n-type semiconductors while MgO and Al_2O_3 are insulators. However, in the temperature range employed all the three mixed oxides behave as insulators except ZnO-TiO_2 catalysts containing higher percentages of ZnO. Since different lots of the catalyst are used for experiments, the relative activities are given. Although for a catalyst showing a steady activity the relative activities of both dehydration and dehydrogenation are taken as unity, the absolute dehydrogenation activity is 1/10 to 1/20 of the dehydration activity.

The possible causes for the enhanced dehydrogenation activity are the following: (i) Dispersion of ZnO in an active form over the supporting oxide, (ii) Formation of an active interface between ZnO and the support, (iii) Formation of a new compound by the interaction between the component oxides which is active for dehydrogenation and (iv) Formation of an active interface between the new compound and ZnO. (i) and (ii) are applicable to all the three systems. In fact it is difficult to separate (ii) completely from (i) because with increased dispersion there will be increased interface also. (iii) and (iv) are not applicable to ZnO-MgO and ZnO-TiO_2 catalysts because X-ray studies give no evidence for the formation of any new compound.

In the ZnO TiO_2 system the highest activity is seen for catalysts containing 10 per cent ZnO and in the ZnO-MgO system for the catalyst containing 14 per cent ZnO. X-ray analysis of this ZnO-MgO catalyst shows that ZnO exists in the active form. No definite conclusion about the nature of ZnO in the ZnO-TiO_2 catalyst can be drawn because the ZnO lines appear very diffuse. However, the fact that ZnO is present in the active form in 64 per cent ZnO-TiO_2 suggests that the same form is stabilized at lower concentrations of ZnO as well. This ZnO in the active form may be responsible for the higher activity. The rise in specific activity with increasing ZnO content may be due to the increased amount of active ZnO which may be dispersed on the surface and the consequent increase in the interface between ZnO and the support. The occurrence of maximum activity at the same composition, whether one considers activity per gram of the catalyst or per gram of ZnO is explained on the basis of the same assumption. If activity is proportional to the concentration of ZnO alone and not the physical state in which it is present, the activity pattern should show a regular increase. But the decrease in activity with increase in ZnO content that is observed must mean that after a certain amount of ZnO has been added the aggregation of ZnO sets in which causes the activity to fall. This aggregation leads to a decrease in the extent of the interface.

In the case of the $ZnO-Al_2O_3$ system the maximum in specific activity occurs at 27 per cent ZnO while for the activity per gram of ZnO it occurs at 6 per cent ZnO. A 6 per cent $ZnO-Al_2O_3$ catalyst prepared by coprecipitation and subsequent decomposition of the hydroxides gives evidence for the formation of $ZnAl_2O_4$. Only a very small quantity of free ZnO is present. This catalyst, however, does not dehydrogenate at all. So neither $ZnAl_2O_4$ nor $ZnAl_2O_4-Al_2O_3$ interface can be the seat of activity. Several workers[4,5)] have reported the dehydrogenation activity in this system to be associated with the formation of the spinel, $ZnAl_2O_4$. If $ZnO-ZnAl_2O_4$ interface acts as the seat of dehydrogenation activity one can understand why the 6 per cent $ZnO-Al_2O_3$ catalyst prepared by coprecipitation is inactive while the impregnated 6 per cent $ZnO-Al_2O_3$ catalyst is active. In the impregnated catalyst some free ZnO is present as can be seen from the X-ray pictures while in the coprecipitated catalyst almost all the ZnO is used for spinel formation because of the better mixing achieved. The consequent presence of a $ZnO-ZnAl_2O_4$ interface confers activity on the impregnated catalyst.

The dehydrogenation activity which is inherent in pure Fe_2O_3 is completely suppressed in $Fe_2O_3-Al_2O_3$ catalysts. A study of $Fe_2O_3-Al_2O_3$ catalysts of different compositions shows that a catalyst containing 8 per cent Fe_2O_3 has a better dehydration activity than even pure Al_2O_3. This seems to be due to an interaction between the two oxides. Considering the nature and extent of modification of the catalytic properties of Fe_2O_3 by Al_2O_3, the interaction between them must be stronger than that in the case of ZnO-MgO and $ZnO-TiO_2$ catalysts. The close values of the radii of Fe^{3+} (0.64 Å) and Al^{3+}(0.50 Å) ions and the similarity in the structures of their sesquioxides suggest that the most facile type of interaction between these two oxides would be the formation of a substitutional solid solution.

All the catalysts used even though prepared at temperatures below 500°C may contain at least a small amount of the solid solution. The X-ray pattern of Fe_2O_3 in the mixed oxides remains unaltered, indicating the presence of large amounts of free Fe_2O_3. In spite of this, the completely altered activity of Fe_2O_3 suggests that the catalytic activity of Fe_2O_3 is strongly influenced by its evnironment. In the absence of X-ray evidence for the presence of Al_2O_3 in a crystalline form in the α-$Fe_2O_3-Al_2O_3$ catalysts, one might conclude that α-Fe_2O_3 exerts an inhibitive effect on the crystallization of Al_2O_3. After sintering under the same conditions pure γ-Al_2O_3 remains unaltered in structure while $Fe_2O_3-Al_2O_3$ catalysts show the presence of α-Al_2O_3. Thus even though a new structure is not formed there is a strong interaction between the two oxides. The absence of a break in the log R vs 1/T plot of the mixed catalysts further shows that the uncombined Fe_2O_3 in $Fe_2O_3-Al_2O_3$ is not similar to pure Fe_2O_3 in all respects. Only the catalysts having a Fe_2O_3 content as high as 85 per cent which have a slight dehydrogenation activity show the break characteristic of pure α-Fe_2O_3. One can, therefore, conclude that Fe_2O_3 even if it is not structurally modified has its properties considerably altered by the presence of Al_2O_3.

The transformation of α-Fe_2O_3 to γ-Fe_2O_3 which takes place with use of the catalyst for decomposing 2-propanol, is complete in the case of catalysts that are finely powdered. It is likely that this transformation is due to the alternate reduction and regenerative oxidation. Pouillard[6)] has observed that the solid solution of α-$Fe_2O_3 \cdot Al_2O_3$ on reduction followed by oxidation forms the solid solution γ-$Fe_2O_3 \cdot Al_2O_3$. The unusual stability of γ-Fe_2O_3 in these catalysts might, therefore, result from an influence of $Fe_2O_3 \cdot Al_2O_3$ solid solution on the entire mass of Fe_2O_3 in contact with it. The solid solution even if present in small quantities is the nucleus of the phase transformation occurring in Fe_2O_3.

Table 2 suggests that a purely dehydrating catalyst acquires some dehydrogenation activity after use for the ketonization of acetic acid or propionic acid. The activity is unaltered when the catalyst is treated with a base like pyridine or an acidic reagent like phenol. Thus the effect of carboxylic acids seems to be very specific and is possibly connected with the formation of iron salts during the ketonization and their subsequent conversion to active Fe_2O_3.

5. CONCLUSIONS

All three oxide supports used for ZnO enhance the dehydrogenation activity. MgO and TiO_2 help the formation and stabilization of ZnO in the active form. The interface between ZnO and the support is the seat of dehydrogenation activity in the ZnO-MgO and ZnO-TiO_2 catalysts. In the ZnO-Al_2O_3 catalysts the appearance of $ZnAl_2O_4$ phase is important since it is the interface between ZnO and $AnAl_2O_4$ that is responsible for the dehydrogenation activity. In Fe_2O_3-Al_2O_3 catalysts an interaction between the two oxides is responsible for the modification of the catalytic activity.

ACKNOWLEDGEMENTS

The authors are grateful to the Department of Metallurgy, Indian Institute of Technology and the Centre for Advanced Study of Physics, Madras University, for facilities provided for the X-ray work. Their thanks are due to Dr. Paul Ratnasamy (Laboratoire de Physico-chimie Minerale, Heverlee-Louvain, Belgium) and Mr. L. Chandrasekhar (Department of Metallurgy, Indian Institute of Technology) for helping with the X-ray analysis. Financial help from the C.S.I.R. (India) is gratefully acknowledged.

REFERENCES

1) V. K. Skarchemko, V. S. Frolova, I. T. Golubchenki, V. P. Musienko, P. N. Galich, Kinetika i Kataliz 5 (1964) 932.
2) J. C. Kuriacose, C. Daniel, R. Swaminathan, J. Catalysis 12 (1968) 19.
3) H. E. Swanson, R. K. Fuyat, Standard X-ray Diffraction Powder Patterns, National Bureau of Standards Circular 539, Vol. II (U.S. Government Printing Office, Washington 25, D.C., 1953).
4) B. C. Alsop, D. A. Dowden, Symposium on the structure and texture of catalysts, Paris, France, 1954.
5) A. Simon, Chr. Oehme, K. Pohl, A. Anorg. Allgem. Chem. 323 (1963) 160.
6) E. Pouillard, Ann. Chim. 5 (1950) 164.

DISCUSSION

F. STEINBACH

I would like to add 3 results concerning the catalytic activity of ZnO.

1) Pure ZnO is an active catalyst for the decomposition of propanol-2. In the temperature range between 180 and 340°C, in UV light and in the dark, propanol-2 vapour is dehydrogenated to acetone and, simultaneously, dehydrated to propylene. The catalyst is a ZnO powder of electrophotographic grade, BET surface 6 m^2/g. Within the pressure range of 2 to 30 Torr propanol-2, the dehydrogenation is a first order reaction with a strong inhibition by water; the dehydration is a zero order reaction. Though the acetone yield is higher than the propylene yield, the activation energy of the dehydrogenation is higher than that of the dehydration.

2) Activation energies and pre-exponential factors of both reactions are decreased by UV illumination and increasing age of the catalyst.

3) By very sensitive neutron activation analysis an oxygen depletion of the catalyst in dark and--more pronounced--in UV light is measured. The decrease of the activation energy is attributed to the loosening of the Zn-O bond in the surface of the catalyst due to promotion of electrons to non-bonding state. This occurs reversibly by excitation by UV light and irreversibly by oxygen depletion of the surface.

K. TANABE

We recently found that TiO_2 showed a large amount of strongly acid character when mixed with 10% ZnO. The acidity decreased with further increase of zinc oxide content. The acidity change is similar to your activity change. In this respect, I would like to hear your opinion about the possibility of the participation of acid sites as a part of active sites for the dehydrogenation of 2-propanol.

J. C. KURIACOSE

The catalysts prepared by heating titania impregnated with zinc nitrate show no dehydration activity. Qualitative tests using a solution of para-ethoxy-chrysoidine indicator in benzene suggest that the zinc oxide-titania catalysts are less acidic than pure titania. Pure titania shows some dehydration activity. On sintering titania, the same qualitative tests show that the acidity is considerably reduced. One also observes that the overall dehydration activity falls even though the energy of activation is lowered. One may conclude from this that acidity in the catalyst results in a certain amount of dehydration activity. Zinc oxide-titania catalysts that have been sintered at 800°C and also catalysts prepared by decomposing zinc nitrate impregnated on sintered titania, all exhibit only dehydrogenation activity. From these observations it seems unlikely that in the zinc oxide-titania catalyst preparations studied, acidity plays any major role in contributing to the dehydrogenation activity.

D. A. DOWDEN

The authors seem to have misunderstood the position taken by Alsop and in their 1954 paper. Stoichiometric, non-defective zinc aluminate spinel was found to be inactive. When the spinel contained dissolved alumina and cation defects, it had dehydration activity. The spinel dissolves little or no zinc oxide. Where there is excess zinc oxide--not enough to be detected by X-ray diffraction but enough to give the typical yellow green fluorescence --the activity is wholly dehydrogenation as one would expect.

J. C. KURIACOSE

Dr. Dowden must be referring to the following sentence in our paper: "Several workers have reported the dehydrogenation activity in this system to be associated with the formation of the spinel, $ZnAl_2O_4$." Since Dr. Dowden is one of the workers we have referred to, no comments are warranted on his clarifications of the work he has done. However, since it is suggested that we seem to have misunderstood the position taken by Alsop and Dowden, a few statements from Alsop and Dowden's 1954 paper are quoted to indicate the basis of our conclusion.

"La déshydrogenation sur le système oxyde de zinc-alumine n'a lieu qu'au voisinage de la composition du spinelle stoechiometrique." (summary) "La déshydratation est la seule réaction observée avec tous les catalyseurs sauf la composition spinelle et il est remarquable que les catalyseurs préchauffés à 500°C diffèrent de tous les autres par leur diminution régulière de la vitesse de déshydratation quand croit le taux de ZnO." (page 4) "La déshydrogenation a lieu pour la composition spinelle." (page 5) "L'extraction d'un catalyseur (ZnO-Al_2O_3 fritté à 500°C, qui ne donnait que la déshydrogenation par l'acide chlorhydrique dilué, suivie d'un lavage soigné jusqu'à élimination des ions chlore, enlève une partie du zinc et produit un catalyseur provoquant la déshydrogenation et la déshydratation en quantités egales." (page 5) "Les catalyseurs chauffés à température non supérieure à 500°C ne contiennent que les phase spinelle de zinc (plus ou moins stoechiométrique) et alumine. La déshydratation est la seule activité jusqu'à et y compris 4 ZnO - 6 Al_2O_3. Les catalyseurs chauffés à 800°C et au-dessus contiennent des solutions solides d'alumine dans le spinelle de zinc, qui doivent donc contenir des réseaux avec vacances de cations. La déshydratation est toujours la seule reaction jusqu'à et y compris 4 ZnO - 6 Al_2O_3, mais les activités spécifiques ne décroissent pas réguilièrement." (page 6) "La deshydrogenation n'apparait qu'à la composition spinelle et est associée à la presence d'une phase extractible qui est surtout ZnO." (page 6)

The above extracts will prove that there is no misunderstanding of the position taken by Alsop and Dowden in the 1954 paper.

Paper Number 12

ADSORPTION SITES ON OXIDES. INFRA-RED STUDIES OF ADSORPTION OF OXIDES OF NITROGEN

N. D. PARKYNS
Gas Council, London Research Station, London SW6 2AD, England

ABSTRACT: Nitrogen dioxide is strongly and irreversibly adsorbed on to transition aluminas. Infra-red spectra show that adsorption takes place with formation of NO_3^- NO^+ species. The NO^+ is relatively weakly held, probably to oxide sites : the NO_3^- ion is very strongly held by co-ordination through its oxygen atoms to Al^{3+} ion sites. Splitting of the absorption bands indicates that different co-ordination sites are present. Adsorption of NO_2 takes place on silica-alumina by a similar mechanism. There is no infra-red evidence for non-dissociative adsorption of nitric oxide on iron-free alumina, on silica or on silica-alumina. Gravimetric adsorption isotherms of NO_2 on alumina enable an estimate to be made of the maximum number of exposed Al^{3+} cations and this is compared to a model of the surface.

1. INTRODUCTION

Infra-red spectroscopy has found considerable application in determining the types of active adsorption sites on metal oxides of interest to catalyst chemists, in particular on alumina, silica, silica-alumina and molecular sieves[1-3].

Previous work in these laboratories showed that freshly prepared alumina in particular had transient oxidising properties for both carbon monoxide[4] and nitrogen[5]. At the same time, Pink and his co-workers[6] had shown that alumina and silica-alumina developed strongly electron-accepting and donating properties on activation by heat. Although Lunsford[7] had demonstrated that nitric oxide gives weak ESR signals when added to alumina, no infra-red spectroscopic work had been done on this or similar systems.

In the present paper, we have looked at the possibility of detecting electron accepting sites by infra-red spectroscopy, using the odd-electron molecules NO and NO_2.

2. EXPERIMENTAL

2.1 Materials

Alumina aerogels were made as described previously[4,8] where the properties were fully described. Alumina type C, made by Degussa, was a gift from Bush, Beach and Segner Bayley Ltd. It had a measured BET surface area of 104 m^2g^{-1} (N_2 gravimetric adsorption): impurity levels (measured by XRF) were :

Fe ~1% Zn ~0.01% Cl ~0.01% Ni,Cu, traces

Alumina labelled "very high purity Gibbsite" was obtained from Peter Spence and Sons Limited. This had a quoted BET surface area of 400 m^2g^{-1} after calcination at 420°C, and the χ crystal structure. Maximum impurity levels were Fe<30 p.p.m., Si<30 p.p.m., Ca<10 p.p.m, Cu<8 p.p.m., other impurities all<10 p.p.m.

Silica was Aerosil (Standard) and had a BET surface area of 175 m^2g^{-1}, the maximum impurities being quoted as HCl<0.025%, Al_2O_3<0.05%, TiO_2<0.03%.

Some samples of Spence alumina were deliberately contaminated by the addition of ferric chloride which was then heated in air at 600°C to form ferric oxide (1% by weight).

Silica-alumina catalyst containing 13% Al_2O_3 was obtained from Joseph Crosfield & Sons Ltd., and had a quoted surface area of 525 m^2g^{-1}.

No and NO_2 were Matheson Research grade gases. The nitric oxide was freed from possible nitrogen dioxide impurity by passing it through a liquid nitrogen/isopentane cold trap. The N_2 impurity (∿1%) was not removed.

2.2 Apparatus

This was similar to that described previously[4], although the spectra are now recorded on a Grubb-Parsons GS2/DB3 grating spectrometer. The cell for holding samples was modified to enable them to be heated to 1000°C.[9]

Spectra were processed by computer to give absorbance-wave number presentation as previously described[10].

Gravimetric adsorption isotherms were obtained on a Cahn RG vacuum microbalance.

2.3 Procedure

In most cases the discs of sample were made by pressing the finely divided powder to a thickness corresponding to 20 mg/cm^2. The sample was degassed in vacuo at 10^{-6} torr for two hours. On cooling the sample, the spectrum was taken, the test gas measured out and added and the spectrum re-run. The degassing time was arbitrarily chosen as being experimentally convenient: the bulk of degassed products had in fact been removed by then as shown by the pressure in the system falling to a low, constant level.

3. RESULTS

3.1 Nitric oxide on alumina

Degussa type C samples showed a characteristic broad band at 1820 ± 10 cm^{-1}. The spectra showed the following characteristics:

(a) The intensity of the 1820 cm^{-1} band increased with pressure up to a maximum of about 10 torr.

(b) The higher the temperature of pre-treatment the more intense the band for a given pressure of nitric oxide.

(c) At higher pressures (>5 torr) of nitric oxide or on leaving to stand for a few hours, a weak broad band appeared at 1960 cm^{-1}.

(d) The 1835 cm^{-1} band was often accompanied by one at 2245 cm^{-1}.

Some aerogel samples showed no bands at all when nitric oxide was added, others showed a band at 1830 cm^{-1} weaker than for Degussa samples, accompanied by one of equal intensity at 1792 cm^{-1}.

The very pure Spence alumina samples showed no absorption at all at about 1800 cm^{-1} at low pressures of nitric oxide (5 torr). At higher pressures (10 torr and above) a very slight general absorption in this region accompanied the sharp gas phase NO band at 1876 cm^{-1}.

Spence alumina containing 1% ferric oxide however gave a very strong band at 1818 cm^{-1} with a pressure of 0.92 torr nitric oxide. At 11.4 torr pressure there was little change in intensity of the 1818 cm^{-1} but there was a suspicion of a shoulder to the main band at about 1740 cm^{-1}. When the sample was reduced with hydrogen at 400°C subsequent addition of nitric oxide gave practically the same spectrum but with a very weak band at 1905 cm^{-1} and no shoulder at 1740 cm^{-1}.

All samples of alumina, whatever the source, which had been degassed at temperatures higher than 400°C showed a broad band at 1220-1240 cm^{-1}, the intensity of which increased with pressure of added nitric oxide and with temperature of degassing. It was not removed by pumping at room temperature.

3.2 Nitric oxide on silica and silica-alumina

Aerosil samples transmitted to about 1280 cm^{-1}. Even after heating to 800°C and exposure to 14 torr of nitric oxide no extra absorption bands were seen except for a weak absorption at about 2230 cm^{-1}, very close to the ν_3 nitrous oxide gas phase band.

The silica-alumina catalyst similarly showed no absorption bands above the cut-off limit of 1300 cm^{-1} after degassing at about 600°C and addition of NO up to 10 torr.

3.3 Nitrous oxide on alumina

Nitrous oxide was added to an alumina aerogel which was degassed at 350°C. No bands which could be attributed to adsorbed species were found and all the gas phase nitrous oxide bands up to a maximum pressure of 10 torr could be removed by pumping. The principal gas phase bands, which were superimposed on the background of the alumina spectrum were found to be at 2242 cm^{-1} with a smaller band at 2218 cm^{-1}.

3.4 Nitrogen dioxide on alumina

All samples of alumina chemisorbed nitrogen dioxide strongly and gave similar spectra of which that of figure 1 is typical. The main features are two very strong absorption bands at about 1600 cm^{-1} and 1240 cm^{-1}: there are less intense bands at 1960 cm^{-1} and 1290 cm^{-1}. At higher coverages (θ = 0.05) the band at about 1600 cm^{-1} (initially at 1597 cm^{-1}) splits into three (at 1570, 1597 and 1615 cm^{-1}), and the band initially at 1225 cm^{-1} moves to 1240 cm^{-1}. At higher coverages still, ($\theta \sim$ 0.25) complete absorption is obtained at these positions and no further detail can be distinguished (Values of θ were obtained by comparison of amounts adsorbed with a gravimetric absorption isotherm on a similar sample).

All three types of alumina show these features. There is a difference in the intensity of the band at 1290 cm^{-1} between the aerogels on the one hand and the Degussa and Spence aluminas on the other, the latter giving much better defined contours.

A very broad absorption occurs at between 2240 - 2270 cm^{-1} at higher coverages (θ>0.15) and is accompanied by bands at 2525, 2830 and 2640 cm^{-1}, all of them being relatively weak and broad.

Evacuation of the system to 10^{-5} torr at room temperature had little effect on most bands although those at 1960 cm^{-1} and 2250 cm^{-1} were markedly reduced. The same effect was produced by allowing the system to stand overnight without evacuation. The bands at ca. 1250 cm^{-1} and 1600 cm^{-1} were so strongly held that evacuation at 500°C for 1 hour was necessary to remove them completely.

The effect of the adsorption of nitrogen dioxide on the -OH groups of the alumina is not shown in figure 1 but a very considerable perturbation occurred at high coverages although there was little effect at low coverages. Of the 3 hydroxyl bands, the highest frequency at 3792 cm^{-1}

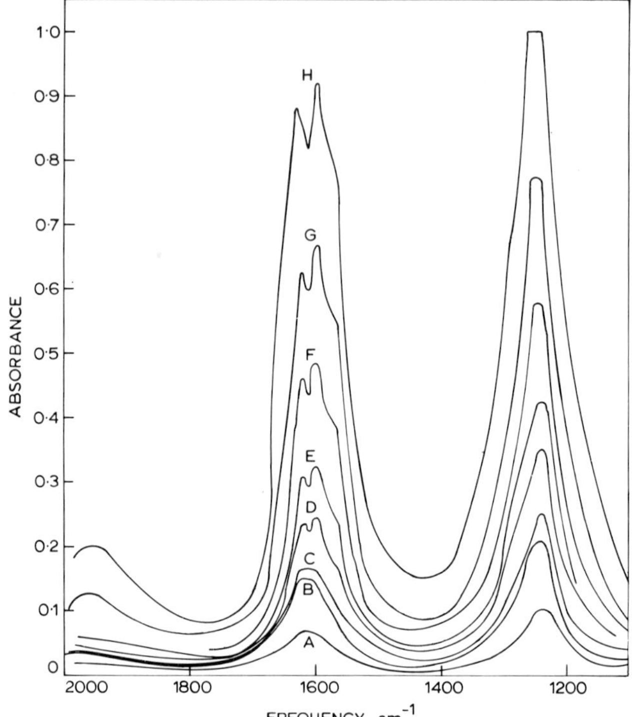

Fig.1. INFRA-RED SPECTRUM OF NITROGEN DIOXIDE ADSORBED ON ALUMINA AEROGEL
A. NO_2 ADSORBED 2·3mg, g^{-1} B. 4·5mg, g^{-1}
C. 55 mg, g^{-1} D. 7·7 mg, g^{-1}
E. 10·8mg, g^{-1} F. 17·4 mg, g^{-1}
G. 25·1mg, g^{-1} H. 41·3mg, g^{-1}
FINAL EQUILIBRIUM PRESSURE, CURVES A-G ~10^{-3} torr, CURVE H, 0·017 torr.

was affected first, the intensity decreasing progressively with increasing coverage until at $\theta = 0.5$, it fell to about 60% of its original size. At higher coverages, detail in the hydroxyl stretching region is lost until a very strong broad band at about 3600 cm^{-1} is the only remaining band. The original alumina bands are not restored until the sample has been evacuated to 500°C.

3.4.1. Effect of degassing temperature on the spectrum of nitrogen dioxide adsorbed on alumina.

The effect of increasing the evacuation temperature of the alumina on the spectrum of adsorbed nitrogen dioxide was investigated in detail : the results will be presented elsewhere[11]. The main effect of increasing temperature of degassing was to increase the maximum intensity of the bands for a given equilibrium pressure of gas. A second effect was to reduce the size of the 1290 cm^{-1} compared to that of the 1220-1240 cm^{-1} band. Finally, the band at ∿ 1960 cm^{-1} was accentuated as the degassing temperature was raised.

3.5 Nitrogen dioxide on silica

Samples degassed at up to 400°C showed no bands when nitrogen dioxide was added up to a pressure of about 4 torr, other than the gas phase bands at ca. 1600 cm^{-1}. Because observation of bands was hampered by the strong silica overtone and combination bands at 1866 and 1634 cm^{-1}, an exactly similar sample to that under study was used in the reference beam and gave a satisfactory background spectrum free of extraneous peaks. On heating aerosil to 600°C, addition of NO_2 gave a small band at 1658 cm^{-1} at 0.09 torr pressure. The intensity of this band increased with pressure levelling out at about 2 torr. It was quite sharp and was unaffected by evacuation at room temperature.

3.6 Nitrogen dioxide on silica-alumina

Silica-alumina degassed at 600°C gave a spectrum very similar to silica with a single-OH stretching frequency at 3740 cm^{-1} and strong silica-like bands at 1855 cm^{-1} and 1640 cm^{-1}. Adsorption of NO_2 up to 0.02 torr pressure gave a very weak band at 1960 cm^{-1} and an increase in intensity at 1640 cm^{-1}. 0.33 torr NO_2 gave considerable increases in both these bands together with an inflection at 1584 cm^{-1}.

4. DISCUSSION

4.1 Nitric oxide adsorbed on oxides

Our spectroscopic results give no convincing evidence for intrinsic undissociated adsorption of nitric oxide on any of the surfaces studied. It seems that the presence of Fe^{3+}ions is necessary in alumina to give the band observed at 1820 cm^{-1}. We did not investigate silica and silica-alumina with added iron impurities but the results of Poling and Eischens[12] and of Blyholder and Allen[13] on silica-supported iron leave little doubt that a similar band would have been obtained had we done so. In both sets of results the principal absorption band was at 1820 cm^{-1} for NO on freshly reduced iron on silica, shifting to 1824 cm^{-1} when pre-oxidised iron was used [12]. Terenin and Roev[14] found that NO on Fe_2O_3 gel had a principal absorption band at 1806 $cm^{-1}$and several smaller bands between 1600 and 2000 cm^{-1}: the whole spectrum was time-dependent. NO absorbed on iron supported on alumina, however, had a dominant band at 2008 cm^{-1} with only a minor band at 1830 cm^{-1}.

All these results combine to throw doubt on the authenticity of any bands observed between 1600-2000 cm^{-1} when nitric oxide is added to alumina or silica and these doubts are confirmed by the absence of any such bands when the purest grade of alumina (Spence) is used. Furthermore, when iron was deliberately added to this alumina, only 1% of iron oxide was sufficient to produce a strong band at 1818 cm^{-1} considerably stronger than any found on the other aluminas alone. This band is identical to that found by Poling and Eischens[12] and by Blyholder and Allen[13] and both papers discuss in detail the nature of the species responsible for the absorption.

Terenin and Roev[14] claim to have obtained additional bands when nitric oxide was added, not only to alumina but to other oxides, at 1625, 1660 and 1698 cm^{-1}. We have found no evidence of absorption at these positions either for alumina or for silica when NO is added, and conclude that these must have been due to some impurity in the Russian experiments.

All samples of alumina including the Spence alumina however, when degassed at temperatures above 400°C, showed an absorption band at 1220-1240 cm^{-1} which was very broad and whose intensity varied with pressure of NO added. It was accompanied by a band at about 2240 cm^{-1} but there was no apparent correlation between the intensities of the two bands. These bands we believe to be inherent to the alumina/nitric oxide system and may arise from decomposition to nitrous oxide and adsorbed oxygen. Gravimetric results support this theory to some extent, because the maximum uptake on Degussa alumina (after degassing at 900°C and at a pressure of 10 torr) of nitric oxide is 5 mg/g of alumina at room temperature. This is an order of magnitude higher than is required for non-dissociative adsorption on to impurity sites.

4.2 Nitrogen dioxide adsorbed on alumina

The absorption bands in figure I can be assigned simply and comprehensively to a complex of the type $NO^+ NO_3^-$, the nitrate ion being co-ordinately bonded by its oxygen atoms to one or more Al^{3+} cations. This follows by analogy to the structure of the complex $SnCl_4.N_2O_4$, isolated by Addison & Simpson[15], which is shown by its infra-red spectrum to have the structure $NO^+(SnCl_4.NO_3^-)$. A complete assignment of the bands in figure I is given in Table I.

Table 1. Frequencies and assignments of the infra-red spectrum of NO_2 adsorbed on alumina.

Species	Band		Frequency (cm^{-1})	
$SnCl_4.N_2O_4$ [15]	1254 (vs. br)		1580 (vs.br)	2216(m)
Free NO_3^- ion			1400	
Unidentate [16] NO_3^-		1254-1346	1464-1560	
Bidentate [17] NO_3^-	1255		1630-1635	
Bridging [18] NO_3^-	1213-1225		1620-1646	
NO_2/alumina	1225-1250	1290 (ms)	ca.1600(vs) (1615,1597, 1570)	1960-1977(m)
Assignment	NO_2 sym.Stretch	NO_2 Sym.Stretch	NO_2 asym.stretch N =O stretch	NO^+

The splitting of the 1600 cm^{-1} band refers to the three possible modes of co-ordination of the nitrate ion;by one oxygen atom, by two, or by bridging between two Al^{3+} ions. The different modes of co-ordination probably reflect different environment of adsorption site. The bands at 1570 cm^{-1} and 1290 cm^{-1} may refer to unidentate co-ordination; the bands at 1597 and 1615 cm^{-1} with the corresponding unresolved band at 1225-1250 cm^{-1} refer to bidentate and bridging co-ordination respectively.

The intensities of the 1970 cm^{-1} in figure 1 (A-H) when plotted against those of the 1600 cm^{-1} and 1250 cm^{-1} bands give very good straight lines right up to the experimental limits (A_{1600} = 1;

corresponding coverage 0.25). This means that NO^+ and NO_3^- ions are produced in constant ratio as expected and also that there is no significant contribution to the 1600 and 1250 cm^{-1} bands apart from the nitrate ions. This is an important point because it enables us to rule out the possibilities of nitrite ions[19,20], both free and co-ordinated, being present in significant amounts.

The frequency of the NO^+ band for adsorbed NO_2 is much lower than for the free ion (2216 cm^{-1} in $SnCl_4.N_2O_4$[15], 2340 cm^{-1} in $NO^+HSO_4^-$[21], 2240 cm^{-1} in $NO^+NO_2^-$[22]). Even for NO adsorbed on aluminium chloride[23], the band appears at 2142 cm^{-1}. This may imply that ionisation is not complete and that the NO^+ is associated with NO_3^- by some partly covalent bond.

The assignment of the small bands of frequency above 2000 cm^{-1} is more speculative. A broad band at about 2250 cm^{-1} is assigned to NO_2^+ although there is no other evidence to back this up. The bands at higher frequencies still are probably due to overtones and combinations of the very strong fundamental nitrate stretching modes.

4.3 NO_2 on silica and silica-alumina.

Although nitric oxide gives no spectra indicative of non-dissociative adsorption on either silica or silica-alumina, adsorption of nitrogen dioxide gives absorption bands on both oxides.

It is somewhat surprising to find that silica has sites capable of adsorbing nitrogen dioxide quite strongly, as the tetrahedral co-ordination of silicon in silica makes for a very stable arrangement in which unsaturated valencies can easily be saturated either by terminal silanol groups or by bridging oxygen atoms. However, it is known that dehydration of silica at higher temperature can produce relatively reactive sites possibly involving surface radicals[24], and a reaction scheme as follows can be envisaged.

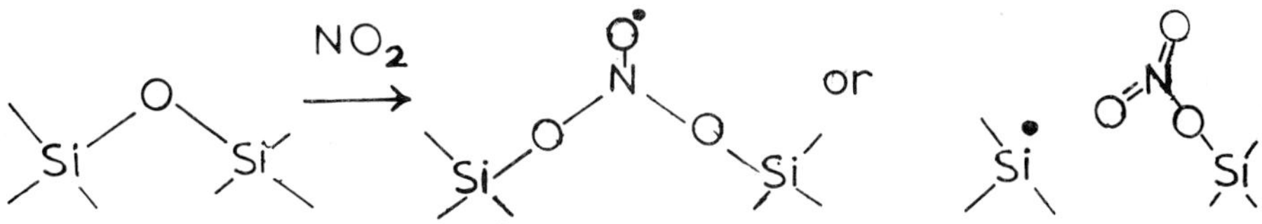

The observed band at 1658 cm^{-1} can therefore be ascribed to formation of nitrate species, the other modes of vibration occurring below 1300 cm^{-1} not being observed because of the absorption by the silica. Support is given to this assignment by the fact that this band does not appear for lower dehydration temperatures (600°C), where siloxane bridge formation is less likely to occur.

Silica-alumina, as might be expected, reacts with nitrogen dioxide through its surface aluminium ions. The bond at 1960 cm^{-1} although weak, is characteristic of the NO^+ ion and the large increase at 1640 cm^{-1} represents one of the stretching modes of co-ordinated NO_3^-, the other modes being hidden by the strong absorption below 1300 cm^{-1} as in the case of silica.

The alumina content was only about 8% on a mole basis but a comparison with an alumina of similar surface area showed that the band intensity of the nitrate ions on the silica-alumina were lower by a factor of only about 2.3. The quantitative comparison is difficult because both samples were highly microporous and as a result the BET areas are not meaningful representations of the surface area. Nevertheless this large discrepancy might suggest that aluminium cations in silica-alumina are preferentially concentrated at the surface.

4.4 Mechanism of adsorption of NO_2 on alumina

The infra-red spectra show that nitrogen dioxide (or tetraoxide) adsorbs dissociatively on alumina to form co-ordinated nitrate ions to aluminium cations and NO^+ ions which presumably are adsorbed near O^{2-} ions. The nitrate ions are very strongly held indeed whereas the nitrosonium ions can be removed by evacuation. We can account for these phenomena by the following reaction scheme:

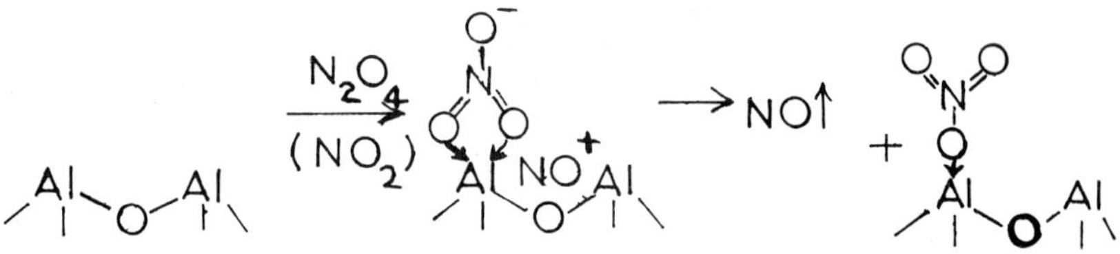

The possible change in co-ordination of the nitrate ion when NO is lost from the surface has not been observed in the infra-red spectrum because the 1600 and 1200 cm^{-1} nitrate bonds are so intense at high coverages as to obliterate fine detail in them.

Further information is obtained from gravimetric isotherms on a type C alumina, degassed at various temperatures. After adsorption of NO_2 and evacuation, there is a loss in weight although it is an order of magnitude higher than the above reaction scheme would predict.

We can also calculate from the adsorption isotherms the maximum capacity for chemisorption of NO_2 on alumina and hence the number of aluminium cation sites, assuming two NO_2 molecules to be associated with each cation. These values are shown in Table 2.

Table 2. Monolayer capacity of alumina 'C' for NO_2 (μmoles.m^{-2}) and no. of adsorption sites ($100Å^{-2}$) after evacuation to temperatures shown.

Degassing temp (^oC)	25	200	400	600	800
Monolayer capacity (μmoles NO_2 per m^2 alumina)	5.0	5.3	6.7	7.1	7.6
No. of adsorption sites per 100 $Å^2$	1.5	1.6	2.0	2.1	2.3

Using as a basis for discussion the model proposed by Peri[25] for alumina aerogels in which the 100 face only is taken into consideration, initially all surface aluminium cations are covered with hydroxyl groups. As the temperature is raised dehydroxylation occurs leaving exposed aluminium cations, oxygen ions and residual isolated hydroxyl groups. The maximum number of aluminium sites on such a model, assuming that all hydroxyl groups are removed, is $6.25/100Å^2$ although they will have slightly different environments.

Peri's model is admittedly an idealised one, although for needle-shaped aerogel crystallites the 100 plane may well be predominant. For the type C alumina which consists of microspheres of 50-300 Å diameter, this cannot be true. It is however possible to make up a rhombicuboctahedron of 26 sides from low index planes of the alumina spinel structure. We can calculate the maximum number of exposed cations on a given plane and these are shown in Table 3.

Table 3. Number of surface aluminium cations (100 $Å^{-2}$) on low index planes of γ-alumina spinel.

Plane Index		No. of octahedral cations	No. of tetrahedral cations
100		5.3	1.4
110	C layer	3.9	3.9
	D	3.9	0.0
111	E	9.1	0.0
	F	1.8	7.2

Unlike those in the model adopted by Peri, these surface cations are formed by a notional cleaving of the crystal whereby stoichiometry is preserved and hence all are potentially capable of acting as adsorption sites. Using these planes to make up the 26-sided solid model gives final figures of 4.7 Octahedral cations and 2.8 tetrahedral ions per 100 $Å^2$ as the maximum number of potential Lewis sites. This compares with the values of 2.3 obtained by gravimetry. Further discussion of these results and the assumptions lying behind them will be discussed elsewhere[11].

4.5 Applications to other oxide systems

The results presented here show that NO_2 may be used to detect, apparently quantitatively, Lewis sites on the surface of alumina and silica-alumina, although a rather crude model would indicate that only one third of the surface aluminium cations react with it. Pyridine is very widely used for determining Lewis sites semi-quantitatively as first proposed by Parry[26], but NO_2 would have certain advantages for this purpose.

Firstly, as it is a much lower boiling point material than pyridine, van der Waals-type interaction at room temperature is reduced, whereas the spectrum of adsorbed pyridine shows considerable hydrogen bonding. Secondly it does not react with Bronsted acids, as does pyridine, giving a clear distinction between the Bronsted and Lewis sites on an oxide having both kinds. Thirdly, it is a much smaller molecule than pyridine (van der Waals radius 4.8 Å as against 6.8 Å) and this would be useful for measuring acidity in very small micropores or in molecular sieves.

5. ACKNOWLEDGEMENTS

I would like to thank Dr. M. Alario-Franco and Mrs. B.J. Gillett for experimental help in determining adsorption isotherms and infra-red spectra respectively, and the Gas Council for permission to publish this work.

REFERENCES

1. L.H. Little, "Infra-red spectra of adsorbed species", Academic Press, London and New York, 1966.
2. M.L. Hair, "Infra-red spectroscopy in surface chemistry", Dekker, New York, 1967.
3. D.J.C. Yates, Catalysis Reviews, 1969, 2, 113.
4. N.D. Parkyns, J. Chem. Soc., A, 1967, 1910.
5. N.D. Parkyns & B.C. Patterson, Chem. Comm., 1965, 530.
6. B.D. Flockhart, J.A.N. Scott & R.C. Pink, Trans. Faraday Soc., 1966, 62, 730.
7. J.H. Lunsford, J. Catalysis, 1969, 14, 379.
8. N.D. Parkyns, paper submitted to J. Catalysis.
9. N.D. Parkyns, in "Laboratory methods in infra-red spectroscopy", 2nd Edition, B.C. Stace & R.G.J. Miller (eds), Heyden, London, 1972.
10. N.D. Parkyns, J. Phys. Chem., 1971, 75, 526.
11. N.D. Parkyns, paper in preparation.
12. G.W. Poling & R.P. Eischens, J. Electrochem. Soc., 1966, 113, 218.
13. G. Blyholder & M.C. Allen, J. Phys. Chem., 1965, 69, 3998.
14. A. Terenin & L. Roev, Spectrochimica Acta, 1959, 15, 946.
15. C.C. Addison & W.B. Simpson, J. Chem. Soc.A, 1966, 775.
16. B.O. Field & C.J. Hardy, Quart. Rev. Chem. Soc., 1964, 18,361.
17. C.C. Addison & W.B. Simpson, J. Chem. Soc., 1965, 598.
18. B.O. Field & C.J. Hardy, J. Chem. Soc., 1964, 4428.
19. K. Nakamoto, "Infra-red Spectra of Inorganic and Co-ordination Compounds", Wiley, New York, 1963.
20. B.J. Hathaway & R.C. Slade, J. Chem. Soc. A, 1966, 1485.
21. W.K. Angus & A.H. Leckie, Proc. Roy. Soc. A, 1935, 149, 327.
22. J.D.S. Goulden & D.J. Millen, J. Chem. Soc., 1950, 2620.
23. A.N. Terenin, W. Filimonov and D. Bystrov, Z. Elektrochem., 1958, 62, 180.
24. R.E. Day, A.V. Kiselev & B.V. Kuznetzov, Trans. Faraday Soc., 1969, 65, 1386.
25. J.B. Peri, J. Phys. Chem., 1965, 69, 220.
26. E.P. Parry, J. Catalysis, 1963, 2, 371.

DISCUSSION

K. OTTO

From a series of gravimetric adsorption studies we have deduced that NO is adsorbed on Al_2O_3, although to a much smaller extent than on iron oxides. Because of the slowness of the adsorption a very weak signal is expected when alumina is exposed to NO for a few hours only. The explanation of the adsorption bands at 1220 - 1240 and 2240 cm^{-1} by decomposition of NO to N_2O and adsorbed oxygen atoms is improbable, as NO decomposition on any oxide catalyst at these temperatures is unknown. The decomposition requires the close proximity of two NO molecules at the catalytic surface to form the necessary nitrogen-nitrogen bond. It appears to be inconsistent that N_2O is produced catalytically while, at the same time, NO is not adsorbed on the surface.

N. D. PARKYNS

Although the presence of trace amounts of transition metals, particularly iron, are necessary for the appearance of bands in the 1800 cm^{-1} region when NO is added to alumina, our gravimetric results, as I have already commented in the paper, show that about 4-5 times as much NO is adsorbed than can be accounted for by adsorption on Fe ion sites alone; furthermore there is a rapid initial adsorption followed by a much slower one, indicating that there are possibly at least two types of adsorption sites. Thus, there is no contradiction between Dr. Otto's remarks and the results given in the present paper. As far as explanation of other infrared bands by invoking a dissociative adsorption is concerned, this is frankly conjectural at this stage but I feel that the 2240 cm^{-1} band is connected with N_2O in some form or other.

H. KNÖZINGER

I agree with Dr. Parkyns' statement that NO_2 has certain advantages as compared to pyridine as a probe for surface acidic sites. To the three reasons mentioned by him, I like to add a fourth one, which is also related to the molecular size of pyridine. This can lead to steric restrictions in the direct interaction even with an easily accessible site in a sufficiently wide pore. These steric effects might always be important where the incompletely coordinated Lewis acidic cations are situated in "holes" in the surface oxide layer as e.g., in the case of alumina. We have shown the influence of such steric effects by means of the competitive adsorption of pyridine and the much stronger base 2,4,6-trimethylpyridine, whose coordination to surface Lewis sites is completely prevented in the presence preadsorbed pyridine. Pyridine itself forms a relatively stable "inner" complex at sufficiently high temperatures through an activated adsorption step. This activated adsorption can also be explained by steric hindrance for the accessibility of the cation sites. (See e.g., H. Knözinger, H. Stolz, Fortschr. Kolloide u. Polymere 55 (1971) 16; and Ber. Bunsenges. Phys. Chem. 75 (1971) 1055).

N. D. PARKYNS

I am glad that Dr. Knözinger has been able to confirm my tentative ideas that pyridines might suffer steric hindrance to adsorption. In our present study, we carried out our quantitative adsorption work on the non-microporous alumina "C," despite its known iron impurity, rather than on the purer aerogel because the latter is extremely microporous, possibly leading to molecular sieving effects whereby pores which admitted N_2 in BET adsorption measurements might not admit the somewhat larger NO_2 and N_2O_4 molecules. On the other hand, pores which would just admit NO_2 would also lead to enhancement of physical adsorption by micropore filling. In any case, it would seem that determining acidity of microporous oxides by adsorption either of pyridine or of nitrogen oxides calls for some care in interpreting the results, because of the difficulty in separating weak chemisorption from strong physical adsorption, whether due to micropore filling or to hydrogen-bonding.

J. A. RABO

I would like to refer you to three publications in which the disproportionation of NO to N_2O and NO_2 is described and established, based on chemical (Barrer) and on electron spin resonance evidence (Lunsford and in a different paper P. H. Kasai). These papers explain the slow irreversible adsorption of NO in the case of zeolites.

N. D. PARKYNS

I have noted Dr. Rabo's comments about disproportionation of NO on zeolites but I do not think that the same mechanism applies in the case of alumina because adsorbed NO_2 would give strong bands at 1600 cm^{-1} and no such bands are observed. On the other hand, when NO is added to TiO_2, as the partial pressure is increased, the infra-red spectra (not yet published) show quite strong bands due to the nitrate ion in this and other regions, indicating that disproportionation may be occurring.

R. C. PITKETHLY

With reference to the discrepancy between the number of surface Lewis sites and the estimated number of surface aluminum cations, I would draw attention to the possibility of making silica-alumina with known amounts of surface aluminum and corresponding number of acid sites which can be converted from Bronsted structure to Lewis by dehydration (and vice versa). The procedure is to treat pure silica gel with an aluminum salt in aqueous solution.[1)] The number of acid sites has been measured by ion exchange and the conversion from Bronsted to Lewis demonstrated through IR spectra of adsorbed pyridine. It would be useful to compare your technique using materials in which the number and state of the sites can be controlled.

1) K. H. Bourne, F. R. Cannings and R. C. Pitkethly, J. Phys. Chem., 74, 2197 (1970).

N. D. PARKYNS

This is a very useful suggestion and I hope that we shall be able to take advantage of the technique outlined by Dr. Pitkethly to establish whether or not all exposed Al^{3+} cations are potential Lewis acids.

Paper Number 13

THE EFFECT OF CATALYST SINTERING ON LOW TEMPERATURE AMMONIA OXIDATION OVER SUPPORTED PLATINUM

J. E. DeLANEY and W. H. MANOGUE
Department of Chemical Engineering, University of Delaware
Newark, Delaware

ABSTRACT: Yields of nitrous oxide of up to 75% were obtained at 3.7 atmospheres pressure when 1% ammonia and 3% oxygen in argon were passed over a fixed bed of 0.5% platinum-on-alumina catalyst between 200–350°C. Sintering the catalyst or lowering the pressure to 1 atm. at a constant mass flow rate reduced the nitrous oxide yield. Ammonia was completely consumed in these experiments in which nitrogen and small amounts of nitric oxide were the other reaction products. These results contrast with the commercial oxidation of ammonia over platinum above 700°C. where nitric oxide is produced in over 95% yield.

1. INTRODUCTION

The effect of crystallite size upon specific catalytic activity and selectivity is of fundamental importance in heterogeneous catalysis and is a phenomenon which is incompletely understood at best[1)]. Although recent work has established that there is a large class of reactions, facile in the classification of Boudart[1)], in which the specific activity of the supported metal is independent of its crystallite size, reactions involving oxygen generally appear to be sensitive to the crystallite size of the metal[1,2,3,4)]. This appears also to be the case with O_2 chemisorption on supported Pt in contrast to H_2 chemisorption[5)].

The oxidation of ammonia over supported platinum is a system for which it is not known if a crystallite size effect exists. The purpose of this work was to determine if there is such an effect in the low temperature (200–350°C.) oxidation of ammonia. In contrast to the situation for reactions such as H_2O_2 decomposition, SO_2 oxidation, and H_2 oxidation in which only the specific catalytic activity can be measured in crystallite size studies, ammonia oxidation can have several reaction products: N_2, N_2O, or NO. A change in selectivity with crystallite size change if observed could offer strong proof of structure sensitivity.

1.1 Ammonia Oxidation

The oxidation of gaseous ammonia over platinum and its alloys is of industrial interest over a wide temperature range.

Above 750°C. ammonia is partially oxidized to nitric oxide, NO, in yields of about 95% as a first step in the production of nitric acid[6]. At lower temperatures the yield of NO falls and that of nitrogen and nitrous oxide, N_2O, increases. The N_2O/N_2 ratio varies markedly with catalyst and reaction conditions[7,8,9,10,11].

Although there has been much research on the oxidation of ammonia[6,12], there is considerable uncertainty as to how various factors affect the course of the reaction over platinum over widely varying conditions. This uncertainty is especially pronounced in the low-temperature region where kinetics control the reaction and where the products are mainly N_2O and N_2 as opposed to the high-temperature region where mass transfer is the rate limiting step[12].

At temperatures below 350°C. ammonia reacts selectively with NO over a catalyst in the presence of oxygen reducing the NO to N_2 and N_2O. This reaction is of direct significance to air pollution abatement because it is a means of selectively reducing NO in effluent gas streams to N_2 and N_2O[13,14,15,16,17,18].

Early investigators[6,12] proposed very complex mechanisms for the oxidation of ammonia over platinum. However recent work utilizing mass spectrometers[7,19,20,21,22] and tracers[23] indicates that NO is the initial product in ammonia oxidation ($NH_3+O_2 \rightarrow NO$) and that the N_2 and N_2O result from subsequent reactions, one of which is the reaction between NO and NH_3 yielding N_2 and the other the reaction of nitric oxide with chemisorbed hydrogen yielding N_2O. Thus the reaction appears to be much simpler mechanistically than was originally thought.

2. EXPERIMENTAL

2.1 Apparatus and Procedure

Two forms of a commercial catalyst containing 0.5 weight per cent platinum deposited on the outside 1/64" of 1/8" cylindrical alumina pellets were used. Material denoted A contained small platinum crystallites which were amorphous to X-ray. Material designated S contained crystallites of about 450Å size by X-ray diffraction. Material similar to S was prepared by heating A catalyst 17 hr. in air at 780°C. according to the procedure of Maat and Mascou[24]. Crystallite size was checked by electron microscopy. The sintered material was found to contain a range of crystallite sizes.

Activity measurements were carried out on a charge of 40 pellets in a fixed bed reactor immersed in a fluidized constant temperature sand bath. A twelve-foot coil of 1/4" O.D. 304

stainless steel tubing served as reactor preheater. Mixtures of 20.9% oxygen in argon, 7.45% ammonia in argon, and argon were metered into a common header to arrive at a desired composition of feed gas. Argon was used as a diluent to permit a nitrogen atom balance to be obtained. The feed gas mixture was then preheated and passed through the reactor, a back pressure regulator, and to a gas chromatograph. Temperatures were measured immediately preceding and 1/4" beyond the catalyst bed. Most runs were made with a constant gas feed rate of 1280 cc/minute measured at 23°C. and 1 atm. This corresponds to a particle Reynolds number of 54 at a reactor temperature of 300°C.

A feed composition of 1% ammonia was chosen as a compromise between minimizing the adiabatic heat up across the catalyst bed and having sufficient reaction products to make an accurate nitrogen balance. For a 1% NH_3 feed the adiabatic temperature rise associated with complete conversion to nitrogen is 150°C. In our work the pellet temperature never rose more than 50°C. above the feed gas temperature. This was determined with a thermocouple placed in the center of a catalyst pellet. In the absence of reaction this thermocouple agreed with the reactor thermocouple within 2 degrees.

In a typical run flows were introduced into the preheated reactor to achieve desired feed composition in the order of argon diluent, 20.9% O_2 in argon, and finally 7.45% NH_3 in argon flow. The ammonia was not introduced until last to avoid any reducing atmosphere about the catalyst. The back pressure regulator was then set for the desired reactor pressure. As soon as steady state was achieved (usually within 5 min.), samples were taken for analysis. The sand bath was then set to the next highest temperature. Following completion of data taken at the highest temperature in a series of runs, the feed gases were turned off in reverse order, the reactor left under low pressure argon, and the sand bath allowed to cool down.

2.2 Product Analysis

A F&M Model 810 dual column gas chromatography unit with thermal conductivity detectors was used for determination of nitrogen and nitrous oxide with helium as a carrier gas. A separate gas sampling valve with a 5 cc sample loop was used for injecting samples into each of the two columns used. Pressure was set at 1 psig $\pm$ 1/20 psi using a bypass valve arrangement to ensure constant mass injection. Nitrogen was separated at room temperature with a retention time of 9 min. using a 12' x 3/16" stainless steel column packed with 60-80 mesh 5A molecular sieve with a minimum detectable level of .004%. Water, ammonia, nitrous oxide, and nitric oxide (in the presence of water) were held up in the column indefinitely. Periodically the 5A column was cleared of these products by passing helium gas through the column at 350°C. overnight. Nitrous oxide was separated at room temperature using a 6' x 3/16" stainless steel column packed with 80–100 mesh Porapak® Q with a detection limit of .002% and a retention time of 2-1/2 min. The column oven was then heated to 110°C. and held 6 min. until water and ammonia were driven out of the column. Calibration was by comparison with responses from a standard mixture purchased from the Matheson Company containing 0.43% NO, 0.39% N_2O, and 0.36% N_2. Peak height was as accurate as peak area for the calculations.

2.3 Nitric Oxide Determination

While the flow reactor was held at one steady state condition during N_2O and N_2 determinations by gas chromatography, the entire effluent gas stream was passed through a 1315 cc gas collecting tube fitted with Teflon® stopcocks on either end and a glass injection port with a rubber septum on the side. Following collecting of a gas sample, 10 cc of a 3% hydrogen peroxide solution was injected into the tube to oxidize any NO or NO_2 gas to nitric acid which was titrated with 0.01 normal sodium hydroxide to a phenolphthalein endpoint. Nitrogen, nitrous oxide, oxygen, or water did not interfere. No ammonia which would interfere was present in samples taken at 260°C. or above.

3. RESULTS

3.1 Conversion Over a Single Pellet

Empty tube tests showed that at the flow rates used in this work 4% of the ammonia was reacted at 300°C. and 31% at 380°C. The alumina carrier was inert up to 300°C. Although this work was not intended to be a kinetic study, some data on conversion over a single catalyst pellet were obtained and are given in table 1. The pellet also contained a central thermocouple which indicated a significant temperature rise over the catalyst. In calculating the rate of reaction corrections were applied for the amount of reaction occurring in the empty tube.

TABLE 1*

Sand Bath Temp. °C.	Pellet Temp. °C.	% NH_3 Consumption			Reaction Rate g. moles/hr., g. catalyst
		Pellet & Reactor	Empty Reactor	Net over Pellet	
200	252	10	0.8	9.2	0.064
250	288	11	1.7	9.3	.065
300	330	16	4.	12.	.084
350	377	25	22.	3.	.021
380	403	49	31.	18.	.126

*TABLE 1. Single Pellet Reactor Data. 0.045 g. Unsintered Catalyst. Reactor Pressure 1 psig. Feed Gas: 1% NH_3, 3% O_2 in Argon at 1280 cc/min. (23°C., 1 atm.) N_{Re} = 54.

3.2 Light-Off Temperature

With a bed of 40 sintered pellets, a 1% ammonia feed and a Reynolds number of 54, the first signs of reaction products, nitrogen and nitrous oxide, appeared at an inlet temperature of 100°C. where 1.5% of the ammonia reacted. At 150°C., 2.5% of the ammonia had reacted. At 175°C. the reaction lit off, with 86% of the ammonia reacting. At 200°C., 98% of the ammonia reacted over both sintered and unsintered catalyst.

3.3 Stability

Following a series of runs on the S catalyst covering some 140 hr. of operation between 150°C. and 550°C., the catalyst was checked for any change that might have occurred by returning to a flow condition used in an earlier run. Similarly fresh A catalyst was checked after 27 hr. of on-stream time at from 150°C. to 400°C. Results given in table 1 indicate that little change in product distribution took place over the period of this study.

TABLE 2
Catalyst Aging Study Results
Feed Conditions: Same as in table 1.

Catalyst	On-Stream Time	Temp.	% N_2	% N_2O	% NO
A (unsintered)	2 hr.	280°C.	0.14	0.29	0.05
"	27 hr.	280°C.	0.14	0.31	0.06
S (sintered)	9 hr.	300°C.	0.25	0.19	0.12
"	138 hr.	300°C.	0.26	0.19	0.10

3.4 Product Distribution

The ammonia oxidation product distribution observed in this work was similar to that reported with platinum gauze catalysts. At the ignition temperature of from 160–180°C. most of the product was nitrogen with some nitrous oxide present and little or no nitric oxide. As the reactor temperature was raised, the nitrous oxide content gradually rose to a maximum at from 275–325°C. and then fell with further increase in temperature. At 550°C. little nitrous oxide was present in the product gas. Nitric oxide formation steadily increased with rising temperatures starting with almost none present at 200°C.

3.5 Influence of Platinum Crystallite Size on Product Distribution

A marked difference was observed between the product distribution with unsintered and sintered catalyst in the temperature range investigated as shown in fig. 1 and 2. (In each of the graphs the lower curve for each catalyst or condition represents the amount of ammonia converted to N_2O; the upper curve is the amount converted to N_2 plus N_2O. Since all data were taken at essentially complete conversion, the difference between the upper curve and 100% represents the amount of ammonia converted to NO.) Yields of N_2O were as much as 50% greater with the unsintered A catalyst as with the same catalyst sintered at 780°C. for 17 hr. While not as significant, lower NO formation was also observed with the unsintered catalyst.

3.6 Effect of Temperature, Pressure, and Contact Time on Product Distribution

A series of runs with both sintered and unsintered catalyst at pressures of 1 and 40 psig at constant Reynolds number with 1% ammonia, 3% oxygen feed showed a marked increase in nitrous oxide and decrease in nitric oxide yield with increasing pressure. In order to determine whether the difference in product distribution with total system pressure at constant mass flow rate was primarily the result of a change in contact time (volumetric flow rate) or a change in system pressure, a run was made on the A-17-780 catalyst using one-half the pressure and one-half the mass flow rate, all other conditions being the same as in the 40 psig run. This represented a reduction in N_{Re} of from 54 to 27. The results summarized in fig. 3 show that contact time rather than pressure is the critical variable since doubling the system pressure while halving the Reynolds number to maintain a constant contact time resulted in little change in product distribution.

3.7 Influence of Oxygen and Ammonia Concentration on Product Distribution

The effect of oxygen concentration on the product distribution at constant temperature, Reynolds number, and ammonia concentration is shown in fig. 4 for 1.25 to 4 times the stoichiometric amount of oxygen required to form nitric oxide. The yield of nitrous oxide rose with increasing oxygen concentration in the feed. The increased nitrous oxide yield was at the expense of nitrogen. Beyond five parts oxygen to one part ammonia in the feed gas, changes in product distribution with oxygen concentration were minimal.

4. DISCUSSION

The purpose of this work was only to determine if there was an effect of crystallite size on the product distribution and to provide some estimate of its extent. No attempt at kinetic analysis was made at this stage. However, significant mass transfer effects, both external to the pellet and within the catalyst layer, were found to be present from the single pellet data of table 1. The tests applied were those given in Satterfield[25].

The explanation for the increased N_2O yields with smaller crystallites is as yet speculative. Previous workers[7,19-23] have shown that NO is the initial oxidation product over platinum and that N_2 and N_2O are the products of subsequent reactions with ammonia. N_2O is also formed in varying yields when ammonia is oxidized over metal oxides[6]. Krauss[26] suggests that metal oxides which contain excess oxygen as defects in the lattice produce N_2O in good yields and he has found a linear relationship between the percentage of N_2O in the product gas and the amount of excess oxygen. Zawadski[11] suggests that the mechanism of nitrous oxide formation on platinum is similar to that on metal oxides in that the yield of nitrous oxide depends on the amount of atomic oxygen on the surface. There is some evidence that platinum crystallite size may affect the amount of oxygen which can adsorb on the surface and this could be associated with a change in the ratios of exposed crystal faces[5].

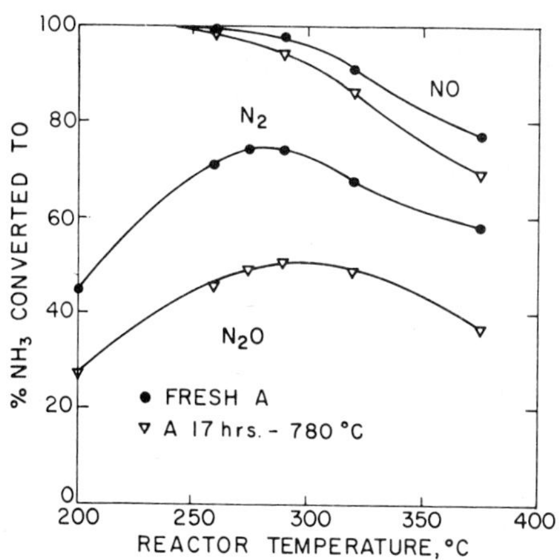

Fig. 1 Effect of sintering catalyst and of operating temperature on product distribution (1% NH_3, 3% O_2, balance argon in feed, 54.6 psia reactor pressure, N_{Re} = 54).

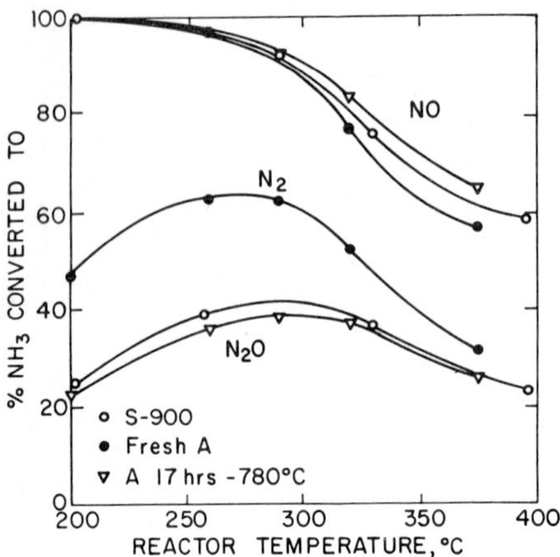

Fig. 2 Effect of sintering catalyst and of operating temperature on product distribution (1% NH_3, 3% O_2, balance argon in feed, 15.6 psia reactor pressure, N_{Re} = 54).

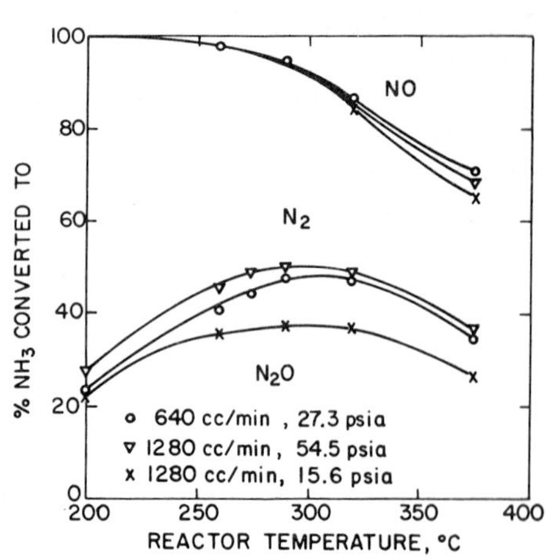

Fig. 3 Influence of pressure, temperature and volumetric flow rate on product distribution (1% NH_3, 3% O_2, balance argon in feed, A-17-780 catalyst).

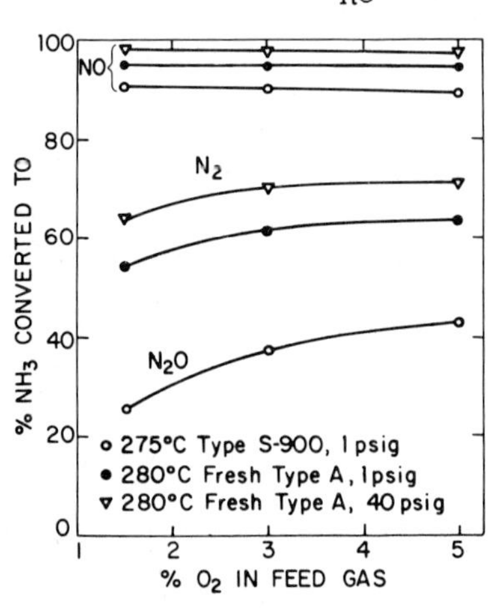

Fig. 4 Influence of oxygen on product distribution at constant temperature and flow rate (1% NH_3 in feed, N_{Re} = 54)

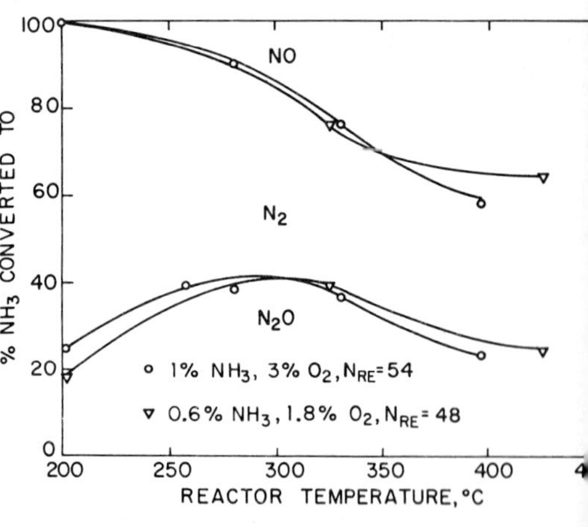

Fig. 5 Influence of ammoni concentration, at constant ammonia to oxygen ratio, on product distributio (S-900 catalyst, 15.6 psia reactor pressure).

REFERENCES

1) M. Boudart, "Catalysis by Supported Metals" in "Advances in Catalysis", Vol. 20, Academic Press, New York, 1969.
2) O. M. Poltorak and V. S. Boronin, Russian J. Phys. Chem. 39 (1965), 781.
3) Ibid., 39 (1965), 1329.
4) Ibid., 40 (1966), 1436.
5) G. R. Wilson and W. K. Hall, J. Catalysis 17 (1970), 190.
6) J. K. Dixon and J. E. Longfield, "Oxidation of Ammonia, Ammonia and Methane, Carbon Monoxide and Sulfur Dioxide", in "Catalysis", Vol. VII, P. H. Emmett, Ed., p. 281, Reinhold, N. Y. (1960).
7) C. W. Nutt and S. Kapur, Nature 220 (1968), 697.
8) T. J. Schriber and G. Parravano, Chem. Eng. Science 22 (1967), 1067.
9) K. Tuszynski, Roczniki. Chem. 23 (1949), 397.
10) K. Marczewska, Roczniki. Chem. 23 (1949), 406.
11) J. Zawadzki, Disc. Faraday Society (1950), 140.
12) E. J. Nowak, Chem. Eng. Science 21 (1966), 19.
13) H. C. Andersen, J. G. E. Cohn and D. R. Steel, U.S. Patent 2,975,025 (March 14, 1961).
14) H. C. Andersen, W. J. Green and D. R. Steele, Ind. Eng. Chem. 53 (1961), 199.
15) H. C. Andersen and C. D. Keith, U.S. Patent 3,008,796 (Nov. 14, 1961).
16) M. Shelef and J. J. Kummer, "The Behavior of Nitric Oxide in Heterogeneous Catalytic Reactions", Paper 13f, A.I.Ch.E. Sixty-Second Annual Meeting, Washington, D. C., Nov. 16-20, 1969.
17) W. Bartok, A. R. Crawford and A. Skopp, Chemical Engineering Progress 67, 2 (1971), 64.
18) National Air Pollution Control Administration, "Control Techniques for Nitrogen Oxide Emissions from Stationary Sources", Pub. No. AP-67, U. S. Government Printing Office (March, 1970).
19) Ya. M. Fogel, B. T. Nadykto, V. F. Rybalko, R. P. Slabospitskii, I. E. Korobchanskaya, and V. I. Shvachko, Kinetika i Kataliz 5 (1964), 154.
20) Ya. M. Fogel, B. T. Nadykto, V. F. Rybalko, V. I. Shvachko and I. E. Korobchanskaya, Kinetika i Kataliz 5 (1964), 496.
21) Ibid., 5 (1964), 942.
22) C. W. Nutt and S. Kapur, Nature 224 (1969), 169.
23) K. Otto, M. Shelef and J. T. Kummer, J. Phys. Chem. 74 (1970), 2690.
24) H. J. Maat and I. Moscou., Proc. Intern. Congr. Catalysis,

3rd, Amsterdam, 2, 1277, Wiley, New York, N. Y., 1965.
25) C. N. Satterfield, Mass Transfer in Heterogeneous Catalysis, MIT Press, Cambridge, Mass., 1970.
26) W. Krauss, Z. Elektrochem. 54 (1950), 264.

DISCUSSION

M. BOUDART

I agree with your expectation that the selectivity of oxidation reactions on supported platinum catalysts may well depend on the particle size of the metal. Indeed, in our paper No. 96, we present evidence to confirm the dependence of the amount of oxygen chemisorbed by platinum on the particle size of the latter. Unfortunately, at the moment, it is too early to decide whether or not you have found the expected effect. My doubts are due largely to the complications introduced by the influence of heat and mass transfer in your work. This could disguise changes of selectivity. To avoid such a difficulty in further work, I recommend that you establish the absence of concentration and temperature gradients by studying your reactions on several catalysts containing various amounts of Pt per g of catalyst, all samples containing metal particles of about the same size. If the selectivity of your reactions is found to be the same on this series of samples, but is different on another similar series of samples containing Pt particles of a different particle size, you would have a much better case for the reality of the expected effect, irrespectively of its ultimate justification.

W. H. MANOGUE

Thank you. We are continuing this work since we believe at this time that we only have evidence of an apparent crystallite size effect.

M. SHELEF

To confirm the conclusion of the authors the determination of structure-sensitivity would have to be made at a constant Pt surface available in both sintered and non-sintered catalysts. The much larger surface area of the non-sintered catalysts may alter the product distribution of the exit even when there is no difference in the reaction paths on the individual sites.

W. H. MANOGUE

Thank you.

C. H. AMBERG

Since the sintering of metals is notorious for changing their surface impurity composition, I should like to suggest this further possibility to explain your variations in product distribution. Have you been able to eliminate this possibility experimentally?

W. H. MANOGUE

We have no estimates of surface impurity composition. However, as shown in Table 2 of the preprint, little change in product distribution was observed over the period of this study. One might expect that the surface concentration of some of the expected impurities, such as sulfur and carbon, would change significantly over this period. It has been noted in discussion of one of the other papers at this Congress that the steady state surface of a working platinum catalyst may be the analogue of a clean surface.

D. L. TRIMM

Your results could be explained by particle size, particle geometry, changes in surface composition or by metal-support effects. Do you have any work dealing with the latter possibility?

W. H. MANOGUE

We plan to examine metal-support effects.

T. P. KOBYLINSKI

Some of your experiments were carried out at temperatures close to 450°C. I would like to know whether or not you observed any NH_3 decomposition, and if so, did you find any difference in ammonia decomposition activity (in the absence of O_2) between fresh and sintered Pt catalysts?

W. H. MANOGUE

We did not perform the experiments you suggest. However, in Section 3.1 of the preprint we point out that significant ammonia consumption in the presence of oxygen was observed about 300°C., both in the empty tube and in the tube packed with support. Therefore, we attach most significance to the data taken at or below 300°C.

K. OTTO

A change of the N_2O/N_2 ratio as a function of crystallite size is interesting in view of a reaction scheme which we have deduced from the reaction between ^{14}NO and $^{15}NH_3$ over Pt (reference 23). This scheme should be applicable here, as the authors assume that NO is the initial product during ammonia oxidation. The minimum of nitrous oxide is roughly determined by the amount of hydrogen liberated during the dissociative chemisorption of ammonia $^{15}NH_3 \rightarrow {}^{15}NH_2 + H$. Thus, if the formation of $^{14}NH_2$ is disregarded, a minimum of 33% of the nitrogen-containing species appears as nitrous oxide ($^{14}N_2O$), or 25%, if $^{14}N_2$ formation is taken into account. This percentage, found by Delaney and Manogue at 200°C, implies that at lower temperatures the only mixed product formed is $^{14}N^{15}N$. Increased nitrous oxide production indicates that $^{15}NH_2$ radicals loose their hydrogen before they can react with ^{14}NO to form $^{14}N^{15}N$ and H_2O. The hydrogen abstraction necessitates formation of $^{15}N^{14}NO$ instead of $^{14}N^{15}N$ and results in additional $^{14}N_2O$, as well. Small crystallites can be expected to favor hydrogen abstraction because of a larger percentage of preferred adsorption sites (corners, edges, strain defects). A similar N_2O increase is also expected as the temperature is moderately increased, and is actually observed.

W. H. MANOGUE

Thank you. We have found your work very helpful in trying to increase our understanding of our observations.

K. OTTO

I would like to point out that we have observed some foreign compounds when adding oxygen to a mixture of nitric oxide and ammonia at room temperature. A white solid was deposited on the glass walls from a stream consisting of 1% of NO, 1% of O_2 and 2% of NH_3 in helium. The deposit is easily volatilized without decomposition when heated moderately with a torch. We have not identified this product and wonder whether you have found any evidence for its formation.

W. H. MANOGUE

We have observed a white deposit in the lines from a sulfuric acid scrubber which we have used to remove ammonia in some oxidation work with a similar stream. Infrared analysis showed that the material was ammonium nitrate.

J. R. KATZER

In comment on this paper and on some of the earlier questions to this paper, I would like to present briefly the results of more recent work which has been carried out under differential conditions. Under these conditions heat and mass limitations are not important, and the kinetics of the reaction are being determined. Supported Pt on η-Al_2O_3 catalysts (~1% Pt) prepared in our laboratory were used. One of these catalysts contained highly dispersed platinum (d_p ~ 20 Å) and the other contained large crystallites (~ 150 Å) as determined by X-ray, electron microscopy, and H_2 chemisorption. The specific catalytic activity of the highly dispersed Pt catalyst was almost one order of magnitude lower than that of the catalyst containing the large Pt crystallites. Selectivity differences were also observed. Thus, a pronounced crystallite size effect was observed in the low temperature ammonia oxidation system. We are not able to define the exact cause of this effect as yet.

W. H. MANOGUE

Thank you.

Paper Number 14

A STUDY OF NITROUS OXIDE DECOMPOSITION ON NICKEL OXIDE BY A DYNAMIC METHOD

CHARLES C. YANG,* MICHAEL B. CUTLIP, and CARROLL O. BENNETT
Department of Chemical Engineering, University of Connecticut,
Storrs, Connecticut 06268

ABSTRACT: The decomposition of N_2O over 14 weight per cent NiO on silica has been studied over a wide range of reactant concentrations in argon at atmospheric pressure and from 332 to 372°C in a completely-mixed or gradientless reactor. The response of the catalytic system to step changes of N_2O concentration in the input stream was measured to steady state conditions. The catalyst activity remained constant during the series of experiments, and various tests showed that the rates measured are not influenced by mass or heat transfer effects.

The results show that the reaction proceeds through the following sequence of steps:

$$N_2O + S \xrightarrow{k_1} N_2 + OS$$

$$2\ OS \underset{k_{-2}}{\overset{k_2}{\rightleftarrows}} O_2 + 2S$$

The first step is irreversible and the second step is reversible but not fast enough to be in quasi equilibrium during reaction.

By comparison of the results predicted by the above sequence of steps and a model of the reactor with the experimental concentration curves, it is possible to estimate the kinetic parameters k_1, k_2, and k_{-2}. The results are

$$k_1 = 7.83 \times 10^6 \exp(-7887/RT)$$

$$k_2 = 2.22 \times 10^{10} \exp(-12{,}977/RT)$$

$$k_{-2} = 3.64 \times 10^{10} \exp(4045/RT)$$

These parameters indicate that equilibrium oxygen adsorption is described by $\Delta H° = -17{,}022$ cal/g mole and $\Delta S° = -22.3$ cal/g mole °K.

1. INTRODUCTION

1.1 The dynamic method

The measurement of rates of a heterogeneous catalytic reaction at steady state usually furnishes insufficient information for the elucidation of the sequence of elementary steps which underlie the observed reaction. Many different experimental methods must be applied to a given reaction system, if we hope to gain some knowledge of the identity and reactivity of the active centers which make up the reaction chain. However, kinetic measurements retain an important role in catalytic studies, and dynamic studies are a much richer source of rate information than are those done at steady state. In our work we send a concentration signal in a steadily flowing stream to a small, well-mixed catalytic reactor. The effluent concentration is continuously monitored by a mass spectrometer; from the measured response of the concentration in the reactor it is possible to calculate

*Present address:
Halcon International, Inc.
Little Ferry, New Jersey 07643

the kinetic parameters for a plausible sequence of elementary steps. The large amount of information furnished by such an experiment improves our ability to distinguish among reaction sequences and values of the appropriate rate constants.

The present method was first described in 1967, although powerful computer methods have now replaced the linearization procedure then proposed for data analysis[1]. The experiments are an extension of the transient methods used by Wagner and Hauffe[2] and by Tamaru[3,4]. Perturbations can be made from steady state, so that catalyst stability is always monitored. By using an internally recycled, gradientless reactor[5], the analysis of the data can be made quantitative and transport falsifications are avoided. Perturbations to the feed of a fixed-bed reactor[6] are more difficult to interpret quantitatively - i.e., it is hard to calculate good rate constants.

1.2 Nitrous oxide decomposition

As a first application of the dynamic method we chose the study of the decomposition of nitrous oxide over nickel oxide supported on silica. Many workers have studied this reaction over the past 20 years[7-19]. The identity of the adsorbed species is known with some assurance, especially from the recent studies of Guilleux et al.[18,19]. However, there is little agreement on which is the rate controlling step[13]. At temperatures above 150°C, where there is negligible adsorbed N_2O[19], basically the following two sequences have been proposed:

$$\text{I} \qquad N_2O + S \rightarrow N_2 + OS$$

$$2OS \rightleftarrows O_2 + 2S$$

$$\text{II} \qquad N_2O + S \rightarrow N_2 + OS$$

$$N_2O + OS \rightleftarrows N_2 + O_2 + S$$

Since the adsorption of oxygen or the decomposition of N_2O increases the electrical conductivity of NiO, a p-type semiconductor, most authors agree with Samaha and Teichner[14] that, above 150°C[18], the adsorbed oxygen (OS) is present as O^- ions. All agree also that N_2 is not adsorbed.

Thus in our work we shall attempt to find out which of the two sequences best agrees with the transient kinetic data, and from the rate constants found we can say something about the rate-determining step, if any, in a given sequence.

2. EXPERIMENTAL

2.1 Materials

The catalyst was made by soaking highly porous silica beads in an ammoniacal solution of nickel formate, according to a published procedure[20]. After drying, the formate was decomposed to NiO in the reactor by heating in a stream of air to about 300°C. The catalyst was then exposed to argon at 400°C for 24 hours. This procedure produced a catalyst of 14.0 percent NiO of total surface area 350 m^2/g. Over several months of use its activity remained essentially constant. The reactor was filled with 6.72 g of 1 mm particles, giving a bed volume of about 17 ml.

Nitrous oxide (Matheson, 98.0%) and oxygen (Matheson, 99.6%) were used directly from their cylinders. Argon (Matheson 99.999%) was used as the inert carrier gas.

2.2 Gradientless reactor

The stainless steel reactor is shown in fig. 1. Its empty volume is 26.5 ml, and the free volume when filled with 6.72 g of catalyst is 23.3 ml, as measured by compression of a known volume of helium into the reactor. As shown, the catalyst is held in an annular space enclosed by upper and lower screens. The absence of internal gradients of temperature and concentration was assured by the operation of the turbine at speeds up to 3600 rpm. A later version of the reactor uses a magnetic drive for the turbine.

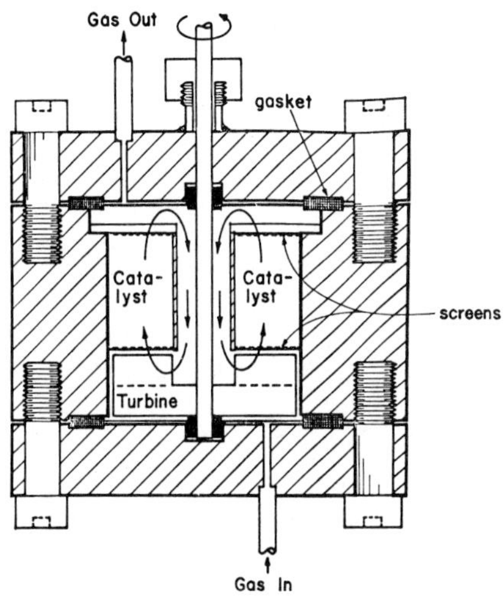

Fig. 1 Gradientless reactor

When a stream of argon flowing to the reactor is suddenly changed to nitrogen, neither of which is adsorbed, the response of the outlet to this step signal shows that the gas space is well mixed at reaction temperature. These tests have already been described in some detail[5]. From these results the volume of the gas space in the reactor can be calculated; it was found to agree with that measured statically (23.3 ml). We also found that the rate of N_2O decomposition at steady state was uninfluenced by turbine rpm over a wide range.

2.3 Experimental flow scheme

The arrangement of the apparatus is shown in fig. 2. For most transient runs, pure argon was stored in vessel A, and an appropriate N_2O-Ar mixture in vessel B. The feed to the reactor was suddenly switched from A to B by the solenoid-operated 4-way valve. The needle valves were carefully adjusted so that this switch did not alter the pressure in the reactor or the flow rate measured by the soap-bubble flowmeter. The second four-way valve permits the feed to by-pass the reactor so that the sharpness of the step function can be tested, and the mass spectrometer can be suitably calibrated by known pure gases and gas mixtures.

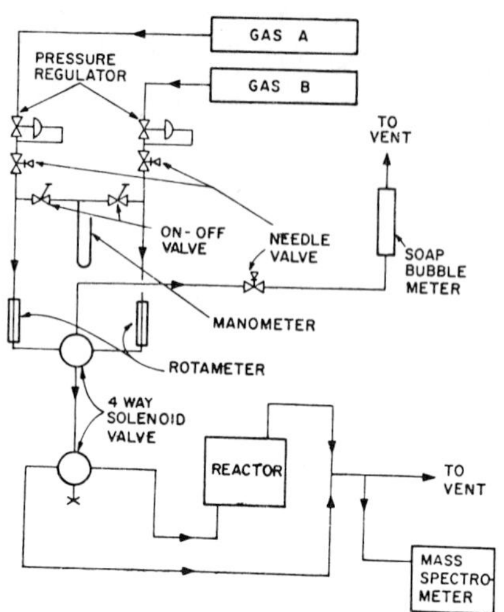

Fig. 2 Experimental flow scheme

As already described[5], the continuous inlet system to the mass spectrometer has a response time sufficiently small to permit good resolution of the inlet step functions. However, automatic repetitive scan and fast response data recording equipment were not available, so the spectrometer was set to follow one mass number during a transient experiment. Thus we had to repeat each run four times, following mass numbers 44, 40, 32, and 28 successively.

3. RESULTS AND INTERPRETATION

3.1 Steady-state data

The rate of the reaction

$$2N_2O \rightarrow 2N_2 + O_2 \tag{1}$$

at steady state is calculated from a simple mass balance on the oxygen produced (no O_2 in the feed):

$$\bar{r} = q\, c_{O_2} \tag{2}$$

where q is the effluent flow rate, ml/sec and c_{O_2} is in g mole/ml.

The rate is thus given in g mole/ml-sec; it can be converted to g mole/g of catalyst-sec by multiplying by 23.3/6.72. The steady-state results are given in fig. 3. A plot of ln $\bar{r}$ vs 1/T at a constant (arbitrary) value of c_{N_2O} gives a straight line yielding an apparent activation energy of 30 kcal/g mole.

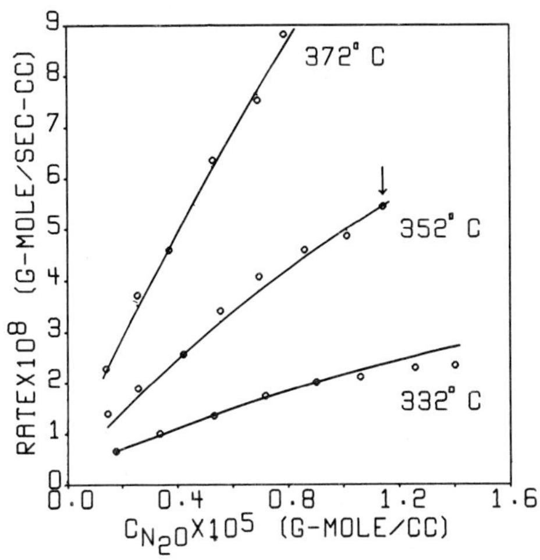

Fig. 3 Steady-state reaction rates

3.2 Transient data

Fig. 4 shows a typical transient run. At time zero the pure argon flowing to the reactor was replaced by a mixture of argon and nitrous oxide, $c_{N_2Of} = 1.382 \times 10^{-5}$ g moles/cc; $c_{O_2f} = c_{N_2f} = 0$. As time proceeds, the concentration of N_2O increases to its steady-state value, $\bar{c}_{N_2O} = 1.144 \times 10^{-5}$, which is less than c_{N_2Of}. This value, taken from the right-hand part of fig. 4, is indicated for this run by the small arrow in fig. 3. Corresponding to every point in fig. 3, there is a preceding transient period like the one shown in fig. 4.

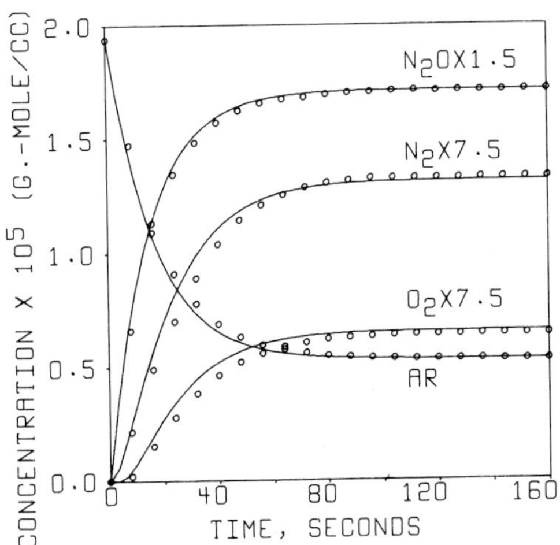

Fig. 4 A transient run

The points in fig. 4 are experimental and the lines are those obtained from a model of the reactor based on the more favored sequence (I) and the best values of its kinetic parameters to be discussed below.

3.3 Method of calculating the kinetic parameters

The mathematical treatment of the data is considered in some detail elsewhere[1,21]. The concentration of a gaseous species c_j in the reacting system obeys the mass balance

$$\frac{dc_j}{dt} = \left(\frac{q_f}{v}\right) c_{jf} - \left(\frac{q}{v}\right) c_j + \sum_{i=1}^{E} Z_{ij} r_i \quad (j=1,\ldots,N) \tag{3}$$

where v is the reactor free volume (23.3 ml) and the subscript f refers to the feed (t > 0) to the reactor. There are E elementary steps of rate r_i described by the chemical equations

$$\sum_{j=1}^{N^*} Z_{ij} c_j = 0 \quad (i = 1,\ldots,E) \tag{4}$$

where Z_{ij} is the matrix of stoichiometric coefficients. There are N gaseous components and N*-N active centers, including vacant sites; thus N* is the total number of components including the surface species. For the latter, the mass balance is

$$\frac{dc_j}{dt} = \sum_{i=1}^{E} Z_{ij} r_i \quad (j = N+1,\ldots,N^*) \tag{5}$$

The rates are assumed to depend on concentration according to the simple Guldberg-Waage law. For sequence I this gives, for example

$$r_1 = k_1 c_{N_2O} c_S \tag{6}$$

$$r_2 = k_2 c_{OS}^2 - k_{-2} c_S^2 c_{O_2} \tag{7}$$

The concentrations must obey the relations

$$\sum_{j=1}^{N} c_j = P/RT \tag{8}$$

and

$$\sum_{N+1}^{N^*} c_j = c_{SO} \tag{9}$$

where c_{SO} is the concentration of total vacant sites at t = 0--i.e., after exposure to argon.

We now can simulate by computer the response of the reactor effluent c_j to an input signal from $c_{N_2O} = 0$ to c_{N_2Of}. The concentration vs time curves thus produced depend on the four parameters c_{SO}, k_1, k_2, and k_{-2}. We now use a computer optimization procedure to find the values of the four parameters which make the curves best fit the experimental transient curves. In mathematical terms, we search for the parameter values which minimize the objective function defined as

$$\text{O.F.} = \sum_{j=1}^{N} \sum_{k=1}^{M} \left[\frac{\hat{c}_{jk} - c_{jk}}{\hat{c}_{jk}} \right]^2 \tag{10}$$

where c_{jk} is c_j at time t_k and there are M equal time intervals. The concentrations calculated from the assumed parameters are $\hat{c}_{jk}$.

3.4 Values of the kinetic parameters

More information on the search procedure will be found in a separate paper[21]. Here we merely give the results of this work: the lowest value of the objective function is obtained for sequence I with the following relations for the parameters:

$$c_{SO} = 19.86 \exp(-21{,}082/RT) \tag{11}$$

$$k_1 = 7.83 \times 10^6 \exp(-7887/RT) \tag{12}$$

$$k_2 = 2.22 \times 10^{10} \exp(-12{,}977/RT) \tag{13}$$

$$k_{-2} = 3.64 \times 10^{10} \exp(4045/RT) \tag{14}$$

The curves of figs. 3 and 4 are based on the above set of parameter values. These values yield for K, the equilibrium constant for oxygen adsorption (the reverse of step 2 in sequence I),

$$K = k_{-2}/k_2 = 1.64 \exp(17{,}022/RT) \tag{15}$$

From this relation we obtain for the dissociative oxygen adsorption $\Delta H° = -17{,}022$ cal/g mole O_2 and $\Delta S° = -22.3$ cal/g mole °K.

4. DISCUSSION

4.1 Nature of adsorbed intermediates

From a material balance on nitrogen atoms during the transient period the quantity of N_2O adsorbed can be computed. We find negligible N_2O adsorption, in agreement with the literature[19]. The nitrogen produced just corresponds to the N_2O which disappears by reaction. Separate tests showed that nitrogen is not adsorbed.

Oxygen, however, is adsorbed during the transient; it appears more slowly than the nitrogen does. Oxygen also reduces the steady-state rate when it is added with N_2O in the feed. As discussed previously, the adsorbed oxygen is probably present on O^- ions, but we designate it merely as OS. The rate of adsorption at 330-370°C is presumed to be influenced by the concentration of vacant sites of suitable energy and not by electron transfer, and is expressed by step 2 of sequence I (see eq. 7).

In the transient experiments, the initial state of the catalyst is that in equilibrium with pure argon at the run temperature. Thus the surface is not denuded of oxygen; this is clearly indicated by the temperature desorption spectra of Gay[14]. Thus the oxygen adsorbed during a transient is that which makes relatively low-energy bonds with the surface. The calorimetric data of El Shobaky et al.[22] show that the differential heat of adsorption falls from the neighborhood of -60 kcal/g mole on clean NiO to less than -20 at high coverage. Our experiments lead to the calculation of a differential ΔH° of adsorption, so that the value of about -17 kcal/g mole does not seem unreasonable.

Our results measure the amount of adsorption during reaction, so that the apparent initial concentration of vacant sites c_{SO} is obtained. Since the amount of irreversible adsorbed oxygen increases with decreasing temperature, we find that the vacant sites which are available for the reversible adsorption of oxygen (our c_{SO}) decrease with decreasing temperature. Note that the quantity $c_{SO}k_1$ has an activation energy of about 28 kcal/g mole. The value of c_{SO} found is of the order of 10^{-7} g mole/ml, or 0.3 ml/g pure NiO (~ 1.0 mg/g) of oxygen at saturation. This quantity represents about 0.6 percent of a monolayer.

4.2 Non equilibrium adsorption during reaction

From the experimental concentrations, the value of the rate parameters (eqs. 11-14), and the rate expressions (eqs. 6 and 7), we can calculate the rate of the first step (irreversible) and the forward and backward rates of the second step of sequence I. The ratio at steady state of the forward rate of step 2 (r_{2f}) to r_1 is a measure of how close the second step is to equilibrium under reaction conditions. If r_{2f}/r_1 is large (say above 20) then the second step is at quasi equilibrium, and the first step is rate-determining. If the ratio is small, neither step is controlling alone.

Fig. 5 shows how r_{2f}/r_1 varies with temperature and nitrous oxide composition (Ar plus N_2O feed). We see that the first step is not rate controlling and thus that the adsorbed oxygen is not in equilibrium with the gaseous oxygen. As the temperature is increased, the first step becomes more nearly rate controlling.

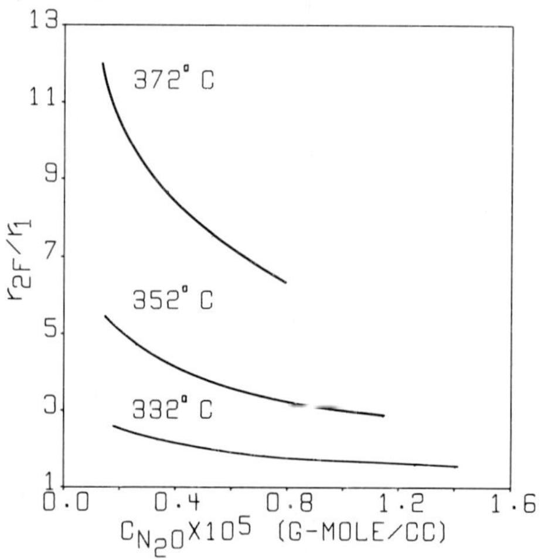

Fig. 5 Ratio of rates of elementary steps

These ideas are illustrated in a different way by fig. 6, where the fractional coverage of oxygen during a transient (same run as fig. 4) is plotted against time. The coverage θ_{eq} which would exist if the surface were in equilibrium with the observed gaseous O_2 concentration is also given. Since r_{2f}/r_1 is not large, the surface is more covered with oxygen than it would be if step 1 controlled. $\theta-\theta_{eq}$ is a driving force for desorption.

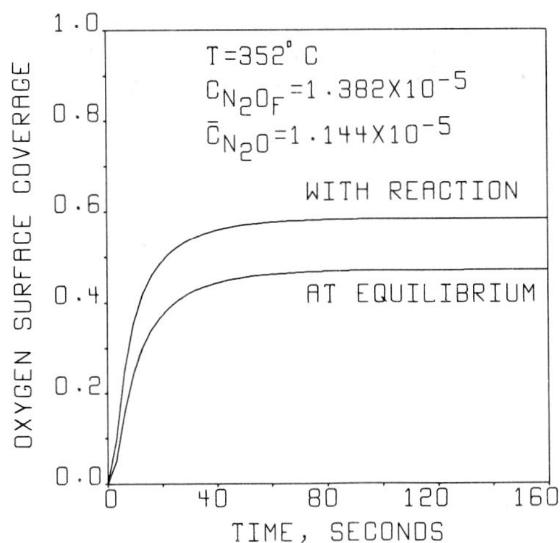

Fig. 6 Non equilibrium adsorption of oxygen

4.3 Steady-state rate expressions

If step 1 of sequence I is rate-controlling, the steady-state rate $\bar{r}$ is given by

$$\bar{r} = r_1 = r_2 = \frac{k_1 c_{SO} c_{N_2O}}{1 + K^{1/2}\, c_{O_2}^{1/2}} \tag{16}$$

From the discussion in the preceding section we know that this equation should not fit our data well, although it is a better approximation at higher temperatures. Other workers have used their kinetic data with eq. (16) to compute $\Delta H°$ for K. For NiO, Gay and Tompkins[15] obtained -40 kcal/g mole and Winter[17] -45 kcal/g mole; for α-manganese sesquioxide, Rheaume and Parravano[23] got -43 kcal/g mole and for chromium oxide Tanaka and Ozaki[24] got -44 kcal/g mole. Our value of - 17 kcal/g mole is much smaller, but we now see that this disagreement is not surprising; eq. (16) probably does not apply accurately to most previous data on N_2O desorption also.

The steady-state rate equation which agrees with sequence I with non equilibrium adsorption of oxygen can be found by solving for c_S by equating r_1 and r_2 (eqs. (6) and (7)) and using eq. (9). This procedure gives

$$\bar{r} = r_2 = r_1 = k_1\, c_{N_2O}\, c_S \tag{17}$$

and

$$\bar{r} = k_1 c_{N_2O}\, \frac{2k_2 c_{SO} + k_1 c_{N_2O} - \sqrt{4k_2 k_{-2} c_{O_2} c_{SO}^2 + 4k_2 k_1 c_{N_2O} c_{SO} + k_1^2 c_{N_2O}^2}}{2\,[k_2 - k_{-2}\, c_{O_2}]} \tag{18}$$

The evaluation of the constants in this redoubtable equation from kinetic data would usually be viewed as a rather foolhardy exercise. However, the transient method used here permits one to consider the equivalent simple model (sequence I and eqs. (6) and (7)). The effect of the various parameters and concentration can be comprehended from this model, whereas their effect in eq. (18) is buried in a maze of algebra. Nevertheless, eq. (18) is perfectly suitable for a computer simulation for steady-state reactor design, and of course it has the virtue of fitting the data better than eq. (16).

This work was supported by NSF grant GK1918.

REFERENCES

1) C. O. Bennett, AIChE Journal 13 (1967), 890.
2) C. Wagner and K. Hauffe, Z. Elektrochem. 45 (1939), 409.
3) K. Tamaru, Advances in Catalysis 15 (1964), 65. Academic Press, New York, N.Y.
4) A. Ueno, T. Onishi, and K. Tamaru, Trans. Faraday Soc. 60 (1970), 756.
5) C. O. Bennett, M. B. Cutlip and C. C. Yang, Chem. Eng. Sci. (in publication process).
6) R. J. Kokes, H. Tobin and P. H. Emmett, J. Am. Chem. Soc. 77 (1965), 5860.
7) K. Hauffe, R. Glang and H. J. Engell, Z. Physik. Chem. (Leipzig) 201 (1952), 223.
8) C. B. Amphlett, Trans. Faraday Soc. 50 (1954), 273.
9) K. Hauffe, Advan. Catal. 7 (1955), 213: 9 (1957), 187.
10) J. Dewing and Cvetanovic, Can. J. Chem. 36 (1958), 678.
11) E. R. S. Winter, Disc. Faraday Soc. 28 (1959), 183.
12) H. B. Charman, R. M. Dell and S. S. Teale, Trans. Faraday Soc. 59 (1963), 453.
13) K. Kuchynka, Coll. Czechoslov. Chem. Comm. 30 (1965), 613, 622.
14) E. Samaha and S. J. Teichner, Bull. Soc. Chim. France 5th Ser. 2 (1966) 650, 667, 672.
15) I. D. Gay and F. C. Tompkins, Proc. Roy. Soc. A293 (1966), 19.
16) I. D. Gay, J. Catal. 17 (1970), 245.
17) F. R. S. Winter, J. Catal. 19(1) (1970), 32.
18) M. F. Guilleux, J. Fraissard and B. Imelik, Bull. Soc. Chim. France 11 (1969), 3787.
19) M. F. Guilleux and B. Imelik, Bull. Soc. Chim. France 4 (1970), 1310.
20) U. S. Patent 3,207,702, Sept. 21, 1965. W. H. Flank, J. E. McEvoy, and H. Shalit, to Air Products and Chemicals, Inc.
21) M. B. Cutlip and C. O. Bennett, AIChE Journal 18 (1972), 1073.
22) G. E. Shobaky, P. C. Gravelle and S. J. Teichner, Advances in Chemistry Series 76 (1968), 292.
23) L. Rheaume and G. Parravano, J. Phys. Chem. 63 (1959), 264.
24) K. Tanaka and A. Ozaki, J. Catal. 8 (1967), 307.

DISCUSSION

D. D. ELEY

It is interesting to compare these results for NiO with earlier detailed work on the N_2O decomposition on Pd and PdAu alloys.[1] We adopted a similar mechanism, but found it necessary to allow for successive stages of desorption 2 O(1 site) $\rightleftarrows$ O_2(2 sites) $\rightleftarrows$ O_2(1 site) $\rightleftarrows$ O_2(gas), the term in the denominator of the rate equation being P_{O_2} rather than $P_{O_2}^{1/2}$. Pd was clearly less active than NiO, our temperature range being 500-900°C. In our case $\Delta H° = -32,200$ cal/g mole and $\Delta S° = -7.4$ cal/g mole °K, so although the oxygen is held more strongly on Pd, it is also more mobile. For Pd, k_1 was $2.56 \times 10^{-1}\ e^{-12,700/RT}\ dm^{-2}\ sec^{-1}$, so the activation energy was higher. The activation entropy (assuming all sites were active) could be compared with NiO, when the latter reaction is expressed in the same units, and this would also be very interesting.

1) D. D. Eley and C. F. Knights, Proc. Roy. Soc. Lond. A 294 (1966) 1.

C. O. BENNETT

The rate of the irreversible dissociative adsorption of N_2O can be written according to transition state theory as

$$r_1 = \nu\ c^{\neq} \tag{1}$$

where the concentration of the activated complex ($c^{\neq}$, cm^{-2}) is in equilibrium with the reactants, N_2O (g) and S. Thus we have also

$$K^{\neq} = \frac{c^{\neq}/c_S^0}{(c_S/c_S^0)\ (c_{N_2O}/n^0)} \tag{2}$$

We choose the same standard state c_S^0 (say $\theta = 0.5$) for both surface species, and set n^0 equal to the concentration of pure N_2O at 1 atm and T, say $10^{19}\ cm^{-3}$. Equation (1) then becomes

$$r_1 = \nu\ K^{\neq}\ c_S\ c_{N_2O}/n^0 \tag{3}$$

or

$$r_1 = \frac{\nu}{n^0} \exp\left(\frac{\Delta S^{0\neq}}{R}\right) \exp\left(-\frac{\Delta H^{0\neq}}{RT}\right) c_S\ c_{N_2O} \tag{4}$$

By comparison with the equation

$$r_1 = k_1 c_S c_{N_2O} = A_1 \exp(-E/RT)\ c_S\ c_{N_2O} \tag{5}$$

we obtain the relation

$$\Delta S^{0\neq} = 2.3 \log_{10} (A_1 n_0/\nu) \quad (6)$$

The value of A_1 (see response to comment No. 4) is 1.29×10^{-17} cm^3 sec^{-1}; for $\nu = 10^{13}$ sec^{-1} we find $\Delta S^{0\neq} = -50$ cal/gmole °K. For the reaction on pure Pd at 650°C, the data of Eley and Knights yield $\Delta S^{0\neq} = -21$ cal/gmole °K, for $c_S = 1.5 \times 10^{15}$ cm^{-2}.

K. TAMARU

In this paper the mechanism of N_2O decomposition on NiO catalyst has been beautifully elucidated by studying the simpler (or elementary) processes of which the overall reaction is composed. In this case the rate constants of the elementary processes were taken to be independent of the surface coverage. The surface properties of the catalyst and the rate constants, in general, depend upon surface coverage, which is one of the characteristic features of heterogeneous catalysis. I wonder if you could examine the dependence of the rate constants upon the oxygen coverage, especially by using a non-stationary technique, and if you could express the rate constants as functions of coverage, you will obtain better and more exact kinetic analysis.

C. O. BENNETT

One could introduce Elovich-type factors such as $e^{-s\theta}$ into each of the parameters k_1, k_2, and k_3; in addition, s might vary with temperature. We did not consider our data extensive enough to justify evaluating even more kinetic parameters than the eight reported. Rather than attempt to find more parameters by fitting the data reported, it would be preferable to use tracers so that rates could be measured at various levels of constant surface coverage. This might be done by suddenly replacing the N_2O in the feed by $N_2{}^{18}O$ and measuring the response of $^{18}O^{16}O$ and $^{18}O_2$ while the total chemical OS concentration is constant.

L. ZANDERIGHI

I am very interested in your treatment of transition state data and I have two questions:

1) Is it possible to calculate the concentrations of vacant sites (S) and occupied sites (OS) during the transient runs?

2) From Figure 3 it appears that many points are in stationary conditions. Do you think that this fact affects analysis of transient runs, that is, the application of transient equations to stationary runs may involve a correlation between c_{SO} and k_1, k_2, and k_3.

C. O. BENNETT

1) The concentrations c_{OS} and c_S during a transient can be obtained from Figure 6 and the relation $c_{OS} = c_{SO}\ \theta$.

2) Of course there are relations among the parameters (constraints) which must hold at steady state, and we use these relations to find a plausible starting place (parameter vector) in the computer search for the best values of the parameters. The procedure is described elsewhere.[1)]

We chose a constant total time of run arbitrarily so as to include, for an average run, about an equal time during the transient and an equal time at steady state. The particular run of Figure 3 has a higher than average steady-state period. A more fundamental method of choosing the length of run is one of the many interesting statistical problems involved in our work.

1) M. B. Cutlip, C. C. Yang, and C. O. Bennett, AIChE Journal 18, 1073 (1972).

M. BOUDART

I wish to emphasize the point made by Professor Eley. It is hoped that workers in catalysis will report their data in units that permit comparison between various investigations. The rate of reaction, in particular, should be reported, whenever possible in molecules $m^{-2}\ s^{-1}$.

C. O. BENNETT

Since preparing the manuscript for this paper we have obtained a measurement of crystallite size of 190 ± 20 Å by X-ray line broadening.[1] This figure, which corresponds to $4.94 \times 10^4\ cm^2$ of NiO/g of supported catalyst, permits the calculation of the kinetic parameters in terms of rates in molecules $cm^{-2}\ sec^{-1}$, gas concentrations in molecules cm^{-3}, and surface concentrations in molecules cm^{-2}.

It is also preferable to replace Equation 7 for the rate of the second step by the more precise expression,[2]

$$r_2 = k_2' c_{OS} \frac{1}{2} Z\theta - k_{-2}' c_{O_2} c_S \frac{1}{2} Z(1-\theta) \qquad (1)$$

or

$$r_2 = k_2 \frac{c_{O_2}^2}{c_{SO}} - k_{-2} \frac{c_S^2 c_{O_2}}{c_{SO}} \qquad (2)$$

where the quantities 1/2 Z are incorporated in k_2 and k_{-2}. The resulting parameters are

$$C_{SO} = 8.38 \times 10^{-17} \exp(-21{,}082/RT)\ cm^{-2} \qquad (3)$$

$$k_1 = 1.29 \times 10^{-17} \exp(-7{,}887/RT)\ cm^3\ sec^{-1} \qquad (4)$$

$$k_2 = 4.41 \times 10^{11} \exp(-34{,}059/RT)\ sec^{-1} \qquad (5)$$

$$k_{-2} = 1.20 \times 10^{-12} \exp(-17{,}037/RT)\ cm^3\ sec^{-1} \qquad (6)$$

The thermodynamic quantities ΔH° and ΔS° are of course unchanged. The new activation energies shown in Equations (5) and (6) seem more realistic than those of Equations (13) and (14) of the paper. The parameter values agree with the criteria given by Boudart.[3]

1) Paul Stonehart, personal communication.
2) M. Boudart, "Kinetics of Chemical Processes," Prentice Hall, 1968, pp. 85-86.
3) M. Boudart, A.I.Ch.E. Journal 18 (1972), 465.

Paper Number 15

AGING OF ALUMINA-SUPPORTED COPPER CATALYSTS

J. C. SUMMERS and R. L. KLIMISCH
General Motors Research, Warren, Michigan, USA

Abstract: The effects of aging on the chemistry of alumina-supported copper catalysts have been investigated. The relationships of surface area and copper concentration on the chemical environment of copper were determined as was the relationship between surface area and the catalyst's oxidation activity.

High surface area blue and green copper catalysts did not contain the spinel copper(II) aluminate but, instead contained a six-coordinate copper(II) species. In oxidizing atmospheres, the spinel formed only upon extended high-temperature treatment (~50 hours at 900°C). However, the spinel tends to form much faster in alternating oxidizing-reducing atmospheres. The spinel is the main copper species present in deactivated low-surface area copper catalysts. A general mechanism of deactivation for copper oxidation catalysts is proposed.

1. INTRODUCTION

Considerable effort is being directed towards the development of catalytic systems that will convert automobile exhaust emissions into harmless products[1,2]. Catalysts for these systems must exhibit high initial activity and retain most of their activity for extended periods under a variety of severe conditions.

Copper-containing catalysts have exhibited high initial activity for oxidizing carbon monoxide and hydrocarbons[1]. However, they often lose much of their activity during extended automotive testing[3]. Recently a number of studies have been performed to elucidate the chemical and physical properties of alumina-supported copper catalysts[4-8]. These studies have generated relatively little information, however, about the effects of aging on these properties or on their relationships to oxidation activity. Therefore, the overall objective of the present study was to determine how the chemical and physical properties of alumina-supported copper catalysts change upon aging and the effect that these changes have on catalytic activity.

2. EXPERIMENTAL

2.1. Catalysts

A series of alumina supports with different surface areas was obtained by heating high-surface area preformed alumina (1/8" spheres of Kaiser KA-302). The heating schedule and resulting surface areas are shown in Table 1.

The alumina supports were impregnated with aqueous copper(II) nitrate, air-dried overnight, and then calcined in air for 4-5 hours at 600°C. Catalysts containing four levels of copper were prepared (Table 2).

Table 1
The Effect of Thermal Treatment on Alumina Surface Area

Surface Area	Heating Schedule
300 m^2/g	untreated
145 m^2/g	850°C - 1 hr
87 m^2/g	850°C - 2 hrs
35 m^2/g	1050°C - 48 hrs

Table 2
Designations of Copper Concentrations

Approx. Concentration (wt %)	Designation
1.6	Cu_1
3.4	Cu_2
6.5	Cu_3
12.3	Cu_4

In a separate preparation, a Cu_3 catalyst (surface area = 264 m^2/g) was heated in a muffle furnace for 48 hours at 950°C. The resulting material had a surface area of 11 m^2/g.

2.2. Catalyst characterization

The surface areas were determined by the standard BET technique, using a Numinco Orr surface area-pore volume analyzer, Model MIC 103. X-ray diffraction powder patterns were obtained from a General Electric XRD-5 unit using the $K\alpha$ line of copper (1.5418Å) as the X-ray source. The electronic absorption spectra were obtained on a Cary Model 14 recording spectrophotometer.

2.3. Aging of the catalysts

Artificial aging was accomplished by heating the catalysts in an environment where the net oxidizing or reducing character approximately simulates that encountered in automobile exhaust. The oxidizing conditions were achieved by heating the catalysts in a muffle furnace in contact with air. Reducing conditions were achieved by flowing a stream of carbon monoxide (4%) and nitrogen (96%) over the catalyst contained in quartz tubes electrically heated by a Lindberg tube furnace.

2.4. Activity measurements

Reactivity measurements were carried out in a continuous flow system. The reactor consisted of a 2 cm I.D. stainless steel pipe containing a catalyst charge of 30 cm^3 heated electrically by a tube furnace[1)].

A feed stream, consisting of 1.0% CO, 0.025% C_3H_6 (propylene), 2% O_2, and 10% steam in nitrogen, was passed over the catalyst at a gas hourly space velocity of 14,200. The hydrocarbon content of the stream was analyzed with a Beckman 109A flame ionization analyzer. Carbon monoxide and carbon dioxide concentrations were measured with Beckman IR315A non-dispersive infrared analyzers.

3. RESULTS

3.1. Catalyst characterization

Blue and green colored "copper oxide" on alumina catalysts have often been thought to contain aluminates since copper(I) and copper(II) oxides are red-brown and black in color. It can be seen from Table 3 that the spinel copper(II) aluminate, $CuAl_2O_4$, was not detected in any of the blue or green catalysts. Instead, the electronic absorption spectra indicated that these materials contained a six-coordinate copper(II) species designated as $[Cu^{2+}(O)_6]$.

Table 3
Characterization of Alumina-Supported Copper Catalysts

Cu Conc.	Surface Area (m^2/g)	Copper and Support Idenity*		Color
		Cu	Al_2O_3	
Cu_1	253	$[Cu^{2+}(O)_6]$	gamma	Light Blue
	145	"	"	Blue-green
	93	"	"	"
	30	$CuO, CuAl_2O_4$	alpha	Gray
Cu_2	263	$[Cu^{2+}(O)_6]$	gamma	Light Blue
	134	"	"	Blue-green
	87	"	"	"
	30	$CuO, CuAl_2O_4$	alpha	Gray
Cu_3	264	$[Cu^{2+}(O)_6]$	gamma	Blue-green
	132	"	"	Dark Green
	83	$CuO, [Cu^{2+}(O)_6]$	"	Green-black
	32	$CuO, CuAl_2O_4$	alpha	Black
	11	$CuAl_2O_4$	"	Tan
Cu_4	221	$[Cu^{2+}(O)_6]$	gamma	Dark Green
	114	CuO	"	Black
	74	CuO	"	"
	34	CuO	alpha	"

* Identified by XRD except for $[Cu^{2+}(O)_6]$ which was identified by its electronic absorption spectra.

The electronic absorption spectra of the blue and green materials were characterized by a single broad assymetric band centered at 6500 to 7500Å (13,000 to 15,000 cm^{-1}). The position and general shape of these bands resemble those obtained when copper(II) is coordinated to six oxygen atoms, e.g., in $CuSO_4 \cdot 5H_2O$[9]) (Figure 1). Presumably,coordination of the copper occurs through oxygen atoms supplied by the alumina support so that this species may also be classified as an aluminate.

Table 3 also shows that the spinel copper(II) aluminate was detected in catalysts having low surface areas (< 50 m^2/g). These catalysts were gray and tan in color. The other copper species that was detected, copper(II) oxide, was favored by low surface areas and high copper concentrations.

Catalysts of surface areas greater than 70 m^2/g contained gamma alumina, whereas only the alpha phase was detected in catalysts of surface areas less than 35 m^2/g.

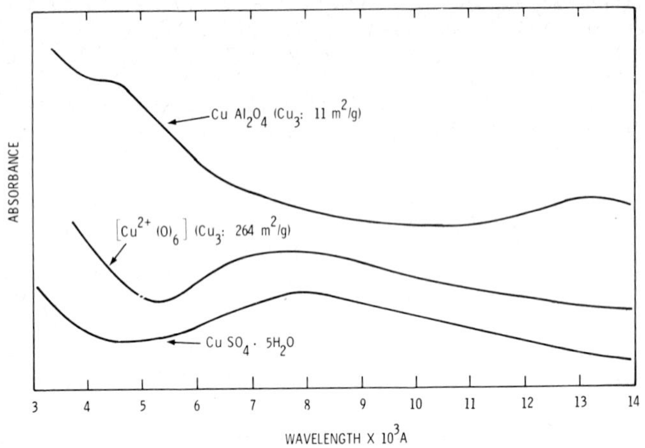

FIGURE 1. ELECTRONIC SPECTRA OF SOME COPPER SPECIES.

3.2. Surface area and activity

The Cu_3 catalyst series was used to investigate the relationship between surface area and catalytic activity. These results are summarized in fig. 2. Catalytic activity was indicated by the temperatures required to oxidize 50% of the propylene or 50% of the carbon monoxide. Decreases in surface area affected hydrocarbon activity more than carbon monoxide activity.

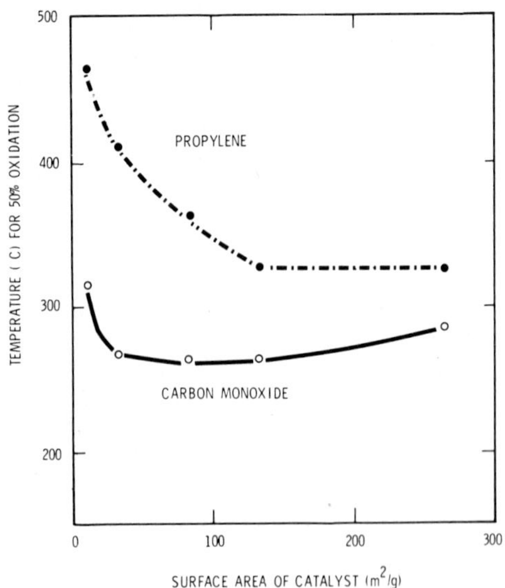

3.3. Aging experiments

Three gamma alumina-supported copper catalysts were heated at 900°C to determine the effects of such treatment on the chemical environment of the copper. The changes in the chemical environment of the catalysts are summarized in Table 4. The chemical composition was determined by electronic absorption spectroscopy and by X-ray diffraction.

Table 4
The Effect of Heating at 900°C on the Chemistry of Copper Catalysts

Time (hr)	Cu_1		Cu_3		Cu_4	
	$CuAl_2O_4$	α-Al_2O_3	$CuAl_2O_4$	α-Al_2O_3	$CuAl_2O_4$	α-Al_2O_3
8	No	No	No	No	Yes	No
24	"	"	"	Yes	"	Yes
48	"	"	Yes	"	"	"
72	"	"	"		"	"
167	"	Yes	"	"	-	-
217	Yes	"	-	-	-	-

It can thus be seen that alpha alumina and the spinel copper(II) aluminate form more readily with increasing copper concentrations.

Under oxidizing conditions, CuO was not observed in any of the aging experiments. Its absence may be caused by a rapid conversion to $CuAl_2O_4$. To test this hypothesis, a previously prepared CuO-containing catalyst (Cu_3 at a surface area of 83 m^2/g) was heated at 900°C in a muffle furnace. The CuO experienced initial conversion to $CuAl_2O_4$ after only one hour, and was completely converted by four hours. Thus, CuO is relatively unstable under oxidizing conditions and converts to $CuAl_2O_4$.

While CuO formation isn't important at 900°C in an oxidizing atmosphere, its importance at lower temperatures could not be ruled out from these studies. Therefore, a Cu_3 catalyst was studied at temperatures from 500 to 800°C (Figure 3). No CuO was observed in this range either, and copper(II) aluminate was detected only after 504 hours of heating.

Despite the fact that CuO is readily converted to the spinel $CuAl_2O_4$ in high-temperature oxidizing atmospheres, CuO has been found in a number

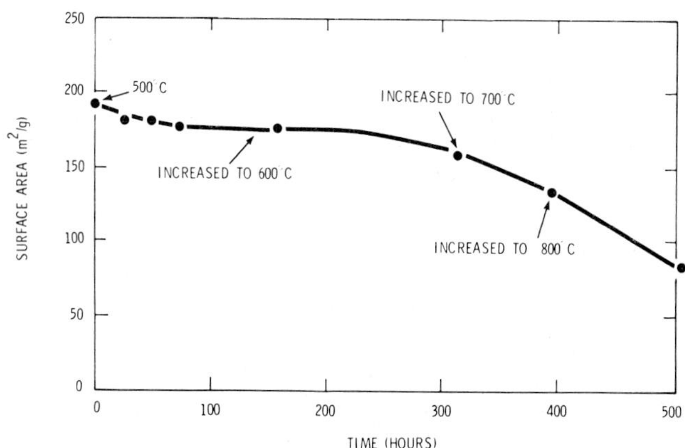

FIGURE 3. SURFACE AREA LOSS OF A Cu_3 ALUMINA-SUPPORTED CATALYST.

of aged automobile exhaust catalysts. Since the spinel and $[Cu^{2+}(O)_6]$ are the only stable high-temperature copper species encountered in this study, a series of oxidation-reduction experiments were performed to determine if CuO could be prepared from these materials.

A $[Cu^{2+}(O)_6]$ catalyst (Cu_3 of surface area 264 m^2/g) was reduced with carbon monoxide overnight at 760°C. The X-ray powder pattern of the resulting black material consisted of broad bands characteristic of finely dispersed metallic copper. Reoxidation with air at 440°C generated both $[Cu^{2+}(O)_6]$ and CuO. Similarly, a $CuAl_2O_4$ catalyst (Cu_3 of surface area 11 m^2/g) was reduced with carbon monoxide overnight at 760°C. The very sharp X-ray bands indicate that metallic copper was present in a highly crystalline state. Reoxidation at 440°C for 5 hours yielded only CuO. Thus, CuO can be formed from both the spinel $CuAl_2O_4$ and $[Cu^{2+}(O)_6]$ by a reduction followed by oxidation.

4. DISCUSSION

It was concluded from the results in Table 3 that copper initally enters the lattice of high-surface alumina as a six-coordinate copper(II) species. Coordination presumably occurs through oxygen atoms supplied by the alumina, and perhaps also by water molecules. When the available six-coordinate sites are filled, copper(II) oxide begins to form.

Decreasing the surface area of the alumina tends to decrease the number of six-coordinate sites available. As a consequence, CuO begins to form at lower and lower copper concentrations and, at temperatures greater than 800°C, it is readily converted to the spinel copper(II) aluminate.

The activity results in fig. 2 indicate that decreasing the surface area from 264 to 132 m^2/g resulted in little deterioration in hydrocarbon activity. Below 132 m^2/g, this activity decreased rapidly with decreasing surface area. On the other hand, the CO activity was relatively independent of the copper environment for catalysts having surface areas of 264 to 35 m^2/g. The copper chemistry in this series varied from six-coordinate copper(II) in $[Cu^{2+}(O)_6]$ to planar copper(II) in CuO[11]. The much less active 11 m^2/g catalyst contains a distorted spinel, $CuAl_2O_4$, in which copper(II) is coordinated both tetrahedrally and octahedrally [10,11].

These results indicate that catalyst deactivation, especially for CO oxidation, is associated with the formation of the spinel $CuAl_2O_4$. Furthermore, these results are in accord with a previous proposal that the alumina support serves not only as a diluent for the copper species in CO oxidation but is actually involved in hydrocarbon oxidation[1]. Thus, surface area changes, as expected, have a minimal effect on CO activity, but have a more significant effect on hydrocarbon activity. In the absence of catalyst poisoning, low-activity copper catalysts are characterized by low surface areas and contain copper(II) aluminate and alpha alumina. Any process that leads to the spinel copper(II) aluminate will lead to catalyst deactivation since copper(II) aluminate catalyzes the conversion of high-surface area gamma alumina to low-surface area alpha alumina[5].

Approximately 50 hours of heating at 900°C under oxidizing conditions are required to convert catalysts containing 6 to 12 wt % copper to alpha alumina and copper(II) aluminate. However, much longer times are required for lower copper concentration (~1.6 wt %), and lower temperatures (500 to 800°C).

Another pathway, involving an oxidizing-reducing cycle can also result in the formation of the spinel copper(II) aluminate:

$$[Cu^{2+}(O)_6] \xrightarrow[\Delta]{CO} Cu \xrightarrow[\Delta]{O_2} CuO \xrightarrow{\Delta} CuAl_2O_4 \quad (1)$$

Even though there are a number of steps involved, milder conditions are required in this oxidizing-reducing cycle than in the one-step oxidizing process. Since CuO is often found in many aged automobile exhaust emission catalysts, either this process or the following must take place:

$$CuAl_2O_4 \xrightarrow[\Delta]{CO} Cu \xrightarrow[\Delta]{O_2} CuO \quad (2)$$

The following equation summarizes the chemical changes that alumina-supported copper automobile exhaust catalysts undergo with aging:

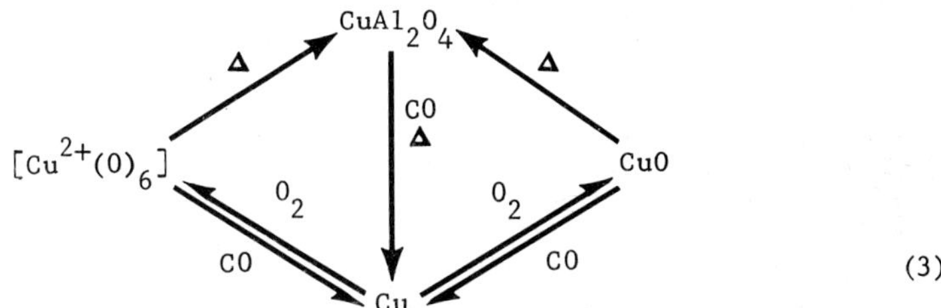

(3)

Up to the point of copper(II) aluminate and alpha-alumina formation, the chemistry of the copper is an equilibrium between $[Cu^{2+}(O)_6]$, Cu, and CuO.

5. CONCLUSIONS

The chemical environment of copper is a function of the catalyst surface area and the amount of copper present. The blue and green high-surface area copper catalysts do not contain the copper(II) aluminate spinel, but contain six-coordinate copper(II). The spinel is formed only after prolonged high-temperature heating in an oxidizing atmosphere or in an alternating oxidizing-reducing atmosphere in which metallic copper and copper(II) oxide are formed as intermediates.

The rate of CO oxidation was nearly independent of surface area except at very low surface areas, but the rate of propylene oxidation decreased markedly with decreasing surface area. The least active catalyst ($\sim$11 m^2/g) examined contained copper(II) aluminate and alpha alumina.

ACKNOWLEDGEMENTS

The authors acknowledge the valuable assistance of Carl M. Nannini, John Melbardis, and Barbara Krieger in obtaining the experimental data.

REFERENCES

1) R. L. Klimisch, Proceedings, First Nat. Sym. on Heterogeneous Catalysis for Control of Air Pollution. Nov. 21-22, 1968, Philadelphia, Pa., pp. 175-198.
2) M. Shelef, Proceedings, First Nat. Sym. on Heterogeneous Catalysis for Control of Air Pollution. Nov. 21-22, 1968, Philadelphia, Pa., pp. 87-112.
3) J. F. Roth, Abstracts of Papers, 161 National Meeting, ACS Petr. 64, Los Angeles, March 1971.
4) J. J. Voge and L. T. Atkins, J. Catalysis 1, 172 (1962).
5) E. D. Pierron, J. A. Rashkin and J. F. Roth, J. Catalysis 9, 38 (1967).
6) A. Wolberg and J. F. Roth, J. Catalysis 15, 250 (1969).
7) P. A. Berger and J. F. Roth, J. Phys. Chem. 71, 4307 (1967).
8) A. Wolberg, J. L. Ogilvie, and J. F. Roth, J. Catalysis 19, 86 (1970).
9) O. G. Holmes and D. S. McClure, J. Chem. Phys. 26, 1686 (1957).
10) A. F. Wells, Structural Inorganic Chemistry, Third Edition (Oxford Universtiy Press, London, 1962).
11) F. Bertant and C. Delorme, Comp. rend. 239, 540 (1954).

DISCUSSION

J. A. RABO

I have one question and one suggestion:

Question: When you refer to the Cu-spinel phase as inactive, do you mean that you establish that the intrinsic activity of the spinel--at equal surface area--is much less than that of the active catalyst phase in the absence of the spinel phase?

Comment: The infrared spectrum of chemisorbed CO could be helpful to establish the valence of the surface copper ions.

J. C. SUMMERS

We do not claim that the copper spinel phase is itself inactive but only that inactive catalysts are characterized as having low surface areas (alumina present in the alpha phase) and as containing the spinel copper (II) aluminate.

R. J. BERTOLACINI

1) In preparing the lower area catalysts was it necessary to use multiple impregnation?

2) Would you predict the same Cu species to be formed if the catalyst area had been reduced by thermal treatment of the $Cu\text{-}Al_2O_3$ catalyst rather than the Al_2O_3?

3) What is the effect of H_2O on the deactivation mechanism of $Cu\text{-}Al_2O_3$?

J. C. SUMMERS

1) Only single impregnations were necessary.

2) One significant difference that we noted was that alumina-supported copper containing catalysts aged from 500-900°C never formed CuO whereas this product was formed on several catalysts prepared from low surface area aluminas.

3) This effect was not studied.

M. SHELEF

There appears to be a discrepancy between this work and that of Roth et al., especially with regard to the absence of the Cu spinel phase at temperatures lower than 900°C. According to Roth, even the fresh catalyst calcined from the nitrate at 600°C for 4 - 5 hours should have contained appreciable Cu spinel, at least in the surface.

J. C. SUMMERS

We have characterized freshly prepared copper catalysts of high surface areas by their electronic absorption spectra and have never seen the low energy band that characterizes the presence of the copper spinel. If the spinel is indeed present, it must be there in small quantities.

W. K. HALL

In the work of Berger and Roth with the same system, Cu^{2+} was found to remain in the surface layer only. What is the difference with your system where you find a discrete $CuAl_2O_4$ phase?

J. C. SUMMERS

The techniques employed to study these catalysts have measured only bulk properties and thus have not been able to distinguish between surface and bulk states. From visual observation, however, it was noted that the spinel $CuAl_2O_4$ can form throughout the catalyst pellet.

M. SCHIAVELLO

In the case of NiO supported on well defined η- and γ-Al_2O_3 the formation of a "surface spinel" occurs, even at 1 atomic percent Ni^{2+} (at 600 C, 24 hrs).[1,2] The cation distribution depends on the atmosphere of firing, on the Ni content and on the type of Al_2O_3 (η- or γ-). Above 6-7 Ni at. percent, a small amount of unreacted NiO was detected. Since it is known that bulk $CuAl_2O_4$ is formed more rapidly than bulk $NiAl_2O_4$, I wish to ask you if there is any special reason why in your case the $CuAl_2O_4$ is formed at higher copper content and not also at low content? A second question regards the possible formation of Cu^{1+} during the reducing cycle. Have you any evidence of the occurrence of Cu^{1+}?

1) LoJacono, M. Schiavello and A. Cimino, J. Phys. Chem. 75, 1044 (1971).
2) M. Schiavello, M. LoJacono and A. Cimino, J. Phys. Chem. 75, 1051 (1971).

J. C. SUMMERS

We did detect the presence of $CuAl_2O_4$ at low surface areas for catalysts containing as little as 1.6 wt % copper (Table 3). In none of these studies was Cu^{1+} observed. However, we have done work in which copper catalysts, after having been exposed to reducing exhaust, were found to contain Cu, Cu_2O, and CuO.

J. F. ROTH

Our results obtained by ESCA and K-absorption edge spectral measurements have indicated that highly dispersed Cu^{+2} ion on gamma alumina exists in a molecular environment resembling that found in crystalline $CuAl_2O_4$ rather than crystalline CuO. Drs. Summers and Klimisch now report that the environment is one in which the Cu^{+2} is principally in octahedral coordination with oxygen. Any apparent discrepancy can be reconciled by assuming that in the crystalline spinel aluminate, Cu^{+2} occupies both octahedral and tetrahedral sites, that the results derived from ESCA and K absorption edge measurements cannot distinguish between Cu^{+2} in octahedral and tetrahedral sites in $CuAl_2O_4$, but that the electronic spectral measurements can.

With regard to Dr. K. Hall's question on the observation of the spinel aluminate in the GM work, we have never observed the presence of crystalline copper aluminate in samples that have been calcined in air at a temperature of 600°C. However, we have not worked much with those samples of the type described by Dr. Summers as yielding the spinel at 600°C, namely, ones in which the alumina support was first thermally sintered to relatively low surface areas prior to impregnation and deposition of Cu^{+2} on the support.

N. D. PARKYNS

γ-Alumina has itself a defective spinel structure, being 2-2/3 cations deficient from the unit cell of 24 cations. In principle, small cations can be accommodated in this structure and Lippens & deBoer have suggested that Na^+ and K^+ can stabilize the lattice against high temperature degradation to α-alumina, the stable form. Our own experience, however, suggests that alkali contents in excess of 1% accelerate the γ- to α-phase change under extreme conditions. I wonder whether it is possible that Cu^{2+} is a sufficiently small ion to be taken up in this way, up to a maximum of a few percent, and if so whether it has a stabilizing effect or the reverse, on your copper-alumina catalysts?

J. C. SUMMERS

We have determined the rate of surface area loss of alumina-supported copper catalysts containing 1.6 to 12.3 wt % copper. In this concentration range, we found that the rate of surface area loss increases with increasing copper content. We have no experience with catalysts containing less than 1.6 wt % copper in this regard.

Paper Number 16

STUDIES OF SURFACE REACTIONS OF *NO* BY ISOTOPE LABELING V. THE REACTION BETWEEN NITRIC OXIDE AND HYDRAZINE AT 25-125°C

K. OTTO and M. SHELEF
Fuel Sciences Department, Scientific Research Staff, Ford Motor Company, Dearborn, Michigan

ABSTRACT: At room temperature the decomposition of hydrazine yields N_2 and H_2 without the splitting of the N-N bond which is somewhat weaker than the N-H bond; above 60°C the decomposition to NH_3 and N_2 becomes predominant. This is a clear indication of the scission of the N-N bond.

Introduction of NO into the system supresses the scission of the N-N bond in the whole investigated temperature range. The products of the surface reaction between ^{15}NO and $H_2{}^{14}N$-$^{14}NH_2$ are H_2O, $^{15}N_2O$, $^{15}N_2$, and $^{14}N_2$, with only ∿ 3% of the nitrogen-containing product molecules being isotopically mixed. This process takes place indiscriminately on many solid surfaces and also in the liquid phase.

From experiments in which the surface to volume ratio was changed and from other runs with pre-adsorbed hydrazine it is deduced that the heterogeneous interaction between nitric oxide and hydrazine remains confined to the surface and proceeds by the adsorption of nitric oxide onto a layer of adsorbed hydrazine. A mechanism of the interaction is proposed.

I. INTRODUCTION

The kinetic resistance to the defixation of the nitrogen atom bound in the thermodynamically unstable nitric oxide is associated with the difficulty of the formation of the N≡N bond.[(1)] This is also the reason why hydrogen reduces NO at low concentrations to NH_3 and not to N_2.[(2)] In contrast, in the catalytic interaction between NO and NH_3 the formation of the N≡N bond is facilitated, as each of the reactants supplies one nitrogen atom.[(3)]

In connection with the above it was of interest to study the heterogeneous reaction of NO with hydrazine, using the ^{15}N-labeling technique. Hydrazine is a reducing agent with a preformed nitrogen-nitrogen bond, although a weak one.

A meaningful difference between the distribution of N-containing products in the ^{15}NO-$^{14}N_2H_4$ reaction and that in the ^{15}NO-$^{14}NH_3$ reaction can be expected only if the formation of NH_3 by thermal decomposition of N_2H_4 is not prevalent.

Above 60°C the heterogeneous decomposition of hydrazine is known to proceed with the formation of nitrogen, ammonia, and sometimes hydrogen in varying proportions depending on the surface and experimental conditions. In the absence of hydrogen formation the stoichiometry can be represented[(4-11)] by

$$3\ N_2H_4 \rightarrow N_2 + 4\ NH_3. \qquad (1)$$

When hydrogen is present in the products the stoichiometry is sometimes represented by the process[(8,9)]

$$2\ N_2H_4 \rightarrow N_2 + 2\ NH_3 + H_2, \qquad (2)$$

or by the combination of (1) and (2).[10]

Equation (2) can, in turn, be represented by a linear combination of (1) and of a decomposition mode leading only to nitrogen and hydrogen[5,6,11]

$$N_2H_4 \rightarrow N_2 + 2\ H_2. \qquad (3)$$

In essence all observed product distributions in heterogeneous decomposition of hydrazine by various means can be expressed by combinations of (1) and (3).[11]

The decomposition proceeding clearly according to (3) without the formation of ammonia, which is a direct evidence of the complete integrity of the nitrogen-nitrogen bond in the hydrazine, has been observed only during the catalytic decomposition in solution.[11]

The room temperature decomposition of gas-phase hydrazine by photolysis,[12] in a glow discharge[13] and by other methods[11] always leads to a certain proportion of products derived from the splitting of the (H_2N-NH_2) bond. In certain instances of photolysis of mixed hydrazine containing $^{15}N_2H_4$ and $^{14}N_2H_4$, the proportion of randomized product molecules was relatively small.[12]

Stripping of hydrogen with preservation of the N-N bond makes hydrazine exceptionally suitable for direct energy conversion in fuel cells. Liebhafsky and Cairns have stated in this connection[14]: "In the anodic oxidation of hydrazine it is probable that one nitrogen-hydrogen bond is broken at a time until N_2 remains. Furthermore, the existence of the nitrogen-nitrogen bond in hydrazine means that hydrazine stands in sharp contrast to ammonia, for which such a bond must be formed before nitrogen evolution can occur . . . the presence or absence of this nitrogen bond may be the principal reason for different anodic behavior of the two fuels."

The reaction of NO and N_2H_4 in the gas phase was also the subject of several studies. In combustion, Grey and Spencer[15] showed that the NO-N_2H_4 system has the lowest ignition point among the six binary fuel-oxidant systems involving the two fuels NH_3 and N_2H_4 and the three oxidants - O_2, NO and N_2O. Stief and DeCarlo[16] demonstrated that the gas-phase reaction between ^{14}NO and $^{15}N_2H_4$ takes place at room temperature under the influence of uv light. They found the non-randomized nitrogen molecules $^{14}N_2$ and $^{15}N_2$ to be the main reaction products. The formation of the randomized $^{15}N^{14}N$ molecule was ascribed to the interaction of the $^{15}NH_2$ radical and ^{14}NO; no analysis for N_2O was made. Production of N_2O during the photolysis of N_2H_4 in the presence of NO was reported by Bamford.[17]

2. EXPERIMENTAL

The vacuum apparatus, the analytical methods, and the supported Pt catalyst have been described earlier.[3,18] An 0.5-g sample of the platinum catalyst was used in some of the experiments; in one case 0.5 g of the alumina support was used without a platinum deposit. Hydrazine (97%, Matheson, Coleman and Bell) was dried by anhydrous NaOH.[13] The hydrazine was vacuum distilled into a storage bulb connected with the circulation loop. The liquid was dosed by distilling it from the storage bulb into a calibrated capillary which was also attached to the circulation loop. The reaction mixture was completed by adding a measured amount of NO and about 100 torr of argon as a diluent. The gases were continuously circulated in the loop and small samples were removed from time to time for later analysis by mass-spectrometer.

Only preliminary data could be obtained in the circulation system due to the tendency of the hydrazine to condense out in certain parts of the system. Therefore the conclusive experiments were carried out in a spherical Pyrex vessel of 300 cc attached directly to a CEC 21-103 mass-spectrometer. A cold finger was attached to the flask through a stopcock. First, a measured amount of nitric oxide was frozen into the cold finger. After closing of the stopcock the reaction vessel was opened for 15 seconds to hydrazine vapor at room temperature. Argon was added as a diluent and the cold finger was quickly warmed up. Small gas samples of 1 cc were removed at given time intervals into the mass-spectrometer for gas analysis.

The analysis by mass-spectrometer involved the peaks at m/e 2, 17, 28, 29, 30, 31, 40, 44, 45 and 46 for the determination of H_2, $^{14}NH_3$, $^{14}N_2$, $^{14}N^{15}N$, $^{15}N_2$ and ^{14}NO, ^{15}NO, Ar, $^{14}N_2O$, $^{14}N^{15}NO$ and $^{15}N^{14}NO$, $^{15}N_2O$, respectively. The contributions of ^{14}NO and $^{15}N^{14}NO$ to peaks at m/e 30 and 45 were assessed from the known isotopic impurity in the ^{15}NO reactant.

In most cases hydrazine and water were removed by freezing before admitting the gas into the mass-spectrometer either by a Dry Ice-acetone trap, or if an accurate analysis for hydrogen was desired, by a liquid nitrogen trap.

3. RESULTS AND INTERPRETATION

3.1 Preliminary experiments in the circulation loop

3.1a Hydrazine decomposition

Above 100°C the decomposition of hydrazine over the platinum catalyst yielded only NH_3 and N_2; hydrogen accounted for less than 0.2%. At 50°C a sizeable peak at m/e 2 showed substantial formation of hydrogen. Finally, at room temperature the H_2/N_2 ratio was 2.04 in correspondence with the stoichiometry of equation (3). The formation of some ammonia which could have been lost by adsorption on the catalyst cannot be excluded. The maximum possible NH_3 formation at room temperature, estimated from the uncertainty in the measured H_2/N_2 ratio, could amount to 20%.

Hence it was ascertained that at least at room temperature there is little formation of NH_3 from hydrazine decomposition and that the interaction of NO with hydrazine apparently will not involve an NH_3 intermediate in the gas-phase. As will be shown below the presence of NO suppresses the formation of NH_3 even at higher temperatures.

3.1b The NO-N_2H_4 reaction

For the experiments in the circulation loop the reaction mixture was prepared by the saturation of the Ar carrier with N_2H_4 vapor at room temperature. Some of the hydrazine underwent observable condensation on the ground glass of the valve seats and balls of the circulation pump. This condensation could not be prevented and therefore the data from reaction in the loop served only for orientation. These are summarized in the form of product distributions in table 1.

TABLE 1
Product distributions in the ^{15}NO-$^{14}N_2H_4$ reaction observed in the circulation system

Run No.	Initial* Amount of Reactants Torr	Passed Over	Product Distribution of Nitrogen-Containing Products and Hydrogen, %							
			$^{14}N_2$	$^{14}N^{15}N$	$^{15}N_2$	$^{14}NH_3$	$^{14}N^{15}NO$	$^{15}N^{14}NO$	$^{15}N_2O$	H_2
1	NO 60	Bypass	37.2	1.0	11.5	3.0	2.2	0.6	44.6	nil
	N_2H_4 70	Catalyst	39.3	4.2	21.0	8.5	nil	nil	nil	26.9
	Ar 100									
2	NO 88	Catalyst	29.4	1.3	6.7	nil	7.1	1.1	54.3	nil
	N_2H_4 29									
	Ar 100									
3	NO 77	Catalyst	35.5	0.9	14.6	nil	1.6	0.2	47.3	nil
	N_2H_4 31	Bypass	39.7	0.7	9.6	nil	1.6	1.0	47.4	nil
	Ar 100	Catalyst	36.9	0.5	10.6	nil	2.8	<0.1	49.1	nil
4	NO 77	Catalyst	35.6	0.1	8.0	nil	2.9	0.5	53.0	nil
	N_2H_4 30	Catalyst	33.4	0.4	8.8	nil	1.7	1.0	54.6	nil
	Ar 100									

*Expressed as pressure (excess liquid hydrazine over the corresponding vapor pressure stored in capillary attached to loop).

Runs 1, 2, 3 were performed at room temperature and run 4 at 125°C. The first observation is that NO and hydrazine react during circulation in the bypass without contacting the catalyst. In run 1 a large excess of hydrazine was employed. The first line gives the product distribution after one hour circulation in the bypass, where ∿ 40% of the NO has reacted, and the second line the distribution after additional circulation for one full day over the catalyst. The main products of the reaction in the bypass are $^{14}N_2$, $^{15}N_2O$, and $^{15}N_2$; mixed products are relatively insignificant. Thus the reduction of NO at room temperature is accomplished without splitting of the nitrogen-nitrogen bond in the hydrazine and results in the pairing of the nitrogen atoms from the nitric oxide. Prolonged circulation over the catalyst leads to complete reduction of all the nitrous oxide, mixed and unmixed, and to the decomposition of excess hydrazine according to equation (3) with the formation of ample molecular hydrogen. Presence of small amounts of ammonia indicates some hydrazine decomposition according to equation (1).

When the reactant ratio was subsequently changed to oxidizing, i.e. to an $NO:N_2H_4$ ratio > 2 in runs 2 and 3, the product distribution remained similar to that observed during the initial stages in run 1, and was essentially not affected by the switch-over from catalyst to bypass. Again, the products contain mainly non-mixed molecules and there is no formation of NH_3 or H_2. It is worth noting that the reaction rate over the catalyst was faster by a factor of ∿ 4 than in the bypass. As discussed below, this increase can be associated with the increase in the overall contact surface rather than with any specific function of the catalyst.

Finally, at 125°C the product distribution is essentially the same as at room temperature. While the decomposition of hydrazine at this temperature proceeds with the scission of the nitrogen-nitrogen bond the presence of NO suppresses the cleavage of this bond almost completely.

Examination of the distribution of mixed nitrous oxides ($^{14}N^{15}NO$ and $^{15}N^{14}NO$) shows that the non-randomized form ($^{14}N^{15}NO$), where the nitrogen-15 remains attached to the oxygen, is prevalent.

The only substantial difference between the behavior of the catalyst and that of the glass surface of the loop was that the reduction of N_2O by hydrazine takes place exclusively over the catalytic surface. The interaction of NO and hydrazine was non-specific with respect to the nature of the surface and from the preliminary experiments it was impossible to ascertain whether the reaction occurs at the walls, in the gas-phase or even in the liquid phase. The presence of condensed hydrazine in the system could not be avoided due to the tendency of this reactant to condense out on the ground glass surfaces of the valve seats and balls of the magnetic circulation pump.[3] Therefore subsequent experiments were performed in a vessel directly attached to a mass-spectrometer as noted in the Experimental Section.

3.2 Experiments in static conditions

3.2a Liquid phase $NO-N_2H_4$ reaction at room temperature

To establish whether the $NO-N_2H_4$ reaction takes place in the liquid phase non-labeled NO was bubbled through liquid hydrazine at contact times of 10-100 seconds. The product stream was diluted with helium and examined continuously in a CEC 21-614 mass-spectrometer with a capillary inlet system for sampling at atmospheric pressure. An exothermic reaction accompanied by a temperature rise of $\sim$ 10°C took place in the liquid even when shielded completely from light. The N_2/N_2O ratio in the products varied between 0.59 and 0.73. This experiment indicated that in the liquid phase the reaction proceeds quite readily at room temperature without the input of external energy.

3.2b $NO-N_2H_4$ reaction with varying surface/volume ratios

The 300 cc Pyrex bulb was filled with 60 torr of ^{15}NO and 14 torr of $^{14}N_2H_4$. The hydrazine pressure was just below the corresponding vapor pressure at 25°C (14.4 torr).[11] In another experiment, carried out at the same initial pressures, segments of glass tubing 4 cm long, 4 mm o.d. and 2 mm i.d. were inserted into the bulb enlarging the geometric surface area from 232 to 2320 cm^2 while lowering the free volume from 302 to 199 cm^3.

Fig. 1 gives the decrease in ^{15}NO and the increases in $^{15}N_2O$ and $^{15}N_2$ as a function of time. In fig. 2 the ^{15}NO decrease is replotted in semilogarithmic co-ordinates and the straight line indicates that the reaction is first-order in NO.

As before, only 2-3% of the ^{15}N-containing products are accounted for by mixed molecules. If the formation of mixed products is neglected, the overall process is described by a linear combination of the two equations

$$^{14}N_2H_4 + 4\ ^{15}NO = {}^{14}N_2 + 2\ ^{15}N_2O + 2\ H_2O \qquad (4)$$

$$^{14}N_2H_4 + 2\ ^{15}NO = {}^{14}N_2 + {}^{15}N_2 + 2\ H_2O. \qquad (5)$$

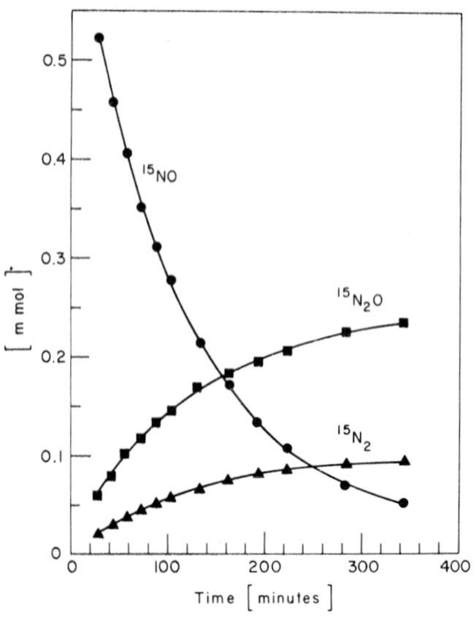

Fig. 1 ^{15}NO consumption and $^{15}N_2$ and $^{15}N_2O$ formation as a function of time in the packed flask.

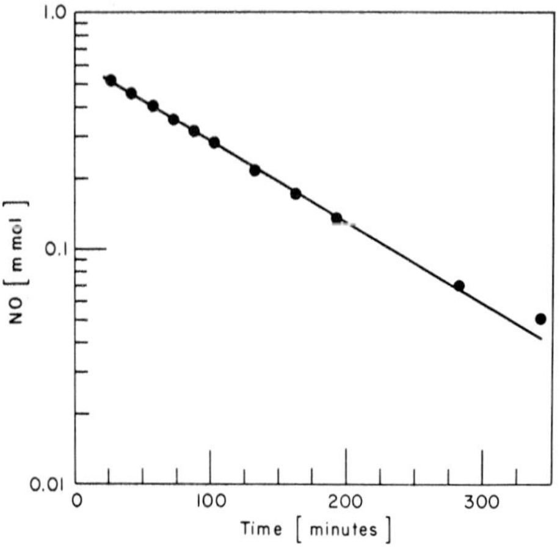

Fig. 2 Semilogarithmic plot of ^{15}NO consumption in the packed flask.

From the material balance it is clear that the 14 torr of N_2H_4 in the gas-phase was not sufficient to reduce the 60 torr NO to the extent shown in fig. 1. The amount of N_2H_4 consumed corresponds, from equations (4) and (5), to an initial pressure of 19.3 torr. This is a direct indication of the sorption of an excess of N_2H_4 during the filling which then participates in the reduction of NO.

This additional amount of 4.9 torr N_2H_4 in a volume of 199 cc can be evaluated in terms of thickness of a liquid film, spread uniformly over the available geometric surface, by the use of the known density[11] of the liquid at room temperature, 1.00 g/cm^3. A film amounting to 18-19 molecular layers is obtained by such an estimate. The roughness factor of glass is generally assumed to be fairly close to unity[19] and therefore some capillary condensation of hydrazine at the points of contact between the glass tubes of the packing must be assumed. The thickness of the hydrazine film on the glass at pressures close to saturation can exceed substantially that represented by a monolayer[20] and the demarcation between multilayer adsorption and capillary condensation is not a sharp one.

Nevertheless, the rate of reaction per unit geometric area was very close in both experiments although the surface/volume ratio differed by more than an order of magnitude and notwithstanding, also, the presence of some condensed N_2H_4 in the packed flask. The values of the rates of NO consumption were 14.0×10^{-12} and 17.0×10^{-12} mol s^{-1} cm^{-2} respectively for the packed and empty flask. As the standard error of the rate determination was 3.3×10^{-12} mol s^{-1} cm^{-2}, these rates can be considered as equal. Thus, the reaction between NO and N_2H_4 takes place at the surface and the contribution due to the presence of condensed phase is minor.

The surface character of the reaction was further confirmed by covering the walls of the empty flask by paraffin wax to prevent adsorption and following the reaction between 60 torr of ^{15}NO and 14 torr of $^{14}N_2H_4$. The slowness of the process required long reaction times.

In fig. 3 the comparison is made between the rates of $^{15}N_2O$ formation for the untreated flask (curve a) and the coated flask (curve b). (Note the difference in the time scale.) Comparison at a pressure of 0.05 mmol N_2O, i.e. at an equal extent of reaction, shows that the rate in the coated flask is only 3% of that in the untreated flask. The slow reaction could well take place on some bare spots at the flask neck not covered by the wax. This experiment confirmed beyond doubt that the NO-N_2H_4 reaction takes place at the surface and not in the gas-phase.

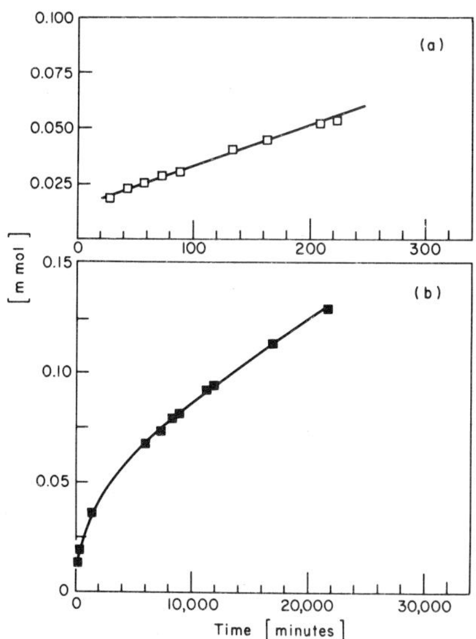

Fig. 3 Comparison of $^{15}N_2O$ formation in plain flask (a) and paraffin coated flask (b).

The final experiment to clarify the importance of the condensed layer of hydrazine, which was found earlier to take part in the reaction, was carried out in the flask filled with the glass tubes (total surface 2320 cm^2). The hydrazine storage vessel was opened to a volume of 200 cc to which the reaction flask was connected. Two minutes elapsed before the pressure in the connecting volume built up to 14 torr. The hydrazine vapor was then allowed to enter the reactor flask for 2 minutes, and then the reaction flask was evacuated for 3 minutes to a residual pressure of less than 0.5 torr, and filled with 25 torr of ^{15}NO and 100 torr of Ar in the usual manner. A gas sample withdrawn immediately into the mass-spectrometer did not show any presence of hydrazine. In spite of the absence of this reactant in the gas the reaction proceeded in a manner similar to that observed before (cf. fig. 1) although at only one-tenth of the rate, as shown in fig. 4.

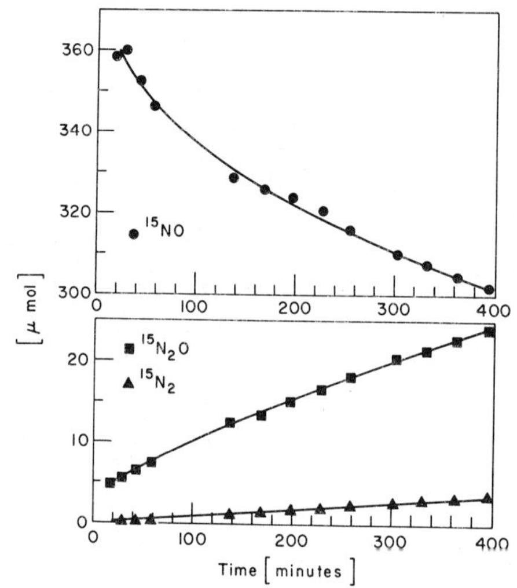

Fig. 4 ^{15}NO consumption and $^{15}N_2$ and $^{15}N_2O$ formation as a function of time in the packed flask with pre-adsorbed hydrazine.

An estimate of the amount of hydrazine in the flask, from the amount of NO consumed, yields an equivalent of 5-6 molecular layers with respect to the geometric area. Most of this hydrazine is conceivably again present as capillary condensate. The slowness of the reaction between this condensate and the NO in the gas-phase is in agreement with its minor contribution to the overall rate in the experiment represented on fig. 1.

4. DISCUSSION

The heterogeneous reaction between hydrazine and nitric oxide is observable at a lower temperature than the reaction between ammonia and nitric oxide in line with the energetically easier hydrogen-atom abstraction from hydrazine as compared with ammonia. Such behavior was observed in other reactions as for instance those of methyl radicals with NH_3 and

N_2H_4[21]; it is also in agreement with the lower value of the bond dissociation energy for $D(H-N_2H_3)$, 76 ± 5 kcal/mol, than for $D(H-NH_2)$, 103 ± 2 kcal/mol.[22]

Within the range of the experimental conditions of this work the heterogeneous reduction of nitric oxide by hydrazine occurs virtually without the scission of the nitrogen-nitrogen bond in hydrazine where the bond dissociation energy is generally accepted as $D(H_2N-NH_2)$, 60 ± 4 kcal/mol.[22] This is somewhat lower than the above quoted bond dissociation energy for $D(H-N_2H_3)$, and therefore recourse must be made to kinetic arguments to explain the preservation of the H_2N-NH_2 bond during the heterogeneous reactions of hydrazine. Such an argument usually presupposes the existence of an activated complex wherein the strength of the hydrogen-nitrogen bond in hydrazine has been considerably diminished.

Before postulating such a complex it must be decided on the basis of the available experimental evidence what is the locus of the heterogeneous reaction between NO and N_2H_4. All the results point to this locus as being the surface film of hydrazine. The thickness of this film cannot be deduced from our results. However, the independence of the rate of hydrazine pressure close to saturation, as shown in the data presented in fig. 2, indicates that under these conditions the film is at least one layer, or maybe even a few layers, thick. On the other hand, when hydrazine is pumped out, as in the experiment of fig. 4, the film occupies an area much smaller than a monolayer, with the exception of the areas covered by condensate pockets. This, in turn, is reflected in the lower reaction rate.

The surface hydrazine layer is replenished from the gas phase and the NO molecules adsorb onto it. The surface coverage by NO must be sparse to reflect the first order dependence in the gas pressure of this reactant. The N_2O is desorbed as the main product and is reduced only to a negligible extent. The ratio of $^{15}N_2/^{15}N_2O$ depends on the reaction conditions, as given in table 2. Under conditions, where the liquid-phase reaction might be important, as under B, C, and D this ratio is lower; when the wall reaction is prevailing, as under A, this ratio tends to be higher.

Owing to the sparsity of the coverage by NO the activated complex in the wall reaction is postulated to contain only one molecule of NO and one of hydrazine.

A complex where one N-H bond has been weakened can be visualized, by analogy with the alkali metal salts of the type Na^+NO^-,[23] such as $[H_3{}^{14}N_2{-}{-}{-}H^+\ {}^{15}NO^-]$. This complex cleaves into a $^{14}N_2H_3$ radical and an $H^{15}NO$ intermediate. The N_2H_3 being unstable is consecutively stripped of the remaining hydrogens, to eliminate molecular nitrogen, either directly by additional NO molecules or by disproportionation into, say, diimide ($^{14}N_2H_2$) and hydrogen atoms. The HNO species recombine pairwise on the surface to water and N_2O,

$$[2\ H^{15}NO] \rightarrow {}^{15}N_2O + H_2O \qquad (6)$$

where the square brackets imply again a surface complex.

TABLE 2
Observed $^{15}N_2/^{15}N_2O$ ratios in the products

Experiment	$^{15}N_2/^{15}N_2O$ ratio in products
A. Static experiments in a. non-coated Pyrex bulb b. glass-packed Pyrex bulb c. paraffin-coated bulb	0.2 - 0.4
B. Static experiment with hydrazine in pre-adsorbed layer	0.1 - 0.15
C. Circulation over Al_2O_3 support or Pt on Al_2O_3 catalyst	0.0 - 0.2
D. In liquid hydrazine*	0.04 - 0.12

*Using non-labeled reactants and calculated from equations (4) and (5):
$^{15}N_2/^{15}N_2O = 0.5\ [N_{2(tot)}/N_2O_{(tot)} - 0.5]$

As pointed out in our work on the surface reaction between NH_3 and NO[3] the formation of N_2O in a system containing hydrogen, oxygen, and nitrogen invariably presupposes the HNO intermediate. The formation of $^{15}N_2$ is best accounted for by further reduction of the pair of HNO intermediates by hydrogen atoms derived from the disproportionation of the N_2H_3 radicals, in a manner essentially analogous to that postulated before[3]

$$[2\ HNO] + 2\ H \rightarrow N_2 + 2\ H_2O. \qquad (7)$$

It is possible also to postulate a direct concerted reaction between two ^{15}NO molecules and one $^{14}N_2H_4$ molecule going directly to products according to equation (5). Such a process seems to be much less likely because it involves a complicated molecular rearrangement.

At least up to 125°C, the surface reaction between hydrazine and nitric oxide does not entail the formation of NH_2 radicals, in marked contrast to the interpretation of Bamford[17] who has studied this reaction, albeit at a higher temperature, without the use of labeled reactants.

REFERENCES

1) M. Shelef, K. Otto and H. Gandhi, J. Catal. 12, 361 (1968).
2) M. Shelef and H. Gandhi, Submitted to IEC, Prod. Res. Develop.
3) K. Otto, M. Shelef and J. T. Kummer, J. Phys. Chem. 74, 2690 (1970).
4) Gmelins Handbuch der anorganischen Chemie, 8. Auflage, Stickstoff, System Nummer 4, p. 317. Verlag Chemie, Weinheim, Germany 1936 (1955).
5) K. Aika, T. Ohhata and A. Ozaki, J. Catal. 19, 140 (1970).
6) J. Völter and G. Lietz, Z. Anorg. Allg. Chem. 366, 191 (1969).

7) G. Schulz-Ekloff and P. Jiru, Collect. Czech. Chem. Commun. 35, 3765 (1970).
8) P. J. Askey, J. Am. Chem. Soc. 52, 970 (1930).
9) C. F. Sayer, US CFSTI AD 1969, No. 710627.
10) M. Szwarc, Proc. Roy. Soc. A 198, 267 (1949).
11) L. F. Audrieth and B. A. Ogg, "The Chemistry of Hydrazine," John Wiley and Sons, New York, N. Y. (1951).
12) L. J. Stief, V. J. DeCarlo and R. J. Mataloni, J. Chem. Phys. 46, 592 (1967).
13) E. F. Logan and J. M. Marchello, J. Chem. Phys. 49, 3929 (1968); 50, 2724 (1968).
14) H. A. Liebhafsky and E. J. Cairns, "Fuel Cells and Fuel Batteries," p. 416, John Wiley and Sons, New York, N. Y. (1968).
15) P. Gray and M. Spencer, Trans. Farad. Soc. 59, 879 (1963).
16) L. J. Stief and V. J. DeCarlo, J. Chem. Phys. 49, 100 (1968).
17) C. H. Bamford, Trans Farad. Soc. 35, 568 and 1239 (1939).
18) K. Otto, M. Shelef and J. T. Kummer, J. Phys. Chem. 75, 875 (1971).
19) W. A. Cannon, Nature 197, 1000 (1963).
20) S. J. Gregg and K. S. W. Sing, "Adsorption, Surface Area and Porosity," Academic Press, N.Y. pp. 149 ff (1967).
21) P. Gray and J. C. J. Thynne, Trans. Farad. Soc. 60, 1047 (1964).
22) V. I. Vedeneev, L. V. Gurvich, V. N. Kondratyev, V. A. Medvedev and Ye. L. Frankevich, "Bond Energies, Ionization Potentials and Electron Affinities," St. Martin's Press, New York, N.Y. (1966).
23) C. C. Addison and J. Lewis, Quart. Rev. (London), 9, 115 (1955).

DISCUSSION

A. LAWSON

We have been studying the low temperature gas-solid interaction of NO on catalysts comprising an organic NH_2-containing compound supported on charcoal and alumina supports. We have found extensive catalytic decomposition of NO to occur on urea/charcoal catalysts at 60°C in the presence of air and moisture. N_2 is formed as a product, but gravimetric and surface area measurements suggest that urea is not removed from the support. The authors' results are therefore of considerable interest to us since we feel we may be observing a similar phenomenon, i.e., the formation and pairing of HNO type complexes from NO/air/H_2O mixtures with the subsequent elimination of N_2. Our proposed reaction mechanism is too complex to discuss here, but clearly in our case adsorption of the NO complex occurs on the solid urea surfaces, whereas in the authors' case adsorption is on a liquid film of N_2H_4. We would like to ask the authors if they have extended their observations to other NH_2 compounds, particularly solid ones, and if they have added O_2 or H_2O vapour to their gas phase reagents. A tracer study of the NO/air/H_2O/urea system such as that described by the authors would be of considerable interest.

K. OTTO

Presently we are not planning to extend our observations to other NH_2 compounds. We have, however, studied extensively the catalytic reaction between $^{14}NH_3$ and ^{15}NO which is currently of special relevance for the reduction of nitric oxide from automotive exhaust. The observed product distribution suggests that the formation of dimerized $H^{15}NO$ is of importance for this reaction, too. We have avoided to add O_2, as nitrate formation may complicate matters; H_2O, of course, is present as a reaction product.

W. J. M. PIETERS

In the discussion of your paper a mechanism was proposed in which the decomposition of NO to N_2O and N_2 proceed as parallel reactions. Figure 1 was used to support your proposed mechanism. The results from Figure 1 can, however, equally well be explained with a first order decomposition reaction of intermediate N_2O.

K. OTTO

Figure 1 represents data which have been obtained at room temperature without a catalyst. Under these conditions N_2O from the gas phase does not decompose and is not reduced by hydrazine. We propose parallel reactions for the formation of $^{15}N_2O$ and $^{15}N_2$ from the same surface complex--a dimer of H^{15} NO[cf. eq. (6) and (7)]. It is, however, not possible to distinguish between the reaction given by eq. (7) and another possibility where $^{15}N_2O$ after its formation on the surface in accordance with eq. (6) is reduced by two more hydrogen atoms before it is released into the gas phase.

Paper Number 17

ADSORBED WATER ON SINGLE CRYSTAL OXIDES

RICHARD W. RICE and GARY L. HALLER
Department of Engineering and Applied Science, Yale University
New Haven, Connecticut, U.S.A. 06520

ABSTRACT: Adsorbed water and surface hydroxyl groups on Al_2O_3, MgO, ZnO and TiO_2 single crystals have been studied by infrared internal reflection spectroscopy (IRS). The results indicate a substantial difference between surface OH on these surfaces and those found on their amorphous counterparts. The heat of adsorption and orientation of adsorbed water were found to be strong functions of both crystal type and plane.

1. INTRODUCTION

Water, whether adsorbed as a molecule or dissociated to form hydroxyl groups, is thought to play an important role in many reactions catalyzed by oxide surfaces[1,2,3]; a fact which has led to many infrared investigations of water adsorbed on amorphous oxides[4,5]. Given planes of the stable crystal are often used as models in the interpretation of such spectra[6,7,8,9], yet infrared observation of adsorbed water on single crystals has not been reported. It is the purpose of this paper to report infrared spectra of adsorbed water and surface OH on the following crystal planes: (0001), $(11\bar{2}3)$, $(4\bar{1}50)$ Al_2O_3; (100) MgO; $(2\bar{1}\bar{1}0)$ TiO_2; (0001), $(2\bar{1}\bar{1}0)$ ZnO; and to compare the results with those from the amorphous forms of these oxides. All spectra were obtained using internal reflection spectroscopy[10] (IRS) with single crystals themselves acting as internal reflection prisms.

2. EXPERIMENTAL

2.1 Apparatus

The equipment used here was essentially the same as that used in a previous study[11], the only major addition being a double-pass vacuum cell which was used with the TiO_2 and ZnO crystals. In this case the cell was a stainless steel tubular design supplied by Harrick Scientific Corp.[12] Since the infrared beam propagates down the length of the plate and is reflected back upon itself by a metallized end-reflecting surface, the beam enters and exits the crystal at the same end and only a single window is required. A zinc sulfide window (Harrick) was used in this work.

2.2 Reagents and crystals

The single crystals, except for ZnO, were supplied, cut and polished by Adolf Meller Co.[13] The Al_2O_3 (sapphire) and TiO_2 (rutile) were flame fusion grown from high purity powder (99.98%) and the MgO crystal (99.9+%) was grown from the melt. The ZnO (wurtzite) crystal was hydrothermally grown, cut and polished by Airtron[14]. Since the growth solution contained both LiOH and KOH, the ZnO single crystal contains Li and K impurities at the several ppm level (about 9 ppm Li).

The major faces of the crystals were oriented to within 1/2° of the specified plane, and were polished optically flat using either diamond dust or fine alumina in oil. The Al_2O_3 and MgO were used in the single pass mode and were 52x20x0.5mm with the ends beveled at 45° to form entrance and exit apertures (trapazoidal configuration). The TiO_2 and ZnO crystals were used in the double-pass mode (parallelopiped configuration) and were 26x20x0.5 and 25x10x0.5mm respectively. For TiO_2 the bevels were 45° and 49.5°, while for ZnO they were 45° and 51°. Both crystals were metallized on the 45° bevel by vacuum deposition of gold. In all cases roughly 100 internal reflections were achieved. The crystals were generally opaque below 2000 cm^{-1}, thus only stretching modes were observable. Water used in these experiments was double ion-exchanged, passed through an organic removal filter, and then distilled. Deuterium oxide (99.7 min. atom % D) was used as received from Merck Sharp and Dohme[15].

2.3 Procedure

Unless otherwise specified, an outgassing at 450-500°C, between 10^{-6} and 10^{-7} torr was used as a standard pre-treatment. Water vapor was admitted and/or exhausted in increments, and allowed to equilibrate 15-20 minutes before the spectrum was recorded. Typically, a X5 ordinate scale expansion was used to enhance detection.

3. RESULTS AND INTERPRETATION

3.1 Surface hydroxyls

The most significant observation arising from this work is the apparent rarity of occurrence of high frequency isolated hydroxyl (3600-3800 cm^{-1}) on these single crystal surfaces. This was not anticipated in light of the multitude of papers reporting such groups on high-area samples[4,5]. High frequency surface OH were noted only in the following instances: (1) at 3750 cm^{-1} on $(41\bar{5}0)$ Al_2O_3 after repolishing; (2) at 3755 cm^{-1} on $(11\bar{2}3)$ Al_2O_3 after heating in flowing H_2 at 1000°C; (3) at 3700 cm^{-1} on MgO. In the first two cases the peak was not evident after outgassing, but for MgO the hydroxyl was only mildly perturbed by either exposure to water or outgassing. Finding these groups apparently absent on most crystals the following attempts were made to produce them[16]: The crystals were heated at 400°C in O_2, then in (a) H_2, (b) D_2, (c) H_2O, and (d) D_2O; all at 350-400°C. Spectra were obtained after each treatment both before and after evacuation, but in no case were free surface OH or OD detected. There were, of course, changes in the low frequency (around 3400 cm^{-1}) hydroxyl and water bands.

In order to assess whether this general absence of surface hydroxyl peaks was merely due to a detectivity problem, an expected peak maximum absorbance, A_{max}, was calculated from the IRS analogue[17] of the Beer-Lambert Law, $A_{max} = \varepsilon c(N\gamma d) = \varepsilon^* c^*(N\gamma)$, eq. (1). Here N is the number of internal reflections; ε (liters/mole-cm) and ε^* (cm^2/mole) are the molar extinction coefficients; c (moles/liter) and c^* (moles/cm^2) are three dimensional and surface OH concentrations respectively; d is the thickness of the surface hydroxyl layer, and γ is a factor arising from amplification of the beam's electric field at the internally reflecting interface[18]. The product (γd) is often termed the "effective film thickness". In this case, as in all described herein, the adsorbed layer thickness is much less than the penetration depth of the evanescent wave in the rarer medium and the energy loss per reflection is so small that this wave is not much perturbed by absorption at the interface[11,18].

The assignment of the value for ε^* in eq. (1) was made from the data of Peri and Hannan[16] for "isolated" OH on γ-alumina. These authors reported an average infrared absorptivity of 8×10^4 cm^2/mole for isolated OH. Peri also estimated that after drying at 400°C, the surface hydroxyl concentration, c^*, was 5 OH/100 Å^2 (or 8.3 μmoles/m^2). The amplification factor, γ, is dependent upon the angle of incidence; the polarization of the beam, and the refractive indices, n_1, n_2 and n_3; of the crystal, adsorbed layer, and gas phase respectively. Using Harrick's equations[18], a 45° angle of reflection, $n_1 = 1.7$ (Al_2O_3 or MgO), $n_3 = 1.0$ and assuming n_2 equal to the refractive index of bulk water, 1.3; we calculate a $\gamma = 3.3$. For N = 103 internal reflections, and using the other values given above, $A_{max} = 0.023$. This is readily detectible. It is noteworthy that in the few cases where free hydroxyls were observed their absorbance was between 0.010 and 0.020.

The preceding calculation has involved some assumptions which might be considered questionable, but which the authors feel are realistic. The extinction coefficient given by Peri is in reasonably good agreement with that measured for monomeric hydroxyl stretching modes of very dilute solutions of alcohol in non-polar solvents[19], a system which may be a good analogue for surface OH. Likewise, the surface hydroxyl concentration of 5 OH/100 $Å^2$ is representative of that found on various oxides[6,7,16,19] and is a conservative estimate for many of the conditions used here. The amplification factor, γ, is not strongly dependent on the highly questionable value of 1.3 assumed for n_2, and has been demonstrated to be correct by experiments with deposited close-packed stearate monolayers[11] where the bulk value of the index of refraction was assumed for n_2.

Considering the above discussion, isolated surface hydroxyls, if they exist on these single crystals in the concentrations reported for high area samples, should have been detected. Mere exposure to water or D_2O, and certainly heating of the crystals to 350°C in these vapors[16], should insure an appreciable surface OH concentration. That the surfaces do contain high concentration of surface hydroxyls is indicated by the strong absorbance in the region of 3400 cm^{-1}, but the absence of sharp bands at higher frequency implies that the overwhelming majority of these hydroxyls on single crystal surfaces are involved in interactions which disqualify them from being properly termed "isolated". Extensive hydrogen bonding is the most obvious possibility, and it is interesting to observe that the crystals on which isolated surface OH were definitely not detected exposed the closest-packed oxygen planes, (0001), of Al_2O_3 and ZnO. At least two authors[7,21] have reported that surface OH peaks can be strongly suppressed by impurities unless the surface is well oxidized. Accordingly, all the crystals studied here were always handled so as to avoid the introduction of surface impurities, and were heated in oxygen to remove such species.

It has been speculated[22] that well-annealed, defect-free surfaces are more likely to have free OH than "rough" surfaces, but the results described here would appear to oppose this view. It would be naive to picture our single crystal surfaces as atomically smooth, but they are probably well described in terms of the terrace-ledge-kink model of Frank[23]. The specified plane would be exposed as large terraces with the principal "roughness" being the step-like irregularity of the surface. In other words these single crystal surfaces would retain considerably more short-range order than their high area counterparts. Thus the results of this study would indicate that isolated surface OH on amorphous oxides may be linked with a defect structure uncommon to single crystals.

The frequent appearance of the 3700 cm^{-1} peak on (100) MgO, in contrast to the general absence of such findings on the other crystals, may be a reflection of a significantly larger extinction coefficient for Mg-OH, resulting from the greater ionic nature of the hydroxyl, but a more likely explanation is a greater reactivity with water to form the hydroxide. Anderson et al.[6] attributed a similar peak, at 3710 cm^{-1}, to a metastable surface OH on MgO formed by decomposition of $Mg(OH)_2$.

Although free surface hydroxyls were only infrequently detected, relatively sharp peaks due to bulk OH were observed in the 3150-3550 cm^{-1} region, as previously described for sapphire[11]. In the case of MgO there were weak or moderately strong peaks at 3298, 3388, 3395, 3412, 3483, 3500, 3519 and 3539 cm^{-1}; and very intense peaks at 3350 and 3557 cm^{-1}. Moderately intense peaks were noted at 3245, 3265, 3290, and 3550 cm^{-1} for ZnO; and rutile exhibited two very intense peaks at 3280 and 3320 cm^{-1}. These internal hydroxyls are probably associated with impurities and/or cation vacancies[24] and are unaffected by evacuation at 500°C. Using a polarized beam[11] the bulk OH in sapphire and rutile were determined to lie perpendicular to the c- axis. A similar finding was made for all except the 3550 cm^{-1} species in ZnO. This OH appears to be oriented perpendicular to the closest-packed oxygen plane, i.e., parallel to the c-axis. No appreciable dependence on beam polarization was found for the internal OH of MgO. Rough estimates of the concentration of these bulk OH, calculated from the observed intensity, ranged from less than one to several hundred ppm; thus if the groups are distributed evenly throughout the crystals, their presence on the surface would be undetectable. Atherton et al.[9] assigned a peak at 3555 cm^{-1} to a surface OH on ZnO prepared by combustion of metallic zinc, and the peak noted here for both ZnO crystals may correspond to this peak. However, the 3550 cm^{-1} peak of (0001) ZnO was not measurably affected when the crystal was heated at 300°C in O_2, H_2, or vacuum, and thus was at least principally due to an internal OH. Attempts to deuterium exchange the bulk OH of (100) MgO and $(41\bar{5}0)$ Al_2O_3 by heating at 1000°C in flowing D_2 proved unsuccessful. After all adsorption experiments were completed, a thin Pt film was deposited on $(41\bar{5}0)$ Al_2O_3 and the deuterium exchange tried again. This time, the Al_2O_3 3313 cm^{-1} peak intensity decreased markedly and an OD peak at 2407 cm^{-1} appeared. The exchange was shown to be largely reversible by re-heating at 1000°C in H_2.

3.2 Water adsorption

Water appears to adsorb in at least two distinct manners on all of the single crystals studied. One type was removable by pumping at moderate temperatures; the other was completely removable only by evacation above 400°C. Invariably, the weakly adsorbed water had a peak maximum just above 3400 cm^{-1} (at higher frequency than water at the same temperature) and a half-band width of 450-500 cm^{-1}. The strongly adsorbed species had a less well-defined absorbance maximum in the 3250-3350 cm^{-1} region, and a half-band width which was a function of the particular crystal, but was generally 550-700 cm^{-1}. Kinetically the strongly bound water (assumed to have reacted to form surface hydroxyls) may be further subdivided into a portion which forms rapidly when exposed to water vapor and a second portion which forms more slowly even when the crystal is in contact with liquid water. The spectrum of water adsorbed on $(41\bar{5}0)$ Al_2O_3, fig. 1, illustrates these features.

Two aspects of water adsorption were studied: the heat of adsorption and the average orientation. A procedure similar to that employed in a previous paper[11] was used to calculate the isosteric heat of adsorption, Q_{iso}, for water at various coverages on (0001) and $(41\bar{5}0)$ Al_2O_3 and on (100) MgO, see table 1. Absorbance at 3400 cm^{-1} was used as a measure of the amount adsorbed because changes in this value were found to parallel those for integrated intensity over a fair range within the accuracy of measurement of the latter. The use of either absorbance at peak maximum or integrated intensity for this purpose is obviously only an approximation

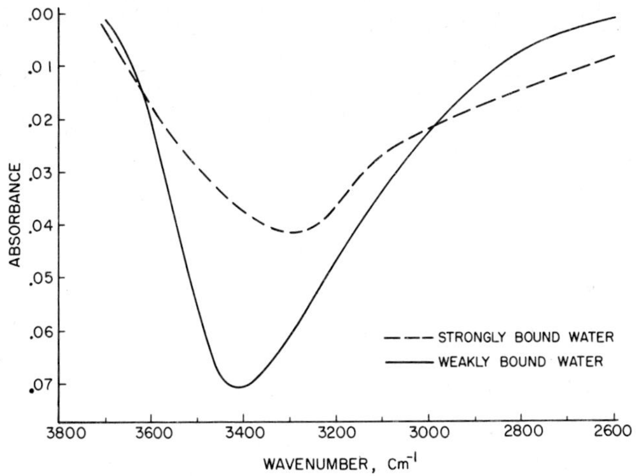

Fig. 1 Spectra of two kinds of adsorbed water on $(41\bar{5}0)$ Al_2O_3; The strongly bound water is at about monolayer coverage and the weakly bound water at greater than monolayer coverage.

since the extinction coefficient undoubtedly changed somewhat with both coverage and temperature. There would appear to be no agreement in the literature[25, 26)] concerning variation of extinction coefficients with experimental parameters. Finch and Lippincott[27)] measured the variation of the integrated intensity of hydrogen bonded alcohol hydroxyl stretching modes with temperature and concentration and found an inverse absolute temperature dependence, but no appreciable concentration dependence. An in-

TABLE 1

Isosteric heats of adsorption for weakly bound water

Crystal Plane	Absorbance at 3400 cm^{-1}	Surface Coverage*	Q_{iso}, Kcal/mole	Q'_{iso}, # Kcal/mole
(0001) Al_2O_3	0.0100	0.25	11±1	9±1
	0.0080	0.20	10	8
	0.0040	0.10	11	9
$(41\bar{5}0)$ Al_2O_3	0.0100	0.25	23±2	19±2
	0.0080	0.20	22	20
	0.0040	0.10	20	19
(100) MgO	0.0120	0.30	16±4	13±4
	0.0100	0.25	17	14
	0.0080	0.20	17	14
	0.0060	0.15	22	19

*The total fractional coverage for water ranged between 0.4 and 0.6 for the section of the isotherms used. #An inverse absolute temperature dependence of the extinction coefficient was assumed and Q_{iso} corrected to Q'_{iso}.

verse absolute temperature dependence of the extinction coefficient uniformly lowers Q_{iso} by 10-20% relative to the uncompensated value, see table 1. Because of the uncertainty of the effect of coverage on the extinction coefficient of adsorbed species[25,28] and because the range of coverage involved in the calculation of Q_{iso} was relatively small, no attempt was made to incorporate this effect. A calculation similar to that presented for isolated OH was made in order to get a rough indication of coverages for the weakly bound water. An ε of 103 liter/mole-cm[29] was used and the water molecule was treated as a sphere of radius 1.9 Å. Assuming close packing of hard spheres and applying $\gamma = 3.3$ and $N = 103$, we calculate an absorbance of 0.04 at 3400 cm^{-1} for monolayer coverage. This value, although based on an admittedly simplistic model, is not unreasonable in light of the experimental results.

The Clausius-Clapeyron equation from which isosteric heats are calculated is valid only for equilibrium adsorption; therefore, Q_{iso} values are presented only for the weakly bound (reversibly adsorbed) water. However, reasonably linear isosteres obtained from total water isotherms suggest that the heat of adsorption for strongly bound water is roughly twice that for the weakly bound water. Typical isotherms for total water adsorption on (41$\bar{5}$0) Al_2O_3 are shown in fig. 2. The absorbance due to irreversibly adsorbed water at any given temperature was subtracted from these isotherms before using them to calculate the heats given in table 1.

The large difference in Q_{iso} between (0001) and (41$\bar{5}$0) Al_2O_3 is nearly the same difference found for n-amyl alcohol on these two crystals[11]. The isosteric heat of adsorption for this alcohol fell between the values reported here for the two types of adsorbed water, a fact which is understandable since the alcohol hydroxyl is undoubtedly the group involved in forming the adsorptive bond. The values given for (0001) Al_2O_3 in table 1 are somewhat lower than those of Yao[30] on microcrystals, but her heats are composite values for the two kinds of adsorbed water and contain a contribution from planes of higher index than (0001). The data for MgO was more limited than for Al_2O_3 as this crystal was ruined after only four isotherms had been measured, but the Q_{iso} values are in good

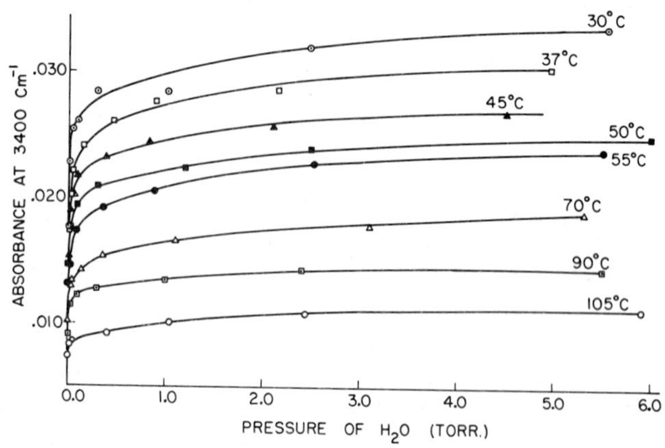

Fig. 2 Isotherms for total water adsorption on (41$\bar{5}$0) Al_2O_3.

agreement with the trend noted for water compared with alcohol[31] and with the findings of Anderson[6].

Dissociation of water into an OH which bonds with a coordinately unsaturated cation and a proton which forms a second OH by bonding to a lattice oxygen would account for the high heats of adsorption estimated for the strongly bound water. A somewhat weaker interaction could involve water adsorbed as a molecule at the vacant ligand position of an exposed cation[32]. The significantly larger heat of adsorption for $(41\bar{5}0)$ compared with (0001) Al_2O_3 is compatable with both possibilities, since Al^{+3} ions on the $(41\bar{5}0)$ plane are more exposed[11]. Outgassing of a hydroxylated surface can be viewed as causing condensation of an OH and the proton of an adjacent OH to form H_2O, which is removed, leaving coordinatively unsaturated pairs of Al^{+3} and O^{-2} ions. The average coordination number for such a pair site is about 9 for (0001) and about 5 for $(41\bar{5}0)$[11].

Magnesium oxide has a cubic (NaCl) structure in contrast to the trigonal corundum structure and is more ionic in nature. The relatively high heat of adsorption estimated for irreversibly adsorbed water on (100) MgO implies, as discussed above, a cationic site for the strongly bound species supplemented by extensive hydrogen bonding. The large band-width and low frequency observed for both Al_2O_3 and MgO indicate that the hydrogen of a cation bound OH or H_2O interacts appreciably with neighboring oxygens. Hydroxyl absorbance below 3000 cm^{-1} has been associated with strong hydrogen bonds involving short O-O distances[33,34]. The adsorption mechanism chosen here agrees with this view. An alternate explanation for the low frequency absorbance involves a labile proton in a very strong hydrogen bond[35].

For the weakly bound water, the primary interaction is assumed to be hydrogen bonding of molecular water with lattice oxygen ions and/or with hydroxyl groups already involved in hydrogen bonding. The heat of adsorption for this species on (0001) Al_2O_3 is, within experimental error, the heat of liquefaction of water; yet, for the other two crystals its heat is noticeably higher. The high Q_{iso} on $(41\bar{5}0)$ Al_2O_3 and the stronger hydrogen bonding evidenced by the more extensive absorbance "tail" to low frequency on this crystal relative to (0001) Al_2O_3 again probably reflect the lower coordination numbers on this surface. For weakly bound water on (100) MgO, Q_{iso} being larger than the latent heat is likely attributable to a polarizing effect of the relatively ionic surface oxygen and OH.

The data on ZnO and TiO_2 is not as extensive as for Al_2O_3 and MgO, but their adsorptive properties appear to be similar. They both show broad bands at low frequency for weakly and strongly bound water much like that shwn in fig. 1, but both their adsorptive capacity and the fraction of the total water which is strongly bound differ somewhat.

3.3 Orientation of surface species

The electric field at an internally reflecting interface varies in the three spatial directions[17], thus allowing calculation of the average orientation of an adsorbed species by measuring absorbance as a function of beam polarization[11]. Using the refractive indices given above in the calculation of γ and assuming unit electric field for the radiation propagating within the crystal, beam polarization perpendicular to the plane defined by the incident and reflected beam will set up an electric field, E_y with amplitude 1.75, across the 20mm width of the prism . Beam polarization parallel to the plane defined above will result in elliptical polarization with amplitude components $E_x = 1.20$ along the length of the crystal, and $E_z = 1.25$ normal to the surface of the crystal. We calculate the absorbance ratios that would result if the absorbing dipole assumed each of three limiting orientations: (1) random orientation, $A_\perp/A_\parallel = E_y^2/(E_x^2 + E_z^2) = 1.07$; (2) random in the x-y plane (in the plane parallel to the surface), $A_\perp/A = E_y^2/E_x^2 = 2.13$; and (3) normal to the surface $A_\perp/A_\parallel = 0/E_z^2 = 0$. Perpendicular and parallel polarization spectra for adsorbed water revealed that the absorbance ratio was roughly constant at 1.2-1.3 for all frequencies over the band for the weakly adsorbed water. The constancy of the absorbance ratio shows that only one species gives rise to this band and the fact that this ratio is greater than 1.07 indicates that it is not randomly orientated. If weakly bound water is in fact molecular water, there are contributions from both the symmetric and the antisymmetric modes (the dipole changes for these modes are both in the plane of the molecule, but normal to each other). This complication does not allow an unambiguous interpretation of the measured absorbance ratio, but this result does suggest that the plane of the water molecule forms an angle of less than 45° to the surface.

The absorbance ratio for the strongly bound water (presumed to be surface hydroxyls) increases with decreasing wavenumber across the band as shown in fig. 3. This variation of the absorbance ratio with frequency, particularly that for MgO, indicated that more than one species contribute to this band. Hydroxyls contributing intensity in the low frequency portion of the bands show an absorbance ratio approaching that calculated for a dipole oriented parallel to the surface. This is consistent with the interpretation given above that these hydroxyls are involved in hydrogen bonds with short O-O distances.

4. CONCLUDING STATEMENT

The results described here indicate that, although single crystal surfaces closely parallel their amorphous counterparts in terms of both the infrared spectrum and heat of adsorption of adsorbed water; they appear to differ intrinsically with regard to concentration of isolated surface hydroxyls. The implications of these findings are: (i) ideal crystal planes may not, in general, be good models for the local structure of amorphous oxide catalysts and (ii) the specific catalytic activity of amorphous and crystalline oxides may be expected to differ for those reactions involving surface hydroxyls.

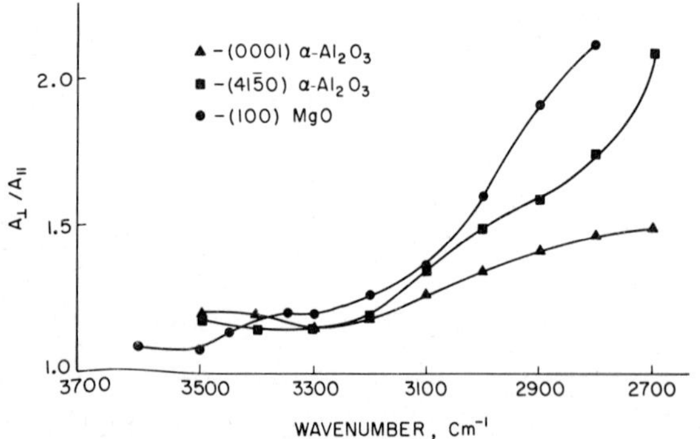

Fig. 3 Variation of absorbance ratio with frequency for the strongly bound water on (0001) and (41$\bar{5}$0) Al_2O_3 and (100) MgO.

ACKNOWLEDGEMENT

Financial support for this research by the National Science Foundation, Grant No. GK 24688, and by the Research Corporation is gratefully acknowledged. R. W. Rice held a NSF Predoctoral Fellowship, June, 1970-May, 1972.

REFERENCES

1) S. G. Hindin and S. W. Weller, Advan. Catal. 9 (1957) 70.
2) Y. Noto, K. Fukuda, T. Onishi and K. Tamaru, Trans. Faraday Soc. 63 (1967) 2300.
3) A. G. Oblad, J. U. Messenger and H. T. Brown, Ind. Eng. Chem. 39 (1947) 1462.
4) L. H. Little, Infrared Spectra of Adsorbed Species (Academic Press, New York, 1966).
5) M. L. Hair, Infrared Spectroscopy in Surface Chemistry (Marcel Dekker, New York, 1967).
6) P. J. Anderson, R. F. Horlock and J. F. Oliver, Trans. Faraday Soc. 61 (1965) 2754.
7) P. Jackson and G. D. Parfitt, Trans. Faraday Soc. 67 (1971) 2469.
8) J. B. Peri, J. Phys. Chem. 69 (1965) 220.
9) K. Atherton, G. Newbold and J. A. Hockey, Discuss. Faraday Soc. 52 (1972).
10) N. J. Harrick, Internal Reflection Spectroscopy (Interscience, New York, 1967).
11) G. L. Haller and R. W. Rice, J. Phys. Chem. 74 (1970) 4386.
12) Harrick Scientific Corporation, Ossining, New York.
13) Adolf Meller Company, Providence, Rhode Island.
14) Airtron Division of Litton Industries, Morris Plains, New Jersey.
15) Merck Sharp and Dohme of Canada Limited, Montreal, Canada.
16) J. B. Peri and R. B. Hannan, J. Phys. Chem. 64 (1960) 1526.
17) N. J. Harrick, in ref. 10), pages 41 and 42.
18) N. J. Harrick, in ref. 10), pages 50 and 51.
19) G. M. Barrow, J. Phys. Chem. 59 (1955) 1129.
20) V. Y. Davydov, A. V. Kiselev and L. T. Zhuravlev, Trans. Faraday Soc. 60 (1964) 2254.
21) K. E. Lewis and G. D. Parfitt, Trans. Faraday Soc. 62 (1966) 204.
22) A. V. Kiselev and V. I. Lygin, Russ. Chem. Rev. 31 (1962) 175.
23) W. K. Burton, N. Cabrera and F. C. Frank, Proc. Roy. Soc. A243 (1951) 299.
24) R. F. Belt, J. Appl. Phys. 38 (1967) 2688.
25) L. H. Little, in ref. 4), pages 382-402.
26) A. S. Wexler, Appl. Spectr. Rev. 1 (1967) 29.
27) J. N. Finch and E. R. Lippincott, J. Phys. Chem. 61 (1957) 894.
28) M. Folman and D. J. C. Yates, J. Phys. Chem. 63 (1959) 183.
29) R. E. Frech, Ph. D. Thesis (University of Minnesota, Minneapolis, Minn., 1968).
30) Y.-F. Y. Yao, J. Phys. Chem. 69 (1965) 3930.
31) R. W. Rice and G. L. Haller, unpublished data (1970).
32) G. Blyholder and E. A. Richardson, J. Phys. Chem. 66 (1962) 2597.
33) R. C. Lord and R. E. Merrifield, J. Chem. Phys. 21 (1953) 166.
34) O. Glemser and E. Hartert, Z. Anorg. Chem. 283 (1956) 111.
35) G. Zundel and G. M. Schwab, J. Phys. Chem. 67 (1963) 771.

DISCUSSION

J. B. PERI

It seems questionable that isolated hydroxyl groups on the surfaces of single crystals of Al_2O_3 at concentrations as high as on γ-alumina would have been detectable in your experiments. The average absorptivity for isolated OH(8.4×10^4 cm^2/mole) used in your calculations was derived from the summed absorbances of 3 isolated OH bands on γ-alumina dried at 800°C. All H-bonded OH groups had been eliminated. For γ-alumina dried at 400°C, at least 5 isolated OH bands, as well as an H-bonded OH band, remain in the spectrum. Probably less than a quarter of the surface OH groups would contribute to any single band in the 3700-3800 cm^{-1} region. This would give, at most, an A_{max} of about 0.006. As an example, consider Figure 4 from your reference 16 (Peri and Hannan). For this alumina aerogel sample the maximum absorbance for OD groups (similar to that for OH) after drying at 400°C would probably be about 1.4. The sample thickness (ca. 40 mg/cm^2) and surface area (ca. 300 m^2/g) put 1.2×10^5 surface layers in the beam path. Assuming 103 reflections and an amplification factor of 3.3 in your experiments, the aerogel sample gave 350 X your sensitivity. I would thus expect A_{max} in your experiments of about 0.004 rather than 0.023 as you calculate. I assume that this would not have been readily detectable.

G. L. HALLER

The point made by Peri is well taken, i.e., implicit in our calculation of expected absorbance is the assumption that isolated hydroxyl groups on a single crystal surface will all adsorb at the same frequency. It should be noted that in three exceptional cases where high frequency OH was detected (see section 3.1), only one band was observed and it was equivalent to 3-5 OH/100 $Å^2$ based on the absorptivity of 8.4×10^4 cm^2/mole reported by Peri and Hannan (Ref. 16). Moreover, the 5 isolated OH bands seen on γ-alumina dried at 400°C are not likely to represent a single crystallographic plane of the polycrystalline sample used. Another qualitative argument that the hydroxyls that remain on the γ-alumina and single crystal α-alumina differ in the extent to which they participate in hydrogen bonding can be made in terms of their behavior upon dehydration. The spectra of both materials are dominated by a broad band with a maximum near that of liquid water when saturated with water. With progressive dehydration of γ-alumina the band shifts to higher frequency and, following heating at 400°C, the relatively sharp bands due to the isolated OH become the prominent feature. Contrast this with single crystal α-alumina where the only band observed remains broad and shifts to lower frequency with increasing dehydration and, following heating to 400°C, it is still detectable but has shifted about 200 cm^{-1} to lower frequency with respect to liquid water. We conclude that the majority of the hydroxyls that remain after dehydration at 400°C are strongly hydrogen bonded on α-alumina but are reasonably isolated on γ-alumina.

N. D. PARKYNS

This very interesting application of IRS to the problem of hydroxylation of single crystal oxides has raised the question of why the observed groups are apparently all hydrogen-bonded as opposed to the "free" hydroxyl groups found on polycrystalline materials. In an attempt to rationalize this finding, I have two comments to make.

1) Samples previously studied, particularly in the case of alumina, have been in crystalline modifications where the coordination of the metal ions is in any case less than perfect. In the stable α-alumina (corundum) structure we might expect to see differences from the transitional γ- and η-forms. It would be interesting to see what the hydroxyl stretching frequency of the transmission spectrum of a polycrystalline α-alumina sample would look like, but to my knowledge this has not been done.

2) If the findings that there are in general no "free" hydroxyls on single crystal faces are correct, then presumably those normally found must be located at defects, at edges and apices of microcrystalline samples.

A further point on the nature of adsorbed water found in the present results is made by considering the models proposed by Parfilt and Jackson and by Hockey and his coworkers for rutile, where it was assumed that the 110 plane was predominant. Some of the hydroxyl bands were ascribed to undissociated water molecules, bonded via the oxygen atom in a ligand-like manner to complete the coordination of exposed Ti^{4+} ions. In some of my own work on rehydroxylation of anatase, a band at 1600-1640 cm^{-1} due to the deformation of (strongly) adsorbed water was found. When the intensity of this band was plotted against the equilibrium pressure of added water vapor, a flat-topped isotherm resulted which reproduced very well that given in Fig. 2 of the present paper, arguing a similarity in origin.

G. L. HALLER

Regarding the first portion of your comment, we agree that α-alumina might be expected to exhibit surface characteristics significantly different from that of γ- and η-alumina on the basis of their respective structures. These differences were emphasized in surface acidity measurements reported by Szabó and Jóvér (see paper No. 57). With respect to transmission spectra of polycrystalline α-alumina, it should be noted that it is difficult to obtain a sample of pure α with surface area greater than a few m^2/g and, therefore, the absorbance would probably not exceed that which we obtained using multiple internal reflection.

Our α-Al_2O_3 crystals did not transmit below 2000 cm^{-1} so it was not possible to confirm the existence of strongly adsorbed molecular water by recording the deformation band. But, we have attributed the difference between water adsorbed reversibly (at room temperature) on (0001) and $(41\bar{5}0)$ Al_2O_3 to the greater coordination unsaturation on the higher index plane (see Ref. 11). Undissociated water is assumed to enter the coordination sphere of Al^{3+} on both planes but the heat of this reaction is expected to be greater for the Al^{3+} of lower coordination, i.e., the $(41\bar{5}0)$ plane.

D. J. C. YATES

I should also like to congratulate Dr. Haller on a very interesting paper. However, as such a discrepancy seems to exist between single crystals and amorphous alumina, I should like to point out that as your crystals were mechanically polished with abrasives, one might expect that such a surface should be essentially amorphous. Under such conditions, isolated OH groups should be found on polished crystals. A further difference might be found in the pretreatment temperatures--most experiments reported with amorphous alumina have used a relatively high pretreatment temperature for the initial cleanup of the surface. Did you pretreat your crystals at high temperatures before adsorbing the water?

G. L. HALLER

For very hard materials like α-Al_2O_3 it is perhaps incorrect to call the mechanically polished surfaces amorphous, but they probably have a roughness comparable to the dimensions of the polishing agents used and the imperfections can be expected to be filled with micro-crystallites of α-Al_2O_3 mis-oriented with respect to the single crystal. Experimentally we find that the characteristic heats of adsorption and saturation coverages (see Ref. 11) are reproducibly obtained for different single crystals of the same orientation. For Al_2O_3 this appears to be the case for heat treatments as low as 500°C, but we have treated some crystals to temperatures in excess of 1000°C in hydrogen and/or oxygen. This is consistent with LEED studies (1) where polished surfaces always give weak and diffuse diffraction before heating above 600°C in vacuum. In the case of (100) MgO, a higher heat treatment is required. For example, a repolished MgO surface treated at 500°C in vacuum resulted in alcohol coverages equivalent to 1.3 monolayers under conditions which formerly had given submonolayer coverage. After heating in oxygen at 850°C the alcohol coverage decreased to the submonolayer value.

1) J. M. Charig, Appl. Phys. Lett., 10, 139 (1967); C. C. Chang, J. Appl. Phys. 39, 5570 (1968), T. M. French and G. A. Somorjai, J. Phys. Chem., 74, 2489 (1970); J. M. Charig and D. K. Skinner, page 34, and C. C. Chang, page 77, in "Proceeding of the Conference on the Structure and Chemistry of Solid Surfaces," G. A. Somorjai, Ed., Wiley, New York, 1969.

Paper Number 18

THE NATURE OF ULTRASTABLE FAUJASITE

J. B. PERI
Research & Development Department, American Oil Company
Whiting, Indiana, USA

ABSTRACT: Evidence on the nature of ultrastable faujasite was sought through potentiometric titrations, through infrared study of structural hydroxyl groups, adsorbed NH_3 and other molecules, and through extraction of zeolites with acetylacetone. The results support Kerr's conclusion that, during formation of ultrastable faujasite, Al migrates from tetrahedral sites in the aluminosilicate framework to cation positions outside the framework. They also indicate that Si replaces the lost Al through recrystallization of the framework. The cation positions appear to hold $Al(OH)^{2+}$, $Al(OH)_2^+$, and AlO^+ ions varying in relative numbers with the extent of dehydration.

1. INTRODUCTION

One of the most **catalytically active** forms of synthetic faujasite is "decationized" Y zeolite [1]. This form is made by cation-exchanging NaY zeolite with an ammonium salt solution to produce NH_4Y, and heating at 350-600°C to first produce HY (which holds H as OH groups), and subsequently to dehydroxylate it. The details of the structural changes caused by the removal of oxide ions from the crystal framework when H_2O is evolved on dehydroxylation are still unknown. Decationized zeolites are usually much less able than cationic forms to retain crystallinity and surface area after contact with water or heating at high temperatures [2].

A few years ago, McDaniel and Maher [3] achieved a remarkable stabilization of decationized Y zeolite. Although they stressed the thermal stability of the "ultrastable faujasite" produced, it is also highly stable in the presence of water. The reasons for the stability are still not entirely clear. The stabilization process includes heating an NH_4Y zeolite in air slightly above 800°C. Ambs and Flank [4] claim that thermal stability depends only on a low Na content and that ultrastable faujasite is not intrinsically different in framework structure from usual NH_4Y. Kerr[5,6], on the other hand, concludes from his studies that stabilization involves removal of Al from the framework to form a new structure in which much of the Al is held as $Al(OH)_2^+$, $Al(OH)^{2+}$, or Al^{3+} ions outside the framework. Evidence for removal of Al and O from the framework has been found in X-ray studies by Maher et al [7]. Scherzer and Bass [8] have recently presented infrared data showing new OH groups on US faujasite and indicating that stabilization involves recrystallization in which Si replaces the Al lost from the aluminosilicate framework. The present study was undertaken to obtain additional evidence on the nature of ultrastable faujasites. As used here, the term ultrastable has been broadened to include zeolites similar to those described by McDaniel and Maher [3], but with a higher Na content.

2. EXPERIMENTAL

2.1 Materials

Linde NaY zeolite (SK 40, Lot 3607-205 -- 12.5% Al, 29.7% Si, and 9.6% Na) was the starting material in NH_4Y preparations, which were made using standard ion-exchange procedures.

Three ultrastable (US) faujasite samples were supplied (in "soda" form) by the Davison Division of W. R. Grace & Company. These held 2.2% (A), 2.5% (B), and 3.0% (C) Na by weight. They were reportedly type B[3] preparations before the final NH_4^+ exchange to reduce Na content.

Reagent-grade or other high-purity chemicals were used unless otherwise specified. For extraction experiments, Eastman (White label) 2,4-pentanedione (acetylacetone) was used without further purification.

2.2 Equipment and Techniques

Data on area loss (as shown by changes in N_2 adsorption) on wetting and redrying were obtained as follows: Roughly 0.3 g of zeolite powder was pressed at 5 tons/in.^{2} in a steel die, 1-1/4 in. in diameter, to form a thin disc. The disc was suspended from a quartz-helix balance, dried by evacuation for 1 hr at 500°C, and cooled to -196°C. Adsorption of N_2 was then measured at pressures between 10 and 200 Torr. After warming to room temperature, and evacuation of the balance for 15 min to remove N_2 the sample was exposed to water vapor (16-18 Torr) for 1 hr. It was then heated to 500°C, dried by evacuation and cooled, and the adsorption of N_2 at -196°C again measured. The treatment was then repeated.

In potentiometric titrations, powder samples (1 g) were added to 50 ml of NaCl solution (200 g/1), and titrated with small additions of 0.1 N NaOH solution with constant stirring. Five minutes was allowed for equilibration before recording pH values.

For infrared study, discs were prepared by pressing roughly 0.15 g of zeolite powder in a steel die. The disc had a "thickness" of 20-25 mg/cm^2. They were then trimmed and mounted in an infrared cell[9] which permitted heating under vacuum. Spectra were recorded using a Beckman IR-9 spectrometer. Conventional procedures were followed in treating samples and recording spectra except that slit widths were wider than normal[9]. Samples were typically dried by evacuation below 10^{-4} Torr for 1 hr and cooled to about 40°C before recording spectra.

Standard X-ray techniques were used in determining zeolite crystallinities.

3. RESULTS

3.1 Stability

Fig. 1, curve a, shows the large loss in area obtained when an NH_4Y (1.9% Na) was heated in vacuum at 500°C, exposed to moisture, and then redried. Heating at 600°C in flowing air improved its hydrolytic stability but, as shown by curve b, the product was still much less stable than US faujasite (curve d). However, as shown by curve c, heating the same NH_4Y at 700°C in a 50 ml crucible gave a product nearly as stable as US faujasite. When similarly heated at 700°C, an NH_4Y that contained 2.9% Na gave a product which (curve e) was as stable as US faujasite and slightly more crystalline.

Fig. 2 shows potentiometric titrations of samples taken after 2 hrs at each indicated temperature when NH_4Y (1.9% Na) was progressively heated in slowly flowing air in a Vycor tube. Marked acidity developed after heating at 300°C, increased after heating at 400 and 500°C, and then after heating at higher temperatures, progressively decreased. The titration curves suggest that two types of acid were present after heating at 400 or 500°C, one with pK_a of about 2.7, the other with pK_a near 3.7. After heating at 700°C, however, two weaker acids, with apparent pK_a values near

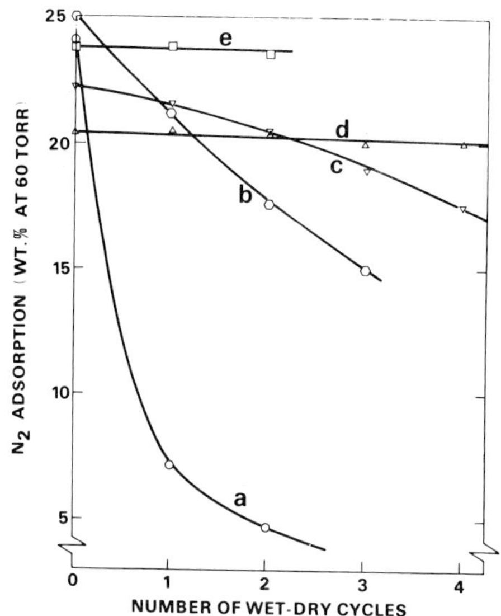

Fig. 1. Hydrolytic stability of decationized zeolites. (All samples were vacuum dried at 500°C). a. NH_4Y (1.9% Na) - predried 2 hr in vacuum at 500°C; b. same - preheated 6 hr in flowing air at 600°C; c. same - preheated 5 hr in static air at 700°C; d. US faujasite (A); e. NH_4Y (2.9% Na) - preheated 5 hr in static air at 700°C.

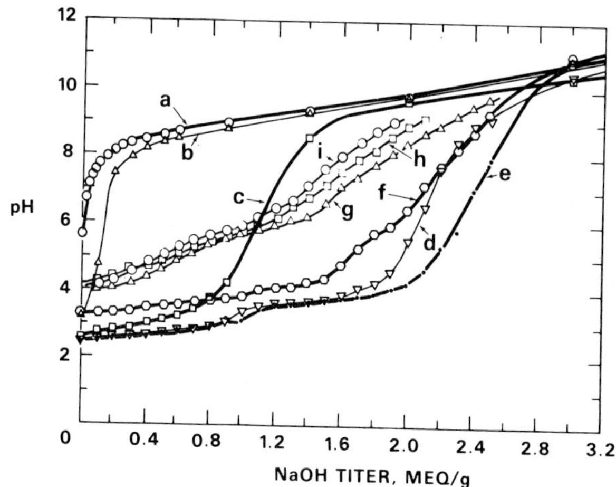

Fig. 2. Potentiometric titrations of zeolite suspensions in NaCl solution. NH_4Y (1.9% Na) samples preheated for 2 hr in flowing air as follows: a. 100°C; b. 200°C; c. 300°C; d. 400°C; e. 500°C; f. 600°C; g. 700°C; h. 800°C. Curve i represents US faujasite (B) predried at 500°C.

4.1 and 5.8, were seen. The maximum acidity was roughly 2.3 meq/g, compared with 3.56 meq/g total acidity expected if one proton was held for each Al ion in the framework not compensated by Na^+. The discrepancy between expected and observed acidity could reflect irreversible condensation of acidic hydroxyl groups on heating, inability of certain protonic sites to exchange H^+ for Na^+, or removal of some Al from the framework so that compensating H^+ was no longer needed.

As also shown in fig. 2, US faujasite and NH_4Y (1.9% Na) heated in air at 800°C gave similar titration curves, showing slightly under half the theoretical acidity based on Al content. The greater hydrolytic stability of these materials may arise, at least in part, from their lower acidity, since high acidity may cause destruction of zeolites on rewetting.

Only limited study of the thermal stability of decationized Y zeolites was made, but none of the samples showed the extremely high thermal stability claimed for US faujasites holding less than 1% Na. Low Na content may be a necessary but is evidently not a sufficient condition for maximum thermal stability. For example, the adsorptions of N_2 observed for samples preheated for 1 hr in vacuum at elevated temperatures were as shown in Table 1.

Table 1
Adsorption of N_2 on decationized zeolites
(wt% at -196°C and 60 Torr)

Sample	Preheat Temperature (°C)			
	500	800	850	900
NH_4Y heated at 700°C in static air-- 3.0% Na	25.8	21.1	18.5	--
US faujasite(B)--2.5% Na	24.2	23.4	--	4.9
US faujasite(B) NH_4Cl exchanged--0.93% Na	21.3	16.5	2.8	--

3.2 Infrared spectra of hydroxyl groups on zeolites

Infrared spectra of HY zeolites made by heating NH_4Y at 300-500°C have shown two major OH-stretching bands near 3550 and 3650 cm^{-1} which represent two types of OH groups, both attached to Si atoms and influenced by neighboring Al atoms[10,11]. These two types of OH groups may represent the two types of acid sites noted in titrations of NH_4Y preheated at 400°-500°C.

Fig. 3-A shows spectra obtained after drying NH_4Y (1.9% Na) in vacuum at 500° and 600°C. The expected OH bands are observed. When the same NH_4Y zeolite was originally heated in static air at 700°C, the OH bands were quite different from those observed after drying in vacuum and resembled those of curve a in fig. 3-B. The band at 3750 cm^{-1} was still present, but in place of the bands at 3550 and 3650 cm^{-1}, two new OH bands were seen near 3625 and 3700 cm^{-1}. Variation of the original Na content up to 4% did not significantly affect the frequencies or the relative intensities of the OH bands on other NH_4Y samples preheated in air at 700°C or higher.

The spectra of the US faujasite samples all showed similar bands near 3700 cm^{-1} (3695 ± 5 cm^{-1}) and 3625 cm^{-1} (± 10 cm^{-1}) in addition to the band near 3750 cm^{-1} (3745 ± 5 cm^{-1}). Compared with the other bands, the 3625 cm^{-1} band appeared to vary more in frequency and intensity with different levels

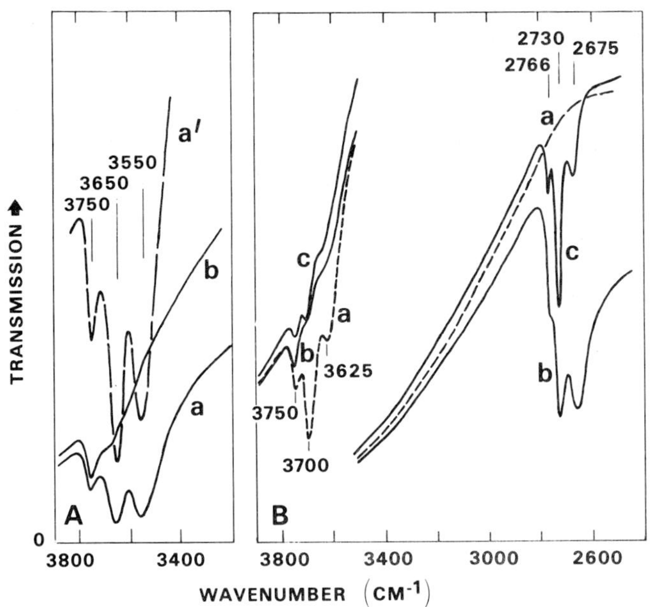

Fig. 3. Hydroxyl and deuteroxyl stretching bands in spectra of heated NH_4Y samples. A.- NH_4Y (1.9% Na) - predried in vacuum for 1 hr at 500° (a and a') and at 600°C (b). Wider slits were used for a'. B.- a. US faujasite(B) after 1 hr evacuation at 600°C; b. after exchange with D_2O and evacuation at 200°C; c. after final evacuation at 600°C.

of hydration. Similar bands have recently been reported by Scherzer and Bass[8] for US faujasite, and roughly similar bands were reported earlier[12] for "deep-bed" calcined NH_4Y samples.

Fig. 3-B shows spectra recorded before and after the OH groups on US faujasite(B) were converted to OD groups by exchange with D_2O. The OH bands at 3750, 3700, and 3625 cm^{-1} were converted to OD bands at 2766, 2730, and 2675 cm^{-1} as expected for isotopic substitution of D for H.

The bands near 3625 and 3700 cm^{-1} can apparently be produced to some extent when NH_4Y is heated in air below 700°C. Spectral evidence of the new bands was seen after heating NH_4Y (1.9% Na) in static air at 500°C. A mixture of the HY and US OH bands was also found after heating in flowing air at 600°C for 6 hr.

Spectral changes resulting from exchange of US faujasite with NH_4^+ are shown in fig. 4. The spectra were obtained under identical conditions. The spectra after drying at 800°C indicate that exchange reduced the intensities of the 3625 and 3700 cm^{-1} bands by about 50%. Some new bands, most evident after drying at 400-600°C, were also apparently produced in the 3550-3650 cm^{-1} region.

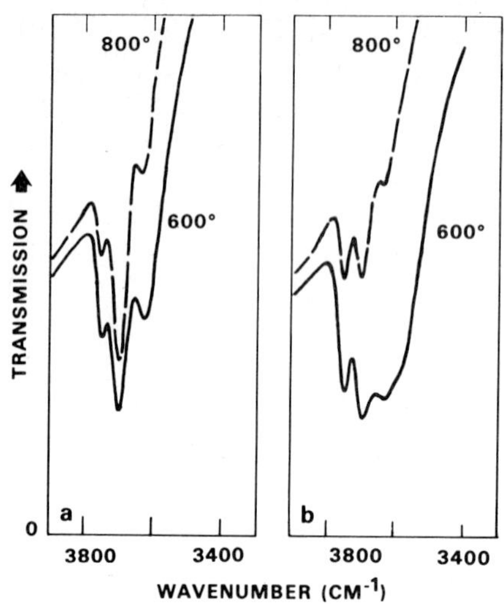

Fig. 4. Effect of NH_4^+ exchange on the hydroxyl bands of US faujasite(C). a. original dried by evacuation at the temperatures indicated; b. after exchange with NH_4Cl solution to reduce Na content to 0.84%, then dried as before.

3.3 Infrared study of rehydration

Heating above 500°C markedly reduced the number of OH groups on US faujasite, but they could be partially restored by subsequent exposure to H_2O vapor. After US faujasite(B) had been dried at 370°C and 800°C, it was heated in H_2O vapor at 370°C for 1 hr. Restoration of the 3625 cm^{-1} band was most evident, although none of the bands was completely restored to its original intensity. Spectra recorded directly at 400°C at varying H_2O vapor pressures, showed that OH content depends mainly on temperature and only slightly on H_2O vapor pressure.

3.4 Gravimetric study of the OH content of US faujasite

A pressed disc (0.3 g) of US faujasite(B) was suspended from a quartz-helix balance, dried by evacuation (1 hr) at a series of temperatures, and was finally heated in vacuum at 875°C for 2 hr to eliminate virtually all remaining OH groups. Weight was recorded at each step. Results indicated that retention of chemisorbed H_2O decreased approximately linearly from about 1.2 wt%, after drying at 300°C, to roughly 0.15%, after drying at 800°C.

3.5 Adsorption of NH_3

Fig. 5 shows spectra of NH_3 adsorbed on US faujasite(B). The bands at 3200-3450 cm^{-1} arise from N-H stretching. Those near 1620 and 1460 cm^{-1} reflect adsorbed NH_3 and NH_4^+ deformation vibrations. As shown by spectrum c, the 3700 cm^{-1} band was eliminated by NH_3 adsorption but the bands at 3750 and 3620 cm^{-1} were relatively unaffected. (Jacobs and Uytterhoeven[12]) reported that NH_3 adsorption did <u>not</u> eliminate a band at 3700 cm^{-1} in

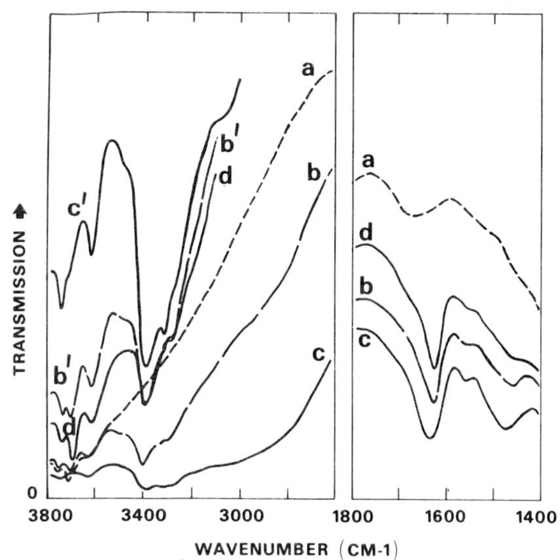

Fig. 5. Adsorbed ammonia on US faujasite(B)
a. after initial evacuation at 600°C; b. after first NH_3 addition (.22 Torr) at 40°C; c. after second NH_3 addition (5.4 Torr) at 40°C; d. after evacuation for 15 min. (Slits were widened for spectra b', c', and d in the 2600-3800 cm^{-1} region).

spectra of deep-bed calcined NH_4Y.) Failure of the OH groups responsible for the two other bands to interact with NH_3 could indicate either inaccessibility or weak acidity.

Adsorption of NH_3 on US faujasite holding OD groups, showed that all three types of OD (or OH) groups exchanged H readily with NH_3 at low temperatures, although the 3700 cm^{-1} OH band (2730 cm^{-1} OD band) showed the most rapid exchange. This suggests that all the groups are accessible to NH_3.

3.6 Adsorption of CO_2, CO, and NO

Some zeolites adsorb CO_2, CO, and NO markedly on exposed metal cations and/or reactive oxide ions. Such sites might be expected on decationized Y zeolite because decationization is thought to expose Al and Si ions[10]. Infrared study showed, however, that very little CO or NO was adsorbed by either US faujasite or vacuum dried HY and, compared to silica-alumina or CaY, relatively little CO_2. US faujasite(B) predried at 500°C showed no adsorption of CO at 40 Torr, and only a small band near 2250 cm^{-1} with NO at 50 Torr. Adsorption of CO_2 was significant, however, giving (at 5 Torr) a band near 2360 cm^{-1} which resembled that usually obtained by adsorption of CO_2 on dry alumina[13].

3.7 Infrared bands from vibrations of the zeolite framework

Spectra of faujasites show that bands in the 200 to 1300 cm^{-1} region generally shift to higher frequencies as the framework silica/alumina ratio increases[14], and suggest that AlO vibrations are responsible for a band (or bands) near 750 cm^{-1}. Spectra of NH_4Y (2.9% Na) and of US faujasite(B) (using KBr pellets containing 6 mg of zeolite) showed that nearly all bands from 400 to 1200 cm^{-1} in the NH_4Y spectrum were 15 to 30 cm^{-1} lower in frequency than those in the US faujasite spectrum. A band at 710 cm^{-1} in the spectrum of NH_4Y was seen at 740 cm^{-1} in the spectrum of US faujasite.

3.8 Extraction with acetylacetone

US faujasite(C) was contacted with refluxing acetylacetone in a Soxhlet extractor until no further removal of aluminum was detectable. Loss of the original Al content was roughly 50%. Subsequent analysis indicated that the residue had lost none of its original crystallinity. Infrared spectra of the residue showed that the OH band at 3700 cm^{-1} and the band at 740 cm^{-1} were almost gone. The 3625 cm^{-1} band remained, however.

As shown in Fig. 6 the crystallinity of residues left after such extraction of calcined NH_4Y samples reached a minimum after preheating at 400°C, but increased sharply after heating above 600°C, except where high Na content (4.4%) caused loss of crystallinity.

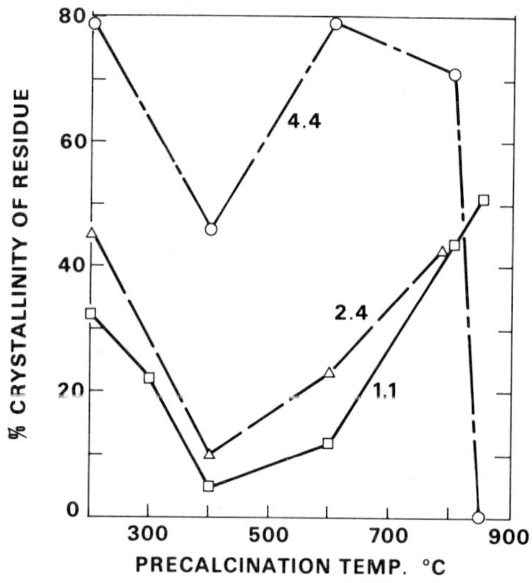

Fig. 6. Crystallinity of residues after acetylacetone extraction of calcined NH_4Y (% Na in original NH_4Y indicated on curves). Samples were preheated for 3 hr in static air.

4. DISCUSSION

Present results support Kerr's conclusion that, during stabilization of decationized Y, Al migrates from sites in the framework to external sites. Results of the potentiometric titrations are consistent with such migration of Al. The infrared spectra show that new hydroxyl groups, giving bands at 3700 cm^{-1} and 3625 cm^{-1} are created during stabilization. That the band at 3700 cm^{-1} arises from OH groups held by Al is supported by the removal of this band when Al is removed by extraction. Assignment of this band to $Al(OH)^{2+}$ ions seems reasonable. A band at 3700 cm^{-1} in the spectrum of γ-alumina has been assigned[15] to OH groups on sites having a net charge of +1. The $Al(OH)^{2+}$ ions on sites otherwise holding a single negative charge, should be electrostatically equivalent to OH groups on sites with a net charge of +1. The 3625 cm^{-1} band could reflect $Al(OH)_2^+$ ions. If so, the low frequency must be attributed to some interaction with the crystal environment. The disappearance on dehydration could reflect $Al(OH)^+_2 \rightarrow AlO^+ + H_2O$.

As observed in the present study, the selective removal of the 3700 cm^{-1} band by adsorption of NH_3 could reflect high accessibility of the $Al(OH)^{2+}$ ions, on sites in the supercage, while $Al(OH)^+_2$ ions might be held in the bridge positions. If so, it is easy to understand why acetylacetone extraction selectively removes the groups causing the 3700 cm^{-1} band. Reduction of the 3625 and 3700 cm^{-1} band intensities by NH_4^+ exchange of US faujasite supports the view that both these bands arise from complex Al cations held on relatively accessible sites.

US faujasite can apparently hold at least 25% of theoretical H for HY of the same composition. If 1/3 of the aluminum were held in cation sites as $Al(OH)^{2+}$, the total retention of hydroxyl groups would be 33% theoretical. Retention of half the aluminum as $Al(OH)_2^+$, however, would require an OH content equal to that of HY, much higher than exists on US faujasite. An average charge of 1.5 for cationic Al could be achieved with 20% of the Al held as $Al(OH)^{2+}$ and 20% as AlO^+, the total OH content being 20% of theoretical for HY. The hydration level varies with drying temperature, however. Some hydration of AlO^+ to $Al(OH)_2^+$ could occur at lower temperatures, while, at higher temperatures, the condensation of two $Al(OH)^{2+}$ ions could form Al^{3+}, AlO^+, and H_2O. Cations such as $Al(OH)^{2+}$, $Al(OH)_2^+$, and AlO^+ could shield the Al atoms so that they would not readily adsorb CO or NO.

In view of the high stability shown by US faujasite the removal of Al and O probably does not simply leave holes in the framework. Instead, as concluded by Scherzer and Bass[8)], SiO_2 probably fills the holes left by removal of $HAlO_2$. That such recrystallization can occur seems clearly established by work of Eberly[16)], showing that pure silica "faujasite" can be made by steaming NH_4Y at 650°C. The crystallinity of residues left after extraction of calcined NH_4Y sieves (fig. 6), and the shifts in IR bands in the 400-1300 cm^{-1} region, indicate that recrystallization does occur. Silica does not necessarily migrate in discrete molecules, e.g., $Si(OH)_4$, to fill holes. The holes may simply migrate through the framework until they reach an external surface of the crystal. This probably occurs readily in the presence of water at 700-850°C. Some amorphous alumina is probably produced by this process, since the silica used to fill the holes must come either from other portions of the crystal or from silica impurities. If 40% of the framework Al is replaced by Si, the amount of alumina produced need, however, be only about 10% of the sample.

As loosely defined here, US faujasites include a range of structures, differing in the relative intensities of characteristic infrared bands, in unit cell dimension, and in other properties. Heat plus moisture can evidently both remove Al (and O) from the framework and cause recrystallization, giving in the limit a silica framework. Changes in the relative rates of removal of Al and recrystallization should produce stable faujasites of differing framework silica/alumina ratios. A "pure-silica" faujasite, although stable, would not show much catalytic activity. Recrystallization should be rapid enough to permit stabilization before removal of nearly all framework Al and also, consequently, of complex Al cations if both stability and activity are essential. This may be achievable at 800° but not at 600°C.

REFERENCES

1) J. A. Rabo, P. E. Pickert, D. N. Stamires, and J. E. Boyle, Actes du Deuxième Congrès International de Catalyse, Paris, 1960 (Editions Technip, Paris, 1961), Vol. 2, p. 2055. (See also U.S. Patent 3,130,006, April 24, 1964.)
2) J. Cattanach, E. L. Wu, and P. B. Venuto, J. Catalysis 11, 342 (1968).
3) C. V. McDaniel and P. K. Maher, "Molecular Sieves," Soc. Chem. Ind. (London), Monograph, 186 (1968).
4) W. J. Ambs and W. H. Flank, J. Catalysis 14, 118 (1969).
5) G. T. Kerr, J. Phys. Chem. 73, 2780 (1969).
6) G. T. Kerr, J. Catalysis 15, 200 (1969).
7) P. K. Maher, F. D. Hunter, and J. Scherzer, "Molecular Sieve Zeolites," Vol. I, Advances in Chemistry Series, American Chemical Society, Washington D.C. (1971), p. 266.
8) J. Scherzer and J. L. Bass, Second North American Meeting of the Catalysis Society, Houston, Texas, February 24-26 (1971).
9) J. B. Peri, Discussions of the Faraday Society 41, 121 (1966).
10) J. B. Uytterhoeven, L. G. Cristner, and W. K. Hall, J. Phys. Chem. 69, 2117 (1965).
11) J. W. Ward, "Molecular Sieve Zeolites," Vol. I, Advances in Chemistry Series, American Chemical Society, Washington D.C. (1971), p. 380.
12) P. Jacobs and J. B. Uytterhoeven, J. Catalysis 22, 193 (1971).
13) J. B. Peri, J. Phys. Chem. 70, 3168 (1966).
14) A. V. Kiselev and V. I. Lygin in "Infrared Spectra of Adsorbed Species," by L. H. Little, Academic Press, N.Y. (1966) pp. 361-363.
15) J. B. Peri, J. Phys. Chem. 69, 220 (1965).
16) P. E. Eberly, Jr., S. M. Laurent, and H. E. Robson, U.S. Patent 3,506,400 April 14, 1970.

DISCUSSION

W. H. FLANK

The introduction to your paper implied that the basis for your work was the resolution of a conflict between sodium removal and aluminum removal from the framework as the reason for stabilization. Ambs and Flank had shown a clear relationship between thermal stability and sodium content, but, as you point out, "this may be a necessary but not sufficient condition for maximum stability." This statement is not supported, however, by your confusing and seemingly contradictory sorption data, and I wonder if this can be clarified. Was the degree of stabilization determined for your samples? The importance of water partial pressure in stabilization, as pointed out by Kerr, should have been noted as well. I also find little to contradict the Ambs and Flank claim that ultrastable faujasite is not intrinsically different in framework structure. This was further discussed in a subsequent publication, where the defect nature of these materials was emphasized and the concept of a continuum of properties was clearly brought out. Your own data seem to support this quite strongly, and you should have pointed out the agreement. In fact, some of your data support our mechanism for aluminosilicate degradation quite nicely also. Finally, could you describe the X-ray method used for determining crystallinity, since I do not believe there are many published standard methods that are adequately reliable?

J. B. PERI

My investigation of thermal stability was very limited, but I believe that the data of Table 1 support my conclusion. Preservation of the adsorption capacity for N_2 was taken as the measure of stability. Comparing the two US faujasites on this basis, it is clear that the one holding 0.93% Na is less stable than the one holding 2.5% Na. To explain these relative stabilities, factors other than the Na content alone must obviously be invoked.

Although the same general framework structure is apparently maintained, displacement of framework Al to cation positions outside the framework, coupled with recrystallization to produce a more-siliceous framework, strikes me as producing some rather important intrinsic differences. The difference between typical US faujasites and usual decationized Y zeolites seems at least as substantial as the difference between X and Y zeolites, for example.

I regret having overlooked the later exchange of comments between Kerr et al. [J. Catalysis 18, 236 (1970)], and Ambs and yourself [ibid. 18, 238 (1970)]. I am happy if we agree that suitable pretreatment of NH_4^+ faujasite can yield an almost continuous range of products.

The relative crystallinity of the zeolites was determined by comparison of the summed intensities of 7 or 8 peaks in the X-ray spectrum of the sample with the summed intensities of corresponding peaks in the spectrum of the original zeolite.

W. M. H. SACHTLER

While the IR bands typical for adsorbed NH_3 and NH_4^+-exchanged faujasite Y are known to disappear upon heating to 290°C,[1] it is quite remarkable that spectrum a in fig. 5 of your paper shows that even after heating to 600°C under vacuum, a band is present at ≈1670 cm^{-1}. How do you assign this band?

1) J. B. Utterhoeven, L. G. Christner, W. K. Hall, J. Phys. Chem. 69 (1965) 2117.

J. B. PERI

This band is characteristic of the zeolite framework and does not reflect adsorbed NH_3. Similar bands are seen with comparably thick samples for X and Y zeolites and, in the 1630-1650 cm^{-1} range, for silica and silica-alumina samples.

W. K. HALL

Most IR measurements are made on thin wafers; most catalytic measurements are made with "deep beds." Since pretreatments of the latter at elevated temperatures lead to IR bands which approximate those which you report for the ultrastable zeolite, I wonder if this means that most catalytic measurements (with decationated zeolites) correspond to the ultrastable material.

J. B. PERI

The exact conditions of pretreatment are crucial in governing the nature of the final product. The presence of similar OH bands in the spectrum is not, in itself, a sufficient indication of the formation of a US faujasite having the extreme hydrolytic stability of material heated as described at 700°C or higher. I would guess that very few, if any of the decationized zeolites used for past catalytic measurements correspond to US faujasites of the type discussed here.

J. W. WARD

With regard to Hall's comment concerning the relationship of infrared and catalytic measurements, it is possible to readily obtain samples in catalytic reactors which are not of the ultrastable type even though the bed is of the "deep bed" type. A critical factor seems to be the careful drying of the sample before treatment at elevated temperature. If suitable care is taken, samples are produced which, at the worst, contain only minor amounts of ultrastable type material, as shown by the usual criterion for detecting such material. Since we were aware of the influence of hydrothermal conditions many years ago, we have taken such precautions and also characterized materials removed from our catalytic reactors.

D. BARTHOMEUF

I would like to make a comment on the acidity results you described in your paper. If one looks at the ratio of the acidity which is measured against the theoretical acidity, a value of 0.63 is obtained for NH_4 zeolite. This value of 0.63 is very close to 0.6 we obtained for an efficiency coefficient for acid sites of Y zeolites. This efficiency coefficient measures the fraction of the total acidity which can be titrated by acid-base reaction (2nd Conf. Zeolites and J. Catalysis). In the short time we have now we can only say that it depends on the framework Al content. Hence we suppose that for U.S. zeolites which have less than 56 Al/uc in framework location, this efficiency coefficient would be higher than 0.63. That is to say with these U.S. samples one would measure a higher fraction of the theoretical acidity, since this theoretical acidity is determined by the framework Al content. Then my question is: did you measure the acidity of U.S. zeolites from which non-framework Al has been extracted by acetylacetone?

J. B. PERI

No, I did not.

P. K. MAHER

I believe this paper is quite conclusive. Low sodium level is a necessary but not a sufficient condition for the high thermal and hydrothermal stability that is characteristic of Ultra-Stable Faujasite. I believe sodium oxide level during the steam dealumination will control the number of aluminum atoms removed from the structure and thus control minimum temperature required for stabilization so that at low sodium levels 700°C is sufficient, but at higher levels 800°C is required.

P. B. WEISZ

You speculate that Si replaces Al in the holes, and thus these holes may migrate ultimately to the surface. It seems that, in a 1 micron crystal, we may have one surface atom per 10^3 internal sites; so the surface roughness can only take care of 0.1% Al extraction, and we must still conclude that a more appreciable portion of the bulk structure must remain disordered.

J. P. PERI

I may not fully understand your comment, but it seems obvious to me that the external crystal surface can accomodate any number of holes without requiring disorder of any of the bulk structure. The crystal surface may, of course, become covered with some amorphous Al_2O_3 "debris."

P. PICHAT

We have studied the vibrations of the framework of aluminum deficient zeolites (Compt. Rend., Ser. C 272 (1971), 612). We observed similar shifts in frequencies as Dr. Peri. This may be a proof of a certain analogy between the aluminum deficient zeolites and the ultrastable zeolites.

You assume that the band at 3625 cm^{-1} is due to $Al(OH)_2^+$ ions. Do you have any idea about the mechanism of the removal of these ions by NH_4Cl? Do you think that the bands around this frequency in the spectra of zeolites are always due to the formation of these ions? We observed the appearance of a band at ca. 3600 cm^{-1} for aluminum deficient zeolites in the absence of aluminum outside the framework.

Paper Number 19

INFRARED STUDY OF ULTRASTABLE FAUJASITE-TYPE ZEOLITES

R. BEAUMONT, P. PICHAT, D. BARTHOMEUF and Y. TRAMBOUZE
Institut de Recherches sur la Catalyse, 69-Villeurbanne, France
et Université Claude-Bernard, Lyon, France

ABSTRACT : Catalysts which show a high thermal stability have been prepared by aluminum extraction of a Y zeolite using E.D.T.A. The faujasite-type structure is preserved. I.R. quantitative measurements of pyridine chemisorption have shown that the concentration of weak Lewis acid sites is lower in the aluminum-deficient zeolites than in Y zeolite, but the concentrations of strong Lewis or Brönsted acid sites are higher. It appears that the protons do not originate only from OH groups. Chemisorption of pyridine made it possible to observe non-acidic OH bands at about 3670 and 3600 cm^{-1} as in ultrastable faujasites prepared by other methods, although this latter band is not always visible before adsorption of the base.

1. INTRODUCTION

Aluminum deficient-zeolites are characterized by a thermal stability which increases with the quantity of aluminum extracted [1-3]. For the same number of aluminums in anionic positions, their properties are alike in most respects to those of zeolites obtained by "deep-bed" calcination [4,5] or by a 870°C-heating [6]. The advantage of the chemical extraction is to remove selectively the aluminic sites catalytically inactive [7]. These sites are also the less acidic as stated by previous measurements of acidity using n-butylamine in benzene [8]. Furthermore, any disturbing effect that might be due to aluminum atoms not situated in anionic positions is obviated in these aluminum-deficient catalysts, which is not the case in the "deep-bed" calcinated [4,5] or heat-treated zeolites [6].

A previous I.R. study in the region 1300-250 cm^{-1} [9] and X rays patterns [10] have shown that the faujasite structure of these zeolites is preserved as long as the number of aluminum atoms per unit cell is not lower than ≃ 25.

The aim of the present work was an investigation of the surface acidity of these aluminum-deficient catalysts by means of infrared spectroscopy using chemisorption of pyridine. This method has the advantage of allowing the distinction between Lewis and Brönsted sites and between acidic and nonacidic OH groups. On the other hand, it seemed of interest to know whether the hydroxyls vibrating at 3680 and 3600 cm^{-1} observed by Jacobs and Uytterhoeven [5] with deep-bed calcined Y zeolites are also found in this type of ultra-stable zeolites.

2. EXPERIMENTAL

2.1 Catalysts

The starting material was a Union Carbide NaY zeolite containing 56 aluminum atoms per unit cell. In a first step, about 85 % of the Na ions were exchanged for NH_4^+. Aluminum was then extracted using ethylenediaminetetraacetic acid (E.D.T.A.). The chemical compositions of the obtained solids are given in table 1. The aluminum-deficient zeolites are designated by Y-x, x being the number of aluminum atoms per unit cell. Before any test, these catalysts were stabilized by heating in a stream of dry air for 15 hr at 380°, then for 15 hr at 550°C so that any "deep-bed" effect was avoided.

TABLE 1

Chemical composition of the catalysts

Catalysts	Al/u.c.	Al_2O_3 loss (%)	SiO_2 loss (%)	Na/u.c.
Y-56	56	0	0	8.7
Y-37.5	37.5	33.0	0	5.6
Y-26.5	26.5	52.7	0	4.2
Y-17.4	17.4	69.0	6.3	2.5

2.2 Infrared measurements

40 mg of powdered zeolite were compressed at 300 kg cm^{-2}. The resulting disk (18 mm diameter) was inserted in a quartz sample-holder which was introduced into an infrared cell as reported previously [11]. After evacuation at room temperature, the samples were heated in O_2 for 4-5 hr at 450°C, the cell being connected to a liquid nitrogen trap, and evacuated overnight at this temperature.

Pyridine (Merck for spectroscopy) was distilled under vacuum into small bulbs. The content of each bulb was thoroughly dried over Linde 5A-molecular sieve just before use. The samples of zeolite were allowed to equilibrate with pyridine at 25°C vapor pressure. In some experiments small increasing amounts of pyridine had been previously admitted. Pyridine was removed at different temperatures from 25 to 450°C for 5 hr for each temperature.

Infrared spectra were recorded on a Perkin-Elmer Model 125 grating spectrophotometer after cooling of the solid at room temperature. This instrument was continuously flushed with air free from H_2O and CO_2. The spectral slit width was less than 2 cm^{-1}.

3. RESULTS

3.1 Pyridine bands

The assignments and frequencies of the infrared bands of pyridine chemisorbed on the studied zeolites are indicated in table 2. These frequencies are close to those reported previously for hydrogen Y zeolites [2,12-16]. Eberly [15] assigned the band at ca.3130 cm^{-1} to a ν(NH) mode, we found that this frequency is a ν (CH) one. Most bands were shifted towards the short wavelengths when the evacuation temperature of pyridine was increased, consequently two limits are listed for each band in table 2. After removing the pyridine at about 350°C, a shoulder appeared on the band at ca.1454 cm^{-1} ; this shoulder became more pronounced at 400 or 450°C and formed in some cases a distinct band at 1460 cm^{-1}.

TABLE 2

Frequencies (in cm^{-1}) and assignments of pyridine chemisorbed on aluminum-deficient zeolites

Modes	Pyridinium ions	Pyridine on Lewis sites
ν (NH)	3245-3235,3170-3165	
ν (CH)	3135-3125,3095-3080,3068-3060, 2977-2970,2885-2880	idem
8a	1632-1630	1624-1616
8b	∿1620	non observed
19a	1489.5-1487.5	idem
19b	1540	1454.5-1452.5,1460

3.2 Acidity measurements

The measurements of the optical densities were carried out by using a background curve drawn according to the method already described [17].

The ratios of the optical densities of the 3240, 3170, 1630 and 1540 cm^{-1} bands corroborate that these bands originate from the same species, i.e. pyridinium ions (PyH^+). The 1540 cm^{-1} band is better isolated than that at 1630 cm^{-1} and more intense than those at 3240 and 3170 cm^{-1}. Therefore, the accuracy is improved by the choice of the measurement of the optical density of the 1540 cm^{-1} band as an indication of the number of PyH^+ ions and consequently of the number of protonic sites that have chemisorbed pyridine.

Coordinately bonded pyridine (Lewis acidity) gives rise to absorptions around 1620 and 1454 cm^{-1}. But the 1620 cm^{-1} band is not isolated and actually contains the 8b mode of PyH^+ ions as proved by the variations of the ratio of optical densities of the bands at 1620 and 1454 cm^{-1}. The 1454 cm^{-1} band is the only one that allows the measurement of the number of molecules of pyridine chemisorbed on Lewis sites. However the accuracy is not so good as in the case of the PyH^+ ions owing to the splitting of this band for the temperatures of evacuation of pyridine ⩾ 350°C.

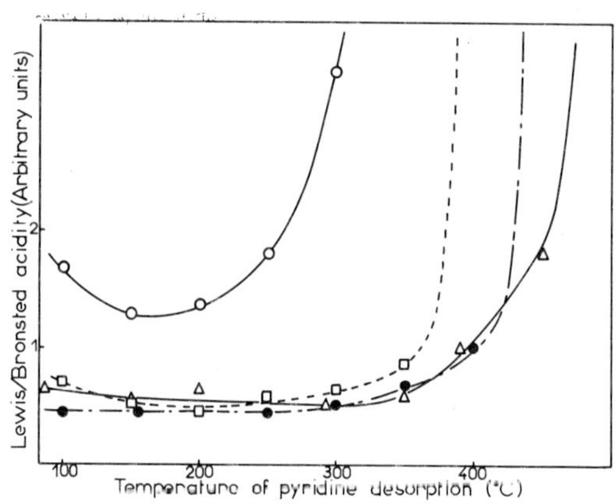

Fig. 1 Ratio of Lewis to Brönsted acidity as a function of the acid strength represented by the temperature of pyridine desorption
O : Y-56 □ : Y-37.5 ● : Y-26.5 Δ : Y-17.

Hence the ratio of optical densities of the bands at 1454 and 1540 cm^{-1} is proportional to the ratio of the number of Lewis sites to the number of Brönsted sites. This ratio (L/B.k) is shown in fig. 1 as a function of the evacuation temperature of pyridine for 4 different catalysts.

To compare the number of Lewis or Brönsted sites in different samples the optical densities have been divided by the number of unit cells contained in the 40 mg-wafers used for the I.R. measurements. The quantities thus obtained are proportional to the number of acid sites per unit cell. They are shown in fig. 2 as a function of the evacuation temperature of pyridine.

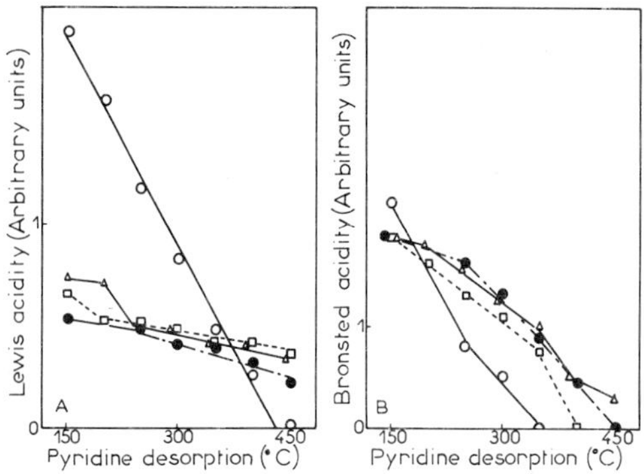

Fig. 2 Number of acid sites (Lewis -A- or Brönsted -B-) as a function of their strength represented by the temperature of pyridine desorption
O : Y-56 □ : Y-37.5 ● : Y-26.5 Δ : Y-17.4

3.3 Interaction of pyridine with OH groups

Besides a band at 3742 cm^{-1}, the infrared spectra of the aluminum-deficient zeolites generally contain bands at ca.3555 and 3630 cm^{-1} and a shoulder more or less pronounced between 3670 and 3695 cm^{-1} (fig. 3B, curve a). Among these spectra whose patterns vary according to the aluminum content of zeolites, that of the Y-26.5 zeolite is noticeable, since it includes only a broad band ($\Delta\nu_{\frac{1}{2}}$ ≃130 cm^{-1}) at 3605 cm^{-1} instead of the 3555 and 3630 cm^{-1} bands (fig. 3A, curve a). It is interesting to note that this sample has an aluminum content close to the limit beyond which the faujasite structure is damaged.

Adsorption of small increasing doses of pyridine at 25°C has shown that the appearance of the bands of PyH^+ ions occurs before changes in the OH bands. When the pressure of adsorption equilibrium reaches the vapor pressure of pyridine at room temperature, the spectra of the aluminum-deficient zeolites always contain a strong and relatively sharp ($\Delta\nu_{\frac{1}{2}}$ ≃70 cm^{-1}) band at ca.3600 cm^{-1}, a shoulder or a band in the region 3650-3685 cm^{-1}, while the intensity of the 3742 cm^{-1} band has considerably decreased and the 3630 and 3555 cm^{-1} bands have disappeared (fig. 3B, curve b). In the particular case of the Y-26.5 sample, the 3600 cm^{-1} band remains at the same position, but is a great deal narrower ($\Delta\nu_{\frac{1}{2}}$ ≃70 cm^{-1}) (fig. 3A, curve b). A similar band never occurs in the spectrum of the Y-56 zeolite.

Except for the 3742 cm^{-1} band, the pattern of the spectra does not change after evacuation of pyridine from 25 to 350°C (fig. 3, curves c).

On fig. 3 (curves d or e), it can be seen that in the temperature range 350-400°C, the spectra return to the original pattern they had before the adsorption of pyridine. Nevertheless, the typical bands of PyH^+ ions are still visible, even when the intensities of the OH bands have reached their maximum. It is noteworthy that this maximum coïncides with the initial spectrum for the Y-26.5 sample and corresponds to a slight decrease in the number of OH groups for the other samples.

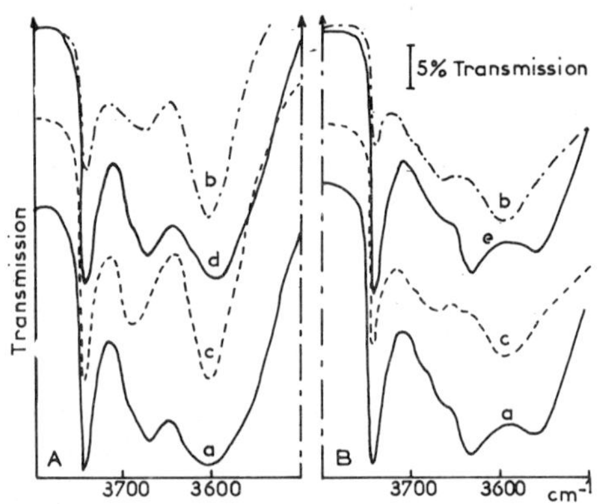

Fig. 3 Spectra of Y-26.5 (A) or Y-37.5 (B) zeolites, after initial treatment (a), adsorption of pyridine vapor pressure at 25°C (b), then evacuation at 150°C (c), 350°C (d) or 400°C (e).

4. DISCUSSION

4.1 Acidity measurements

In what follows, we shall consider that the acid sites are the stronge[st] as they retain pyridine at a higher temperature. Fig. 1 shows that the nature of acidity is deeply changed for the aluminium-deficient samples as com[-]pared with the initial material. Their Lewis acidity appears to be less pronounced or their Brönsted acidity seems to be enhanced.

Fig. 2 allows one to specify this point since the variations of the protonic and nonprotonic acid sites are distinguished. The three aluminum-deficient zeolites considered have many fewer Lewis sites than the initial zeolite (60-70 % less, roughly), but this difference comes principally from the quasi-disappearance of the sites of weak or medium strength, whereas on the contrary the number of very strong sites has increased (fig. 2 A). About half of the Lewis sites of the aluminum-deficient solids still retains pyridine after evacuation at 450°C. It appears that elimination of part of the aluminum atoms causes an increase in the acid strength of the nonprotonic sites related to the remaining aluminum atoms. This difference between the acid strengths of the examined catalysts does not lead however to a variation in the wavenumber of the 19b mode of coordinately bonded pyridine.

Extraction of aluminum does not change the total number of Brönsted sites, but increases their strength (fig. 2 B). Whilst for the Y-56 zeolite the PyH^+ ions have all disappeared after evacuation at 350°C, they are removed only between 400 and 450°C for the aluminum-deficient samples. The Y-17.4 zeolite even retains PyH^+ ions at 450°C, but the faujasite structure is not enterily preserved. The samples of fig. 2 and two other samples (Y44.5 and Y-30.5) allow one to state that the lower the aluminum content is, the stronger the Brönsted acidity. The strength of this acidity, as that of Lewis acidity, is raised by the dealumination method. Topchieva and al. [2] have previously pointed out that PyH^+ ions are more stable on a half-dealuminated sample than on the initial Y zeolite.

Acidity measurements by means of n-butylamine have been achieved on zeolites held in suspension in benzene [7]. They have shown that the solids obtained by extracting aluminum have a much more homogeneous acid strength. In particular, the number of sites of weak or medium acid strength (the reference being 72 % H_2SO_4) drops to zero when the aluminum content is lower than approximatøly 35 Al per unit cell. The present I.R. study enables one to specify on one hand that the weak acid sites removed are chiefly Lewis sites and on the other hand that the acid strength of both the protonic sites and especially the Lewis sites varies within a smaller range when aluminum is withdrawn.

4.2 OH bands at ca.3600 and 3670 cm^{-1}

Evacuation of pyridine in the temperature range 150-350°C leaves in the spectra of the aluminum-deficient zeolites a better defined band at ca. 3670 cm^{-1} and makes a 3600 cm^{-1} band appear, or in the case of the Y-26.5 sample sharpens the preexisting 3600 cm^{-1} band. Pyridine adsorption at 25°C has also shown that at least part of these OH groups are not hydrogen-bonded to pyridine either. Therefore they are nonacidic to pyridine.

Bands at 3673-3700 and 3580-3610 cm^{-1} insensitive to pyridine have been previously observed for divalent exchanged zeolites [18,19,20]. The assignments of these bands are still under discussion, but in any case they have been attributed to the presence of the divalent cations. Ward [21] has pointed out that the I.R. spectra of Y zeolites prepared according to the procedure of Mc Daniel and Maher [6] contains a band or shoulder near 3660-

3670 cm^{-1} and a strong and broad band in the 3570-3610 cm^{-1} region. When pyridine was reacted with one of the samples, a sharp band was observed near 3680 cm^{-1} and the structureless absorption between 3670 and 3600 cm^{-1} was eleminated. The behavior of the 3600 cm^{-1} band was not reported. Very recently, Jacobs and Uytterhoeven [5)] have stated that bands around 3675-3700 and 3600-3620 cm^{-1} occur in the spectra of deep-bed calcined Y zeolites and that they are nonacidic to ammonia and pyridine. They did not find these bands for a slightly dealuminated Y zeolite containing 49.4 Al per unit cell.

The present study indicates that if the 3600 cm^{-1} band is not visible in the spectra of most of the aluminum-deficient zeolites, except for the sample with an aluminum content close to the lower limit of the faujasite structure (Y-26.5). On the other hand, this band appears in presence of pyridine for the other samples. Therefore, it seems that there is a certain analogy between the OH groups of the "deep-bed" and aluminum-deficient zeolites Since in the latter catalysts the extracted aluminum does not form deposits in some form, the conclusion of Jacobs and Uytterhoeven [5)] according to which these nonacidic OH groups are not associated with hydroxyaluminum ions is corroborated. Furthermore the present work tends to show that these particular hydroxyls are formed because of the aluminum deficiency of the lattice However, Ward's results [19)] have brought evidence that they could also be generated by divalent cation exchange.

It would be interesting to know whether, with the exception of the Y-26.5 sample, the 3600 cm^{-1} band does not exist before adsorption of pyridine or is blurred in the other OH absorptions. Attempts to decompose the OH bands by means of a "Dupont curve resolver, type 310" would incline us to reject the latter hypothesis. Pyridine is known to induce displacements of transition cations in Y zeolites [22)]. Consequently it is likely that it causes the protons of aluminum-deficient zeolites to migrate. So it is not unexpected that the environment of OH groups not bonded to pyridine is modified and / or that a new distribution of the hydrogens over the oxygen atoms takes place. An evacuation at a temperature higher than 450°C brings about a similar shift in the OH bands [3)].

4.3 Stability of the OH groups

Complete desorption of pyridine does not dehydroxylate the Y-26.5 sample and causes the OH bands of all the other aluminum-deficient zeolites to decrease only slightly. This result proves the stability of the OH groups of these dealuminated catalysts compared with the initial material whose distribution in OH groups is very noticeably changed after the desorption of pyridine as pointed out by Jacobs [23]. This increased stability was also observed towards heating [3].

4.4 Origin of the protons

After evacuation at 150°C, all the hydrogen-bonded pyridine has been removed. Therefore, the decline in the OH absorption near 3630 and 3555 cm^{-1} or the decrease in the broadness of the 3600 cm^{-1} band for the Y-26.5 sample is produced only by the formation of PyH^+ ions. In this latter case, owing to the strong intensity of the 3600 cm^{-1} band, the acidic OH groups do not cause definite maxima of absorption but a broadening of this band.

However, progressive adsorption of pyridine at 25°C has shown that the PyH^+ appear before the OH bands have been modified. This tends to prove that some of the protons are not characterized by infrared bands in the OH region. It does not seem that this fact has been pointed out for hydrogen zeolites [24], whereas it is admitted for amorphous silica-alumina catalysts.

5. CONCLUSION

The diversity of Y zeolites whose spectra contain OH bands around 3600 and 3680 cm^{-1} leads us to assume that different treatments -various heating conditions [4-6,21], exchange with divalent cations [18,19], aluminum extraction- may be at the origin of these bands. Even in the zeolites thus treated, their occurence and frequencies depend on the cations (mono-and divalent) and molecules (for instance, pyridine) introduced in the catalyst. Also, a connection between these nonacidic OH bands and the increased thermal stability of these zeolites is corroborated by all the investigations.

With regard to acidity, the aluminum extraction results chiefly in catalysts whose acid strength varies within a smaller range. All the sites associated with aluminum do not have catalytic properties for isooctane cracking as reported in a previous work [7]. The present results tend to show that the inactive sites in this catalysis are the weak Lewis acid sites.

ACKNOWLEDGEMENT

The authors are indebted to Mrs M. C. Bertrand for excellent experimental assistance.

REFERENCES

1) G. T. Kerr, J. Phys. Chem. 72 (1968), 2594.

2) K. V. Topchieva and Huo Shi T'Huoang, Kinetika i Kataliz 11 (1970), 490.

3) P. Pichat, R. Beaumont and D. Barthomeuf, to be published.

4) G. T. Kerr, J. Catalysis 15 (1969), 200.

5) P. Jacobs and J. B. Uytterhoeven, J. Catalysis 22 (1971), 193.

6) C. V. Mc Daniel and P. K. Maher, "Molecular Sieves", Soc. Chem. Ind. London Monogr. (1968), 168.

7) R. Beaumont and D. Barthomeuf, Compt. Rend., Ser. C 272 (1971), 363.

8) R. Beaumont and D. Barthomeuf, to be published.

9) P. Pichat, R. Beaumont and D. Barthomeuf, Compt. Rend., Ser C 272 (1971) 612.

10) P. Gallezot, R. Beaumont and D. Barthomeuf, to be published.

11) M. V. Mathieu and P. Pichat in "La Catalyse au Laboratoire et dans l'Industrie", Masson et Cie (1967), 319-323.

12) S. P. Zhdanov, A. V. Kiselev, V. I. Lygin and T. I. Titova, Zhur. Fiz. Khim. 40 (1966), 1041.

13) B. V. Liengme and W. K. Hall, Trans. Faraday Soc. 62 (1966), 3229.

14) T. R. Hughes and H. M. White, J. Phys. Chem. 71 (1967), 2192.

15) P. E. Eberly, J. Phys. Chem. 72, (1968), 1042.

16) J. W. Ward, J. Catalysis 9 (1967), 225.

17) P. Pichat, M. V. Mathieu and B. Imelik, Bull. Soc. Chim. Fr. (1969), 2611.

18) J. B. Uytterhoeven, R. Schoonheydt, B. V. Liengme and W. K. Hall, J. Catalysis 13 (1969), 425.

19) J. W. Ward, J. Phys. Chem. 72 (1968), 4211.

20) J. Kermarec, J. F. Tempere and B. Imelik, Bull. Soc. Chim. Fr. (1969), 3792.

21) J. W. Ward, J. Catalysis 19 (1970), 348.

22) P. Gallezot, Y. Ben Taarit and B. Imelik, Compt. Rend. Ser C 272 (1971), 261.

23) P. Jacobs, Thesis, Louvain (1971).

24) J. W. Ward, Advan. Chem. Ser. 101 (Molecular Sieve Zeolites - I) (1971), 380.

DISCUSSION

W. H. FLANK

Were your samples characterized in any other way besides IR and chemical analysis to determine their properties? In particular, how much damage was done in exchanging 90% of the sodium ions with protons? I believe it is rather difficult to achieve this proton exchange level without damage. Your Table 1, however, indicates closer to 84% proton exchange, and it is interesting that the percentage of sodium loss during extraction closely parallels the percentage of alumina loss, suggesting that a structural unit is involved in the extraction reaction.

P. PICHAT

Our samples were also characterized by X-ray analysis, catalytic reactions, amine titration and adsorption of hydrocarbons. With respect to the exchange of Na^+ ions with protons, there was a mistake in our preprint. Sodium ions were exchanged for NH_4^+ ions and not for protons before the dealumination process. This method prevents any damage in the faujasite structure as evidenced by X-ray diffraction patterns.

It is interesting to note in Table 1 that the percentage of sodium loss from NH_4NaY zeolites (84% NH_4^+ exchange) closely parallels the percentage of alumina loss. Generally it is accepted that the number of cations involved in the extraction reaction and that of aluminum atoms are similar. Hence, the equivalent losses in Al and Na suggest an homogeneous removal of the NH_4^+ and Na^+ ions. The structural unit assumed by Dr. Flank might account for this homogeneous removal.

J. SCHERZER

Concerning the effect of aluminum extraction on the acidity:

a) Can you explain the increase in the strength of Bronsted type acidity with partial dealumination of the framework?

b) If I understand correctly, you suggest that the groups responsible for the strong acid sites absorb at the same frequency as those responsible for the weaker acidity. Wouldn't we expect different absorption frequencies for OH groups with different acid strength?

c) Your spectra in Fig. 3 indicate that pyridine adsorption results in the disappearance of the band at ca. 3550 cm^{-1}. Since it is generally accepted that pyridine does not affect this band because the corresponding OH groups are not accessible, how do you explain the disappearance of this band in your spectra?

d) Do you have any suggestions about the position of the OH groups corresponding to the 3600 and 3700 cm^{-1} bands?

P. PICHAT

a) As a matter of fact, up to now, we have no real coherent explanation to suggest for the increase in the strength of Brönsted acidity.

b) It would be attractive to utilize the stretching frequency of the

O-H groups by itself as an indication of the acid strength. But the experimental facts preclude this use. For instance, amorphous silica is generally considered as showing no protonic acidity, and, however, this material has OH groups vibrating at the same frequency as those of silica-alumina surfaces with a high Brönsted acid strength.

c) The interaction of the band at ca. 3550 cm^{-1} with pyridine depends on the pressure. In HY zeolites, this band can be strongly reduced with 10 torr (Ward, J. Phys. Chem., 71 (1967) 3106) or completely removed with 25 torr. In the present work, it is restored only after evacuation at 350°C, probably because of the higher strength of Brönsted acidity.

d) The determination of the position of the OH groups is a difficult problem; it needs more work and is still open to discussion.

P. K. MAHER

What is the hydrothermal or thermal stability of these dealuminated structures? What is the X-ray unit cell size?

P. PICHAT

These dealuminated zeolites still show a good catalytic activity after being treated at 900°C in a stream of dry air. Comparatively to that of the starting material, their hydrothermal stability is also improved.

The unit cell size decreases with dealumination. The u.c. sizes measured by X-ray are analogous to those of the ultrastable zeolites prepared by other methods.

H. KNÖZINGER

I come back to the former discussion on the constancy of the pyridine vibration frequencies. I can understand the constancy of frequencies of the pyridinium ion (PyH^+), in which the proton is really added to the pyridine, if the interaction of PyH^+ with the surface does not influence the adsorbed species too much. As to the coordinately bonded pyridine, however, I cannot agree with Dr. Pichat's statement that the vibration frequencies are constant irrespective of the desorption temperature. On the contrary, there is evidence in the literature for an increase in the ring vibration frequencies with increasing desorption temperature, which indicates the stronger coordination bond of the respective chemisorbed species.

P. PICHAT

I agree that in most cases an increase in the ring vibration frequencies of the coordinated pyridine indicates a stronger coordination. In particular, this has been elegantly proved in Dr. Knözinger's work and it was also my opinion when I wrote the paper of reference 17. But it seems that in some cases the 19b vibration may occur at high frequencies (1460-62 cm^{-1}) without implying a stronger coordination than for the band remaining at lower frequencies (1455 cm^{-1}), perhaps because of steric hindrance. This phenomena has been described in J. Chem. Soc., Faraday I, 68 (9), 1972, p. 1712. On the other hand, in the present work, the variation of the Lewis acidity strength corresponds only to a 2 to 3 cm^{-1} variation in the 19b mode (Table 2, 3rd column).

W. K. HALL

In our experience, the 19b vibration for the Lewis species shifts from about 1440 cm^{-1} to 1455 cm^{-1} as the polarity of the interaction increases, for example from Na^+ to Mg^{2+} as the acceptor site. Usually the strength of the adsorption increases concommitantly. These phenomena are described in the work of Liengme and Hall.

P. PICHAT

Same answer as given to Dr. Knözinger.

J. W. WARD

We have studied zeolites prepared by hydrothermal treatment of zeolites containing various amounts of sodium ions (see Fig. 1). Samples containing about 7% sodium contain only one band which is located near 3700 cm^{-1}. More extensive ion exchange to about 0.1% sodium results in the most intense band appearing near 3590 cm^{-1} and no absorption near 3700 cm^{-1}. Exchange to intermediate sodium levels before hydrothermal treatment results in spectra exhibiting both bands. Hence the type of bands appearing depends strongly on the sodium level. The observation of only the 3700 cm^{-1} band for the 7% sodium level is interesting since at this level of exchange only a single band near 3650 cm^{-1} is observed in the spectrum of hydrogen Y. Hence it is possible that the two types of hydroxyl groups are formed from the same type of site. The band near 3600 cm^{-1} increases with decreasing sodium level and hence appears to be associated with removal of aluminum from the lattice.

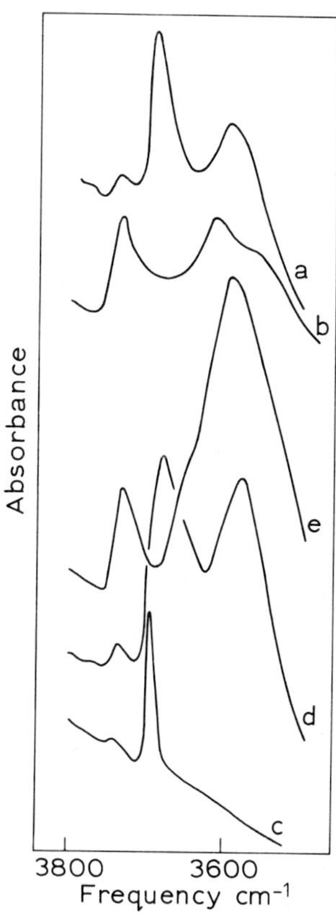

Fig. 1

a. Ultrastable Y precursor (3.4% Na)
b. Ultrastable Y (0.3% Na)
c. Hydrothermally heated Y (8.2% Na)
d. Hydrothermally heated Y (5.8% Na)
e. Hydrothermally heated Y (0.1% Na)

P. PITCHAT

Dr. Ward's results prove that the Na level affects the OH bands in ultrastable zeolites. But the work which we have presented here, as well as unpublished results, show that the aluminum level has an influence on the OH bands independently of the Na level. For instance, after evacuation of the samples at 450°C, the 3600 cm^{-1} band occurs in the spectrum of the Y-26.5 sample with 4.2 Na^+ per unit cell and does not appear in that of the Y-17.4 sample with 2.5 Na^+ per unit cell, although the faujasite structure of this latter sample is only slightly damaged.

Paper Number 20

INFRARED STUDY OF WORKING SULFONIC ACID RESIN MEMBRANE CATALYST--THE DEHYDRATION OF METHANOL

R. THORNTON and B. C. GATES
Department of Chemical Engineering, University of Delaware, Newark, Delaware, USA

ABSTRACT: Methanol dehydration to dimethyl ether in a differential flow reactor at 97°C and atmospheric pressure was catalyzed by a 5 μ-thick polystyrenesulfonic acid membrane. Infrared absorption spectra of the functioning catalyst were measured simultaneously with reaction rates for CH_3OH, CH_3OD, and CD_3OD feeds diluted with nitrogen and CH_3OH feeds containing water. The catalyst approached saturation at 0.6 atm methanol partial pressure; water inhibited reaction. The results indicate the presence of a hydrogen-bonded network of undissociated $-SO_3H$ groups in the catalyst, as well as a reaction intermediate identified as two methanol molecules doubly hydrogen-bridged between a pair of $-SO_3H$ groups. A concerted reaction mechanism is suggested in which the C-O bond-breaking and -forming are rate determining.

1. INTRODUCTION

Infrared spectroscopy is a powerful method for identifying adsorbed reaction intermediates, especially when spectra of functioning catalysts are obtained simultaneously with reaction rates and amounts of adsorption, allowing distinction between reactive and unreactive species[1-6]. In the research described here infrared absorption spectra have been measured simultaneously with dehydration reaction rates in a steady-state flow reactor as methanol vapors contacted sulfonic acid groups in the matrix of a 5 μ-thick polymer membrane. Structures of the sulfonic acid ion-exchange resin at various degrees of hydration are known from the thorough investigations of Zundel[7], whose techniques for determining membrane spectra have been adapted to this work. Zundel's results demonstrate the importance of hydrogen bonds which form between undissociated sulfonic acid groups and either adjacent sulfonic acid groups or polar molecules such as water and methanol.

2. EXPERIMENTAL

2.1 Catalyst Preparation

Membranes of polystyrenesulfonic acid crosslinked with eight mole percent divinylbenzene were synthesized between glass plates separated by a 5 μ-thick Ta foil spacer[7]. A typical membrane was 5 cm in diameter, weighed 0.04 gm in an air-dried state (containing about 20 wt% water), and contained 0.17 meq $-SO_3H$ groups as determined by titration against standard base. A single membrane was used in all reported experiments.

2.2 Apparatus

The flow reactor system of stainless steel and glass is shown in fig. 1. Liquid feed from a syringe pump (Sage model 249) flowed into a vaporizing column packed with glass beads

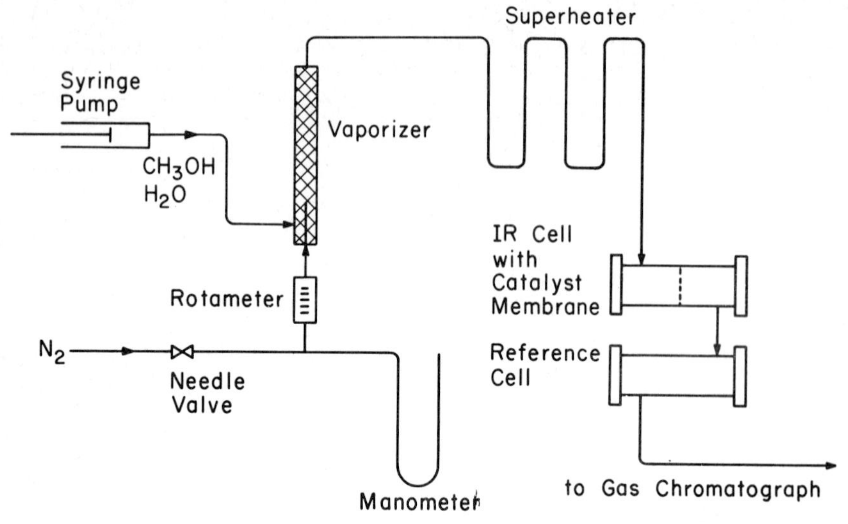

Fig. 1. Schematic diagram of reactor system.

where it was mixed with a metered stream of nitrogen carrier gas. The vapor stream flowed through electrically heated lines to the reactor, which was an infrared gas cell designed to hold a single resin membrane perpendicular to the infrared beam and to the direction of vapor flow. The electrically heated stainless steel cell had NaCl windows clamped to the ends and sealed by silicone rubber gaskets. Product vapors flowed from the reactor through the reference cell to the heated gas sampling valve of a Hewlett-Packard 5750 gas chromatograph. The reactor and reference cell were mounted in a Perkin-Elmer 247 infrared grating spectrophotometer. Temperatures in the reactor and at other points in the system were held constant and monitored with thermocouples.

2.3 Measurement of Reaction Rates and Infrared Spectra

Liquid feeds were CH_3OH (Fisher, 99.9% pure), occasionally mixed with distilled water; CH_3OD; or CD_3OD (both from Stohler Isotope Chemicals, 99% D). Liquid feeds contained about 50 ppm HNO_3, which had no effect on spectra or reaction rates, yet prevented the slow catalyst activity loss observed in preliminary runs, presumably caused by accumulation in the catalyst of a basic feed impurity. Feeds also contained one percent by weight diethyl ether, serving as an internal standard in the product analysis for dimethyl ether.

During a run methanol and nitrogen flowed at constant rates, and steady-state conversions were measured repeatedly with the gas chromatograph. Products were separated in a column packed with polystyrene beads (Porapak N, 80-100 mesh) held at 140°C. Helium carrier gas flow rate was 30 ml/min; a flame ionization detector was used. Since the detector responses for the three methanols were indistinguishable, they were assumed to be the same for CH_3-O-CH_3 and CD_3-O-CD_3.

Infrared absorption spectra of the catalyst were obtained during each run at a scanning speed of 1.4 cm^{-1}/sec.

Methanol flow rate was varied from 0.01 to 0.08 moles/hr, and methanol partial pressure from 0.02 to 0.6 atm. Pressure was 1.09±0.04 atm, and reaction temperature was 97.0±0.3°C. Conversions of methanol were usually about 0.1%.

3. RESULTS

The observed conversions, obtained in the absence of mass transfer influence[8)], side reactions, and catalyst activity loss, were easily low enough to be in the differential conversion range, determining reaction rates directly[8)]; the rate data have a precision of about ±5%.

Rates for CH_3OH, CH_3OD, and CD_3OD, respectively diluted with nitrogen, are shown in fig. 2. Saturation kinetics were observed for each alcohol, with both a primary and a secondary isotope effect indicated by the distinct curves. The rate data of fig. 3, obtained with CH_3OH feeds containing water, demonstrate inhibition of reaction by water.

Infrared spectra of the catalyst with CH_3OH and CH_3OD feeds are shown in figs. 4 and 5, respectively. Spectra of resins containing CH_3OD and CD_3OD were indistinguishable. Spectra obtained as mixtures of nitrogen and either H_2O or D_2O contacted the catalyst at the reaction temperature are shown in figs. 5 and 7, respectively. Spectra indicated that the $-SO_3H$ groups were deuterated by contacting with either CH_3OD or D_2O.

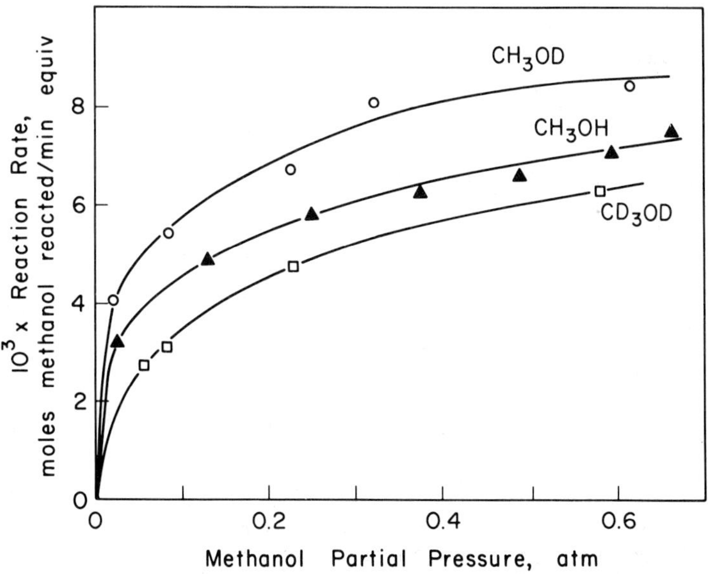

Fig. 2. Rate of dehydration of methanol diluted with nitrogen.

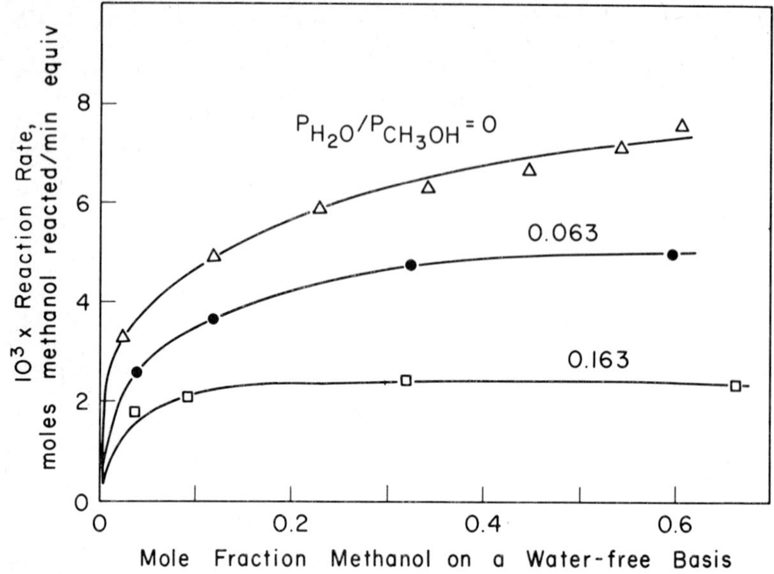

Fig. 3. Rate of dehydration of methanol mixed with water and nitrogen.

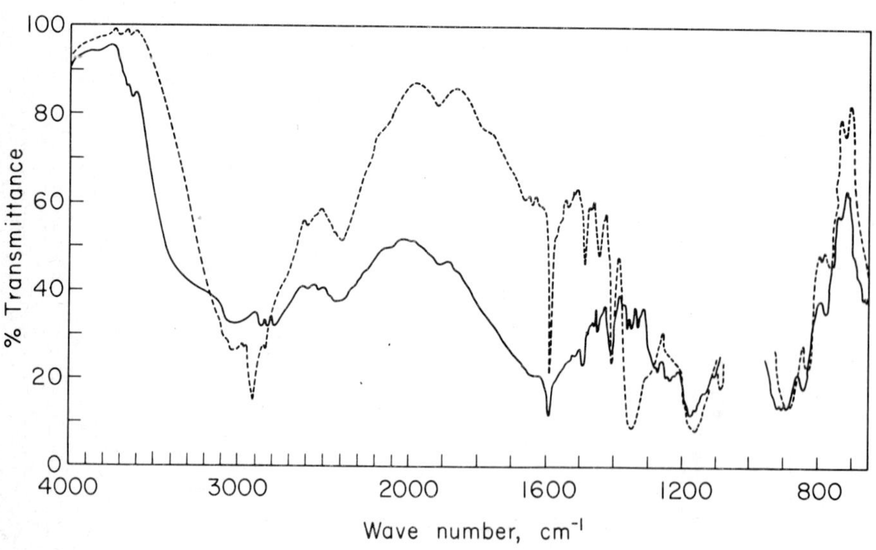

Fig. 4. Spectrum of working catalyst containing CH_3OH. ——P_{CH_3OH} = 0.60 atm; ---P_{CH_3OH} = 0.026 atm.

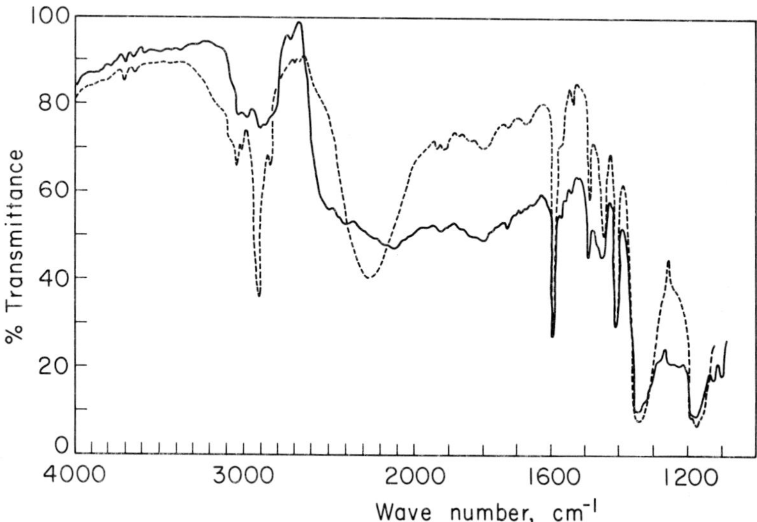

Fig. 5. Spectrum of working catalyst containing CH_3OD.
—— P_{CH_3OD} = 0.60 atm; --- P_{CH_3OD} = 0.026 atm.

4. DISCUSSION

The infrared spectra are similar to those observed by Zundel[4,9] and by Knözinger[10] for water and methanol combined at 20°C with $-SO_3H$ groups in the resin and in solutions containing p-toluenesulfonic acid. The assignments to be discussed are based on their interpretations; the small differences from the wave numbers of bands shown in figs. 4-7 are believed to result

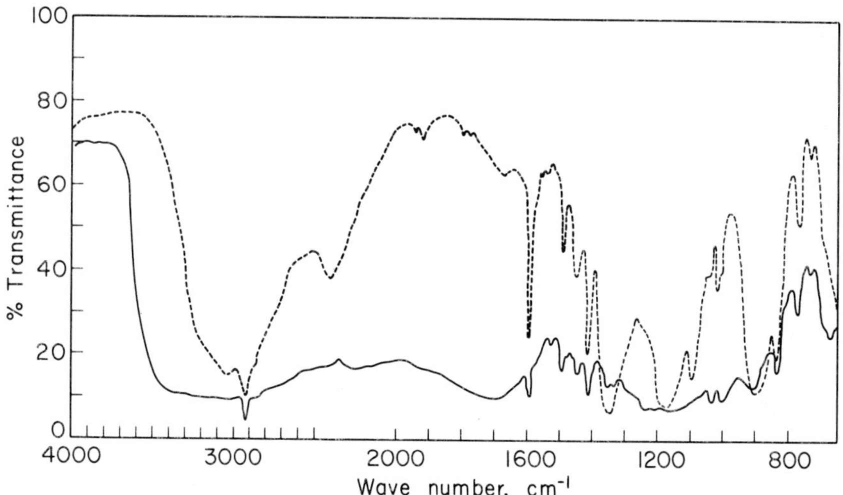

Fig. 6. Spectrum of catalyst in contact with H_2O.
—— P_{H_2O} = 0.22 atm; --- P_{H_2O} = 0.051 atm.

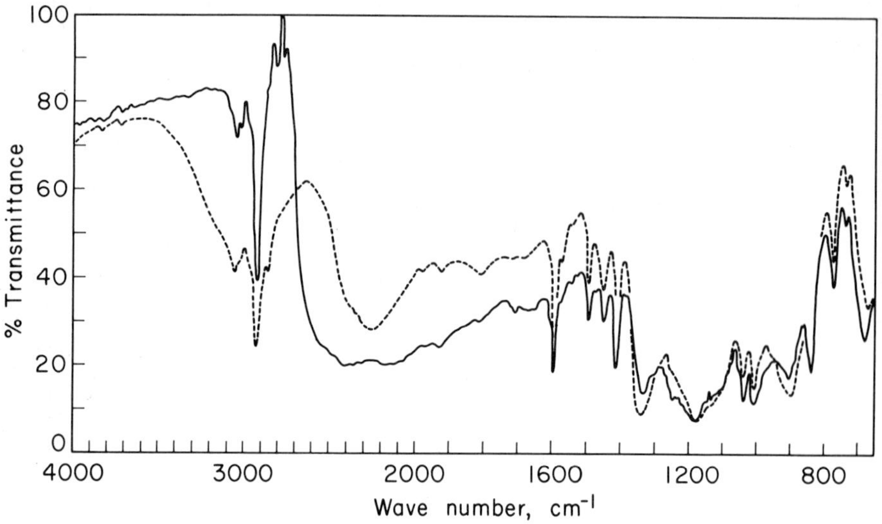

Fig. 7. Spectrum of catalyst in contact with D_2O.
—— P_{D_2O} = 0.24 atm; - - - P_{D_2O} = 0.11 atm.

primarily from the difference in temperatures of the respective experiments. The gaps in the spectra between 1100 and 950 cm^{-1} with CH_3OH and below 1100 cm^{-1} with CH_3OD are attributed to black-body radiation from the catalyst, which became an important fraction of the radiation incident on the detector in the regions where the methanol vapor in the cells was absorbing strongly.

Zundel's interpretation of resin hydration in terms of a network containing hydrogen-bonded water and $-SO_3H$ groups (fig. 8) was derived from identification of infrared absorption bands of undissociated and dissociated $-SO_3H$ groups and of hydrogen bonds involving the undissociated groups both as strong proton acceptors and donors. As water[7] or methanol[10] was added to the anhydrous resin at 20°C, there was a decrease in the concentration of S-O and S=O bonds, indicated by disappearance of the 907 and 1350 cm^{-1} bands, respectively, while there was an increase in $\left(-S\begin{smallmatrix}O\\O\\O\end{smallmatrix}\right)^-$ ion concentration indicated by bands at 1034 and 1200 cm^{-1}.

The spectra of figs. 6 and 7, obtained at 97°C, indicate that the same pattern of hydration prevails at the higher temperature. The spectra of fig. 4 demonstrate similar behavior of $-SO_3H$ groups when methanol contacts the catalyst. The clearest data are provided by the disappearance of the 1350 cm^{-1} band, indicating the antisymmetric stretching vibration of the S=O bond. The decrease of the logarithm of the inverse transmittance with increasing methanol partial pressure is shown in fig. 9. The results show that the bonding of methanol to the catalyst leads to dissociation of $-SO_3H$ groups.

The spectra of fig. 4 further demonstrate the formation

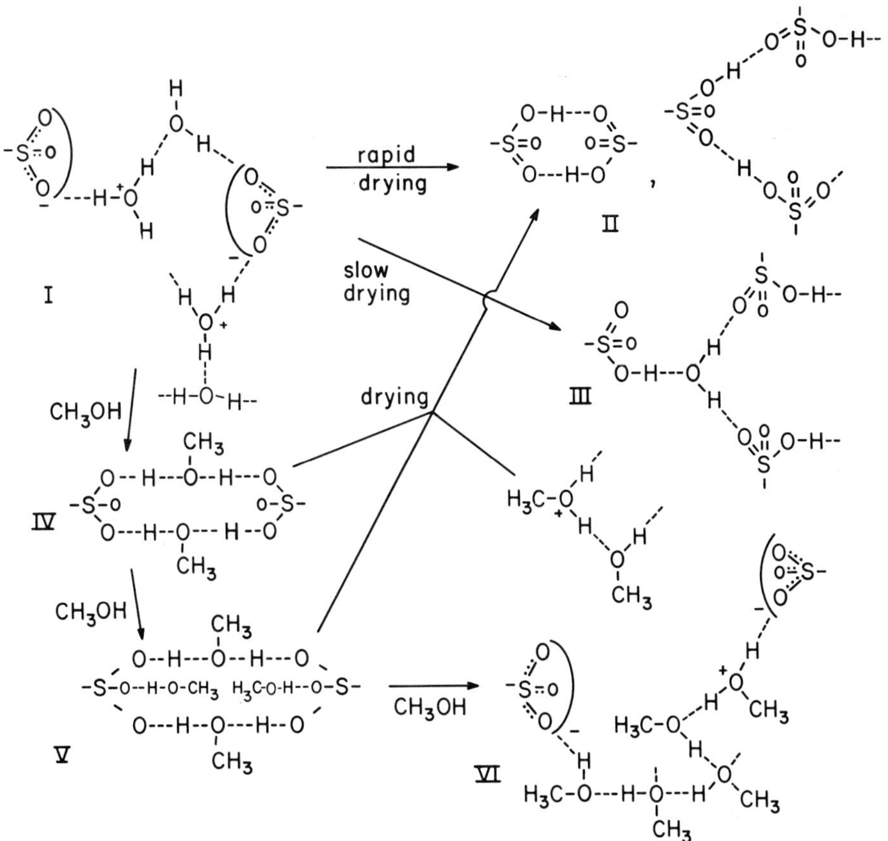

Fig. 8. Hydrogen-bonded structures involving $-SO_3H$ groups. I-III[7]; V[10]; VI[9].

of hydrogen bonds involving methanol and $-SO_3H$ groups. As has been found for combinations of water[7], methanol[10], and formic acid[11] with the groups, there is no significant band in the spectrum indicating the presence of free -OH groups (to be expected in the range 3800-3400 cm^{-1}), a result consistent with the formation of hydrogen bonds involving all the -OH groups of both methanol and the resin.

With increasing resin methanol content at 20°C, Knözinger[10] found increasing absorption between 3600 and 2000 cm^{-1}. Such a broad band is also present in the spectrum of hydrated resin (structure I, fig. 8) as shown by fig. 6, and is identified with a proton hydrogen-bonded between two oxygen atoms[7,10]. The band disappeared rapidly with decreasing resin methanol concentration at 20°C (at higher concentrations than observed in this work), and Knözinger ascribed it to hydrogen bonds involving -OH groups of methanol (structure V). The band is also evident in the spectrum of the working catalyst con-

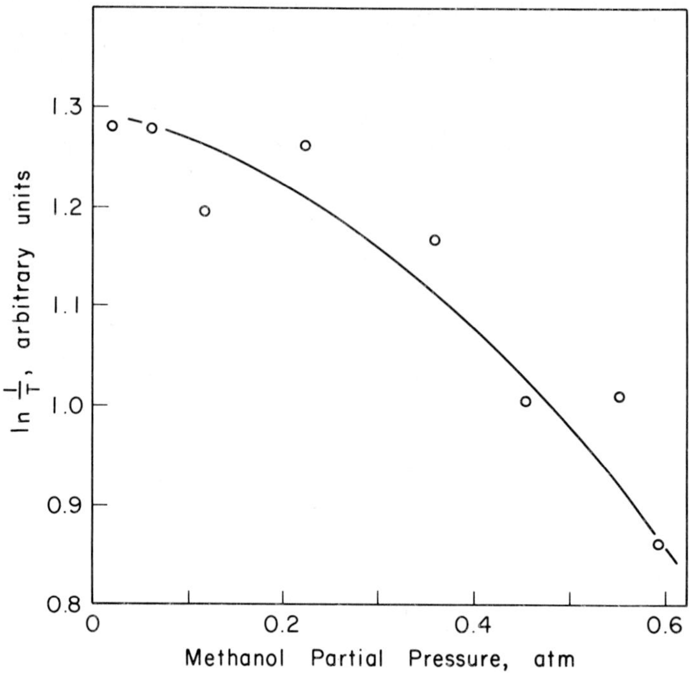

Fig. 9. Extinction coefficient of 1350 cm^{-1} band in working catalyst containing CH_3OH.

taining CH_3OH (fig. 4); it increased in intensity with increasing methanol partial pressure, accompanied by a decrease in the intensities of bands at 2950 and 2405 cm^{-1}. The latter bands are associated with protons in the hydrogen bridges of $-SO_2OH$ groups (structure II); evidently the network of these groups is incompletely broken up by methanol at reaction conditions.

The corresponding broad band for the working catalyst containing CH_3OD and $-SO_3D$ groups is observed between 2600 and about 1500 cm^{-1} (fig. 5). This band similarly increased in intensity with increasing methanol partial pressure, while the bands at 3260 and 1800 cm^{-1}, indicating deuterated structure II, decreased correspondingly.

A second broad band between 3000 and 2000 cm^{-1} has also been identified with a hydrogen-bonded proton[10]. Since at 20°C this band disappeared less rapidly with decreasing methanol concentration than the band at 3600-2000 cm^{-1}, it was ascribed[10] to a hydrogen bond involving a more strongly combined methanol, suggesting structure V (previously proposed by Ulbricht[12]) and identifying the 3000-2000 cm^{-1} band with hydrogen bonds involving protons donated by $-SO_3H$ groups.

While the data for the working catalyst cover too small a range in methanol concentration to allow discrimination of the two broad absorption bands, the spectral results are clearly consistent with those of Knözinger and with structure IV, which involves two doubly hydrogen-bonded methanol molecules per pair of $-SO_3H$ groups.

Substantiating evidence of the simultaneous occurrence of structures II and IV in the working catalyst is provided by the adsorption equilibrium data of Kabel and Johanson[13] for ethanol in the resin at 100°C. Though methanol adsorption data are lacking, the spectra of methanol and ethanol in the resin are indistinguishable at 20°C[10], and the two alcohols are expected to have nearly equal adsorption equilibrium constants[14]. The ethanol adsorption data fit the Langmuir isotherm over the observed range to 0.34 atm, and, extrapolated, correspond to combination of 0.6 alcohol molecules per $-SO_3H$ group at 0.6 atm alcohol partial pressure. This result suggests that at the lower temperature of reaction (97°C), there was slightly less than one methanol molecule per $-SO_3H$ group at 0.6 atm. This conclusion is supported by the rate data, which indicate a nearly saturated catalyst at this pressure (fig. 2). Combined with the evidence that the resin offers a nearly homogeneous array of catalytic sites[14], the results imply that structure IV is uniquely the reaction intermediate.

The hydrogen-bonded intermediate structure is consistent with the observed inhibition by water, which competitively forms hydrogen bonds with $-SO_3H$ groups. It also leads directly to the suggestion of a bimolecular reaction mechanism (fig. 10) which is indicated by the Langmuir-Hinshelwood kinetics observed for methanol[14] and ethanol[13] dehydrations at 120°C, and by the second-order dependence of dehydration reaction rate on resin alcohol concentration at 120°C, measured during reaction by Herlihy[15].

The spectra provide further information related to the reaction mechanism. The broad infrared absorption bands observed in hydrated resin with structure I are associated with tunneling protons; the motion of protons between oxygen atoms is described by a more or less symmetrical double minimum potential with a low intervening barrier, as inferred from the lack of temperature dependence of the continuous absorption spectrum[7].

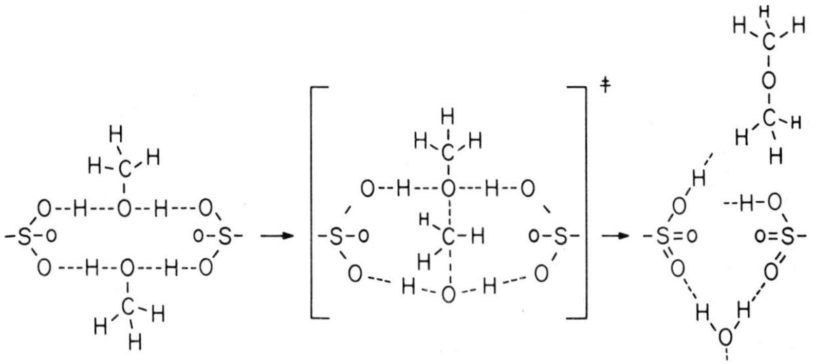

Fig. 10. Proposed reaction mechanism

Comparison of Knözinger's spectra obtained for methanol in the resin at 20°C[10]) with those of fig. 4 indicates little if any dependence of the broad absorption bands on temperature, imply-that protons in structure IV similarly require very little activation energy to move between minima in the potential. This result suggests that the O-H bond-forming and -breaking steps depicted in fig. 10 have little influence on the reaction rate. The suggestion is consistent with the unusual primary isotope effect (fig. 2), as CH_3OD reacts about 20% more rapidly than CH_3OH; evidently the activation energy is lower for reaction of the deuterium-containing intermediate.

It is inferred then that the rate determining step is the concerted C-O bond-forming and -breaking represented in fig. 10. The mechanism is an example of tautomeric catalysis, relying on the simultaneous proton-donating and -accepting property of the $-SO_3H$ groups. The process requires inversion of a methyl group configuration. It is consistent with the fourfold lower rate constant for ethanol than for methanol at 120°C[14]), which is explained by the added steric hindrance when hydrogen of a methyl group is replaced by a methyl group. While the unusual secondary isotope effect (fig. 2) is unexplained, it is perhaps best accomodated by a highly structured transition state, as has been suggested.

ACKNOWLEDGMENT

This research was supported by grants from E. I. du Pont de Nemours and Company and the Petroleum Research Fund, administered by the American Chemical Society.

REFERENCES

1) Y. Noto, K. Fukuda, T. Onishi and K. Tamaru, Trans. Faraday Soc. 63 (1967), 2300.
2) Y. Soma, T. Onishi and K. Tamaru, Trans. Faraday Soc. 65 (1969), 2215.
3) A. L. Dent and R. J. Kokes, J. Phys. Chem. 73 (1969), 3772.
4) A. L. Dent and R. J. Kokes, J. Phys. Chem. 73 (1969), 3781.
5) R. F. Baddour, M. Modell and U. K. Heusser, J. Phys. Chem. 72 (1968), 3621.
6) R. F. Baddour, M. Modell and R. L. Goldsmith, J. Phys. Chem 74 (1970), 1787.
7) G. Zundel, "Hydration and Intermolecular Interaction," Academic Press, New York, 1969.
8) B. C. Gates and L. N. Johanson, J. Catal. 14 (1969), 69.
9) I. Kampschulte-Scheuing and G. Zundel, J. Phys. Chem. 74 (1970), 2363.
10) E. Knözinger, Dissertation, University of Munich, 1966.
11) E. Knözinger and H. Noller, Z. Physik. Chem. (Frankfurt) 55 (1967), 59.
12) K. A. Ulbricht, Dissertation, University of Munich, 1964.
13) R. L. Kabel and L. N. Johanson, Amer. Inst. Chem. Engrs. J. 8 (1962), 621.
14) B. C. Gates and L. N. Johanson, Amer. Inst. Chem. Engrs. J. 17 (1971), 981.
15) J. C. Herlihy, Ph.D. Thesis, University of Washington, Seattle, 1968.

DISCUSSION

H. NOLLER

After Zundel,[7] the broad band in Figs. 4-7 in the low pressure region must be assigned to a hydrogen bridge between different species (SO_3H and H_2O for example), whereas the continuous absorption at higher pressure is due to a proton between two equal species, for example, two methanol molecules. This might be taken into account when discussing structures IV, V, and VI. Of course, this statement does not affect your mechanistic conclusions since several species should exist at the same time.

As to the anomalous isotope effect, it may be worthwhile to mention that E. Knözinger[10] has expected such an effect for formic acid dehydration after his IR studies of formic acid adsorption. Unfortunately, he did not find it. Your results show that he was not so wrong with his prediction.

B. C. GATES

Thank you for your comment.

J. MANASSEN

You call your isotope effect abnormal. From the point of view of an organic chemist it seems to be normal:

1) Rate $CH_3OD > CH_3OH$
   this suggests a preequilibrium
   $CH_3OH \rightleftarrows CH_3O^- + H^+$
2) Rate $CD_3OD < CH_3OD$

This is in accoradnce with your mechanism, which suggests a change in hybridization in the rate determining step. A Walden inversion of $-CD_3$ can be expected to go at a lower rate than that of $-CH_3$.

B. C. GATES

Thank you for your comment.

P. B. WEISZ

In view of the mechanism involving pairs of sulfonate sites, I wonder about the background in this field of research, as to the dependence of rate on the sulfonate population density. It would seem then that much rich information of the type you have demonstrated could be gleaned from performing these measurements at greatly different sulfonate concentrations.

B. C. GATES

We agree that experiments with catalysts of various $-SO_3H$ group concentrations can be interesting--it is a unique advantage of the resin catalyst that this concentration can be varied systematically by ion exchange.

Experiments were reported[8] which determined initial rates of methanol dehydration catalyzed by a series of resins containing $-SO_3H$ and $-SO_3Na$ groups by a series containing $-SO_3H$ and $-SO_3Li$ groups. The salt forms of the resins lacked catalytic activity, and activity increased with increasing

$-SO_3H$ group concentration. Resins containing $-SO_3H$ and $-SO_3Li$ groups were more active than resins containing equivalent numbers of $-SO_3H$ and $-SO_3Na$ groups. We infer that while $-SO_3H$ groups are necessary for catalysis, $-SO_3Li$ groups may participate in the reaction; this inference itself suggests a mechanism involving two groups. It is clear that if only one of the groups in Fig. 10 were replaced by $-SO_3Li$, the postulated proton transfers could still occur.

Related experiments were reported for an olefin-forming elimination reaction, the dehydration of t-butyl alcohol (B. C. Gates, J. S. Wisnouskas and H. W. Heath, Jr., J. Catal. 24 (1972) 320). Here the salt groups simply diluted the catalytically active acid groups. The rate of reaction was proportional to about the fourth power of $-SO_3H$ group concentration at high values, which again suggests a concerted mechanism involving a network of hydrogen-bonded $-SO_3H$ groups.

H. W. HABGOOD

I am surprised that you did not postulate a network-type of hydrogen-bonded methanol (similar to that found with water) but give only the highly regular structure of species IV.

B. C. GATES

We agree that a network involving methanol similar to the hydrate structures would be formed. Species IV is only the simplest of the structures involving hydrogen-bonded $-SO_3H$ groups and methanol.

K. KOCHLOEFL

1) Have you observed the OD-OH exchange in the unreacted alcohol? How large was the extent of this reaction?

2) The lower reactivity of ethanol in comparison with methanol was explained on the basis of steric effects. Is it not possible to consider the action of electronic effects also?

B. C. GATES

1) We observed that when $-SO_3H$ resin was contacted with CH_3OD or CD_3OD, there was a rapid exchange to give $-SO_3D$ resin. All of our steady state data are either for $-SO_3H$ resin with CH_3OH or for $-SO_3D$ resin with CH_3OD or CD_3OD.

2) The lower reactivity of ethanol was explained as a steric effect because the direction and magnitude both seem to be appropriate. Some confirmation for the explanation might be inferred from the result that isopropanol also forms ether at roughly the same rate (Gottifredi, J. C., Yeramian, A. A. and Cunningham, R. E., J. Catal. 12 (1968), 245). We do not rule out alternative explanations.

J. W. HIGHTOWER

Kinetic isotope effects are being used more frequently to provide significant information about the mechanisms of catalytic reactions and the exact nature of the transition states involved. Since the effects are frequently small, highly accurate measurements are required in order for the results to be meaningful. Even though the catalytic activity appears to be "stable," the possibility always exists that an impurity present in only one of the reactants may influence the reaction when independent rate measurements are made with each isotopically labeled compound. It would be much more convincing if both isotopically labeled compounds were used simultaneously and a mass spectrometer (or other appropriate analytical device) were used to determine to what extent each had reacted under identical conditions.

I have two questions: 1) Is there H-D scrambling in the methyl groups? and 2) could some impurities have caused the somewhat unusual inverse isotope effect reflected in your Fig. 2?

B. C. GATES

1) We do not know whether there was H-D scrambling in the methyl groups; we have no data from a mass spectrometer.

2) We cannot rule out the possibility that impurities were responsible for our isotope effect. Considering the reproducibility of our results and the purities of feeds, we regard it as unlikely. We agree that use of the isotopically labeled and unlabeled compounds simultaneously is generally a better experiment. We would mention that in the case of methanol dehydration such an experiment is expected to lead to formation of mixed ethers, thereby adding an interesting complication.

Paper Number 21

INFRARED STUDY OF THE EFFECT OF PREADSORBED OXYGEN FOR HYDROGEN CHEMISORPTION ON PLATINUM

D. J. DARENSBOURG and R. P. EISCHENS
Texaco Research Center, Beacon, New York

Abstract: Previous reports which indicate that an enhancement of room temperature hydrogen chemisorption on alumina-supported platinum is produced by preadsorption of oxygen are confirmed. However, the hypothesis which attributes the enhancement to hydrogen adsorbed on platinum oxide is not verified. Instead, it is concluded that the enhancement is due to an increase in platinum area which is produced by a low temperature reduction. This conclusion is based on observation of a parallel enhancement of carbon monoxide chemisorption without evidence for carbon monoxide adsorbed on oxide, and on reducibility studies which show that preadsorbed oxygen is removed from platinum as water by exposure to hydrogen at room temperature. In related experiments, efforts were made to test the concept that platinum-hydrogen species can be obtained by dissociative chemisorption of water. No valid evidence for this concept was observed and it was found that there is danger of misleading results due to impurity carbon monoxide.

1. INTRODUCTION

Eley and co-workers[1)] have recently published an infrared study of hydrogen chemisorption on supported platinum. They observed bands at 2120 and 2040 cm^{-1}. These experimental results are in general agreement with earlier work[2)] where bands were observed at 2115 and 2060 cm^{-1}. In the earlier work the two bands were attributed to two forms of hydrogen adsorbed on metallic platinum, $\overset{H}{\overset{|}{Pt}}$ and $\overset{H}{Pt\diagup\diagdown Pt}$. However, Eley et al found that the 2120 cm^{-1} band was enhanced by preadsorption of oxygen. They therefore attribute this band to hydrogen atoms adsorbed on Pt^{+4} ions of platinum oxide. They also found that their 2040 cm^{-1} band was observed after addition of water and concluded that this band indicated dissociative chemisorption of the water on oxide patches. The 2040 cm^{-1} band was also attributed to H atoms adsorbed on Pt^{+4} ions. The lower frequency of this band, as compared to the 2120 cm^{-1} band, was ascribed to the effect of hydrogen bonding with neighboring hydroxyl groups.

Calorimetric studies of Aston and co-workers[3)] have shown that preadsorption of oxygen enhances hydrogen adsorption on carbon-supported platinum. They attributed the enhancement to a rearrangement of platinum atoms.

The work to be described here is concerned with the effect of preadsorbed oxygen on the 2120 cm^{-1} band and a study of the 2040 cm^{-1} band produced by addition of water. Studies of the reducibility of platinum are included to determine whether it is reasonable to expect that oxide patches will remain on the platinum surface after exposure to hydrogen at room temperature. The reducibility studies are also of interest with regard to removal of oxygen in the hydrogen titration method which is currently used in measuring the surface area of supported platinum[4,5,6].

2. EXPERIMENTAL

The infrared spectra were measured with a Cary-White Model 90 grating spectrophotometer. All spectra were obtained at 25°C. The samples were suspended from a Cahn microbalance so the weight of adsorbed gas could be determined. This gravimetric apparatus is satisfactory for determining quantities of adsorbed water, oxygen, and carbon monoxide. It is not satisfactory for the smaller weight of adsorbed hydrogen. Details of the experimental apparatus have been published[7].

A 5.0 wt % platinum-on-silica gel sample was prepared by the ion exchange procedure described by Benesi et al[8]. The sample was reduced in flowing hydrogen for 16 hours at 360°C. It was then evacuated at 380°C for one hour to remove adsorbed hydrogen. This is the standard reduction-evacuation procedure for the present work. Chemisorption of carbon monoxide at room temperature gave a CO/Pt ratio of 0.65 indicating that 65% of the total platinum atoms were exposed.

A 9.0 wt % platinum-on-alumina (Alon-C) was prepared by impregnation with chloroplatinic acid. It was dried and reduced by the standard procedure. Portions of this 9.0% platinum/alumina were used in a number of experiments. For each portion the reduction was carried out in the infrared cell. The CO/Pt ratios for the various portions ranged between 0.3 and 0.4. Thus, the platinum was not as highly dispersed as in the case of the platinum-on-silica sample. One portion of the 9.0% platinum/alumina was diluted with alumina to a platinum concentration of 2.2% in order to provide a sample suitable for observation of the peaks of the strong bands in the spectrum of chemisorbed carbon monoxide.

3. RESULTS AND DISCUSSION

3.1 Effects of preadsorbed oxygen

3.1a Hydrogen on platinum/alumina: Fig. 1 presents evidence for the enhancement of platinum-hydrogen and platinum-deuterium bands by preadsorption of oxygen on the 9 wt % platinum/alumina. In fig. 1 there is one series of spectra based on hydrogen experiments and one series based on deuterium experiments. In both series the dashed lines represent the background which has been corrected for the frequency independent changes which occur when oxygen is adsorbed. The initial background was measured after the sample had been subjected to the standard reduction and evacuation and then cooled to 25°C. The negative band in the background near 2060 cm^{-1} is due to impurity carbon monoxide on the reference beam sample. The same sample disc (weight 102 mg after reduction) was used for the hydrogen and deuterium experiments. The hydrogen and deuterium pressures were 760 Torr.

Fig. 1 a, Hydrogen and Deuterium Chemisorbed on Reduced Platinum;

b, on Oxygen Treated Platinum

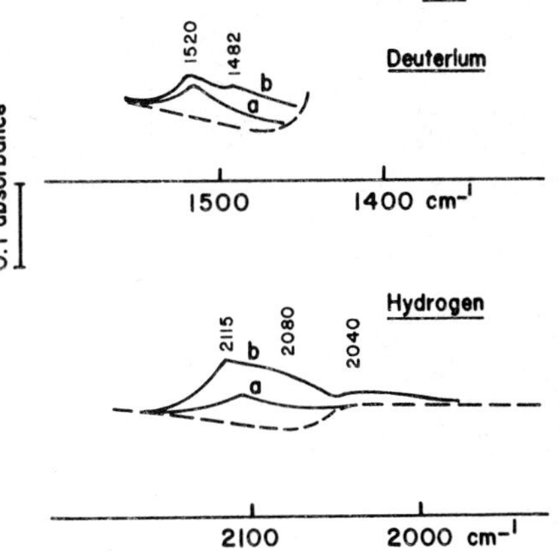

In the hydrogen series the area between spectrum a and the background represents the infrared absorption due to hydrogen chemisorbed on reduced platinum. Spectrum b was observed after hydrogen was admitted to the sample which had been pretreated with oxygen. The total area between spectrum b and the background represents hydrogen chemisorbed on preoxidized platinum. The oxygen pretreatment was at room temperature with an oxygen pressure of 300 Torr. Constant weight was attained in less than 10 minutes. The oxygen uptake was 0.26 mg which corresponds to an O/Pt ratio of 0.35. This ratio is approximately the same as the CO/Pt ratios measured for this sample. Room temperature oxygen uptakes showed no difference attributable to pressure in the range of 35 to 300 Torr.

The deuterium series in fig. 1 follows the general pattern of the hydrogen series. The oxygen exposure was started at room temperature and raised to 200°C with a pressure of 300 Torr. After several hours the weight uptake leveled off at 0.55 mg. This is equivalent to an O/Pt ratio of 0.74. It is evident that oxygen pretreatment also produces an enhancement in the platinum-deuterium bands. In the deuterium experiments the total area under spectrum b is approximately twice as large as the area under spectrum a.

In previous work from this laboratory[2)] platinum-deuterium bands were observed at 1520 and 1485 cm^{-1}. Iridium-deuterium bands[9)] have been found at 1520 and 1490 cm^{-1}. The bands in spectrum b are near these frequencies although the lower frequency band is more poorly defined than in the previous work. Eley reported platinum-deuterium bands at 1465 and 1460 cm^{-1} (both $\pm$ 5 cm^{-1}). For the high frequency Pt-D band the difference between 1520 cm^{-1} and Eley's 1465 cm^{-1} is larger than would be expected from ordinary experimental variations. There is also a significant difference in band shape since Eley's platinum-deuterium bands must be extremely narrow to allow resolution into two well defined peaks.

Eley and his co-workers emphasized the 2120 cm^{-1} band in their discussion of the effect of preadsorbed oxygen and emphasized the 2040 cm^{-1} band in their studies of the adsorption of water. The spectra in fig. 1 indicate that the enhancement applies to both the high and low frequency bands. Since these bands are not well resolved, it is not feasible to consider them separately. The significant finding illustrated by fig. 1 is the confirmation of Eley's observation of platinum-hydrogen band enhancement caused by oxygen treatment.

3.1b Carbon monoxide on platinum/alumina: Although the spectra of fig. 1 show the enhancement caused by preadsorbed oxygen, the mechanism of this effect is not established by those experiments. Use of carbon monoxide provides information on whether the supplementary adsorption is on oxide patches or on an increased surface area of metallic platinum. This advantage is gained from the fact that the carbon-oxygen stretching frequency of carbon monoxide adsorbed on cations of metal oxides is generally found above 2100 cm^{-1}. A discussion of the adsorption on oxidized silica-supported platinum has been presented by Heyne and Tompkins[10)] who attribute bands at 2120 and 2170 cm^{-1} to CO on platinum oxide. An additional advantage of using carbon monoxide is that the gravimetric methods can provide an independent measure of the amount adsorbed. This eliminates questions of spurious results due to changes in the specific extinction coefficient of the adsorbed gas.

Fig. 2 shows how preadsorption of oxygen affects carbon monoxide chemisorption on the 2.2 wt % platinum/alumina sample (disc weight 107 mg). Spectrum A was observed after carbon monoxide (pressure one Torr) was chemisorbed on the reduced sample and Spectrum B was observed after carbon monoxide was chemisorbed after preadsorption of oxygen.

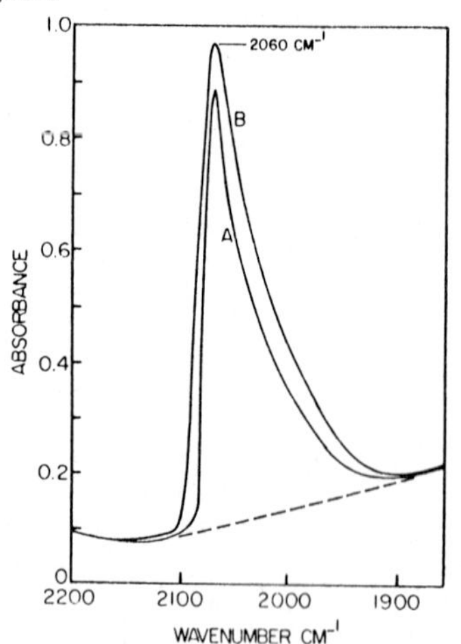

Fig. 2 A, Carbon Monoxide Chemisorbed on Reduced Platinum;

B, on Oxygen Treated Platinum

The oxygen exposure at 200°C gave an O/Pt ratio of 0.68. After oxygen treatment the sample was evacuated, cooled to room temperature, exposed to hydrogen and reevacuated prior to adsorption of carbon monoxide. Although carbon monoxide converts preadsorbed oxygen to carbon dioxide at room temperature, the final stages of the reaction are slow. Moreover, the gravimetric measurements are difficult to interpret because some of the carbon dioxide is desorbed into the gaseous phase and some remains adsorbed on the alumina.

Since the carbon monoxide of Spectrum B had been added to the platinum after evacuation of hydrogen at room temperature, the standard reduction-evacuation procedure was modified for Spectrum A by cooling in hydrogen before evacuation so both could be observed under comparable conditions. The weights of adsorbed carbon monoxide were 0.13 mg and 0.19 mg for Spectra A and B, respectively. These weights correspond to CO/Pt ratios of 0.39 and 0.58. The integrated areas for the bands in Figure 2 are 44 and 63 cm^{-1}.

Spectra A and B have bands in the 1750-1850 cm^{-1} region which are not included in the integrated areas. Since they are small and broad, they are difficult to integrate. The 1750-1850 cm^{-1} bands may be seen in fig. 3.

The gravimetric measurements indicate that oxygen pretreatment produces an enhancement of 49% in carbon monoxide adsorption. The integrated intensities indicate an enhancement of 43%. The spectra of fig. 2 supplement the gravimetric measurements of the amount of carbon monoxide adsorbed. In addition, the spectra show that there is no band above 2100 cm^{-1} on the oxygen treated sample. This indicates that none of the carbon monoxide is bonded to Pt^{+4} cations of the type which Eley visualizes as adsorption sites on oxide patches.

3.1c <u>Hydrogen on platinum/silica</u>: Preadsorption of oxygen on 9 wt % platinum/silica did not produce a detectable enhancement of hydrogen or carbon monoxide adsorption. The work with platinum/silica was carried out prior to the platinum/alumina studies described above. The failure to detect enhancement appeared to be in conflict with Eley's experimental observations for platinum/silica. However, the conflict is probably not real. Our platinum/silica sample was initially highly dispersed with a CO/Pt ratio of 0.65. Thus, a further dispersion would be difficult to achieve. It is noteworthy that the platinum rearrangement mechanism is consistent with the failure to detect enhancement for the highly dispersed platinum/silica while the oxide patch hypothesis is not.

3.2 Study of the 2040 cm^{-1} band

3.2a <u>Water on platinum/alumina</u>: In fig. 1 the lower frequency platinum-hydrogen band appears as a broad absorption below 2080 cm^{-1}. Eley et al observed a relatively narrow lower frequency band at 2040 cm^{-1}. The 2040 cm^{-1} was observed after addition of water or after long waiting periods. This band was attributed to a platinum-hydrogen species produced by dissociative chemisorption of water. The 2120 cm^{-1} band did not appear under these conditions. Eley et al also observed a moderately strong band at 2040 cm^{-1} which they attributed to impurities but the nature of the impurity was not determined. However, arguments were advanced to deny the possibility that a band at 2040 cm^{-1} could be due to chemisorbed carbon monoxide. These arguments were based mainly on experiments which indicated that deliberate addition of carbon monoxide produced a band at 2060

cm^{-1} rather than at 2040 cm^{-1}.

Because of the inherent interest of the water dissociation concept, attempts were made to confirm Eley's findings by adding water to our 9 wt % platinum/alumina sample. No bands were observed which could be attributed to a platinum-hydrogen species, even though small bands (absorbance 0.03-0.05) were detectable in the 2040-2060 cm^{-1} region. Similar bands were observed after addition of D_2O. Dissociative chemisorption would be expected to produce platinum-deuterium bands near 1500 cm^{-1} and no bands were observed in this region. The D_2O experiments involved a sequence in which the sample was pre-deuterized to avoid misleading results due to dilution of the deuterium by exchange with hydrogen from surface hydroxyls.

The similarity of results observed after addition of water and D_2O suggests that the small bands in the 2040-2060 cm^{-1} region are not produced by dissociative chemisorption of water. Instead, the bands appear to be due to impurity carbon monoxide which is either introduced with the water or displaced from the cell walls. The absorbances of 0.03-0.05 would correspond to a carbon monoxide surface coverage in the range of 1-2%.

The 2040-2060 cm^{-1} bands may be obtained even with the most carefully purified water or D_2O. In these cases desorption from the cell walls is probably the source of the impurity carbon monoxide. The cell can be cleansed of carbon monoxide by a series of water addition experiments. After each cycle the impurity carbon monoxide bands in the 2040-2060 cm^{-1} region are smaller until after the fourth cycle they are barely detectable.

3.2b Carbon monoxide as an impurity: In the preceding discussion the presence of impurity carbon monoxide was attributed to desorption from the cell walls. Carbon monoxide may also be introduced with the water. The water and D_2O discussed above were purified by bubbling a stream of argon through the liquids as they were heated to boiling. The spectra of fig. 3 were observed after improperly purified water was admitted to the 9 wt % platinum/alumina. This was distilled water which had been put through a series of freeze-pump-thaw cycles using liquid nitrogen as a freezing agent. The bands at 2050 cm^{-1} and near 1800 cm^{-1} are attributed to carbon monoxide chemisorbed in the linear, Pt-C≡O, and bridged forms $\substack{Pt \\ Pt}\!>\!C{=}O$. The band at 1630 cm^{-1} is due to the HOH bending vibration. The OH stretching region near 3500 cm^{-1} was not studied because of interference from strong AlOH bands. Spectrum C of fig. 3 was observed after addition of 2.6 mg of water. The 2050 cm^{-1} band of Spectrum C corresponds to 0.05 mg of chemisorbed carbon monoxide or a surface coverage of 10-15 per cent. It is likely that the freeze-pump-thaw purification procedure is unsatisfactory because it does not eliminate carbon dioxide which is known to dissociate to carbon monoxide on platinum[11]).

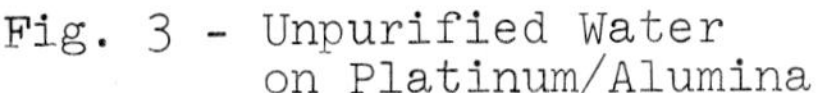

Fig. 3 - Unpurified Water on Platinum/Alumina

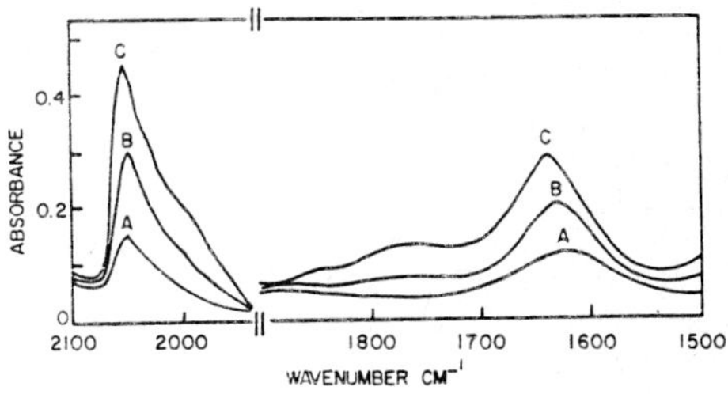

Impurity carbon monoxide is a serious problem because its specific extinction coefficient is large relative to that of chemisorbed hydrogen. The integrated area of the bands in fig. 2, and related gravimetric data, indicate an extinction coefficient of $16x10^{-17}$ cm/molecule for chemisorbed carbon monoxide. It is difficult to obtain precise measurements of the amount of adsorbed hydrogen in apparatus designed for infrared work. Gravimetric and volumetric techniques are limited by the small weight and the pressure sensitivity of the adsorption. Extinction coefficients for chemisorbed hydrogen, measured here and in previous work,[2)] range from 0.15 to $0.50x10^{-17}$ cm/atom. These values are not adequate to answer questions of current interest as to whether there are one or two hydrogens adsorbed on each surface platinum. However, the small values account for low intensity of bands attributable to chemisorbed hydrogen compared to those observed for chemisorbed carbon monoxide.

Eley et al exclude carbon monoxide as a source of their 2040 cm^{-1} band because the frequency is 20 cm^{-1} lower than the band they observed after deliberate adsorption of carbon monoxide. However, it has been shown that the frequency of the carbon-oxygen band varies with surface coverage on platinum[12)]. Although several variables may affect the carbon-oxygen frequency, it is commonly observed to shift from 2040 to 2060 cm^{-1} as the surface coverage is increased. This shift can be explained on the basis of dipole-dipole interactions between neighboring adsorbate molecules.

It has also been shown[12)] that the change of frequency with coverage may be obscured by stratification when carbon monoxide is adsorbed at room temperature. The sample discs contain the equivalent of several thousand layers of platinum particles. When small doses of gas are added an immobile adsorption at low temperature allows particles near the outer surfaces of the disc to be completely covered with carbon monoxide while particles deep in the interior of the disc remain bare. Subsequent doses merely adsorb on strata of particles further inside the disc. Thus, dosing at room temperature produces a band at 2060 cm^{-1} which does not change in frequency as more carbon monoxide is added.

Stratification may be avoided by adding the carbon monoxide at higher temperatures. Uniform distribution is attained rapidly at 200°C. When the amount of chemisorbed carbon monoxide is small compared to the total adsorptive capacity of the sample, the redistribution of the carbon monoxide is detected by a change in frequency from 2060 cm^{-1} to 2040 cm^{-1}.

In discussion of the experiments on the adsorption of satisfactorily purified water the impurity band frequency was referred to as 2040-2060 cm^{-1}. This range was specified because a number of experiments were conducted and there were examples of the band being observed at 2040 cm^{-1}, at 2060 cm^{-1}, and at intermediate frequencies. It is reasonable to assume that the different frequencies merely reflect the temperature to which the sample had been heated by the exothermic adsorption of water.

Attempts to detect a dissociative chemisorption of water on platinum/silica yielded similar negative results. Since the present work does not confirm Eley's hypothesis concerning the origin of the 2040 cm^{-1} band, it appears that further work will be needed before the concept of dissociative chemisorption of water on platinum can be accepted.

3.3 Reducibility of supported platinum

3.3a Calibration for the 1630 cm^{-1} band: This concept of hydrogen adsorption on patches of platinum oxide implies that the surface oxide would not be reduced by hydrogen at room temperature. There is evidence that the reduction easily goes to completion in the case of unsupported platinum[4]). However, the reducibility is less well defined for supported platinum. It is often observed that metals are more difficult to reduce when they are supported on alumina or silica. Thus a direct measure of reducibility is needed when the question of reducibility of any supported metal is being considered.

For unsupported metal the weight change due to the removal of oxygen may be used to measure the extent of reduction. However, this method is not satisfactory for supported metals because the water formed by the reduction may be re-adsorbed on the support. In the present work the 1630 cm^{-1} hydrogen-oxygen-hydrogen band was used to measure the water produced by the reduction. This approach involves the assumption that all of the water would be adsorbed by the support. For present purposes, this is not a dangerous assumption because it could not lead to an erroneously high estimate of platinum reducibility.

Although dissociative chemisorption of water in platinum was not detected in the present work, it is reasonable to expect dissociation of water on the alumina support. This dissociation would produce surface hydroxyl groups, HOH + (Al_2O) $\longrightarrow$ 2 AlOH. Since these surface hydroxyls do not contribute to the 1630 cm^{-1} band, it is necessary to have a calibration curve for the adsorbance of the 1630 cm^{-1} band as a function of the weight of water taken up by the alumina. Fig. 4 represents this calibration curve for the 9 wt % platinum/alumina sample. The points represented by squares were obtained by adding water at room temperature after the 9% platinum/alumina had been exposed to oxygen. The points represented by circles were obtained by adding water to a sample after it had been reduced by the standard procedure. The change in slope of the calibration curve at low coverages is attributable to dissociation of water on the alumina as discussed above.

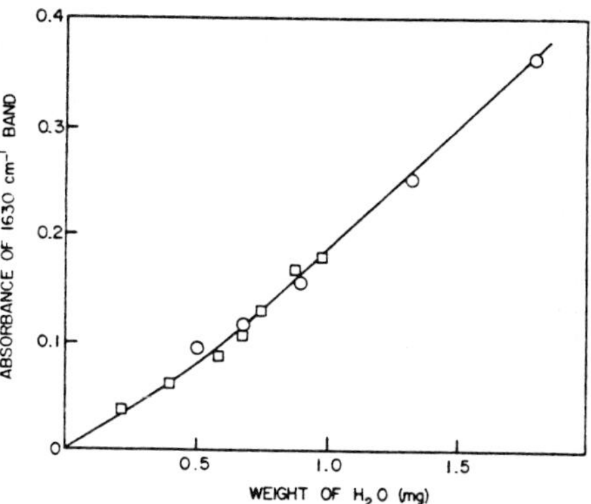

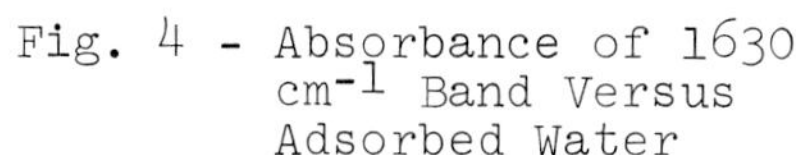
Fig. 4 - Absorbance of 1630 cm^{-1} Band Versus Adsorbed Water

3.3b <u>Removal of oxygen from platinum</u>: The calibration data of fig. 4 provide a means to measure the amount of water produced when oxygen-covered platinum is exposed to hydrogen. Table 1 compares the amount of water produced with that expected from complete conversion of the oxide. The first column indicates the weight of oxygen taken up by the sample. The 0.38 mg value in the first row resulted from exposure to oxygen at room temperature. The two higher values were obtained after exposure at 200°C. The second column gives the weight of water calculated on the basis of conversion of all of the preadsorbed oxygen. The third column indicates the absorbance of the 1630 cm^{-1} band observed after exposure to hydrogen. The fourth column shows the amount produced based on the absorbance of the 1630 cm^{-1} band and the calibration curve, fig. 4.

Table 1

Reduction of Preadsorbed Oxygen at Room Temperature

O_2 mg		H_2O mg Calculated	1630 cm^{-1} Absorbance	H_2O mg Observed
0.38		0.43	0.055	0.40
0.74		0.83	0.160	0.90
0.74		0.83	0.120	0.70
	Total	2.09		2.00

On the basis of the comparison of the fourth and second columns of Table 1, we conclude that the reduction goes at least to 90% completion, and perhaps is 100% complete, under experimental conditions equivalent to those of fig. 1. This implies that oxide patches do not play a significant role in the enhancement of hydrogen or carbon monoxide adsorption.

The carbon monoxide chemisorption and the reducibility studies show that the enhanced hydrogen adsorption of fig. 1 can best be attributed to a rearrangement of the platinum. This rearrangement may produce a greater exposure of crystal faces favorable to chemisorption of hydrogen or carbon monoxide. Another possibility is a larger effective area which is the result of a smaller particle size. It is not possible to differentiate between these possibilities on the basis of the evidence presented here. However, the question of the mechanism of the rearrangement is amenable to physical and chemical study and merits further attention.

REFERENCES

1) D. D. Eley, D. M. Moran and C. H. Rochester, Trans. Far. Soc. 64 (1968), 2168.
2) W. A. Pliskin and R. P. Eischens, Z physik Chem. 24 (1960), 11.
3) J. G. Aston, E. S. J. Tomezsko and R. A. Fisher, J. Am. Chem. Soc. 86 (1964), 2097.
4) J. E. Benson and M. Boudart, J. Catalysis 4 (1965), 704.
5) D. E. Mears and R. C. Hansford, J. Catalysis 9 (1967), 125.
6) G. R. Wilson and W. K. Hall, J. Catalysis 17 (1970), 190.
7) F. P. Mertens and R. P. Eischens "The Structure and Chemistry of Solid Surfaces" edited by G. A. Somorjai, John Wiley and Sons Inc. (1969).
8) H. A. Benesi, R. M. Curtis, and H. P. Studes, J. Catalysis 10 (1968), 328.
9) F. Bozon-Verduraz, J-P Contour, and G. Pannetier, CR Acad. Sc. Paris 269C (1969), 1436.
10) H. Heyne and F. C. Tompkins, Trans. Far. Soc. 63 (1967), 1274.
11) R. P. Eischens and W. A. Pliskin, Proc. Second Int. Congress on Catalysis (1961) 789.
12) R. M. Hammaker, S. A. Francis and R. P. Eischens, Spectrochemica Acta 21 (1965), 1295.

Present address of D.J.D. - State University of New York, Buffalo, New York

DISCUSSION

R. C. HANSFORD

The hydrogen titration of oxygen chemisorbed on platinum, of course, involves the oxidation and reduction of the surface at room temperature. In some cases, where subsequent measurements have been made on the same sample, we have also observed an enhancement of hydrogen adsorption. We employ a flow system using nominally 2% H_2 in argon for the hydrogen adsorption and titration measurements and 1% O_2 in helium for the oxygen adsorption measurement, all at atmospheric pressure. A catalyst consisting of 1.86% Pt on high area silica gel showed a H/Pt (total) ratio of 0.81 after reduction at 600° for 125 minutes and outgassing at 400° for 120 minutes. After a hydrogen titration of adsorbed oxygen, the sample was mildly reduced for 40 minutes at 200° and outgassed for 155 minutes at 400°. Remeasurement of hydrogen adsorption at room temperature gave a H/Pt (total) ratio of 1.01. A similar result was found for 0.99% Pt on eta-alumina, the H/Pt (total) ratio increasing from 0.52 to 0.65 with the same pretreatment except for reduction at 400° after the titration.

Since the hydrogen adsorption was measured at an effective hydrogen pressure of about 15 torr at room temperature, we believe a H/Pt (total) ratio of unity represents a dispersion of about 50%. Thus, the effect, while not large, is significant even with fairly well dispersed platinum. There may be other reasons for the enhancement besides increased dispersion of the metal, but the latter may well occur due to alternate oxidation and reduction under very mild conditions.

J. A. BETT

The authors' conclusion that neither of the infrared bands associated with hydrogen chemisorption can be ascribed to adsorption of hydrogen on the surface platinum oxide is similar to that arrived at by Tsuchiya et al.[1] for the corresponding chemisorbed species observed in Temperature Programmed Desorption. In added support of this conclusion the electrochemical measurement of hydrogen adsorption on platinum by potentiodynamic sweep techniques also shows strongly and weakly chemisorbed hydrogen at potentials where no surface platinum oxide can exist.

Both TPD and the potentiodynamic sweep technique show the weakly adsorbed hydrogen species to be predominant on platinum black. The infrared data also indicate that the weakly bonded species is predominant.

It therefore appears likely that the relationship between the strongly and weakly adsorbed species and the various crystal faces on platinum which has been demonstrated elastrochemically by Will,[2] also applies in the gas phase measurements.

1) S. Tsuchiya, Y. Amenomiya and R. J. Cvetanović, T. Cat. 19, 245 (1970).
2) F. Will, J. Electrochem. Soc. 112, 451 (1965).

D. D. ELEY

Dr. Rochester and I consider there are facts in our paper which cannot be explained by the present communication. Thus in our paper, p. 2177, we stated "when deuterium was added to a sample which had been fully deuterated (as shown by negligible absorption around 3700 cm^{-1}) no band at 2040 cm^{-1} was observed." Therefore the 2040 cm^{-1} band is due to Pt-H and not CO on Pt. On p. 2179 "In our system neither addition of hydrogen to samples treated with CO, nor addition of hydrogen mixed with a small amount of CO have the main CO absorption as low as 2040 cm^{-1}." We observed the band at 2060-2090 cm^{-1} for CO adsorbed at low coverage in the presence of excess hydrogen. Thus, Eischen's comment on the middle of p. 2177 is not a valid criticism of our work. Also, our results on the treatment of a fully deuterated surface with hydrogen (top paragraph, p. 2179) are incompatible with the 2040 band being CO.

R. P. EISCHENS

The difficulties due to impurity carbon monoxide are not always encountered. The extent of the problem will depend on the prior history of the experimental system. I do not believe that "clean" experiments can be extrapolated to rule out the possibility of impurity carbon monoxide when it is suspected.

The question of the 2040 cm^{-1} versus 2060 cm^{-1} frequency has been discussed in our Section 3.2b. The 2060 cm^{-1} band, indicative of high coverage, is often observed after limited dosing at room temperature. This is due to stratification whereby metal particles near the outer edges of the disk attain high coverage even though the overall coverage is low.

If I correctly interpret the last sentence of Professor Eley's statement, he implies that the Freundlich relationship between hydrogen pressure and band intensity is evidence that the 2040 cm^{-1} band is due to chemisorbed hydrogen. It is not clear to me whether this relationship refers to a band superimposed on the 2040 cm^{-1} "impurity" band which he discusses on page 2176. Despite this uncertainty, I would accept the premise that chemisorbed hydrogen may produce a 2040 cm^{-1} band, indeed, Dr. Pliskin and I observed this band in our previous work. In Eley's paper the 2040 cm^{-1} band is presented mainly as evidence for the dissociative chemisorption of water. Our question is related to whether his 2040 cm^{-1} band has been correctly interpreted in that context.

M. BOUDART

That water leaves the surface of Pt black which has been exposed to O_2 and then to H_2 both at room temperature, has been demonstrated by M. A. Vannice et al. (J. Cat., 1971) in a gravimetric study. More recently, in his Ph.D. Dissertation (Stanford, 1971), J. C. Schlatter has observed that the role of hydrogenation of ethylene on a Pt/SiO_2 sample after a low temperature reduction of oxidized samples, was higher than that measured under identical conditions on the same sample reduced at higher temperatures. This effect, which was of the same order of magnitude as the increase of surface area reported in your paper, was attributed to a surface reconstruction of the Pt surface by oxygen chemisorption, this reconstruction being erased by high temperature treatment in hydrogen.

R. C. PITKETHLY

The use of the phrase "rearrangement of platinum" to account for increased absorption is interesting. Rearrangement of the surface of platinum single crystal faces has been observed by LEED during treatment with oxygen. Without going into detail, one can say that the (100) face changes, and additional diffraction spots appear giving the so-called (5x1) structure which can be explained by a rearrangement of the top layer of metal to a close-packed hexagonal (111)-like layer. The greater compactness of the (111) layer can only account for ca. 14% increase compared with the values about 50% reported in the paper.

However, dispersion of platinum into smaller crystallites on supported catalysts and reorganization of platinum black (as just mentioned by Professor Boudart) by oxygen treatment are known.

Has Dr. Eischens evidence for a surface rearrangement or does he favor the more robust second process?

R. P. EISCHENS

We did not intend to imply a specific mechanism for the retention of high surface area after the low temperature reduction. The preservation of extremely small particles is one possibility. A second possibility is the preservation of unstable shapes such as platelets or whiskers. I believe there is a need for a further physical study of this problem because simple measurement of adsorptive capacity can provide only limited information on the nature of the particles. The physical studies should be supplemented by catalytic work such as that described by Professor Boudart.

M. M. BHASIN

I would like to know if the H_2 chemisorption and H_2-titration experiments were done on a catalyst after CO desorption.

Have you analyzed for CO_2 in the desorbed gases? We have found that some disproportionation of CO to CO_2 occurs during desorption of CO. Such CO disproportionation can give rise to carbon on the surface and thus complicate the chemisorption results.

R. P. EISCHENS

In most cases the hydrogen experiments were done without prior deliberate exposure to CO. In cases where CO was known to be adsorbed it was removed by hydrogen reduction rather than by simple desorption. In general, we try to avoid exposure of the sample to CO at high temperatures. We did not analyze for CO_2 in the experiments described in the present paper. In other situations we did detect small amounts of CO_2 but the results were not sufficiently quantitative to distinguish between CO disproportionation and CO_2 in the normal background.

J. M. BASSET

You mention in your paper that there is danger of misleading results due to impurity carbon monoxide. In a recent work on Pt/Al_2O_3[1)] we have also observed the presence of CO as an impurity on platinum. The extent of the pollution by CO increases with the time of evacuation of the catalyst at room temperature (see Fig. 1). My question is the following: In the following figure, how can you eliminate the band due to CO in order to observe the band of irreversibly adsorbed hydrogen?

1) M. Primet, J. M. Basset, M. V. Mathieu and M. Prettre, J. of Catalysis under press (1973).

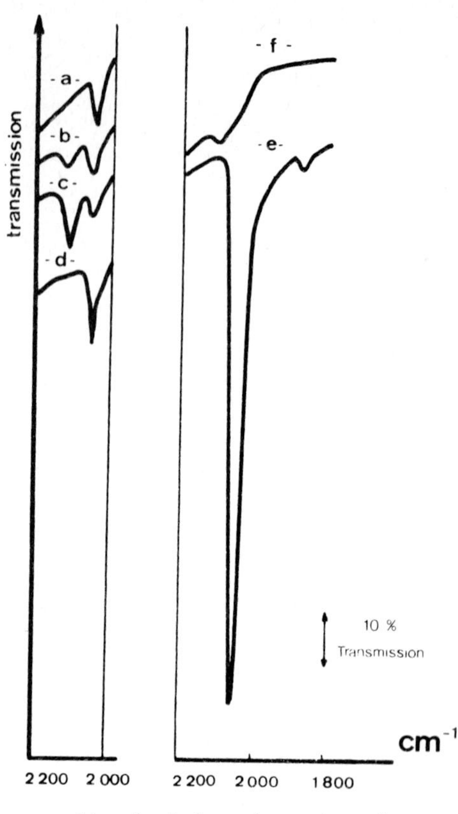

Fig. 1 Infrared spectra of Pt/Al_2O_3
a. initial solid (solid A),
b. under 20 torr of hydrogen at 30°C,
c. under 600 torr of hydrogen at 30°C,
d. after evacuation at room temperature for 10 mn,
e. after evacuation at room temperature for 13 h,
f. introduction of oxygen (pressure: 20 torr) at room temperature.

R. P. EISCHENS

The situation represented by the figure is worse than one would be forced to work with. However, if faced with spectrum a, we would rely on using deuterium instead of hydrogen. The platinum-deuterium bands are found in the 1500 cm^{-1} region where there is no significant interference from bands due to chemisorbed CO. An extension of your question would be whether the frequency of the platinum-deuterium bands could be affected by an interaction between chemisorbed carbon monoxide and chemisorbed deuterium. Dr. Pliskin and I made a study of this and found no evidence of such an effect when the coverage with CO was relatively small. Platinum-deuterium bands were found at 1520 and at 1485 cm^{-1} in cases where no chemisorbed carbon monoxide was present. Thus, it was not possible to interpret one of the bands as evidence of a shift caused by do-adsorption of CO.

Paper Number 22

AN I.R. AND VOLUMETRIC STUDY OF HYDROGEN CHEMISORPTION ON ZINC OXIDE

J. J. F. SCHOLTEN and A. van MONTFOORT
Central Laboratory, DSM, Geleen, and Laboratory of Chemical Technology, the Technological University of Delft, the Netherlands

Abstract: The infrared transmission of zinc oxide samples from various sources was studied. An explanation is given of the influence of hydrogen adsorption on the infrared transmission.

A new type of hydrogen adsorption on zinc oxide was discovered.

1. INTRODUCTION

Taylor and Strother[1)] were the first to discover two maxima in the isobars of hydrogen adsorbed on zinc oxide. The first maximum lies in the vicinity of 80 °C and was called type A adsorption. Eischens, Pliskin and Low[2)], using the infrared technique, demonstrated that type A produces two absorption bands in the vibrational part of the spectrum. The band at 1709 cm^{-1} was attributed to the formation of Zn-H with a covalent bond character at the surface, and the one at 3496 cm^{-1} to the oxygen-hydrogen stretching vibration of OH-groups. As the ratio of their intensities remains constant during adsorption and desorption of hydrogen, it was concluded that these bands are simultaneously formed by the dissociative adsorption of a hydrogen molecule.

According to Dent and Kokes[3)] we are dealing with a type A I adsorption, which is characterized by fastness and high reversibility and is associated with the occurrence of the I.R. bands mentioned above. Type A II, on the other hand, is partly rapid, partly slow at room temperature, and reversible only at temperatures above 150 °C.

It is the purpose of the present study to further clarify the complicated way in which hydrogen is adsorbed on zinc oxide, and to discover the location of the various adsorption sites, by combining observations made in the infrared with measurements of adsorption isotherms.

2. EXPERIMENTAL

2.1. Materials

Hydrogen and deuterium were purified by diffusion through a palladium thimble. Oxygen was purified by condensation at -196 °C, and subsequent distillation. The zinc oxide samples were obtained from various sources. Sample no. 1 came from B.D.H., England; the mode of preparation is unknown. Samples no. 2 and 3 from the New Jersey Zinc Company, were prepared by burning zinc vapour in air. Sample no. 4 was made from a batch of $ZnC_2O_4 \cdot 2\ H_2O$. The zinc oxalate was decomposed in air for 21 h at

350 °C. Sample no. 5 was prepared after Derrough c.s., i.e. by decomposition of zinc hydroxide ammoniacate[4]. The data of the sample texture are gathered in table 1.

TABLE 1
Texture data of zinc oxide samples

Sample no.	Origin	Surface area (m^2/g)	Mean crystallite size $\bar{d}_{v.s.}$ (Å)	Mean crystallite size $\bar{d}_w$ (Å)	Analysis
1	B.D.H.	3.6	3050	5000	reagent grade C: 0.13 %
2	N.J. Zinc Comp. SP 500	2.8	3920	5000	C: 0.08 %
3	N.J. Zinc Comp. Kadox 25	9.4	1170	1000	C: 0.10 %
4	ex oxalate	22.0	500	240	C: 0.27 %
5	ex ammoniacate	11.0	1000	215	reagent grade C: 0.17 %

The BET surface areas were measured after the samples had been degassed for 2 hr at 350 °C. The mean-volume-surface diameter $\bar{d}_{v.s.}$ was calculated from the BET surface area, and the weight mean diameter $\bar{d}_w$ from Röntgen line broadening. Plotting by means of Klug and Alexander's method[6], $\beta \cos \theta$ vs $\sin \theta$, for six diffraction lines, where θ is the Bragg angle and β is the integral line breadth, we found that the line broadening observed in all five samples is due only to the smallness of the crystallite size. In no case did we find indications of lattice distortions in the samples.

2.2. Combined infrared and volumetric adsorption studies

Samples of zinc oxide were pressed to circular disks of 21 mm diameter, under a pressure of 19,000 psi in a stainless steel ring. The disk was centered in a cylindrical gas cell with rocksalt windows. The central part of the cell could be heated by Kanthal coils, while water jackets kept the windows at room temperature. The sample temperature was measured by means of a Pt-Pt/Rh thermocouple; the configuration was such that the junction formed a pressure contact with a flat platinum ring firmly attached to the zinc oxide disk. The cell communicated via tubular connections with a conventional gas-dosing system, and also with a flask which, placed in a furnace, contained 50 g of an auxiliary sample of zinc oxide to increase the amount of hydrogen adsorption, the amount of zinc oxide in the infrared cell being only a few hundred milligrams. By means of an electronic device the temperature of the auxiliary sample was kept equal to that of the disk within one degree centigrade. Pressures in the system were measured by means of a sensitive dibutylphthalate manometer.

Spectra of the disks were taken by means of a model GS 4 Grubb Parsons grating spectrophotometer.

The particle size distribution of the zinc oxides in the 1-50 micron range was determined by the photosedimentographic method of Fortuin and Prop[5]. In some experiments excitation of electrons from the donor levels to the conduction band was brought about by concentrating the light of a 125 Watt type HPL Philips U.V. lamp on the disk by means of an aluminium reflector. The U.V. lamp was mounted very close to the infrared source of the photospectrometer.

3. RESULTS AND DISCUSSION

3.1. The infrared spectrum of pressed disks of zinc oxide powders

There being no qualitative differences on this point between samples no. 1 to 5, we shall deal with sample no. 4 only. A pressed disk of sample no. 4 was heated in the infrared cell at 350 °C for two hrs, and after that for three hrs in 5 cm oxygen at 350 °C to ensure good transmission. After cooling to room temperature the infrared spectrum was measured. Low intensity stretching vibration bands of hydroxyl groups not removed by activation at 350 °C, were found at 3660 cm^{-1}, 3610 cm^{-1}, 3550 cm^{-1} and 3439 cm^{-1}. From an exchange experiment with D_2 between 25 °C and 100 °C, the total OH-coverage was found to be at most 20 %. In conformity with Bozon-Verduraz[7)], we found strong absorption bands in the 1600-1200 cm^{-1} and the 1100-800 cm^{-1} ranges. The absorption maxima at 1525 cm^{-1} and 1330 cm^{-1} may, according to Bozon-Verduraz[7)], be ascribed to a "carboxylate" species. We observed these bands in all samples investigated, and noted that their absorption intensity increased with the carbon content.

It is not at all easy to remove this contamination; after 3 hrs heating in vacuo at 350 °C and subsequent cooling to room temperature the bands disappeared, but upon heat treatment at 350 °C they became visible again. This shows that we are dealing with adsorbed carboxylate groups which, after decomposition at the surface, are replenished by diffusion from the bulk. Heating in vacuo at 450 °C for at least 120 hrs proved necessary to remove the carboxylate contamination; all bands in the 1600-1200 cm^{-1} range disappeared then and the carbon content of the sample dropped from 0.27 to 0.04 %.

In the 1100-800 cm^{-1} range bands are observed at 1080 cm^{-1}, 1030 cm^{-1}, 1020 cm^{-1}, 980 cm^{-1}, 875 cm^{-1}, 720 cm^{-1}, 700 cm^{-1} and 685 cm^{-1}. These occur in all ZnO spectra whatever the carbon content, pretreatment or origin of the sample. We therefore conclude in accordance with Bozon-Verduraz[7)], that these bands are due to the lattice vibrations of ZnO.

3.2. Factors determining the I.R. transmission of zinc oxide powder disks

3.2.1. Influences of texture

For adsorbed molecules to be detected, their number per unit of pellet-thickness must be high. If the sample has a low surface-concentration of adsorption sites this requirement is difficult to fulfil. This explains why in the detection of adsorbed hydrogen on zinc oxide (θ is only $\sim$ 0.1 %) use must be made of relatively thick pellets, thickness: $>$ 0.5 mm, which implies that the material must have a high specific transmission. Therefore, a careful study was made of the factors influencing this characteristic.

In fig. 1 the transmissions of our samples are given as a function of the wavelength (absorption bands have been omitted).

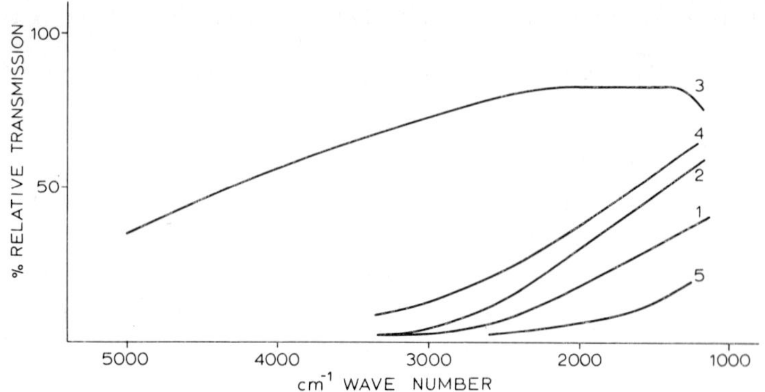

Fig. 1. Infrared transmission of various zinc oxide preparations. Pretreatment of all samples: 1 hr heating in oxygen at 350 °C, followed by 1 hr evacuation at room temp. Numbers refer to table 1.

The large differences in transmission after the same pretreatment can in no way be correlated with the texture data in table 1, the carbon content or with differences in lattice distortion; in fact, the latter were absent as appeared from our X-ray analysis. There exists, however, a correlation with the particle size distribution, as was established sedimentographically on suspensions of the powders in a saturated caprolactam solution.

TABLE 2
Photosedimentographic particle size analysis of various zinc oxide samples

fraction	percentage by weight				
μ	Nr. 1 B.D.H.	Nr. 2 SP 500	Nr. 3 Kadox 25	Nr. 4 Zn ox.	Nr. 5 ex ammoniacate
< 2	25	31	51	47	13
2-3	10	30	7	5	7
3-4	15	11	9	3	14
4-6	3	6	4	3	14
6-8	4	5	3	9	10
8-11	10	6	8	10	13
11-40	27	11	18	23	26

Comparison of the results presented in table 2 with the transmission data (fig. 1) shows that the higher the percentage of zinc oxide granules smaller than 2 microns diameter, so the higher the infrared transmission.

That the Kadox 25 sample lends itself better for detection of adsorbed hydrogen in the infrared than the other types of zinc oxide is due to the fact that it combines a relatively large surface area with a high percentage of particles with diameters much smaller than the infrared wavelength. It is clear now why only very weak absorption bands of adsorbed hydrogen could be detected in samples nr. 2 and nr. 4.

3.2.2. Influence of interstitial zinc

When zinc oxide is heated in vacuo, interstitial zinc atoms are formed:

$$ZnO \longrightarrow ZnO(Zn_i) + O_2\uparrow \qquad (1)$$

Kesavulu and Taylor[8] as well as Bozon-Verduraz c.s.[9] pointed out that, after heating at temperatures below 400 °C, the majority of interstitial zinc atoms appear to have concentrated in the surface. This is due to the fact that diffusion of zinc in zinc oxide calls for a very high activation energy (74 Kcal/mol Zn[10]). Owing to the presence of interstitial zinc, the surface exhibits n-type semiconductivity. The ionization energy of the interstitial zinc atoms is small; at room temperature most of them are singly ionized[11], while their electrons are thermally excited to the conduction band. In the case of bulk zinc oxide we have:

$$Zn_i \rightleftharpoons Zn_i^+ + ⓔ \qquad (2)$$

The gap between the Zn_i level and the conduction band corresponds to 0.9 Kcal/mole Zn. At higher temperatures double ionization can occur:

$$Zn_i^+ \rightleftharpoons Zn_i^{++} + ⓔ \qquad (3)$$

The Zn_i^+-level is situated 50 Kcals/mole Zn below the conduction band.

At the surface the semiconductivity must differ from that in the bulk. Little is known on this point: Heiland[10] found that the semiconductivity at the surface is less temperature dependent than that in the bulk.

Let us now turn to the influence ionization processes like (2) and/or (3) have on the infrared transmission of zinc oxide. Fig. 2 shows the relative transmission of a disk of sample nr. 3 (thickness: 0.5 mm), at an infrared wave number of 1176 cm^{-1}.

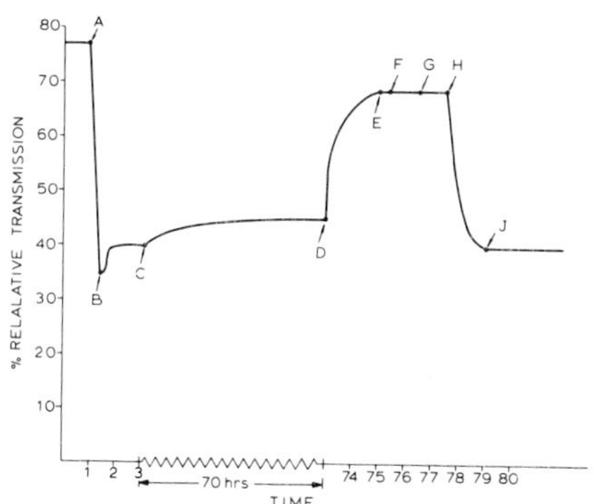

Fig. 2. Relative infrared transmission of a Kadox 25 disk, pre-evacuated overnight at 150 °C, as influenced by thermal excitation, cooling, and adsorption of hydrogen

Upon heating the disk (pre-evacuated overnight at 150 °C) in vacuo from 150 °C to 300 °C, interstitial Zn atoms are formed according to equation (1), while at the same time equilibria (2) and/or (3) are caused to shift to the right. This promotion of electrons from the Zn_i and/or Zn_i^+ donor levels to the conduction band enormously increases the electrical conductivity. Owing to this, the infrared absorption also increases, and this increase is reflected in a strong decrease of the infrared transmission (from point A to point B in fig. 2 and 3).

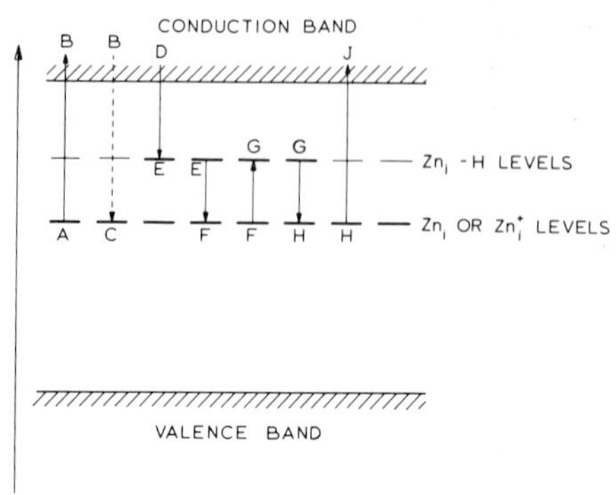

Fig. 3. Schematic representation of the energy-level diagram of a surface zinc oxide semiconductor. The energy-levels being unknown, the distance between them have been chosen arbitrarily.

This is in agreement with the results of D.G. Thomas[12)] who found that at any given wavelength, the infrared absorption of zinc oxide crystals varies directly with the electron concentration in the conduction band.

It appeared that the reverse process, i.e. capturing of promoted electrons from the conduction band in the donor levels, is extremely slow. On cooling from 300 °C to room temperature in vacuo (point B to C in fig. 2 and fig. 3) the samples showed only a 4 % change in relative transmission, while if it was kept at room temperature in vacuo for 70 hrs (point C to D in fig. 2) the change in transmission was hardly noticeable. Probably the kinetic energy of the electrons in the conduction band cannot easily be transformed into vibrational lattice energy owing to the absence (or very low concentration) of contaminations and lattice distortions; this accounts for the extremely slow change of the electron concentration in the conduction band (and the percentage of relative infrared transmission). As far as we know, such a very slow relaxation process has never been mentioned for zinc oxide. Perhaps it occurs only in samples with a low concentration of interstitial zinc in the surface, as is the case here. At higher temperatures, for instance at 300 °C (from point B to C), the rate of capture is higher than at room temperature, owing to the "thermal crystal imperfection" introduced under these conditions.

The following experiment shows that adsorption of hydrogen (mainly type A II hydrogen, see paragraph 3.3) may rapidly bring electrons to a level below the conduction band. Hydrogen was introduced into the infrared cell at room temperature and 6 cm Hg pressure (point D in fig. 2). This raised the relative transmission in about two hrs from 45 % to 70 % (point D to E in fig. 2 and 3); about one third of this increase took place within a few minutes.

This influence of hydrogen adsorption on transmission can be made understandable in the following way. The hydrogen molecules are adsorbed dissociatively, partly in a few minutes, partly slowly in a few hrs (see also the next paragraph). One hydrogen atom of each hydrogen molecule is bound to an interstitial Zn-atom; hence we have:

$$H + Zn_i^+ + \text{electron in the conductivity band} \longrightarrow Zn_i\text{-}H \quad (4)$$

This means that the electron is transported from the conductivity band to a new Zn_i-H donor level (as indicated in the energy-level diagram in fig. 3), and that the conductivity decreases (increase in transmission).

On desorption of the hydrogen, effected by pumping for ten minutes at room temperature, no change in relative transmission was observed (from point E to F in fig. 2 and 3). In this stage the Zn_i-H donor level disappeared, and the electrons returned into the Zn_i^+ or Zn_i donor level. The latter phenomenon does not influence the electron concentration in the conductivity band; hence, no change in transmission was observed. Since all electrons were in equilibrium now, readsorption of hydrogen at 6 cm Hg pressure and room temperature (point F to G in fig. 2 and 3) had no influence on the transmission, neither had desorption of hydrogen (point G to H). Finally, when the sample was heated in vacuo at 300 °C, the electrons became re-excited to the conduction band (from point H to point J in figs. 2 and 3). Fig. 3 also explains Kubokawa and Toyama's observation that hydrogen adsorption leaves the conductivity of zinc oxide unaffected.

Finally, we remark that other zinc oxide preparations exhibited an analogous behaviour as Kadox 25. The Kadox 25 experiment (fig. 2) was performed at a wave number of 1176 cm^{-1}, but the results obtained between 830 and 3330 cm^{-1} were qualitatively the same.

Promoting the electrons to the conduction band by U.V. radiation (see fig. 2, from point A to B) instead of by thermal excitation, did not make any difference as to the further course of the experiment.

3.3. Hydrogen adsorption on zinc oxide

3.3.1. Type A I-a hydrogen adsorption

In fig. 4 the amount of hydrogen adsorbed on freshly activated zinc oxide at room temperature is plotted as a function of pressure.

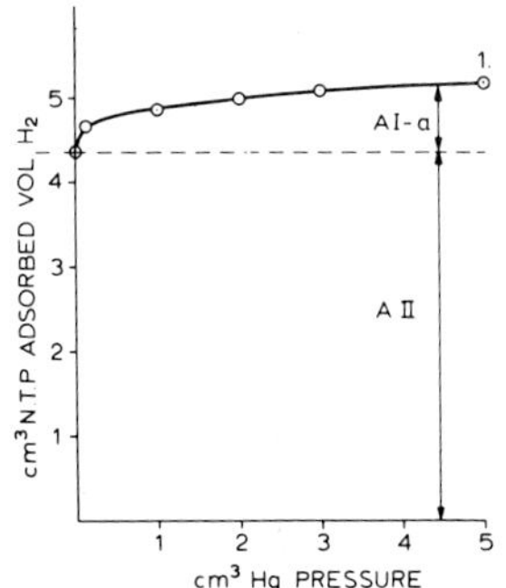

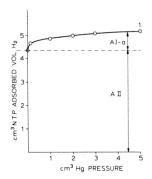

Fig. 4. Hydrogen adsorption isotherm at 25 °C on Kadox 25 zinc oxide upon heating in vacuo at 325 °C for 16 hrs. Total surface area 475 m^2

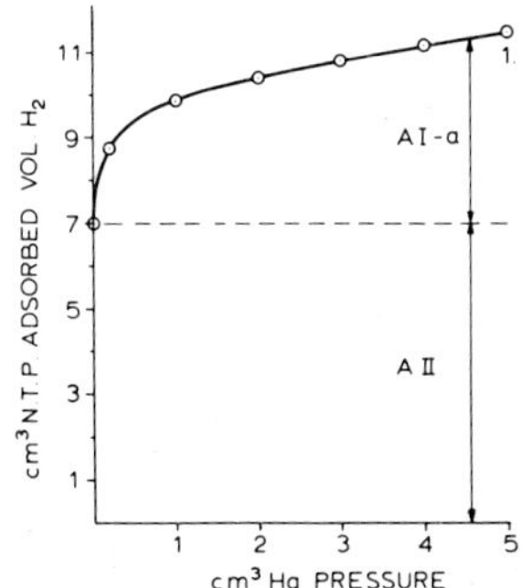

Fig. 5. Hydrogen adsorption isotherm measured at 25 °C on zinc oxide Kadox 25. Five fold cleaning with oxygen and heating in vacuo at 325 °C.

The Kadox 25 zinc oxide was activated in vacuo at 325 °C for 16 hrs.

Point 1 in the plot of fig. 4 denotes the situation after 20 hrs of equilibration. The part of the isotherm above the dotted line in fig. 4 (0.8 cm^3) is rapidly reversible. The isotherm section irreversible at room temperature (4.37 cm^3, below the dotted line in fig. 4) refers to type A II adsorption[3].

A second experiment was performed with a Kadox 25 sample which, to clean the surface thoroughly, was five times evacuated around 300 °C, heated in oxygen around 300 °C and re-evacuated around 300 °C. After that, it was heated in vacuo at 325 °C for one hr, whereupon the hydrogen adsorption as a function of pressure was measured at room temperature (see fig. 5). The overall behaviour was the same as that of the sample in fig. 4, but the degree of A I-a adsorption was five to six times, and that of A II adsorption nearly two times higher than in the first experiment.

Rapidly reversible adsorption on zinc oxide was previously reported i.a. by Dent and Kokes[3], but these investigators dealt with a mode of adsorption producing infrared absorption bands at 3510 cm^{-1} (OH) and at 1710 cm^{-1} (Zn-H). The reversible adsorption described in this paragraph is a new form of hydrogen adsorption which does not produce bands detectable in the infrared, and can be distinguished from the infrared-active form, only after relatively mild activation of the zinc oxide (pumping at 325 °C).

The nature of A I-a adsorption is not fully clear. It corresponds to a coverage θ, of about 0.03, at 25 °C and p = 5 cm Hg, which is much too high to be accounted for by physical adsorption. Since samples without interstitial zinc do not exhibit A I-a adsorption, it is likely that Zn_i sites in the surface are the adsorption sites. The absence of the I.R. absorption bands of dissociatively adsorbed hydrogen (or deuterium) may point to a molecular form of chemisorption.

3.3.2. Type A I-b hydrogen adsorption

After the experiment described in fig. 5, the catalyst was treated at 350 °C in oxygen for two hrs, and subsequently heated in vacuo at 350 °C for 16 hrs, so 25 °C above the temperature used in the experiments described in paragraph 3.3.1. Now 12 cm^3 of A II chemisorption was found, plus 8.7 cm^3 of rapidly reversible adsorption at 5 cm Hg pressure and room temperature.

The rapidly reversible adsorption was partly A I-a adsorption, partly a mode producing I.R. absorption bands at 2580 cm^{-1} and 1225 cm^{-1}: type A I-b. A quantitative separation between A I-a and A I-b adsorption could not be made.

Infrared-active A I-b adsorption was earlier reported by Eischens[2]. The good reversibility and the nature of the attendant infrared bands as regards their intensity ratio and pressure dependency reported by this author, could fully be confirmed.

Similarly as A I-a adsorption, A I-b adsorption is found only if Zn_i atoms are present in the sample. However, for A I-b to be observed the zinc oxide surface must be de-oxygenated more drastically, for instance by heating the sample for at least 16 hrs in vacuo at 350 °C or higher. Hence we have:

$$ZnO(Zn_i)_s + H_2 \longrightarrow (Zn_i)_sH + ZnOH \qquad (5)$$

where $(Zn_i)_s$ is an interstitial surface atom. This conclusion is in accordance with that of Kesavulu and Taylor[8].

3.3.3. Type A II hydrogen adsorption

This type of adsorption, incidentally indicated already in paragraphs 3.3.1 and 3.3.2, is described in detail by Dent and Kokes[3]. In agreement with these authors we noted that it can be distinguished from the A I-a and A I-b types, since:

1. it is not reversible at room temperature
2. it proceeds rapidly for a small part, whereas the main part consists of slow chemisorption.

TABLE 3

The extent of the three forms of A type hydrogen (c.q. deuterium) adsorption on 51 grams of zinc oxide Kadox 25, as influenced by various pretreatment conditions, at a pressure of 5 cm Hg

exp.nr.	pretreatment	A I-a cm^3 N.T.P.	A I-b cm^3 N.T.P.	A II (after 20 hrs equilibration) cm^3 N.T.P.
1	1 hr heated in vacuo at 150 °C	0	0	0
2	16 hrs heated in vacuo at 300 °C	0.1	0	1.1
3	16 hrs heated in vacuo at 325 °C	0.7	0	4.4
4	16 hrs heated in vacuo at 375 °C	8.2		6.9

From table 3 it is seen that the adsorption sites needed for type II (as also for A I-a and A I-b) arise upon introduction of interstitial zinc into zinc oxide by heating in vacuo. However, the Zn_i sites necessary for the A II-adsorption are formed already at relatively low temperatures, that is to say under conditions where no or insufficient Zn_i sites are available for type A I-a and A I-b adsorption to take place. If moreover, we take into account that Dent and Kokes[3] did not observe participation of type II in the hydrogenation of ethylene at room temperature, we arrive at the conclusion that:

a. the A II interstitial zinc sites are situated in sub-surface layers not accessible to ethylene
b. interstitial zinc in such sub-surface layers is formed more rapidly and at lower temperatures and, therefore, in a larger quantity than in the surface itself.

In one of our experiments with Kadox 25, we measured 3 cm^3 of rapid A II-adsorption, 9 cm^3 of slow A II-adsorption and 8 cm^3 of A I-a and A I-b adsorption. On pumping at room temperature, the infrared bands at 2580 and 1225 cm^{-1} rapidly disappeared, except for a small part (about 10 % of the total extinction) which persisted as long as the A II-adsorption. The wave-numbers of the slowly disappearing bands were at 2580 and 1231 cm^{-1}. These observations could be duplicated in a second experiment. We conclude that type A II-adsorption also gives rise to I.R. detectable bands, but that the extinction of these bands is clearly weaker (at about the same degree of adsorption) than for type A I-b.

In the following experiment we succeeded in demonstrating that A II-adsorption indeed gives rise to formation of deuteroxyl groups and, hence, is dissociative in nature. Hydrogen in a Kadox 25 sample was exchanged for D_2 at 5 cm Hg pressure until the concentrations of hydroxyl and deuteroxyl groups were at equilibrium. After activation overnight at 300 °C, D_2 was adsorbed at the same pressure and at room temperature (type A II-adsorption), under conditions where A I-b was practically negligible. After 20 hrs the total (OH + OD) extinction appeared to have increased by 9 %; after 66 hrs the increase was 25 %.

Hence, for type II adsorption we have:

$$ZnO(Zn_i)^* + H_2 \longrightarrow (Zn_i)^* - H + ZnOH \qquad (6)$$

where Zn_i^* is an interstitial zinc atom in a subsurface layer.

Some kinetic experiments (not fully reported here) point to zero activation energy of adsorption (in conformity with the results of Kesavulu and Taylor[8]), and to a first-order pressure dependency in hydrogen. The slowness of type A II-adsorption is probably due to the fact that hydrogen has to penetrate under the surface, perhaps as atomic hydrogen, via a very small concentration of A I-sites in the surface.

The weakness of the infrared bands of type A II with respect to those of type A I-b adsorption at the same concentration, is another argument in favour of the subsurface position of type A II.

REFERENCES

1) H.S. Taylor and C.O. Strother, J. Am. Chem. Soc. 56 (1934) 589

2) R.P. Eischens, W.A. Pliskin and M.J.D. Low, Journ. of Catalysis 1 (1962) 180

3) A.L. Dent and R.J. Kokes, Journ. of Physical Chemistry 73 (11) (1969) 3772

4) M. Derrough, R. Bardet, P. Turlier and Y. Trambouze, Comptes Rendus des journées d'études sur les solides finement divisés, Saclay, 27-28-29 september 1967, p. 47

5) J.M.H. Fortuin and J.M.G. Prop, Staub 22 (11) (1962) 469

6) H.P. Klug and L.E. Alexander, "X-ray diffraction procedures", N. York, John Wiley and Son Inc. 1954, chapter 9

7) F. Bozon-Verduraz, Journ. of Catalysis 18 (1970) 12-18

8) V. Kesavulu and H.A. Taylor, J. Phys. Chem. 64 (1960) 1124

9) F. Bozon-Verduraz, B. Arghiropoulos and S.J. Teichner, Bull. Soc. Chim. France 8 (1967) 2854

10) G. Heiland, E. Mollwo and F. Stöckmann, in "Solid State Physics", edited by F. Seitz and D. Turnbull, Vol. 8, p. 208, New York, Academic Press Inc., 1959

11) S. Roy Morrison, Advances in Catalysis VII (1955) 259, Academic Press Inc. Publishers, New York, N.Y.

12) D.G. Thomas, J. Phys. Chem. Solids 10 (1959) 47

Notes added by author at Meeting

1) Table a shows that an electronmicrosocpic study of particle size distributions furnishes a much better explanation of the superior transmission of Kadox 25 than do the photosedimentographic results in Table 2 of the paper.

Table a

Sample	Particle diam. (d), from E.M. (in microns)	Agglomeration	I.R. transmission at	
			$3000\ cm^{-1}$	$2000\ cm^{-1}$
No. 3, Kadox 25	$0.05 < d < 0.3$	No	70%	80%
No. 2, SP 500	$0.25 < d < 1$	No	5%	30%
No. 4, ex oxalate	$d \simeq 0.03$	Strong, in strings of 10 microns or more	15%	40%

This is because the photosedimentographic method itself leads to agglomeration of very small particles and hence does not give a fully clear picture of the real particle size distribution.

It is seen from Table a that with particle sizes below 0.25 micron an enormous improvement in transmission is obtained. Agglomeration, as with the oxalate sample, offsets this again (see Table a). It is not possible to improve the transmission further by grinding and/or sedimentation - fractionation, because this gives rise to agglomeration as appears from Table b.

Table b

Kadox 25	Transmission at $2000\ cm^{-1}$
1. Not ground	80%
2. Ground in ball mill in MEK, 20 hours	9%
3. After sedimentation-fractionation in water (particles < 1.5 microns)	14%

2) S. R. Morrison and P. H. Miller (J. Chem. Phys. 25, 1062, 1965) found that oxygen, like hydrogen is adsorbed on interstitial Zn atoms. The coverage is, however, much lower than for hydrogen, and at 325°C and 10 mm Hg it is only 0.4%. We, too, found only 0.53 cm^3 chemisorbed O_2 at 25°C and 540 mm Hg, as against 20 cm^3 H_2 under the same conditions. After poisoning our smaple at 300°C with oxygen, cooling to room temperature and evacuation, the I.R. bands of H_2 could be reproduced, and their intensity was only slightly lower. Hence, only part of the Zn_i was covered with oxygen.

3) Calculation shows that changes in I.R. transmission as depicted in Fig. 2 can largely be described by Eq. 2:

$$Zn_i \rightleftarrows Zn_i^+ + e$$

it being assumed that the energy-gap for the surface semiconductor equals 0.9 kcal/mole.

4) I.R. bands at 1231 and 2580 cm^{-1} given in this paper refer to Zn_i-D and Zn_i - OD.

DISCUSSION

A. L. DENT

In our experiments dealing with ZnH and OH on zinc oxide we activated our catalysts as follows: the catalyst was treated with dried oxygen at high temperatures, cooled in dried oxygen to room temperature and evacuated briefly at room temperature. The infrared band intensities under these conditions were at least as strong as for samples activated by high temperature degassing alone. In view of this, we do not believe that the sites for this kind of hydrogen chemisorption (i.e. designated as Type I) involve interstitial zinc or, if you prefer, non-stoichiometry due to oxygen deficiency.

In addition, we have observed that catalysts pretreated with hydrogen and subsequently exposed to oxygen do, indeed, show a poisoning effect. This poisoning appears to be due to the formation of water caused by the interaction between the adsorbed hydrogen and gaseous oxygen rather than to non-stoichiometry.

R. J. KOKES

I agree that it is difficult to remove interstitial zinc 100% by roasting zinc oxide in oxygen and cooling in oxygen. Nevertheless, it seems logical to conclude that such treatments at up to 500°C will substantially reduce the number of interstitial zinc atoms at the surface. The fact that such pretreatment does not reduce the intensity of the Zn-H and O-H bonds convinces me that the active sites for this adsorption do not involve interstitial zinc atoms.

J. J. F. SCHOLTEN

Removal of interstitial zinc by roasting and cooling in oxygen is indeed a slow process, even under the conditions used by Dent and Kokes (Adv. in Cat. 22, 9, 1972). Whereas a H_2 molecule can easily dissociate on the surface by binding the first H atom to Zn_i and the second H atom to an adjacent O-ion, this is much more difficult for oxygen which needs two Zn_i atoms for dissociative bonding.

This slow sorption of oxygen, which does not attain equilibrium below 450°C and, above this temperature, is further retarted owing to dissociation of ZnO, has been demonstrated by S. R. Morrison (Adv. in Cat. VII, 1955 and diss. Pensylvania 1953); see Fig. 1.

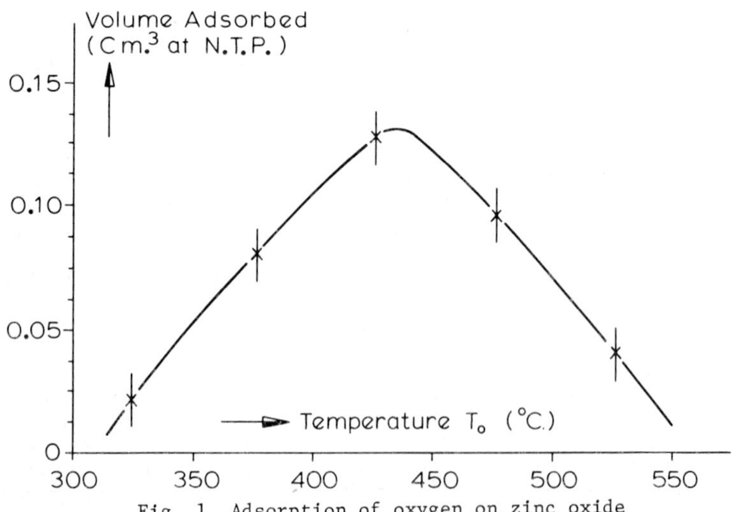

Fig. 1 Adsorption of oxygen on zinc oxide

The maximum O_2 uptake in Fig. 1 is 6.8 cm^3 (N.T.P.) per 1000 m^2ZnO at 450°C. This is only 6% of the H_2 type A adsorption found by us on Kadox 25 at room temperature!

Our view that Zn_i atoms are the adsorption sites for H_2 is in accordance with that of V. Kesavalu and H. A. Taylor (J. Phys. Chem. 64, 1124, 1960), who found that at 1 atm. O_2 pressure, temperatures above 500°C are necessary to achieve partial reoxidation of Zn_i. After this procedure type A hydrogen adsorption still persists.

Final and decisive evidence that Zn_i is involved in type A hydrogen adsorption is afforded by its influence on the semiconductivity of ZnO (Fig. 2 of this paper) and by the fact that the degree of adsorption is strongly dependent on the amount of Zn_i (Table 3).

Z. G. SZABÓ

In paragraph 3.2.1 the authors point to the fact that the number of the adsorbed molecules must be high, since, otherwise, they cannot be detected. To take transmission spectra, thick pellets of high transmission are required. From a catalytic point of view, only the surface atoms/or ions and the bonds between them are significant. Therefore I should like to ask if reflectance spectrometry could not afford further, and more direct information? I admit that the technique is not very easy and demands very great precision.

Secondly, I should like to remark that the investigations of this paper could be considerably refined, if the hydrogen to be adsorbed were added in extremely small, successive amounts (pulse method). It would be a real art to titrate the active centers.

J. J. F. SCHOLTEN

I agree with Professor Szabó's views.

J. HABER

I would like to ask Professor Scholten what is his opinion about the physical meaning of the distinction between surface and subsurface interstitial zinc ions. Trying to obtain zinc oxide surfaces with reporducible interstitial zinc concentrations and characterize them using oxygen absorption, we found that when consecutive runs of oxygen adsorption are carried out in the temperature range 100-800°C on samples outgassed at 800°C, the surface rapidly loses its adsorption activity; reproducible results being obtained only after 5-10 runs. A plot of the adsorption isobars shows that the maximum coverage in the pressure range 10^{-1} - 10^{-4} torr increases continuously with temperature up to 800°, without a maximum being observed in the isobars. These results indicate that:

1) Interstitial zinc ions in the surface layer are very mobile even at low temperatures, showing that incorporation of adsorbed oxygen takes place rapidly. This rapid reconstruction of the surface may render the surface and subsurface interstitial ions undistinguishable from each other.

2) Adsorption isobars cannot be used for characterizing the material, as the position of the maximum on the isobar critically depends on the pretreatment of the sample.

3) It seems to me that different modes of hydrogen adsorption might be related to the presence of different crystal planes in which zinc ions are situated in different arrangements.

J. J. F. SCHOLTEN

Three facts led me to the supposition that type A II adsorption occurs on subsurface interstitial zinc atoms. 1) type A II does not participate in the hydrogenation of ethylene. 2) A II is largely a slow type of adsorption, though the activation energy is zero, which points to a diffusion process. 3) the I.R. bands are weaker than those of type A I at the same concentration.

The possibility that the occurrence of various A types of H_2 adsorption is due to plane specificity cannot fully be excluded, however.

Dr. Haber's results with samples outgassed at 800°C may not be compared with the results reported in this paper, as the highest activation temperature was only 375°C. In the temperature range studied by us a rapid dissociative oxygen uptake has not been found, and there are no reasons to believe that Zn_i in the surface is very mobile, the activation energy for bulk migration being 75 Kcal/mole.

T. ONISHI

Grain size and surface area are important factors in the detection of adsorbed species by the infrared technique. In the case of Kadox 25, we observed the strong infrared bands of adsorbed hydrogen when the sample was heated at 350°C for 3 hrs and subsequently heated at 350°C for 1 hr in oxygen of 10^{-3} mm Hg. In other ZnO samples, very strong hydrogen adsorbed bands were also detected after a similar pretreatment.

These results show that the surface condition of zinc oxide is one of the most important factors in the detection of hydrogen adsorbed bands and that a drastically de-oxygenated zinc oxide surface does not always provide the best condition for detection of type A I-b adsorption.

J. J. F. SCHOLTEN

Table 3 of my paper clearly demonstrates the great influence the intensity of de-oxygenation has on the extent of the A types H_2 adsorption. As to the influence of oxygen, I refer to my reply to the comments by Dent and Kokes.

Paper Number 23

COMPARATIVE STUDY OF THE MECHANISM OF OLEFIN POLYMERIZATIONS IN THE PRESENCE OF ONE- AND TWO-COMPONENT SYSTEMS BASED ON $TiCl_2$

Yu. I. ERMAKOV, V. A. ZAKHAROV and G. D. BUKATOV
Institute of Catalysis, Novosibirsk, USSR

ABSTRACT: The polymerization of ethylene and propylene in the presence of two catalytic systems - $TiCl_2$ and $TiCl_2 + AlEt_2Cl$ - was studied. For the determination of the number of propagation centers and of the propagation rate constant (K_p) the method of stopping the polymerization by CO^{14} and CO_2^{14} was used. With $TiCl_2$ the number of propagation centers was a small fraction (< 0.1%) of the total number of titanium ions on the surface of catalyst. The value of K_p was about 1.10^4 l/mole·sec (75°) for ethylene polymerization and about 1.10^2 l/mole·sec (70°) for propylene polymerization, the values of K_p being approximately the same for isotactic and atactic additions. On adding $AlEt_2Cl$, the rate of polymerization in the presence of $TiCl_2$ decreases for both monomers due to the decrease of the number of propagation centers at a constant value of K_p. The isotacticity of propylene obtained increases in the presence of $AlEt_2Cl$ because of the reduced rate of atactic polymerization. These results suggest that the centers active in isotactic polymerization are formed without participation of an aluminum-organic compound.

1. INTRODUCTION

The investigation of Ziegler-Natta catalysts has played an important role in developing the general features of the mechanisms of processes catalyzed by transition metal ions. However there is no agreement yet concerning such important points of the Ziegler-Natta catalysts as the concentration and nature of active centers on a catalyst surface,[1-7] the values of propagation rate constants and the role of a metalorganic component in forming stereoregular polymer (hypothesis of "monometallic" and "bimetallic" stereospecific active centers).

The catalyst used in this work was $TiCl_2$ which causes olefin polymerization even in the absence of metalorganic compounds.[9,10]

2. EXPERIMENTAL PROCEDURE

$TiCl_2$ was obtained by the vacuum decomposition of $TiCl_3$ at 500°; its specific surface was about 25 m^2/g (by BET-method). The procedure of preparing reactants, performing polymerization and its stopping by introducing radioactive inhibitors was analogous to that used by us earlier.[11] The inhibitor was introduced in the amount necessary for the complete stopping of polymerization. Examples of kinetic curves obtained while introducing an inhibitor are given in Figure 1. The polymer was freed from the catalyst and radioactive impurities by treatment with a 1:1 mixture of methanol and isopropanol at 75° for an hour. The radioactivity of the polymer film was determined by a top-counter in case of C^{14} and by a flow 2π-

counter in case of H^3. The number of labelled polymer molecules (C^*) was calculated from the relation:

$$C^* = \frac{A\ G}{a\ Q} \text{ mole/mole}_{TiCl_2},$$

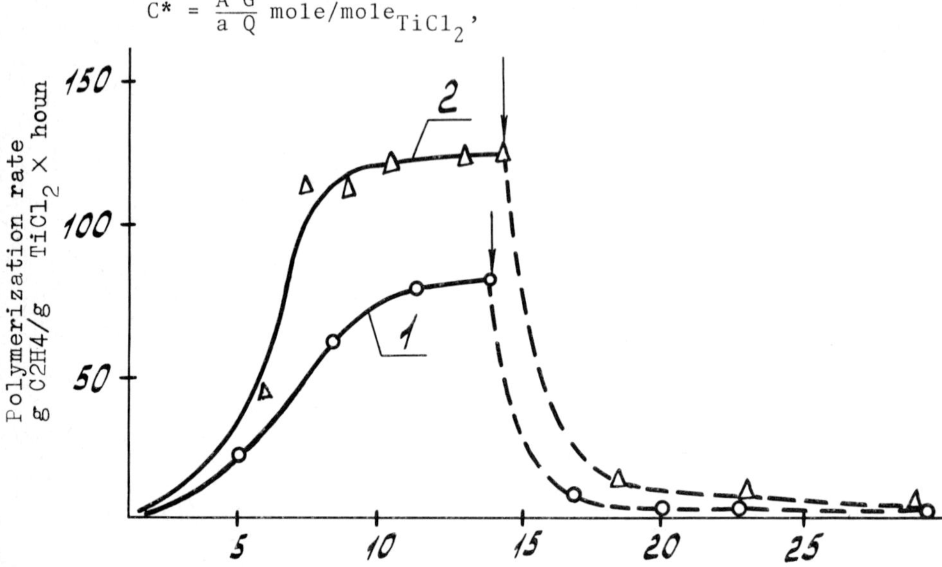

Fig. 1 Kinetic curves obtained at 75° and ethylene pressure of 5.5 kg/cm^2. Arrows show the moment of inhibitor introduction.
1 - $TiCl_2$ + $AlEt_2Cl$ (see Table 2, Experiment 3);
2 - $TiCl_2$ (see Table 1, Experiment 8).

where A = polymer radioactivity, mcuric/g; G = polymer yield, g; a = quantity of $TiCl_2$, mole; Q = inhibitor radioactivity, mcurie/mole. The value of K_p was determined from the equation:

$$V = K_p\ C^*\ C_M,$$

where V = polymerization rate at the moment of inhibitor injection, mole of monomer/mole$_{TiCl_2}$ x sec; C_M = monomer concentration, mole/l. The polymer fractionation was performed in a Kumagawa extractor first by boiling ether and then by boiling n-heptane. The intrinsic viscosity was measured in decalin at 135°. The crystallinity of polypropylene was determined by an X-ray method.[12)]

3. EXPERIMENTAL RESULTS

3.1 Ethylene polymerization

The results obtained stopping the $TiCl_2$-catalyzed ethylene polymerization by different radioactive compounds are given in Table 1. The absence of radioactivity when using $CH_3{}^3OH^3$ is related to the low sensitivity of a 2π-counter used which cannot detect tritium at levels less than 5.10^{-5} mcurie/g.

Using different inhibitors the same values of K_p were obtained. Using $C^{14}O_2$ it was found that the value of K_p does not depend on polymerization time, polymer quantity at the moment of introducting an inhibitor, polymerization rate, concentration of $C^{14}O_2$ and its specific radioactivity nor on the contact time of the inhibitor with the reaction medium.

On stopping the polymerization with $C^{14}O$ the radioactivity found in the polymer increased with the contact time (τ) of the inhibitor with the reaction medium. As a result, the calculated value of K_p decreases with increasing τ (Table 1, Experiments 4 and 5).

Table 2 shows the results of stopping the polymerization of C_2H_4 on $TiCl_2$ + $AlEt_2Cl$. With $C^{14}O$ and $C^{14}O_2$ as the inhibitors, the values of C* and K_p were similar, but with CH_3OH^3 the number of labelled molecules was about 100 times higher. Methanol treatment of the CO-stopped reaction mix gave also a high C* value (see Experiments 1 and 4). Slow accumulation of polymer radioactivity with increasing contact time was also observed.

The addition of $AlEt_3$, $AlEt_2Cl$ or $AlEtCl_2$ to $TiCl_2$ decreased the polymerization rate (see Fig. 1). In this case the propagation rate constant remained unchanged and therefore the lower polymerization rate must be ascribed to a decrease of the number of propagation centers.

3.2 Propylene polymerization

The results of two typical experiments on propylene polymerization are given in Table 3. Propylene is polymerized by $TiCl_2$ alone to a high molecular weight polymer containing about 25% of a fraction insoluble in boiling n-heptane (isotactic polypropylene) and about 60% of an atactic amorphous polymer soluble in boiling ether. The fraction insoluble in boiling ether but soluble in boiling n-heptane (about 15% in case of $TiCl_2$) seems to be a stereoblock polymer.[13] On adding diethylaluminiumchloride to $TiCl_2$, the isotactic fraction increases (up to 55%) and the polymerization rate decreases.

Table 3 shows that the values of K_p are similar for the atactic and isotactic additions with both catalytic systems. On adding $AlEt_2Cl$ there is a sharp decrease in the number of propagation centers for the atactic, but not for the isotactic, polymer.

On stopping the polymerization by CH_3OH^3 the number of labelled polymer chains exceeds the corresponding number with $C^{14}O_2$ by two orders of magnitude.

4. DISCUSSION

4.1 Determination of number of propagation rate constants for one-component and two-component systems

The use of the radioactive quenching technique for studying propagation reactions in catalytic polymerization is most advantageous for one-component heterogeneous catalysts. In this case the effect of the inhibitor on the process of polymerization is determined only by the interaction of the inhibitor molecule with the propagation center and is not complicated by side reactions. Using radioactive inhibitors in the $TiCl_2$-catalyzed polymerization the quantitative determination of the active bond numbers is possible. It is to be noted

TABLE 1

Stoppage of the ethylene polymerization on $TiCl_2$ by different inhibitors at 75°C

No. of Experiment	1	2	3	4	5	6	7	8
Inhibitor Type	$CH_3^3OH^3$	$C^{14}H_3CH_2CH_2I$		$C^{14}O$		$C^{14}O_2$		$C^{14}O_2$
Q, mcurie/mole	575	452		1210		2070		2760
Concentration of inhibitor, $mole/mole_{TiCl_2}$	0.1	0.24	0.85	0.01	0.02	0.04	0.04	0.04
V, $\frac{G_{C_2H_4}}{G_{TiCl_2}\cdot hour\cdot atm}$	19.4	17.6	11.2	14.0	4.7	5.0	17.8	21.4
$C^*\cdot 10^5$, $\frac{mole}{mole_{TiCl_2}}$	-	4.4	2.2	2.3	2.6	1.0	2.9	3.4
$K_p\cdot 10^4$, $\frac{1}{mole\cdot sec}$	-	0.8	1.0	1.2	0.36	1.0	1.2	1.25
τ, min	15	20	15	20	165	20	140	20

TABLE 2

Stoppage of the ethylene polymerization on $TiCl_2$ + $AlEt_2Cl$ by different inhibitors at 75°

No. of Experiment	1	2	3	4[1)]	5
Inhibitor Type	$C^{14}O$		$C^{14}O_2$	CH_3OH^3	
Q, mcurie/mole	1210		2070	575	
Inhibitor concentration, mole/mole	0.09	0.1	0.08	7.0	7.7
V, $\frac{G\ C_2H_4}{G\cdot hour\cdot atm}$	9.4	5.0	14.7	-	9.0
$C^*\cdot 10^5$, mole/mole	2.0	3.0	2.2	270[2)]	450[2)]
$K_p\cdot 10^{-4}$, 1/mole·sec	1.15	0.33	1.25	-	-
τ, min	4.5	25	20	30	30

[1)] Alcohol was introduced into the reactor after stopping polymerization in Experiment 1.

[2)] γ-kinetic isotopic effect

Note: Al/Ti = 1, except for Expt. 3 for which Al/Ti = 3.

that: a) the values found for C* and K_p do not depend on the type of an inhibitor used (see Table 1); and that b) using $C^{14}O_2$ as the inhibitor, the change of polymerization conditions (its duration, polymer yield and the introduced portion of $C^{14}O_2$) did not affect the value of K_p.

With $TiCl_2$ + $AlEt_2Cl$ quantitative measurements are only possible when using "specific" inhibitors which interact with transition metal-carbon bonds in active centers, but not with aluminum-polymer bonds, by transfer reactions with an aluminum-organic co-catalyst. Such inhibitors are $C^{14}O$ and $C^{14}O_2$. The introduction of CO into a polymer chain during interaction with

TABLE 3

Number of labelled molecules (C*) and value of propagation rate constant (K_p) on stopping propylene polymerization by $C^{14}O_2$ and CH_3OH^3; polymerization in liquid propylene at 70°C; Al/Ti = 1

Catalytic System	$TiCl_2$					$TiCl_2 + AlEt_2Cl$					
Average Rate of Polymerization,[1] $\frac{G_{C_3H_6}}{G_{TiCl_2} x hour}$	5.6					2.2					
Inhibitor	$C^{14}O_2$										CH_3OH^3
Inhibitor Concentration, $\frac{mole}{mole_{TiCl_2}}$	0.04					0.05					1.0
Results of Fractionating	Portion of Fraction, % weight	$[\eta]^2$ $\frac{dl}{g}$	Crystallinity	C* 10^7, $\frac{mole}{mole_{TiCl_2}}$	K_p, $\frac{l}{mole\ sec}$	Portion of Fraction, % weight	$[\eta]^2$ $\frac{dl}{g}$	Crystallinity	C* 10^7, $\frac{mole}{mole_{TiCl_2}}$	K_p, $\frac{l}{mole\ sec}$	C* 10^7, $\frac{mole}{mole_{TiCl_2}}$
Ether Fraction	60	2.6	nil	36	63	30	1.6	nil	5.5	80	
Heptane Fraction	15	4.2	medium	4.5	126	15	3.8	medium	2.5	87	240 γ^3
Polymer Insoluble in n-Heptane	25	12.0	high	12.6	76	55	8.6	high	8.5	94	1700 γ^3

[1] Polymerization rate was calculated from the polymer yield.
[2] Intrinsic viscosity.
[3] Kinetic isotopic effect.

an active center is caused by the coordination capability of CO.[14] The accumulation of radioactivity in the polymer with increasing contact time of the CO is due to slow co-polymerization of CO with ethylene.

A labelled carbon dioxide can be also used for the determination of the number of active bonds the the presence of $AlEt_2Cl$. CO_2 does not interact with Al-C-bonds provided that at least one alkyl group on the Al atom is substituted by a halogen.[15] The absence of interaction between CO or CO_2 and the aluminumorganic co-catalyst is also confirmed by the fact that with $C^{14}O$ and $C^{14}O_2$ polymerization stops at one-tenth of the inhibitor concentration necessary with diethylaluminum-chloride (Table 2).

With the two-component system the introduction of ROH^3 leads to a radioactive polymer due to the interaction with non-active metal-polymer bonds (see Tables 2 and 3). It is characteristic that on ROH^3 addition high radioactivity is also observed after polymerization has been completely stopped by CO or CO_2. These data cast some doubt on the C* and K_p values obtained for Ziegler-Natta systems with labelled alcohols.

Our K_p values for $TiCl_2$ systems are considerably higher than the corresponding values obtained earlier by others.[16,17,18,19]

In $TiCl_2$ systems the portion of propagation centers active for ethylene polymerization is not more than 0.1% of the total number of titanium ions on the surface. In a one-component system, $TiCl_2$ propagation centers seem to form from titanium ions with high coordinative unsaturation. An active bond is apparently formed by the alkylation of titanium ions by the oxidative addition of olefin molecules to coordinatively unsaturated titanium ions in a low oxidation state. The low number of propagation centers during the $TiCl_2$-catalyzed propylene polymerization compared to the corresponding number for ethylene polymerization is worth noting. It may be connected with a lower output of propagation centers during alkylation of titanium ions by propylene.

4.2 Role of metalorganic co-catalyst

With $TiCl_2$ the presence of a metalorganic co-catalyst is not necessary for forming an active metal-carbon bond. According to Tables 2 and 3 the alkylation of $TiCl_2$ by $AlEt_2Cl$ does not increase the number of propagation centers. Actually the addition of $AlEt_2Cl$ before or during polymerization reduces the rate by decreasing the number of propagation centers. Thus, for the investigated catalytic systems propagation centers are formed by the alkylation of coordinatively unsaturated titanium ions.

In $TiCl_2$-catalyzed polymerization in the absence of added aluminumorganic compounds active centers are "monometallic," i.e., do not contain other metals. The results obtained on propylene polymerization by this catalyst (Table 3) show that on $TiCl_2$ there are at least two types of active centers - "atactic" and "isotactic." Both types of centers have nearly the same activity for the polymerization of propylene and ethylene. The difference between them resides in the different steric arrangement of chloride ions surrounding the titanium ion. In an atactic center the surrounding of every chloride ion does not provide the same steric position for propylene in the π-complex before insertion in an active bond. An atactic center seems to be able to acquire temporarily the ability for isotactic stereoregulation by the reversible adsorption of a monomer (or impurities present in the system) in the immediate vicinity of a transition metal ion. Thus the formation of a polymer consisting of blocks of atactic and isotactic polymer can be explained.

An aluminumorganic compound added to $TiCl_2$ is reversibly adsorbed primarily on atactic centers. As a result the total polymerization rate decreases mainly because of the reduced formation of the atactic polymer (see Table 3). In this case neither number or activity of isotactic centers change significantly, indicating that the introduction of an aluminumorganic compound does not change the structure of the center. On the addition of the aluminumorganic compound the polymerization rate decreases for ethylene also because of decrease in the number of propagation centers.

With both $TiCl_2$-based catalysts stereoregulation is dependent on the arrangement of chloride ions in an active center. Metallorganic co-catalysts prevent non-stereospecific polymerization by being reversibly adsorbed on active centers which are responsible for atactic polymerization.

ACKNOWLEDGEMENT

The authors wish to thank Dr. Emirova for the determination of polymer properties.

REFERENCES

1) H. W. Coover, J. Guillet, R. Combs, F. B. Joyner, J. Polymer Sci., A-I, 4, 2583 (1966).
2) L. A. M. Rodriguez, H. M. van Looy, J. Polymer Sci., A-I, 4, 1971 (1966).
3) P. Cossee, J. Catalysis, 3, 80 (1964).
4) E. J. Arlman, P. Cossee, J. Catalysis, 3, 99 (1964).
5) G. Allegra, Makromol. Chem., 145, 235 (1971).
6) Yu. V. Kissin, S. N. Mezhikovsky, N. M. Chirkov, Europ. Pol. J., 6, No. 2, 267 (1970).
7) Yu. V. Kissin, N. M. Chirkov, Europ. Pol. J., 6, No. 3, 525 (1970).
8) Yu. I. Ermakov, V. A. Zakharov, Uspekhi chimii, in press (to be published in No. 2, 1972).
9) F. X. Werber, C. J. Benning, W. R. Wszolek, G. E. Ashby, J. Polymer Sci., A-I, 6, 743 (1968).
10) G. D. Bukatov, V. A. Zakharov, Yu. I. Ermakov, Kinetika i kataliz, 12, 505 (1971).
11) Yu. I. Ermakov, V. A. Zakharov, Principles of Prediction of Catalytic Activity. Proceedings of IV International Congress on Catalysis, 1, 200, M, 1970.
12) G. Natta, P. Corradini, M. Cesari, Accad. naz. Lincei, Rend. Cl. Sci., fisiche, mat. natur. Sez., II, 22 (8) (1957) II.
13) J. Pasquon, G. Natta, A. Zambelli, A. Marinangeli, A. Surico, J. Polymer Sci., C, 16, 2501 (1967).
14) J. P. Candlin, K. A. Taylor, D. T. Thompson, Reactions of Transition Metal-Complexes, Elsevier Publishing Company, Amsterdam-London-New York, 1968.

15) Organometallic Chemistry, Edited by Zeiss, Reinhold Publishing Corporation, New York, 1960.
16) C. F. Feldman, E. Perry, J. Polymer Sci., 46, 217 (1960).
17) G. Natta, J. Pasquon, Adv. Catalysis 11, 1 (1959).
18) E. Kohn, H. Schnurmans, J. V. Cavender, R. A. Mendelson, J. Polymer Sci., 58, 681 (1962).
19) J. C. W. Chien, J. Polymer Sci., A, 1, 425 (1963).

DISCUSSION

D. F. HOEG

I have several comments about these interesting results based on our past work describing the use of $TiCl_2$ as a single component catalyst for the polymerization of ethylene and propylene. First, it was observed that the addition of Et_2AlCl to $TiCl_2$ reduced the polymerization rate for propylene. With trialkylaluminum, as I recall, we observed a marked increase in the rate of polymerization compared to pure $TiCl_2$. This would seem consistent if, as has been proposed, Et_2AlCl coordinates accessible exposed Ti centers, but does not create new centers. R_3Al would appear to create more centers than originally present on the surface of the $TiCl_2$. These tracer techniques should be useful in clarifying this. Have you any results along these lines?

Secondly, the work done by my former coworkers Werber, et al. established that the structure of the polyethylene formed by $TiCl_2$ was essentially the same as with the Phillips catalyst but of higher molecular weight. It was a linear polyethylene with the growing chain end a saturated alkane structure. Where then did the H atom come from to form this saturated end? Is a dissociative adsorption of ethylene involved in the initiation stage? This question is still unresolved on $TiCl_2$. The difference in the initiation reaction between ethylene and propylene may, in part, account for the large difference in their overall rates of polymerization with $TiCl_2$.

YU. I. YERMAKOV

In our experiments with highly active samples of $TiCl_2$ the addition of Al Et_3 as well as of Al Et_2Cl reduced the polymerization rate for ethylene and propylene. That is why we consider Al Et_3 not to create new centers on the surface of the $TiCl_2$. Using the obtained values of propagation rate constant and data of molecular weight of polyethylene, it is easy to show that hundreds of polymer molecules appear in one center in a short space of time (several minutes). In this case the structure of polyethylene is not determined by the initiation reaction but by the chain transfer, which must lead to the appearance of one methyl-group and one vinyl-group. Correspondingly polyethylene obtained on $TiCl_2$ contains only methyl and vinyl end groups.

The mechanism of the initiation reaction on $TiCl_2$ is not clear at present. But we have the evident data that at the polymerization of propylene on this catalyst the number of propagation centers is approximately by an order lower than at the polymerization of ethylene. We suppose that the difference is accounted by various yields of active centers in the initiation stage.

J. L. GARNETT

The fact that the present authors have used CH_3OT to quench the reaction is interesting since the principle can be used as a new tritium labeling technique. We have found that if one adds Et_2AlCl to an aromatic compound (or an olefin, even some monomers) then quenches the reaction with T_2O (or CH_3OT), immediate tritium exchange occurs on the parent compound at room temperature (Long, Garnett, Vining and Mole, JACS, in press). While Et_2AlCl is an active exchange catalyst, lower valence $TiCl_2$ is inactive. However, higher valence titanium chlorides are very rapid exchange catalysts for this tritium work. This procedure constitutes a remarkably rapid tritium labeling method which should be of value in the tritiation of a wide range of compounds.

YU. I. YERMAKOV

The use of alcohols labeled by tritium and water for the determination of number of propagation centers in catalytic polymerization is often complicated by the side effects, including the effect of isotope exchange mentioned in the present comment. The absence of this effect for catalyst used by us was checked by special experiments. Besides for the determination of the number of propagation centers we mainly used the compounds labeled by the radioactive carbon ($C^{14}O$, $C^{14}O_2$). In this case it is possible not to take into consideration the kinetic isotope effect while calculating the propagation rate constant.

L. M. SAJUS

New titanium catalysts for ethylene polymerization are supported titanium catalysts. In that case, the productivity of the catalyst, expressed in pounds of polymer produced per gram of titanium is largely higher, by a factor of 50 to 100, than with your $TiCl_2$ catalysts.

One can think that this increase of productivity involves a better use of titanium and that about 10% of the total content of titanium of those catalysts are active centers, compared with 0.1% in the case you discussed.

Do you agree with that consideration, and did you make some experiments with those supported titanium catalysts?

YU. I. YERMAKOV

We investigated the supported titanium catalysts on silica and found that their activity was by two orders higher than that of the samples of $TiCl_2$ (till 10,000 g polyethylene/g Ti·hour·atm for supported catalyst in comparison of to 70 g polyethylene/g Ti·hour·atm for $TiCl_2$ in the most active samples). We consider it is due to the participation of greater part of titanium ions in the active center formation. While for $TiCl_2$ the number of propagation centers does not exceed 0.01% titanium ions, for the supported titanium catalyst this number is about 1.0%.

Paper Number 24

INVESTIGATION OF THE CATALYTIC OLIGOMERIZATION OF BUTADIENE IN THE PRESENCE OF π-ALLYL COMPLEXES OF PLATINUM

A. M. LAZUTKIN, A. I. LAZUTKINA, I. A. OVSYANNIKOVA, E. N. YURTCHENKO, and V. M. MASTIKHIN
Institute of Catalysis, Novosibirsk, USSR

ABSTRACT: For the first time the π-allyl complexes of platinum $(C_3H_5)_2PtX$ and $[C_3H_5PtX]_2$ were obtained (X = Cl, Br, I). With the help of infrared, NMR and X-ray spectroscopy the π-nature of a Pt-allyl group bond was proved and it was shown that the electron structures of Cl and Br-derivatives are similar and differ significantly from that of the I-derivatives. It is suggested that the stability of a Pt-X bond increases in the Cl<Br<I series and that the C_3H_5-group has acceptor properties. The complexes $(C_3H_5)_2PtX$ are thought to be associated in solution.

In the presence of these and related complexes the butadiene oligomerization was studied. Halide-containing complexes cause the formation of low-molecular polybutadiene, a monomeric complex being active. The polymerization is accompanied by the formation of trimers and tetramers due to the conversion of some $(C_3H_5)_2PtX$ into $Pt(C_3H_5)_2$. This conversion tendency decreases in the Cl>Br>I series in accordance with the stability of a Pt-X bond. Correspondingly, a bis-(π-allyl)-platinum causes mainly the process of forming trimers and tetramers. The mechanism of the oligomerization process is discussed on the basis of kinetic data and the composition of products.

1. INTRODUCTION

Recently, various types of π-allyl compounds of transition metals have gained widespread use as homogeneous catalysts for reactions of unsaturated hydrocarbons. Thus, π-allylnickel halides and π-allylpalladium halides[1,2] are successfully used as catalysts of sterospecific polymerization of butadiene. Much attention was given to the study of π-allyl bond nature and the structure of these compounds.[3-8] But for a more fundamental understanding of the mechanism of the reactions catalyzed by these compounds, it is necessary to be able to compare their properties not only in the ligand series but also in the isoelectronic metal series. For this reason we prepared π-allylplatinum halides and studied their catalytic activity. At the beginning of our work the inactivity of a bis (π-allyl) platinum[3] and the synthesis of allylplatinum chloride,[3] but not its structure or properties, was reported. Therefore we included into our study the complexes obtained by us as well as those known earlier.

2. SYNTHESIS AND INVESTIGATION OF π-ALLYL COMPLEXES OF PLATINUM

By treating a solution of bis-(π-allyl)-platinum in pentane by allyl halides (C_3H_5X, X = Cl, Br, I) diallylplatinum halides of the composition $(C_3H_5)_2$ PtX (X = Cl(I), Br(II) I(III)) were obtained quantitatively. These are converted by dissolution in halogenated solvents into monoallylplatinum halides C_3H_5PtX (X = Cl(IV), Br(V), I(VI)). The structures of the complexes obtained were deduced from infrared (Fig. 1), NMR (Table 1) and X-ray spectra and their comparison with well-known spectra of π-allyl-palladium and -nickel halides.[4-9] The infrared spectra of π-allyl-palladium halides[7] in the Cl, Br, I series are virtually identical and hardly differ from those of the corresponding nickel derivatives.[4,5] This is due to the nature of infrared spectra in the range of 400-4000 cm^{-1} for these compounds determined by local symmetry of a π-allyl group.[6]

A comparison of analogous infrared spectra shows that only the spectrum of iodide (VI) bears resemblance to the spectra of π-allylpalladium and π-allylnickel halides. The difference between them consists only in the displacement of the δ_{CCC} band up to 552 cm^{-1} and in the appearance of bands at 805 cm^{-1} which we assign to δ_{CH}, and at 758 cm^{-1}. The band 1018 cm^{-1} is assigned to $\nu(ccc)s$, 1382 cm^{-1} to $\nu_{(ccc)as}$, 1460 cm^{-1} to δ_{CH_2}. It suggests for complex VI the structure of the π-allyl complex $[C_3H_5PtI]_2$ with a bridge Pt $\begin{smallmatrix} I \\ I \end{smallmatrix}$ Pt bond and a small deformation of the local symmetry of the π-allyl group. The spectra of the chloride (IV) and the bromide (V) are identical but differ significantly from that of (VI). The displacement of $\nu_{(ccc)as}$ up to 1485-1490 cm^{-1}, the decrease of the intensity of δ_{CCC}, the changes in the region $\nu_{(ccc)s}$ and the absence of bands above 1490 cm^{-1} suggest that IV and V are dimeric bridge π-allyl complexes in which the C_3H_5-groups have electron configurations different from those in VI. Because of the poor solubility of the complexes of (IV-VI), their NMR-spectra were studied in pyridine solutions. In this solution the initial complexes were converted to $C_3H_5Pt\begin{smallmatrix} X \\ Py \end{smallmatrix}$ (X = Cl(VII), I(IX)), shown by the comparison of the areas under the lines of O-protons in pyridine and H_1 of the allyl group (the relation 1:1) in the spectrum of (VII) dissolved in chloroform.

The parameters of the NMR-spectra of (VII) and (IX) (Table 1) are close to the corresponding parameters of the π-allyl complexes studied earlier.[5] The difference of τ_{H_3} and J^{Pt-H_2} for (VII) and (IX) seems to be connected with higher non-complanarity of the C_3H_5-group in (VII) as compared with that in (IX). In comparing infrared spectra of the chloride (VII), bromide (VIII) and iodide (IX) derivatives an interesting feature was found: the spectra of (VII) and (VIII) are identical while that of (IX) differs again the the region δ_{CCC}, $\nu_{(ccc)s}$ and $\nu_{(ccc)as}$. This may indicate the stable nature of the C_3H_5 - Pt bond in complexes formed from (IV-VI) with pyridine. The same phenomenon has been recently shown for π-allylpalladium halides.[6,8]

The results obtained confirm the assumption that (IV-VI) are π-allyl complexes and that the deviation of the electron structure of a C_3H_5-group in the halide complexes from that typical for π-allyl complexes increases in the series I<Br<Cl. This circumstance will no doubt influence the ability of (IV-VI) to substitute the C_3H_5-group for other ligands.

Infrared spectra (I-II) have the same pecularity as (IV-VI): spectra (I) and (II) are analogous and differ from spectrum (III). This, we believe, may point to higher equivalence in the electron structure of a C_3H_5-group in (I) and (II) than in (III). The consideration of typical C_3H_5-frequencies in

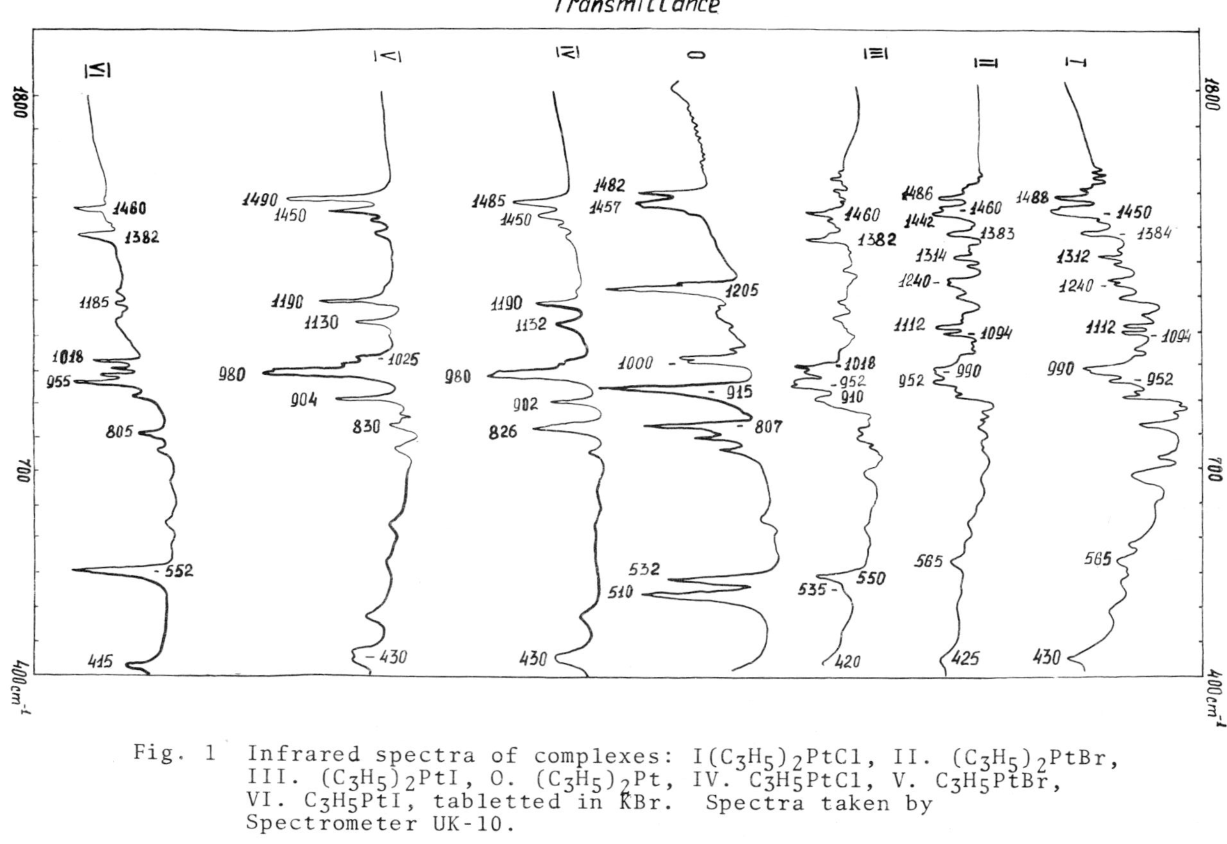

Fig. 1 Infrared spectra of complexes: I$(C_3H_5)_2PtCl$, II. $(C_3H_5)_2PtBr$, III. $(C_3H_5)_2PtI$, O. $(C_3H_5)_2Pt$, IV. C_3H_5PtCl, V. C_3H_5PtBr, VI. C_3H_5PtI, tabletted in KBr. Spectra taken by Spectrometer UK-10.

spectra of (I-III) shows that all allyl groups in the complex are π-bound. The results of the catalytic studies with I-IX indicate that these complexes can be regarded to be in the same series with π-allylnickel and π-allylpalladium halides.

The comparison of spectra of (III) and (VI) on the basis of the location of δ_{CCC}(550 cm^{-1}), $\nu(ccc)_s$(1018 cm^{-1}), $\nu(ccc)_{as}$ (1382 cm^{-1}) suggests that at least one C_3H_5-group in the complex is bound as π-allyl (VI). However we can not draw definite conclusions on the structure of complexes (I-III) from the NMR-spectra since these complexes are only soluble in those solvents with which they react forming monoallyl $(C_3H_5PtX)_2$ complexes (in chloroform) or $C_3H_5Pt\begin{smallmatrix}X\\Py\end{smallmatrix}$ (in pyridine) with elimination of a C_3H_5-group. NMR-spectra of (I) and (III) taken immediately after dissolution in chloroform at 0°C have two groups of broad signals at τ = 7.90; τ = 5.50 and τ = 7.70; τ = 5.85, respectively. The reason for broadening may be association of several molecules $(C_3H_5)_2PtX$.

To confirm the assumption that the changes in infrared spectra in the conversion of (I) and (II) to (III), of (IV) and (V) to (VI) and of (VII) and (VIII) to (IX) are connected with a change of the electronic structure of the complexes, X-ray L_{III} absorption spectra of platinum in these compounds were studied. The displacement of L_{III}-edges of absorption of metal atoms (ΔE ev) from a certain chosen start reflects the change of electron density around the central atom and hence the influence of a ligand on the electron condition of a metal in the complex and vice versa.[10] The results for ΔE are given in

TABLE 1
Parameters of H'NMR-spectra (spectrometer INM-4H-100)

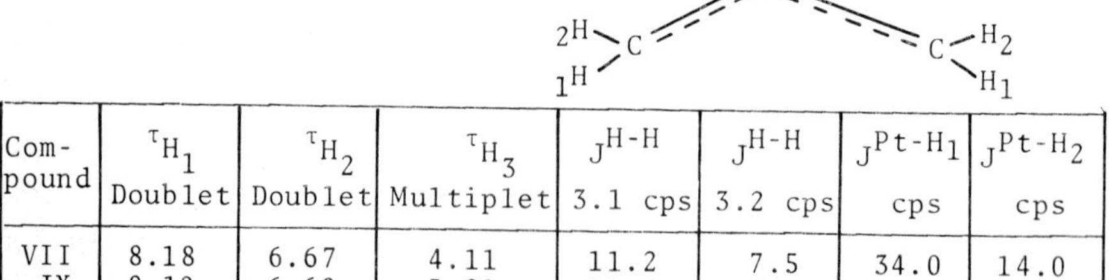

Compound	τ_{H_1} Doublet	τ_{H_2} Doublet	τ_{H_3} Multiplet	J^{H-H} 3.1 cps	J^{H-H} 3.2 cps	J^{Pt-H_1} cps	J^{Pt-H_2} cps
VII	8.18	6.67	4.11	11.2	7.5	34.0	14.0
IX	8.10	6.60	5.81	11.2	7.5	33.5	11.0

Fig. 2. For all complexes ΔE decreases in the Cl>Br>I series (the electron density around Pt increases, the positive charge of platinum decreases), and the most abrupt change is observed in the conversion to iodine derivatives. This can be explained by the formation of π-donor bonds due to d-orbits of I which becomes significant on forming the Pt-I bond. It appears that because of this additional stabilization the iodine-derivatives will be more stable and less reactive. The comparison of the ΔE values for diallyl, monoallyl and monoallyl-pyridine derivatives shows that the C_3H_5-group is a definite acceptor of electrons. The electron density on Pt in diallyl complexes proves to be less than in monoallyl complexes.

The comparison of the results of using infrared and X-ray spectroscopy for the investigation of (I-IX) shows that the change of the electron condition of Pt in the X = Cl, Br, I-series also influences on the electron condition of the C_3H_5-groups and this distinguishes the corresponding platinum complexes from those of nickel and palladium.

3. BUTADIENE OLIGOMERIZATION IN THE PRESENCE OF BIS-(π-ALLYL) PLATINUM

Butadiene oligomerization in the presence of a bis-(π-allyl)-platinum was performed in sealed ampoules. After a predetermined time one of the ampoules was opened and evacuated at room temperature up to constant weight to remove butadiene and volatile additives. The residue (yellow oily oligomers of butadiene) was then separated by sublimation into sublimable (90%) and non-sublimable (10%) fractions. Chromatographically it was shown that a sublimable fraction was a mixture of

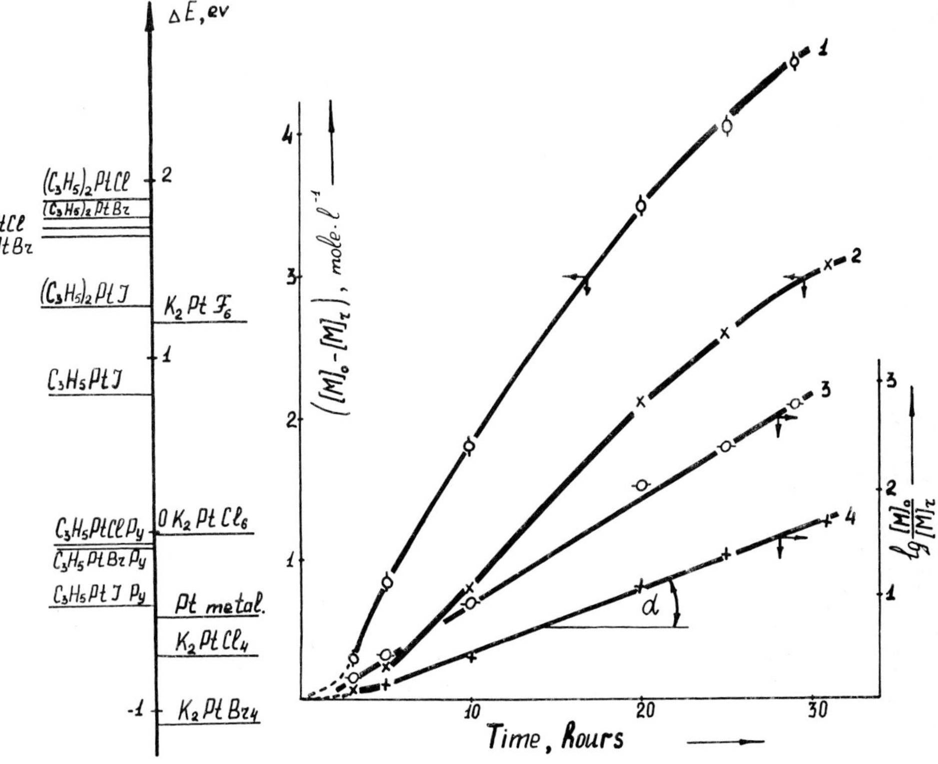

Fig.2 Displacements of X-ray L_{III} edge of adsorption of platinum

Fig. 3 Relationship between $([M]_0-[M]_\tau)$ and lg $\frac{[M]_0}{[M]_\tau}$ and time.
$[(C_3H_5)_2Pt]$ = 18.10^{-3} mole/l,
1.3 $[M]_0$ = 9.6 mole/l, 98°;
2.4 $[M]_0$ = 9.8 mole/l, 90°.

trimers (95%) and tetramers (5%) of butadiene. According to NMR and infrared spectra the trimers have a linear structure and have no CH_3-groups.

From the total weight of oligomers the number of moles of butadiene reacted per liter ($[M]_0-[M]_\tau$) was determined ($[M]_0$ = initial concentration of butadiene, $[M]_\tau$ = concentration of butadiene at time τ). The kinetic curves of the formation of oligomers have an induction period (Fig. 3, relationships 1,2).

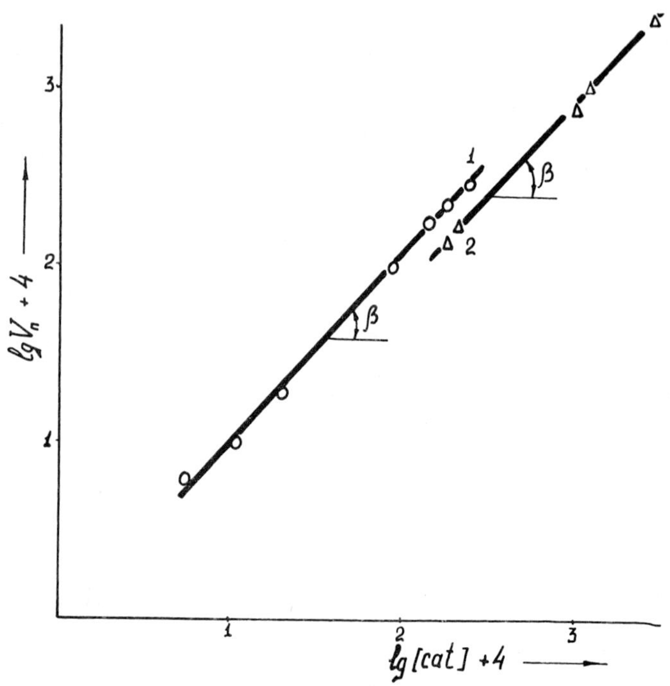

Fig. 4 Relationship between lg V_n and lg[cat], $(C_3H_5)_2Pt$. 1. $[M]_0$ = 9.6 mole/1, 98°; 2. $[M]_0$ = 9.8 mole/1, 90°.

With the help of NMR it was found that the induction period is related to the conversions of bis-(π-allyl)-platinum to another unidentified complex. The conversion of the kinetic curves (1,2) into a plot of $[M]_0$ vs. time gives a straight line (3,4) with tgα, first order rate constant, equal to 0.06 and 0.10 $hour^{-1}$, respectively. Fig. 4 is a plot of the logarithm of the rate expressed in butadiene concentration vs. the logarithm of the catalyst concentration. tgβ=1 indicates first order for the catalyst concentration. The values of rate constants found from the plot are $6.3\cdot10^{-5}$ 1/mole hour and $8.7\cdot10^{-5}$ 1/mole hour, respectively. The presence of the induction period, the absence of CH_3-groups in oligomers and their linear structure by analogy with[11] can be explained on the basis of the following scheme for the reaction

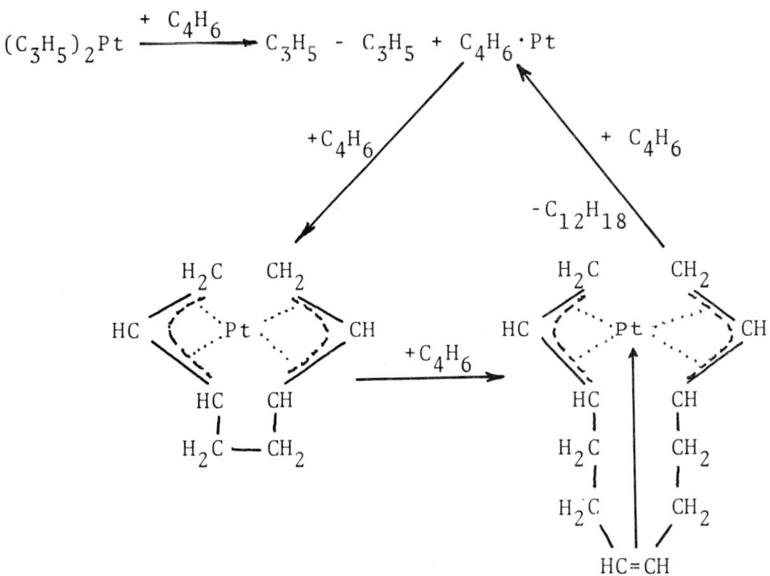

4. BUTADIENE OLIGOMERIZATION IN THE PRESENCE OF HALIDE-CONTAINING π-ALLYL COMPLEXES OF PLATINUM

The investigation of butadiene oligomerization in the presence of halide-containing complexes of platinum was performed in the same way as with bis-(π-allyl)-platinum. Oligomerization proceeds in the presence of (I-III) and (VI). (IV) and (V) are insoluble under the conditions of the experiment.

Oligomers represent a mixture of sublimable (1-25%), and non-sublimable (75-99%) fractions. The portion of the sublimable fraction of the total quantity of oligomers decreases in the Cl>Br>I series and depends on the reaction conditions (catalyst and butadiene concentrations and reaction temperature). Under the optimal conditions in the presence of diallylplatinum chloride the portion of a sublimable fraction reaches 25%. The infrared and NRM-spectra of the sublimable fractions obtained in the presence of a bis-(π-allyl)-platinum and halide-containing π-allyl complexes of platinum are analogous. The formation of a sublimable fraction (trimers and tetramers) in both cases is catalyzed by the same complex. It is possible that diallylplatinum halides undergo the conversion

$$[(C_3H_5)_2PtX]_n \longrightarrow n(C_3H_5)_2Pt + nX^{\ominus}, \quad (1)$$

as a consequence of a scission of the Pt-X bond. As it was shown in the first section the degree of this conversion should decrease in the Cl>Br>I series. In the same series the quantity of bis (π-allyl) platinum formed by (1) should decrease and hence the yield of sublimable oligomers. The tentative estimate indicates that the degree of such conversion under our conditions is within the limits of 2-15% of diallylplatinum halide introduced.

A non-sublimable fraction (75-99%) of the product is a low molecular polybutadiene with the intrinsic viscosity [η]=

0.1 - 0.5 and contains 65-75% 1.4-trans-, 9-17% cis- and 12-23% 1.2-butadiene links. The frequency 1380 cm^{-1} is not present in the infrared spectra, showing the absence of methyl groups in the polybutadiene. In the oligomerization on a bis-(π-allyl)-platinum the yield of a non-sublimable fraction is negligible, and the kinetics is determined by the formation of sublimable trimers and tetramers. With a π-allylplatinum halide as the catalyst the relative amounts of the fractions change. The microstructure of polybutadiene remains essentially unchanged with changing temperature.

While plotting kinetic curves for diallylplatinum halide-catalyzed reaction the total rate of oligomerization (the formation of sublimable and non-sublimable fractions) was taken into account.

Fig. 5 shows that butadiene oligomerization proceeds with a constant rate for a considerable time. Control experiments showed that under these conditions spontaneous butadiene polymerization does not occur at a significant rate. Fig. 6 represents the relationships between the oligomerization rate and initial concentration of a complex (Curves 1, 2, 3). The reaction order with respect to the catalyst decreases to zero with increasing catalyst concentration. Zero order kinetics is possible in case of radical polymerization at a high rate of the initiation (generation of radicals). But a number of indications provide evidence against a radical mechanism. First of all, the dependence of the oligomerization rate on the nature of the catalyst and the microstructure of oligomers are not characteristic of radical polymerization products and do not change significantly with the reaction temperature.

As it was shown in the first section, the decrease of the reaction order with increasing concentration of the complex can be attributed to the tendency of complexes to the associate and thus to reduce the concentration of the catalytically active monomer form.

The rates of oligomerization in the presence of (IV) and (VI) are the same. On the basis of the abovementioned data on the structure of (III) and (IV) it seems likely that (I-III) in a butadiene solution lose one allyl group and change into a monoallyl derivative.

5. INFLUENCE OF ELECTRON ACCEPTOR AND ELECTRON DONOR ADDITIVES ON THE CATALYTIC PROPERTIES OF π-ALLYL COMPLEXES OF PLATINUM

The addition of electron acceptors or electron donors is one of the methods for regulating the catalytic properties of π-allyl complexes.[1)] In our experiments the additions of electron acceptors ($AlCl_3$, $TiCl_4$, etc.) did not influence on the catalytic properties of complexes (I-III, VI). Electron donors (pyridine, diethylamine) inhibited the conversion of butadiene in the presence of (I-III, IV) but were without effect in the presence of a bis-(π-allyl)-platinum. A stronger donor - triphenyl phosphine- inhibited the reaction on all complexes. The stable complex $(C_3H_5)_2Pt.(PPh_3)_2$ was separated and studied by us. When a stable coordinative bond is formed with an electron donor ligand[12)] the butadiene polymerization is hindered. This is the reason for the low activity of complexes (VIII-IX).

Thus it is justified according to our results to consider

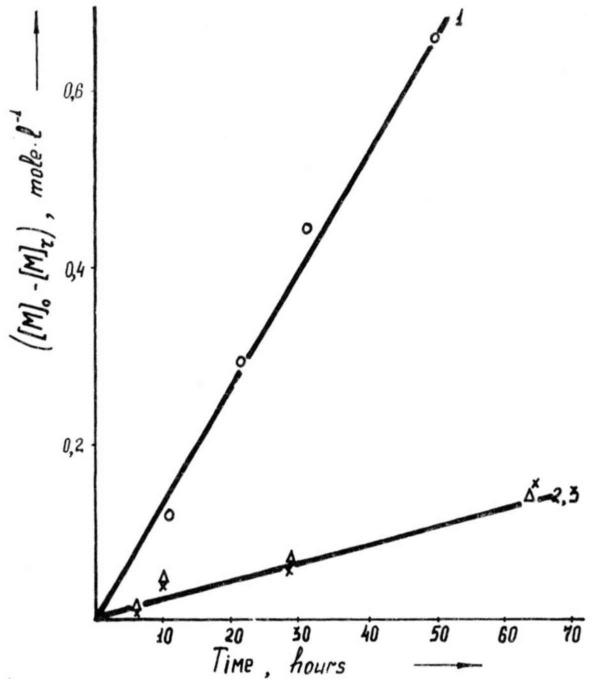

Fig. 5 Relationship between $([M]_0-[M]_\tau)$ and time. $[M]_0$ = 9.8 mole/l, 90°; · $[(C_3H_5)_2PtCl]$ $1\cdot10^{-4}$ mole/l; 2. · $[(C_3H_5)_2PtBr]$ $1.05\cdot10^{-4}$ mole/l; 3. ×$[(C_3H_5)_2PtI]$ $1.35\cdot10^{-3}$ mole/l.

platinum complexes $(C_3H_5)_2PtX$, and particularly C_3H_5PtX, to belong in the same series with their nickel and palladium analogs. All these complexes cause catalytic oligomerization (polymerization) of butadiene. Iodine derivatives of nickel[1)], palladium,[2)] and platinum are characterized by the same mechanism of stereoregulation and are probably involved in the same mechanism of polymer chain propagation.

Also, the change of the electronic properties in the 3d-4d-5d series of the metals results in regular changes in composition and structure of complexes and in the reaction mechanism. In particular, with the change of the platinum complexes the mechanism limiting the polybutadiene chains changes and the tendency of complexes to associate increases.

Fig. 6 Relationship between oligomerization rate and lg[cat]. $[M]_0$ = 9.6 mole/l, 98°; 1. $(C_3H_5)_2PtCl$; 2. $(C_3H_5)_2PtBr$; 3. ×$(C_3H_5)_2PtI$; + C_3H_5PtI. 4. Relationship between oligomerization rate and $[M]_0$, $[(C_3H_5)_2PtCl] = 10^{-3}$ mole/l, 98°.

REFERENCES

1) B. A. Dolgoplosk *et al.*, "Diene Polymerization by π-Allyl Complexes," M., Nauka, 1968.
2) A. I. Lazutkina, L. Ya. Alt, T. L. Matveeva, A. M. Lazutkin, "Kinetika i Kataliz," XI, 1591 (1970).
3) W. Keim, Dissertation, Technische Hochschule, Aachen, 1963.

4) H. P. Fritz, Chem. Ber., 94, 1217 (1961).
5) E. O. Fischer, H. Werner, Metal π-Complexes. Complexes with Di- and Oligo-Olefinic Ligands. Elsevier Publishing Company. Amsterdam-London-New York, 1966.
6) L. A. Leites, V. T. Aleksanyan, S. S. Bukalov, A. Z. Rubezhov, Chem. Commun., 265 (1971).
7) K. Shobatake, K. Nakamoto, JACS, 92, II, 3339 (1970).
8) L. A. Leites, Preprints of International Congress on Metalloorganic Chemistry, 302, Moskwa, 1971.
9) D. Adams, A. Squire, J. Chem. Soc., A, 10, 1808 (1970).
10) K. I. Matveev, L. N. Ratchkovskaya, N. K. Eremenko, I. A. Ovsyannikova, Proceedings of III All-Union Conference on Catalytic Reactions in Liquid Phase, Alma-Ata, 1971.
11) G. Wilke, B. Bogdanovic, P. Hard, P. Heimbach, W. Keim, M. Kröner, W. Oberkirch, K. Tanaka, E. Steinrücke, D. Walter, H. Zimmermann, Angew, Chem., 78, 157 (1966).
12) F. D. Mango, Advances in Catalysis, v. 20 (1969).

DISCUSSION

R. UGO

Complexes of type II (bis π-allyl Pt X)$_y$ could be Pt(III) or Pt(I). Are they diamagnetic or paramagnetic? If they are diamagnetic, do you suppose they are dimers with Pt-Pt interactions or are they a mixture of (bis π-allyl Pt) and Pt Cl_2? Have these complexes been characterized in the solid state?

A. M. LAZUTKIN

The (bis π-allyl Pt X) cannot be in principle the mixture of (bis π-allyl Pt) and Pt Cl_2 since the first is instable in the air according to our data and the second is insoluble in organic solvents. The (bis π-allyl Pt X) is stable in the air and very well soluble in such organic solvents as pyridine, dietylamine, chloroform, etc. (see ref. A. I. Lasutkina, T. P. Lasarenko, E. N. Yurchenko, W. M. Mastichin, L. J. Alt, A. M. Lasutkin, Journal of General Chemistry, USSR, 42, 7, 1583 (1972)). During the synthesis at the various conditions it has constant composition. IR-spectrum of the (bis π-allyl Pt) differs regularly from that of (bis π-allyl Pt) (see Fig. 1, sp. o). In solid state it is diamagnetic. It is possible that the reason this phenomenon is the existence of the Pt - Pt bonding, but we have no direct evidence of such bond. The question is being studied.

Paper Number 25

THE MECHANISM OF HYDROFORMYLATION

V. Ju. GANKIN
All-Union Scientific Research Institute of Petrochemical Processes, Leningrad (USSR)

ABSTRACT: In the reaction between cobalt hydrocarconyl and hexene-1 the concentrations of isomeric alkylcobalt carbonyls, acylcobalt carbonyls, aldehydes, olefins, cobalt hydrocarbonyl and dicobalt octacarbonyl have been determined as the reaction progressed. The results show that the reaction proceeds autocatalytically, alkylcobalt carbonyls are the intermediate products, cobalt hydrocarbonyl is added according to Markovinikov rule and the products of normal composition are formed by the isomerization of acylcobalt carbonyl. The olefin isomerization is observed only when alkylcarbonyls exist.

The autocatalytic character of hydroformylation reaction can be satisfactorily explained by the concept that the reaction is catalyzed by coordination compounds and that it proceeds by a chain mechanism in which coordinatively unsaturated particles (so called "conenses") are active.

The same mechanism is suggested also for heterogeneous catalysis but in this case the chain mechanism is due to a partial or full transfer of the electron cloud from a coordinatively saturated compound to the free orbit of a coordinatively unsaturated compound together with a ligand whereby a new vacant position forms.

1. INTRODUCTION

It is known that in the reaction of cobalt hydrocarbonyl with olefins, olefin isomerization and hydroformylation take place.[1,2] This stoichiometric hydroformylation (3), and the related catalytic hydroformylation have been extensively studied, and the experimental results and various hypotheses on the hydroformylation mechanism, isomeric aldehyde formation and isomerization reaction have been summarized in recent reviews.[1,2] It appears that while the mechanism proposed by Breslow and Heck[4] for the hydroformylation reaction has received wide acceptance, the hypotheses concerned with the mechanism of isomeric aldehyde formation and olefin isomerization are still being debated.

2. RESULTS AND DISCUSSION

In the reaction of cobalt hydrocarbonyl (HCC) with hexene-1 cobalt acylcarbonyls, cobalt hydrocarbonyl, aldehydes and hexenes have been analyzed by procedures[4),5)] and[6)] respectively. It has been found that cobalt acylcarbonyls are intermediate products obtained before aldehyde formation (Fig. 1).

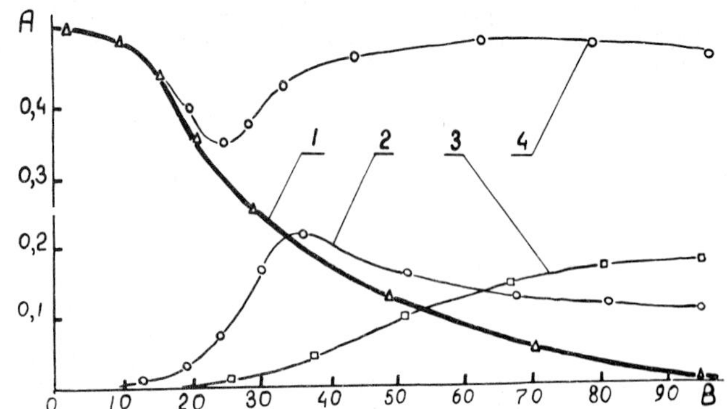

Fig. 1 Changes in $HCo(CO)_4$ (Curve 1), acylcobalt carbonyl (2), and aldehyde concentration (3).
A=concentration in mmol/g, B=reaction time (minutes).
4= sum of cobalt-containing compounds.
P_{CO}=1.1 at., t=+10°C, initial hexene-1 concentration =3,5 mmol/g, solvent=n-heptane.

A total cobalt curve (one mole of aldehyde obtained equals two moles of cobalt) has a minimum corresponding to the formation of an intermediate product which appears before the acylcarbonyls (Fig. 1).

According to Breslow and Heck[4] we have proposed that cobalt alkylcarbonyls form before acylcobalt carbonyls do in the region of minimum. Because of their chemical properties, alkylcarbonyls form iodohexanes on treatment with iodine, but iodohexanes can form also by the following reactions:

$$2HCo(CO)_4 + 3I_2 \rightarrow 2CoI_2 + 2HI + 8CO$$

$$RCH = CH_2 + HI \rightarrow RCHI - CH_3$$

For this reason it seemed necessary to remove cobalt hydrocarbonyl from the product.

The treatment of a sample with 40% alkali at 50° and then with iodine resulted in the appearance of 2- and 3-iodohexanes in the eighteenth minute of the reaction showing the presence of free cobalt hydrocarbonyls in the product before the iodine treatment.

Infra-red spectrum of the sample (Fig. 2) taken in the twelfth minute (minimum of acylcarbonyls) and treated with alkali to remove cobalt hydrocarbonyl is similar to that of methylcobalt carbonyl described in literature.[7]

The quantitative analysis of 2-iodohexane carried out during the reaction has shown:

1. Alkylcarbonyls are products preceding acylcobaltcarbonyls.

2. Cobalt hydrocarbonyl, alkylcobalt carbonyls and acylcobalt carbonyls are the sole intermediates while the total cobalt curve remains constant (Fig. 2).

In experiments carried out under elevated pressure (Fig. 4) and temperature the character of products remained unchanged. There is a smooth transfer into the region of

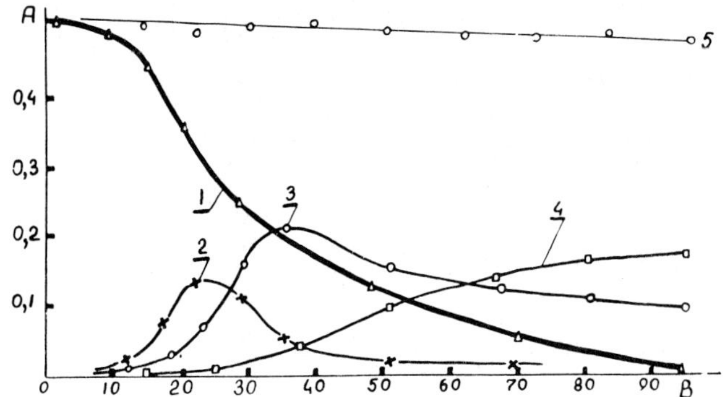

Fig. 2 Isomeric hexene content of the reaction mixture
A = isomer per cent content in the mixture.
B = reaction time in minutes.
1: hexene-1; 3: trans-hexene-2, 2: cis-hexene-2,
4: hexene-3.
P_{CO} = 1.1 at. t = +10°C, initial hexene-1 concentration = 3,5 mmol/g, solvent = n-heptane.

catalytic synthesis during which the concentration of intermediates changes only slightly. In the catalytic region the formation of the aldehydes (Fig. 4) is practically independent of the cobalt hydrocarbonyl concentration and proportional to the acylcarbonyl concentration. This proves that the acylcarbonyls are cleaned by hydrogen. The infra-red spectrum of a sample taken from the catalytic region was similar to that of the sample produced by stoichiometric synthesis at the same initial concentrations of cobalt carbonyls.

It is thus apparent that under the conditions of both stochiometric and catalytic synthesis, cobalt is present mainly in the form of hydrocarbonyl, alkylcarbonyl, acylcarbonyl and dicobalt octacarbonyl.

It was also shown that alkylcobalt carbonyl is not formed in the reaction (the introduction of 1-iodohexane has proved that the analytical method used allows its determination at a concentration one hundred times lower than that of 2-iodohexane).

The linear products, acylcarbonyl and aldehyde, increase in the course of the reaction (Fig. 5, 6).

Thus it appears that the initial addition of cobalt hydrocarbonyl to olefins follows the Markovinikov rule and that normal products are obtained in the isomerization of acylcobalt carbonyl.

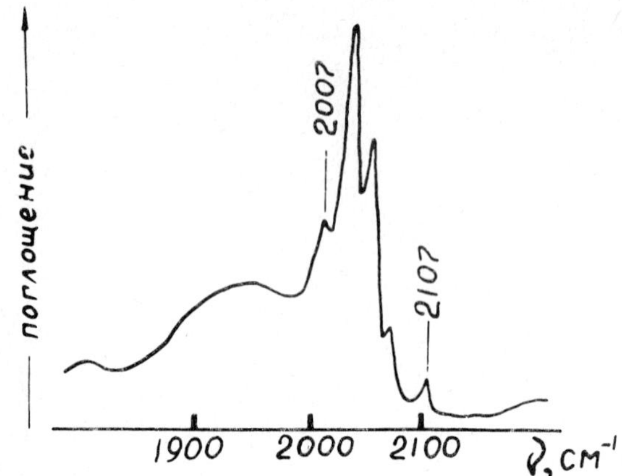

Fig. 3 The infra-red spectrum of the sample taken in the 12-th minute and treated with 40% KOH.

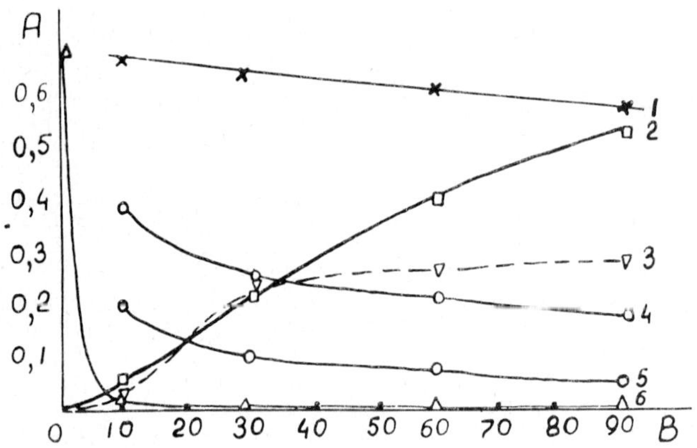

Fig. 4 Changes in product concentrations in the reaction mixture.
A = concentration, mmol/g; B = reaction time (minutes)
1 = sum of cobalt-containing compounds,
2 = enanthic aldehyde, 3 = $Co_2(CO)_8$,
4 = acylcarbonyl, 5 = alkylcarbonyl, 6 = $HCo(CO)_4$,
H_2= CO (1:1) mixture pressure = 50 at., temperature = 70°C, initial hexene-1 concentration = 0,3 mmol/g, solvent = n-heptane.

The analyses of the hexene isomers in the course of the reaction shows (Fig. 7) that the isomerization occurs only while alkylcobaltcarbonyl are present. (Compare Fig. 7 and Fig. 2).

The autocatalytic character of the hydroformylation reaction can be (Fig. 2) satisfactorily interpreted by a reaction mechanism which involves catalysis by coordination compounds.

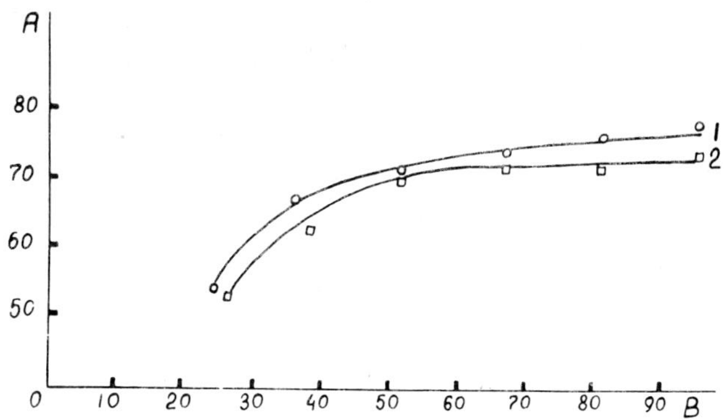

Fig. 5 Normal acylcarbonyls and aldehydes in the reaction mixture
A = content, per cent,
B = reaction time (minutes).
1 = normal acylcarbonyl, 2 = normal aldehyde.

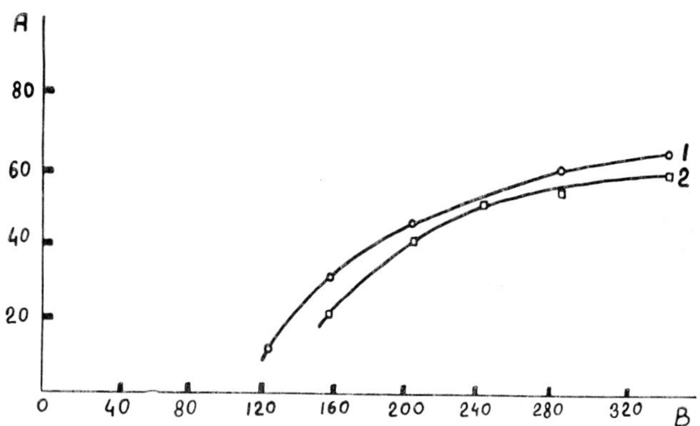

Fig. 6 Normal acylcarbonyls and aldehydes in the reaction mixture.
A = content, per cent,
B = reaction time, minutes.
1 = normal acylcarbonyl, 2 = normal aldehyde.

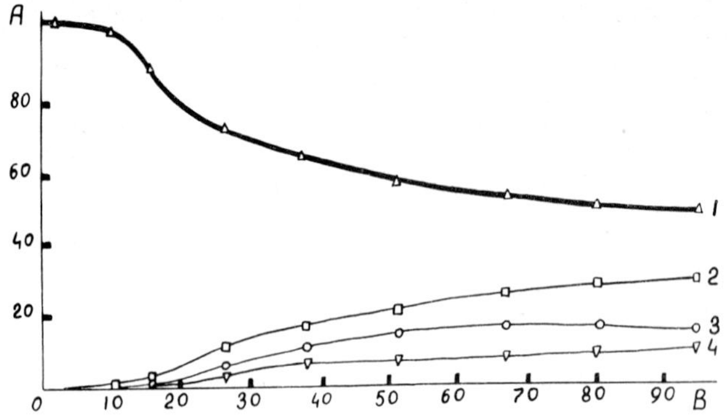

Fig. 7 The isomeric hexene composition of the reaction mixture.
A = content,
B = reaction time (minutes).
1 = hexene-1, 2 = trans-hexene-2, 3 = cis-hexene-1
concentration = 3,5 mmol/g, solvent = n-heptane.

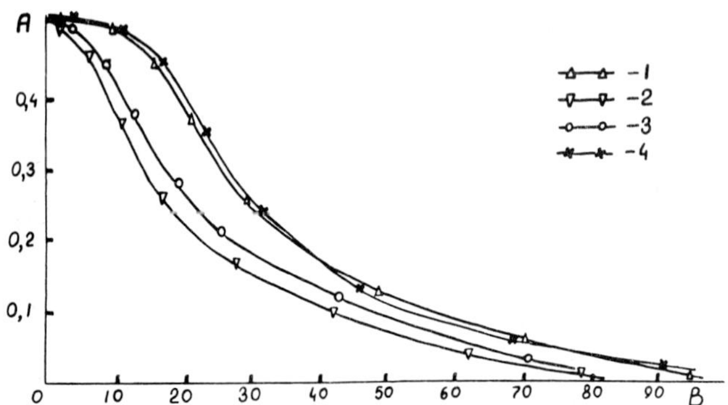

Fig. 8. $HCo(CO)_4$ disappearance in the stoichiometric synthesis.
A = concentration in mmol/g, B = reaction time, minutes.
1 = without initiating additive,
2 = addition of products taken at the stage of alkylcarbonyl formation; 3 = addition of products taken at the end of the reaction, 4 = addition of products treated with carbon monoxide.

The reactions of coordination compounds in the active intermediate is a coordinatively unsaturated compound which we shall call "conense."

In applying this concept [8] to the hydroformylation reaction and the autocatalytic conversion of dicobalt octacarbonyl into hydrocarbonyl we have postulated that conense reactions can proceed by a chain-branching mechanism.

This chain conense mechanism satisfactorily interprets the main kinetic features of reactions catalyzed by coordination compounds (ligand suppression, autocatalytic character) and enables us to deduce their pathway. Thus, the mechanism of the stoichiometric interaction of cobalt hydrocarbonyl with an olefin can be presented as follows:

$$CH_3(CH_2)_4CH=CH_2 + HCo(CO)_4 \rightarrow CH_3(CH_2)_4\underset{\downarrow\, HCo(CO)_3}{CH=CH_2} + CO \qquad \text{i}$$

$$\underset{\downarrow\, HCo(CO)_3}{CH_3(CH_2)_4CH=CH_2} \rightarrow CH_3(CH_2)_4\underset{|\, Co^*(CO)_3}{CH}-CH_3 \qquad \text{ii}$$

$$CH_3(CH_2)_4\underset{|\, Co^*(CO)_3}{CH}-CH_3 + HCo(CO)_4 \rightarrow CH_3(CH_2)_4-\underset{|\, Co(CO)_4}{CH}-CH_3 + HCo^*(CO)_3 \qquad \text{iii}$$

$$HCo^*(CO)_3 + CH_3(CH_2)_4CH=CH_2 \rightarrow CH_3(CH_2)_4\underset{\downarrow\, HCo(CO)_3}{CH=CH_2} \qquad \text{iv}$$

$$CH_3(CH_2)_4-\underset{|\, Co(CO)_4}{CH}-CH_3 \rightarrow CH_3(CH_2)_4\underset{|\, CO\, |\, Co^*(CO)_3}{CH}-CH_3 \qquad \text{v}$$

$$CH_3(CH_2)_4\underset{|\, CO\, |\, Co^*(CO)_3}{CH}-CH_3 + HCo(CO)_4 \rightarrow CH_3(CH_2)_4\underset{|\, CO\, |\, Co(CO)_4}{CH}-CH_3 + HC\overset{*}{o}(CO)_3 \qquad \text{vi}$$

$$CH_3(CH_2)_4\underset{|\, CO\, |\, Co^*(CO)_3}{CH}-CH_3 + HCo(CO)_4 \rightarrow CH_3(CH_2)_4\underset{\backslash\, CHO}{CH}-CH_3 + Co_2^*(CO)_7 \qquad \text{vii}$$

$$Co_2^*(CO)_7 + HCo(CO)_4 \rightarrow HCo^*(CO)_3 + Co_2(CO)_8 \qquad \text{viii}$$

$$CH_3(CH_2)_4\underset{|\, CO\, |\, Co(CO)_4}{CH}-CH_3 \rightarrow CH_3(CH_2)_4CH_2-CH_2COCo(CO)_4 \qquad \text{ix}$$

$$CH_3(CH_2)_5CH_2COCo(CO)_4 + HCo^*(CO)_3 \rightarrow CH_3(CH_2)_5CH_2C{\overset{\displaystyle O}{\underset{\displaystyle H}{\lessgtr}}} \qquad \text{x}$$

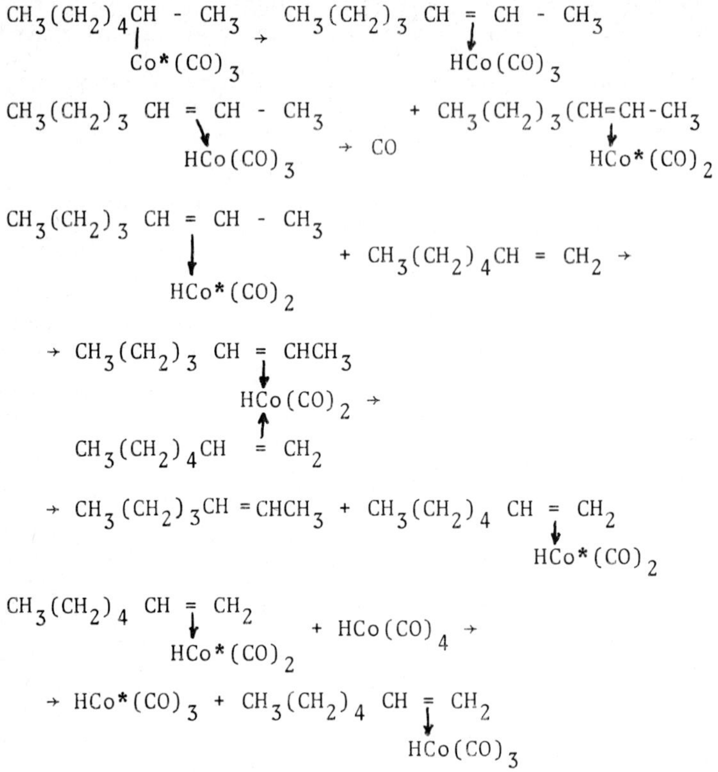

$CH_3(CH_2)_4CH(Co^*(CO)_3) - CH_3 \rightarrow CH_3(CH_2)_3 CH = CH - CH_3 (\downarrow HCo(CO)_3)$ xi

$CH_3(CH_2)_3 CH = CH - CH_3 (\downarrow HCo(CO)_3) \rightarrow CO + CH_3(CH_2)_3(CH=CH-CH_3 (\downarrow HCo^*(CO)_2)$ xii

$CH_3(CH_2)_3 CH = CH - CH_3 (\downarrow HCo^*(CO)_2) + CH_3(CH_2)_4CH = CH_2 \rightarrow$ xiii

$\rightarrow CH_3(CH_2)_3 CH = CHCH_3 \downarrow HCo(CO)_2 \uparrow CH_3(CH_2)_4CH = CH_2 \rightarrow$ xiv

$\rightarrow CH_3(CH_2)_3CH = CHCH_3 + CH_3(CH_2)_4 CH = CH_2 (\downarrow HCo^*(CO)_2)$

$CH_3(CH_2)_4 CH = CH_2 (\downarrow HCo^*(CO)_2) + HCo(CO)_4 \rightarrow$

$\rightarrow HCo^*(CO)_3 + CH_3(CH_2)_4 CH = CH_2 (\downarrow HCo(CO)_3)$ xv

In addition to the above reactions leading to final products chains are broken as a result of conense reaction with CO:

$$HCo^*(CO)_3 + CO \rightarrow HCo(CO)_4$$

This reaction explains the inhibition of the reaction rate (chain breaking) by CO and such kinetic features as auto-catalytic character of escaping cobalt hydrocarbonyl; the order of formation and consumption of intermediate products and the coincidence of the curves of isomerization and escape of cobalt hydrocarbonyl at the stage of alkylcobalt carbonyls existence).

The isomerization of 2-acylcobalt carbonyl into 1-acylcobaltcarbonyl (stage IX) explains satisfactorily the formation of normal acylcobalt carbonyls in the absence of 1-alkylcobalt carbonyls and the increase of n-acylcobalt carbonyl in the course of the reaction.

The portion of normal products is defined by the ratio of rates in stages VII and IX; the relative rate of reaction VII decreases with the concentration of the hydrocarbonyl.

The greater influence of the factors hindering hydro-formylation (increase of CO pressure, decrease of temperature, introduction of phosphine ligands, replacement of Rh with Co)

---

*active coordinatively-unsaturated compound (conense).

on the stage VII than on the stage IX has been suggested as an explanation of the well-known yield increase of normal products with the decreasing rate.

Similarly, the greater influence of higher temperatures and of lower CO pressure on route XI-XV leading to isomeric hexenes than on stage V explains the often observed rise of initial olefin isomerization[2] with the rise of temperature and fall of CO pressure.

In the proposed mechanism route only a few modes of developing chains have been presented, and alternative mechanisms could be suggested to account for the observed autocatalysis. However, Figure 8 shows that the addition of 3% of the product taken after the completion of the induction period and kept at -50° under argon to the initial mixture eliminates the induction period while the addition of the same sample kept under CO has no influence on the induction period.

In general conenses are the compounds which can form coordination bonds with compounds having unpaired electrons or π-electrons.

Thus, the square planar coordination compounds can be designated as conenses or considered as sites for chain initiation by associative mechanism.

In the case of octahedral coordination compounds the chain initiation occurs by a dissociative mechanism.

In a solution conenses are in a solvated state, a fact that explains the influence of solvents on the reaction rate.

Lewis bases can form coordination bonds and have therefore the greatest influence on the reaction.

The same conense chain mechanism could be proposed for heterogeneous catalysis, in which case the chain mechanism occurs but by the partial or full transfer of the electron cloud from a coordinatively saturated compound to the free orbit of a coordinatively unsaturated compound together with a ligand whereby a new vacant position forms.

REFERENCES

1) A. Chalk, J. Harrod, Adv. Organomet. Chem. 6, 142, 1968.
2) Yn. Falbe "Synthesis based on carbon oxide," "Khimia" Publishing House, 1971, p. 11.
3) J. Takegami, J. Watanabe, H. Masada, T. Mitsuda, Bull. Chem. Soc. Japan 42, 2920, 1969.
4) D. Breslow, R. Heck, J. Am. Chem. Soc. 83, 4023, 1961.
5) R. Jwanaga, Bull. Chem. Soc. Japan 35, 247, 1962.
6) G. Karapinka, M. Orchin, J. Org. Chem. 26, N11, 4187-4190 (1961).
7) L. Marko, G. Bor, G. Almasy and P. Szabo, Brennsfoff-Chemie, 6, 44, 184-187 (1963).
8) V. Yn. Gankin, D. M. Rudkovsky, "Carbonylation of unsaturated compo," "Khimia" Publishing House, 1968, p. 25.
9) V. Yn. Gankin, "Theses of the Reports of the Third All-Union Conference on Liquid Phase Catalysis," Alma-Ata, 1970, p. 211.

DISCUSSION

P. D. TAYLOR

The suggestion that cobalt hydrocarbonyl adds exclusively to an olefin according to the Markovnikov rule is contrary to recently published[1] results wherein cobalt hydrocarbonyl was allowed to react with deuterated propene and exchange H for D atoms. This study indicated that approximately 70% of the cobalt hydrocarbonyl proceeds according to the Markovnikov rules. The significance of this discovery becomes important when applied to the hydroformylation mechanism. One need not include the rearrangement of branched acyl cobalt carbonyl to linear acyl cobalt-carbonyl, a very difficult reaction to envisage, in order to account for the observed aldehydes. Orchin and Rupilius[2] have written an excellent review which incorporates some of our most recent experimental results as they apply to the hydroformylation mechanism.

1) P. Taylor and M. Orchin, J. Am. Chem. Soc., 93, 6504 (1971).
2) M. Orchin and W. Rupilius, Catalysis Reviews 6, 85 (1972), Marcel Dekker, Inc., N. Y., Edited by H. Heinemann.

Paper Number 26

STUDIES OF THE MECHANISM OF THE PROPYLENE DEUTEROGENATION WITH HYDRIDOCARBONYLTRIS(TRIPHENYLPHOSPHINE)RHODIUM AND RELATED IRIDIUM COMPLEXES BY MEANS OF MICROWAVE SPECTROSCOPY

TOMIKO UEDA
The Institute of Physical and Chemical Research,
Wako-Shi Saitama, 351, Japan

ABSTRACT: The propylene deuterogenation with $RhH(CO)(PPh_3)_3$, $IrH(CO)(PPh_3)_3$, trans-$IrCl(CO)(PPh_3)_2$ and β-$IrHCl_2(PPh_3)_3$ have been studied by analyzing quantitatively the isotopic isomers of the two monodeuterated propanes and the four monodeuterated propylenes by means of microwave spectroscopy. In the case with $RhH(CO)(PPh_3)_3$ in N,N-dimethylformamide at 30° C, the relative reaction rates were 23.6 (propylene-2-d_1); 3.5 (propylene-3-d_1); 1 (cis-propylene-1-d_1); 1 (trans-propylene-1-d_1). The propane-2-d_1 was produced greater than propane-1-d_1. Similar results were found also with $IrH(CO)(PPh_3)_3$ and trans-$IrCl(CO)(PPh_3)_2$ at 60°C. With β-$IrHCl_2(PPh_3)_3$ in toluene at 60°C, the relative initial rates were 2.8 (propylene-2-d_1); 3.6 (propylene-3-d_1); 1 (cis-propylene-1-d_1); 1 (trans-propylene-1-d_1). It is concluded that the hydrogenation and isotopic exchange occur with the two paths which are characterized by the metal-normalpropyl and metal-isopropyl intermediates, respectively, and the former path is more important than the latter with $RhH(CO)(PPh_3)_3$, $IrH(CO)(PPh_3)_3$ and trans-$IrCl(CO)(PPh_3)_2$. In the case with β-$IrHCl_2(PPh_3)_3$—SiO_2 at 0°C, propylene-3-d_1 was produced predominantly. An intermediate of π-allyl type was confirmed in the exchange reaction.

1. INTRODUCTION

Some years ago, Vaska and co-workers [1,2] found that hydridocarbonyltris(triphenylphosphine)iridium, trans-chlorocarbonylbis(triphenylphosphine)iridium, and the corresponding complexes with rhodium act as catalysts for the hydrogenation of ethylene or propylene under ambient conditions. Since then, several workers have proposed the catalytic mechanism for the hydrogenation of olefins with these complexes [3-6]. The deuterium tracer study on the distributions of deuterium in the deuterated species of olefin and paraffin has been made about the ethylene hydrogenation with trans-$IrCl(CO)(PPh_3)_2$ by means of mass spectrometry [3]. However, there has been no report on the propylene hydrogenation with the complexes as described above. The mechanisms [1] proposed by Vaska et al. for the ethylene hydrogenations with $IrH(CO)(PPh_3)_3$ and with $OsHCl(CO)(PPh_3)_3$ are analogous to those proposed by Horiuti-Polanyi [7] and by Ridial [8] for the heterogeneous hydrogenations of ethylene on metals. It is interesting to see such close conceptual interrelation between homogeneous and heterogeneous catalysis. On the other hand, Wilkinson and co-workers [9] considered that addition of the two hydrogen atoms of the catalytically active species $RhCl(H_2)(PPh_3)_2$ to an olefin was synchronous on the basis that tris(triphenylphosphine)chlororhodium is an effective homogeneous hydrogenation catalyst. However, Hassey et al. reported recently the formation of the σ-bonded rhodium-alkyl species as an intermediate in the olefin hydrogenation with the Wilkinson catalyst [10].

In this work, the propylene-deuterium exchange and the **deuterogenation** of propylene were studied with $RhH(CO)(PPh_3)_3$, $IrH(CO)(PPh_3)_3$ and trans-$IrCl(CO)(PPh_3)_2$ in N,N-dimethylformamide (DMF) and/or benzene, and with β-$IrHCl_2(PPh_3)_3$ in toluene and benzene. In addition, these reactions were carried out with $IrHCl_2(PPh_3)_3$ supported on silica gel in the absence of solvent. Analyzing the deuterated products in detail,the intermediates and the mechanisms of these reactions have been proposed. The microwave spectroscopy [11,12] was applied to analyse not only the four different isomers of monodeuterated propylenes, but also the two different isomers of monodeuterated propanes, since neither IR nor NMR spectroscopy could distinguish easily cis- and trans-propylene-1-d_1, propane-1-d_1 and propane-2-d_1 in the presence of dideuterated propane.

2. EXPERIMENTAL

2.1 Material

Trans-$IrCl(CO)(PPh_3)_2$ and β-$IrHCl_2(PPh_3)_3$ were prepared by refluxing a mixture of iridium tetrachloride and triphenylphosphine under nitrogen atmosphere in N,N-dimethylformamide [13], and in ethanolic water [14,15], respectively. β-$IrHCl_2(PPh_3)_3$ supported on silica gel was prepared in extremely air free condition as follows. 3.90 g of silica gel (Shimazu Seisakusho LTD.) was added to a solution of 50 mg of β-$IrHCl_2(PPh_3)_3$ in 20.00 g of toluene and the toluene in the mixture was removed by distillation. $RhH(CO)(PPh_3)_3$ or $IrH(CO)(PPh_3)_3$ was prepared by treating trans-$RhCl(CO)(PPh_3)_2$ or $IrCl(CO)(PPh_3)_2$ with sodium borohydride in ethanol [16]. Propylene and D_2 (99.5 %) obtained from the Takachiho Chemical Industrial Co. were used without any further purification.

2.2 Procedure

A required amount of complex ($RhH(CO)(PPh_3)_3$ = 90 mg, $IrH(CO)(PPh_3)_3$ = 100 mg, trans-$IrCl(CO)(PPh_3)_2$ = 100 mg, β-$IrHCl_2(PPh_3)_3$ = 50 mg or β-$IrHCl_2(PPh_3)_3—SiO_2$ = 395 mg) was placed on the bottom of a reaction vessel of volume 150 ml. The vessel was connected to a conventional vacuum apparatus which was maintained at 10^{-4} Torr by means of a mercury diffusion pump backed by an oil rotary pump. In the reaction vessel charged with one of the complexes except β-$IrHCl_2(PPh_3)_3—SiO_2$, 20.00 g of degassed solvent (DMF, benzene or toluene) was condensed. 0.8 mM of propylene and a required amount of deuterium were admitted to the reaction vessel. The mixture was stirred and reacted with $RhH(CO)(PPh_3)_3$ at 30°C, with β-$IrHCl_2(PPh_3)_3—SiO_2$ at 0°C and with the other complexes at 60°C. At a required reaction time, propylene and propane were sampled by expanding into an evacuated vessel, and transferred to the analysis.

2.3 Analysis

Composition of the normal propylene and the four kinds of monodeuterated propylenes and that of the **non-deut.propane** and the two kinds of monodeuterated propanes were determined quantitatively by measuring the intensities of the rotational spectra of the $1_{01} - 0_{00}$ transition for the former species[11, 12,17,18] and those of the $3_{12} - 3_{03}$ and $4_{13} - 4_{04}$ transitions for the latter species[19,20], respectively. Their frequencies for propane are listed in Table 1. The $1_{01} - 0_{00}$ transition frequencies for propylene have been already described in reference 17. The rotational spectra of propylene and propane were measured with a conventional 100-kHz sinusoidal Stark modulation microwave spectrometer by the method of Morino and Hirota[11,12]. The measurements were made at Dry Ice temperature by using a 3-m absorption cell. The absorption lines of propane were recorded under the condition of Stark dc voltage gradient of about 500 V/cm with sinusoidal modulation amplitude of 1000 V/cm. The correction coefficients [11,12] of intensities of absorption lines used to estimated the relative abundance of deuterated propane from the appearance intensities of absorption lines were also shown in Table 1.

TABLE 1

Rotational spectra of propane and its monodeuterated species used in the analysis

Species	Transition	Frequency (MHz)	Correction Coefficient
$CH_3 \cdot CH_2 \cdot CH_3$	$3_{12} - 3_{03}$	24365.47	1.226
sym-$CH_2D \cdot CH_2 \cdot CH_3$	$3_{12} - 3_{03}$	24324.96	1.000
asym-$CH_2D \cdot CH_2 \cdot CH_3$	$4_{13} - 4_{04}$	24283.11	1.000
$CH_3 \cdot CHD \cdot CH_3$	$4_{13} - 4_{04}$	23971.76	1.106

3. RESULTS AND INTERPRETATION

3.1 $RhH(CO)(PPh_3)_3$, $IrH(CO)(PPh_3)_3$ and trans-$IrCl(CO)(PPh_3)_2$

The observed change in the composition of the propylenes during the propylene deuterogenation with $RhH(CO)(PPh_3)_3$ in DMF at 30°C is shown in Fig. 1, which shows that the exchanges to the respective deuterated propylenes occur simultaneously. The formation rate of propylene-2-d_1 can be expressed approximately a first order kinetics, i.e.,

$$k = \frac{2.303}{t} \log \frac{C_e}{C_e - C_t}$$

where C_t and C_e are the concentrations of propylene-2-d_1 at time t and equilibrium, respectively. The rate constant k was estimated as 0.050 (min^{-1}). It was also confirmed that the kinetic orders of the isotopic exchange reactions of propylene-3-d_1, cis-propylene-1-d_1 and trans-propylene-1-d_1 are also approximately first. The relative rates of the formation are 23.6 (propylene-2-d_1): 3.5 (propylene-3-d_1); 1 (cis-propylene-1-d_1); 1 (trans-propylene-1-d_1). It was found that

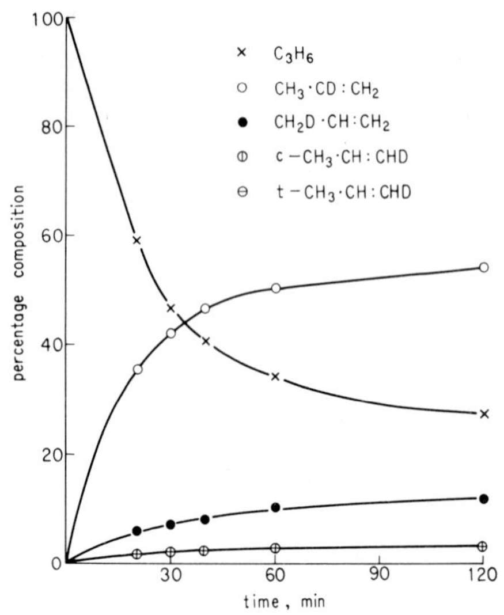

Fig. 1. Composition-time curves for the propylene-deuterium exchange with $RhH(CO)(PPh_3)_3$ in DMF at 30°C. D_2 / C_3H_6 = 1.

TABLE 2

Compositions of the monodeuterated propanes for the propylene deuterogenations (D_2 / C_3H_6 = 1) with $RhH(CO)(PPh_3)_3$ and trans-$IrCl(CO)(PPh_3)_2$ in DMF.

Catalyst	Temp., °C	Hydrogenation, %	Propane-d_1, %	
			$CH_2D\cdot CH_2\cdot CH_3$	$CH_3\cdot CHD\cdot CH_3$
$RhH(CO)(PPh_3)_3$	30	9	34.2 ± 0.8	65.8 ± 0.8
$IrCl(CO)(PPh_3)_2$	60	14	23.6 ± 0.5	76.4 ± 0.5
$IrCl(CO)(PPh_3)_2$	60	66	27.9 ± 2.9	72.1 ± 2.9

the orders of deuterogenation are first with respect to propylene and deuterium, respectively. The initial rate of deuterogenation was 0.37 % (min^{-1}) which is comparable with that of the isotopic exchange into propylene-3-d_1. Observed compositions of the monodeuterated propanes in the course of hydrogenation is shown in Table 2. This result indicates that the amount of propane-2-d_1 (65.8 %) at the initial stage of the hydrogenation is greater than thermodynamic equilibrium value of 50 %, which is coincided with the predominant formation of propylene-2-d_1 shown in Fig. 1. Similar results were observed in the case with $RhH(CO)(PPh_3)_3$ in benzene, excepting that the deuterogenation rate in benzene was somewhat faster than that in DMF. In Fig. 2, the composition of the four species of monodeuterated propylenes versus conversion in the case of the propylene-deuterium exchange with $RhH(CO)(PPh_3)_3$ in DMF or benzene is shown. It shows that the catalytic selectivity is regardless of the solvent being polar or nonpolar. The results of the propylene-deuterium exchanges and the propylene deuterogenations with $IrH(CO)(PPh_3)_3$ and $IrCl(CO)(PPh_3)_2$ were also similar to the ones obtained with $RhH(CO)(PPh_3)_3$.

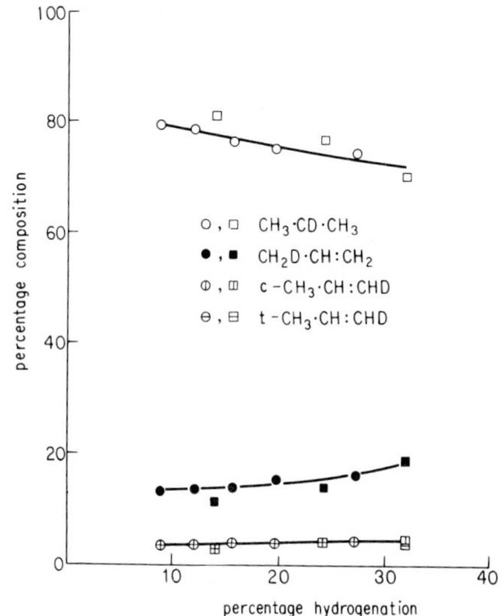

Fig. 2. Composition-conversion curves for the propylene-deuterium exchange with $RhH(CO)(PPh_3)_3$ in DMF (circles) and benzene (squares) at 30°C. D_2 / C_3H_6 = 1.

3.2 β-$IrHCl_2(PPh_3)_3$

In order to elucidate the catalytic behavior of the six-coordinated complex of iridium, β-$IrHCl_2(PPh_3)_3$ has been chosen, as an example. Although the preparation and structure of β-$IrHCl_2(PPh_3)_3$ had been studied [14], there have been no published work concerning to its catalytic properties, yet. The propylene deuterogenation with $IrHCl_2(PPh_3)_3$ has been carried out in toluene and benzene. The relative initial formation rates obtained in toluene at 60°C are 2.8 (propylene-2-d_1); 3.6 (propylene-3-d_1); 1 (cis-propylene-1-d_1); 1 (trans-propylene-1-d_1). In the case with benzene, similar relations could be observed.

Here, it should be noted that the hydrogenation rate is very much enhanced with the coexistence of small amounts of O_2. Measuring the infrared spectra of $IrHCl_2(PPh_3)_3$ treated with oxygen in toluene, we observed the absorption bands of triphenylphosphine oxide at 1122 cm^{-1} and 1198 cm^{-1}, which are in accordance with the results obtained from $RhCl(PPh_3)_3$ [22]. The formation of triphenylphosphine oxide was also confirmed by thin-layer chromatography analysis.

3.3 β-$IrHCl_2(PPh_3)_3$—SiO_2

In order to examine the catalytic behavior of the complexes in the solid state, the system of β-$IrHCl_2(PPh_3)_3$—SiO_2 was chosen, as an example. The compositions of the produced monodeuterated propylenes and propanes in the deuterogenation with the system have been observed. The results of the propylene-deuterium exchange are shown in Fig. 3 (D_2 / C_3H_6 = R = 1) and Fig. 4 (R = 2 and 3). The produced amounts of the monodeuterated propylenes were independent of R and are

propylene-3-d_1 ≫ cis-propylene-1-d_1
≈ trans-propylene-1-d_1 > propylene-2-d_1.
The both experimental results that the deuterogenation rate was faster than the isotopic exchange rate, and propylene-2-d_1 was scarcely produced at the initial stage are the remarkable differences with the reaction with $RhH(CO)(PPh_3)_3$ in solvent. Variation in the composition of the monodeuterated

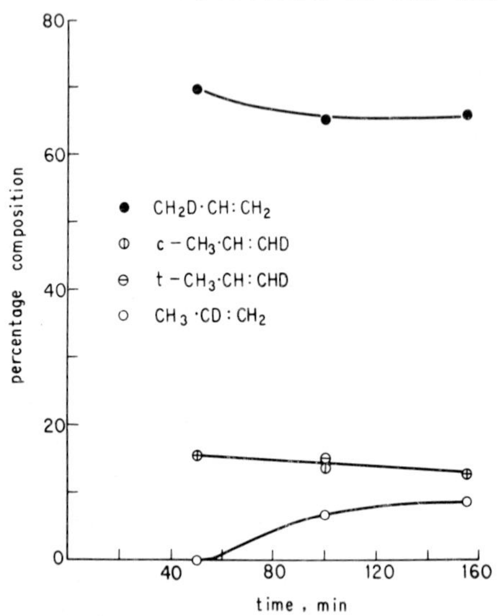

Fig. 3. Composition-time curves for the propylene-deuterium exchange with β-$IrHCl_2(PPh_3)_3$ — SiO_2 at 0°C. D_2 / C_3H_6 = 1.

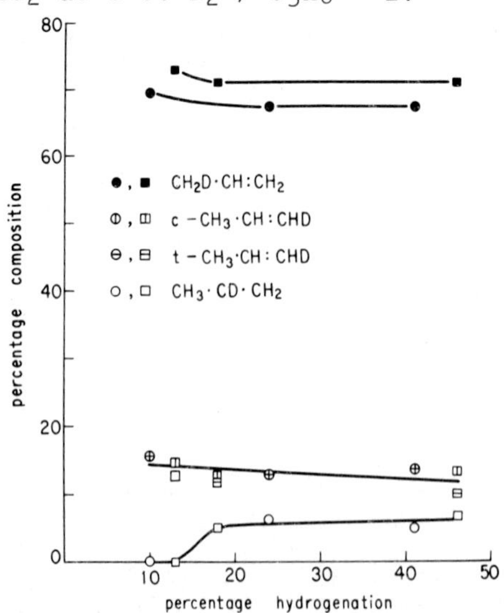

Fig. 4. Composition-conversion curves for the propylene-deuterium exchange with β-$IrHCl_2(PPh_3)_3$— SiO_2 at 0°C. D_2 / C_3H_6 = 2 (circles), 3 (squares).

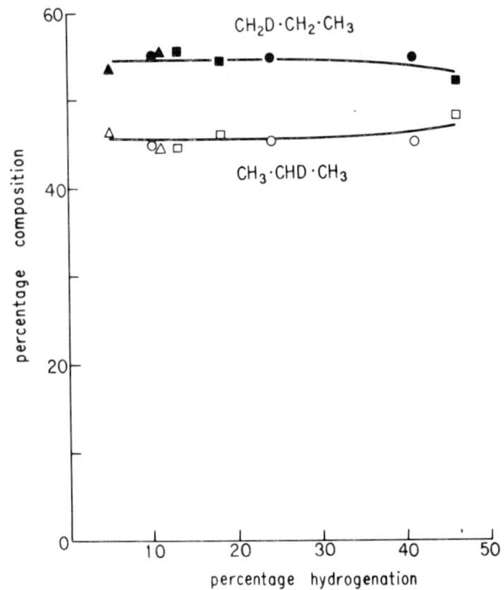

Fig. 5. Composition-conversion curves for the propylene deuterogenation with β-$IrHCl_2(PPh_3)_3$ — SiO_2 at 0°C. D_2 / C_3H_6 = 1 (triangles), 2 (circles), 3 (squares).

propanes during the propylene deuterogenation is shown in Fig. 5, which shows that the formation rates of propane-2-d_1 and propane-1-d_1 are comparable. No influence of deuterium/propylene ratio on the composition has been also confirmed.

4. DISCUSION

The experimental fact that the appreciable amount of various subspecies of propylene-d_1 and propane-d_1 are produced during the propylene deuterogenation shows that the addition of hydrogen atoms is stepwise process. The Wilkinson's mechanism [9], the simultaneous addition of two hydrogen atoms to an olefin, can not be applied to this experimental result. It can be considered that the catalytic activity is due to the dissociation of triphenylphosphine ligand from the complex, because the formation of triphenylphosphine oxide was confirmed in treating $IrHCl_2(PPh_3)_3$ in solution with oxygen, of which traces in the system enhanced the hydrogenation rates very much with this complex and $IrCl(CO)(PPh_3)_2$ [5]. Such a dissociation has been also realized in the case of $RhH(CO)(PPh_3)_3$ by Wilkinson et al. [4] who found the catalytically active species $RhH(CO)(PPh_3)_2$. With the hydrogenation of maleic acid with $IrCl(CO)(PPh_3)_2$, it has been shown that after the coordination of olefin by π-bonding to metal had taken place, hydride formation was accompanied by formal oxidation of metal [5]. If the availability of the formation of the metal-ligand bond depends on the nucleophilic nature of the reactant, hydride formation should occur before coordination of a π-bonded propylene with central metal because of the order of the nucleophilic nature, maleic acid > hydrogen > propylene.
Thus, for instance, in the isotopic exchange it is expected

that the following mechanism to be operative,

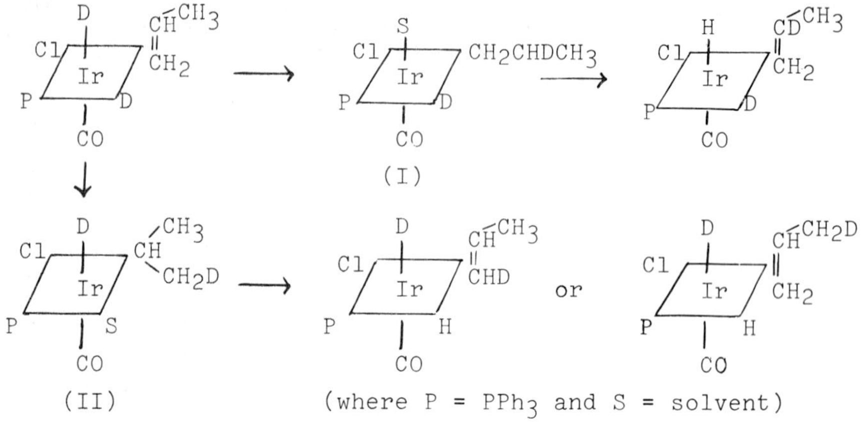

D CH$_3$–CH=CH$_2$ / Cl–Ir–D, P, CO ⟶ S / Cl–Ir–CH$_2$CHDCH$_3$, P, D, CO (I) ⟶ H / Cl–Ir–CD(CH$_3$)=CH$_2$, P, D, CO

↓

D / Cl–Ir–CH(CH$_3$)CH$_2$D, P, S, CO (II) ⟶ D / Cl–Ir–CH(CH$_3$)=CHD, P, H, CO or D / Cl–Ir–CH(CH$_2$D)=CH$_2$, P, H, CO

(where P = PPh_3 and S = solvent)

Although the σ-bonded iridium-normalpropyl (I) and -isopropyl (II) intermediates are formed in the systems, most of the isotopic exchange and hydrogenation may be taken place fast via the normalpropyl species (I) in the early stage, because propylene-2-d_1 and propane-2-d_1 were in the main products.

In the case with $RhH(CO)(PPh_3)_3$, the relative rates of formation of the σ-bonded metal-normalpropyl and isopropyl intermediates are estimated as v_n = 7.7 and v_i = 1.0, from the observed relative formation rates of various subspecies of propylene-d_1,

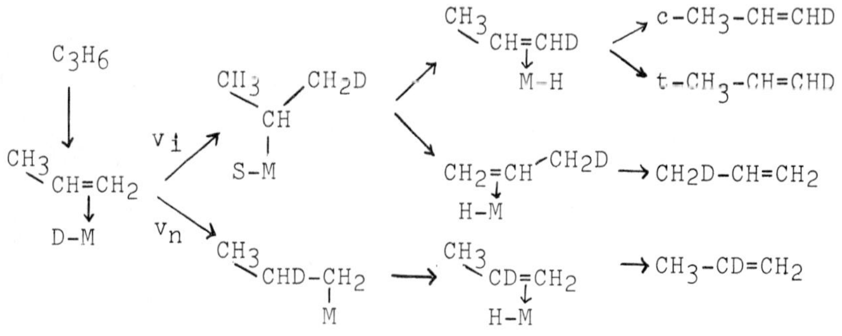

C_3H_6 ⟶ CH_3–CH=CH_2 (↓ D–M)

$\xrightarrow{v_i}$ CH_3–CH(CH_2D)–(S–M) ⟶ CH_3–CH=CHD (↓ M–H) ⟶ c-CH_3-CH=CHD, t-CH_3-CH=CHD

⟶ CH_2=CH–CH_2D (↓ H–M) ⟶ CH_2D-CH=CH_2

$\xrightarrow{v_n}$ CH_3–CHD–CH_2–M ⟶ CH_3–CD=CH_2 (↓ H–M) ⟶ CH_3-CD=CH_2

where M is the central metal atom of a complex. The fact that the hydrogenation rate is less than that of the exchange into propylene-2-d_1 indicates that the addition of H atom to the σ-bonded metal-normalpropyl intermediate is the rate-determining step in the hydrogenation.

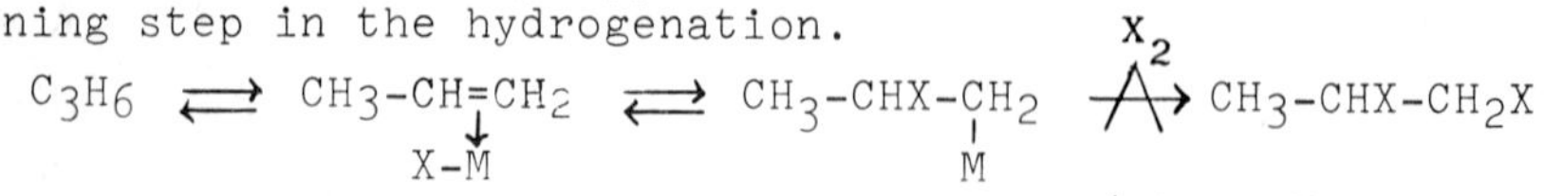

C_3H_6 ⇄ CH_3-CH=CH_2 (↓ X–M) ⇄ CH_3-CHX-CH_2–M $\overset{X_2}{\nrightarrow}$ CH_3-CHX-CH_2X

(where X = H or D)

In the case of $IrH(CO)(PPh_3)_3$, since the similar results were obtained with $IrCl(CO)(PPh_3)_2$ and $RhH(CO)(PPh_3)_3$, the σ-bonded metal-normalpropyl intermediate may be involved in the main steps of the hydrogenation.

However, in the case with $IrHCl_2(PPh_3)_3$ in solvent, it is considered that the formation rate of an intermediate of metal-isopropyl type is comparable with that of metal-normal-propyl type, on the basis of the good agreement between the calculated relative rates with the same v_n and v_i and the experimental values, as below.

	propylene-2-d_1,	propylene-3-d_1,	cis-propylene-1-d_1,	trans-propylene-1-d_1
calc.	3.0	3.0	1.0	1.0
obs.	2.8	3.6	1.0	1.0

Since $IrHCl_2(PPh_3)_3$ molecule has two chlorine atoms of the large electronegativity, it may be explained that the intermediate of secondary alkyl type, which is more **nucleophilic** than that of primary alkyl type, is dominant with this complex.Therefore, the isotopic exchange must proceed via propylene molecule coordination to the vacant trans position against triphenylphosphine ligand where has been filled by the other triphenylphosphine ligand previously.

In the case with $IrHCl_2(PPh_3)_3—SiO_2$, it was found the different behavior of the exchangeable hydrogens in propylene; i. e., propylene-3-d_1 was produced predominantly, as described. The results can be explained on the basis of an intermediate of π-allyl type, as follows.

$$\begin{array}{ccccccc} \overset{C_3H_6}{\downarrow} & & \text{CH} & \xrightarrow{D_2} & \text{CH} & & \\ CH_3\text{-}CH{=}CH_2 & \longrightarrow & CH_2 \cdots CH_2 & \longrightarrow & CH_2 \cdots CH_2 & \longrightarrow & CH_2D\text{-}CH{=}CH_2 \\ \downarrow & & \downarrow & & \downarrow & & \downarrow \\ M & & H\text{-}M & & D\text{-}M & & M \end{array}$$

However, since the observed amounts of produced propane-2-d_1 and -1-d_1 are reasonably equal, both intermediates of metal-normalpropyl and -isopropyl types can be considered as comparably operative in the hydrogenation mechanism, which is essentially the same with that in the case with solvent. Considering the fact that the hydrogenation rate is greater than the isotopic exchange rate in the absence of solvent, it appears that the amount of the intermediate of π-allyl type is less than those of normalpropyl and isopropyl types. In addition, it is reasonable to assume that the active center of the complex for the intermediate of π-allyl type is the coordination position **created by losing a hydrogen ligand.**

Finally, it should be emphasized that although the information about the exchange products is helpful to discuss the hydrogenation mechanism, the intermediate species in exchange reaction are not always the same as the hydrogenation ones, therefore the quantitative analysis of the deuterated paraffins is important for the studies of the hydrogenation mechanism.

REFERENCES

1) L. Vaska, Inorg. Nucl. Chem. Letters 1 (1965), 89.
2) L. Vaska and R. E. Rhodes, J. Am. Chem. Soc. 87 (1965), 4970.
3) G. G. Eberhardt and L. Vaska, J. Catalysis 8 (1967), 183.
4) C. O'Connor and G. Wilkinson, J. Chem. Soc. (A) (1968), 2665.
5) B. R. James and N. A. Memon, Can. J. Chem. 46 (1968), 217.
6) M. Yamaguchi, Kogyo Kagaku Zasshi 70 (1967), 675.
7) M. Polanyi and J. Horiuti, Trans. Faraday Soc. 30 (1934), 1164.
8) G. C. Bond, "Catalysis by Metal", Academic Press, London (1962) Chapter 11.
9) J. A. Osborn, F. H. Jardine, J. F. Young and G. Wilkinson, J. Chem. Soc. (A) (1966), 1711.
10) A. S. Hassey, Y. Takeuchi, J. Am. Chem. Soc. 91 (1969), 672.
11) Y. Morino and E. Hirota, Nippon Kagaku Zasshi 85 (1964), 535.
12) E. Hirota, Shokubai (Catalyst), 13 (1971), 31.
13) J. P. Collman and J. W. Kang, J. Am. Chem. Soc. 89 (1967), 849.
14) R. C. Taylor, J. F. Young and G. Wilkinson, Inorg. Chem. 5(1966), 20.
15) L. Vaska and J. W. DiLuzio, J. Am. Chem. Soc. 84 (1962), 4989.
16) D. Evans, G. Yagupsky and G. Wilkinson, J. Chem. Soc. (A) (1968), 2660.
17) T. Ueda and K. Hirota, J. Phys. Chem. 74 (1970), 4216.
18) D. R. Lide, Jr., and D. E. Mann, Ibid. 27 (1957), 868; D. R. Lide, Jr., ibid. 35 (1961), 1374.
19) D. R. Lide, Jr., Ibid. 33 (1960), 1514.
20) E. Hirota, C. Matsumura and Y. Morino, Bull. Chem. Soc. Jap., 40 (1967), 1124.
21) C. H. Townes and A. L. Schawlow, "Microwave Spectroscopy" McGraw-Hill, New York, N. Y. (1955), 102.
22) H. van Bekkun, F. van Rantwijk, and T. van de Putte, Tetrahedron Lett. No. 1 (1969), 1.

ACKNOWLEDGMENT: The author wishs to express her gratitude to Professor Eizi Hirota, the University of Kyushu, Dr. Shuji Saito, the Sagami Central Chemical Laboratory, Professor Yozaburo Yoshikawa, Osaka University, and Mr. Masao Mineo, for their advice and help concerning the construction of the microwave spectrometer, and to Dr. Hiroshi Yamazaki, Professor Tominaga Keii and Dr. Kiyoshi Otsuka, Tokyo Institute of Technology, for helpful discussions of the present work.

DISCUSSION

P. B. WELLS

This paper has focussed attention upon the relative rates of formation of normal- and iso-intermediates when hydrogen atom addition to coordinated olefin occurs.

We have examined the isomerization of $CHD{:}CDCH_2CH_2CH_3$ to cis- and trans-2-pentene catalyzed by a wide range of Group VIII metal complexes in benzene solution in the range 35 to 80°C. The relative rates of formation of n-pentyl complexes (v_n) and of 2-pentyl complexes (v_i) have been calculated from the observed redistribution of deuterium in the reactants and the products at low conversions. Some values are compared with those quoted in Paper 26 in the Table. The ratio $v_n = v_i$ is clearly a variable quantity even for a given complex in solution, and this is puzzling. I should like to ask Dr. Ueda whether she can interpret the preferential formation of n-propyl complexes that she has presented in her paper.

Olefin	Complex or Catalyst	$v_n{:}v_i$	Reference
Propylene	$RhH(CO)(PPh_3)_3$	7·7:1.0	Paper 26
	$IrH(CO)(PPh_3)_3$		
1-Pentane	$RhH(CO)(PPh_3)_3$	0.4 to 1.5:1.0	1
1-Pentene	$IrH(CO)(PPh_3)_3$	0.3:1.0	2
	$Ni[P(OEt)_3]_4$	0.1:1.0	2
	Alumina-supported Ni	9.0:1.0	3

1) B. Hudson, D. E. Webster, and P. B. Wells, unpublished work.
2) D. Bingham, D. E. Webster, and P. B. Wells, J. Chem. Soc. in press.
3) D. McMann and P. B. Wells, unpublished work.

T. UEDA

I also think these variations are interesting. I do not have a good explanation offhand. The ratio may be very sensitive to many conditions such as the conversion or the steric requirement.

H. van BEKKUM

Regarding the suggestion on the ninth page of your interesting paper that an allyl complex might be involved in the exchange of propene over $IrHCl_2(PPh_3)_3$ on silica, I have some questions:

1) Which ligands are dissociated or substituted in the complex in order to accomodate a hydride as well as an allyl ligand in the intermediate?

2) Is there any evidence that the next step, H-M complex reacting with molecular D_2 yielding D-M complex, can occur? Generally, iridium-hydride complexes do not exchange readily either with molecular deuterium or with hydroxylic solvents.

3) Do you have any ideas about the role of the silica?

T. UEDA

1) A plausible mechanism is shown in the following scheme.

$Ir^{III}HCl_2(PPh_3)_3 \rightarrow Ir^{I}Cl(PPh_3)_3 + HCl$

$Ir^{I}Cl(PPh_3)_3 + C_3H_6 \rightarrow \pi\text{-allyl } Ir^{III}HCl(PPh_3)_3.$

2) I can offer the following observations. In the n.m.r. spectrum of β-$IrHCl_2(PPh_3)_3$, the hydride resonance is split into a symmetrical quartet with relative intensities of 1:3:3:1 at $\tau = 29.2$. After the reaction, this hydride resonance was found to be weak.

For the propylene hydrogenation, β-$IrHCl_2(PPh_3)_3$--SiO_2 catalyst is as active as metal rhodium catalyst, though $IrH(CO)(PPh_3)_3$ and $IrCl(CO)(PPh_3)_2$ are poor catalysts.

3) The distribution of D atom in deuterated propylene species produced by the propylene-deuterium exchange with β-$IrHCl_2(PPh_3)_3$ in the absence of SiO_2 and solvent was similar to that in the presence of SiO_2.

K. TAMARU

Suppose we obtain 100% propylene-3-d_1 as the initial reaction product. Your interpretation is that a π-allyl species is the reaction intermediate. However, if one of the hydrogen atoms of methyl group would be dissociated to form σ-allyl species, we still obtain 100% propylene-3-d_1 in the same way. If the push-pull mechanism is valid, and deuterium atoms attacks the $=CH_2$ end pulling one of the methyl hydrogens,still we obtain 100% propylene-3-d_1. Accordingly, as far as the d_1-species is concerned, those three mechanisms give the same reaction products, and we cannot tell which is the correct. However, if we analyze the d_2-species together with the d_1-species, those three mechanisms give different distributions of geometrical isomers. In other words, the σ-allyl intermediate gives double exchange only at methyl hydrogen in its d_2-species, whereas in the push-pull mechanism the two deuterium atoms come at each end carbon to form 100% propylene-3, 1-d_2. The π-allyl intermediate, on the other hand, gives 1:1 mixture of these two d_2-isomers. In this way, we can determine the reaction intermediate by analyzing not only the d_1-species, but also the d_2-species, which has been accomplished for the first time by us (Kondo, Ichikawa, Saito and Tamaru) for elucidating the mechanism of catalytic exchange reaction.

T. UEDA

I think that is a good approach.

I have analyzed the isotopic isomers of dideuterated propylene by the microwave spectroscopy in the case of the self-exchange of deuterium in propylene-3-d_1 on platinum catalyst and discussed the mechanism of the reaction.[17] In the present case, the reaction rate of propylene deuterogenation is extremely faster than that of propylene-deuterium exchange, and propylene-3-d_2, cis-propylene-1,3-d_2 and trans-propylene-1,3-d_2 cannot be detected. (The measurable limit on concentration is 0.1% for each species.)

K. TAMARU

The results given in Fig. 3 are interpreted to demonstrate that a π-allyl species is the intermediate in the exchange reaction between propylene and deuterium (ninth page). If a π-allyl species is the reaction intermediate, the d_1-species initially formed should be 100% propylene-3-d_1, whereas, if iso-propyl type is the intermediate, it should be propylene-3-d_1 (60%), cis-propylene-1-d_1 (20%) and trans-propylene-1-d_1 (20%).

The results given in Fig. 3 appear to demonstrate mainly the latter intermediate rather than the former one, as cis-propylene-1-d_1 and trans-propylene-1-d_1 both do not drop to zero, but approach to nearly 20% at zero reaction time.

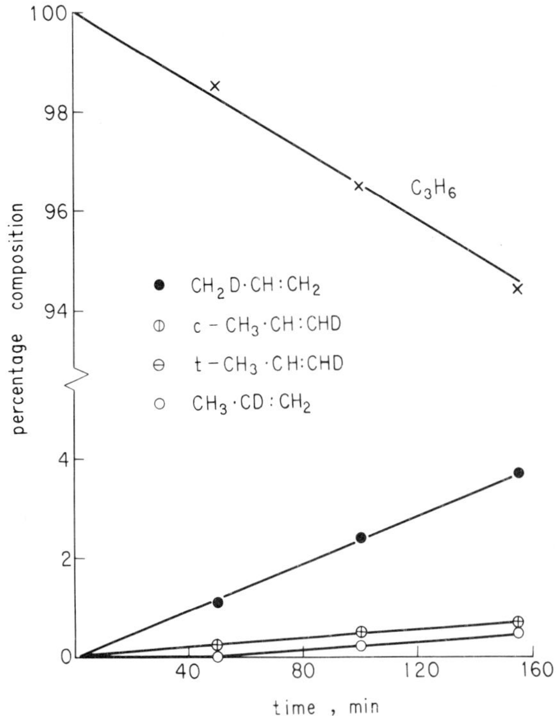

Fig. 3b Composition-time curves for the propylene-deuterium exchange with β-$IrHCl_2(PPh_3)$ --SiO_2 at 0°C. $D_2/C_3H_6=1$

T. UEDA

The initial rates of formation of three isomers can be obtained from Fig. 3(b) more accurately than from Fig. 3. The relative rates are 4.6 for propylene-3-d_1, 1 for cis propylene-1-d_1 and 1 for trans propylene-1-d_1. Therefore, it can be pointed out that an intermediate of π-allyl type is fairly produced besides isopropyl type in the isotopic exchange, though the former species may play a minor role for the hydrogenation.

Paper Number 27

OXIDATION OF OLEFINS OVER PALLADIUM SALTS - ACTIVE CHARCOAL CATALYSTS

K. FUJIMOTO and T. KUNUGI
Department of Synthetic Chemistry, University of Tokyo
Hongo, Bunkyo-ku, Tokyo, Japan

ABSTRACT: Palladium salts adsorbed on active charcoal are excellent catalysts for the oxidation of olefins without any co-catalyst such as cupric chloride. Ethylene is oxidized to form acetaldehyde in the vapor phase with constant activity and high selectivity (about 99%). The reaction is accelerated markedly by an increase in steam pressure and it is suppressed by a rise in temperature. The reoxidation of palladium to palladium ion is more rapid than carbonylation. Active charcoal seems to catalyze the reoxidation reaction through the activation of oxygen by dissociative adsorption.

1. INTRODUCTION

The oxidation of olefins by acidic solutions of palladium (II) salts is a typical reaction involving coordination complexes. It proceeds through the intermediate formation of palladium-olefin π-complexes and the redox decomposition of the complexes. Smidt et al. discovered a catalyst system for olefin oxidation which uses cupric chloride and oxygen as oxidation agents for reduced palladium.[1)]

We have discovered that palladium salts adsorbed on active charcoal with no redox reagent are excellent catalysts for the oxidation of olefins by oxygen in the presence of steam. The catalytic behavior of such solid catalysts and the reoxidation of palladium catalyzed by charcoal are described in this paper.

2. EXPERIMENTAL

2.1 Apparatus and procedure

The apparatus was a conventional tubular flow reactor equipped with devices for measuring and controlling the temperature and flow rates. The reactor was a glass tube (500 mm long, 16 mm i.d.), with a thermocouple sheath along the central axis. It was heated by electric furnace or oil bath. The catalyst charged in the reactor was less than 20 g. Glass spheres of 1 - 2 mm diameter were packed above and below the catalyst. Water from a micro-feeder was vaporized in the upper part of the reactor. The mixture of gaseous reactants was then passed through the catalyst bed. The liquid products were collected in traps cooled with ice water.

An apparatus with gas recirculation was used to measure the adsorption rate of oxygen. In this, the decrease in pressure of the system through reaction was balanced by the addition of the gas mixture comprising ethylene and nitrogen.

2.2 Analytical

Acetaldehyde was determined by volumetric titration. Other organic products and the gaseous mixture comprising oxygen, carbon monoxide, carbon dioxide and ethylene were analyzed by gas chromatography.

2.3 Materials

Ethylene and propylene (99.9% purity) were fed to the reactor after passing through calcium chloride and active charcoal. Commercially available oxygen and nitrogen were also passed over calcium chloride and active charcoal. Ion exchanged water was employed.

A commercially available active charcoal was used. It had been prepared from wood and activated by steam treatment. The specific surface area was 1280 m^2/g (BET).

2.4 Catalyst

$PdCl_2$-Active Charcoal Catalyst:

Palladium dichloride (0.84 g) was dissolved in 300 ml of 1 N hydrochloric acid (Solution A). Active charcoal, which had been boiled with dilute nitric acid and then washed with pure water, was mixed with 200 ml of 1 N hydrochloric acid (Mixture A) and boiled for 10 minutes. After cooling to room temperature, Solution A was added with stirring (Mixture B). After 24-48 hours at room temperature, the palladium chloride had been adsorbed completely on the charcoal. The charcoal containing palladium chloride was then washed with 200 ml of pure water and dried at 150°C for 6 hours *in vacuo*.

The palladium chloride-active charcoal catalyst so prepared contained some free hydrochloric acid and the atomic ratio of chlorine to palladium was about 2.0 to 2.2.

Other palladium salts were adsorbed on active charcoal from the solution in the acid corresponding to the anion of the salt.

3. RESULTS AND DISCUSSION

3.1 Catalysis by palladium salt-charcoal catalysts

Oxidation of ethylene was examined with $PdCl_2$-charcoal, $PdSO_4$-charcoal, $PdCl_2$-γ-alumina and $PdCl_2$-silica gel. As shown in Fig. 1, palladium salts are effective catalysts only when supported on active charcoal. The activity and selectivity of palladium chloride-charcoal catalysts are substantially constant over runs of 50 hours. The number of moles of acetaldehyde formed during 50 hours is more than 2000 times the number hour is more than 40 moles per mole of palladium. In later studies, the rate was increased to about 200 even under normal pressure by selecting the proper conditions for catalyst preparation and for reaction. It is apparent that active charcoal has a unique function as a co-catalyst in the oxidation of ethylene.

Other palladium salts supported on charcoal were also found to have catalytic activity as shown in Fig. 2. It is not clear why palladium chloride has the highest activity nor why its temperature dependence differs from those of other palladium salts.

Smidt *et al.* have shown[1] that the unit reactions in the $PdCl_2$ - $CuCl_2$ catalyst system are as follows:

$$PdCl_2 + C_2H_4 + H_2O \rightarrow CH_3CHO + Pd + 2HCl \quad (1)$$

$$Pd + 2HCl + 1/2\ O_2 \rightarrow PdCl_2 + H_2O \quad (2)$$

where the reaction (2) is divided into two steps.

$$Pd + 2CuCl_2 \rightarrow PdCl_2 + 2CuCl \quad (3)$$

$$2CuCl + 2HCl + 1/2\ O_2 \rightarrow 2CuCl_2 + H_2O \quad (4)$$

In the present reaction system, it seems possible that the reaction (1) proceeds through a route similar to that postulated by Henry[2] or Jira[3] and that reaction (2) is catalyzed by active charcoal.

3.2 Reactivity of various olefins

Olefins other than ethylene are also oxidized by oxygen with this catalyst to give saturated carbonyl compounds. Reactivities and product distributions are shown in Table 1. The order of reactivity is ethylene > propylene > cyclohexene > butadiene. Except with ethylene, the main products are ketones but small amounts of aldehydes are also formed. The similarity of these phenomena to those in the homogeneous reaction[1,4] suggests that the mechanism of carbonylation is much the same in both reactions.

3.3 Influence of temperature and steam pressure

In the previous work,[2] more than one water molecule has been considered as involved in the formation of one molecule of carbonyl compound. However, the influence of water on the

TABLE 1
Oxidation of olefins over $PdCl_2$-A.C. catalysts

Olefin	Conversion of Olefin (mol %)	Selectivity of Product (mol %)		
		Aldehyde	Ketone	Carbon dioxide
Ethylene	59.9	98	-	0.9
Propylene	27.2	10	89	0.5
1-Butene	7.7	4	93	0.9
Butadiene	1.5	25	20	23
Cyclohexene	0.5	-	84	13

$PdCl_2$-A.C (Pd 3 wt%), Temperature 105°C,
Olefin:O_2:H_2O = 5:2:15, W/F = 25 g-cat·hr/mol.

TABLE 2
Re-oxidation of Palladium

Anion	Olefin	Reagent	Temp.(°C)	Pressure of Olefin (atm)	$K_2\sqrt{K}$*
Cl^-	C_2H_4	H_2O	95	0.23	29.4
Cl^-	C_2H_4	H_2O	105	0.17	18.0
Cl^-	C_2H_4	H_2O	105	0.23	12.0
Cl^-	C_2H_4	H_2O	115	0.24	9.5
SO_4^{--}	C_2H_4	H_2O	105	0.23	8.0
NO_3^-	C_2H_4	H_2O	105	0.24	17.5
OAc^-	C_2H_4	CH_3CHOH	130	0.29	1.1
Cl^-	CO	H_2O	100	0.20	193

PdX_2-A.C. (Pd 0.5 wt%), P_{H_2} 0.50 atm
*mole/g-cat·hr·atm$^{1/2}$ x 10^3

reaction rate has not been clear in the homogeneous systems because of the presence of excess water. In the present work, steam pressure has a marked effect on the rate of reaction and the rate decreases remarkably with increasing temperature as shown in Fig. 2 and Fig. 3.

If the mechanism of ethylene carbonylation follows equations (5) - (8) and if the concentrations of intermediate complexes are very low, the rate of reaction (1) is given by equation (9).

$$PdCl_2 + H_2O \underset{}{\overset{K_1}{\rightleftharpoons}} PdCl_2(H_2O) \tag{5}$$

$$PdCl_2(H_2O) + C_2H_4 \overset{K_2}{\rightleftharpoons} PdCl_2(H_2O)(C_2H_4) \tag{6}$$

$$PdCl_2(H_2O)(C_2H_4)+H_2O \overset{K_3}{\rightleftharpoons} PdCl_2(OH^-)(C_2H_4)+H_3O^+ \tag{7}$$

$$PdCl_2(OH^-)(C_2H_4)+H_2O \xrightarrow{k} Pd+2Cl^-+H_3O^++CH_3CHO \tag{8}$$

$$v = kK_1K_2K_3[PdCl_2][C_2H_4][H_2O]^n \quad (n = 2\sim3) \tag{9}$$

Thus, the above equations well interpret the effect of steam pressure shown in Fig. 2.

The apparent activation energy of this reaction is calculated from Fig. 3 as about -18 kcal/mol. The value seems strange from the standpoint of usual solid-catalyzed oxidation reactions. We interpret this as indicating that the surface concentration of water raised to the power n (equation (9)) decreases with rise in the temperature more rapidly than the rate constant increases and, therefore, that the overall reaction rate decreases. This explanation is supported by the result obtained from reaction in the liquid phase catalyzed by a dispersion of the finely divided charcoal catalyst, which provides an apparent activation energy of about 11 kcal/mol.

3.4 Re-oxidation of the reduced palladium compound

Overall reaction rates can be represented by equation (10),

$$v = \frac{k_1k_2f_1(P_{C_2H_4}, P_{H_2O})f_2(P_{O_2})}{k_1f_1(P_{C_2H_4}, P_{H_2O}) + k_2f_2(P_{O_2})}[Pd_T] \tag{10}$$

where $[Pd_T]$ is the total amount of palladium. The results shown in Fig. 4, which were obtained by the use of an apparatus at constant pressure and volume, show that the effect of oxygen pressure on the reaction rate is appreciable only when its value is very small. This means that the rate of reaction (2) is larger than that of reaction (1).

Trial and error methods based on equation (10) show that $k_2f_2(P_{O_2})[Pd^o]$, the rate of re-oxidation of several palladium salts are represented by equation (11),

$$k_2f_2(P_{O_2})[Pd^o] = \frac{k_2\sqrt{K}\sqrt{P_{O_2}}}{1+\sqrt{K}\sqrt{P_{O_2}}}[Pd^o] \tag{11}$$

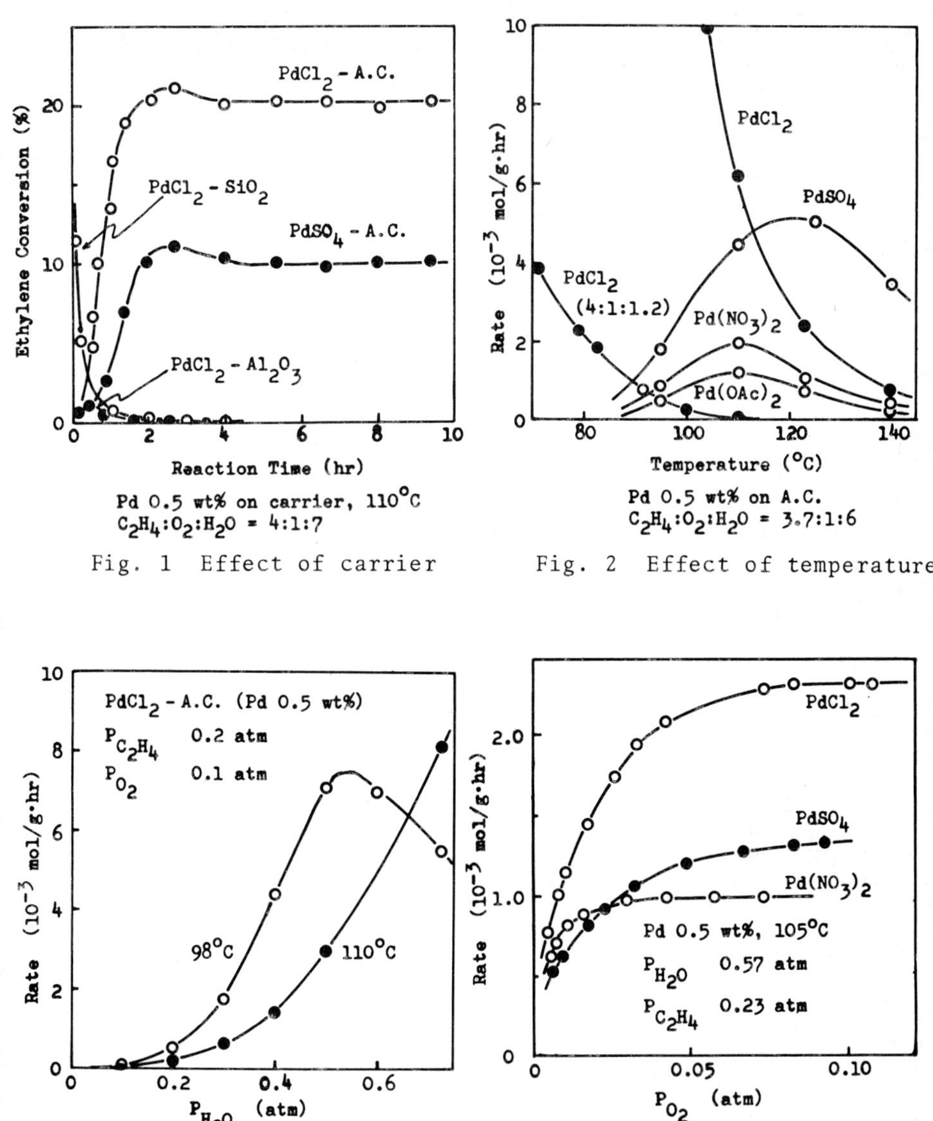

Fig. 1 Effect of carrier

Fig. 2 Effect of temperature

Fig. 3 Effect of steam pressure

Fig. 4 Effect of oxygen pressure

where K is equilibrium constant for the dissociative adsorption of oxygen. Equation (11) indicates that oxygen takes part in re-oxidation after being dissociatively adsorbed on the catalyst. The term, $k_2\sqrt{K}$, the measure of re-oxidation rate, varies with the partial pressures of ethylene and of steam, the temperature, the kind of active charcoal, and palladium salts as shown in Table 2.

REFERENCES

1) J. Smidt, W. Hafner, R. Jira, J. Sedlmeier, R. Sieber, R. Rüttinger, and H. Kojer, Angew. Chem., 71, 176 (1959).
2) P. M. Henry, J. Am. Chem. Soc., 86, 3246 (1964).
3) R. Jira, J. Sedlmeier, and J. Smidt, Ann. Chem., 693, 99 (1966).
4) P. M. Henry, J. Am. Chem. Soc., 88, 1595 (1966).

DISCUSSION

H. HEINEMANN

Active charcoal is sometimes prepared by coalification of wood in the presence of sublimable metal salts, such as $ZnCl_2$. Is it possible that your charcoal contained $ZnCl_1$ or the like, which could be reduced by hydrocarbon to Zn and then reoxidized by HCl and oxygen? This would provide a redox mechanism similar to that provided by Cu in the Wacker chemistry.

T. KUNUGI

Almost all of the active charcoals used in our study were made of wood by steam activation. Contents of inorganic substances were as follows: SiO_2 0.73, Fe_2O_3 0.08, Al_2O_3 0.08, CaO 0.08, MgO 0.03, K trace, Cu trace (wt %). The catalytic reaction could also be observed over $PdCl_2$-charcoal which was prepared from beet sugar of more than 99.9% purity by steam activation in our laboratory. The catalytic activity of $ZnCl_2$-activated charcoal was considerably less than that of steam-activated charcoal.

J. W. ESPY

In your paper you state that the oxidation of ethylene was examined with $PdCl_2$-active charcoal, $PdCl_2$-γ-alumina, and $PdCl_2$-silica gel. Furthermore, you give some evidence to show that palladium salts are effective catalysts only when they are supported on active charcoal. Do you feel there is some interaction between $PdCl_2$ and the other supports which makes them inactive?

T. KUNUGI

In general, there is no evidence of the interaction between $PdCl_2$ and the supports other than active charcoal. In the case of γ-Al_2O_3, silica gel etc., the rate of acetaldehyde formation decreased markedly and a lot of HCl flowed out from the catalyst bed and many crystals of metallic palladium were found on the catalyst surface after the reaction. From these results it is reasonable to consider that the reaction, $PdCl_2 \xrightarrow{(C_2H_4)} Pd^\circ + HCl$, proceeds easily, but the reverse reaction cannot take place over the supports other than active charcoal.

J. J. F. SCHOLTEN

It might be that we are dealing here with the normal Wacker catalyst. Various impurities are present in active charcoal which go into solution when steam is capillarily condensed in the pores. This is supported by the fact that coals of various sources with about the same surface area give different activities. Hence, I support Dr. Heinemann's remark.

T. KUNUGI

In our experiment, the capillary condensation was of course observed. At the reaction temperature lower than dew point of feed gas the reaction did not occur practically. At the temperature above dew point some of micro-pores of smaller diameters are supposed to be filled with water. Such pores filled with water can scarcely contribute to the reaction rate because of the extremely slow diffusion of feed gas and reaction products.

Active charcoals used were boiled with 5 - 10% nitric acid and washed with water before the preparation of catalysts. Hence, there is little chance of dissolution of impurities into water in the pores. The fact that coals of various sources with about the same surface area give different activities seems to depend upon the difference of dispersion of $PdCl_2$ on the surface, the extent of capillary condensation etc. caused by the difference of surface structure of charcoals.

C. E. FRANK

1) The paper is entitled "oxidation over Pd salts." Have you analyzed the catalyst after some hours of operation to determine if Pd is actually still present as Pd^{++}, or it is possibly converted largely to Pd°?

2) The pronounced activity of the C supports compared with alumina and silica may be due to a much greater surface area. Have you determined the relative surface areas of these supports?

T. KUNUGI

1) After 3-4 hours of operation (about 1,000 moles of acetaldehyde were produced from 1 mole of palladium chloride), the catalyst was reduced with hydrogen at 300 - 400°C. About 2 moles of hydrogen chloride was formed from 1 mole of supported palladium compound and the rate of acetaldehyde formation became nearly zero. This means that almost all of palladium is still present as Pd^{2+} after several hours of operation.

2) The surface areas of the supports used in Fig. 1 are as follows: active charcoal 1,280 m^2/g, alumina 230 m^2/g and silica gel 350 m^2/g. The rate of acetaldehyde formation is proportional to the surface area in the case of active charcoal of same source. The constant catalytic activity is observed only in charcoal support.

H.-J. ARPE

What kind of information do you have that the oxidation state of palladium in a working catalyst is still Pd^{2+}?

There are several indications that liquid and gas phase oxidation of ethylene and also benzene in the presence of acetic acid--which seems to follow quite the same mechanism as the oxidation in the presence of water--can occur on metallic palladium.

T. KUNUGI

The working state of palladium is Pd^{2+} as mentioned in the preceeding comment.

Oxidation rate of ethylene over metallic palladium-active charcoal catalyst in the presence of steam is only 1/10 - 1/100 times as much as our palladium chloride-active charcoal and main product is acetic acid.

Paper Number 28

CATALYTIC ACTIVATION OF CARBON MONOXIDE IN HOMOGENEOUS AND HETEROGENEOUS SYSTEMS

V. A. GOLODOV, A. B. FASMAN, and D. V. SOKOLSKY
The Institute of Organic Catalysis and Electrochemistry
The Kazakh SSR Academy of Sciences, Alma-Ata, USSR

ABSTRACT: Low temperature conversion of carbon monoxide in aqueous solutions, viz

$$CO + H_2O \rightleftarrows CO_2 + \underbrace{2H^+ + 2e^-}_{H_2},$$

proceeds on both the surface of the Pt-Group metals and in solutions of their complex compounds. The complex center ion, oxidizer molecule (in homogeneous system), as well as hydronium ions, reduceable into molecular hydrogen (heterogeneous catalysis), may act as electron acceptors depending on reactant and catalyst redox potential values. During the first step of the heterogeneous and homogeneous processes, the CO molecule penetrates into the interior of the complex, or adsorbs on the metal surface, thereby displacing the less strongly associated ligands, especially H_2O. In the catalytic process, the linear form of coordinated CO is the principle intermediate. Increased concentration of the complexing particles and the center ion charge leads to screening of available sites and stable compound formation, thus impeding water penetration and activation, causing a sharp decrease in the rate of the catalytic process. Similar results are obtained from the increase in concentration of strongly adsorbed bridge structures, followed by H_2O molecule displacement from the monolayer of the metal surface. Thus heterogeneous and homogeneous CO activation mechanisms have many common properties, due to the similarity of the activated complex structure in solution and on the surface.

I. INTRODUCTION

Recently a number of papers were published, giving a comparative analysis of homogeneous and heterogeneous catalysis.[1,2] E.g., Halpern found that in the most active hydrogenation catalysts the metals and metal ions are isoelectronic;[3] Fleed does not observe any fundamental differences in catalytic acetylene conversion processes over the same metals in homogeneous and heterogeneous systems.[4] Other investigators report, that there are common steps to the homogeneous and heterogeneous isomerization of olefins and in their deuterium exchange reaction.[5,6]

Attempts to find out the principal similarity between the homogeneous and heterogeneous molecule activation mechanism are most successful when studying comparatively simple reactions, yielding identical or structurally similar products.

Low temperature conversion of carbon monoxide in aqueous solutions, as follows from our experimental data, proceeds both on the surface of Pt-Group metals and in the presence of their

$$CO + H_2O \overset{k}{\rightleftharpoons} CO_2 + \underbrace{2H^+ + 2e^-}_{H_2} \qquad (1)$$

complex compounds.[7,8]

In the homogeneous system the central atom reduces to the zerovalent state according to the reaction

$$M^{n+} + CO + \frac{n}{2} H_2O \rightarrow M^0 + CO_2 + nH^+ , \qquad (2)$$

where M designates ions of the subgroups I and II and of platinum metals.[9]

In the presence of oxidizers whose redox potential (ψ_0) is higher than that for the M^{n+}-M^0 couple, the reaction

$$Ox + CO + H_2O \rightarrow Red + CO_2 + 2H^+ \qquad (3)$$

is realized, where Ox = quinones of various structure, $Cr_2O_7^{2-}$, Cu^{2+}, Fe^{3+}, BrO_3^- and other strong oxidizers. For example, for Ox = p-benzoquinone, the reaction proceeds as follows:

$$C_6H_4O_2 + CO + H_2O \xrightarrow{k} C_6H_4(OH)_2 + CO_2 \qquad (4)$$

On a solid surface the reaction is completed by reduction of the hydronium ions to molecular hydrogen (reaction 1).

During the first step of homogeneous and heterogeneous processes the CO molecule penetrates into the interior of the complex or adsorbs on the metal surface displacing less strongly bound ligands

$$\text{In solution: } [MX_n]^{2-n} + CO \underset{}{\overset{K}{\rightleftharpoons}} [(CO)MX_{n-1}]^{3-n} + X^- \qquad (5)$$

$$\text{On the surface: } \text{-M-M-} + CO \overset{K_1}{\rightleftharpoons} \text{-M-}\overset{\overset{O}{\|}}{\underset{\downarrow}{C}}\text{M-} \text{ or } \text{-M-M-} \text{ (bridged C=O)} , \qquad (6)$$

where X = ligands.

A necessary condition for a catalytic process is the activation of the two reaction components: CO and H_2O by the oxygen donor. Otherwise the process results in the formation of mixed carbonylic compounds in homogeneous systems (step 5) and stable carbonyls on the surface (step 6).

The mixed carbonyls of Pt, Rh and Ru are relatively stable even in aqueous solution,[10-12] whereas the voluminal carbonyls of palladium are stable only in the absence of water.[13-18]

The nature of further conversions of voluminal and surface carbonyls is determined by the energy and form of CO and H_2O bonds with the metal atom (ion).

The investigation of CO chemisorption on Pt-Group metals using spectral,[19] potentiometric and galvanostatic,[20,21] conductometric,[22] thermodesorptive and electronographical[23] methods led to the conclusion that there are two structures on the surface--a linear one (CO associated with one M atom) and a bridge-type (CO associated with two M atoms). The heats of adsorption for the two forms are 20-27 and 40-50 kcal/mol respectively;[24,25] i.e., the bridge-type bond is much stronger.

In the sequence Ru < Os ~ Ir < Rh ~ Pt << Pd the relative portion of the bridge-type structures increases, and on

Pd it predominates. The relation between optical densities of linear and bridge-type structures decreases from 4 to 0.1-0.2 over the transition from Ru, Os and Ir to Pd. The sequence mentioned above correlates with the paramagnetic susceptibility (χ) and specific electronic density (ρ) of metals. The width of the metal d-band decreases in the same sequence.[26)]

Calculations show that even in the presence of Cl^- ions the Pt surface coverage with carbon monoxide is near to the monolayer.[27)] In solutions of chloride and bromide complexes of Pd (II) the equilibrium (5) is almost completely shifted to the mixed carbonyl formation. The reaction equilibrium constant K is $2.5\text{-}3.2 \cdot 10^3$, as determined experimentally and derived from the kinetic equation describing p-benzoquinone reduction.[28,29)]

However, in homogeneous catalysts there are some sites available for the H_2O molecule after the formation of M:CO = 1:1 complexes, whereas during heterogeneous catalysis the whole surface is blocked with carbon monoxide. In this case it is the activation of H_2O molecule that determines the possibility for the reaction to proceed and its rate.

Water adsorbs on metals more weakly than CO. According to data from Schulz-Ekloff <u>et al.</u>,[30)] the heat of adsorption Q(Pt-H_2O) is only 4 kcal/mol. H_2O adsorption on Pt does not actually change the electronic work function.[31)]

During carbon monoxide adsorption or carbonyl formation in solution, a simultaneous transfer of CO ρ-electrons onto the metal vacant d-orbital (donor-acceptor bond), and that of metal d-electrons onto the loosening CO orbitals (dative bond) take place.

Both localized and collective electrons take part in the formation of dative and donor-acceptor bonds of M-CO. Changes in the electronic structure of transition metals lead to changes in the relative contribution of different bonds to the total bond energy M-CO.[32)] The C - O bond order increases with the metal d-band coverage but the M - C bond energy decreases.[33)] The CO neutral molecule gets into an excited state, its reactivity increasing considerably.

However, not all of the adsorbed carbon monoxide can take part in the conversion reaction. Experimental data show a well-defined correlation between structure of energy bands of Pt metals (paramagnetic susceptibility, specific electron concentration and d-band width ε), form and structure of adsorption bond and reactivity of surface carbonyl compounds.

Low-temperature CO conversion proceeds only over metals with average values of χ, ρ and ε, i.e., within the optimum interval of the bond energies between reactant and catalyst. Equally unfavorable for catalysis is a too strong metal-adsorbate (Pd) bond as well as too weak one (Os, Ir, Cu, Au) --see Table.

There is a linear relationship (Fig. 1) between linearly adsorbed CO and reaction rate (1). This is direct evidence that the linear form participates in the catalytic process. It is essential to note that the linear form concentration was determined by two independent methods--spectral[34)] and potentiodynamic.[35)]

At present there are no data available for an analogous comparison in the homogeneous process, but it may be stated, that halogen salts of Pd (II) form only mononuclear, very active linear, carbonyls in aqueous solutions. Di- and polynuclear carbonyls of other Pt- Group metals are low in activity.[36]

If the linear form is responsible for the course of conversion, then the equilibrium shift to the side of its formation should lead to an increase of the reaction rate. This assumption is confirmed by the work of Binder et al.,[37] showing that blocking of sites having a high adsorption potential with sulphur is followed by increase in the interaction rate of carbon monoxide with water by two orders of magnitude. Similar effects are exerted by platinum alloyed with a small portion of nickel[38] or by reactions on a Pt-plate in the presence of another component--Pt-powder or tungsten carbide powder. Brummer[39] assumes that in this case the concentration of weakly-bonded CO increases and the reaction of H_2O with CO is facilitated.

Using IR-spectroscopy methods, it was found, that due to platinum metal sulphuration, bridge-type adsorption complexes disappeared.[40] Haman believes that CO molecules, located in the second adsorption layer, may oxidize on the surface in the presence of sulphur.[41] The addition of nickel as proved by the thermodesorption method[38] increases the concentration of weakly-bonded CO from 25 to 70%.

During one of the surface CO - H_2O reaction steps deprotonation of the water molecule is possible similar to homogeneous catalysis. Over Pt, for instance, the degree of surface coverage for CO decreases in alkaline solution to half that in acidic solution. Taking this into account,[35] one might expect an essential promotion of the process. Indeed, the authors of the present paper as well as other authors[42] observed an increase in Pd and Pt activity with substitution of acidic by alkaline media. By analogy to hydroxy and aqueous complexes in solutions the M - OH bond strength should be greater than Q_{M-H_2O}.

Since substitution of ligands in the inside region of complexes in aqueous solutions proceeds, as a rule, through a preliminary step,[43] step (5) may be represented as the displacement of H_2O

$$[(H_2O)\ MX_{n-1}]^{3-n} + CO \rightleftarrows [CO)\ MX_{n-1}]^{3-n} + H_2O \qquad (7)$$

The resulting complex has a strong transactive CO group. This facilitates the subsequent penetration and deprotonation of the H_2O molecule considerably, viz;

$$[(CO)MX_{n-1}]^{3-n} + H_2O \rightleftarrows [(CO)MX_{n-2}H_2O]^{4-n} + X^- \qquad (8)$$

$$[(CO)MX_{n-2}H_2O]^{4-n} \underset{}{\overset{K_2}{\rightleftarrows}} [(CO)MX_{n-2}(OH)]^{3-n} + H_3O^+ \qquad (9)$$

The increase of the concentration of complexing particles (Fig. 2--right section of the curves) and of the central atom charge (Fig. 3), followed by the formation of stable complexes and screening of the third coordination of axis hinders the intermediate mixed carbonyl formation (steps 5,7 and 8) and the water activation (step 9).

Similar results are obtained by increasing the concentration of strongly-adsorbed bridge structures, thus hindering the H_2O molecule penetration into the monolayer on a solid metal surface.

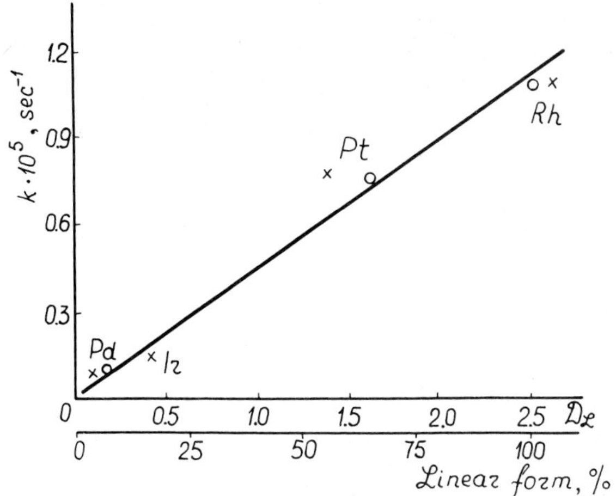

Fig. 1 Dependence of carbon monoxide conversion rate on CO linear surface concentration (o) and on optical density of linear-structure carbonyls (x)

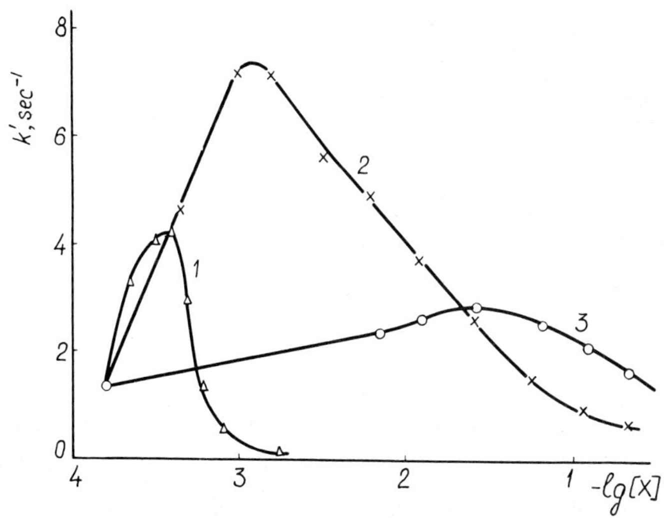

Fig. 2 Effect of concentration of J^-(1), Br^-(2) and Cl^-(3) on rate of bensoquinone reduction with carbon monoxide catalyst K_2PdCl_4 - $1.57 \cdot 10^{-4}$ mol/l, solvent-water: dioxane (3:7), temperature 20°C, $HCLO_4$ - $2.5 \cdot 10^{-3}$ mol/l.

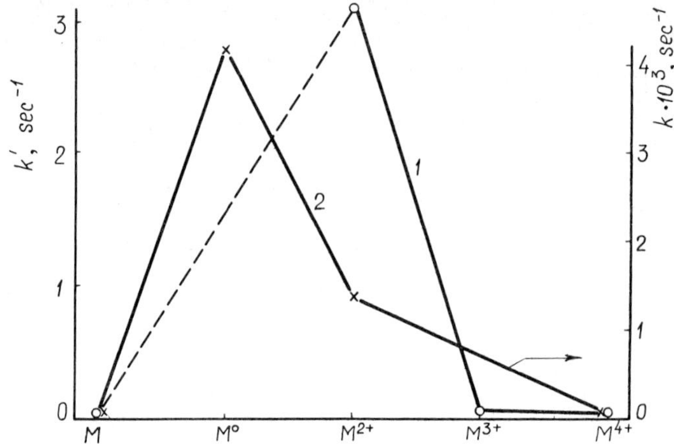

Fig. 3 Effect of the central atom charge on the activity of complex compounds of platinum metals in reactions 4(curve 1) and 1(curve 2)
M = surface of metals
1 = Pd(II), Rh(III), Pt(IV);
2 = Pt(0), Pt(II), Pt(IV).[10]

In homogeneous catalysis, as in the heterogeneous systems[44,45] an optimum value for interaction energies between reactant and the catalyst's central ion should exist. Such an interaction can be obtained by varying the nature of the ligand. Thus a well-defined maximum is observed at pK = 13.6 for Pd (II) and 25 for Pt (II)[46] on the curve for CO oxidation rate constants against pK or ϕ of complexes.

Changing the concentration of complexing particles gives a maximum portion of complexes responsible for the catalysis or those possessing the optimum catalytic properties.

It should be kept in mind that the catalytic activity of complexes is determined not only by thermodynamic but by kinetic parameters as well. From Grinsberg's work ,[43] it follows that the ligand displacement rate in the inside region increases in going from less to more stable complexes. Thus, the catalytic activity of complexes is determined by the relation between thermodynamic stability and kinetic lability values.

The main advantages of homogeneous catalysis over heterogeneous CO activation reactions consist to a great extent, in the more favorable conditions for the formation of the intermediate active complex. This is apparent from all the reaction steps: CO and H_2O penetration into the inside region, activation of H_2O molecules located on the opposite sites (to deprotonation, particularly), subsequent rearrangement of π-into σ-complex and its decomposition.

Thus, the acidity of the H_2O molecule increases considerably, if it is located on sites opposite to such ligands as C_2H_4 or CO.[43] Indeed, a comparison of values for equlibrium constants K_3 found experimentally for the reaction

$$[(H_2O)PdX_3]^- + H_2O \overset{K_3}{\rightleftharpoons} [(OH)Pd\,X_3]^{2-} + H_3O^+ \qquad (10)$$

with those estimated for step (9) shows, that $K_3 : K_2 =$ 5-10:1.[47,29]

The subsequent internal rearrangement of OH^- into the cis-position to CO or the transition to the σ-complex with the formation of a Pd-COOH particle leads to an unstable intermediate decomposing into the final products Pd°, CO_2 and H^+.

On the surface the number of available sites is substantially smaller. This hinders the simulultaneous activation of all the reactants considerably. Consequently, the availability of sites for the activation of the weakly-bound H_2O molecule is of great importance to heterogeneous catalysis. As in homogeneous systems, the metal functions as an electron-transfer bridge (donor to acceptor). Moreover, the CO and H_2O adsorbed molecules may be located either on neighboring sites of the crystal lattice (a) or on the same metal atom (b)

```
       O    H   H                  O   H   H
       ||    \ /                   ||   \ /
a)     C      O              b)    C     O
       ↓      ↑                     ↘   ↗
   —  M  —   M  —                  — M —
       |      |                       |
```

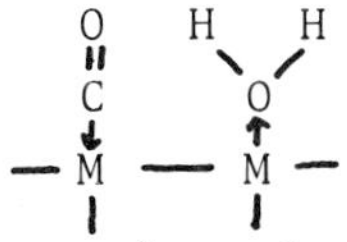

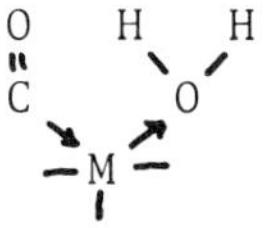

The portion of the dative and donor-acceptor CO bond component varies according to how the metal d-band changes structure. For a single metal it varies as a function of degree of oxidation. No well proved interpretation is known as yet, relating to the electronic structure of linear and bridge-type forms; though it may be assumed from available calculations of CO chemisorption states,[48] that the contribution of the π-component of the M-CO bond increases as the metal electronegativity decreases, attaining a maximum on Pd. In homogeneous systems this maximum is achieved at the change from charged to zerovalence complexes.

A highly oxidized central ion is unfavroable for catalysis. It was shown in paper[10] that the relationship between the activity of Pt(IV), Pt(II) and Pt(0) complexes in reaction (1) is 1:27:100. Fig. 3 shows also that the CO oxidation rate decreases sharply at the change to tri- and tetravalent metal complexes.

In Table 1 the activity of the platinum group elements in homogeneous and heterogeneous CO oxidation processes is given (decreasing order of activity).

As seen from the table, the activity sequence of the homogeneous system does not coincide with those of the heterogeneous one and the first to convert is Pd-position. This is due to the high concentration of the nonreactive bridge form of the adsorbed CO on a solid metal, as well as to the formation of an extremely unstable (in aqueous solution) carbonyl complex with the Pd^{2+} ion. A square-planar configuration mostly favors the elementary reaction step and the simultaneous activation of several reactants.

TABLE 1
Activity sequence of homogeneous and heterogeneous catalysts in reaction (1) and (3)

Catalyst	State densities on the Fermi level $N(\varepsilon f)$[26]	Atomic paramagnetic susceptibility,[49] $\chi \cdot 10^6$	Electron density[50] $\rho \cdot 10^{-23}$, e/cm^3	Activity in reaction(1), relative units	Catalyst	Electron structure of the external quantum level of the ion	Activity in reaction (3), 2 $O\chi=Cr_2O_7$, relative units
Rh	18.7	101	6.6	31	Pd^{2+}	d^8	330
Pt	23.2	189	6.6	27	Rh^{3+}	d^6	6
Ir	13.8	35	6.4	3	Ru^{4+}	d^4	4
Os	-	9	5.7	2	Pt^{4+}	d^6	3
Ru	-	44	5.9	1	Ir^{4+}	d^5	1
Pd	32.7	558	6.8	very low	Os^{4+}	d	1

From the analysis of experimental and published data it can be seen that there are many common features to homogeneous and heterogeneous CO activation, due primarily to the similar structure for the activated complex both in solution and on the surface. The observed differences in the absolute activity are due to the sharp decrease for the probability of simultaneous activation of both reactants within the limits of one active center on the surface as compared to the analogous center in the volume. A similar conclusion was reached by Heinemann.[51]

REFERENCES

1) G. C. Bond, P. B. Wells, Adv. in Cat., 15, N.Y.-L., Acad. Press, 1964, p. 91.
2) K. Tarama, Kagaky Kogyo, 19, 995 (1968).
3) J. Halpern, Adv. in Cat., 11, N.Y.-L., Acad. Press, 1959, p. 301.
4) R. M. Fleed, Vest. Acad. Nauk SSSR, N 6, 104 (1971).
5) J. L. Garnett, R. J. Hodges, W. A. Sollich-Baumgartner, Proceedings of IV International Congress on Catalysis, 1, "Nauka," Moscow, 1970, p. 62.
6) V. Sh. Feldblum, ibid., p. 192.
7) A. B. Fasman, G. L. Padyukova, D. V. Sokolsky, Dokl. Acad. Nauk, SSSR, 150, 856 (1963).
8) V. A. Golodov, A. B. Fasman, D. V. Sokolsky, Dokl. Acad. Nauk SSSR, 151, 98 (1963).
9) V. D. Marcov, A. B. Fasman, Zh. Fis. Khim. (USSR), 40, 1564 (1966).
10) K. I. Matveev, L. N. Rachkovskaya, N. K. Eremenko, Isv. Sib. Otd. Acad. Nauk SSSR, Ser. Khim. Nauk, N 7, 113 (1968).
11) J. Stanko, G. Petrov, C. K. Thomas, Chem. Comm., 1969, 1100.
12) J. Halpern, B. R. James, A. L. Kemp, J. Am. Chem. Soc., 88, 5142 (1966).
13) W. Manschot, J. Konig, Ber., 59, 883 (1926).
14) V. A. Golodov, C. G. Kutykov, A. B. Fasman, D. V. Sokolksy, Zh. Neorg. Khim. (USSR), 9, 2319 (1964).

15) E. O. Fischer, A. Vogler, J. Organomet. Chem., 3, 161 (1965).
16) A. Treiber, Tetrahedron Letters, 1966, 2831.
17) V. F. Vozdvizhensky, Yu. A. Kushnikov, G. G. Kutyukov, A. B. Fasman, Zh. Neorg. Khim. (USSR), 12, 1518 (1967).
18) W. Schnable, E. Kober, J. Organomet, Chem. 19, 455 (1969).
19) N. N. Kavtaradze, N. P. Sokolova, J. Res. Inst. Catalysis, Hokkaido Univ., 13, 496 (1966).
20) S. Gilman, J. Phys. Chem., 66, 2657 (1962); 67, 1898 (1963); 71, 4339 (1967).
21) M. W. Breiter, J. Phys. Chem., 72, 1305 (1968).
22) T. Sugita, S. Ebisawa, K. Kawasaki, Surface Sci., 11, 159 (1968).
23) A. E. Morgan, G. A. Somorjai, Surface Sci., 12, 405 (1968).
24) N. N. Kavtaradze, N. P. Sokolova, Zh. Fis. Khim (USSR), 44, 171 (1970).
25) G. Ertl, P. Rau, Surface Sci., 15, 443 (1969).
26) O. K. Andersen, Phys. Rev., B. Solid State, 2, 883 (1970).
27) S. Gilman, J. Phys. Chem., 70, 2880 (1966).
28) V. A. Golodov, A. B. Fasman, G. G. Kutyukov, V. D. Markov, "Probl. of Kin. and Catal." 12, "Nauka," Moscow, 1968, p. 69.
29) D. V. Sokolsky, Ya. A. Dorfman, Ligand Catalysis in Aqueous Solutions, "Nauka," Alma-Ata, 1972.
30) G. Schulz-Ekloff, D. Baresel, J. Heidemeyer, Coll. Czech. Chem. Commun., 36, 928 (1971).
31) A. A. Fokina, N. A. Shurmovskaya, R. H. Burstein, Electrokhimjya, 5, 225 (1969).
32) A. Andreev, Comp. rend. Acad. Bulgare Sci., 20, 1309 (1967).
33) G. Blyholder, M. C. Allen, J. Amer. Chem. Soc., 91, 3158 (1969).
34) N. N. Kavtaradze, N. P. Sokolova, Zh. Fis. Khim (USSR), 42, 1286 (1968).
35) P. Stonehart, "Power Sources, 1966," Oxford-London et al., Pergamon Press, 1967, p. 507.
36) E. W. Abel, Quart. Rev., 17, 133 (1963).
37) H. Binder, A. Kohling, G. Sanstede, Angew. Chem. 79, 477 (1967).
38) D. W. McKee, M. S. Pak, J. Electrochem. Soc., 116, 516 (1969).
39) K. D. Brummer, J. Catalysis, 9, 207 (1967).
40) C. R. Guerra, J. Colloid. Interface Sci., 29, 229 (1969).
41) C. H. Hamann, Ber. Buns. Ges. physik. Chem., 75, 542 (1971).
42) O. G. Tyurikova, N. B. Miller, V. I. Veselovsky, Electrokhimjya, 5, 55 (1969); 6, 468 (1970).
43) A. A. Grinberg, Introduction to Chemistry of Complex Compounds, "Chimiya," Moscow-Leningrad, 1966.
44) A. A. Balandin, Multiplet Theory of Catalysis, p. 2, Moscow University, Moscow, 1964.
45) G. K. Boreskov, Catalysis, pp. I and II, "Nauka," Novosibirsk, 1971.
46) A. B. Fasman, V. A. Golodov, S. S. Kutyukov, V. D. Markov, in "Homogeneous catalysis," "Illim," Frunze, 1970, p. 160.
47) V. A. Golodov, A. B. Fasman, G. G. Kutyukov, V. V. Roganov, Zh. Fis. Khim. (USSR), 41, 1085 (1967).

48) H. H. Dunken, H. Opitz, Proceedings of IV International Congress on Catalysis, I, "Nauka," Moscow, 1970, p. 73.
49) P. W. Selwood, Magnetochemistry, N.Y.-L., Inters. Publ. 1956.
50) M. A. Jensen, B. T. Mathias, R. Anders, Science, 150, 1448 (1965).
51) H. Heinemann, Chem. Tech. 4 (1971).

DISCUSSION

I. V. KALECHITS

The similarity of homogeneous and heterogeneous catalysts with respect to the course and mechanism of reactions has been shown in the very interesting papers presented by Drs. Golodov and Garnett. There are also some additional important evidences of such similarities, which stem from energetical and geometrical factors. Reaction of the stable radical 2,2,6,6-tetramethyl-pyperidone-4-oxyl-1 with 2,6-dimethyl-3,5-dicarbethory 1,4 dihydropyridine

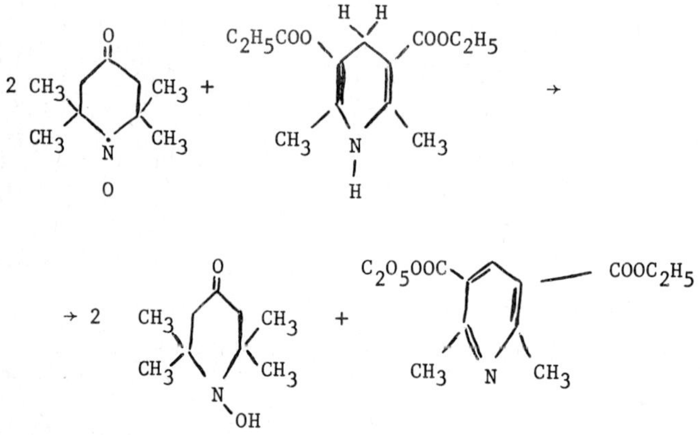

O
2 CH_3 CH_3 + C_2H_5COO H H $COOC_2H_5$ →
CH_3 N CH_3 CH_3 N CH_3
O H

O
→ 2 CH_3 CH_3 + C_2O_5OOC $COOC_2H_5$
CH_3 CH_3 CH_3 N CH_3
N
OH

is catalyzed by many dyes and quinones. The catalysts must be hydrogen acceptors at first and hydrogen donors later. Indeed, the plot at activity versus oxydation--reduction potential of catalyst gives maxima, for example, in the series of quinones:

9,10-anthraquinone < 1,2-benzanthraquinone < 1,4-napthaquinone > 1,4-benzoquinone > tetramethoxydiphenoquinone > diphenoquinone

or in the line of complex catalysts containing metal ions:

$Ni^{+2} < Cu^{+2} > Fe^{+3} > Co^{+2}$

and quinone ligands:

alizarine < O-naphtaquinone > 3,5-ditertbutyl-O-benzoquinone.

These curves are completely similar to the Balandin's "volcano-like" curves, the physical sense of which is well known.

Geometrical factors in homogeneous catalysts were investigated in the case of binuclear complexes

$Cl(CO)_2RhH_2N-(CH_2)_n-NH_2Rh(CO)_2Cl$

for the hydrogenation of decene-1 and cyclohexene. The rate of hydrogenation is higher the longer the bridge chart is, but in the same sequence the selectivity is decreased.

	n=2	n=6	n=10
Rate of hydrogenation of decene-1		Increased	
Ratio of conversions decene-1 and cyclohexene	∞	85	14

Such difference may be caused by steric hinderance.

V. A. GOLODOV

The results by Professor Kalechits have given further positive support to our point of view developed herein.

J.-Y. RYU

1) It is not clear if the excited state means electronic excitation or not. However, if it means the electronic state, what will be the excited electronic state of CO? Will the electronic excitation effect, the bond order or the stretching vibration frequency between carbon and oxygen atoms?

2) It has been reported that the carbon monoxide can adsorb on metals which do not have any valence d-orbital electrons, for example, Cu, Ag, and Au. In these cases, will d-orbital participation in the chemisorption bondings be possible? In other words, will the electron backdonation from metal to the adsorbed CO be possible?

V. A. GOLODOV

1) Indeed, the excited state appears, firstly, as a result of electron excitation. It is most typical for the CO molecule, that the excited state manifests itself through variation in the stretching vibration frequency of the C-O bond. Thus, in the case of free CO ν = 2150 cm^{-1}, during formation of a palladium carbonyl complex, the absorption band shifts into the long-wave region up to 2000-1850 cm^{-1}, depending on the charge of the central atom. The bond order of CO thereby decreases correspondingly.[1,2]

2) In our paper it has been reported, that the M-CO adsorption bond is too weak for metals of Group I Subgroup (Au, Ag). Thus, on their surface no conversion reaction occurs. In this case the bond between metal and carbon monoxide molecule may be formed, theoretically, by transfer of the metal d-electrons on to the π-orbital of CO; however, data concerning the quantitative description of such donation bondings are not available in the literature.

1) M. Orchin, H. H. Jaffe, The Importance of Antibonding Orbitals, N.Y.-Atlanta, 1967.
2) R. P. Eischens, Accounts. Chem. Res. 5, 74 (1972).

Paper Number 29

NEW HETEROGENEOUS OXO CATALYSTS FOR LIQUID PHASE PROCESSES

W. O. HAAG and D. D. WHITEHURST
Mobil Research and Development Corporation
Central Research Division, Princeton, New Jersey

ABSTRACT: Transition metal compounds chemically bound to organic polymers through coordination represent a new class of solid catalysts. A wide variety of organic reactions formerly catalyzed by soluble transition metal complexes can now be conducted under heterogeneous reaction conditions. The liquid phase heterogeneous catalysis of the oxo reaction, presented here, is an illustration.

A kinetic description of the conversion of olefins to aldehydes was developed. This provides a basis for the comparison of the activities and selectivities of various heterogeneous and homogeneous catalysts. Analogous homogeneous and heterogeneous catalysts have nearly identical selectivities for aldehyde formation.

Rhodium-phosphine polymer complexes exhibit extremely high selectivity for aldehyde production. In contrast to this, rhodium-amine polymer complexes exhibit selectivities for alcohol production which were heretofore unknown. With these catalysts, both aldehyde formation and hydrogenation occur under quite mild conditions. Olefin hydrogenation, however, does not take place. Analogous homogeneous catalysts have much lower selectivities than heterogeneous catalysts for alcohol production.

1. INTRODUCTION

In recent years a number of commercial processes have been developed which use soluble compounds of transition metals as catalysts in the production of basic chemicals. Because of a number of processing difficulties associated with catalysts such as these, especially with expensive metals, an increasing number of attempts have been made to synthesize analogous transition metal catalysts in solid form[1].

We have previously reported that transition metal compounds may be chemically bound to organic polymers through coordination; the resultant solid represents a new class of heterogeneous catalysts[2]. Recently a number of other workers have described similar catalysts[3]. We have shown that these catalysts are useful in a wide variety of organic reactions in both liquid and gas phase. In addition, it was found that catalysts produced in this way could exhibit activities identical to those of analogous soluble compounds[2].

The present work is intended as an illustration of the principles involved in the synthesis and use of these novel catalysts for liquid phase reactions. The conversion of olefins to aldehydes and/or alcohols (oxo reaction) with rhodium catalysts is used as a test reaction because of its known sensitivity to changes in selectivity with catalyst structure[4a)]. In addition, it provides an extremely good model for a reaction in which retention of the active metal compound may be difficult in liquid phase processes. Some of the coordinating polymers which have been used and a typical catalyst preparation are shown below.

$$\left[-CH_2-CH(C_6H_4X^*)-CH_2-CH(C_6H_4X)-CH_2-CH(C_6H_4X)- \right]_n \xrightarrow{[Rh(CO)_2Cl]_2} \left[-CH_2-CH(C_6H_4X \rightarrow Rh(CO)_2Cl)-CH_2-CH(C_6H_4X)-CH_2-CH(C_6H_4X)- \right]_n$$

*$X = -P(C_4H_9)_2$, $-CH_2-P\phi_2$, $-N(CH_3)_2$, $-CH_2-N(CH_3)_2$, -SH

The results of these studies will show that a transition metal complex, when confined to the surface of a solid through coordination, in many instances can act as a catalyst with identical activities and selectivities to those of catalysts composed of analogous metal complexes which are freely soluble in the reaction medium. Thus, many differences previously thought to exist between "homogeneous" and "heterogeneous" catalysis no longer apply, and the only real difference between them is the physical form of the active species. In other instances, a transition metal complex, when confined to the surface of a solid, can exhibit activities and selectivities which are not achievable with analogous soluble complexes. In this case, the solid environment can impose restrictions in the nature of the catalytic specie or the assemblage of reacting molecules.

2. EXPERIMENTAL

2.1 Synthesis of heterogeneous catalysts

The catalysts used in this work are solid insoluble rhodium-organic polymer complexes. Their synthesis has been previously described[2)]. Briefly, a porous poly (styrene-divinylbenzene) copolymer, in bead form, is functionalized by chemically incorporating pendant groups, for example: tertiary amines, tertiary phosphines, thiols that are capable of coordinating with transition metals. Rhodium compounds were then introduced by ligand exchange or bridge splitting reactions.

2.2 Batch oxo reactions

Batch experiments were conducted in 300 ml stirred autoclaves equipped with automatic pressure and temperature recording and with provision for gas and liquid sample withdrawal and introduction under pressure. The reactants and catalysts were charged to the autoclave; the apparatus was flushed with N_2 or CO and heated to the reaction temperature. Gas of the desired composition was then introduced and the reaction begun. Samples were periodically withdrawn and analyzed by glc using Carbowax 1000 and tricresylphosphate columns.

2.3 Continuous flow oxo reactions

Continuous flow experiments were conducted in tubular stainless steel reactors approximately 1 cm x 50 cm. Heating was done by circulating hot oil around the exterior of the reactor tube. Gas consumption was continually monitored with gas mass flow meters, and the liquid effluent was periodically analyzed by glc as above.

3. RESULTS AND INTERPRETATION

3.1 Reaction kinetics

We have found that under conditions of constant pressure, temperature, and solvent composition, the overall reaction kinetics for the production of aldehydes may be adequately accounted for in terms of five pseudo first-order rate constants as outlined in the following model.

$$\text{terminal olefins} \xrightarrow{k_1} \text{linear aldehydes}; \quad \text{terminal olefins} \xrightarrow{k_2} \text{branched aldehydes}; \quad \text{terminal olefins} \underset{k_R}{\overset{k_F}{\rightleftharpoons}} \text{internal olefins}; \quad \text{internal olefins} \xrightarrow{k_3} \text{branched aldehydes}$$

The validity of this model for a wide variety of different catalysts is indicated by fig. 1 in which the points are experimentally derived and the lines are the calculated curves using the indicated rate constants which are presented here relative to k_3.

These relative rate constants, as well as their absolute values, are subject to change as operating conditions are varied.

3.2 Selectivities, definition

The rate constants can be used to define several selectivities which provide a basis for comparison of various catalysts under standard conditions. Selectivities which are characteristic parameters are defined as follows:

Linear Selectivity

$$S_L = \frac{k_1 \times 100}{k_1 + k_2}$$

Isomerization Selectivity

$$S_I = \frac{k_F \times 100}{k_1 + k_2 + k_F}$$

3.3 Selectivity versus structure

A number of rhodium-organic polymer complexes were prepared and catalytically tested for the conversion of 1-hexene to oxo aldehyde. The results of these tests, as well as comparisons with analogous soluble rhodium complex catalysts, are presented in table 1.

One interesting result from these studies is that the linear selectivity (S_L) is quite insensitive to the nature of the ligands which are attached to rhodium. Significantly, for comparable homogeneous

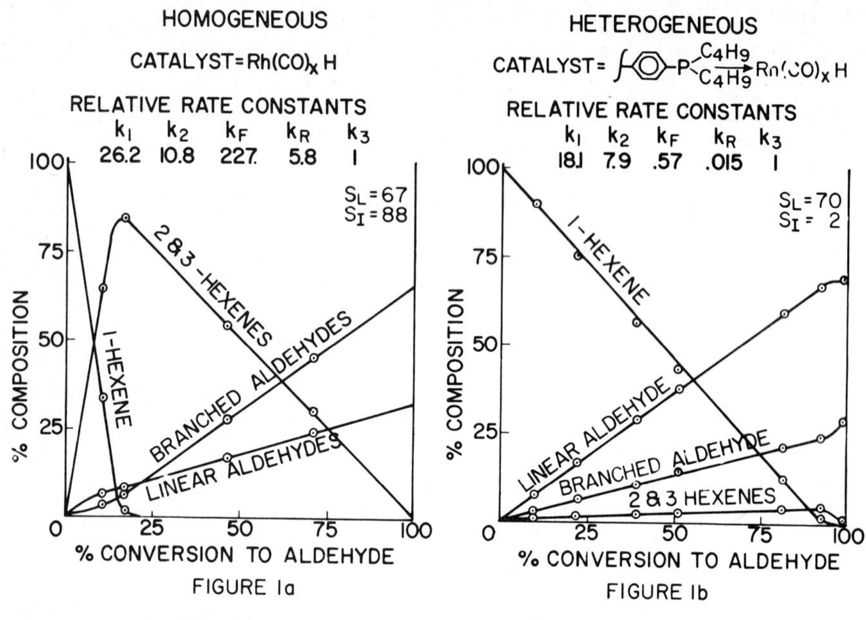

FIGURE 1a FIGURE 1b

CONVERSIONS OF 1-HEXENE TO OXO ALDEHYDES
100°C 1000psi 1/1 CO/H_2

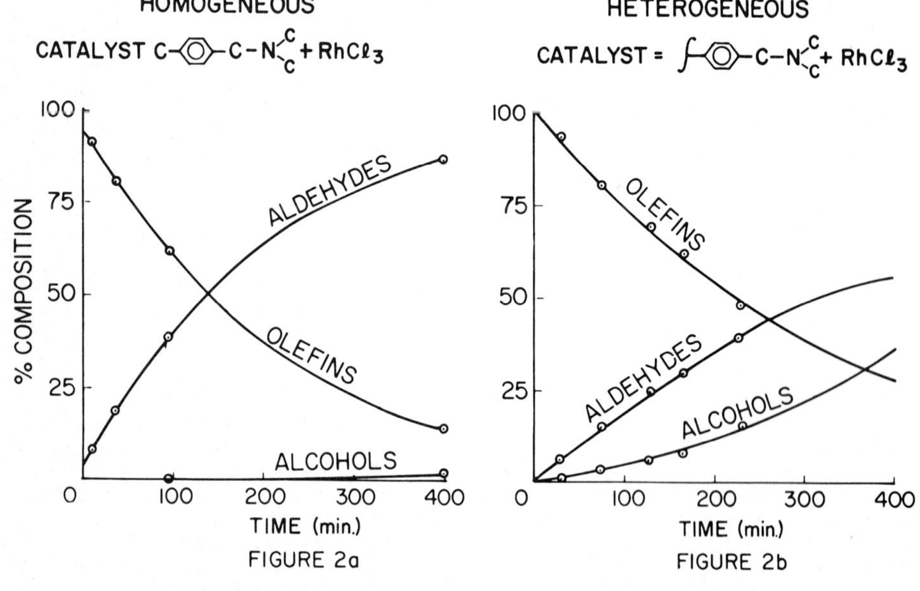

FIGURE 2a FIGURE 2b

CONVERSIONS OF 2-HEXENE TO OXO ALCOHOLS
100°C 1000psi 1/1 CO/H_2

and heterogeneous catalysts, the values obtained were identical.

TABLE 1

Selectivities in oxo aldehyde formation from 1-hexene[a)]

Ligand	Homogeneous		Heterogeneous	
	S_L	S_I	S_L	S_I
Carbon monoxide	67	88		
Benzyldimethylamine	76	69	76	67
Phenylthiol			73	57
Phenyldibutylphosphine			70	2.
Triphenylphosphine[b)]	71	23		
Phenyldimethoxyphosphine			71	14

a) All reactions were run at 90-100°C (except for b) with 1000 psi of 1/1 carbon monoxide/hydrogen in pure 1-hexene.

b) This reaction was run at 76°C.

The isomerization selectivity (S_I), however, was found to vary greatly for different complexes. In general, it was found to decrease as rhodium was coordinated with the following sequence of ligands ($CO > NR_3 > SR > PR_3$). The values obtained with comparable homogeneous and heterogeneous amine-rhodium complexes were essentially identical.

Variations of S_I among the catalysts may be accounted for by order of magnitude differences in the specific activity (g prod. / g Rh/hr.) the catalysts have for promoting isomerization (k_F).

The specific activities for aldehyde production from terminal olefins ($k_1 + k_2$), however, showed much less variation. In general, the heterogeneous catalysts were slightly less active than their homogeneous analogs. We believe this effect results primarily from some diffusional restriction within the polymer matrix. For the same reason, metal-polymer complexes made from intrinsically porous polymer compositions showed greater activities than those made from non-porous varieties.

3.4 Selective production of aldehydes

It is generally known that hydrido rhodium carbonyl phosphine complexes are highly active and promote the conversion of olefins to aldehydes under very mild reaction conditions. Selective conversions, not possible with conventional cobalt catalysts, can thus be achieved. For example, olefin isomerization or hydrogenation, skeletal rearrangements of highly strained systems, or aldehyde hydrogenation can be nearly eliminated and extremely high yields of linear aldehydes can be realized[4b)]. These known high selectivites and activities observed with soluble catalysts are directly translatable to insoluble rhodium-phosphine polymer complex catalysts. In addition, the physical form of these solid catalysts allows the economical use of the expensive metal as well as the ligand. Rhodium and phosphine losses are extremely low, and the catalysts can be used in fixed-bed continuous flow liquid phase reactions for extended periods of time. A few examples of conversions and yields which are achievable with these new catalyst systems are presented in table 2.

TABLE 2

Olefins converted to oxo aldehydes in continuous flow reactions

Olefin	Temp. (°C)	Pressure (psig)	LHSV[a)]	Conversion %	Selectivity Aldehyde %
Propylene	118	500	1.1	96	97.8
1-Hexene	88	1500	2	87	99
1-Octene	128	1500	2.5	90	98.5
1-Dodecene	149	2000	2.5	92	-

a) Volumes of charge/volume of catalyst/hr.

3.5 Formation of alcohols

It is well known that in the practice of the oxo reaction, aldehyde hydrogenation often occurs as a side reaction simultaneously with aldehyde formation. Under conditions necessary to achieve high conversions to alcohols with cobalt catalysts, such as cobalt carbonyls or cobalt carbonyl phosphine complexes[5)], olefin hydrogenation to paraffin cannot be avoided. With soluble rhodium catalysts, alcohol formation has been observed only under reaction conditions that are much more severe than are commonly employed for the synthesis of aldehydes[6)].

As mentioned above, heterogeneous rhodium-phosphine polymer complexes do not readily catalyze aldehyde hydrogenation. However, we have found that even under very mild oxo conditions, rhodium-amine polymer complexes are exceptional aldehyde hydrogenation catalysts. Thus olefins can be converted directly to alcohols in high yield. Soluble rhodium-amine complex catalysts are far less active in this respect than their heterogeneous counterparts. Fig. 2 illustrates this difference in the conversion of 2-hexene to its oxo alcohol.

3.6 Kinetics of alcohol formation

For rhodium-amine complexes, the following rate law was established for aldehyde hydrogenation:

$$r = k' [\text{Aldehyde}] \times P_{H_2} \times P_{CO}.$$

The first order dependence on carbon monoxide is surprising and does not seem to be general for all rhodium complexes.

The kinetic scheme described above is applicable for describing the formation of linear and branched products even when aldehyde hydrogenation to alcohols occurs. The overall conversion of olefins to alcohols can be approximately described by sequential first order reaction kinetics illustrated below for linear products:

$$\text{olefin} \xrightarrow{k_1} \text{linear aldehyde} \xrightarrow{k_H} \text{linear alcohol}.$$

3.7 Selectivities in alcohol formation

The ratio of the rate constants in the above scheme, $\left(\frac{k_H}{k_1}\right)$, defines a selectivity term (S_H), which allows estimation of the amount of alcohol product in the total product mixture at various degrees of conversion of the olefin. This selectivity, as well as the previously described selectivities, are presented in table 3 for a number of rhodium-amine complex catalysts, both homogeneous and heterogeneous.

TABLE 3

Selectivities in oxo alcohol formation from 1-hexene[a)]

Amine Ligand	Homogeneous[b)]				Heterogeneous		
	pK_B	S_L	S_I	$S_H \times 10^2$	S_L	S_I	$S_H \times 10^2$
Pyridine	8.77	72	43	< .5	73	42	< .4
N,N-Dimethylaniline	8.94	74	58	< .5	74	50	3.
Benzyldimethylamine	5.07	74	53	3.	73	58	18.
Aliphatic 3° amine[c)]	3.24	72	57	3.6	74	62	1.
1,4-Diazabicyclo[2.2.2] octane	2.91	71	32	< .5			

a) All reactions were run at 100°C with 1000 psi of 1/1 carbon monoxide/hydrogen in pure 1-hexene.

b) Soluble amine concentrations were .11 M.

c) The ligands for homogeneous and heterogeneous catalysts were not directly comparable.

It can be seen that heterogeneous catalysts are superior to their homogeneous analogs, for aldehyde hydrogenation, and that the basicity of the amine has a marked influence on the catalysts selectivity.

One other distinction between homogeneous and heterogeneous rhodium-amine complex catalysts lies in their ability to promote olefin hydrogenation under oxo reaction conditions. In general, the rhodium-amine polymer complexes are more selective for aldehyde formation than their soluble analogs and less product is lost because of paraffin by-product formation.

3.8 Continuous flow alcohol production

The rhodium-amine polymer complex catalysts have been used in continuous flow liquid phase conversions of olefins directly to oxo alcohols. Illustrative examples are presented in table 4. Retention of the rhodium was found to be quite good and the catalysts exhibited extended lifetimes.

TABLE 4
Olefins converted to oxo alcohols in continuous flow reactions

Olefin	Temp. °C	Pressure psig	LHSV a)	Conversion %	Selectivity Alcohol %
Propylene	96	1500	.33	89	98.5
1-Hexene	93	2000	.33	91.4	99.8
Hexenes (equil).	119	2000	.33	98.8	99+
1-Octene	104	2000	1.	83.4	98+
1-Dodecene	103	2000	1.2	75	87

a) Volumes of charge/volume of catalyst/hr.

REFERENCES

1) a. G. C. Bond et al., J. Cat. 6 (1966), 139.
b. P. R. Rony, J. Cat. 14 (1969), 142.
c. K. K. Robinson, F. E. Paulik, A. Hershman, and J. F. Roth, J. Cat. 15 (1965), 245.

2) a. W. O. Haag and D. D. Whitehurst, Belg. Patent #721,686 (1969).
b. W. O. Haag and D. D. Whitehurst, 2nd N. Am. Mtg. Cat. Soc., Houston, Texas, February 24, 1971.

3) a. J. Manassen, Chim. Ind. (Milan), (1969), 1058.
b. R. H. Grubbs and L. C. Kroll, J.A.C.S. 93, (1971), 3062.
c. British Petrol. Co. Ltd., Neth. Appl. 7006740 (1970).
d. C. U. Pittman, Jr., Chem. Tech. (1971), 1116.

4) a. J. A. Osborn, J. F. Young and G. Wilkinson, Chem. Comm. (1965), 17.
b. J. H. Craddock, A. Hershman, F. E. Paulik, and J. F. Roth, I&E Prod. Res. and Develop. 8, (1969), 291.
c. C.K. Brown and G. Wilkinson, J. Chem. Soc. (A) (1970), 2753.

5) a. Enjay Co. Inc. "Higher Oxo Alcohols," New York, N.Y. (1957).
b. L. H. Slaugh and R. D. Mullineaux, J. Orgmet. Chem. 13 (1968), 469.

6) a. J. F. Falbe, U. S. Patent 3509221, (1970).
b. W. H. Brader, Jr., S. B. Cavitt, and R. M. Gipson, U.S. Patent 3594425 (1971).

DISCUSSION

H. BREMER

It is stated in your paper (page 5) that the losses of rhodium and phosphine are "extremely low" when using the carrier-catalysts in a fixed bed. How many moles of substrate per mole of rhodium can be converted before the activity has decreased to 50%?

The same question arises for the rhodium-amine-carrier catalyst (page 7).

D. D. WHITEHURST

With a rhodium-phosphine complex catalyst in a fixed bed flow reactor, we have produced over 400,000 moles of aldehyde per mole of rhodium before the experiment was terminated arbitrarily. At this point, the activity was still above 50% of the fresh activity.

Some of the rhodium-amine complex catalysts behave very similarly to the rhodium phosphine catalysts with regard to catalyst stability.

P. R. RONY

I would like to compliment the authors of this paper on their work, the implications of which are very exciting. My questions are as follows:

You mentioned in your paper that "these catalysts are useful in a wide variety of organic reactions in both the liquid and gas phase." May I inquire whether a liquid phase present within the pores is required to make a gas phase reaction proceed? If your answer is yes, why would you need to anchor the metal complex?

D. D. WHITEHURST

Our calculations indicate that during a gas phase reaction the majority of the catalyst pores do not contain a liquid phase although it is possible that some of the micropores contain a condensed phase; therefore, we do not believe that a liquid phase is a requirement for our catalysts.

Under the same conditions, non-anchored, impregnated catalysts have a considerably lower activity. We believe this results from the low surface area of the solid rhodium complexes. If a liquid phase were present capable of forming a supported liquid phase catalyst of the type you have studied so successfully, then we would have to conclude that it has a very low effectiveness factor.

R. C. PITKETHLY

I wish also to compliment Drs. Haag and Whitehurst on their work. The formula for "linear selectivity" ignores the reverse reaction (k_r) and does not always give a good measure of the linear to branched ratio of the product. The data of Fig. 1a show this. However, the proper fitting of the data to the mechanisms is effective and gives the correct picture.

There are several questions I would like to ask:

1) It is stated in the paper (paragraph 3.4) that rhodium and phosphine

losses were extremely low. Were phosphine losses in fact observed and if so, could they be accounted for by extraction of occluded reagent used in preparation of the catalyst?

2) In the continuous flow alcohol production (paragraph 3.8) the dependence on carbon monoxide pressure is certainly puzzling. Could this be a displacement of ligand by CO to give a more active hydrogenation catalyst? How good was the retention of rhodium in the amine catalysts and was it affected by oxygen in the feed?

D. D. WHITEHURST

We agree that the "linear selectivity" does not define the ratio of linear to branched products. The latter is a function of the linear selectivity, the isomerization selectivity and the degree of conversion. This is apparent from Figures 1a and 1b, and is readily calculated for any degree of conversion from the selectivities given.

With regard to your question about phosphine losses, we have noted that the very small losses that do occur decrease with time on stream during continuous processing. This is not inconsistent with your suggestion that they could be due to occluded reagents, although we have no other evidence.

We have, of course, also puzzled about the carbon monoxide dependence of the aldehyde hydrogenation rate. While displacement of some ligand by CO might be displaced, since a) amine free-systems have orders of magnitude lower hydrogenation rates, and b) displacement of amine ligand by CO would lead to rapid rhodium elution from the catalyst which is not observed.

Under the conditions described in the text, the retention of rhodium by the amine polymers was very good and comparable to that of the phosphine polymers. The reactor effluent from continuous flow experiments contained much less than 1 ppm rhodium, which is similar to the results that you reported.

We made no attempt to exclude or remove dissolved oxygen from our reaction feeds but other than this observation, we cannot comment on oxygen sensitivity.

H. S. BLOCH

In Table 4, conversion is shown of both hexene-1 and an equilibrium hexene mixture. What is the amount of linear alcohol (heptanol) made from each of these feeds?

D. D. WHITEHURST

The product linearities from 1-hexene and equilibrated hexenes under the reaction conditions described in Table 4 were 62% and 16% respectively.

However, as mentioned in the text, the product linearity is a function of the starting olefin isomer as well as the conditions of the reaction and can be varied greatly.

J. MANASSEN

You seem to assume that only one phosphine ligand is bound to the rhodium in the polymers. My question is twofold:

1) What grounds do you have for believing this?

2) For a good comparison with homogeneous catalysts, they must also contain only one phosphine ligand. How do you arrange that?

D. D. WHITEHURST

1) In the preparation of rhodium-phosphine polymer complexes from $Rh_2(CO)_4Cl_2$ we have good evidence based on infrared studies for the formation of a cis- $Rh(CO)_2PR_3,Cl$ species within the polymer. The spectrum of this polymer complex changes however under reaction conditions. The interpretation of the resultant spectrum is also believed to be consistent with Rh con-

taining one phosphine and two carbon monoxide ligands (L. D. Rollmann, Inorg. Chim. Acta, 6 137 (1972).

2) Homogeneously it is quite difficult to obtain a monophosphine complex as the only rhodium species. This is primarily due to the establishment of an equilibrium of rhodium species containing 0,1,2, or 3 phosphine ligands as reported by Wilkinson *et al*. (G. Wilkinson *et al*. J. Chem. Soc. A 2660 (1968), *ibid*., 937, 2753 (1970). Unless large excesses of phosphine are used, we have observed under reaction conditions significant catalytic contributions of rhodium species containing no phosphine ligands as judged from olefin isomerization.

Thus, an exact comparison of homogeneous and heterogeneous catalysts is problematic in this case.

V. HAENSEL

In the excellent paper just presented, the authors show variations in selectivity (in certain cases) and a substantial maintenance of activity or conversion for the primary reaction. If one assumes a usual conversion-reciprocal space velocity relationship for the homogeneous and heterogeneous reactions, a different selectivity ratio for the various reactions would be obtained depending upon the specific reciprocal space velocity value. Thus, it would not be surprising to obtain either similar or different selectivities for the heterogeneous reaction as compared to the homogeneous reaction. A question of rather critical interest is the way in which the catalyst activity finally declines for the heterogeneous system. The authors indicate a usually slow decline for the first 1000 hours or so--is this followed by an extremely rapid decline in a few hours?

D. D. WHITEHURST

The selectivity ratios as defined are ratios of *rate constants* and as such are *not* dependent on space velocity or conversion.

As far as the nature of the decline in catalytic activity is concerned, this varies with catalyst composition and operating conditions. We have observed both types of aging behavior, i.e., slow uniform aging as well as faster aging following a period of long constant activity.

Paper Number 30

SUPPORTED TRANSITION METAL COMPLEXES AS HETEROGENEOUS CATALYSTS

K. G. ALLUM, R. D. HANCOCK, S. McKENZIE and R. C. PITKETHLY
British Petroleum Co. Ltd., Sunbury-on-Thames,
Middlesex, England

ABSTRACT: New classes of heterogeneous catalysts have been developed which combine the potential versatility and selectivity of homogeneous catalysts with the practical advantages associated with use of a solid material. Techniques are described for chemically linking transition metal complexes to the surfaces of organic polymers and inorganic oxides by way of trivalent phosphorus groups and evidence is adduced for the resulting structures. The metal-containing solids so prepared have been used as heterogeneous catalysts for the hydroformylation and hydrogenation of olefins and the cyclo-oligomerization of phenyl acetylene.

INTRODUCTION

Conventional heterogeneous catalysts, although widely used industrially, generally suffer from the disadvantages that (i) design and improvement are difficult because the active sites are not well-defined, (ii) only a small percentage of the active components are accessible and effective and (iii) the control that can be exerted over the composition and structure of the active sites is relatively limited. In contrast, homogeneous transition metal catalysts are not only better defined, but the steric and electronic environment of the catalytically active site can often be varied widely and in a systematic manner so that the course and rate of reaction may be precisely controlled (1)(2)(3).

However, the use of homogeneous catalysts on an industrial scale can lead to a number of practical problems, eg difficulties associated with corrosion, plating out on the reactor walls and catalyst recovery. If the catalyst is not to be left in the product (a wasteful, and often uneconomic, procedure which results in product contamination), it is necessary to adopt some special approach in order to (i) recover the catalyst from the reaction products eg by distillation, or ion-exchange (Such techniques can be expensive and inefficient), (ii) retain a non-volatile catalyst in the reactor while taking off the products as vapour (The products must clearly be volatile but in addition the build-up of higher boiling side-products must not be excessive and must not poison the catalyst), (iii) impregnate the catalyst on a solid support, (Catalysts used in this way can only be applied in vapour phase reaction. Moreover, impregnation often leads to an uneven surface distribution and poor catalyst utilization), or (iv) link the catalyst chemically to the surface of a solid support. The introduction of phosphine groups into polystyrene and the use of the resulting polymer as a support for metal complexes has been claimed in patents (4)(5) and subsequently briefly described in the literature (6)(7).

This paper is concerned with the last approach and describes the synthesis and use of several phosphorus-containing polymers and a new class of catalyst in which a metal complex is chemically linked to an inorganic oxide support.

EXPERIMENTAL

1. Preparation of (2-diphenyl phosphino ethyl) triethoxysilane (Reagent I)

This compound was prepared by the method of Niebergall (22).

2. Preparation of $RhCl(CO)(\emptyset_2PCH_2CH_2Si(OEt)_3)_2$ (Reagent II)

$[RhCl(CO)_2]_2$ (6.4 mmole) was treated with reagent I (25.5 mmole) dissolved in benzene. The colour of the solution changed from red to yellow. The reaction mixture was stirred overnight, evaporated to small bulk and diluted with pentane to precipitate chloro-carbonyl bis-(2-triethoxysilyl ethyl diphenyl phosphine) rhodium (I). (Found: C, 53.7, 53.9; H, 6.6, 6.4; P, 7.1, 7.1; Cl, 3.6, 3.7; Rh, 11.1, $C_{41}H_{58}O_7P_2ClRhSi_2$ requires, C, 53.6; H, 6.3; P, 6.7; Cl, 3.9; Rh, 11.2 %. ν(CO), 1966 cm^{-1} (nujol).

3. Preparation of $RhH(CO)(\emptyset_2PCH_2CH_2Si(OEt)_3)_3$ (Reagent III)

Reagent II (5g) and reagent I (8.3g) were reacted with $NaBH_4$ (2g) in refluxing ethanol. The product separated as a yellow precipitate and after recrystallization from benzene/ethanol and benzene/pentane showed IR absorption bands at 1990 and 1915 cm^{-1}. (Found: Rh, 7.5; P, 6.8; $C_{61}H_{88}O_{10}P_3Si_3Rh$ requires Rh, 8.2; P, 7.3).

4. Preparation of Phosphorus-containing Solids

Chloromethylation (20) of pre-dried Amberlite XAD-2 and reaction of the product with $\emptyset_2$PK yielded Polymer A (4, 21). Cross-linking chloromethylation (21) of a linear polystyrene (MW 250 000) and reaction with $\emptyset_2$PK (4, 21) gave Polymer B. Polymer C was prepared by reaction of polyvinyl chloride in tetrahydrofuran with $\emptyset_2$PK (4, 21). Dry U40 Grade silica (J. Crosfield and Co) was converted to Silica-P$\emptyset_2$ by treatment with an xylene solution of Reagent I and azeotropic removal of the ethanol formed.

5. Preparation of Metal-containing Solid Catalysts

The catalysts were made from Polymers A, B, C, silica and silica-P$\emptyset_2$ by treatment with solutions of metal complex reagents as listed in Table 1. Unless otherwise stated, all reactions were carried out using dry degassed solvents under oxygen-free nitrogen. Products were either Soxhlet extracted under nitrogen or washed with solvent until the washings were colourless. Finally, products were vacuum dried and analyzed.

6. Catalytic Studies

(a) Hydroformylation

Details of autoclave experiments are given in Table 2. The conditions employed were Temperature 79-83°C; Pressure 41.3 bar (ga) of H_2/CO (1 : 1); Duration 4 hours. Feedstock was hexene-1 in n-heptane. With the catalyst from Polymer A, 84g of olefin were used; with the other catalysts, 68g. Pilot plant tests were carried out using 20 ml of catalyst in a trickle-phase reactor. The catalysts were loaded under nitrogen. The feedstock (hexene-1, 1 vol, in n-heptane, 2 vol) was freed from peroxide before use. Gaseous and liquid feeds were deoxygenated when the catalyst based on Polymer A was tested.

(b) Hydrogenation

(i) The catalyst (0.81g) prepared from $[RhCl(COD)]_2$ and Polymer A, was heated at 50-55°C and 13.8 bar (ga) of hydrogen with hexene-1 (14 ml), cyclohexene (14 ml) and benzene (42 ml). After 4 hours, 71% wt of the hexene and 37% wt of the cyclohexene had been hydrogenated.

(ii) The catalyst (0.45g) made from silica and Reagent II was heated at 140°C and 13.8 bar (ga) of hydrogen with a similar test feed

TABLE 1

CATALYST PREPARATIONS

Support	Phosphorus Content % wt	Solvent	Temp °C	Reagent	Metal Content % wt
Polymer A	2.0	Hexane	25	$Rh(acac)(CO)_2$	1.9
Polymer C	4.8	THF	25	"	7.3[a]
Silica-PØ$_2$	1.0	Toluene	25	"[b]	2.0
Polymer A	2.2	Benzene	25	$[RhCl(COD)]_2$	2.7
Silica-PØ$_2$	0.9	CH_2Cl_2	Reflux	"	1.0
Silica	-	Toluene	Reflux[c]	Reagent II	0.9
Silica	-	Toluene	Reflux[c]	Reagent III	1.2
Polymer B	4.4	B/B[d]	Reflux	$CoCl_2$	2.8
Polymer C	10.8	THF	Reflux	$CoCl_2$	4.5
Silica-PØ$_2$	1.4	EtOH	Reflux	$CoCl_2$	0.6
Polymer B	5.3	B/B[d]	Reflux	$NiCl_2$	3.4
Polymer C	10.8	THF	Reflux	$NiCl_2$	6.1
Silica-PØ$_2$	1.4	EtOH	Reflux	$NiCl_2$	0.5
Polymer A	2.2	Toluene	0	$Ni(COD)_2$	0.3
Silica-PØ$_2$	0.9	Toluene	0	$Ni(COD)_2$	0.8

a IR band at 1980 cm^{-1} (nujol); b Under hydrogen;
c Alcohol formed was azeotroped out; d Benzene/butanol (1:1 vol)

TABLE 2

HYDROFORMYLATION OF HEXENE-1 USING RHODIUM CATALYSTS

Autoclave experiments under conditions given in text, section 6 (a)

Catalyst	Weight of Catalyst Used g	Conversion % wt	Selectivity to Aldehydes % wt	Normal:Branched Aldehyde Ratio
Solid from Polymer A/ $Rh(acac)(CO)_2$	0.620	60	69	2.5
Solid from Polymer C/ $Rh(acac)(CO)_2$	0.223	43	100	2.5
Solid from silica-PØ$_2$/ $Rh(acac)(CO)_2$	0.200	99	39	2.0
$Rh(acac)(CO)_2$	0.0209	99	83	1.2
$Rh(acac)(CO)$ PØ$_3$	0.0398	88	76	2.9

mixture to that used in (i). After 2 hours, 96% wt of the hexene and 95% wt of the cyclohexene had been hydrogenated.

(iii) Reaction (ii) was repeated in the presence of 700 ppm of mercaptan sulphur (as n-C_4H_9SH) using 0.55g of catalyst. 74% wt of the hexene and 3% wt of the cyclohexene were hydrogenated.

(c) Oligomerization of Phenylacetylene

$NaBH_4$ (0.1g) in THF was added to a suspension of the nickel catalyst (0.12g) made from Polymer C in a solution of phenylacetylene (5g) in THF and the reaction mixture stirred at 20°C for four hours. Working up yielded 3.5g (70% wt) of a mixture which contained predominantly 1,2,4- and 1,3,5-triphenyl benzenes.

RESULTS AND DISCUSSION

A wide range of catalytically active transition metal complexes contain phosphine (R_3P) or phosphite ($(RO)_3P$) groups as ligands. For example, complexes containing trivalent phosphorus groups have been shown to be catalysts for the hydroformylation, hydrogenation, dismutation, oligomerization of olefins, the oligomerization and hydrogenation of diolefins. Two methods for chemically linking such transition metal complexes to the surface of solid supports are described in this paper

Method A: Trivalent phosphorus groups are chemically linked to the surface of the support and then reacted with a transition metal compound to form the desired surface complex.

Method B: A bifunctional molecule, containing a trivalent phosphorus group and a group which will react with the support surface, is synthesized. The phosphorus group in the molecule is coordinated to a transition metal and the resulting complex then linked to the support via the second group.

1. Supported Transition Metal Complexes prepared by Method A

We have found that trivalent phosphorus groups can be chemically linked to the surface of polymers using a range of techniques (4) but the most useful of these involves reacting a halogen-containing polymer with an alkali-metal phosphide (4a). Thus chloromethylated, cross-linked polystyrene reacts as follows :-

$$-CH_2CH(-\langle O \rangle-CH_2Cl)- \quad + \varnothing_2PK \longrightarrow -CH_2CH(-\langle O \rangle-CH_2P\varnothing_2)- \quad + \quad KCl$$

Two different polystyrenes have been used as starting materials, viz a macroreticular polystyrene for Polymer A and a conventional polystyrene for Polymer B. Polyvinyl chloride reacts in a similar manner to yield polymer C :-

$$-CH_2CH(-Cl)- \quad + \quad \varnothing_2PK \longrightarrow -CH_2CH(-P\varnothing_2)- \quad + \quad KCl$$

In the subsequent discussion the formulae of Polymers A,B and C will be abbreviated to polymer-$P\varnothing_2$.

For viable commercial operation, it is desirable that a heterogeneous catalyst should be a physically and chemically robust material with an adequate surface area and a neglible solubility in the reaction medium. In addition, the majority of the reaction sites should be readily accessible. The polymer prepared from the high surface area, macroreticular polystyrene (Polymer A) most closely fulfils these requirements.

However, it is often difficult to obtain polymers in a satisfactory physical form. For this reason techniques have been developed for synthesizing transition metal complexes on the surface of inorganic oxides.

The Si-OEt group readily reacts with the surface of silica to form the very stable Si-O-Si linkage. We have utilized this fact to link trivalent phosphorus groups to the surface of silica. For example, the compound $\phi_2PCH_2CH_2Si(OEt)_3$ reacts with surface silanol groups as follows:

$$(\text{-O-})_3\text{Si-OH} + (\text{EtO})_3\text{SiCH}_2\text{CH}_2\text{P}\phi_2 \xrightarrow{-\text{EtOH}} (\text{-O-})_3\text{SiO}\overset{|}{\underset{|}{\text{S}}}\text{iCH}_2\text{CH}_2\text{P}\phi_2$$

to yield a phosphorus-containing silica (designated as silica-$P\phi_2$). This technique is widely applicable and may be used to link a range of phosphorus-containing compounds to the surface of solids provided the solid contains surface hydroxyl groups (4e).

Trivalent phosphorus compounds form complexes primarily with those transition metals which are in low oxidation states and particularly with those in Group VIII. In general, complex formation occurs by either ligand displacement or by expansion of the coordination shell of the transition metal.

A study of the reactions of the phosphorus-containing solids with a range of transition metal compounds and an examination of the properties of the resulting materials has demonstrated that an analogous series of reactions can take place on solid surfaces.

(a) Rhodium-containing Solids

(i) The green complex $[Rh(acac)(CO)_2]$ reacts with phosphines with the liberation of carbon monoxide and the formation of the yellow, square planar, complex $[Rh(acac)(CO)PR_3]$ (8)(9). Polymers A and C and the silica-$P\phi_2$ also react with $[Rh(acac)(CO)_2]$ with the elimination of carbon monoxide and the formation of yellow, rhodium-containing solids. The infra-red spectrum of the rhodium-containing polymer prepared from Polymer C contains a single absorption in the region 1900-2000 cm^{-1} attributable to ν(CO).

Both $[Rh(acac)(CO)_2]$ and $[Rh(acac)(CO)PR_3]$ are olefin hydroformylation catalysts. However, the former complex catalyzes the hydroformylation of hexene-1 at 79-83°C and 41.3 bar (ga) H_2/CO (1 : 1) to yield a product with a normal-to-branched aldehydes ratio of 1.2 : 1 whereas the latter complex, under similar conditions, yields a product with a normal-to-branched aldehydes ratio of greater than 2 : 1 (Table 2).

All three rhodium-containing solids (Table 2) are olefin hydroformylation catalysts. Moreover all yield products with normal-to-branched aldehydes ratios of at least 2 : 1 indicating that the active species contains a phosphorus-rhodium bond. It seems reasonable to conclude, therefore, that $[Rh(acac)(CO)_2]$ reacts with polymers A and C and with the silica-$P\phi_2$ to yield square planar surface species of the type - (polymer-$P\phi_2$)Rh(acac)(CO) and (silica-$P\phi_2$)Rh(acac)(CO).

The polymer-supported rhodium complex prepared from Polymer A has also been examined for the hydroformylation of hexene-1 in a pilot plant and shown to have substantial and sustained activity. Thus over 90 HOS at 80-85°C, 41.3 bar (ga) H_2/CO (1 : 1), GHSV = 1250 and a hexene-1 LHSV = 1.3, the following results were obtained :-

rhodium content of the product $<$ 1 ppm,
conversion $>$ 97% wt; selectivity to aldehydes $\geqslant$ 95% wt;
normal-to-branched aldehydes ratio $\geqslant$ 2 : 1.

Clearly this type of catalyst can be used for liquid phase reactions in a manner similar to conventional heterogeneous catalysts. It is necessary to deoxygenate the feedstocks to avoid the elution of rhodium from the catalyst surface.

(ii) Bis (chloro cyclo-octa-1,5-diene rhodium (I)) reacts with trivalent phosphorus compounds to yield complexes containing between one and five phosphorus groups per rhodium atom depending on the reaction conditions and the nature of the phosphorus compound employed (10). For example, in benzene complexes of the type $[RhCl(COD)PR_3]$, $[RhCl(PR_3)_3]$ and, in the presence of hydrogen, $[RhClH_2(P\emptyset_3)_2(C_6H_6)]$ are obtained whereas cationic complexes (eg $[Rh(COD)(PR_3)_2]^+$) are often obtained in alcoholic solvents.

Polymer A reacts with a benzene solution of $[RhCl(COD)]_2$ in the presence of hydrogen to yield a yellow, rhodium-containing polymer. For steric reasons, it is unlikely that the surface species contains more than two phosphorus groups bound to each rhodium atom and therefore the species most likely to be present under the conditions employed is either (polymer-$P\emptyset_2)_2RhClH_2(C_6H_6)$, or (polymer-$P\emptyset_2$) RhCl(COD). The rhodium-containing polymer is a catalyst for the hydrogenation of both terminal and internal olefins. Since the complex $[RhClH_2(P\emptyset_3)_2(C_6H_6)]$ is formed under the conditions used above and is a catalyst for the hydrogenation of olefins (2), it is possible that the heterogeneous catalyst contains surface species of the first type. A more detailed study of this catalyst will be required to elucidate the exact nature of the surface species. The silica-$P\emptyset_2$ also reacts with $[RhCl(COD)]_2$ to yield a rhodium-containing solid.

(b) Cobalt-containing Solids

Cobalt (II) phosphine complexes usually occur as tetrahedral species of the type $[CoX_2(PR_3)_2]$ although 5-coordinate complexes are also known (11)(12). The tetrahedral-complex $[CoCl_2(P\emptyset_3)_2]$ is a bright blue solid with a group of intense electronic absorption bands in the region 600-760 nm. The splitting of the bands in this region is dependent on the nature of the ligands attached to the cobalt (12). In contrast, the 5-coordinate complexes (eg $[CoCl(\emptyset_2PCH_2CH_2P\emptyset_2)_2]^+$ are generally greenish yellow and have absorption bands around 400 and 600 nm (11).

Polymers B and C react with anhydrous cobalt chloride to form blue cobalt-containing polymers. The diffuse reflectance spectra of these polymers (Figure 1) show strong absorptions in the region 600-760 nm but have no significant absorption around 500 nm. It is also evident from the spectra that the polymers do not contain significant quantities of unreacted cobalt (II) chloride. These results indicate that the polymers contain tetrahedral surface species of the type (polymer-$P\emptyset_2)_2$ $CoCl_2$.

Cobalt chloride also reacts with silica-$P\emptyset_2$ to yield a blue cobalt-containing solid. The diffuse reflectance spectrum (Figure 1) indicates the presence of a tetrahedral cobalt (II) phosphine complex.

Examination of molecular models indicates that coordination to the same metal atom can readily occur, (i) between neighbouring groups in Polymer C, (ii) between neighbouring groups and alternant groups in Polymer B, although the latter seems more likely for steric reasons and (iii) between phosphine groups on different loops of the same polymer chain or on different polymer chains (when they approach sufficiently closely to allow coordination to occur). The solubility of the cobalt-containing polymer prepared from Polymer C is considerably less than that of the parent

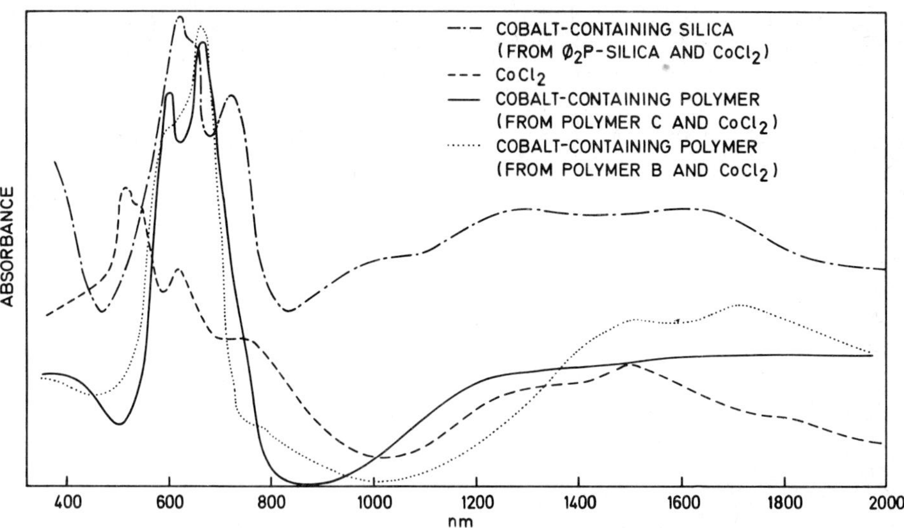

FIG 1 ELECTRONIC ABSORPTION SPECTRA

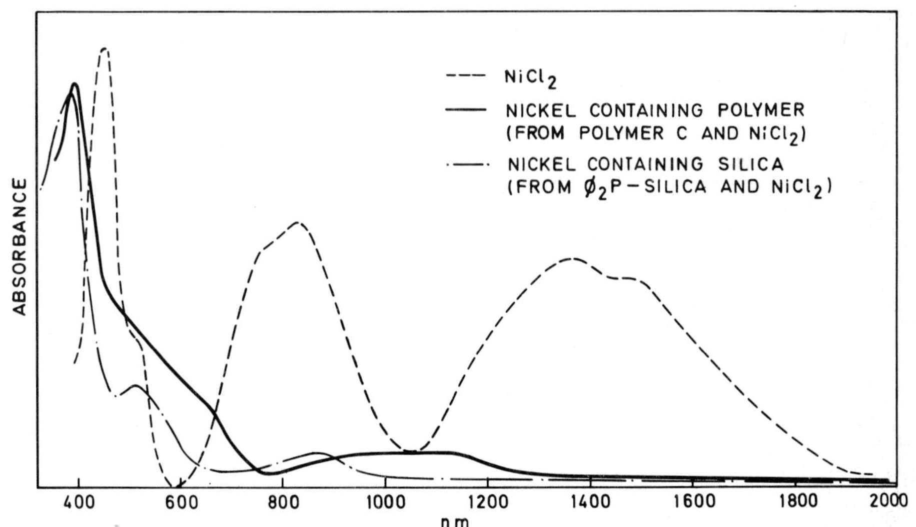

FIG 2 ELECTRONIC ABSORPTION SPECTRA

polymer. This may be the result of cross-linking or possibly of a shortening and stiffening of the polymer chain on coordination.

(c) Nickel-containing Solids

(i) Nickel halides react with monophosphines to form either square planar or tetrahedral complexes of the type $[NiX_2(PR_3)_2]$. The stereochemistry of the complex is determined by the nature of the phosphine and the halide (13). For example, the complexes $[NiX_2(P\phi R_2)_2]$ (where R = alkyl) are square planar whereas $NiCl_2(P\phi_3)_2$ is tetrahedral. Complexes of the type $[NiX_2(P\phi_2R)_2]$ may exist in both square planar and tetrahedral forms with the tetrahedral form becoming relatively more stable as the bulkiness of R increases. In general, the tetrahedral complexes are green or blue and the square planar complexes are red. Although the two forms have fairly similar electronic absorption spectra, the tetrahedral complex has an additional absorption band in the region 800-1000 nm (13).

Nickel (II) chloride reacts with Polymer C to form a red/brown nickel-containing polymer. The diffuse reflectance spectrum of this material (Figure 2) indicates that the bulk of the nickel is present as a square planar complex of the type (polymer-$P\phi_2)_2NiCl_2$. However, the weak absorption band at 960-1100 nm suggests that a small concentration of the tetrahedral species may also be present.

The complex $[NiCl_2(P\phi_3)_2]$, in the presence of sodium borohydride, is an active catalyst for the cyclo-oligomerization of acetylenes (14). For example, phenyl acetylene is converted to 1,2,4 and 1,3,5 triphenyl benzenes. The same reaction is catalyzed when the complex $[NiCl_2(P\phi_3)_2]$ is replaced by the nickel-containing polymer prepared from Polymer C.

Nickel (II) chloride also reacts with Polymer B and the silica-$P\phi_2$ to yield mauve and red nickel-containing solids respectively. An examination of the electronic absorption spectrum of the nickel-containing silica indicates that the nickel is primarily present as a square planar species with relatively small amounts of the tetrahedral species (Figure 2).

(ii) Phosphines react with bis (cyclo-octa-1,5-diene) nickel (0) with the displacement of either one or two cyclo-octadiene molecules and the formation of either $[Ni(COD)(PR_3)_2]$ or $[Ni(PR_3)_4]$ (15).

Both Polymer A and the silica-$P\phi_2$ react with bis (cyclo-octa-1,5-diene) nickel (0) with the formation of red, air-sensitive, nickel-containing solids. For steric reasons, it is likely that only one of the cyclo-octadiene groups on $[Ni(COD)_2]$ is displaced and that the surface species formed are (polymer-$P\phi_2)_2$ Ni(COD) and (silica-$P\phi_2)_2$ Ni(COD). However, in contrast to the analogous homogeneous complex (3), the supported nickel species have little activity for the cyclo-oligomerization of butadiene. It is possible that pairs of the phosphine groups on the polymer form a chelating ligand and thereby inhibit the formation of an active species analogous to $R_3P\text{-}Ni(C_8H_{12})$, (3).

2. Supported Transition Metal Complexes prepared by Method B

The technique outlined above in many cases provides a convenient method for supporting transition metal complexes. However, as the complex must be synthesized in situ on the solid surface, the nature of the complex formed is determined by the arrangement and concentration of the surface phosphine groups. Certain complexes (eg $[RhH(CO)(PØ_3)_3]$) are therefore difficult, if not impossible, to support by this technique unless a ligand exchange reaction is used. For this reason, a series of phosphine complexes have been synthesized containing the ligand $Ø_2PCH_2CH_2Si(OEt)_3$. These complexes can be prepared, purified and characterized by conventional techniques and subsequently reacted with the surface of a hydroxyl-containing solid eg :-

$$M(CO)_n + Ø_2PCH_2CH_2Si(OEt)_3 \longrightarrow \underset{\text{Complex A}}{[M(CO)_{n-1}\,Ø_2PCH_2CH_2Si(OEt)_3]} + CO$$

$$\text{surface-OH} + \text{complex (A)} \longrightarrow (\text{surface-O-}\overset{|}{\underset{|}{Si}}CH_2CH_2PØ_2)M(CO)_{n-1} + EtOH$$

The complex $[RhH(CO)(PØ_3)_3]$ is a yellow, crystalline solid with a characteristic infra-red spectrum in the region 1900-2050 cm^{-1} (16)(17). In solution one phosphine group readily dissociates to yield a species which is an olefin hydroformylation catalyst (17). The complex $[RhH(CO)(Ø_2PCH_2CH_2Si(OEt)_3)_3]$ has now been synthesized. It is a yellow, crystalline solid with an infra-red spectrum in the region 1900-2050 cm^{-1} which is very similar to that of $[RhH(CO)(PØ_3)_3]$. Moreover, the complex readily reacts with the surface of silica to yield a rhodium-containing solid which is an olefin hydroformylation catalyst of considerable potential. For example, over 75 HOS in a pilot plant at 100-120°C, 41.3 bar (ga) H_2/CO (1 : 1), GHSV = 1250 and a hexene-1 LHSV$\simeq$1, the following results were obtained :-

conversion $>$ 98% wt; selectivity to aldehydes $\geqslant$ 97% wt;
normal-to-branched aldehyde ratio $>$ 2 : 1;
rhodium content of product after 75 HOS $<$ 0.07 ppm.

The complex $[RhCl(CO)(PØ_3)_2]$ is a yellow crystalline solid with a single CO stretching frequency in the region of 1960 cm^{-1} (18). It is also a catalyst for the hydrogenation of olefins (19). The analogous complex $[RhCl(CO)(Ø_2PCH_2CH_2Si(OEt)_3)_2]$ has now been prepared. It is also a yellow crystalline solid with a single CO stretching frequency in the region of 1960 cm^{-1}. Moreover, it reacts with silica to form a rhodium-containing solid which is a catalyst for the hydrogenation of olefins. It is also active in the presence of considerable quantities of mercaptan sulphur.

CONCLUSIONS

It is evident from the foregoing discussion that not only can transition metal complexes be readily chemically linked to solid surfaces and used as heterogeneous catalysts for a variety of reactions but also that it should be possible to control the electronic and steric environment of the coordinated metal with considerable precision. For example, the nature of the polymer backbone and the nature and size of the R groups can be varied. Thus the electron density and accessibility of the metal atom, M, will be different for R = phenyl from that where R = methyl or tertiary butyl. Moreover, in those cases where the metal atom is coordinated to the surface of a polymer by more than one group, it may be possible to vary the geometry and steric strain about the metal atom by altering the distance between, and relative positions of, the phosphine groups.

REFERENCES

(1) Bogdanovic B, and Wilke, G. Brennstoff-Chemie, 1968, 49 (11, 323

(2) Jardine, F.H., Osborn, J.A., Young J.F., Wilkinson, G., JCS, 1966 (A), 1711.

(3) Wilke, G., Brenner, W., Heimback, P., Hey, H., Muller, E., Liebigs Ann Chem, 1969, 727, 161.

(4) a. Belgian Patent 739 607,
b. Dutch Patent 70 06586,
c. Belgian Patent 745 273,
d. Belgian Patent 739 609,
e. Belgian Patent 760 556.

(5) Belgian Patent 721 686.

(6) Grubbs, R.H., Kroll, L.C., JACS, 1971, 93, 3062.

(7) Manassen, J., Platinum Metals Review, 1971, 15, 142.

(8) Bonati, F., Wilkinson, G., JCS, 1964, 3156.

(9) Bailey, N.A., Coates, E., Chem Comm, 1967, 1041.

(10) a. Haines, L.M., Inorganic Nucl Chem Letters, 1969, 5, 399,
b. Ohno, K., Tsuji, J., JACS, 1968, 90, 99,
c. Osborn, J.A., Schrock, R., JACS, 1971, 93, 2397,
d. Haines, L.M., J Organometallic Chem, 1970, 25, C85.

(11) Horrocks, W., Hecke van G., Hall, D., Inorganic Chem, 1967, 6, 694.

(12) Cotton, F.A., Faut, O.D., Goodgame, D., Holm R., JACS, 1961, 83, 1780.

(13) Hayter, R.G., Humiec, F.S., Inorganic Chem, 1965, 4, 1701.

(14) Donda, A.F., Moretti, G., J Org Chem, 1966, 31, 985.

(15) British Patent 935 716; Canadian Patent 779 362.

(16) Bath, S., Vaska, L., JACS, 1963, 85, 3500.

(17) Evans, D., Yagupsky, G., Wilkinson, G., JCS, 1968, A, 2660.

(18) Chatt, J., Shaw, B.L., JCS, 1966, A, 1437.

(19) Strohmeier, W., Rehder-Stirnweiss W., Z. Naturforsch, 1971, 26B, 61.

(20) Pepper, J., Paisley, H.M., Young, M.A., JCS, 1953, 4097.

(21) Issleib, K., Pure and Appl Chem, 1964, 9, 205.

(22) Niebergall, H., Makromal Chem, 1962, 52, 218.

ABBREVIATIONS

Ø = C_6H_5 Et = C_2H_5

acac = $CH_3COCHCOCH_3$ COD = cyclo-octa-1,5-diene

ACKNOWLEDGEMENT

Permission to publish this paper has been given by The British Petroleum Company Limited.

DISCUSSION

M. G. NOAK

The general applicability of your approach to converting homogeneous catalysts to heterogeneous ones for a wide variety of catalytic processes should, among other factors, depend on the thermal stability of the polymer backbone. What are the upper temperature limits of applicability for styrene and other suitable polymeric catalyst carriers?

R. C. PITKETHLY

Although the decomposition temperatures of polymers are known, e.g. polystyrene is about 230-250°C, it is not possible to give a simple straightforward answer to this question because the stability of polymer may be increased by the presence of phosphorus or decreased by the process gases. However, we have not found this to be a limiting factor since the objective of our work has been to develop low temperature processes and the problem does not arise with the silica base.

P. R. RONY

You mentioned that you anchored metal complexes to four different types of supports--polystyrene, crosslinked polystyrene, polyvinyl chloride, and silica. Which of these supports would you generally prefer for use in industrial processes, and why?

R. C. PITKETHLY

The choice of support depends on several factors. For example the degree of cross-linking which is acceptable will depend to some extent on the velocity of the catalyzed reaction and the importance of diffusion limitations. If a high active site density is required PVC could be preferred to chlormethylated polystyrene whereas for the attachment of bulky complexes a degree of flexibility may be more desirable. A requirement for thermal and mechanical stability would direct attention to the inorganic supports. In short, one chooses a support to suit the particular circumstances.

R. L. BURWELL, JR.

One can probably rather widely extend the use of silica as a peg-board for the attachment of catalytically active groups. We have attached the group $-CH_2CH_2CH_2Cl$ to silica gel by similar use of the triethyoxysilane and have converted this product to the N-alkylimidazole. This heterogeneous catalyst gives about the same rate for the hydrolysis of p-nitrophenyl acetate in aqueous solution as does the homogeneous analog, N-ethylimidazole.

R. C. PITKETHLY

It is gratifying to hear of this confirmation of the effectiveness of our silica technique. I am convinced that we are on the threshold of far-reaching developments in this work and can see the possibility of preparing multifunctional catalysts with sites controlled not only with respect to their activity and selectivity as suggested in the paper but also with respect to their relative location and orientation. From this it is an admittedly long but logical step to modelling of enzymes.

M. M. BHASIN

The first question is whether you have observed rhodium losses lower than those reported for silica? The second question is whether the loss of rhodium varies during the duration of runs?

R. C. PITKETHLY

The rhodium losses reported for silica are below the limit of the analytical method even when the product sample was concentrated 30-fold by distillation before analysis.

In answer to the second question: it is usual to observe some loss of metal at the start of a run and, under constant conditions, this decreases to negligible amounts as the run proceeds.

H. BREMER

To the remarks on page 4: Is it proved that the cyclotrimerization of phenylacetylene takes a heterogeneous course when using the Ni-polymer C-catalyst or is the reaction effected by nickel which has separated from the catalyst? If this is proved, the question arises, why the cyclooligomerization of butadiene on a similar catalyst (see page 8) takes a very bad course, though the demands for free sites of coordination should be similar to those of the oligomerization of acetylene?

R. C. PITKETHLY

With regard to the first question: the experiments were carried out batch-wise in autoclaves and so it is difficult to be certain that the amount of dissolved nickel was insufficient to account for the catalysis. However, the catalysts were exhaustively extracted before testing and the following evidence indicates that the nickel was firmly bonded.

On the second question, it must be remembered that the catalyst systems are quite different; an activating agent, $NaBH_4$ was used for cyclooligomerization of phenylacetylene but none in the case of butadiene.

It is known from Wilke's work that butadiene is converted over nickel cyclopentadienyl complexes to cyclododecatriene; over nickel complexes containing one phosphine ligand, the product is cyclooctadiene and the complexes containing two phosphine ligands are inactive or produce open chain products. Thus, ignoring the possibility of other causes such as steric hindrance in the polymer, one might argue that the absence of catalysis of the butadiene reaction indicates the presence of nickel chelated to two phosphine ligands and therefore firmly attached to the support.

In the presence of phenylacetylene and sodium borohydride, coordination of three acetylene molecules leading to the formation of triphenyl benzene requires exchange of one phosphine for the third acetylene. One phosphine remains attached to the metal, otherwise the unblocked homogeneous nickel catalyst would produce tetraphenyl-cyclooctatetraenes, which were not detected and could be only minor products.

J. R. OWEN

This is an interesting paper, and could be even more so if the data were presented in such a way that comparison of the activity of homogeneous catalyst, polymer-supported catalyst, and silica-supported catalyst were possible. For example, in the hydroformylation tests given in Table 2, the same amount of rhodium per gram of olefin is used only in the last two (homogeneous) tests. Similarly, it is not possible to compare the activity of the catalysts for hydrogenation. Nor is any comparison made with a conventionally-prepared rhodium/silica catalyst. Do the authors have data for such comparisons? Have they tried tying the transition metals to molecular sieve substrates? I have the feeling that these catalysts and their close relatives are the next generation of industrial catalysts, but I would certainly like to see more definitive comparative data.

R. C. PITKETHLY

We do not have data for comparison of the complexes with a conventional rhodium-on-silica catalyst. On the other hand, we have succeeded in bonding these catalysts to molecular sieves. If these catalysts are to become the next generation of industrial catalysts as we hope, many more definitive comparisons will certainly be made.

Paper Number 31

SIMILARITIES IN THE ISOTOPIC HYDROGEN EXCHANGE OF POLYCYCLIC AROMATIC HYDROCARBONS, LONG-CHAIN ALKYLBENZENES AND ALKANES WITH HOMOGENEOUS AND HETEROGENEOUS PLATINUM

K. DAVIS, J. L. GARNETT, K. HOA, R. S. KENYON and M. LONG
School of Chemistry, The University of New South Wales, N.S.W., Australia

ABSTRACT: A range of polycyclic aromatic hydrocarbons, long-chain alkylbenzenes and alkanes have been exchanged in the presence of homogeneous sodium chloroplatinate (II) and heterogeneous platinum. Remarkable similarities in isotope orientation patterns are observed, particularly during initial rates of exchange, for these compounds using both catalytic systems. The condensed polycyclics such as naphthalene initially exchange in the β position with an M value of unity; the polyphenyls exchange with severe ortho deactivation and M values of approximately 3, while the order of exchange in the hydrogens of the long-chain alkylbenzenes such as n-butylbenzene, is ring > α side-chain > terminal side-chain > methylene side-chain. The pattern of exchange in the butyl group of butylbenzene has also been related to exchange in n-butane. The results have been interpreted in terms of π-complex mechanisms common to both catalytic systems.

1. INTRODUCTION

Correlations between homogeneous and heterogeneous processes are important to fundamental catalytic theory. Isotopic hydrogen exchange reactions are particularly useful in this respect since the manner in which specific positions of a molecule exchange can be directly related to the type of interaction between the molecule and the metal atom whether this be in a homogeneous or heterogeneous system.

In preliminary studies, a new homogeneous platinum catalyst[1,2] has been reported which is capable of catalysing exchange in organic compounds such as polycyclic aromatics, simple alkylbenzenes (toluene etc) and the monosubstituted benzenes. These compounds had previously been deuterated with a heterogeneous catalyst, π-complex mechanisms being proposed for these exchange reactions.[3]

In this earlier work, no isotope orientation studies in both heterogeneous and homogeneous systems had been reported with the polycyclic aromatic hydrocarbons, especially during initial rates of exchange. This is very important in catalytic theory since π-bond localisation concepts have been invoked to explain the observed isotope orientation in this series when the compounds have exchanged for a considerable period of time with a homogeneous platinum catalyst.[2,4] Thus only isotope orientation data with these hydrocarbons are available for the homogeneous system and then only for systems that have extensively exchanged. No equivalent systematic heterogeneous study with the polycyclics has been undertaken because of the difficulty of separating the exchange reaction from the competing randomization process. Randomization has only been observed with heterogeneous catalysts, presumably because exchange rates are very slow under these conditions and to observe finite isotope incorporation in the minimum of time, higher temperatures than are used homogeneously, are necessary.

For a direct comparison between homogeneous and heterogeneous catalytic exchange it is thus essential to be able to observe initial deuteration rates and orientation in both systems. An epr method has been developed for this purpose, since nmr is not sufficiently sensitive at low deuterium incorporation.[5)] Also, mass spectrometry gives the various deuterated species present during initial exchange rates, but does not distinguish positional substitution such as the difference between naphthalene-α-d_1 and naphthalene-β-d_1.

The data to be discussed in the present paper show that there are identical deuteration patterns between heterogeneous platinum catalysed reactions and the corresponding homogeneous platinum (II) systems for a wide range of polycyclic aromatic hydrocarbons. The correlation between homogeneous and heterogeneous platinum systems has also been extended to the long-chain alkylbenzes using n-butylbenzene as representative molecule. The long chain alkylbenzenes are important since preliminary studies with these compounds under homogeneous conditions show that the pattern of isotope incorporation in the side-chain is similar to the pattern observed in the homogeneous platinum catalysed exchange of the corresponding saturated hydrocarbons (n-butylbenzene vs n-butane.) Thus it is now possible to directly relate the manner of bond formation on the surface of a metal catalyst with the chemistry of inorganic complexes in solution for both aromatic and aliphatic molecules.

2. EXPERIMENTAL

2.1 Exchange Procedures

The deuteration method using homogeneous platinum (II) has previously been described.[2)] For the heterogeneous runs, the published method[3)] was followed, the catalyst being dry, pre-reduced platinum black prepared by $NaBH_4$ reduction of the oxide. For multiple equilibration experiments, D_2O(or H_2O) was recharged at the appropriate time.

2.2 Analytical Procedures

Nmr were obtained on both Varian A-60 and JEOL (JNM-4H-100) instruments, using deuterochloroform as solvent. For infrared, samples were analysed as KBr discs on an Hitachi-Perkin Elmer grating instrument (EPI-G2). Mass spectra were obtained at both high and low voltages on Metropolitan Vickers MS-12, MS-9 and Hitachi-Perkin Elmer RMS-4 instruments. Epr measurements were made at room temperature on a Varian V-4502 X band spectrometer, spectra being obtained with a 2-3 milliwatts of power. Several low temperature runs showed no improvement in resolution.

2.3 Radical Anion Preparation

The anion epr method for isotope orientation was only used where the resulting hyperfine spectral pattern was not complex. Occasionally, attempts to obtain radical ions failed. In order to obtain well resolved spectra, the exchanged material required careful purification, since traces of organic impurities or metal catalyst severely inhibited radical ion formation. For example, labelled naphthalene samples which were representative of the more reactive polycyclics gave well resolved spectra one to three hours after preparation, whereas some labelled pyrene samples required to be shaken for two months at room temperature before satisfactory spectra were obtained. The radical monoanions were prepared by reaction of the aromatic with sodium or potassium metal in 1,2-dimethoxyethane at 10^{-3} torr.

3. RESULTS

3.1 Homogeneous exchange of condensed polycyclics

Exchange results of the condensed polycyclics and polyphenyls are shown in tables 1-5, the long-chain alkylbenzenes and alkanes in table 6.

TABLE 1

Homogeneous platinum catalyzed exchange of condensed polycyclics

Run	Compound	Time (hr)	% D (theor)	% D (found)	Predominant[b] Substitution Position[c]	Predominant[b] Orientation[d] d-species
1	Naphthalene	0.25	96.3	3.7	β	d_1
2	Naphthalene	0.5	96.3	1.8	β	d_1
3	Naphthalene	1.0	96.3	5.2	β	d_1
4	Naphthalene	2.25	96.3	13.3	β	$d_1 >> d_2$
5	Naphthalene	168	96.3	44.2	β	d_4
6	Naphthalene	336	96.3	54.8	β>>α	$d_4 >> d_5$
7	Anthracene	1.0	98.8	1.50	β	d_1
8	Anthracene	2.25	99.3	4.2	β	d_1
9	Anthracene	168	99.5	37.7	β	d_4
10	Anthracene	336	99.4	33.4	β>>α	$d_4 >> d_5$
11	Anthracene	5	99.4	29.2	-	[e]
12	Pyrene	0.7	99.7	5.2	β	d_1
13	Pyrene	12	99.7	17.6	β>>α	$d_2 >> d_3$
14	Pyrene	24	99.7	16.8	β>>α	$d_2 >> d_3$
15	Pyrene	36	99.7	17.0	β>α	$d_2 > d_3$
16	Pyrene	1.5,2.5	99.9	41.5	β>α>γ	$d_6 > d_7$
17	Pyrene	6	99.8	67.3	β>α>γ	$d_6 > d_7$

a Exchange temperature 75°C except runs 11 (100°C), 16 (107°C) and 17 (116°C). Typical procedure[2)] involved compound (0.2g) in CH_3COOD (50 mole % in D_2O) containing Na_2PtCl_4 (0.02M) and HCl (0.02M).

b Numbering system[2)] for naphthalene, anthracene, α = position 1, β = 2; for pyrene α= 1, β= 2, γ= 4.

c By epr and/or nmr

d By low voltage mass spectrometry and/or epr

e No acid used, thus no acid catalysed exchange in γ position.

With the homogeneous platinum (II) catalyst, naphthalene and anthracene initially exchange in the β position with an M value of unity (tables 1,7). Prolonged reaction at 75°C with naphthalene and anthracene gives the β-d_4 species with a small degree of alpha scrambling. Exchange in the alpha position is accentuated at high temperatures (130°C); however, platinum precipitation is severe under these conditions. Pyrene (tables 1,5) also exchanges initially with an M of unity (table 7) in the β position so that with careful temperature control, the 2,7-d_2 species can be obtained with reasonable selectivity.

TABLE 2

Heterogeneous platinum catalyzed exchange of condensed polycyclics [a]

Run	Compound	Time (hr)	% D (theor)	% D (found)	Predominant [b]	
					Substitution position	Orientation species
1	Naphthalene	168	95.7	17.9	β	$d_1 >> d_2$
2	Naphthalene	336	95.7	28.0	β	$d_1 \approx d_2$
3	Naphthalene	2x168	99.8	42.3	β>>α	$d_4 >> d_5$
4	Naphthalene	4x168	100.0	98.5	β,α	d_8
5	Anthracene [c]	6	87.3	-	β	d_1
6	Anthracene [c]	48	86.8	25.2	β>>α>γ	d_4
7	Anthracene [c]	168	85.9	73.0	β>α>γ	d_9
8	Anthracene [c]	2x336	97.8	35.7	β,α,γ	d_{10}
9	Pyrene	6	96.8	5.6	β	d_1
10	Pyrene	24	96.8	19.9	β>>α	$d_2 >> d_3$
11	Pyrene	72	96.8	-	β>>α	$d_2 >> d_3$
12	Pyrene	168	96.8	-	β>α>γ	$d_2 > d_3 > d_4$
13	Pyrene	336	96.8	62.6	β>α>γ	$d_8 > d_9 > d_{10}$
14	Pyrene	672	96.8	-	β,α>γ	d_{10}
15	Pyrene	3x336	100.0	94.7	β,α,γ	d_{10}

a Temperature of exchange 150°C. Typical reaction procedure[3)] involved compound (0.4g) with D_2O (5.0ml)and PtO_2 (0.05g).

b See footnotes b-d in Table 1.

c Benzene (0.5ml) added to improve solubility.

TABLE 3

Homogeneous platinum catalyzed exchange of polyphenyls*

Run	Compound	Time (hr)	% D (theor)	% D (found)	Predominant	
					Substitution position	Orientation species
1	Diphenyl	32	92.4	-	m,p	d_6
2	Diphenyl	39	92.4	37.6	m,p	d_6
3	p-Terphenyl	3x2.5	100.0	41.2	m,p	d_6
4	m-Terphenyl	3x2.5	100.0	48.4	m,p	d_7
5	o-Terphenyl	3x2.5	100.0	56.6	m,p	d_8

* Temperature of exchange, 116°C except runs 1,2 (75°C). See footnotes in table 1.

3.2 Heterogeneous exchange of condensed polycyclics

Using heterogeneous platinum (tables 2,5) remarkably similar exchange results to the homogeneous catalyst are observed with naphthalene, anthracene and pyrene. M values of approximately unity are observed and initially exchange is to the β position. Prolonged reaction (with naphthalene for example), gives the β-d_4 species, then finally the fully

TABLE 4

Heterogeneous platinum catalyzed exchange of polyphenyls*

Run	Compound	Time (hr)	% D (theor)	% D (found)	Predominant Substitution position	Predominant Orientation species
1	Diphenyl	24	92.7	-	m,p	d_6
2	Diphenyl	72	92.7	82.2	o,m,p	d_{10}
3	p-Terphenyl	336	89.6	37.3	m,p	d_6
4	p-Terphenyl	3x336	100.0	93.9	o,m,p	d_{14}
5	p-Terphenyl	5x336	100.0	-	o,m,p	d_{14}
6	m-Terphenyl	336	94.4	42.8	m,p	d_7
7	m-Terphenyl	4x336	100.0	99.0	o,m,p	d_{14}
8	o-Terphenyl	336	97.3	58.9	o,m,p	d_{14}

* See footnotes in table 2.

TABLE 5

Back protonation of deuterated polycyclics with homogeneous and heterogeneous platinum

Compound	Catalyst type	Temp. (°C)	Time (hr)	% D (theor)	% D (found)	Predominant Substitution position	Predominant Orientation species
Naphthalene-d_8	homo.	75	504	6.7	-	α	d_4
Anthracene-d_{10}	homo.	75	4	0.7	78.6	α,γ	d_9
Anthracene-d_{10}	homo.	75	336	0.7	79.3	α,γ	d_6
Pyrene-d_{10}	homo.	75	48	0.8	78.5	α,γ	d_8
m-Terphenyl-d_{14}	homo.	116	24	1.3	43.3	0	d_7
p-Terphenyl-d_{14}	homo.	116	24	1.3	52.6	0	d_8
p-Terphenyl-d_{14}	homo.	75	2	0.8	84.6	o,m,p	d_{13}
Naphthalene-d_8	hetero.	150	2x336	0.0	0.0	nil	d_0
Anthracene-d_{10}	hetero.	150	96	1.4	9.3	α,γ	$d_0>d_1>d_6$
Anthracene-d_{10}	hetero.	160	168	1.4	6.3	nil	d_0
Pyrene-d_{10}	hetero.	150	6	1.0	95.4	α,γ	d_8
Pyrene-d_{10}	hetero.	150	48	2.1	-	effectively nil	d_0
Pyrene-d_{10}	hetero.	150	168	2.1	62.1	nil	$d_8 \simeq d_0$

* See footnotes in tables 1 and 2.

TABLE 6

Homogeneous and heterogeneous platinum catalyzed exchange of pentane, n-butylbenzene, n-pentylbenzene

Run	Compound	Catalyst system	D found %	Deuterium Orientation (%) Aromatic	α	Other CH_2	Terminal CH_3
1	n-Butylbenzene	homo.	23	47.2	15.0	2.2	13.7
2	n-Pentylbenzene	homo.	18	44.8	12.0	1.6	10.7
3	Pentane	homo.	75		92	57	
4	n-Butylbenzene	hetero.	20	55.0	6.0	0	1.6

TABLE 7

Approximate M values for exchange reactions*

Compound Type	Type of Catalysis	M
Condensed polycyclics	Homogeneous	1
	Heterogeneous	1
Polyphenyls	Homogeneous	3
	Heterogeneous	3

*From mass spectra during initial exchange.

---

deuterated d_8 compound.

3.3 Homogeneous and heterogeneous exchange of polyphenyls

For the polyphenyls, exchange again is remarkably similar with both homogeneous and heterogeneous platinum catalysts (especially in terms of initial rates(tables 3-5,7). Both systems exhibit multiple M values (≈3) and very severe ortho deactivation (except heterogeneous o-terphenyl which is anomalous and will be discussed elsewhere). Prolonged exchange heterogeneously at 150°C gives the fully deuterated compounds.

As an independent confirmation of the results in tables 1-4, the fully deuterated aromatics have been back-exchanged with water (table 5). The advantage of this procedure is that nmr can now be used to check the initial orientation since the appearance of the proton will now be observed where epr only could be used for this purpose for the forward exchanges.

3.4 Homogeneous and heterogeneous exchange of alkylbenzenes

When the long-chain alkylbenzenes are used as representative systems (n-butylbenzene and n-pentylbenzene in table 6), again remarkably similar data for isotope orientation are observed with both homogeneous and heterogeneous platinum, even during the course of the reaction from initial to equilibrium conditions. Homogeneously, rates of exchange of protons are ring > α side-chain > terminal side-chain > methylene side-chain during initial deuteration. After prolonged exchange, the α position incorporates slightly more deuterium than the terminal position. This pattern of exchange in the side-chain of the long-chain alkylbenzenes is also similar to the orientation in the corresponding alkane (n-pentylbenzene versus pentane) under homogeneous conditions. With heterogeneous platinum, the isotope orientation in the long-chain alkylbenzenes resembles the homogeneous data except that the activity of the methylene hydrogens is enhanced.

4. DISCUSSION

The significant feature of the results is the remarkable similarity in isotope orientation patterns observed in both platinum systems for all the polycyclics studied(except o-terphenyl), the long-chain alkylbenzenes and alkanes. These results are thus consistent with preliminary studies for the monosubstituted benzenes and simple alkylbenzenes such as toluene.[2,4] In terms of π-complex theory the present observed relationships are further evidence to support the concept that the manner of bond formation involving adsorbed molecules and the chemistry of inorganic co-ordination complexes are intimately related.[3] In this respect, the exchange of the polycyclic aromatic hydrocarbons is particularly important since the data show that mechanistically it is convenient to divide this class of compound into two groups, namely the polyphenyls and the condensed polycyclics. This has previously been done only for the homogeneous platinum(II) catalysed exchange of this series of compounds.[2]

4.1 Homogeneous and heterogeneous exchange of polyphenyls

Both homogeneously and heterogeneously, the polyphenyls (except for heterogeneous o-terphenyl) exchange initially with ortho deactivation and multiple M values. In terms of the mechanism of the exchange, classical and π-complex associative and dissociative processes have been proposed for aromatic and aliphatic molecules.[2,3,6,7] For aromatic exchange the present data are consistent with the predominant participation of the dissociative π-complex exchange mechanisms in both heterogeneous (eq.(1)) and homogeneous (eq.(2)) systems for reasons which have already been discussed elsewhere by the present authors.[2,3] It is felt that if the

Heterogeneous π-complex exchange mechanism

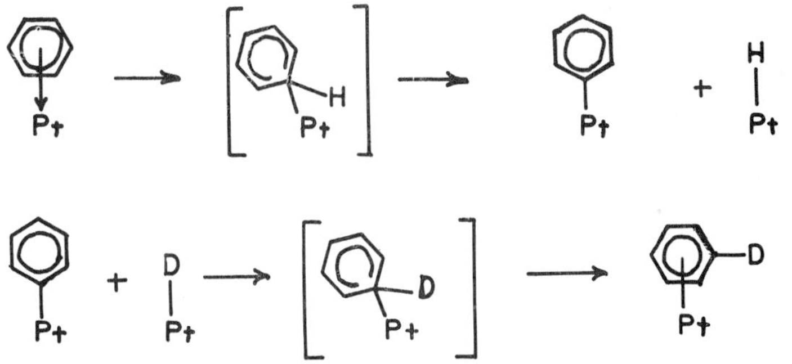

(1)

Homogeneous π-complex exchange mechanism

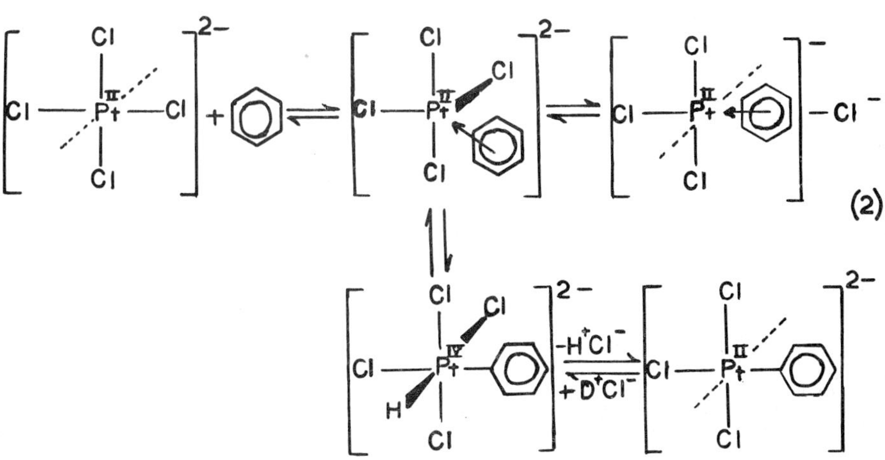

(2)

corresponding homogeneous[2] and heterogeneous[3] π-associative mechanisms are involved, their contribution to the overall exchange process is small, although in the heterogeneous system this conclusion remains controversial.[8,9] The important feature of this work with respect to those mechanisms is that the π-complex with the polyphenyls appears to be essentially with each aromatic ring in turn and it is formed as a delocalized complex with the whole ring. Exchange then occurs by π-σ conversions from this generalised π-bonded state.

4.2 Homogeneous and heterogeneous exchange of condensed polycyclics

By contrast, the exchange characteristics of the condensed polycyclics can be best interpreted if the bond order concept is considered as previously invoked for the homogeneous exchange of this series. Using naphthalene as representative compound, the necessity to consider π-bond localisation is associated with an M value of unity and selective β-orientation during initial exchange. This implies that each time naphthalene forms a π-olefin type complex (species I), the π-σ conversion can only proceed to the β position because of steric hindrance to the α-position. Thus for one naphthalene molecule to deuterate in two β positions and still be consistent with an M value of unity, the π-olefin complex must break and be reformed at another "olefinic" bond. This theory is further supported by work with the substituted polycyclics[10] which show that as the bond order of the 2-3 bond (in say naphthalene) is increased, the M value increases above unity and dideuteration in the β- positions is observed during initial rates of exchange.

The fact that the same isotope orientation pattern (M=1, β-orientation initially) is observed for the condensed polycyclics both homogeneously and heterogeneously again suggests that analogous π-complex dissociative mechanisms as mentioned above, operate in both systems. However, the data do indicate that a refinement in the original heterogeneous dissociative π-complex mechanism is in order. Thus it is now proposed that condensed polycyclics such as naphthalene adsorb as π-olefin complexes (species II) then exchange through π-σ conversion processes. This interpretation is consistent with the strong toxicity of olefins in heterogeneous exchange with D_2O due to solvent displacement effects and also with the observed toxicity of naphthalene in these heterogeneous reactions. These data are also consistent with the earlier heterogeneous work which showed that the condensed polycyclics are relatively toxic in these exchanges, whereas the polyphenyls are not.[3] Thus, when the condensed polycyclics were exchanged in the presence of benzene on heterogeneous platinum, both the deuteration of the parent condensed polycyclic and benzene were severely retarded whereas with the polyphenyls, only little retardation in either benzene or the parent compound was observed.

4.3 Exchange in long-chain alkylbenzenes and alkanes

The salient features of the homogeneous platinum (II) exchange of the long-chain alkylbenzenes from n-propylbenzene onwards are that the ring

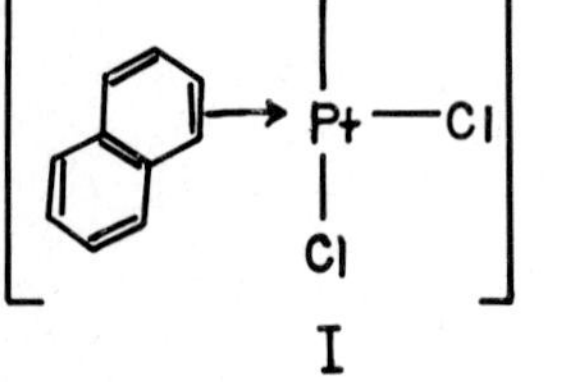

I 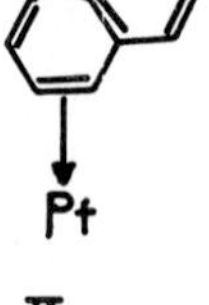 II 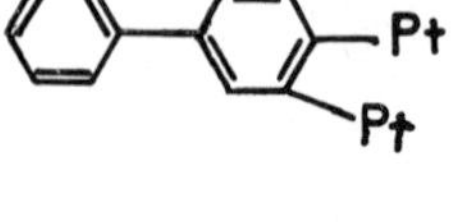 IV

hydrogens deuterate faster than the side-chain.[7] From the n-butylbenzene and n-pentylbenzene results (table 6) exchange in the α group of the side-chain is slightly faster than in the terminal group after appreciable deuteration. By comparison, the methylene protons are relatively slow. To account for this orientation, a species (III) involving simultaneous complexing to the ring and the terminal side-chain position was proposed as the intermediate in exchange at the terminal position, whereas π-allylic species were invoked to explain α-deuteration. The mechanism involving species (III) was used to explain terminal exchange in preference to one where the alkylbenzene was initially complexed to the platinum through the ring, followed by α-β rearrangements to the terminal position.[7]

Using n-butylbenzene as representative long-chain alkylbenzene (table 6) the pattern of exchange on heterogeneous platinum is similar to the result for homogeneous platinum (II), except that deuteration of the methylene protons is slightly more extensive in the heterogeneous system.[11] Preliminary experiments with n-pentyl, n-hexyl and n-nonylbenzenes show similar trends in isotope orientation.[7,11] Thus the implication is that a similar mechanism operates again in both homogeneous and heterogeneous systems, although detailed kinetics are not yet available for the alkylbenzenes and thus no refinement in mechanism is yet possible.

The other interesting feature of these results is that the pattern of exchange in the side-chain of the long-chain alkylbenzenes (table 6) is similar to that observed for the orientation in the simple alkanes[6,11,12] using the homogeneous platinum(II) catalyst. The results of preliminary studies for the corresponding reactions on heterogeneous platinum are also similar to the homogeneous runs with common mechanistic implications in both catalytic systems.[11]

4.4 Significance of exchange results

There is an important general conclusion from the fact that (i) analogous isotope orientation patterns are observed for a wide variety of aromatic and aliphatic compounds in both homogeneous and heterogeneous platinum systems and (ii) the kinetics of homogeneous platinum (II) catalysed exchange with benzene are first order in catalyst concentration. These results suggest that only <u>one</u> platinum atom need be selected by a molecule in a heterogeneous surface array of atoms for reaction to proceed. After adsorption has occurred, the full exchange cycle can then take place on this one atom in a manner similar to the processes that occur with one $PtCl_4^{2-}$ species homogeneously. Because of steric considerations with certain molecules, such platinum atoms which catalyse the reactions in a surface must be favorably placed, i.e. presumably they are easily accessible surface sites which either protrude from the surface, or are located at edges or corners such that molecules will possess a certain degree of mobility especially to enable π-σ conversion processes to occur. Epr measurements for the determination of active surface areas in these platinum catalysts confirm this conclusion since these epr surface areas (active sites only) are lower than surface areas obtained from BET measurements (all sites)[13,14]

A further important conclusion related to the preceding discussion concerns the significance of the M value in homogeneous and heterogeneous exchanges. The fact that molecules such as diphenyl and toluene exhibit M values of approximately 3 in both the homogeneous and heterogeneous platinum systems reported in the current work indicates that it is not necessary to postulate the occurrence of species such as (IV) as previously suggested to explain high M values in heterogeneous processes. This does not mean that such species do not exist on surfaces or do not contribute to the overall exchange. The homogeneous kinetics simply show that high M values can be obtained without invoking the formation of such multiple bonded metal species.

By comparing homogeneous and heterogeneous processes, it is also now possible to chemically define the well-known site effect (sites of differing activities) concept which has often been proposed in heterogeneous catalysis to explain observed phenomena. In the homogeneous system, if chlorine atoms are replaced by bromine atoms in the platinum salt ($PtCl_4^{2-}$ vs $PtBr_4^{2-}$) there is a significant reduction in rate of exchange with benzene.[15,16] In like manner neglecting steric effects, the activity of a platinum atom in a surface should depend markedly upon the nature of the ligands to which it is attached. Since, in the present instance, this may be other platinum atoms, chlorine atoms, water molecules etc., it is not difficult to envisage why site effects may occur in heterogeneous catalysis. This suggestion may also explain why in the present work, temperatures of 150°C and exchange periods of 48 hours are required with heterogeneous platinum to obtain equivalent isotope incorporation to that achieved homogeneously after several hours at 80-100°C.

ACKNOWLEDGEMENT

The authors thank the Australian Research Grants Committee and the Australian Institute of Nuclear Science and Engineering for the support of this research.

REFERENCES

1) J.L.Garnett and R.J.Hodges, J.Am.Chem.Soc. 89 (1967), 4546.
2) R.J. Hodges and J.L.Garnett, J.Phys Chem. 73 (1969), 1525.
3) J.L.Garnett and W.A.Sollich-Baumgartner, Adv. Catalysis 16 (1966), 95.
4) R.J. Hodges and J.L. Garnett, J. Catalysis, 13(1969), 83.
5) K.P. Davis, J.L.Garnett and J.H. O'Keefe, Chem. Comm.(1970), 1672.
6) R.J. Hodges, D.E. Webster and P.B. Wells, Chem.Comm.(1971), 462
7) J.L. Garnett and R.S.Kenyon, Chem. Comm. (1971), 1227.
8) R.J.Harper, S. Siegel and C.Kemball, J. Catalysis 6 (1966), 72.
9) J.L.Garnett, Catalysis Rev. 5 (1971), 229
10) J.L. Garnett and R.S. Kenyon, to be published.
11) J.L.Garnett, W. Hannan, K. Hoa, R.S.Kenyon and M.Long, to be published.
12) N.F. Gol'dshleger, M.B.Tyabin, A.E.Shilov and A.A.Shteinman, Zhur. fiz Khim. 43 (1969), 2174.
13) J.L. Garnett, A.T.T. Oei and W.A.Sollich-Baumgartner, J. Catalysis 7(1967), 305.
14) R.B. Noyes, P.B. Wells, K. Baron, K. Compson, J. Grant and R. Heselden, J. Catalysis 18 (1970), 224.
15) J.L. Garnett and J.C. West, to be published.
16) J.L. Garnett, R.J. Hodges and W.A. Sollich-Baumgartner, Fourth International Congress on Catalysis Preprints 1(1969), 1.

DISCUSSION

J. R. ANDERSON

The well-known work of Professor Garnett and his group on the deuterium exchange of aromatic molecules over homogeneous and heterogeneous platinum catalysts provides strong evidence that a very important reaction pathway--that involving a single platinum atom as the catalytic site--is common to both systems. In the present paper the authors also claim to have shown that the same can be said for the exchange of the alkanes. However, I am unable to see how this last claim can be substantiated from the data provided in the paper itself or from the literature cited therein. My main point is that there exist quite substantial differences in the distributions of deuterated reaction products from homogeneous and heterogeneous platinum catalysts. For instance, Hodges et al.[1] studied the exchange of a number of alkanes and cycloalkanes with a homogeneous Pt (II) catalyst at 100°C, and found the following values for M, the average number of deuterium atoms entering a hydrocarbon molecule during a single period of interaction with the catalyst: ethane, 1.7; cyclohexane, 1.4; neohexane, 1.4. On the other hand, over evaporated platinum film catalysts the M values are: ethane, 3.5 (150°C);[2] cyclohexane, 2.2 (0°C),[3] and neohexane, 1.5 (20°C).[4] Moreover, the data given in the present paper for the distribution of exchange products from η-butylbenzene over the two types of catalyst show quite important differences in the way the deuterium atoms are located in the side-chain.

I do not see how one can escape the conclusion that alkane (or cycloalkane) and alkyl exchange can differ quite markedly over homogeneous and heterogeneous platinum catalysts. There can be little doubt that the main source of difference lies in the ability of a heterogeneous catalyst to offer adsorption modes in which more than one platinum atom is involved, as in 1-2 or 1-3 adsorbed hydrocarbon. Rather than try to blink these differences away, I believe it would be much more profitable to explore the way in which exchange is dependent on the surface topography of the heterogeneous catalysts. For instance, one might well expect to find that exchange over a highly dispersed platinum catalyst tends towards the behavior of an homogeneous catalyst.

1) Hodges, R. J., Webster, D. E. and Wells, P. B., Chem. Comm. p. 462 (1971).
2) Anderson, J. R. and Kemball, C. Proc. Roy. Soc. A223, 361 (1954).
3) Anderson, J. R. and Kemball, C. Proc. Roy. Soc. A226, 472 (1954).
4) Prudhomme, J-C. and Gault, F. G. Bull. Soc. Chim. France, p. 832 (1966).

J. L. GARNETT

This comment by Dr. Anderson is opportune. We have always been concerned about our rationalization of M values in both heterogeneous and homogeneous systems. We have thus stressed in the current Congress paper and elsewhere,[1] that (i) since aromatic and aliphatic molecules can exhibit multiple M values with both homogeneous and heterogeneous platinum, it is not necessary to postulate the occurrence of multiple bonded metal species analogous to (IV) to explain such M values in heterogeneous processes. This observation is not meant to imply that species such as (IV) do not exist heterogeneously; it simply indicates, from homogeneous kinetics with both aromatic[2)] aliphatic[3,4] systems, that multiple exchange can be explained by the formation of a complex with one platinum atom; (i.e. a π-complex for the aromatic and an analogous intermediate complex for the aliphatic).

The relevant additional alkane material which Dr. Anderson seeks and which we presented at this Congress is shown in the attached table (see response to C. Kemball, following). These data show M values greater than unity in both homogeneous and heterogeneous alkane exchange. In the heterogeneous runs, there appears to be some exchange of low multiple character (as in the homogeneous system) superimposed on exchange of high multiplicity which could occur by di-adsorbed species. Further data from the homogeneous kinetics of alkane exchange have been published[3] and it is interesting to note that these workers, as well as ourselves,[1,4] have commented in a similar manner on the mechanistic significance of M values in excess of unity in homogeneous alkane exchange.

1) J. L. Garnett, Catalysis Rev. 5 (1971) 229.
2) R. J. Hodges and J. L. Garnett, J. Phys. Chem., 73 (1969), 1525.
3) R. J. Hodges, D. E. Webster and P. B. Wells, J. Chem. Soc. (A) (1971), 3230.
4) J. L. Garnett, R. S. Kenyon, M. Long and A. McLaren, to be published.

P. B. WELLS

The results in Table 6 for the homogeneously catalyzed exchange of hydrogen for deuterium in pentylbenzene and in pentane reproduce closely the values reported in our study[1)] in which we used a similar catalyst solution developed from your earlier publications. Have you any further comment on the reasons why terminal methyl groups in, say, pentane undergo exchange more rapidly than the interval methylene groups? It seems unlikely to me that this is simply attributable to steric factors.

1) R. J. Hodges, D. E. Webster, and P. B. Wells, J. Chem. Soc. (A) 3230 (1971).

J. L. GARNETT

Our reason for invoking steric considerations to explain the orientation in alkanes is based on the fact that in the exchange of polycyclic aromatics[1)] and alkylbenzenes,[1)] steric effects are always much more severe in the homogeneous $PtCl_4^{2-}$ system than with heterogeneous platinum and it is reasonable to extend this conclusion to the simple alkanes. The remaining alternative involving an electronic effect to explain preferential terminal methyl exchange seems less plausible from our preliminary studies of electronic effects of substituent groups in homogeneous exchange.[2)] These effects are marginal.

1) J. L. Garnett, Catalysis Rev. 5 (1971) 229.
2) J. L. Garnett, R. S. Kenyon and J. C. West, to be published.

C. KEMBALL

It is important to stress that the patterns of behavior for exchange reactions of hydrocarbons in heterogeneous catalysis can be influenced by the form in which the labeling isotope is supplied. Patterns observed using D_2O are often different from those using D_2. I presume that the additional results which Professor Garnett presented for the exchange of alkanes, reacting preferentially in the methyl groups and to a smaller extent in the methylene groups, were obtained using heavy water. Dr. Jaggers and I found no evidence for such an effect on exchanging n-butane with deuterium on platinum films. Both primary and secondary C-H bonds appeared to react at similar rates, and there was also a multiple exchange process which yielded some perdeutero-butane as an initial product.

J. L. GARNETT

Professor Kemball's remarks are most important and we have always been cognisant of possible differences between heterogeneous D_2O and D_2 systems.[1)] The additional heterogeneous data presented at this Congress (Table 8, column 1) were for D_2O systems and show a significant preference for terminal exchange in hexane. We have since done further work with D_2 (Table 8, column 2) and the results also show a preference for terminal methyl positions using nmr and mass spectrometric analyses. Our catalyst was prereduced platinum oxide or chloride, whereas Professor Kemball's results were obtained on platinum films. Such changes in catalytic conditions may well alter orientation patterns in the alkanes although there is little effect with the alkylbenzenes.[1)] The differences in rate between D_2O and D_2 runs is probably due to water activation of the catalyst, a phenomenon observed previously with aromatic exchange.[1)]

1) J. L. Garnett, Catalysis Rev. 5 (1971) 229.

Table 8
Homogeneous and heterogeneous platinum catalyzed exchange of hexane

Deuterium Distribution	Heterogeneous		Homogeneous
	D_2O System	D_2 System	
	(90h at 150°)	(90h at 150°)	(120h at 120°
D_0	57.8	75.6	0
D_1	6.7	17.9	2.0
D_2	10.5	6.5	14.4
D_3	2.6		28.1
D_4	1.5		28.1
D_5	1.1		15.1
D_6	1.3		6.8
D_7	1.3		3.4
D_8	1.9		1.4
D_9	2.1		0.7
D_{10}	3.0		
D_{11}	3.5		
D_{12}	3.9		
D_{13}	2.1		
D_{14}	0.8		
Atom % D	17.9	2.21	27.5
Equil.% D	48.0	3.3	90
M	5.5	1.26	
Orientation (α/β)	11/7	$\alpha>\beta$	4/1

Paper Number 32

CATALYTIC ISOMERIZATION OF BUTENES ON $PdCl_2$. HOMOGENEOUS AND HETEROGENEOUS REACTIONS

E. TIJERO, F. CASTAÑO and E. HERMANA
Department of Catalysis, Instituto "Rocasolano," C.S.I.C.
Madrid-6, Spain

ABSTRACT: The isomerization of butenes on $PdCl_2$ has been studied by UV spectrophotometry, using the catalyst in two different phases, supported on silica gel and dissolved in diglyme.

The spectra of the system butene -$PdCl_2$ are different for both phases. The three butenes used, 1-, 2-cis and 2-transbutene, show a principal band, at 240 nm for the dissolved $PdCl_2$ and at 250 nm for the supported $PdCl_2$, and a secondary one, at 310 nm and 268 nm respectively for each phase.

The difference in energy atributed to these secondary bands, 13 kcal/mol, is similar to the difference in the activation energies kinetically determined for these systems, and published elsewhere.

The existence of two different intermediate compounds, one for each catalytic phase, are discussed.

1. INTRODUCTION

The growing interest for the practical and fundamental implications of homogeneous catalysis lies mainly on its high specificity, as shown in many publications [1-6]. Liquid catalysts show a capacity for coordinating several types of ligands to a specially reactive state, allowing thus the realization of reactions at low temperatures, whith low probability of ocurrence of succesive or paralell reactions.

This special reactivity has been attributed to the existence of a intermediate compounds with π orbitals [3,7] and speculations have been raised [11] on the possibility of explaining heterogeneous catalytic activities by the formation of such compounds on the surface of solid catalysts.

With the aim of studying this hypothesis, a suitable catalytic reaction, butenes isomerization on $PdCl_2$, has been studied in two different systems, homogeneous and heterogeneous, using catalyst in solved and supported form.

Previously, a kinetic study of this reaction, for both systems, has been published [8]. In it, different activation energies were determinated for each one, with a difference of 13 kcal/mol between them.

In this paper, a paralell study, UV spectrophotometry of the evolution of the catalyst during the reactions, is presented for both systems.

2. EXPERIMENTAL

2.1 Reagents

The following chemical products were used:
Butenes from the Philips Petroleum Co. "Research" grade, 99.9%.
Palladium Chloride, from Fluka A.G. Pure.
Diglyme, (DG), dimethyldiglykol ether, from Fluka A.G. 99.5%.
Tetraethyl orthosilicate, $(C_2H_5)_4SiO_4$, from B.D.H., 95%.
Ethyl alcohol, from the Merck Co.

2.2 Catalyst samples preparation

Solid catalyst. The $PdCl_2$ ought be supported on a carrier suitable for UV transmission spectrophotometry. The chosen support, silica gel, is inactive for butene isomerization.

Samples of support were prepared by hydrolisis of tetraethyl orthosilicate in the presence of ethyl alcohol, common solvent for silicate and water, keeping the ratio, in weight %, of

$$15\ H_2O,\ 36\ C_2H_5OH,\ 49\ (C_2H_5)_4SiO_4$$

A minimum of 25 h should be allowed for time of gelification. The gel, dried at 110°C for 3 hours and activated at 600°C for 24 h, has a final specific surface of 236 m^2/g.

The sample must be cleaned of organic adsorbates by heating at 600°C for 3 h in a slow oxygen stream.

Small plaques of this gel, with 5 x 10 mm section, are impregnated with a solution 0.005 M of $PdCl_2$ in 1 M $HClO_4$ with HCl. The concentration of Pd in the final catalyst is 4×10^{-6} mols of $PdCl_2$/g of catalyst.

Liquid catalyst. The $PdCl_2$ is dissolved in diglyme, chosen by its low vapor pressure, lack of nucleophility and great molecular volume, that allow to work under vacuum and avoid the possibility of nucleophylic attack.

2.3 Apparatus and procedure

Optical spectra were determined using the UNICAM spectrophotometer model SP700A, double beam. A reference sample was employed in every run: Silica gel for the case of solid catalyst and diglyme for the liquid one.

A standard vacuum line was used for the addition of butenes to the samples. Solid catalyst samples were mounted in special frames as previously published [9-10].

A sealed recint was built on the top of standard silica 1 cm cells, where the catalyst could be handled at pressures down to 10^{-6} torr. Every experiment was carried out at room temperature.

The initial pressure of the butenes, for each experiment, was 500 torr.

3. RESULTS AND INTERPRETATION

3.1 Spectra of the catalyst phases

Figure 1 shows the spectra of dry silica gel (A) and supported $PdCl_2$ (B). A band at 250 nm appears for the catalyst, against a blank spectrum for the support.

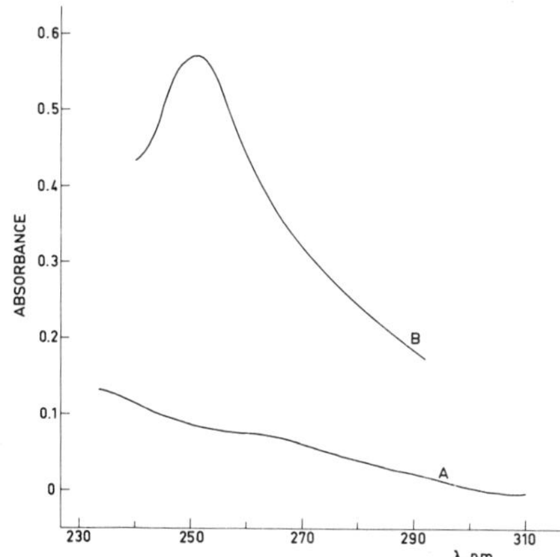

Fig. 1 Spectra of support and supported $PdCl_2$.

The spectrum of $PdCl_2$ in diglyme, curve 1 of Figure 2, shows a peak a 240 nm, with a extinction coefficient of 16, 500 mol^{-1} cm^{-1} liter. The solution follows the Beer's Law up to 10^{-3} mols/1.

According to the molecular orbital calculations of Gray and Ballhausen 12), the band at 240 and 250 nm could be attributed to square planar tetracoordinated complexes of palladium, particularly to transitions (ligand) $\rightarrow$ $d_{x^2-y^2}$. The particular energy of the transitions will depend of the type of ligand. The more stabilized the electrons in these ligands, the highest will be the energy of the transition.

The surface of silica gel has water enough to give the complex $PdCl_2(H_2O)_2$. The oxygens of this water will be at the surface, at a distance that, pressumably, does not coincides with the normal distance in the square complex. This distortion will produce a decrease of the energy of the orbital $d_{x^2-y^2}$. Therefore, the transition will display a bathochromic shift, as we see from 240 to 250 nm.

3.2 Effect of butenes

The effect of linear butenes on the spectra of both catalyst phases is shown in Figs. 2 and 3.

In both cases, the intensity of the original band increases with the addition of butenes, and new bands appear, with smaller intensity, at higher wavelenght. The new bands appear at 268, 270 and 272 nm for the solid catalyst and 1-, cis- and trans-butene, respectively. For the liquid catalyst, the bands appear at 306, 310 and 310 nm for the same sequence of butenes.

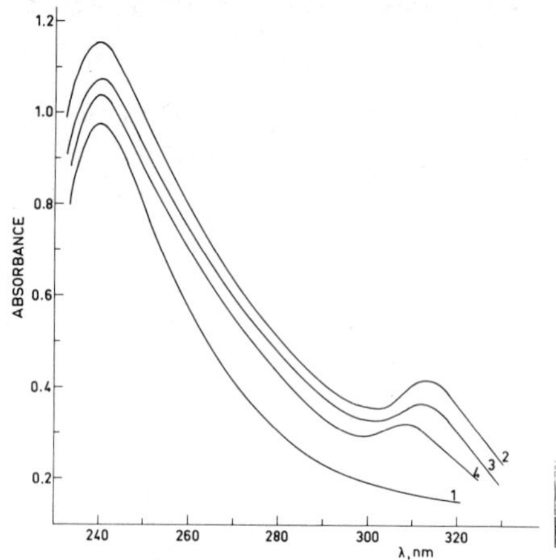

Fig. 2 Effect of butenes on $PdCl_2$ solution: 1, $PdCl_2$ in diglyme solution. 2, 1-butene. 3, cis-2-butene. 4, trans-2-butene.

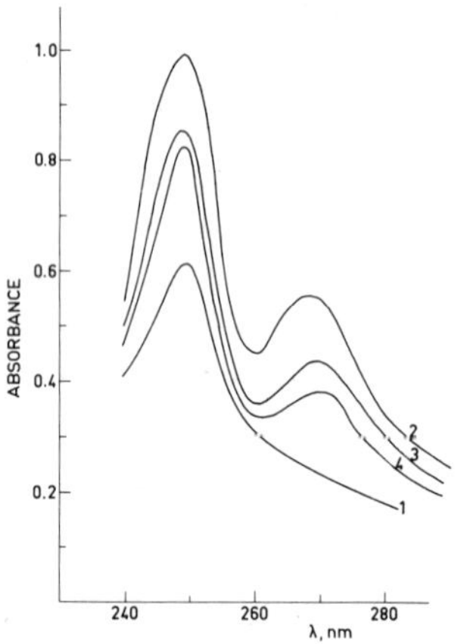

Fig. 3 Effect of butenes on supported $PdCl_2$: 1, $PdCl_2$ on SiO_2. 2, 1-butene. 3, cis-2-butene. 4, trans-2-butene.

These bands suggest the existence of a pentacoordinated charge transfer complex for both phases. For the liquid catalyst, the approximation of butenes should occur through the z axis of the $PdCl_2DG_2$, giving rise to a σ bonding between the π-electrons of butenes and the mixing of 5s and $5p_z$ orbitals of the central atom [13].

The symmetry of the penta coordinated complex is C_g. Therefore, the empty orbital $d_{x^2-y^2}$ decreases its symmetry. As it is well known, this change explains the increase of the extinction coefficient at 240 nm. The transition π(butenes) $\rightarrow$ $d_{x^2-y^2}$ will imply less energy than the π (chlorine) $\rightarrow$ $d_{x^2-y^2}$, since the π electrons in butenes are bound with less energy. Besides, as the interaction butene-complex is weaker for the trans- than for 1- and cis-butene, the trans-butene will show a higher transition energy, explaining the small wavelenght difference, 306 to 310 nm.

For the solid catalyst, the hypsochromic shift of the transition π (butene) $\rightarrow d_{x^2-y^2}$, in relation with the same band in diglyme solution, indicates a higher interaction of the butenes in the penta coordinated charge transfer complex. This interaction is highly sensitive to steric effects, as shown by the inversion of transition energies of the butenes in relation with the effect in solution.

The behaviour with temperature of these penta coordinated complexes must be different for each phase. The influence of temperature in a loose butene (liquid) should be much smaller than on a complex in which one of the ligands (H_2O) must be separated from its fixed position at the surface, changing the symmetry of the complex and leading to a smaller interaction of the butene with the surface. From a cualitative point of view, it could be said that the activation energy in the solution will be smaller.

3.1 Effect of time

Every band develops rapidly, and grows to its maximun intensity in less time than the minimun allowed by the experimental procedure. No attempt has been made, consequently, to measure its rate of formation.

The evolution with longer times has been determined. The effect of time on the liquid phase is shown in Fig. 4.

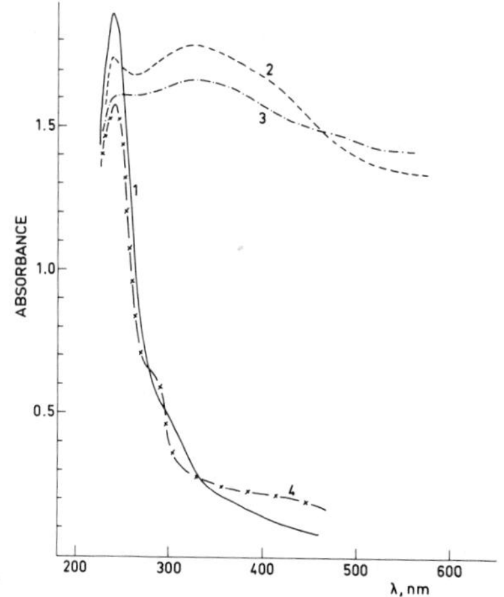

Fig. 4 Evolution of liquid phase with time: 1, initial time. 2, one hour. 3, two hours. 4, after centrifugation.

Curve 1 shows the initial spectrum of the liquid phase. Curves 2 and 3 show the evolution of the spectrum during experiment after one and two hours of sampling. By this time, the rate of reaction was already zero. A broad absorbance develops over the whole visible region. The same liquid catalyst, after centrifugation, has recovered the original form of the spectrum, but with a lower band at 240 nm. The absorbance in the visible region has been eliminated. Thus, it is attributed to light scattering from metallic palladium suspended in the liquid.

The solid phase spectrum, on the contrary, remains constant for the whole two hours.

4. DISCUSSION

As mentioned in the Introduction, this work must be discussed in relation with the fact of the difference in activation energies for isomerization of 1-butene found in a previous work. The difference was as high as 13 kcal/mol.

The energies corresponding to the secondary bands that appear with the addition of butenes, developed from the general equation $\Delta E = h\nu$, are

Catalyst phase	butene	λ, nm	ΔE, kcal/mol
solid	1-	268	105
solid	c-2-	270	104.2
solid	t-2-	272	103.4
liquid	t-2-	306	92
liquid	c-2-	310	91
liquid	1-	310	91

The difference between the energies for 1-butene in both phases is 14kcal/mol. This value is fairly close to the value of 13 found in kinetic studies.

Evidently, the compounds responsible for the 310 and 268 nm bands can not be the activated complex of the catalytic mechanism, as it could been deduced from this similarity of values. There are several reasons for this negation: The bands develop quickly, after the addition of butenes, and, assuming a normal extinction coefficient, represent a concentration too big for any activated complex. Besides, we have proposed [8) for this isomerization a mechanism in which the rate controlling step is the formation of the activated complex.

Thus, these bands must be attributed to some form of absorbed or adsorbed butene. Previously, we have proposed a pentacoordinated charge transfer complex for that compounds, differing in the ligands employed in the square plane for each phase.

However, these complexes have energies very close to a carbene, that, due to the mobility of the bond along the chain, can be proposed as the real activated complex. The difference could not be bigger than 1-2 kcal/mol. The difference in activation energies from kinetic studies may be correlated, then, to the difference in energies found in this study.

Consequently, we propose the existence of the same type of activated complex for both phases, a carbene, with different ligands for each one.

The experimental fact [8) of the rapid decay of the catalytic activity for the liquid catalyst finds an explanation with the results shown in Fig. 4. There is a disminution of the available $PdCl_2$ in solution, due to its reduction to metallic palladium by nucleophylic attack. This decay is not present, however, in the solid phase. As a matter of fact, we assume that molecules of water are part of the composition of the pentacoordinated complex. The fact that this water does not oxidize the butenes, as easily as it does in the liquid phase, must be attributed to the strong bonding of its molecules to the silica gel, that unables them to lose electrons so easily.

ACKNOWLEDGEMENT: The authors wish to express their gratitude to Dr. J. F. Garcia de la Banda and Dr. A. Gamero for their continual collaboration and fruitful discussions on this work.

REFERENCES

1) J. Smidt et al., Angew. Chem. intern. Edit. 1 (2) (1962) 80.
2) E. Stern and M. Spector, Proc. Chem. Soc. (1961) 370.
3) J.J. Rooney and G. Webb, J. of Catalysis 3 (1964) 488.
4) R.W. Schaftlein and T.W. Fraser, Ind. Eng. Chem. 60 (5) (1968) 12.
5) L. Hatch, Hydrocarb. Process. (3) (1970) 101.
6) E. Stern, Catalysis Rev. 1 (1967) 73.
7) E. Crawford and C. Kemball, Trans. Faraday Soc. 58 (1962) 2452.
8) E. Hermana, A. Gamero, E. Tijero and J. Blanco, An. Quím. 67 (1971) 1051.
9) H.P. Leftin, J. Phy. Chem. 64 (1960) 1714.
10) H.P. Leftin and E. Hermana, Proc. 3rd Int. Congr. Catalysis, North Holland Publ. Co. Amsterdam (1965) 1064.
11) H. Heinemann, Chem. Tech. 1 (1971) 286.
12) H.B. Gray and C.J. Ballhausen, J. Am. Chem. Soc. 85 (1963) 260.
13) E. Orgel, "Introducción a la Química de los metales de transición", Edit. Reverte, Barcelona (1964).

DISCUSSION

H. HEINEMANN

This paper presents an interesting attempt to compare active species in homogeneous and heterogeneous reactions. I wonder though, whether the catalyst preparation permitted similar active complexes to exist. Have the authors added a polar material (water, alcohol) to the soluble system and then compared the spectra?

There is literature indicating that a platinum-olefin complex dichloride reacts with a SiO_2 surface under elimination of HCl. Have the authors analyzed the solid catalyst for chloride concentration?

Finally, there is a possibility that the homogeneous catalyst may be dimeric with Cl bridges for the Pd complex, while the heterogeneous catalyst is probably monomeric. Is there evidence for either dimeric or monomeric species?

E. HERMANA

We think there are two similar complexes which are by no means the same. In solution we assume the complexing is made by diglyme, whilst on silica surface the complexing is made by water. There is no possibility of adding polar compounds to the liquid system, because an oxidation of the olefin would take place.

Concerning with the second part of your comment, no deactivation of the solid catalyst was observed at any case. We conclude that reaction with SiO_2 does not occur, at least in a significant extension.

Finally, a dimeric species of olefin-$PdCl_2$ complexes has been proposed by N. R. Davies (Nature 201, 490 (1964)), explaining isomerization in acetic solution. However, such a dimer should stabilize the electrons of Cl in the bridge. Consequently the band at 310 nm, should split, giving rise to a band at shorter wavelength, and decreasing the absorbance by a factor of about 2. None of these effects have been observed when the spectra of solution and solids are compared.

G. C. BOND

Has Dr. Hermana tried to use solid palladium chloride as a catalyst for butene isomerization? The reason for my question is that some years ago we found that solid rhodium trichloride trihydrate effectively catalyzed this reaction: its activity declined with time, but could be regenerated by HCl gas.

E. HERMANA

No attempt has been made in that sense. It looks difficult to get solid $PdCl_2$ with adequate high specific surface, in order to get detectable activity. Besides, the fall in activity could have been very fast, and the possibility of HCl regeneration would introduce a new complicating factor we tried to avoid.

Paper Number 33

THE INITIAL STAGES OF THE OXIDATION OF PLATINUM: AN ULTRA-HIGH VACUUM STUDY OF A PLATINUM (111) SINGLE CRYSTAL

W. H. WEINBERG,* R. M. LAMBERT, C. M. COMRIE and J. W. LINNETT
Department of Physical Chemistry, University of Cambridge, Cambridge CB2 1EP, England

*Permanent address: Department of Chemical Engineering, California Institute of Technology, Pasadena, California 91109

ABSTRACT: The initial stages of the oxidation of platinum have been studied using low-energy electron diffraction (LEED) and Auger electron spectroscopy (AES). The measured initial sticking probability of molecular oxygen on a clean platinum (111) surface is shown to be approximately 7 x 10^{-7}. Previous literature values which indicate a much larger value are believed to represent experiments conducted on contaminated surfaces. The initial sticking probability on clean platinum was found to be independent of oxygen pressure and surface temperature for the experimental conditions investigated ($10^{-5} \leq p_{O_2} \leq 10^{-3}$ torr and $575 \leq T_s \leq 775^{\circ}K$). All of the adsorbed oxygen was removed from the surface either by heating for a few seconds at about 1300°K or by reaction with hydrogen at an elevated temperature to form water. The present results are important not only from the point of view of platinum oxidation but also from the point of view of oxidation catalysis in those cases where the reacting oxygen is absorbed on the surface.

1. INTRODUCTION

The main aim of this work was to investigate the chemisorption of molecular oxygen on clean platinum and thus to help elucidate the initial stages of the oxidation of platinum. An additional reason for the study was to explain and understand the differences between the values obtained for the initial sticking probability of molecular oxygen on platinum by a number of different experimenters.

Bond[1)] has reported that oxygen is readily adsorbed on all metals with the exception of gold. Experiments using powdered platinum black have yielded initial heats of adsorption of 53 kcal/mole[2)] and 60 kcal/mole[3)]. A more recent investigation using an evaporated platinum film has given 68 kcal/mole[4)]. Because of the methods used in their preparation and use, the metal powders must have been contaminated, and the evaporated films cannot have provided clean surfaces either. A very common contaminant of platinum surfaces is carbon, and all the early experiments must have used platinum surfaces contaminated with carbon. This will be discussed in more detail later.

Some of the more recent ultra-high vacuum (UHV) experiments of oxygen chemisorption on platinum have seemed to confirm the earlier result that oxygen adsorption is relatively fast, but others have tended to show that oxygen does not adsorb on platinum. Tucker[5)], who cleaned his surface only by heating, has reported that a (2x2) LEED pattern is formed when a platinum (111) surface is heated to about $575^{o}K$ and cooled in oxygen at a pressure of 2×10^{-6} torr. This corresponds to an initial sticking probability of not less than about 5×10^{-3}. New LEED patterns due to oxygen adsorption were also observed by Tucker[5)] on both platinum (110) and (100) faces with a sticking probability approximately equal to that on the (111) surface.

Procop and Völter[6)] have investigated the adsorption and desorption of oxygen on a platinum foil in an UHV system using the flash filament method. They also only used heating to clean the surface. Oxygen pressures between 9×10^{-8} and 2×10^{-6} torr, and surface temperatures between 300 and $900^{o}K$ were used. The initial sticking probability was found to be 0.14 at $323^{o}K$ and to be nearly independent of surface temperature.

Wood *et al.*[7)] have also recently reported UHV experiments which indicate that oxygen adsorption on polycrystalline platinum is unactivated with a sticking probability of about 0.2. They cleaned the metal in the UHV system by heating.

These UHV experiments[5-7)] agree therefore with the older results and indicate that oxygen is adsorbed on platinum with a high sticking probability which is greater than 0.005 and may be as large as 0.1 or 0.2.

On the other hand, Morgan and Somorjai[8)] have found that oxygen does not seem to adsorb at all on a platinum (100) surface. The investigation was conducted in a UHV LEED apparatus. The platinum was subjected to argon ion bombardment and was reconstructed into a (5x1) LEED structure which is believed by Somorjai and his co-workers[8-10)] to be the equilibrium clean surface configuration of the platinum (100) surface. Morgan and Somorjai exposed the platinum to oxygen at both room and elevated temperatures. They observed no new LEED features, no change in intensity of the substrate LEED spots, and no mass spectrometric evidence of oxygen desorption when the platinum was heated. From the experimental conditions it may be inferred that the sticking probability is less than about 10^{-3}.

Also Lampton[11)] has recently attempted to adsorb oxygen on a platinum (111) surface which had been cleaned by rare gas ion bombardment. An exposure of 0.072 torr-sec of oxygen 2×10^{-5} torr for one hour) was given at a surface temperature of 575°K. The higher temperature was used to prevent the adsorption of background gases (hydrogen and/or carbon monoxide). There was no detectable adsorption, i.e. no new LEED features, no change in the intensity vs. voltage curve for the (00) diffraction spot, and no change in the half-width of a scattered helium atomic beam from that observed from a clean surface. The latter probe has been used effectively to study adsorption on clean surfaces[12)] and is sensitive to surface coverages at least down to 0.1 monolayer. Lampton's results imply that the sticking probability on the (111) surface is less than about 6×10^{-6}.

The above discussion shows clearly that two sets of conflicting data exist in the literature concerning oxygen chemisorption on platinum. One set gives a high sticking probability while the other a much lower one.

2. EXPERIMENTAL

The oxygen adsorption experiments were conducted in a Vacuum Generators post-acceleration and display type 3-grid LEED system which also included a glancing incidence electron gun for use in studying Auger electron emission. Oxygen pressure was monitored by a nude ionization gauge although a quadrupole mass spectrometer was also occasionally used to check oxygen purity.

The platinum single crystal was prepared using standard metallurgical procedures, i.e. by orienting it to within $\leq \frac{1}{4}^{\circ}$ of the (111) plane by back-reflection Laue X-ray methods, and then spark cutting and polishing both sides of the crystal with various grades of abrasive. The final abrasive was 0.05 μ Al_2O_3, and prior to insertion in the LEED chamber the crystal was given a light chemical etch (15 seconds in 50% diluted aqua regia).

After the crystal was inserted in the UHV system (background pressure $\sim 10^{-10}$ torr), both sides were cleaned *in situ* using argon ion bombardment. It has previously been found that platinum cannot be cleaned by thermal treatment alone[13-15)], and ion bombardment has proved effective in producing a (111) platinum surface which is well-ordered on a microscopic scale as judged by atomic beam scattering, a very sensitive surface probe [12,16-19)]. Further details regarding the experimental apparatus and technique may be found elsewhere[15,20)].

A photograph of the clean platinum (111) LEED pattern at an electron energy of 116 volts is shown in figure 1. The Auger electron spectrum corresponding to this clean surface is shown in figures 2 and 3.

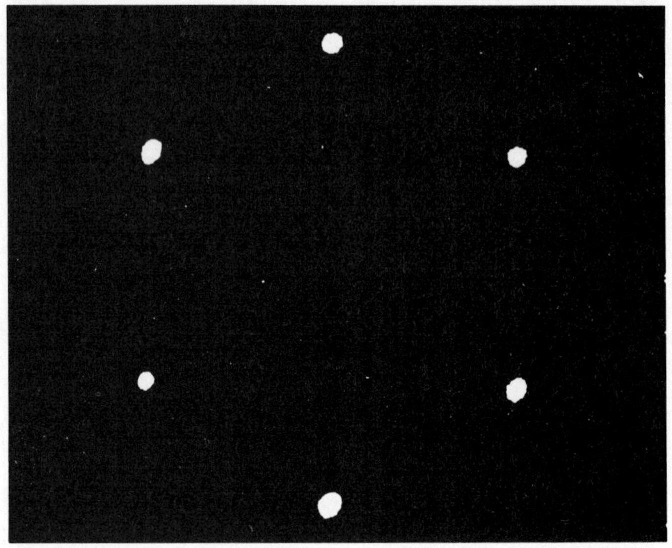

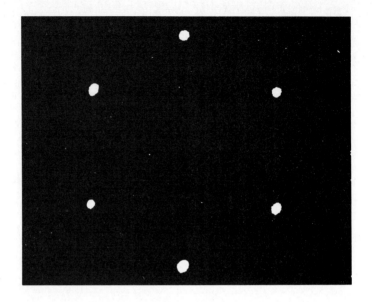

Fig. 1 LEED pattern of Pt(111) - (1x1) clean surface. Photograph taken at 116 volts.

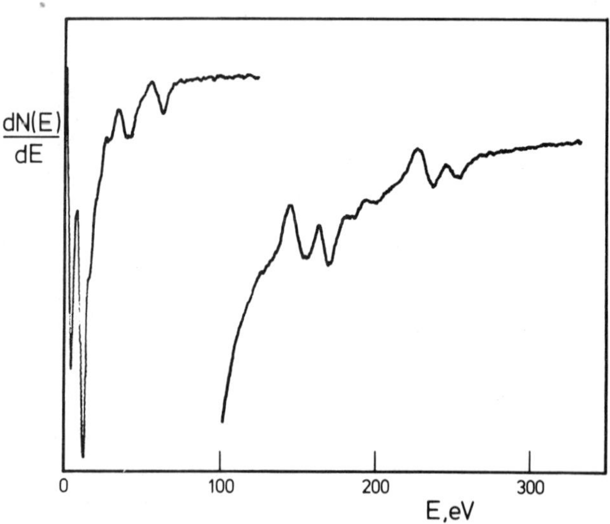

Fig. 2 Auger spectrum of clean Pt(111) surface using 1500 volt primary electron beam, 44 μA at 20° angle to surface. x 1 : 0.28 volt RMS suppressor grid modulation; and x 180 : 4 volts RMS suppressor grid modulation.

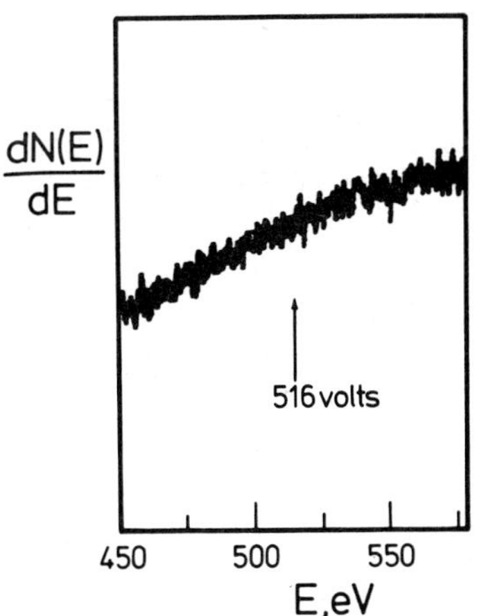

Fig. 3 Auger spectrum of clean Pt(111) surface using 1500 volt primary electron beam, 44 μA at 20° angle to surface. x 2000 (relative to x 1 in figure 2): 4 volts RMS suppressor grid modulation.

All the electronic transitions observed may be associated with platinum, and, in particular, there is no evidence of carbon (278 volts) sulphur (150 volts), silicon (90 volts) or oxygen (516 volts) as may be seen in figures 2 and 3. A (00) LEED spot spectrum of the clean platinum (111) surface is shown in figure 4a with the nth order Bragg peaks marked by arrows after an inner potential correction of 13 volts. The position of the Bragg peaks (analogous to those found in X-ray diffraction) may be found from the relation

$$2h \cos\Psi = n\lambda \qquad (1)$$

where h is the crystal plane spacing, Ψ is the incident angle from the crystal normal, n is the order of the Bragg peak, and λ is the wavelength of the incident electrons. The latter is given by

$$\lambda = \frac{h}{p} = \frac{h}{(2meV)^{\frac{1}{2}}} \qquad (2)$$

where h is the Planck constant, p is the electronic momentum, m is the electronic mass, e is the fundamental unit of electrical charge and V is the voltage of the electron beam. For V in volts and λ in Å, equation (2) may be written approximately as

$$\lambda \simeq \left[\frac{150}{V}\right]^{\frac{1}{2}} \qquad (3)$$

The inner potential correction must be applied since the incident electrons penetrate somewhat into the bulk solid.

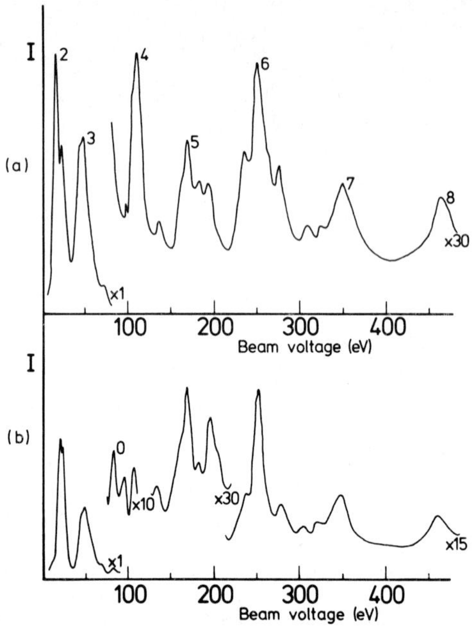

Fig. 4a Intensity-voltage curve of (00) diffraction spot of clean Pt(111) at 6.5° angle of incidence to surface normal. Calculated Bragg peaks are shown through 8th order with a 13 volt inner potential correction.

Fig. 4b Intensity-voltage curve of (00) diffraction spot of Pt(111) after exposure to O_2. Incidence angle is 6.5° relative to surface normal. The new peak at 82 volts is due to adsorbed oxygen.

The clean platinum surface was exposed to oxygen with the surface held between 575 and 775°K. The elevated temperature was used to prevent the adsorption of contaminants from the background gases (hydrogen and carbon monoxide). The oxygen partial pressure used varied between 10^{-5} and 10^{-3} torr for the several different experimental runs and the platinum was usually exposed to it for about one hour. To decrease the concentration of impurities, the oxygen was continuously pumped with the sputter ion pump which was throttled in order to prevent back-streaming from the pump.

3. RESULTS

After the oxygen treatment, no new spots were visible in the LEED pattern. However, the Auger electron spectrum showed that some oxygen had been adsorbed. The oxygen signal at 516 volts after oxygen adsorption is shown in figure 5 for the particular case of an oxygen pressure of

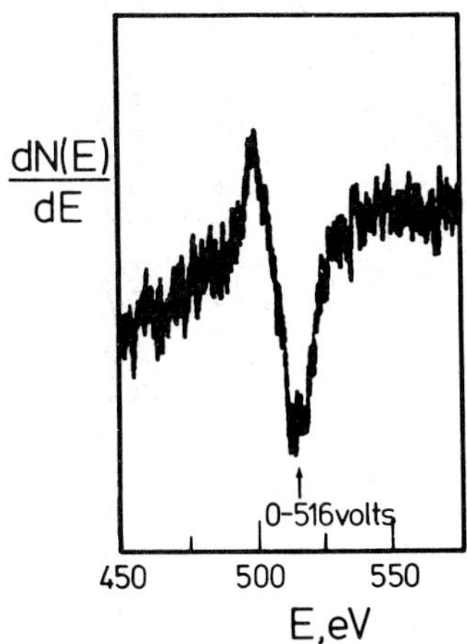

Fig. 5 Auger spectrum of Pt(111) after O_2 adsorption using 1500 volt primary electron beam, 44 μA at 20° angle to surface. x 2000 (relative to x 1 in figure 2): 4 volts RMS suppressor grid modulation.

7×10^{-5} torr, surface temperature of 675°K, and an exposure time of one hour.

The variation of intensity with electron voltage for the (00) LEED spot also confirms that oxygen has been adsorbed, as may be seen in figure 4b. Although most of the diffraction maxima are not shifted in voltage, their intensities are substantially changed; and, in addition, there is a new, fairly intense, 82 volt peak after oxygen adsorption which was not present before adsorption.

The amplitude of the oxygen Auger peak shown in figure 5 may be related approximately to an absolute oxygen coverage because the oxygen Auger signal has previously been calibrated using LEED in the same apparatus in experiments on a molybdenum (111) surface[20]. The signal from the molybdenum surface with an oxygen coverage of between 0.3 and 0.5 of a monolayer was about ten times as strong as that from this platinum surface after exposure to oxygen. If it is assumed that amplitudes in the Auger spectra are proportional to surface coverage, then the oxygen coverage (θ) on the platinum (111) surface is between 0.03 and 0.05 of a monolayer. Because oxygen is adsorbed dissociatively on both platinum and molybdenum, the above procedure for calculating the oxygen coverage on platinum should be reliable.

The sticking probability with the gas at room temperature may now be calculated from

$$S = \frac{4.2 \times 10^{-6}\ \theta}{\tau\ p_{O_2}} \tag{4}$$

where S is the sticking probability, p_{O_2} the oxygen partial pressure in torr, and τ is the exposure time in seconds. Using $p_{O_2} = 7 \times 10^{-5}$ torr, $\tau = 3600$ sec. and θ between 0.03 and 0.05, S is calculated to be between 5×10^{-7} and 8×10^{-7}. This same experimental value was obtained for all conditions of oxygen partial pressure, surface temperature and exposure time which were used. These conditions were mentioned previously. The platinum was restored to its clean state either by treating witn hydrogen at 5×10^{-8} torr at an elevated temperature or by heating to ca. $1300\,^{0}$K for a few seconds. It is known that platinum is a very efficient catalyst for the oxidati of hydrogen[1,21]. Also, Tucker noted the immediate disappearance on exposure to hydrogen of the (2x2) LEED pattern, wich had been formed by exposing a platinum (111) surface to oxygen[5].

4. DISCUSSION

The present results agree with those of Morgan and Somorjai[8] and Lampton[11] whose results showed that the sticking probability of oxygen on platinum is less than 10^{-3} on the (100) surface and less than 6×10^{-6} on the (111) surface respectively. They are in clear disagreement with the results indicating a high sticking probability[1-7].

It is suggested that the sticking probability on a clean platinum surface is low, while that on a surface containing carbon is high. In the earlier work, in which UHV equipment was not used[2-4], it is highly probable that the surface was contaminated with carbon because it is widely prevalent in transition metals[12-15]. Also, in those experiments using UHV in which high sticking coefficients were reported[5-7], the platinum surfaces were given a thermal treatment to clean them, rather than being subjected, for instance, to bombardment by inert gas ions. Platinum cannot be cleaned by heating alone[12-15] and so it is to be expected that carbon remained in the surfaces used by Tucker[5], by Procop and Völter[6], and by Wood *et al.*[7].

A theoretical investigation of the oxygen-platinum system using the absolute rate theory of gas phase kinetics and the crystal field surface orbital-bond energy bond order (CFSO-BEBO) model of Weinberg and Merrill[22-27] has been made and will be presented elsewhere[28]. The model calculations can account for the low value of the sticking probability of oxygen on clean platinum and also the higher sticking probability on carbon contaminated platinum. They therefore provide support for the interpretation given above.

It is thus concluded that those experimenters who have reported rapid adsorption of oxygen on platinum have been dealing with contaminated surfaces[1-7]. It is possible that rougher crystal planes than the (111) adsorb oxygen more efficiently, but the results of Morgan and Somorjai[8] on platinum (100) show that this is unlikely. Also Tucker[5] found that the sticking probability on the (100), (110) and (111) planes were approximately equal; but there was probably a small amount of carbon on each of these surfaces based on the observed value of the sticking probability as was noted previously.

It is finally interesting to speculate that perhaps a carbon contaminated platinum surface, and not clean platinum, is the actual effective oxidation catalyst in industrial use. It has been reported elsewhere that a carbon covered platinum surface, and not clean platinum, is the efficient catalyst for the hydrogenation of unsaturated hydrocarbons[24, 26].

ACKNOWLEDGEMENT: W.H.W., R.M.L., and C.M.C. wish to thank the National Science Foundation (NATO Post-Doctoral Fellowship), I.C.I. Ltd., and the Elsie Ballot Foundation, respectively, for financial support.

REFERENCES

1) G.C. Bond, "Catalysis by Metals", Academic Press, London, 1962.
2) O.D. Gonzalez and G. Parravano, J. Am. Chem. Soc. 78, 4533 (1956).
3) E.B. Maxted and N.J. Hassid, Trans. Faraday Soc. 29, 698 (1933).
4) D. Brennan, D.O. Hayward and B.M.W. Trapnell, Proc. Roy. Soc. A256, 81 (1960).
5) C.W. Tucker, Jr., J. Appl. Phys. 35, 1897 (1964).
6) M. Procop and J. Völter, Paper presented at the IInd International Symposium on Adsorption-Desorption Phenomena, Florence, Italy, 1971 (to be published by Academic Press).
7) B.J. Wood, N. Endow and H. Wise, J. Catalysis 18, 70 (1970)
8) A.E. Morgan and G.A. Somorjai, Surface Science 12, 405 (1968).
9) A.E. Morgan and G.A. Somorjai, J. Chem. Phys. 51, 3309 (1969).
10) L.A. West, E.I. Kozak and G.A. Somorjai, J. Vac. Sci. Tech. 8, 430 (1971).
11) V. Lampton, Master's Thesis, Department of Chemical Engineering, University of California, Berkeley, 1971.
12) D.L. Smith and R.P. Merrill, J. Chem. Phys. 52, 5861 (1970)
13) T.W. Haas, J.T. Grant and G.J. Dooley, J. Vac. Sci. Tech. 7, 43 (1970).
14) T.W. Haas, J.T. Grant and G.J. Dooley, Phys. Rev. B1, 1449 (1970).
15) R.M. Lambert, W.H. Weinberg, C.M. Comrie and J.W. Linnett, Surface Sci. 27, 653 (1971).

16) R.P. Merrill and D.L. Smith, Surface Science 21, 203 (1970)
17) D.L. Smith and R.P. Merrill, J. Chem. Phys. 53, 3588 (1970)
18) A.G. Stoll, D.L. Smith and R.P. Merrill, J. Chem. Phys. 54, 163 (1971).
19) W.H. Weinberg and R.P. Merrill, paper presented at the IInd International Symposium on adsorption-desorption phenomena, Florence, Italy, 1971 Academic Press, London, 1972, pp.151-169.
20) R.M. Lambert, J.W. Linnett and J.A. Schwarz, Surface Science 26, 572 (1971).
21) R.L. Palmer, paper presented at the California Catalysis Society, Santa Barbara, 1970.
22) W.H. Weinberg and R.P. Merrill, "BEBO Model Calculations for Surface Reactions on Platinum. II. NO + CO, O_2 + CO and H_2 + O_2", to be submitted to J. Catalysis (1972).
23) W.H. Weinberg and R.P. Merrill, "CFSO-BEBO Model Calculations for Chemisorption. II. CO, O_2 and CO_2 on Pt(111) and Ni(111)", to be submitted to surface Sci.(1972).
24) W.H. Weinberg and R.P. Merrill, "CFSO-BEBO Model Calculations for the Chemisorption of Hydrogen on Platinum (111)", (accepted for publication in Surface Sci.).
25) W.H. Weinberg and R.P. Merrill, "CFSO-BEBO Model Calculations for Chemisorption. III. N_2, NO and N_2O on Pt(111) and Ni(111)", to be submitted to surface Sci. (1972).
26) R.P. Merrill and W.H. Weinberg, "BEBO Model Calculations for Ethylene Hydrogenation on Platinum(111)", Paper presented at the Second National Meeting of the Catalysis Society, Houston, Texas, 1971 and to be submitted to J. Catalysis (1972).
27) W.H. Weinberg, Ph.D. Thesis, Department of Chemical Engineering, University of California, Berkeley, 1971.
28) W.H. Weinberg, R.M. Lambert, C.M. Comrie and J.W. Linnett, Surface Sci. 30, 299(1972).

DISCUSSION

M. BOUDART

It is not clear from your paper whether you removed chemisorbed oxygen by hydrogen at room temperature or at a higher temperature. Was the removal of oxygen quantitative?

H. WEINBERG

On a Pt(111) surface oxygen is only removed by heating in the presence of H_2 to a surface temperature at least > 800°K. The removal appeared to be quantitative according to Auger Spectroscopy; but most of the experiments reported here were conducted on surfaces which were Ar^+ bombarded and annealed as a final step in producing a clean, well-ordered, reproducible surface.

G. OHLMANN

May I ask, did you make some experiments on a carbon contaminated surface so you could prove whether C promotes the O_2 adsorption? We have determined the sticking probability, too, but on polycrystalline foil[1] (Figure 1). We have cleaned the surface not by thermal treatment in vacuum,

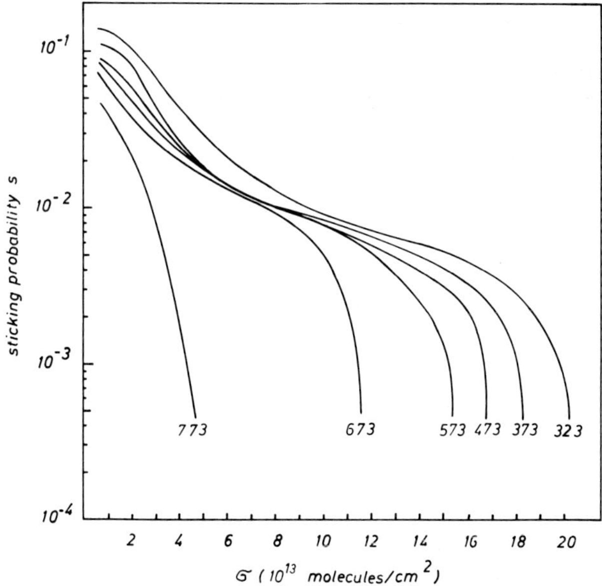

Fig. 1 Sticking probability of O_2 on polycrystalline foil

as you quoted, but by heating in oxygen. The evolving CO indicated that C was removed from the surface. During this pretreatment the CO desorption decreased, whereas the O_2 adsorption increased. Finally we have found a high and stable oxygen adsorption with an initial sticking probability of 0.14. This can only be explained if C is a poison and not a promoter of oxygen adsorption. Moreover, the same oxygen pretreatment results in a surface which very rapidly adsorbs a large amount (4.8×10^{14} molecules/cm^2) of hydrogen.[2] One can hardly imagine that hydrogen is adsorbed easier on carbon than on Pt. This is a further hint that a rapid adsorption occurs on a clean and not on a carbon contaminated Pt surface.

1) M. Procop and J. Volter, Z. physik. Chem. (Leipzig) (1972) in press.
2) M. Procop and J. Volter, Surface Science (1972) in press.

H. WEINBERG

We obviously conducted experiments on a carbon contaminated surface since we produced a clean Pt(111) surface by heating in oxygen. We found a sticking probability (or reaction probability to form gaseous CO and adsorbed oxygen) several orders of magnitude higher than that for adsorption on the clean surface. We have formulated a model which predicts that this effect is due both to a smaller activation energy and most especially to an enhanced pre-exponential factor of the rate coefficient (see ref. 28). We reject the notion that C is a poison for the adsorption of O_2 on Pt(111), and we see no a priori reason why that should be so on polycrystalline Pt foil.

We find relatively rapid adsorption (sticking probability ~ 0.1) of H_2 on both clean and carbon contaminated Pt(111). Although we realize that the H_2 chemisorption is a rate phenomenon, may we remind you that the single order C-H energy is significantly greater than the Pt-H bond energy (~ 101 kcal/mol as compared with ~ 67 kcal/mol (see ref. 24). The comment of Professor Eley and our response below is pertinent to this question as well.

V. I. SAVCHENKO

We have investigated the surface of Pt ribbons by AES method after heating at temperatures ~1000°C. Ba and Mg impurities were found on the surface. I should like to know if you found Ba impurities on the surface of your single crystal Pt(111)? The Auger spectrum on your slides limits the energy to about 520 eV. The characteristic energy of Auger electrons for Ba is about 605 eV.

We have also shown that carbon is removed very easily from the Pt surface by high temperature treatment in oxygen for 1 hour. Such treated Pt foil adsorbs oxygen rapidly with a sticking probability of about 0.1.

H. WEINBERG

We have periodically scanned the Auger energy up to ~1500 eV. At no time have we observed alkaline impurities. If present they were removed by the original 1300°K anneal (which left only carbon as a surface impurity). It might also be mentioned that the bulk concentration of Ba, Mg, etc. must have been small since at no time, even upon subsequent heating, did we observe their segregation to the surface.

The differences between clean Pt foil and Pt(111) vis-a-vis O_2 chemisorption are analyzed in the response to Dr. Wise's comment below.

J. A. JOEBSTL

I investigated the chemisorption of oxygen on platinum (111) by LEED and Auger spectroscopy and obtained similar results as Dr. Weinberg. I observed no (or very little) oxygen adsorption under experimental conditions as described by Dr. Weinberg. However, in some of my experiments I heated the crystal to 900°C in 10^{-6} torr oxygen and kept the crystal in the oxygen atmosphere during the cooling process. In these cases I observed considerable amounts of oxygen on the platinum (111) surface by Auger spectroscopy; furthermore, LEED revealed a ring pattern with equal spacings as the (111) diffraction spots. Apparently, platinum passes during the cooling period a temperature zone where oxygen can be removed easily by heating the crystal to 900°C. These results indicate that differences in the experimental conditions may be the reason for the variety of results which were reported concerning the chemisorption of oxygen on platinum.

H. WEINBERG

We have also performed the experiment described by Dr. Joebstl, but we observed no enhanced chemisorption of O_2 in contradistinction to his results. We conducted this experiment since it is essentially equivalent to the one described by Tucker (Ref. 5) in which he observed significant O_2 chemisorption. We have presented compelling arguments both in this paper as well as the paper of Ref. 28 for believing that Tucker's surface was contaminated by C. Lampton has also conducted this same experiment (cooling Pt(111) in an O_2 ambient with $P_{O_2} \gtrsim 10^{-6}$ torr) and has reproduced our result, i.e. little or no O_2 chemisorption. See Ref. 11 for details.

R. KLEIN

It is well known that it is most difficult to avoid carbon monoxide when introducing oxygen into a system, and if so, is there a possibility that it could have lowered the sticking probability for oxygen on platinum (111)?

H. WEINBERG

In general no hot filaments were operated during the adsorption experiments described in this paper. We intermittently checked the O_2 and CO partial pressures with a quadrupole mass spectrometer equipped with a (low work function) thoriated iridium filament. The CO partial pressure was found to never exceed 1% of the O_2 partial pressure, but this could nonetheless approach 10^{-6} torr due to the very high O_2 pressures we were obliged to use. Thus, we maintained the Pt at a surface temperature greater than 575°K. In other experiments we have shown that under these conditions the equilibrium coverage of CO on the Pt(111) is immeasurably small. Thus, adsorption of background gases did not interfere with the measurement of the O_2 sticking probability.

I might also add the reason we kept the surface temperature below 775°K was to ensure that no oxygen was removed from the surface, e.g. according to the following chemical reactions:

$$Pt = O + CO \rightarrow Pt + CO_2$$

$$Pt = O + H_2 \rightarrow Pt + H_2O$$

We confirmed the reactions of the adsorbed oxygen with the background gases were not important under our experimental conditions. In addition, no oxygen was desorbed from the surface at 775°K. Therefore, we report a reliable value of the sticking probability of O_2 on Pt(111).

D. D. ELEY

In work in press in J. Catalysis, Breakspere, Norton, and I found a sticking coefficient of hydrogen on a polycrystalline Pt at 77K and 5 x 10^{-8} torr which was about 0.1 over a protracted coverage estimated from 0 to 0.3, after which it fell off in value. We found this result when wire was cleaned either a) by simple heating to 1400K in 10^{-10} torr vacuum or b) by successive heating in oxygen, hydrogen, and vacuum to remove carbon.

Dr. Weinberg's paper brings home the need for further experimental work on cleaning procedures for Pt.

H. WEINBERG

As mentioned above in the response to the comment of Ohlmann, we tend to confirm these interesting results reported by Professor Eley. We find the sticking probability of H_2 on Pt is not particularly sensitive to the cleanliness of the Pt vis-a-vis C contamination , and we also establish that the initial sticking probability is ~ 0.1. It is very likely that those surfaces cleaned by simple heating to 1400°K in 10^{-10} torr vacuum are contaminated by C, whereas those cleaned chemically with O_2 and H_2 are presumably quite clean.

H. WISE

The low sticking coefficient deduced by the authors for oxygen on Pt(111) does not seem to apply to a polycrystalline surface of platinum. Recent ultra high vacuum studies in our laboratory have demonstrated that the removal of a carbon impurity from the polycrystalline Pt surface, by treatment in oxygen at 500°C, causes an increase in the sticking coefficient to a value of $S_0 = 0.25$, rather than a decrease as would be expected on the basis of the hypothesis contained in the present paper. We must conclude therefore that the smooth Pt (111) face exhibits entirely different sorption properties from those encountered with a Pt-foil which on its surface will exhibit numerous other crystalline faces as well as surface roughness due to steps and dislocations.[1)]

1) Lang, Joyner, and Somorjai, Surface Science 30, 454 (1972).

H. WEINBERG

It may well be that polycrystalline Pt foil and the smooth (111) crystallographic orientation of a Pt single crystal are fundamentally different. This difference may be understood using our model givin in Ref. 28 if the out-of-plane C atom described in the reference is replaced by a Pt atom, a situation which would obtain, for example, on a rough polycrystalline surface. Then the same free energy arguments used in Ref. 28 may be applied with the following conclusions: (1) The pre-exponential of the sticking probability, i.e., the entropy factor, would be expected to have a "normal" value for polycrystalline Pt, i.e., a value close to unity; and (2) the activation energy for adsorption of O_2 would not be expected to vary significantly from one orientation of Pt to another, i.e., the activation energy to adsorption on polycrystalline Pt should be ~ 2.1 kcal/mol. Assuming multiple thermalizing collisions of O_2 with the rough surface, the sticking probability of O_2 on polycrystalline Pt at 500°C is then given approximately by

$$S_0 \sim \exp\left[- \frac{2100}{(1.987)(773)} \right] = 0.25$$

as quoted by Dr. Wise. Our model is evidently successful in explaining the observed differences in the experimental data on the various Pt surfaces. The entropy of activation plays a crucial role in this explanation.

Paper Number 34

INVESTIGATION OF HYDROGEN-OXYGEN INTERACTIONS ON PLATINUM SURFACES BY FIELD EMISSION MICROSCOPY

V. V. GORODETSKI and V. I. SAVCHENKO
Institute of Catalysis, Novosibirsk, USSR

ABSTRACT: The adsorption of oxygen on platinum and the interaction of hydrogen with adsorbed oxygen have been studied by field emission microscopy.

1. INTRODUCTION

In recent years reports have appeared of significant specificity of different crystal planes of metals for the chemisorption of simple gases,[1,2] and for catalytic activity. In works of Farnsworth[3] and Gwathmey[4] the rates of H_2-D_2 exchange and ethylene hydrogenation were measured on nickel monocrystals. Unfortunately, measurement of a reaction rate on monocrystals is complex experimentally, laborious and limited to planes of low indexes.

Field emission microscopy permits us to establish the relation between catalytic activity of different planes and the structure of metal atoms at a surface. On the emitter surface practically all crystallographic planes are present and direct observation of changes in an adsorbed layer at the surface of each plane is possible.

It has been shown[5,6] that the work function sharply decreases during interaction of hydrogen with oxygen adsorbed on nickel films. In the field emission microscope, this should result in increased emission from those planes on which the interaction takes place.

In the present work, oxygen adsorption and hydrogen-oxygen interaction on platinum have been studied by field emission microscopy.

2. EXPERIMENTAL PROCEDURE

Platinum emitters were made of pure (99.9%), 0.1 mm Pt wire by electrolytic etching in molten 50% NaCl + 50% KCl at a constant voltage of about 3 V. The emitter was fastened to a tungsten wire with a diameter of 0.20 mm and a length of 10 cm. The temperature of the emitter in the range 78 - 1700°K was measured with a chromel-alumel thermocouple having a wire diameter about 40μ. Platinum emitters were cleaned by pulse heating to 1750°K in an ultra-high vacuum of about 10^{-11} torr. Gas composition was analyzed with an omegatron IPDO-I. Hydrogen and oxygen were introduced into the system by diffusion through the walls of palladium and silver capillaries, respectively; water vapor, via an ultra-high vacuum leak-valve.

Obtaining a clean surface of Pt emitters by simple heating in ultra-high vacuum is difficult because around the (100) planes intensely emitting rings appear which seem to result from impurities on the surface.[7,8] Even temperatures close to the Pt melting point fail to remove them. A clean surface can be obtained by heating in hydrogen to about 1700°K. Our platinum emitters had a radius after cleaning of about 0.5μ.

The average work function was calculated from the change of voltage which is necessary to maintain a constant value of the emission currently by the simplified Fowler-Nordheim equation:[9]

$$\phi_{ads} = \phi_{Pt}(V_2/V_1)^{2/3}$$

where ϕ_{Pt} is the work function of clean Pt taken equal to 5.32 eV[10]; ϕ_{ads} is the work function of Pt covered with adsorbed gas; and (V_2/V_1) is the ratio of voltages which give the same value of the emission current (10μa) with covered and clean Pt surfaces.

To eliminate ion etching the field emission microscope was switched on only for the short period of time necessary for measurement of the emission current and photographing a picture on film of high sensitivity.

3. EXPERIMENTAL RESULTS

3.1 Oxygen adsorption

Oxygen adsorption causes an increase in the work function of Pt of ~1.40 eV. Fig. 1a presents the results of the change in the Pt work function during oxygen adsorption. The emission patterns of a Pt emitter with orientation (111) after adsorption of oxygen to saturation at 78°K is given in Fig. 2c. No sharp specificity in the properties of different Pt planes is observed.

Increasing the temperature of platinum with adsorbed oxygen ($\Delta\phi$ = 1.40 eV) from 78 to 1700°K results in reasonable changes of the work function and the following emission patterns:

T ~ 150°K - $\Delta\phi$ increases slightly possibly because of desorption of "electropositive" oxygen,[10] Fig. 1b, Curve 1.

T ~ 650°K - intensely emitting regions appear around planes (110), Fig. 2d.

T ~ 900°K - intensely emitting regions decompose into four regions symmetrically located around planes (110), and at 950°K their complete disappearance is observed, Fig. 2e.

T ~ 970°K - on plane (111) a number of intensely emitting centers of small sizes appear, Fig. 2f.

T ~ 1270°K - the displacement of bright points to a plane center and the formation of specific emission rings around planes (111), Fig. 2g; the appearance of dark regions on planes (102), Fig. 2g.

T ~ 1700°K - the platinum surface is completely free of oxygen; but even such high temperatures fail to eliminate impurities from the (110) planes. Heating in hydrogen at P_{H_2} = 5 x 10^{-6} torr and ~1200°K results in the disappearance of impurities on these planes.

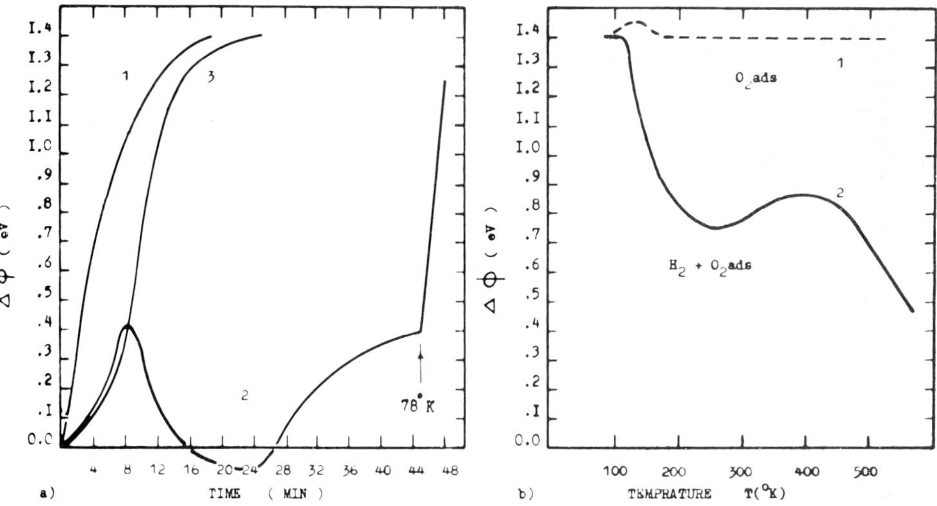

Fig. 1 a) Change of the work function of platinum upon oxygen adsorption: curve 1, 300°K, P_{O_2} = 5 x 10^{-8} torr; curve 2, 650°K, P_{O_2} = 5 x 10^{-8} torr; curve 3, 78°K, oxygen adsorption from a stoichiometric mixture of H_2+O_2 at total pressure of 1 x 10^{-7} torr, b) interaction of hydrogen with adsorbed oxygen.

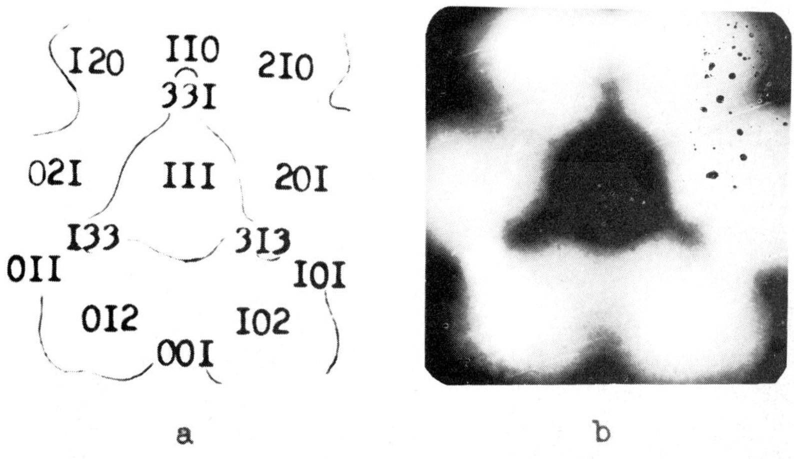

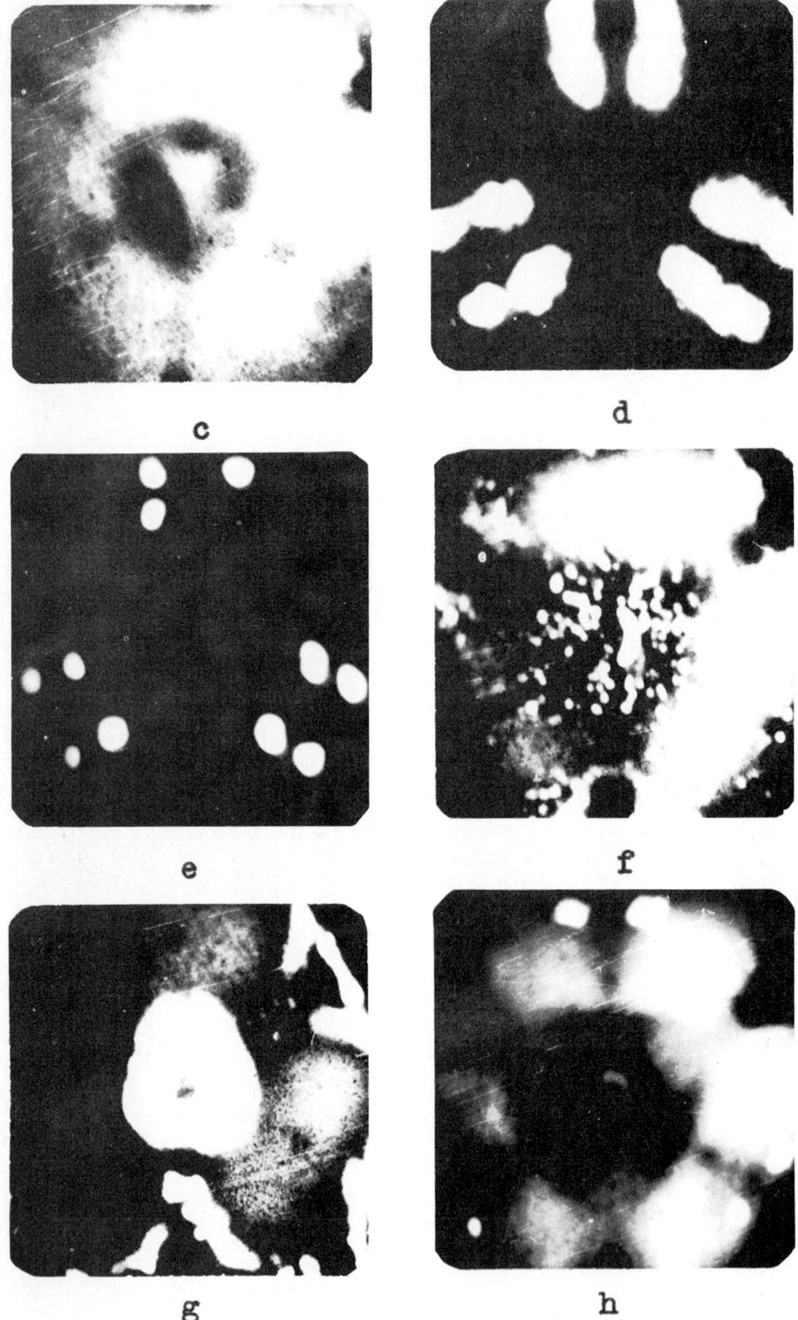
c
d
e
f
g
h

Fig. 2 Oxygen adsorption on platinum. a) Stereographic projection of planes of Pt emitter with orientation (111); b) clean platinum with orientation (111); c) adsorption of O_2 at 78°K, $\Delta\phi$=1.40 eV; d) T=650°K, $\Delta\phi$=0.60 eV; e) T=950°K, $\Delta\phi$=0.36 eV; f) T=970°K, $\Delta\phi$=0.36 eV; g) T=1270°K, $\Delta\phi$=0.53 eV; h) adsorption of O_2 at 78°K($\Delta\phi$= 0.9 eV) and heating to 680°K, $\Delta\phi$=0.36 eV.

Oxygen adsorption at lower surface coverage ($\Delta\phi$=0.90 eV) followed by heating from 78°K to 650°K also results in the appearance of intense emission on the regions around planes (110), Fig. 2h, but the size of these regions is much smaller than that in Fig. 2d. Thus the process of platinum surface reconstruction depends on the oxygen concentration on the surface.

Oxygen adsorption at T = 650°K differs greatly from that at lower temperatures (Fig. 1a, Curve 2). After the initial increase of $\Delta\phi$ to 0.42 eV the work function sharply decreases to $\Delta\phi$ = -0.06 eV and then slowly increases. The decrease is connected with the development of specific regions of intense emission on surface areas around planes (110)(Figs. 3a-d) similar to those obtained upon heating Pt with adsorbed oxygen to 650°K. If the temperature is now decreased from 650°K to 78°K (Fig. 1a, Curve 2) additional adsorption of O_2 is observed, $\Delta\phi$ increases to 1.27 eV, and the emission pattern in Fig. 3d is obtained.

To test the assumption that the appearance of these specific formations (Fig. 3c) is connected with the growth of a platinum oxide phase around (110), the behavior in a hydrogen atmosphere was studied. The introduction of H_2 into the system at 650°K results in a sharp decrease of $\Delta\phi$ from 0.37 to -0.84 eV, although the emission pattern obtained (Fig. 3e) is analogous to the initial one (Fig. 3c). The picture of the Pt emitter with $\Delta\phi$ = -0.84 eV is unstable and for a short period of time surface reconstruction is observed which results in the disappearance of intense emission regions around (110) and the formation of a clean surface of the Pt emitter (Fig. 3f).

3.2 Interaction of hydrogen with adsorbed oxygen

The interaction of H_2 with adsorbed oxygen was studied at 78-700°K at a hydrogen pressure of 5 x 10^{-6} torr. After adsorption of oxygen to saturation at 78°K, hydrogen was introduced. As can be seen from Fig. 1b, Curve 2, $\Delta\phi$ begins to decrease from 1.40 eV at about 120°K and it reaches 0.75 eV at 250°K.

The change of emission patterns (Figs. 4a-b) makes it possible to relate the decrease of $\Delta\phi$ with the course of the processes on different planes. In the temperature range of 100-130°K (Fig. 4b) the interaction of H_2 with adsorbed O_{ads} results in increased emission around plane (111) and on planes (331). Increase of temperature to 135°K (Fig. 4c) results in widening the intense emission region around plane (111). Within the range of 145-160°K (Figs. 4d-h) the process proceeds on planes (102). During the interaction with hydrogen the regions of these planes occupied by adsorbed oxygen decrease in size along the boundary of an adsorbed layer on the side of planes (111). Removal of hydrogen followed by addition of oxygen results in rapid adsorption of O_2 (Fig. 4i) on plane (111).

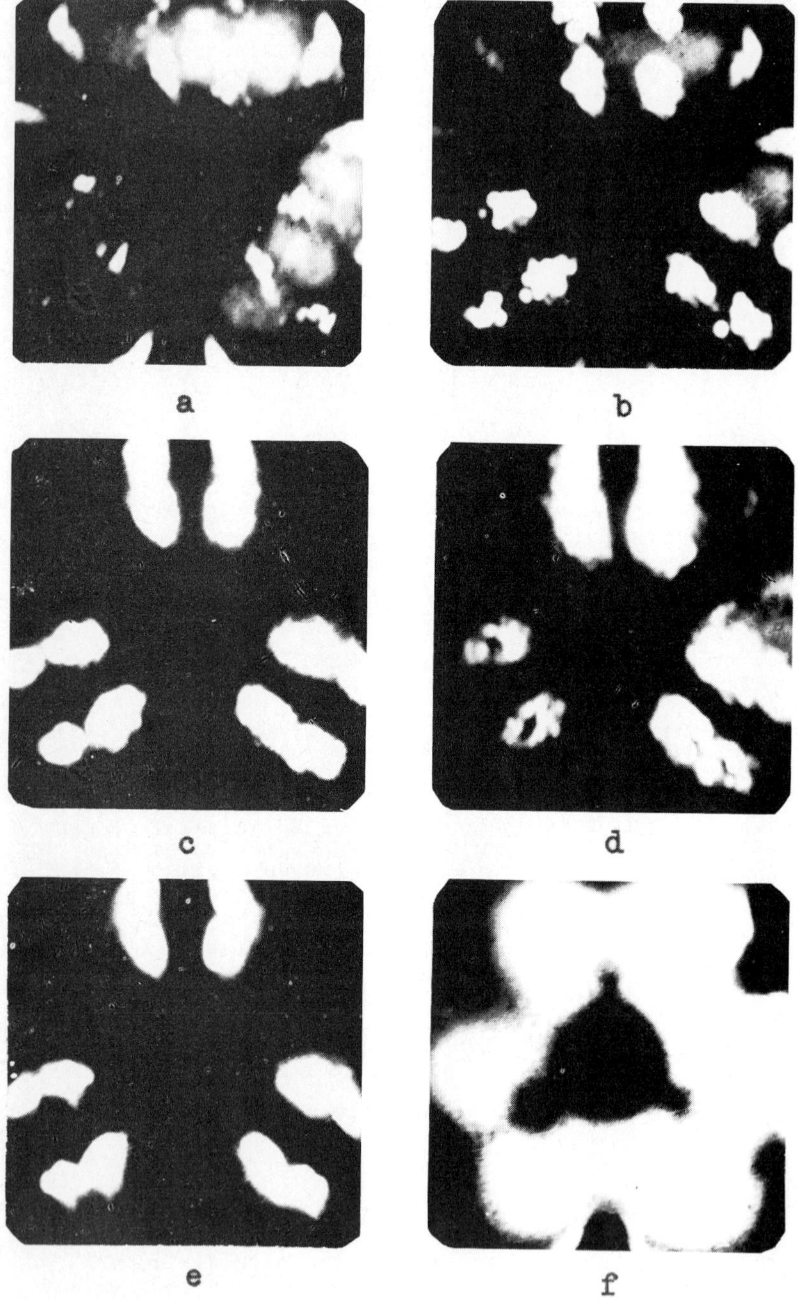

a b c d e f

3.3 Stoichiometric mixture of H_2 and O_2

As can be seen from Fig. 1a (Curve 3), adsorption of oxygen at 78°K from a stoichiometric mixture of $H_2 + O_2$ at a total pressure of 10^{-7} torr results in an increase of the work function by 1.40 eV.

Increasing the temperature of the platinum emitter to about 125°K in $H_2 + O_2$ results in a small decrease of $\Delta\phi$ to 1.35 eV and the appearance of intense emission around plane (111) and on planes (331) (Fig. 4j) as was also observed in the interaction of H_2 with preadsorbed oxygen on clean Pt. At 350°K (Fig. 4k) $\Delta\phi$ decreases to the 0.92 eV value associated with the decrease of the concentration of O_{ads}, but in this case the emission pattern is close to that of the clean surface. At this temperature the concentration of the reacting gases on the Pt surface is very small and $\Delta\phi$ is small (0.37 eV).

It should be noted that $H_2 + O_2$ at 600-700°K does not lead to the formation of the intensely emitting regions around planes (110) which occurs as a result of adsorption of pure oxygen.

4. DISCUSSION

4.1 Adsorption of oxygen

After the adsorption of oxygen at 78°K the picture of a platinum emitter differs little from that of clean platinum although $\Delta\phi = 1.4$ eV. Planes (110) do have slightly more emission. No distinct specificity of the adsorption properties of different planes is observed even though low-energy electron diffraction has shown that oxygen is not adsorbed on Pt plane (100),[12,13] but is adsorbed on Pt (111).[17]

At about 120°K "electropositive" oxygen is desorbed from the surface[10] and then neither the pattern of the emitter nor $\Delta\phi$ changes until a temperature of about 650°K is reached. At this point, bright regions appear around planes (110). Two explanations for the appearance of these regions are possible: either the presence of adsorbed oxygen accelerates the diffusion of electropositive impurities from the bulk of the platinum to the surface, or microcrystals of platinum oxide are formed that may result in a local increase of the field intensity and, consequently, in an increase of emission. Some data point to the second explanation:

1) The size of the regions around plane (110) depends on oxygen concentration on the surface.

2) Heating in hydrogen results in the disappearance of the bright regions accompanied by a sharp decrease in the apparent $\Delta\phi$ (-0.84 eV). The initial clean surface of the Pt emitter ($\Delta\phi = 0$) appears almost instantly.

Fig. 3 Adsorption of O_2 on Pt at 650°K and subsequent interaction with hydrogen: a) after 8 min., $\Delta\phi = 0.42$ eV; b) after 12 min., $\Delta\phi = 0.18$ eV; c) after 42 min., $\Delta\phi = 0.37$ eV; d) Temperature decrease from 650°K to 78°K, adsorption of O_2, $\Delta\phi = 1.27$ eV; e) interaction with hydrogen upon temperature increase from 78 to 650°K, $P_{H_2} = 5 \times 10^{-6}$ torr, $\Delta\phi = -0.84$ eV; f) in 1 min., $\Delta\phi = 0$.

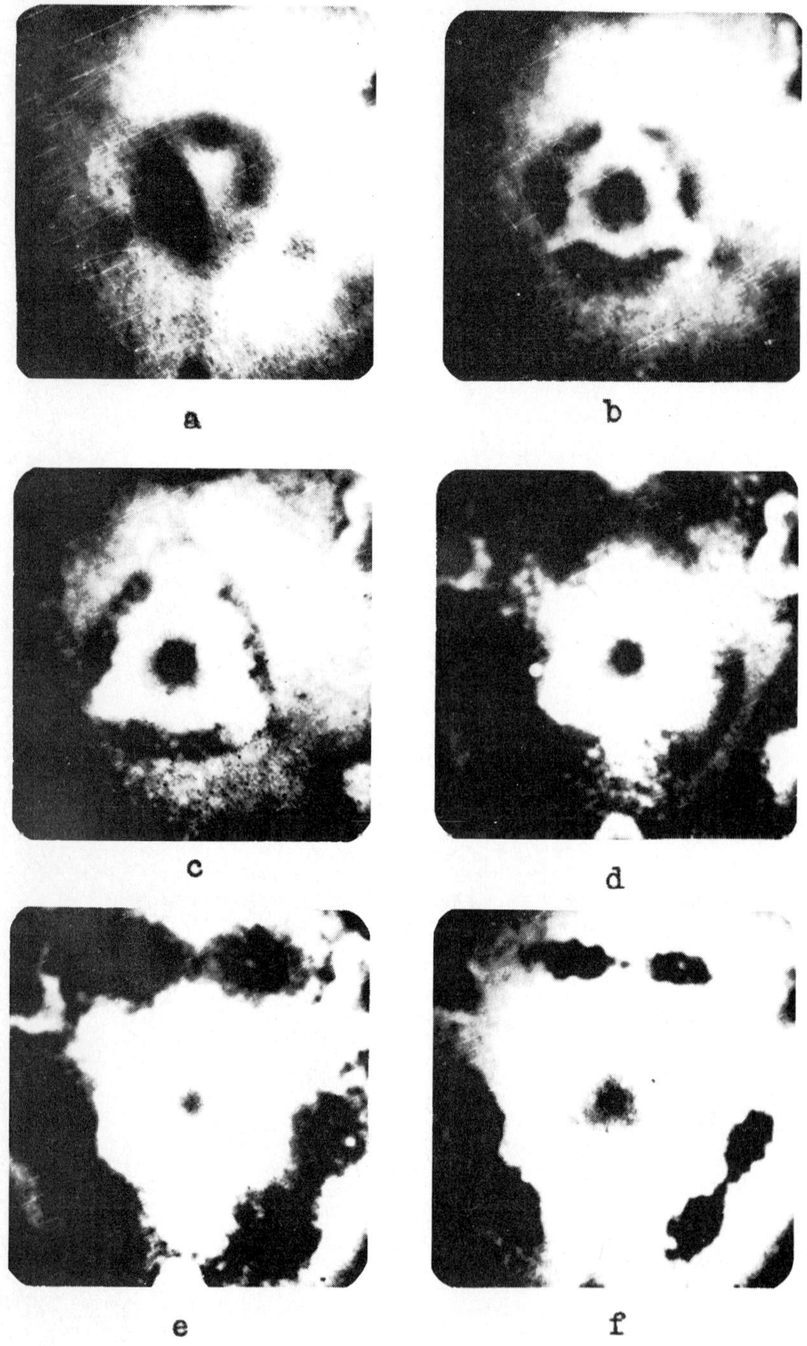

a b c d e f

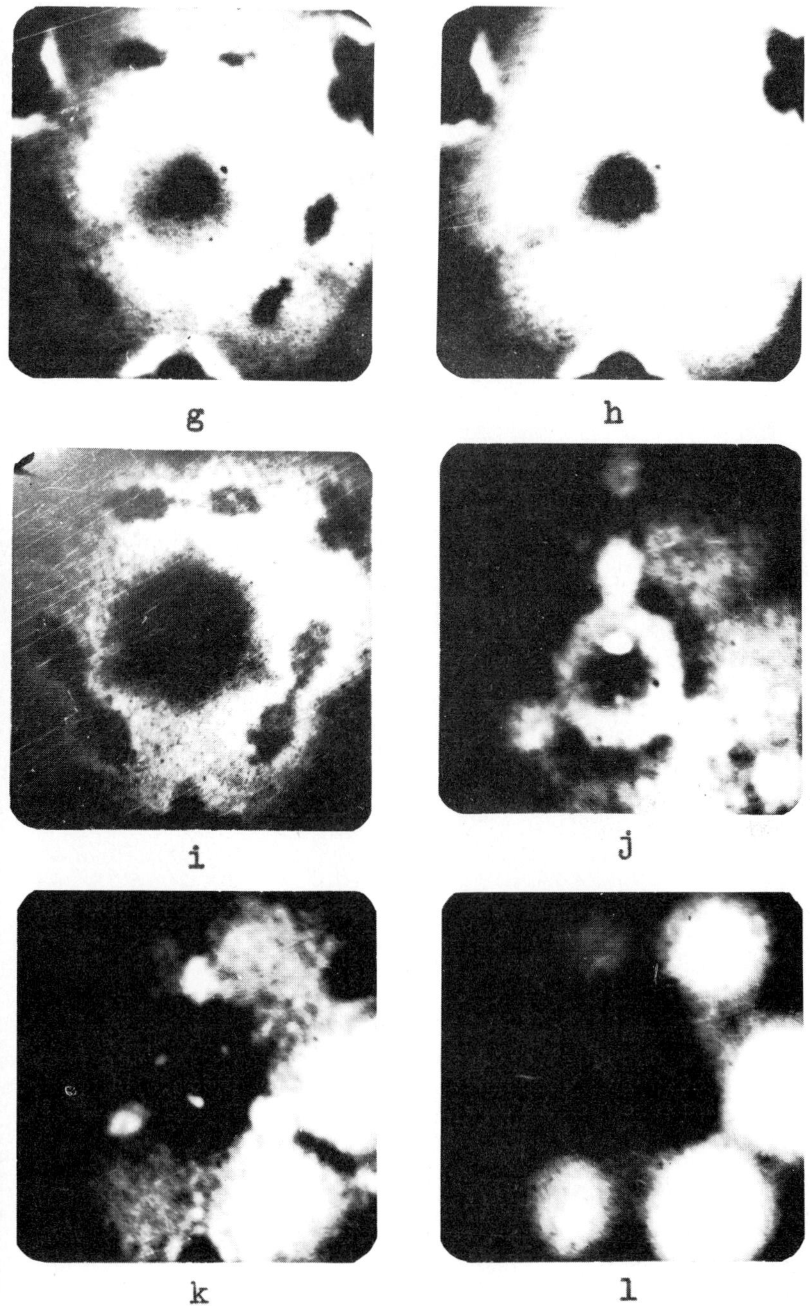

g h i j k l

3) The study of Pt foil of a purity of 99.9% by Auger spectroscopy showed that impurities of alkali-earth metals diffuse from the bulk to the surface beginning at about 1200°K. Most probably the appearance of these impurities can be related to the appearance of the intensely emitting regions around planes (100) observed in the present work and in the earlier work of Melmed[7)] and of Vanselow[8)] and with the points of intense emission on plane (111) (Fig. 2f).

4) Many reports indicate the possibility of formation of platinum oxides at increased temperatures. Thus, Khasin and Boreskov[14)] found that several monolayers of oxygen can be dissolved in platinum films at T = 525°K. Vanselow[15)] and Weber et al.[16)] who studied oxygen adsorption on platinum by a flash-filament method showed that not all adsorbed oxygen is desorbed to form O_2 molecules in the gas phase. Field desorption coupled with direct mass-spectrometric analysis showed the presence of PtO_2^+, PtO^+ etc.[8)]

It is important to note that a study of oxygen adsorption by low-energy electron diffraction on Pt planes (100), (111) and (110) by Tucker[17)] led to the conclusion that the reconstruction of surface atoms is observed particularly on plane (110). The whole complex of the data given leads to the conclusion that at temperatures of 600-700°K an oxide phase is formed in the region of the Pt plane (110).

Roginski and Krylov[18)] studied the rate of oxidation of hydrogen at 300°K in relation to the temperature of pretreatment of platinum in oxygen. A maximum in activity was found for heating in oxygen at 600-700°K. At this temperature (Fig. 2d) microcrystals of platinum oxide appear on the surface. This should result in an increase in the specific surface of platinum and, therefore, in the rate of the hydrogen oxidation reaction.

4.2 Interaction of hydrogen with adsorbed oxygen on platinum

A step-wise mechanism for the oxidation of hydrogen has been proposed[14,19)]

$$O_{ads} + H_{ads} \text{ (or } H_2 \text{ gas)} \rightarrow OH_{ads} \qquad \text{Stage I}$$

$$OH_{ads} + H_{ads} \rightarrow H_2O \text{ gas} \qquad \text{Stage II}$$

Fig. 4 Interaction of hydrogen with adsorbed oxygen: a) adsorption of O_2 at 78°K, $\Delta\phi$ = 1.40 eV; b) P_{H_2} = 5 x 10^{-6} torr, T = 125°K, $\Delta\phi$ = 1.40 eV; c) T = 135°K, $\Delta\phi$ = 1.18 eV; d) T = 145°K, $\Delta\phi$ = 1.10 eV; e) T = 150°K, $\Delta\phi$ = 1.06 eV; f) T = 155°K, $\Delta\phi$ = 1.00 eV; g) T = 157°K, $\Delta\phi$ = 0.96 eV; h) T = 160°K, $\Delta\phi$ = 0.85 eV; i) adsorption of O_2 at T = 125°K and P = 5 x 10^{-7} torr after replacing hydrogen in the gas phase by oxygen at T = 155°K - see Fig. 4e; j) Stoichiometric mixture of H_2 + O_2 at P_{total} = 1 x 10^{-7} torr; T = 135°K, $\Delta\phi$ = 1.35 eV; k) T = 350°K, $\Delta\phi$ = 0.92 eV; l) T = 650°K, $\Delta\phi$ = 0.37 eV.

Naturally the question arises whether the observed decrease of ϕ is connected with the formation of intermediate compounds (OH_{ads}) or with the formation and desorption of water.

Hopkins et al.[20] have shown that dissociative adsorption of water on the Pt plane (110) at a room temperature increases ϕ by 1.13 eV; oxygen adsorption, by 1 eV. The electronegativity of OH-groups is 3.9. This is approximately equal to the electronegativity of O_{ads}, 3.5. Therefore, the formation of hydroxyl groups during the interaction of H_2 with O_{ads} should not result in appreciable changes in ϕ unless some change of the orientation of OH-groups relative to the platinum surface occurs. An analogous conclusion was reached by Tompkins and Siddigi[21] when studying the interaction of atomic hydrogen with preadsorbed oxygen on nickel films at 78°K.

If, nevertheless, we assume that the observed $\Delta\phi$ is connected with the first step in the reaction, then after the interaction of hydrogen with adsorbed oxygen at 120°K the region around plane (111) will be covered with adsorbed (OH)-groups. It is difficult to imagine rapid and facile oxygen adsorption on a platinum surface covered with hydroxyl groups. However, reintroduction of oxygen results in rapid adsorption and an increase of ϕ in the area of plane (111) (Fig. 4i).

From the data given it follows that the observed change of ϕ during the interaction of hydrogen with adsorbed oxygen is connected with the formation of water and desorption of H_2O from the platinum surface. The adsorption of hydrogen on the platinum surface subsequent to the desorption of water cannot significantly influence the emission pattern since the maximum change of a work function during hydrogen adsorption on platinum is small $\Delta\phi$ = 0.2 eV.[22]

It is interesting to note that preliminary data indicate that water adsorbed on platinum at 78°K is first desorbed at a temperature of about 150°K and from the planes (111).

Consequently, the most reactive oxygen is that adsorbed on planes (331) and in the region of plane (111). On these planes the reaction occurs even at 120°K. With the increase of temperature to 140°K, planes (102) are involved. As shown in Figs. 4d-h, the decrease in the dumbbell-like dark regions occurs on the side of planes (111). Once water has been formed in the region of planes (111), (221), and (331) and hydrogen has been adsorbed on these planes, the reaction proceeds in an adsorbed layer at the boundary between immobile O_{ads} and adsorbed hydrogen at the edge of planes (111).

A different mechanism could be proposed: O_{ads} (201) migrates to the central regions of the emitter where it interacts with hydrogen. But the first proposal is more probable since migration of O_{ads} on planes (201) begins at about 450°K.[10] As has been already mentioned, the surfaces of platinum oxides are reduced by hydrogen at higher temperatures, ~700°K. In H_2 + O_2 mixtures, three intensely emitting planes (331) can be distinctly seen at 120-350°K and hydrogen oxidation takes place on them. At 600-700°K the pattern of the emitter hardly differs from the pattern of a clean tip and the formation of an oxidized phase is not observed, i.e. the rate of oxide reduction exceeds the rate of its formation.

It would be interesting to know what causes the increased activity of oxygen adsorbed on planes (331). Unfortunately, it is as yet impossible precisely to measure the heat of oxygen adsorption on these planes.

The increased activity of oxygen adsorbed on planes (331) and on the surface region around plane (111) may be connected with a terrace structure of atom packing on these planes which could promote the transfer of H atoms to oxygen atoms. Similar views have been given by Bond (23) and by Moss et al. (24).

REFERENCES

1) G. Ehrlich, Advances in Catalysis, 14, 255 (1963).
2) N. P. Vasko, Yu. G. Ptushinski, B. A. Tchuikov, Surface Science, 14, 448 (1969).
3) H. E. Farnsworth, Advances in Catalysis, 14, 31 (1963).
4) R. E. Cunningham, A. T. Gwathmey, Advances in Catalysis, 9, 25 (1957).
5) C. M. Quinn, M. W. Roberts, Trans. Faraday Soc., 60, 899 (1964).
6) M. W. Roberts, B. R. Wells, Trans. Faraday Soc., 62, 1068 (1966).
7) A. J. Melmed, J. Appl. Phys., 36, 3691 (1965).
8) R. Vanselow, Phys. Status. Solidi, 21, 69 (1967); Z. Naturforsch., 21a, 1190 (1966).
9) R. Klein, J. Chem. Phys., 21, 1177 (1953).
10) R. Lewis, R. Gomer, Surf. Sci., 12, 157 (1968).
11) E. W. Müller, Ergeb. exact. Naturwiss., 27, 290 (1953).
12) A. E. Morgan, G. A. Somorjai, Surf. Sci., 12, 405 (1968).
13) J. T. Grant, T. W. Hass, Surf. Sci., 18, 457 (1969).
14) A. V. Khasin, E. K. Boreskov, DAN, 152, 1387 (1963).
15) R. Vanselow, W. A. Schmidt, Z. Naturforsch., 22a, 717 (1967).
16) B. Weber, J. Fusy, A. Cassuto, J. Chim. Phys. et Chim. Biol., 66, 708 (1969).
17) C. W. Tucker, J. Appl. Phys., 35, 1897 (1964).
18) O. V. Krylov, S. Z. Roginskii, DAN, 88, 293 (1953).
19) V. Ponec, A. Knor, S. Cerny, Proc. Third Intern. Congr. on Catalysis, 1964, p. 353.
20) C. W. Jowett, P. J. Dodson, B. J. Hopkins, Surf. Sci., 17, 474 (1969).
21) F. C. Tompkins, M. M. Siddigi, Proc. Roy. Soc., A268, 452 (1962).
22) R. Lewis, R. Gomer, Surf. Sci., 17, 333 (1969).
23) G. C. Bond, 4th Intern. Congr. on Catalysis, Moscow, 1968, paper 67.
24) L. Whalley, B. J. Davis, R. L. Moss, Trans. Faraday Soc., 66, 576, 3143 (1970).

J. H. BLOCK

1) The evaluation of work function data by the simplified Fowler-Nordheim equation involves uncertainties, since the pre-exponential may be of importance. If you compare Figure 2b and Figure 2e, the emitting area has changed considerably. Furthermore, the interpretation of these field ion images will face difficulties, since in Figure 2e effects of oxygen adsorption are attributed to small emitting spots (with low work function?) around (011), although the integral work function increased relatively to Figure 2b.

2) We have tried to measure the $H_2 + O_2$ reaction in a field ion mass spectrometer. These experiments failed on account of water formation on metal parts of the UHV chamber at room temperature. How did you control this reaction possibility?

V. I. SAVCHEKO

1) We consider that the adsorption of oxygen at the ~650°K results in reconstruction of Pt surface in region of planes (110) (Figs. 2 d-e, Figs. 3a-c). It is difficult to determine the exact value of work function by using the probe-hole technique. Therefore, in this work we were limited to a qualitative estimation of $\Delta\phi$ by using the simplified Fowler-Nordheim equation. The following facts influence the $\Delta\phi = f(t)$ curves (Fig. 1 (curves 1,2)): a) Change of work function at the adsorption of oxygen. b) Reconstruction of the surface emitter and, connected with this, a local change of the field intensity. c) The O_2 dissolution at ~650°K at the supersurface layers of Pt.

Consequently, the resulting effect (Fig. 1 (curve 2)) is determined by different deposits of this process.

2) The field emission microscope and ultrahighvacuum installation were made of glass. Practically, FEM was completely covered by liquid nitrogen. Gas composition was analyzed by an omegatron IPDO-1; residual pressures of water vapour and CO were about 1.10^{-10} torr when the unit was filled with H_2, O_2, or $H_2 + O_2$.

G. OHLMANN

Your $\Delta\phi$ results exhibit great differences between oxygen adsorption at 650°K and adsorption at lower temperatures. We have made flash-desorption from a Pt foil.[1] The desorption spectrum (Figure 1) shows great differences between adsorption at higher and lower temperature, too. They are represented by these two peaks, the α and the β state with binding energies of 10 and 60 kcal/mole, respectively.

Moreover, you discuss the problem whether the patterns at temperatures higher than 1200 °K may be caused by oxygen or by impurities. In our manometric adsorption and desorption experiments we could not detect any adsorbed oxygen at temperatures higher than 1120°K.

1) M. Procop and J. Volter, ZI. physik. Chem. (Leipzig) (1972) in press.

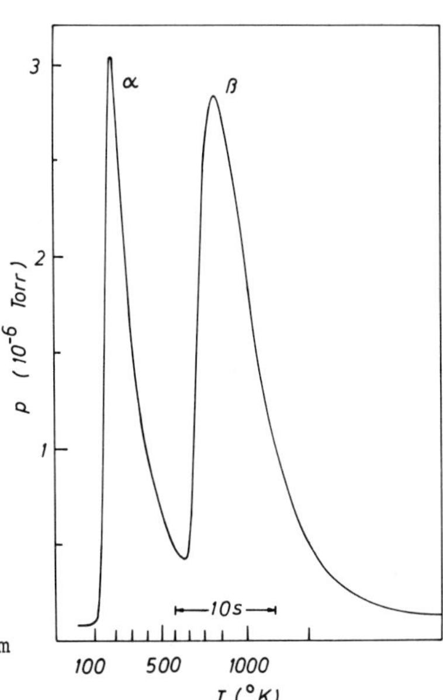

Fig. 1 Flash desorption from Pt foil

V. I. SAVCHENKO

We also consider that the appearance of intensity emitting centers of small sizes on plane (111) (Figs. 2f-g) is connected with diffusion of impurities of alkali-earth metals from the bulk to the surface at high temperatures. This was studied by the Auger spectroscopy on the Pt-foils. (Kin. and Cat. 1972, in press).

R. KLEIN

The appearance of a "pseudo-clean" platinum pattern with a monolayer of oxygen adsorbed on the platinum emitter would indicate that the coverage over all planes including the 111, is essentially equivalent. This does not seem to be in agreement with the results of Weinberg, et al. (preceeding paper) with regard to the low sticking probability observed by him for this plane and the considerably higher value obtained by others for polycrystalline surfaces.

V. I. SAVCHENKO

As can be seen from comparison of Figs. 2b and 2e, a distinct specificity of the emission pattern of Pt during adsorption of oxygen at 78°K is not observed. Possibly this is due to the coverage and sticking probability over all planes including the (111) and (100) being higher than at T = 300°K. On the other hand, the work functions of planes (111) and (100) are higher than the average work function Pt, which was taken equal to 5.32 eV. Therefore, even if the O_2 coverage of these planes is not considerable, we can expect that the emission from these planes would also be negligible, and the general emission should be connected with high-index planes. The approximate estimation of sticking probability for O_2 adsorption at 78°K (Fig. 1, curve 1) gives a value about 10^{-1} - 10^{-2} if one assumes a monolayer capacity of ~ 1.10^{15} atoms.

Paper Number 35

ROLE OF DISLOCATIONS AND FAULTING IN THE CATALYTIC ACTIVITY OF CERTAIN Fe-Cr-Ni ALLOYS

J. M. CRIADO, E. J. HERRERA AND J. M. TRILLO
Department of Inorganic Chemistry, University of Seville, Spain

ABSTRACT: In the present investigation the para-ortho hydrogen conversion reaction was used to study the catalytic activity of two austenitic Fe-Cr-Ni alloys of different stacking-fault energy at various degrees of cold work,previous and after several annealing treatments. The results were correlated with structural studies carried out on these samples by both optical and transmission electron microscopy and hardness measurements.

It has been found that the catalytic activity increases with the degree of cold work. A marked decrease in activity occurs at recrystallization temperature,the minimum activity coinciding with the disappearance of all main lattice defects. When comparing the activity of both alloys at just the same degree of reduction in thickness-50%-,a greater activity is shown by the alloy of lower stacking-fault energy. This fact is consistent with the different work-hardening behaviour of these two alloys.

1. INTRODUCTION

When a metal is plastically deformed by cold working,a small fraction of the mechanical energy remains stored in the material in the form of dislocations and other types of lattice defects,like vacancies,stacking faults and twins. The energy,thus,stored can be released by proper annealing.

Whereas lattice point defects and their chemical consequences are well-known to researchers on heterogeneous catalysis,dislocations and their associated effects have not been till recently systematically explored (1). Little,if any attempt has been made so far to relate catalytic activity to more complex lattice imperfections connected with faulted material.

Studies about the influence of cold working on the activity of metal catalysts have been mainly carried out by Uhara and coworkers (2-7). These authors investigated the activity of a series -Ni,Cu,Pt- generally in the form of wire specimens previously cold worked by twisting. The deformed samples were then annealed at different temperatures before finally being used as catalysts in several heterogeneous systems. They observed that a marked decrease in catalytic activity took place after annealing at a temperature,which according to other sources (8-10),is known to be the recrystallization temperature. The variation of catalytic activity with annealing treatment was related to changes of some physical properties of the specimens-resistivity,EMF,hardness and others-. On the basis of these experiments,the relative influence of vacancies and dislocations on activity was tried to be assessed. No reference, nevertheless, was made to the possible role of internal-surface imperfections of the type of stacking faults, twins and grain boundaries.

It must also be mentioned that in most cases the samples used for catalytic studies and for measurements of the physical property, though supposed to be similar in nature, had a different origin. Since recrystallization temperature and annealing behaviour of metals is remarkably affected by impurity content and distribution, degree, method and rate of coldworking, and previous grain size, it would have been very desirable to make these comparisons with just the same type of specimens.

Another difficulty in that sort of studies arises from the fact that catalyst structural changes, supposed to be responsible for the enhanced or decreased activity, have not been observed directly, but taken for granted. It is also difficult to separate the relative part played by dislocations and other lattice imperfections in the increase of activity. This has been cause of disagreement in the interpretation of results. Thus, Kishimoto (5) has censured the Cratty and Granato's assumption (11) on the type of defect responsible for the loss of catalytic activity of Eckell's cold-worked nickel samples (12), when annealed in the region from 200°to 300°C.

In the present investigation, in addition to hardness tests, both optical and transmission electron microscopy were used to study the defected structure of cold-worked metal catalysts and to follow the disappearance of lattice imperfections, as a consequence of appropriate annealing treatments.

The para-ortho hydrogen conversion reaction was employed for examination of catalytic activity. The reason of this choice being its simple mechanism and the high sensibility of gas-mixture analysis by a proper gauge.

Two high-purity oxidation-resistant Fe-Cr-Ni alloys of different nickel content and of the same original hardness and grain size were used as catalysts. These materials really were carbon-free austenitic stainless steels of different stacking fault energy. Annealing behaviour can best be studied in carbon-free alloys, since the complicating effect due to carbide precipitation in commercial austenitic steels (13) is eliminated. The alloy compositions (wt %) were Fe-20Cr-25Ni (designated Alloy A) and Fe-20Cr-13Ni (designated Alloy B); from published data on related materials, it is expected that the stacking fault energies of the austenite are approximately 30 and 15 erg/cm^2, respectively (14-17).

The catalytic activity as a function of the amount of cold rolling and of annealing temperature of previously deformed samples was studied in the present work. A comparison was also made between the activities of both alloys at just the same degree-50%-of cold rolling.

2. EXPERIMENTAL METHOD

2.1 Material

The alloys were prepared as 200 g. ingots in a non-consumable tungsten-electrode arc furnace, using high purity basis materials(electrolytic iron and chromium and carbonyl nickel).After homogenisation,sheet of 1,2 mm thickness was prepared by cold rolling with appropriate interstage annealing.Samples, protected against oxidation by sealing in silica capsules under 1/3 atmosphere of argon,were heated 1 hour at 1100°C and water-quenched.The protection was a standard operation for all heat treatments.The hardness after quenching was about 104 HV,and the grain size about 8.10^{-2} mm for both alloys.These samples were then cold rolled 25,50,75 and 95%,respectively. Some of the samples,deformed in this way,were isochronally annealed for 1 hour in the range 300-800°C.Previously to catalytic studies,the samples were wet-ground using different grades of silicon carbide abrasive paper and carefully polished.The specimens employed as catalysts were disks of 3 mm. diameter and 0.3 mm thick.

2.2 Metallographic Procedure

Hardness measurements were carried out on polished samples using a Vickers hardness tester with a 1 Kg load.Mean values were obtained from about ten impressions.

The grain size was measured by a lineal intercept method using the number of grain boundaries crossed in a known distance to calculate the average grain diameter.The grain size reported in mm is the average of 40 readings.

For optical microscopy examination,the polished specimens were,thereafter,electropolished and finally etched electrolytically at 6 V for 5 seconds in a solution of 20% HCl and 80% of 10% aqueous CrO_3 solution.

Samples for transmission electron microscopy were prepared from 3 mm diameter disks.The specimens were first thinned down to about 0.15 mm thick by mechanical fine-grinding.Thin-film regions for electron microscopy were then obtained by jet -polishing both sides of the 3 mm disks with a 20% hydrocloric acid solution at 80 V with a current of 0.1 A,followed by electrolytic polishing in a solution containing 54% orthophosphoric acid,36% sulfuric acid,and 10% ethyl alcohol at 6 V.The foils were examined in an A.E.I. EM 6G microscope operating at 100 KV.

2.3 Catalytic Activity

The apparatus for the study of the para-hydrogen conversion was essentially that used in an earlier work (19).A cylindrical vessel containing the catalyst was connected via liquid-nitrogen trap and mercury cutoff to the usual vacuum line and gas storage bulbs.The reactin system could be pumped out to 10^{-6} mm Hg or better.The para-ortho hydrogen mixtures were analyzed by a micro-Pirani gauge.

3. RESULTS AND DISCUSSION

The time course of para-hydrogen conversion followed the usual first order law

$$k_e = (1/t)\ \ln(C_o/C_t)\ \text{min}^{-1}$$

where C_o (C_t) is the excess concentration of species at time zero (t) over the equilibrium concentration at the experimental temperature.

In the experimental range of temperature 200-400°C,in which activity was measured,k_e was pressure independent.The simplest explanation is that the conversion reaction proceeds via a recombination of hydrogen atoms in a chemisorbed layer (Bonhoefer-Farkas´mechanism) (19).

The geometric area of the catalysts,in the form of 3 mm disks,was made used of for referring the rate constant to unit area of catalyst surface.The difference in values of rate constant were large enough to neglect the variation of rugosity,taking into account the method of sample preparation.

The effect of outgassing time of samples at 400°C and 10^{-6} mm Hg on their activity for the para-hydrogen conversion at 200°C was examined.At a lower outgassing temperature reproducible results could not be obtained.All activity measurements were,therefore,carried out after sample outgassing at 400°C for eight hours.This temperature is higher than the one employed by Uhara et al. (6) for degassing of nickel catalysts -150°C and 10^{-5} mm Hg-.The reason for the choice of that long outgassing time can be deduced upon examination of Fig.1. In this figure, data concerning the activity at 200 °C of -

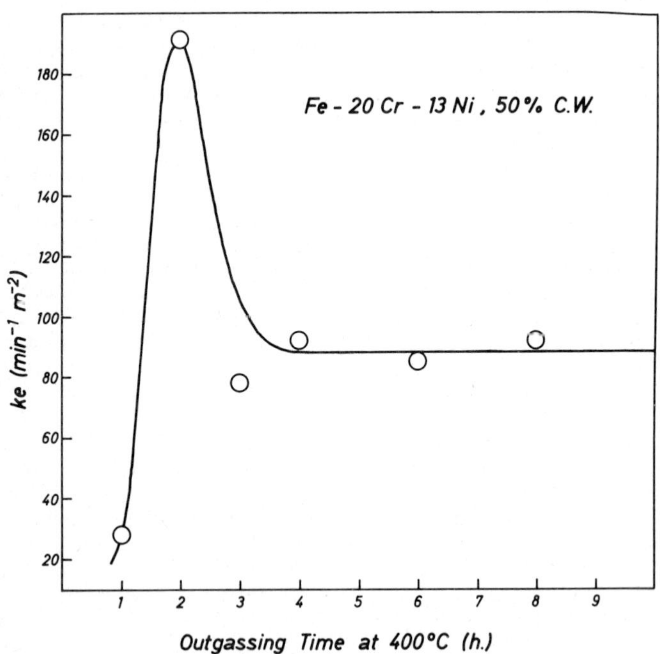

Fig.1 Specific activity of alloy B,50% cold rolled,for the parahydrogen conversion after outgassing for t hours at 400°C,measured at 200°C.

50 % cold-rolled samples of alloy B as a function of outgassing time at 400 °C are plotted;it is observed that the ac-

tivity stabilizes after six to eight hours outgassing treatment. The maximun of activity in figure 1 appeared in all studied samples before reaching reproducibility of results. It may be interpreted by the occurrence of two simultaneous processes:catalyst surface cleaning and a partial recovery of the material.

It is today well known that when a metal is cold worked, a small fraction of the mechanical energy spent in the process remains stored in the material in the form of lattice defects, mainly dislocations. The most important result of cold working is strengthening of the metal. With increasing amounts of deformation the resistance of the metal to further deformation constantly increases-strain or work hardening-,but other changes in both its physical and chemical properties also occur. The strain energy stored in the lattice renders it thermodinamically unstable relative to the undeformed one. Consequently,the cold-worked material will try to return to a state of lower energy,i.e. a more perfect state. In general,this return to a more equilibrium structure cannot occur spontaneously,but has to be thermally activated so as to allow sufficient atomic mobility.

The removal of the cold-worked condition occurs by a combination of three processes:recovery,recrystallization and grain growth -the recrystallization process being the most important one-. The most significant changes in the structure-sensitive properties- for instance,hardness-occur during the recystallization stage. In very general terms it can be defined as the nucleation and growth of strain-free grains out of the matrix of cold- worked metal. Recrystallization is preceded by some degree of recovery.

It must be remembered that the structure of a deformed metal consists,mainly,of dense dislocation networks. The recovery stage is chiefly concerned with the annealing out of vacancies and the re-arrangement of these dislocations to reduce the lattice energy. Recovery is accompanied by only a slight,if any,lowering of hardness. One of the most important recovery processes which leads to a re-arrangement of the dislocations is polygonization. This is the mechanism whereby dislocations all of one sign align themselves into walls to form small-angle,polygonization,or sub-grain boundaries. The basic movement during polygonization is a climb process whereby the edge dislocations change their arrangement from a horizontal to a vertical grouping. It is,therefore,a process which involves the migration of vacancies to or from the edge of the half-planes of the dislocations. In addition to polygonization another basic mechanism of recovery is the running together and mutual annihilation of two dislocations of opposite sign. When the two dislocations are on the same slip plane,the activation energy for annihilation is,generally,small.

Electron microscopical examination of samples cold-rolled 50% showed no obvious structural differences between alloys A and B,both showing a tangled dislocations structure, containing banded features -referred to as "deformation bands" -assumed to be deformation twins and associations of stacking fault and twin. The deformation structure was still maintained after annealing for 1 hour at 300°C. Figure 2.a shows an as-

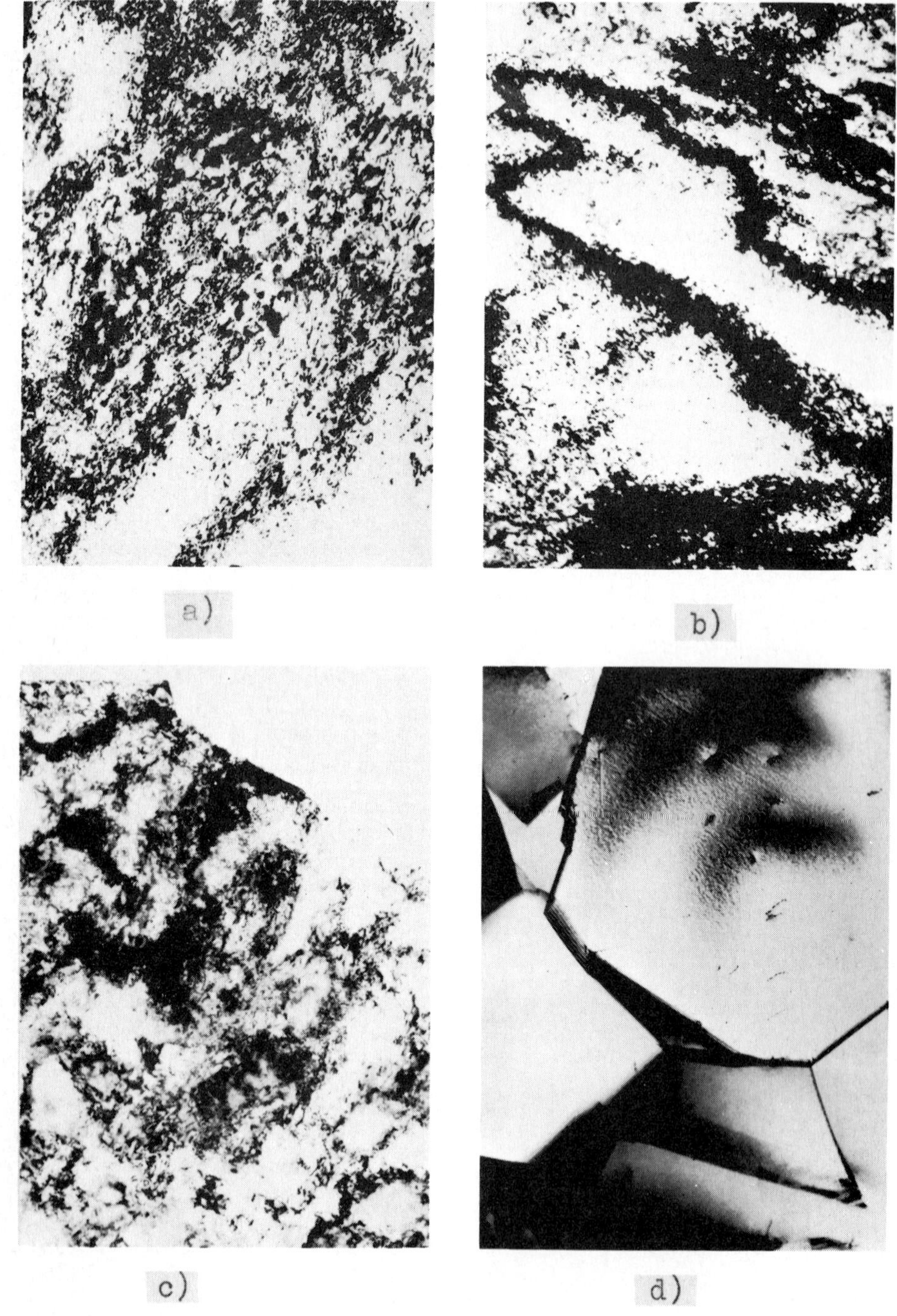
a)
b)
c)
d)

pect of the 50% cold-worked structure of alloy B after this annealing treatment;a high density of dislocations can be observed. Annealing at 500°C only gave rise to some dislocation re-arrangement,but complete recovery did not take place. Figure 2.b represents a sample of alloy A annealed for 1 hour at 500°C,and shows a high density of tangled dislocations,and although some dislocation rearrangement has occurred no subgrain structure is formed.

The maximun of activity of Fig.1.above mentioned,really suggests that simultaneously to surface cleaning a partial recovery of the material is taking place. No complete recovery occurs at 400°C,but removal of vacancies and a certain dislocation rearrangement and annihilation can be expected.

Figure 3 shows the changes of catalytic activity,measured at 400°C,of alloy A 50% cold worked after 1 hour annealing

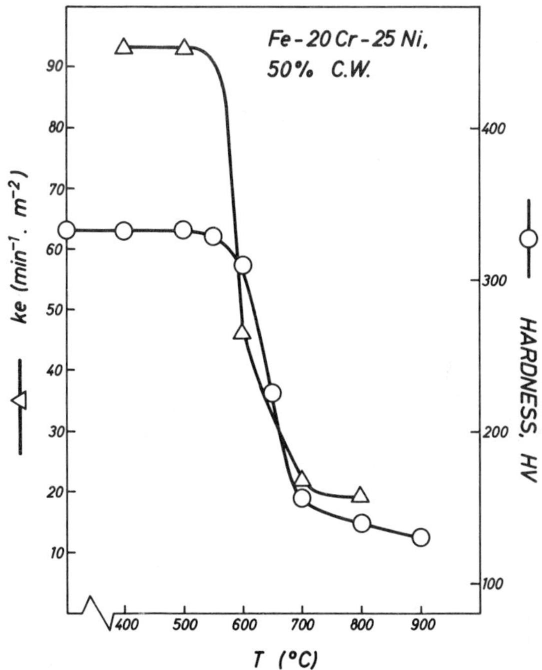

Fig.3 Catalytic activity at 400°C (Δ) and hardness variation (O) versus annealing temperature for alloy A after 50% cold rolling.

Fig.2 Transmission electron microscopy micrographs after annealing 50% cold-rolled samples of alloys A and B for 1 h at various temperatures.
a) Alloy B annealed at 300°C.
b) Alloy A annealed at 500°C.
c) Alloy B annealed at 600°C.
d) Alloy A annealed at 800°C.

treatment at various temperatures. A sudden decrease in activity is observed between 500 and 600°C. In the same figure, the change of cold-worked hardness with annealing temperature can also be followed. The hardness begins to decrease at about 600°C, too, and both the catalytic activity and the softening curves have the same change course. Optical microscopy observations showed that recrystallyzation commences at this temperature and that at 800°C the material is completely recrystallized. This was confirmed by electron microscopy. Thin-film examination of samples of both alloys annealed after 50% cold work did not provide evidence of the formation of a distinct sub-grain structure throughout the material, but recovery and recrystallization were observed to occur as overlapping and competing processes. For examp-, in Fig.2.c corresponding to the structure of alloy B after annealing for 1 hour at 600°C, some recovered areas co-exist with a deformed matrix and a clear recrystallized grain. Both optical and electron microscopy showed that deformation bands were favoured sites for recrystallization and that at 800°C the two alloys were fully recrystallized. Figure 2.d shows the recrystallization of a sample of alloy A annealed for 1 hour at 800°C. The concurrent change and course of catalytic activity and hardness in Fig.3, and the microscopical observations may lead to the conclusion that this decrease in activity is due to the disappearance of dislocations and internal-surface defects.

The variation of catalytic activity, measured at 200°C, of alloy A samples as a function of the amount of cold work is shown in Fig.4. On passing from specimens cold rolled 25 to 50 and to 75%, an increase of the activity is observed. The low value of catalytic activity shown by the 95% cold-rolled sample is explained by the fact that activity decreases upon 400°C treatment end even at 300°C in this case.

The enhanced activity of samples of alloy A with the degree of cold rolling is consistent with the expected higher lattice strain energy. This leads to a greater cold-work hardness. In Table I, the increased values of hardness as a result of the greater reduction in thickness are shown. It is interesting to observe that after 50% cold-rolling alloy B has a hardness of 371 HV against 327 for alloy A.

Samples of alloy A, deformed at different degre, were isochronally annealed for 1 hour at various temperatures and the catalytic activity, measured at 200°C, was studied. In Fig.4 it is observed that the temperature at which the activity starts to decrease is lower as the deformation degree is higher. These

TABLE I

Vickers hardness of samples of alloys A and B at various degree of cold work

Degree of C.W., %	25	50	75	90
Alloy A, HV	260	327	373	407
Alloy B, HV	..	371	..	..

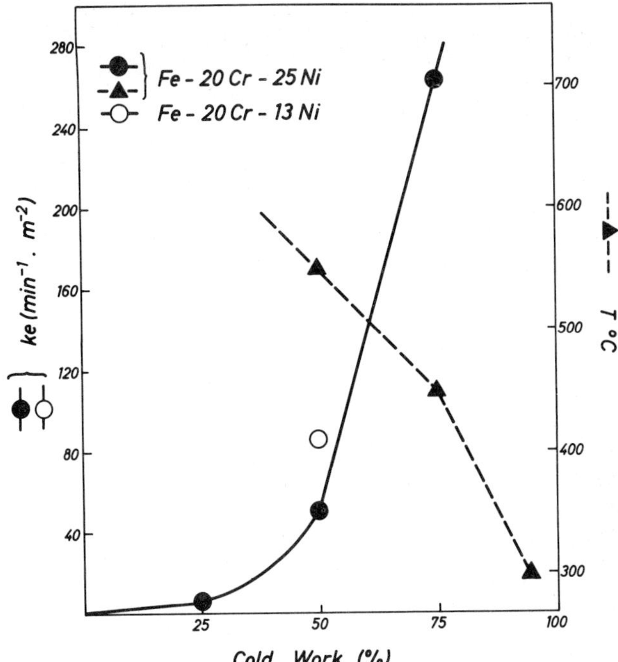

Fig.4 Ortho para conversion activity (Alloys A,●,and B,O) and initial temperature of activity decrease (Alloy A,▲) vs.degree of cold work.

phenomena have an easy explanation,since it is well known to metallurgists that the rate of recrystallization depends very deeply on the amount of prior deformation (the greater the degree of cold work the lower the recrystallization temperature)

The catalytic activity of alloy B 50% cold rolled is also plotted in Fig.4. It is observed that its value is greater than that one pertaining to alloy A at the same degree of reduction in thickness. This result is consistent with the higher hardness of alloy B,as shown in Table I,as a consequence of the lower stacking-fault energy of this alloy,as compared with alloy A. Low values of stacking-fault energy give rise to wide stacking-fault ribbons (a "ribbon" being a sheet of stacking fault between the partial dislocations). The width of the stacking-fault ribbon is of importance in many aspects of plasticity. For instance,in some stage of deformation,dislocations must intersect each other and the difficulty in doing so is a source of work-hardening. The width of the ribbon is also important to cross-slip in which a dislocation changes from one slip plane to another intersecting slip plane. If the dislocations is extended,the partials must first associate and remove the stacking fault to form an unextended dislocation before cross-slip takes place. In materials of low SFE,cross--slip is,therefore,difficult at reasonable stress levels and the screw dislocations,thus,cannot escape from their slip plane and dislocation density is high.

The apparent lack of difference between the 50% cold-worked structure of alloys A and B,when observed by electron microscopy is due to the high density of lattice imperfections present in both materials. Examination of low deformed samples -2.5 and 5%,respectively- showed the expected difference in structure.

From the previous results and discussions,the general conclusion may be drawn that dislocations act as the main active sites for the studied reaction,but other lattice defects, like vacancies,stacking faults and twins also play a role.

REFERENCES

1) J.M. Thomas.Adv.in Catalysis 19 (1969),293.
2) I. Uhara,S. Yanagimoto,K. Tani and G. Adachi,Natur 192, (1961),867.
3) I. Uhara,T. Hikino,Y. Numata,H. Hamada and Y. Kageyama, 66 (1962),1374.
4) I. Uhara,S. Yanagimoto,K. Tani,G. Adachi and S. Teratani, J.Phys.Chem. 66 (1962),2691.
5) S. Kishimoto,J.Phys. Chem.661(1962),2694.
6) I. Uhara,T. Hikino,Y. Numata,H. Hamada and Y. Kageyama,
7) S. Kishimoto,J.Phys.Chem. 67 (1963),1161.
8) L.M. Clarebrough.M.E. Hargreaves and G.H. West,Proc.Roy. Soc.A232 (1955),252.
9) L.M. Clarebrough,M.E. Hargreaves and G.W. West,Phil.Mag. 8 (1956),1528.
10) L.M. Clarebrough,M.E. Hargreaves,M.H. Loretto and G.W.West Acta Met. 8 (1960),797.
11) L.E. Cratty and A.V. Granato,J.Chem.Phys.26(1957),96.
12) J. Eckell,Z.Electrochem.39 (1933),433.
13) V. Ramaswamy and D.R.F. West,J.I.S.I. 208 (1970),395.
14) D.L. Douglas and G. Thomas,Corrosion,20 (1966),1,15t.
15) R. Sumerling and J. Nutting,J.I.S.I. 203 (1965),398.
16) M.A.P. Dewey,G. Sumner and I.S. Brammer,J.I.S.I.203(1965), 938.
17) D. Dulieu and J. Nutting,"Metallurgical Develpments in High Alloy Steels",ISI Special Report 86 (1964),140.
18) G.J.K. Acres,D.D. Eley and J.M. Trillo,J.of Catalysis 4 (1965),12.
19) D.D. Eley and D. Shooter,J.of Catalysis 2(1963),259.

ACKNOWLEDGMENTS

The authors wish to thank Prof.F.Gonzalez García for his interest in this investigation . The collaboration of Dr.D.R. F. West of Imperial College,London,during the metallographic work is also gratefully acknowledged.

DISCUSSION

G.-M. SCHWAB

In recent work Dr. R. Schmidt and I have tried to examine the effect of plastic deformation on catalytic activity by a new method. While all previous workers have studied the catalyst before and after cold working and have it exposed to air in between, we have carried out the deformation (torsion of sheets and mainly wires) in situ during catalysis and have compared activity and activation energy of the formic acid dehydrogenation before and after deformation. The catalysts used in our case were pure metals: copper, silver, iron, nickel, molybdenum. In contradiction to the work of the present authors on austenitic alloys, the effects found were small, especially in the activity, whereas the activation energies showed systematic trends, depending on the metal used. A compensation effect was always observed. No systematic relation to the position in the Periodic table or to lattice structure could be stated.

J. M. TRILLO

The authors are aware of the big problem of maintaining a really clean metal surface exposed to air. One of the reasons of choosing the austenitic alloys was their high oxidation resistance. But the problem of oxidation of metal samples also subsists during the catalytic reaction, due to the possible interaction of the surface with the reactant or with its decomposition products. Precisely, in one of the systems investigated by Professor Schwab --the decomposition of formic acid on nickel--the oxidation of nickel at certain temperatures by formic acid decomposition products has been stated.[1)]

The small changes in activity found by Schmidt and Schwab in the study of their deformed samples could be related with a chemical attack of the specimen, which would mask other effects.

Concerning the last point of the commentor, we would like to mention that the cold-work behavior of metals is remarkably affected by impurity content and distribution, grain size, degree, method and rate of cold deformation, and difference in stacking-fault energy.

1) R. E. Eischens and W. A. Pliskin, Actes due Ceuxième Congrès International de Catalyse, Vol. 1, Editions Technip, Paris (1961), p. 789.

D. D. ELEY

I have made some rough calculations from Reference 19 and find if we assume a roughness factor of 4, an Fe film at 200°C has a K_e of 700 $min^{-1}m^{-2}$, and with Ni and Cr films probably somewhat less. These values are of the same order as found by Professor Trillo. I think both Trillo's sheets and our films may have some carbon in the surface, but possibly not much since Ni films from Ni wires cleaned from carbon have similar absolute rates of 293K.[1)] The problem I find here is how to eliminate the possibility that cold rolling induces orientation of crystallites and exposure of specially active lattice planes, which are lost on annealing, i.e., the extra activity is associated with a special lattice spacing in a normal plane, rather than a dislocation.[2)]

1) D. D. Eley and P. R. Norton, Disc. Faraday Soc. 41 (1966) 135.
2) D. D. Eley and D. M. MacMahon, J. Catalysis 14 (1969) 193.

J. M. TRILLO

We do not disregard the possibility that the samples could have some small amounts of carbon on their surface.

The point raised by Professor Eley concerning cold-rolling texture is a very interesting one. In fact, during the deformation of a polycrystal, the rotations of the individual grains lead to a preferred orientation that develops gradually with strain. The exposure of specially active lattice planes could contribute to the observed extra activity. But, when a heavily deformed metal is annealed, the grains of the recrystallized metal may not develop a random orientation. After recrystallization, metals usually exhibit annealing textures that are either similar to their deformation texture or quite different. Therefore, on the same grounds, the special lattice spacing could appear upon recrystallization.

The parallel change course of catalytic activity and hardness curves on annealing temperature (Fig. 3), as a consequence of the decrease of dislocations and other lattice defects, proved by electron microscopy observations (Fig. 2), strongly supports our interpretation of the phenomena.

C. G. HARKINS

It is well known that the solubility of hydrogen in iron-based alloys is enhanced by stress, such as residual stresses introduced by cold working. Furthermore, the mechanical or electropolishing of metals in contact with hydrogen-containing liquids is an efficient method of introducing hydrogen into the metal phase. Thirdly, the introduction of interstitial hydrogen would tend to enhance the brittleness or hardness measurements observed.

The pressure of interstitial hydrogen at the catalyst surface as a reactant in the ortho-para-hydrogen conversion is a simple and apparently sufficient explanation of the preprint results. Figure 3 showing the decrease of activity above 600°C corresponds also to the temperatures necessary for analytical determination of hydrogen by vacuum extraction.

Did you perform hydrogen analyses of your sample to show that interstitial hydrogen was not a factor, especially to show that mechanical polishing and electropolishing in hydrogen containing liquids did not introduce more hydrogen into cold-working metals?

J. M. TRILLO

It is really well known that hydrogen enters ferrous metals causing hydrogen embrittlement (loss of ductility) and even, in some cases, spontaneous cracking. But, it is also well known that cold rolling enhances metal hardness, mainly due to the fact that multiplication of dislocations during plastic deformation increases the stress necessary for dislocation motion.

The explanation of the results suggested by Dr. Harkins does not apply in any way to our case, since the samples were neither electropolished nor washed in any liquid which liberates hydrogen.

J. W. HALL

It has been shown that rapid cooling causes movement and multiplication of dislocations. Were the samples used for the electronmicrographs in Figure 2 the same as those used in the catalytic activity tests?

My work, reported in 1964, shows a linear relation between dislocation density and catalytic activity for ionic crystals.

We have recently measured the effect of interatomic spacing on the surface by growing silver epitaxially. As the interatomic spacing increased, the catalytic activity increased. This confirms the correlation proposed by Beeck.

J. M. TRILLO

The specimens employed as catalysts and for electron microscopy observations were 3 mm. disks cut from the same sheet of material, prepared as described in section 2.1 of the paper, and submitted to just the same thermal cycle. We do not believe that the dislocation density and distribution of our samples have been significantly altered by cooling. For instance, the recrystallized sample cooled from the highest temperature (Fig. 2-d) shows practically dislocation-free grains.

Paper Number 36

EFFECTS OF RADIATION DAMAGE ON THE CATALYTIC ACTIVITIES OF LITHIUM-DOPED COPPER, NICKEL, AND COPPER-NICKEL

T. TAKEUCHI, D. MIYATANI, K. OKAMOTO, Y. TAKADA and O. TAKAYASU
Faculty of Literature and Science, Toyama University,
Toyama, Japan

ABSTRACT: The effects of radiation damage caused by the nuclear reaction of $^6Li(n,\alpha)^3H$ on the catalytic hydrogenation of ethylene were studied, using two types of copper, nickel, and copper-nickel(1:1). The powdered lithium-doped catalysts were used for the test of the catalytic activity and the sheets of the same metals covered with lithium were used for the autoradiographical observation of the behavior of the tritium in the course of the hydrogenation reaction.

The irradiation of neutrons increased about 3 times in the catalytic activities of the nickel and the copper-nickel catalysts, but did not in the activity of the copper catalyst. The autoradiographs given by the nickel and the copper-nickel indicated that the tritium is concentrated in the grain boundary and this tritium is highly active for the hydrogenation reaction. The autoradiograph of the copper indicated that the tritium was trapped in the depleted zone of the metal in the state of molecule.

1. INTRODUCTION

The effects of lattice imperfections on the catalytic activity of metals have been studied mainly by means of cold working or by the quenching of metals.[1)2)] This study is concerned with the effects of radiation damage on the catalytic hydrogenation reaction. Copper, nickel and copper-nickel, which have previously been doped with lithium and irradiated with thermal neutrons were employed for this purpose. The irradiation of those metals with thermal neutrons causes the nuclear reaction of $^6Li(n,\alpha)^3H$. The energy of ^3H and that of the ^4He produced as the fission products are 2.7 and 2.1 MeV respectively, and the threshold energy of copper and that of nickel are ca. 24 eV. Therefore, a large amount of fresh lattice imperfections can be expected to be produced by the impacts of these fission products with metal atoms of the catalyst. A comparison of the change in catalytic activities upon the irradiation by neutrons was made. In addition, the behavior of the tritium produced by the irradiation in the course of the hydrogenation was observed by means of autoradiography.

2. EXPERIMENTAL

Two types of catalysts were used in this study. Powdered copper, nickel and copper-nickel were used for the study of the catalytic activity, and sheets of the same metals were used for the study by means of autoradiography.

The powdered catalysts were prepared from mixed solutions of the nitrates of each metal and lithium nitrate; each mixture was gently dried on a water bath and finally calcinated at 1000°C in air. The atomic ratio of copper and nickel in the alloy catalyst was 1:1. These nitrates were extra pure grade and were used without further purification. 0.1 gram of the oxide prepared from the mixture of nitrates was placed in a quartz ampul (1cc in volume) and then reduced with hydrogen at 350°C. After the reduction, the catalyst was evacuated under a pressure of 10^{-6} Torr. The ampul was sealded after the evacuation and was then subjected to the neutron irradiation. The irradiation was performed in the atomic pile, JRR-2 of the Japan Atomic Energy Research Institute. A liquid-nitrogen-temperature loop was installed in this atomic pile in order to cool the sample during the irradiation. The radiation dose of the neutrons was 10^{19}nvt.

The catalytic activity test was carried out after two months, in the meanwhile, the induced radioactivity of the catalyst had decayed below the prescribed safe limit. The catalyst was kept this period in a liquid nitrogen bath. However, it was exposed at room temperature when the catalyst was placed in the reaction vessel. The catalytic activity tests were carried out by means of the hydrogenation reaction of ethylene. The ampule of the catalyst was crushed by the aid of a magnet after the reaction vessel had been evacuated to 10^{-6} Torr. A mixture of ethylene and hydrogen (1:1) at a pressure of 20 Torr was introduced. The rate of the reaction was measured by observing the decrease in the pressure.

In the experiments by means of autoradiography, sheets of copper, nickel and copper-nickel (1:1) 4.5 mm wide, 18 mm in length, and 0.4 mm in thickness were used. The purity of these sheets was 99.99%. They were heated to 1100°C, then gradually cooled, polished with fine emery paper, and then subjected to electrolytic polishing. The sheet thus treated was washed with distilled water and then placed in the B portion of the quartz vessel of Fig. 1 after the sheet had been dried. 0.5 grams of normal lithium was placed in the A portion of the vessel, after which the vessel was evacuated to 10^{-6} Torr; then the A portion was heated in order to deposit the lithium on the sheet. The sheet was transferred by the aid of a magnet to the C portion, which was then sealed by fusing and disconnected in order to subject it to irradiation. The radiation dose was the same as in the case of the powdered catalysts.

The electrolytic polishing of the sheets was carried out at 60°C with a current density of 2.5 A/cm^2, employing the sheets as cathodes in the usual solution which had been prepared by mixing 50 cc of phosphoric acid, 1 gram of agaragar, and 0.5 grams of sodium hydroxide.

The assay of the tritium produced in the powdered catalysts by the nuclear reaction was carried out by means of a gas proportional counter. The radioactivity of metal sheets after the irradiation was measured by means of a 2π counter, using methane as the filling gas.

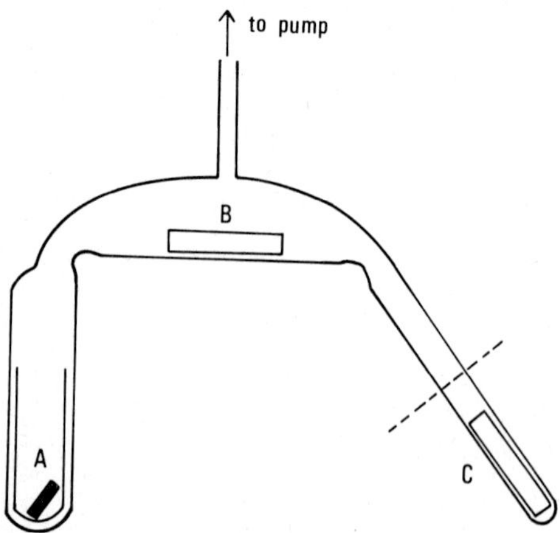

Fig. 1 The vessel for the preparation of the sheet covered with lithium.

The autoradiograph was obtained by the stripping method using Fuji-ET-2F, which was developed by the FD-111 method. Prior to the autoradiographical procedure, a quartz ampule containing irradiated sheet was broken in running water in order to remove the lithium on the surface; the sheet was taken out of water, dried by means of an electric fan, and then subjected to study. The contact time of the stripping film was ca. 20 hours.

3. RESULTS

The hydrogenation reaction of ethylene on each powdered catalyst was expressed by first-order kinetics. Fig. 2 shows the doping effects of lithium on the specific activity and the BET area of the powdered nickel catalysts. This figure indicates that the addition of lithium up to 2% markedly decreased both the activity and the BET area. Fig. 3 shows the Arrhenius plots of the hydrogenation reaction on 4.3%-lithium-doped nickel and on pure nickel. Both catalysts were irradiated with neutrons. The aim of the study on pure nickel was to distinguish the effect of damage caused by the nuclear fission of ^{6}Li and by the neutrons, because ca. 1/100 of the total amount of neutrons radiated in this study were fast neutrons. Fig. 4 shows the Arrhenius plots of the hydrogenation reaction on 4.3%-lithium-doped copper-nickel and those on non-lithiun-doped copper-nickel catalysts. Figs. 3 and 4 reveal that the catalytic activity increased upon the neutron irradiation and that the increase in the activity of lithium-doped catalysts is greater than that in the case of non-lithium-doped catalyst. These results are summarized together with the activation energies of the reaction in Table 1.

It is particular interest that no radiation effect was found in the catalytic activity of lithium-doped copper. The temperatures of the hydrogenation reactions on this catalyst were 19° and 26.8°C. The decrease in the pressure due to the reaction was not observed even after 18 hours, as has been recognized in the catalytic hydrogenation reaction on copper which was not irradiated.

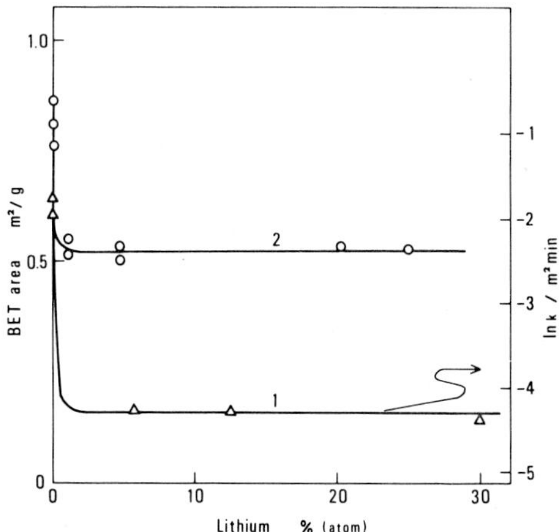

Fig. 2 Doping effects of lithium on specific activity and BET area.
1 specific activity at 0°C, 2 BET area.

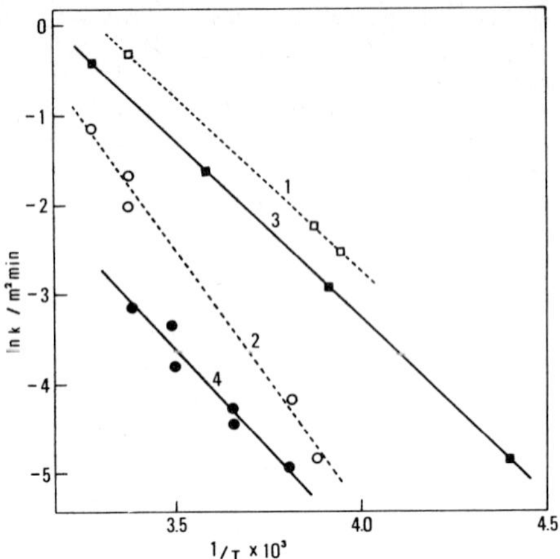

Fig. 3 Arrhenius plots for the hydrogenation reaction on nickel.
1 Ni irradiated, 2 Ni-Li irradiated, 3 Ni unirradiated,
4 Ni-Li unirradiated.

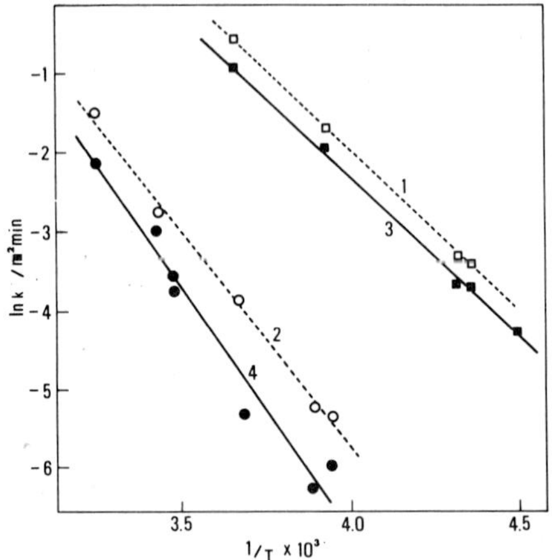

Fig. 4 Arrhenius prots for the hydrogenation reaction on copper-nickel.
1 Cu-Ni irradiated, 2 Cu-Ni-Li irradiated,
3 Cu-Ni unirradiated, 4 Cu-Ni-Li unirradiated.

Fig. 5 shows the dependency of the temperature of the treatment of lithium-doped catalysts upon the amount of tritium taken out by the contact with normal hydrogen, 20 Torr for 14 hours. The dotted line in the figure indicates the amount of tritium obtained theoretically by means of the following equation:

$$(^3H) = W/M\lambda \cdot N\phi\delta_{act}(1 - e^{-\lambda t})e^{-\lambda T}$$

where W is the amount (gr) of ^{6}Li in the sample. M is the atomic weight of ^{6}Li, N is the Avogadro number, and ϕ is the dose of irradiation (cm^{-2}, sec^{-1}). δ_{act} is a cross section of the neutron (barn). λ is the disintegration constant, and t is the irradiation time. T is the cooling time.

Fig. 5 indicates that ca. 50% of the total tritium was taken out from nickel and copper-nickel at 150°C, but that only 30% was taken out from copper at 300°C.

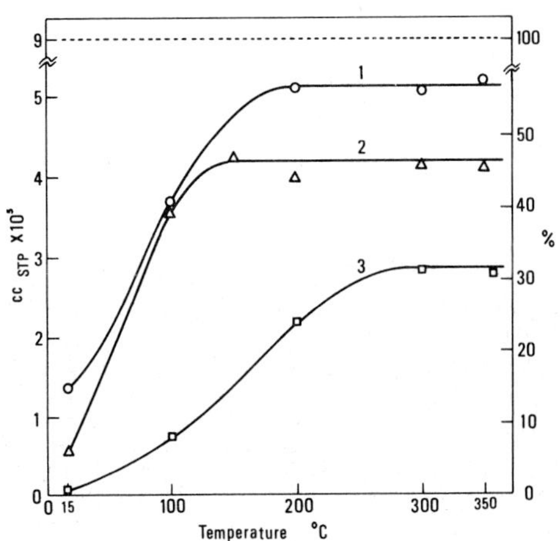

Fig. 5 Dependency of the temperature of the contact with hydrogen on the amount of tritium taken out.
1 Ni-Li, 2 Cu-Ni-Li, 3 Cu-Li

TABLE 1
Activation energies and ratios of the activity of irradiated catalyst to that of unirradiated catalyst

Catalyst	Activation energy K cal/mol	Ratio of activity k_{irr}/k_{unirr}
Ni irr.	8.0	2
Ni unirr.	7.8	
Ni-Li irr.	11.2	3 - 4
Ni-Li unirr.	9.2	
Cu-Ni irr.	8.3	1.5
Cu-Ni unirr.	8.3	
Cu-Ni-Li irr.	12.3	3
Cu-Ni-Li unirr.	12.0	
Cu unirr.	-	
Cu-Li irr.	-	
Cu-Li unirr.	-	

Fig. 6 shows the autoradiograph obtained from the nickel sheet which was covered with lithium and irradiated with neutrons. This figure is composed of numerous spots and indicates a somewhat stronger radioactivity of the grain boundary than that of the inside of the grain, suggesting that the tritium was concentrated in the grain boundary. Fig. 7 shows the autoradiograph of a nickel sheet with no lithium deposit taken after neutron irradiation. This figure indicates that the radioactivity of the grain itself is much stronger than the grain boundary. The radioactivity of this sheet is probably caused by ^{65}Ni produced by the nuclear reaction.

Suspecting the presence of lithium on the sheets, even though they had been washed very well with water, its surface was polished once more with emery paper. The amount of shaved metal correspond to 16 - 20μ, ca. 10^4 atoms layers. The sheets thus shaved were divided into two pieces, and only one half of it was placed in contact with 300 Torr of a mixture of hydrogen and ethylene (1:1) at 20°C for 65 hours in a vessel with a volume of 150 cc.

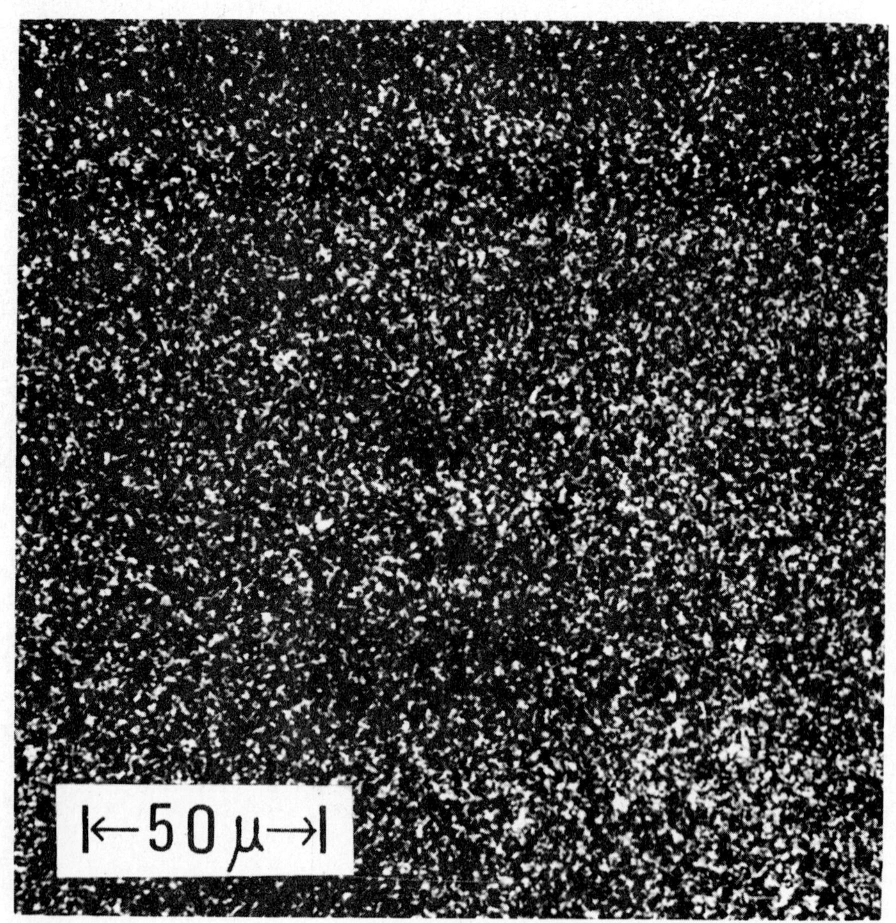

Fig. 6 Autoradiograph of Ni.

Fig. 7 Autoradiograph of Ni with no Li deposit.

Then the gas was evacuated and the autoradiograph was taken. The sensitivity of the half used for the hydrogenation reaction was much lower than that of the half which was not used for the hydrogenation reaction in the cases of both nickel and copper-nickel catalysts. However, the autoradiographs of these sheets were masked by the streaks left by the polishing by the emery paper; therefore, these sheets were again subjected to the electrolytic polishing, and then their autoradiographs were taken and compared with each other. The results of the autoradiograph on nickel are shown in Fig. 8, together with a photograph of the surface of the metal observed by means of a metallurgical microscope. The autoradiograph of one half is just the reverse of the other. That is, the grain boundary of the sheet which was not used for the hydrogenation reaction shows a much stronger sensitivity than the inside of the grain. On the other hand, the other half, which was used for the reaction, gave just the opposite autoradiograph.

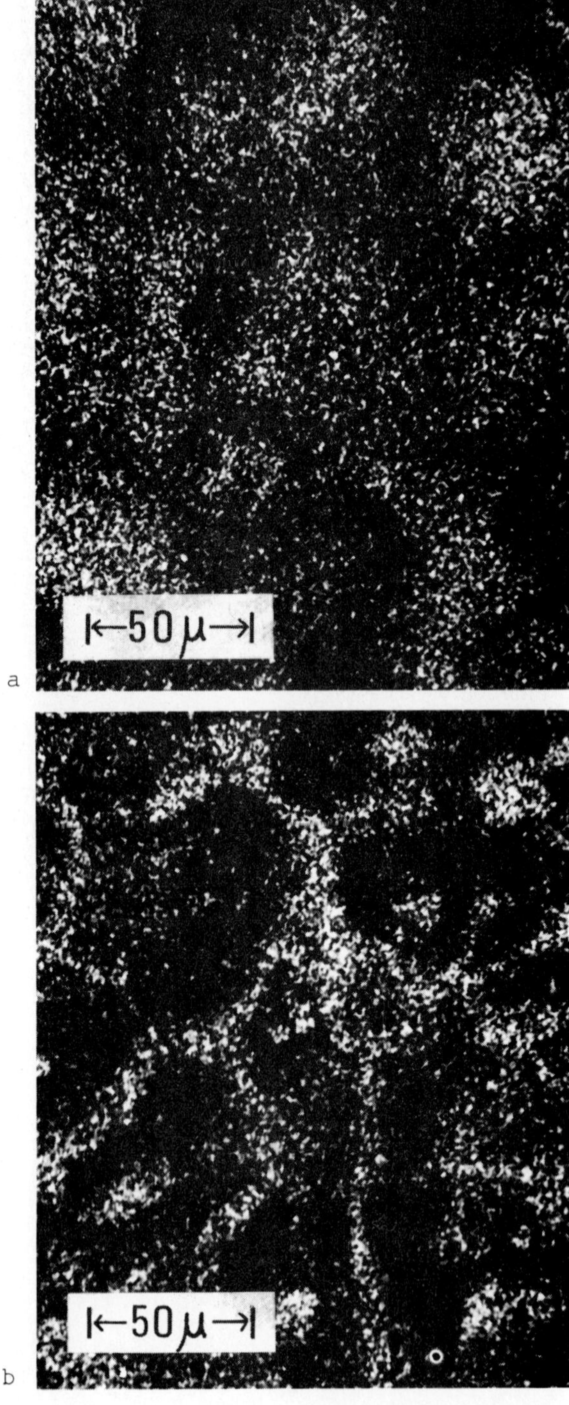
|←50μ→|
a
|←50μ→|
b

c

Fig. 8 Autoradiographs before and after the catalytic reaction, and the metallurgical micrograph of Ni.
a: Autoradiograph of Ni before the catalytic reaction.
b: Autoradiograph of Ni after the catalytic reaction.
c: Metallurgical micrograph of Ni.

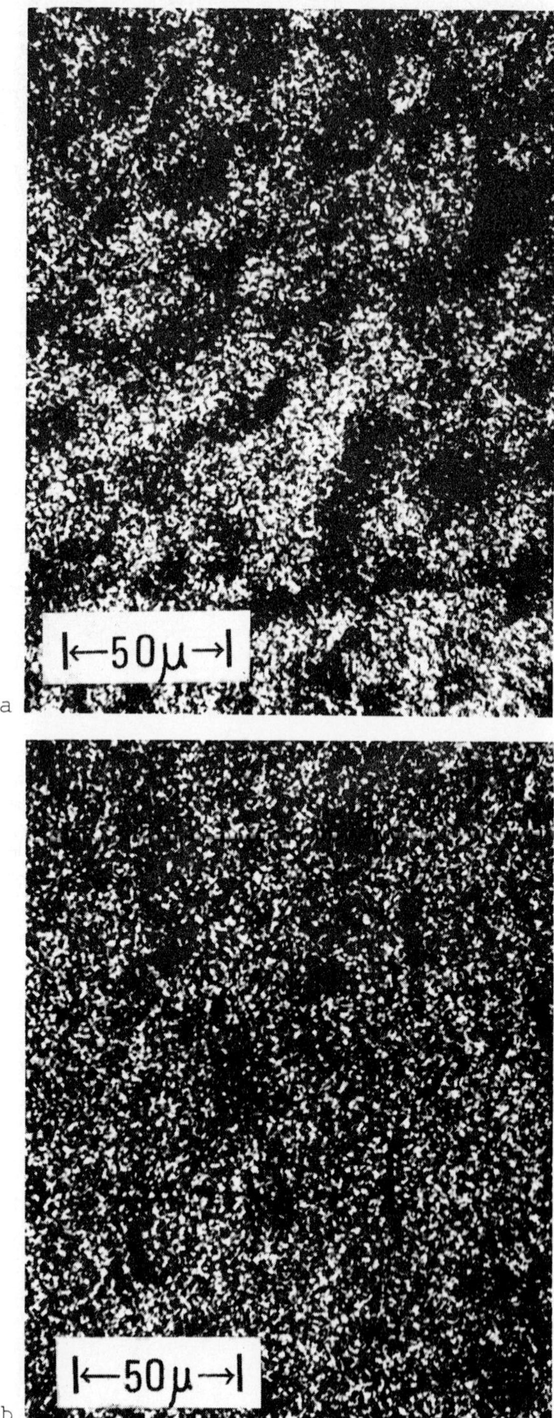
|←50μ→|
a
|←50μ→|
b

c

Fig. 9 Autoradiographs before and after the catalytic reaction, and the metallurgical micrograph of Cu-Ni.
a: Autoradiograph of Cu-Ni before the catalytic reaction.
b: Autoradiograph of Cu-Ni after the catalytic reaction.
c: Metallurgical micrograph of Cu-Ni.

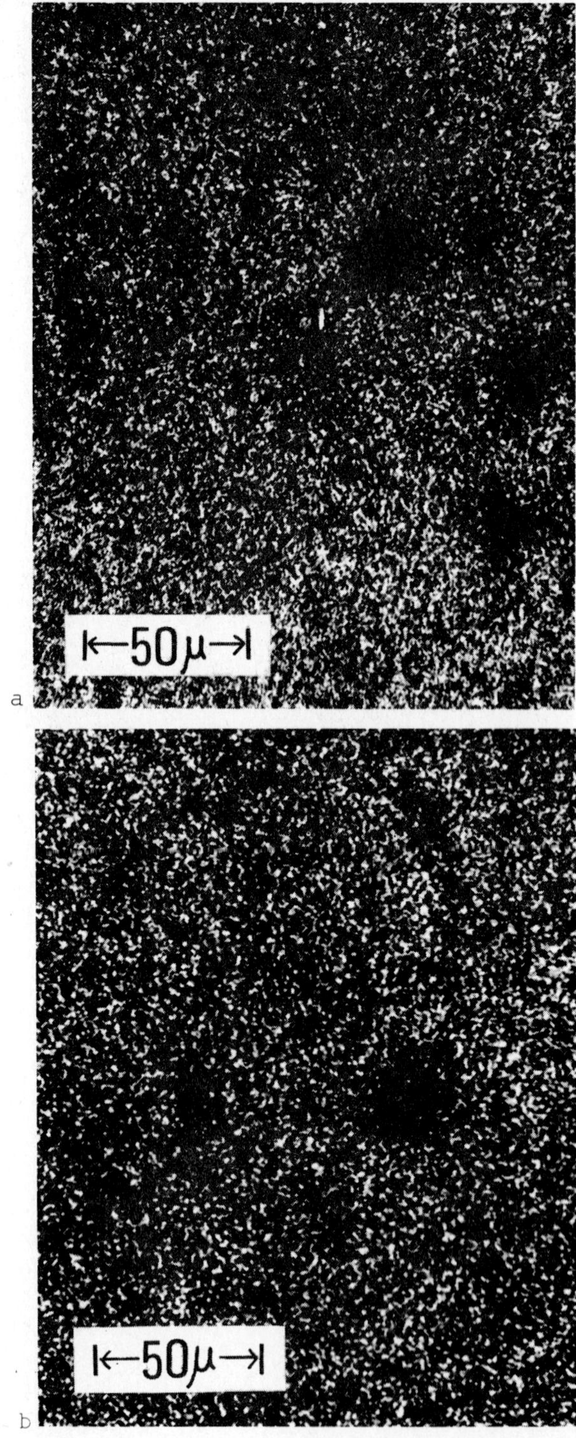

|←50μ→|
a
|←50μ→|
b

c

Fig. 10 Autoradiographs before and after the catalytic reaction, and the metallurgical micrograph of cu.
a: Autoradiograph of Cu before the catalytic reaction.
b: Autoradiograph of Cu after the catalytic reaction.
c: Metallurgical micrograph of Cu.

Fig. 9 consists of the autoradiographs and the photograph of the surface on the copper-nickel sheet. These results indicate that tritium was concentrated in the grain boundary and was consumed preferentially by the hydrogenation reaction, as in the case of the nickel. Fig. 10 consists of the autoradiographs and the metallurgical microphotograph of copper. The stripes which correspond to the grain boundary could not be found in the autoradiograph, which did not change upon the hydrogenation reaction.

Fig. 11 consists of the electron micrograph of nickel which was covered with lithium and irradiated with neutrons and that not doped with lithium and irradiated with neutrons. The surface was shaved by electrolytic polishing to a depth of ca. 16μ before the photograph was taken. Numerous micro-holes ca. 500Å in diameter can be observed in the photograph of the former. However, the micro-hole of this type can not be found in the photograph of the latter.

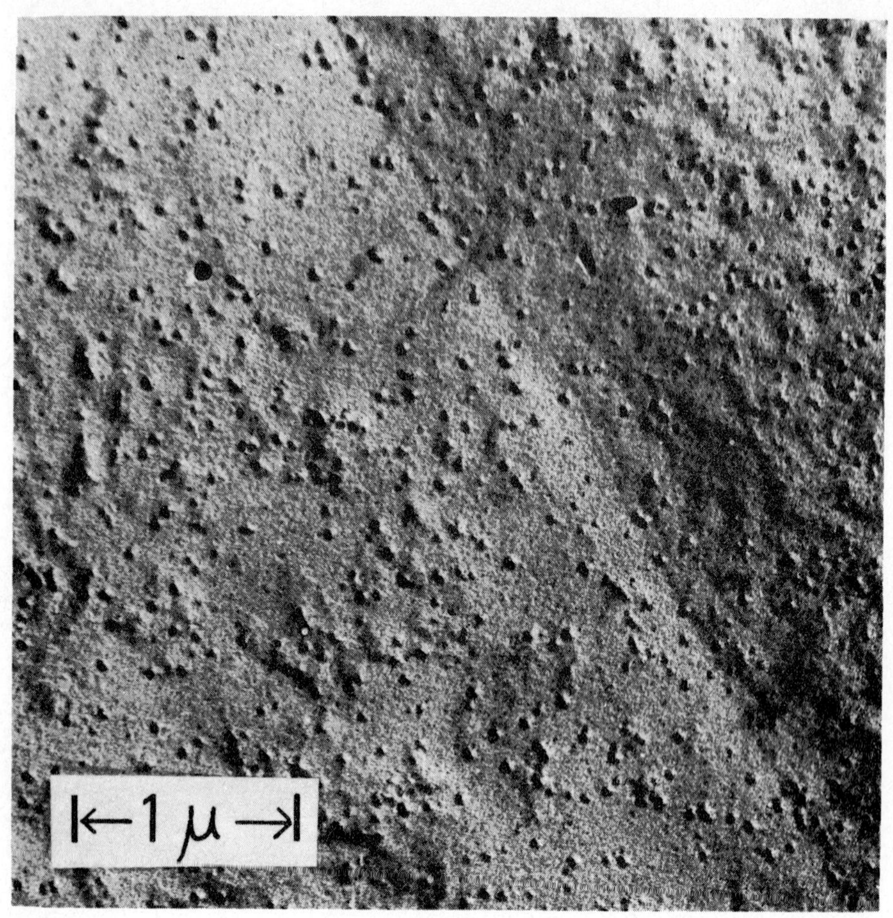

a

b

Fig. 11 Electron micrographs of the surface of nickel, after the electrolytic polishing.
a: Ni-Li irradiated with neutrons.
b: Ni irradiated with neutrons.

4. DISCUSSION

The lithium in the lithium nitrate used for the preparation of powdered catalysts and that used for the deposition on the sheets were normal. It contains 7.4% of ^{6}Li, which is associated with the nuclear reaction. A large portion of the lithium doped in the powdered catalysts or deposited on the sheets remains without change after the irradiation. According to metallurgical conceptions, lithium does not alloy with nickel and copper, and the lithium oxide produced by the calcination of lithium-doped catalysts is difficult to reduce by means of hydrogen. The observation of the X-ray diffraction pattern did not show the difference between powdered lithium-doped catalysts and those with no lithium. This finding suggests that lithium was dispersed in the form of fine crystallites of oxide. The decrease in the catalytic activity upon the doping of lithium would be due to the lithium oxide on the surface. That is, the effective area of the surface would be diminished.

The increase in the catalytic activity upon the irradiation suggests the occurrence of structural change in the metal - that is, the formation of fresh lattice imperfections. The increase in the irradiated catalysts which were not doped with lithium is probably due to the increase in the lattice imperfections produced by the fast neutrons included in the flux. It may also be concluded that the greater increase in the catalytic activity of lithium-doped nickel and lithium-doped copper-nickel is attributable to the additional increase in lattice imperfections produced by the fission products.

Copper has been regarded as intrinsically an inactive catalyst for hydrogenation reaction.[3)4)] If the lattice imperfections can always act as effective sites for the catalytic reaction, copper should become active upon the neutron irradiation. An unexpected result of this study is that the increase in the activity by the increase the amount of lattice imperfections arises only in the metal which is satisfied with the electronic factor for the reaction, as was proposed by Dowden and Reynolds.[5)]

The autoradiographs of the sheets of nickel and copper-nickel indicate clearly that a large amount of tritium exists in the grain boundary, that this tritium is highly active for the hydrogenation reaction, and that it reacts preferentially with ethylene. The accumulation of tritium at the grain boundary would take place quickly along lattice imperfections, such as dislocation, as was reported in connection with the diffusion of impurities.[6)7)]

Taking Brinkman's theory[8)] in consideration, The numerous micro-holes found on the sheet of nickel by means of electron microscope (Fig. 11a) might be the result of the thermal spike which is produced by the penetration of fission products. The depth of the penetration of tritium in the nuclear reaction is estimated to be ca. 20µ.[9)] The micro-holes are produced by the electrolytic polishing of the area of the thermal spike, whose lattice imperfections are very active for the reaction with the solution of electrolytic polishing and which are shaved preferentially.

The autoradiograph of copper did not give a distinct pattern of stripes corresponding to the grain boundary, unlike the figure given by the metallurgical microscope. This finding indicates that the tritium produced as the nuclear fission product was dispersed equally, without distinction between the grain boundary and the inside of the grain. This finding indicates that tritium was trapped in the so-called depleted zone in the state of the molecule. Therefore, a higher temperature than in the cases of nickel and copper-nickel was necessary to taken tritium from copper, as may be seen in Fig. 5.

REFERENCES

1) I. Uhara, S. Yanagimoto, K. Tani, G. Adachi, and S. Teratani, J. Phys. Chem., 66 (1962), 2691.
2) E. M. A. Willhoft, Chem. Comm., (1968), 146.
3) P. H. Emmett and N. Skau, J. Am. Chem. Soc., 65 (1943), 1029.
4) G. C. Bond, Catalysis by Metals (Acad. Press, London, 1962) p. 244.
5) D. A. Dowden and P. W. Reynolds, Disc. Faraday Soc., 8 (1950), 184.
6) D. S. Billington and J. H. Crawford, Radiation Damage in Solids (Princeton Univ. Press, Princeton, 1961) p. 72.
7) F. R. N. Nabarro, Theory of Crystal Dislocation (Oxford Press, 1967), p. 464.
8) J. Brinkman, J. Appl. Phys., 25 (1956), 961.
9) J. B. Marion and F. C. Young, Nuclear Reaction Analysis Graphs and Tables (North Holland Pub. Co., 1968), p. 2.

DISCUSSION

G.-M. SCHWAB

In former work in Munich, catalysts consisting of copper, nickel, aluminum and zinc oxides were irradiated by external sources of alpha, beta, neutron and X-rays before catalysis of ethylene hydrogenation, deuterium exchange and parahydrogen conversions (orthodeuterium conversion). In all cases a decrease of activity by irradiation was observed, and in all cases the activity could be regenerated by annealing in hydrogen. It was concluded that the ionizing radiation removes chemically bound hydrogen from the solid and that this results in a loss of activity which can be repaired by the hydrogen treatment. It might be that also in the present authors' work the presence of tritium (being a hydrogen isotope) is more important than its radiation.

T. TAKEUCHI

In this study only 3×10^{-2} of the total tritium was consumed by the hydrogenation reaction, and the tritium which was shifted to ethane by the reaction constituted 10^{-6} of the total number of ethane molecules. These facts show that the effect of tritium upon the rate of the reaction can be disregarded.

W. M. H. SACHTLER

How did you check, that upon reducing at 350°C the Li is reduced from Li^{1+} to Li^0?

How did you check that Li was homogeneously distributed through the bulk of the alloys?

Is it not possible that Li or Li_2O was accumulating on grain boundaries thus causing tritium to be formed at the grain boundaries?

T. T. TAKEUCHI

We did not check the state of lithium in the alloy. It is possible that lithium accumulates preferentially in grain boundaries rather than in grains, as occurs in the cases of metallic impurities.

P. TETENYI

I would like to suggest the following:

i) It would be preferable to check the effect of irradiation upon the catalytic activity with some other catalytic reactions also, before making some general conclusions.

ii) To be sure the changes in catalytic activity were caused by lattice imperfections, you can dissolve the catalyst, prepare it newly and check the activity. It is possible by such manner to distinguish between the effects caused by chemical changes and those caused by lattice imperfections.

iii) There are some papers dealing with the effect of very low radioactivity on the catalytic activity. The question of the effect of the remaining radioactivity on the catalytic activity has to be raised in this case also.

T. TAKEUCHI

The catalytic activities of nickel and copper-nickel increased when the irradiation of neutrons was carried out at liquid nitrogen temperature but did not when the irradiation was carried out without liquid nitrogen. These findings indicate that the increase in the catalytic activity does not depend upon the remaining radioactivity of the catalysts.

C. G. HARKINS

Your tritium autoradiography shows directly the introduction atoms into the metal phase. Observation of tritium concentrations at the phase boundaries, the thermodynamically stable location for impurity atoms, indicates that Ni and Ni-Cu alloys are more permeable to hydrogen migration than pure copper, and this is confirmed by the exchange data of Figure 5.

The authors also indicate that metal phase tritium is preferentially consumed in ethylene hydrogenation. This suggests a Sabatier-type mechanism is involved in the activity enhancement, and that the catalysts intrinsically permeable to hydrogen show enhancement. Copper is intrinsically non-permeable to hydrogen and radiation damage does not improve the permeability.

My question concerns the necessity of invoking lattice defects or electronic factors when hydrogen mobility variation is the least complicated explanation of the enhancements observed.

T. TAKEUCHI

As mentioned in my talk, the important fact which leads to the conclusion of our study is that the increase in the catalytic activity during irradiation of neutrons was found only in the metals having an incomplete d-shell. Therefore, it is necessary to discuss our results from the standpoint of lattice imperfections or electronic factors.

Paper Number 37

METHOD OF MONOLAYER CAPACITY DETERMINATION OF CATALYSTS

JURO HORIUTI* and CHAO-FEANG LIN**
*Hokkaido University, Sapporo, Japan; **Nippon Medical School, Kawasaki, Japan

ABSTRACT: The previous statistical-mechanical formulation[2)3)] of isotherm is extended by leaving out one of its premises that either of the two congruent lattices of sites, provided by an underlying compact layer of adsorbate, is exclusively available, which holds precisely only when fully occupied. Conclusions thus arrived at, admitting the uniform potential of adsorbate due to adsorbent over each layer, are that critical value ξ_c of Boltzmann factor ξ of attraction potential between two contacting adsorptive molecules, such that only for $\xi \geqq \xi_c$ two-dimensional gas and its condensate coexist stably, is found $\xi_c = 3.406$ in place of previous 2.48[2)] and that the uniform potential due to adsorbent rises stepwise from lower to upper layers. Monolayer capacity is deduced on these bases from stepwise as well as smoothly ascending isotherms. The monolayer capacities thus found are greater than the respective BET-values by 5～69 %.

1. INTRODUCTION

The BET-method attributes to each molecule adsorbed on the top of kindred molecule or molecules a potential drop equal to the heat of vapourization of the condensate of adsorptive per molecule, irrespective of the number of molecules surrounding the latter; hence there may grow "theoretically" one-dimensional molecular rosaries each consisting of molecules equally numerous as adsorbate layers then formed[1)].

Consider the condensate of adsorptive consisting of molecules behaving like so many rigid spheres of definite radius attracting one another, the potential of attraction between them being significant only in their direct contact. Any molecule within the condensate is thus surrounded by twelve kindred ones forming either a cubic or hexagonal closest packing. It follows that the heat of condensation of the adsorptive per molecule amounts to the total attraction potential of six contacting pairs, hence the BET-method overestimates the adsorption potential and in consequence the adsorbed amount per site. It was tried in previous works[2)3)] to obviate this defect by counting the potential of each adsorbate molecule with the number of those contacting with the latter relying upon the following simplifying assumption.

Suppose the adsorption of molecules each seated on a site provided by three adjacent ones in the underlying layer of kindred molecules. Another compact layer of the adsorbate can now occupy only one of the two congruent lattices of sites as indicated respectively by ●'s or ○'s each marked at the center of a site in fig.1, whereas those of both the lattices are available for sparse adsorbate. The simplifying assumption was that either lattice of sites is exclusively available irrespective of the density of adsorbate.

The adsorption isotherm is now formulated leaving out the latter assumption, but postulating as before[2)] that (i) adsorptive molecules behave like rigid spheres of definite radius attracting one another, potential of the attraction being significant only in their contact, that (ii) a site of adsorption is provided by three kindred molecules in contact with one another, that (iii) the adsorbent is thoroughly covered with a compact first layer prior to adsorption in upper layers and that (iv) the adsorptive is deposited on adsorbent, only by adsorption and not otherwise, e.g. by capillary condensation. An inference based on these postulates is, qualitatively in accordance with that in the previous works[2)], that, for sufficient magnitude of attraction among adsorptive molecules, the adsorbate splits spontaneously for a certain range of its density into two-dimensional gas and condensate, denoted by I and II, which stably coexist.

2. ADSORPTION ISOTHERM

The isotherm is formulated for adsorption on a compact underlying layer and the appropriate coexistent state of I and II determined.

Two congruent lattices of sites provided by the underlying compact layer are shown in fig.1 respectively by ●'s and ○'s. Σ denotes the set of ten sites particularly noticed in the formulation of isotherms as indicated by a hexagon circumscribing them*. We see that the total number

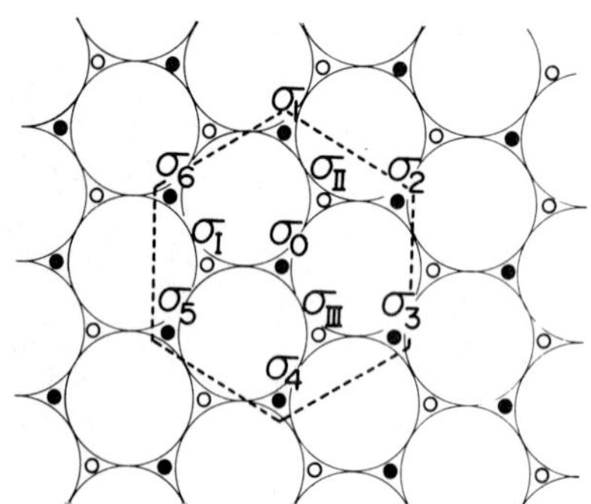

Fig.1 Sites of adsorption provided by underlying compact layer. The constituent molecules of the compact layer are indicated by circles of their orthographs. The lattice of sites, each provided by three adsorbate molecules in the underlying layer, is represented by that of ●'s or of ○'s drawn respectively at centers of sites. We see that either lattice is exclusively occupied by another compact layer of kindred molecules. The set of ten sites, $\sigma_0, \dots \sigma_6$, σ_I, σ_{II} and σ_{III}, circumscribed by the hexagon is denoted by Σ.

of sites provided by a compact layer of adsorbate molecules is twice the number of its constituent molecules for its infinite extension. It follows from the present premise which allows sites of both the lattices accessible, that the occupation of peripheral sites of Σ may possibly be obstructed by molecules occupying proximate ○-sites just outside the hexagon in fig.1, which intrude the peripheral sites. A site is termed occupied, if an adsorbate molecule is seated right above the center of the site, unoccupied, if only not occupied, i.e. even if just intruded, and vacant, if clear even of the intrusion.

* Only seven sites, $\sigma_0, \dots,$ and σ_6, were thus treated in the previous works[2)3)].

The symbols used in the present formulation are as follows:
δ, δ_a: adsorptive or adsorbate molecule respectively.
σ: site provided according to postulate (ii).
N, N_1, N_n: total number of δ_a's present, that in a compact layer or that in the n-th layer respectively. N_1 is the monolayer capacity or the total number of σ's on one of the congruent lattices*.
v, v_m: amount of gas adsorbed or the monolayer capacity on unit of v.
$\theta_n \equiv N_n/2N_1$: fraction of σ's occupied by δ_a's in the n-th layer, its maximum being at $N_n = N_1$, i.e. 1/2.
$\theta_{n,I}$, $\theta_{n,II}$: particular values of θ_n respectively of I and II.
$\theta \equiv N/2N_1$.
C: assembly to be dealt with, which consists of adsorbent, adsorbate and adsorptive gas.
$\mathfrak{P}C$: canonical partition function of C.
γ_n: factor by which $\mathfrak{P}C$ is multiplied as δ is transferred within C from gas to a definite, vacant σ to create δ_a in the n-th layer apart from the potential due to δ_a's in the same and upper layers.
$\gamma_{n,e}$: particular value of γ_n at which I and II coexist in equilibrium.
σ_o, σ_1,... σ_6 or σ_I, σ_{II}, σ_{III}: ●-sites or O-sites of Σ shown in fig.1.
$\mathfrak{P}C_{\Sigma(0)}$, $\mathfrak{P}C_{\sigma_0(\delta)}$: $\mathfrak{P}C$ for the particular state of C, where Σ are all unoccupied, or σ_o is occupied by a δ_a respectively.
$\mathfrak{P}C_{\sigma_0(0)}$, $\mathfrak{P}C_{\sigma_1(0)}$, $\mathfrak{P}C_{\sigma_I(0)}$: $\mathfrak{P}C$ for the particular state of C, where σ_o, σ_1 or σ_I is respectively unoccupied.
η_1, η_2: Boltzmann factor of the sum of potential of δ_a occupying one of σ's of Σ, due to δ_a's outside Σ and the work required, if any, to wedge out δ_a intruding the site; η_1 is relevant to a peripheral site, σ_1, σ_2, ..., σ_5 or σ_6, and η_2 to σ_I, σ_{II} or σ_{III}.
p_G: factor by which $\mathfrak{P}C$ is multiplied on addition of a δ from outside C to its gas phase; $-RT\ln p_G$ is the chemical potential of δ in the gas[4)].
ξ: Boltzmann factor of attraction potential between two δ's in contact.
ξ_c: critical value of ξ such that $\gamma_n = \gamma_n(\theta_n,\xi)$ increases monotonously with θ_n for $\xi < \xi_c$, but follows loops for $\xi \geqq \xi_c$ as seen in fig.2.
p, $p_{n,e}$, p_o: pressure of adsorptive gas in C, its value at $\gamma_n = \gamma_{n,e}$ or that in equilibrium with the condensate of adsorptive respectively.
$\Delta\varepsilon_n$: potential due to the adsorbent, of δ_a in the n-th layer per mole.

The θ_n is given in terms of $\mathfrak{P}C_{\sigma_0(\delta)}$ and $\mathfrak{P}C_{\sigma_0(0)}$, which are respectively proportional to the probabilities of relevant states, as

$$\theta_n/(1 - \theta_n) = \mathfrak{P}C_{\sigma_0(\delta)}/\mathfrak{P}C_{\sigma_0(0)}\,.$$

$\mathfrak{P}C_{\sigma_0(\delta)}$ and $\mathfrak{P}C_{\sigma_0(0)}$ are developed as

$$\mathfrak{P}C_{\sigma_0(\delta)}/\mathfrak{P}C_{\Sigma(0)} = \gamma_n[1 + 6\xi\eta_1\gamma_n + 3(3 + 2\xi)\xi^2\eta_1^2\gamma_n^2 +$$

$$+ 2(1 + 6\xi + 3\xi^2)\xi^3\eta_1^3\gamma_n^3 + 3(3 + 2\xi)\xi^6\eta_1^4\gamma_n^4 + 6\xi^9\eta_1^5\gamma_n^5 + \xi^{12}\eta_1^6\gamma_n^6]$$

and

* cf. foregoing paragraph.

$$\mathfrak{R}C_{\sigma_0(0)}/\mathfrak{R}C_{\Sigma(0)} = 1 + 6\eta_1\gamma_n + 3(3 + 2\xi)\eta_1^2\gamma_n^2 + 2(1 + 6\xi + 3\xi^2)\eta_1^3\gamma_n^3 +$$
$$+ 3(3 + 2\xi)\xi^2\eta_1^4\gamma_n^4 + 6\xi^4\eta_1^5\gamma_n^5 + \xi^6\eta_1^6\gamma_n^6 + 3[1 + 4\eta_1\gamma_n + 3(1 + \xi)\eta_1^2\gamma_n^2 +$$
$$+ 2(1 + \xi)\xi\eta_1^3\gamma_n^3 + \xi^3\eta_1^4\gamma_n^4]\eta_2\gamma_n + 3(1 + 2\eta_1\gamma_n + \xi\eta_1^2\gamma_n^2)\xi\eta_2^2\gamma_n^2 + \xi^3\eta_2^3\gamma_n^3.$$

The above expression of $\mathfrak{R}C_{\sigma_0(\delta)}$ is identical with that of Eq.(3), ref.2) with γ_0 and η respectively replaced by γ_n and η_1, inasmuch as inclusion of σ_I, σ_{II} and σ_{III} in Σ makes no difference, any occupation of them being excluded as incompatible with the prescribed condition of $\mathfrak{R}C_{\sigma_0(\delta)}$ that σ_0 is occupied by a δ_a. The development of $\mathfrak{R}C_{\sigma_0(0)}$ differs, on the other hand, from Eq.(1.a), ref.2) with γ_0 and η respectively replaced by γ_n and η_1, by the last three terms of the above expression, which allow for the occupation of O-sites in Σ respectively by one, two and three δ_a's, it being now compatible with the prescribed condition of $\mathfrak{R}C_{\sigma_0(0)}$ that σ_0 is unoccupied.

Parameters η_1 and η_2 are evaluated so as to find θ_n as a function of γ_n with single parameter ξ by the equation, $\mathfrak{R}C_{\sigma_0(0)} = \mathfrak{R}C_{\sigma_1(0)} = \mathfrak{R}C_{\sigma_I(0)}$, based on the physical identity of sites σ_0, σ_1 and σ_I in accordance with Bethe and Peierls[5]. $\mathfrak{R}C_{\sigma_1(0)}$ and $\mathfrak{R}C_{\sigma_I(0)}$ are developed as

$$\mathfrak{R}C_{\sigma_1(0)}/\mathfrak{R}C_{\Sigma(0)} = 1 + \gamma_n + 5(1 + \xi\gamma_n)\eta_1\gamma_n + 2(3 + 2\xi)(1 + \xi^2\gamma_n)\eta_1^2\gamma_n^2 +$$
$$+ (1 + 6\xi + 3\xi^2)(1 + \xi^3\gamma_n)\eta_1^3\gamma_n^3 + (3 + 2\xi)(1 + \xi^4\gamma_n)\xi^2\eta_1^4\gamma_n^4 +$$
$$+ (1 + \xi^5\gamma_n)\xi^4\eta_1^5\gamma_n^5 + [3 + 10\eta_1\gamma_n + 6(1 + \xi)\eta_1^2\gamma_n^2 + 3(1 + \xi)\xi\eta_1^3\gamma_n^3 +$$
$$+ \xi^3\eta_1^4\gamma_n^4]\eta_2\gamma_n + (3 + 5\eta_1\gamma_n + 2\xi\eta_1^2\gamma_n^2)\xi\eta_2^2\gamma_n^2 + \xi^3\eta_2^3\gamma_n^3$$

and

$$\mathfrak{R}C_{\sigma_I(0)}/\mathfrak{R}C_{\Sigma(0)} = 1 + \gamma_n + 6(1 + \xi\gamma_n)\eta_1\gamma_n + 3(3 + 2\xi)(1 + \xi^2\gamma_n)\eta_1^2\gamma_n^2 +$$
$$+ 2(1 + 6\xi + 3\xi^2)(1 + \xi^3\gamma_n)\eta_1^3\gamma_n^3 + 3(3 + 2\xi)(1 + \xi^4\gamma_n)\xi^2\eta_1^4\gamma_n^4 +$$
$$+ 6(1 + \xi^5\gamma_n)\xi^4\eta_1^5\gamma_n^5 + (1 + \xi^6\gamma_n)\xi^6\eta_1^6\gamma_n^6 + 2[1 + 4\eta_1\gamma_n + 3(1 + \xi)\eta_1^2\gamma_n^2 +$$
$$+ 2(1 + \xi)\xi\eta_1^3\gamma_n^3 + \xi^3\eta_1^4\gamma_n^4]\eta_2\gamma_n + (1 + 2\eta_1\gamma_n + \xi\eta_1^2\gamma_n^2)\xi\eta_2^2\gamma_n^2 .$$

The γ_n is thus solved as a function of θ_n with single parameter ξ. The numerical calculation has been conducted with electronic computer of the Hokkaido University Computing Center, FACOM 230-6 manufactured by Fujitsu CO., Ltd., Tokyo.

2.1 Results of calculation

Fig.2 shows $\gamma_n$vs.θ_n-curves thus worked out. The maximum of θ_n is 1/2 as mentioned in 2. The ξ_c is found at 3.406 in place of the previous value, 2.48, obtained on exclusion of δ_a's from O-sites. Function $\gamma_n = \gamma_n(\theta_n,\xi)$ is universal among n, as shown in 4.1., ref.3), inclusive of n = 0 for the surface layer of condensate of the adsorptive, hence conclusions drawn for $\gamma_0 = \gamma_0(\theta_0,\xi)$[2] apply generally to $\gamma_n(\theta_n,\xi)$. It thus follows from conclusions in 3.2., ref.2) that I and II coexist at $\gamma_n = \gamma_{n,e}$, which satisfy the condition, $\log_{10}\gamma_{n,e} = \int_{\theta_{n,I}}^{\theta_{n,II}} \log_{10}\gamma_n d\theta_n/(\theta_{n,II} - \theta_{n,I})$, that $\gamma_{n,e}$, $\theta_{n,I}$ and $\theta_{n,II}$ are respectively of universal values among n and that

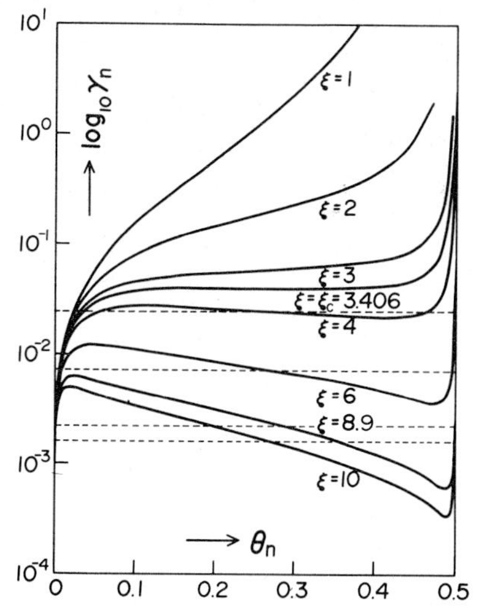

Fig.2 $\log_{10}\gamma_n$ vs. θ_n-curves for different values of ξ. Broken horizontal drawn to each curve for $\xi \geqq \xi_c$ is of such height, $\gamma_{n,e}$, as the area encircled by loops above it just equals that below it, i.e. $\log_{10}\gamma_{n,e} = \int_{\theta_{n,I}}^{\theta_{n,II}} \log_{10}\gamma_n d\theta_n/(\theta_{n,II}-\theta_{n,I})$. The θ_n-values, $\theta_{n,I}$ and $\theta_{n,II}$, at its left and right ends are respectively those of I and II.

two-dimensionally homogeneous adsorbate corresponding to a point on $\log_{10}\gamma_n$ vs. θ_n-curve between $\theta_{n,I}$ and $\theta_{n,II}$ splits spontaneously into I and II.

3. ISOTHERMS AND MONOLAYER CAPACITY

Isotherms are now investigated on the basis of the theory of $\gamma_n = \gamma_n(\theta_n,\xi)$ developed since the previous works[2)3)], then the monolayer capacity derived. The latter development is based, however, on a tacit assumption, that $\Delta\varepsilon_n$ is uniform over each layer of adsorbate, which assumption is termed that of uniform potential layers in what follows. We will throw it to experimental tests in subsequent sections for $\xi > \xi_c$ *, which is the case with most of isotherms observed.

3.1 Uniform potential layers

As follows from conclusions in 2.1, γ_n increases monotonously from naught up to $\gamma_{n,e}$ as θ_n increases similarly up to $\theta_{n,I}$, where the coexistence, at constant $\gamma_n = \gamma_{n,e}$, of I and II begins; with increase of adsorbate the latter gains upon the former until n-th layer is packed completely with II; the n-th layer is then termed completed or compact as in what follows. We have from Eqs.(6.b) and (7.a), ref.3).

$$\gamma_n/\gamma_{n,e} = p/p_{n,e}, \tag{1}$$

where the ratio of concentration of adsorptive in gas to its particular value in equilibrium with coexistent I and II is replaced with $p/p_{n,e}$. We have from Eqs.(7.b) and (8), ref.3) similarly

* The ξ is calculated as the Boltzmann factor of -1/6 times the internal heat of vaporization at the relevant temperature according to Eq.(2.b) and (21), ref.2).

$$p_{n,e}/p_o = \exp(\Delta\varepsilon_n/kT), \qquad (2)$$

hence referring to Eq.(1)

$$\gamma_n/\gamma_{n,e} = (p/p_o)\exp(-\Delta\varepsilon_n/RT). \qquad (3)$$

3.1.1 $\Delta\varepsilon_{n(>1)} = 0$. This postulate underlying the BET-theory leads by Eq.(3) to the equation, $\gamma_n/\gamma_{n,e} = p/p_o$, for n(>1) or that γ_n exceeds $\gamma_{n,e}$ for all n(>1)'s at once, hence the second and upper layers are completed simultaneously resulting in the bulk condensation, as soon as p exceeds p_o even infinitesimally, whereas none of them is completed so long as p keeps below p_o; even in the latter case δ_a's in the second and upper layers may be seated upon sites provided by the compact first layer according to postulate (ii), which count among N. The lower limit N_L to N is given, inclusive of δ_a's in the first layer, as

$$N_L = N_1 + 2N_1\theta_2, \qquad (4.L)$$

the right-hand side being short of δ_a's in the third and upper layers possibly heaped in various lots. Suppose, on the other hand, that δ_a's in the second layer are joined together to make an island consisting of $2\theta_2N_1$ pieces of δ_a's. The number of δ_a's in the third layer upon the island is now $2\theta_2N_1\cdot2\theta_3$ at most, inasmuch as sites are fictitiously formed by the supposed junction, which are originally incapable of seating δ_a's above. We have thus the upper limit, N_U, to N, i.e.

$$N_U = N_1 + (2\theta_2)N_1 + (2\theta_2)(2\theta_3)N_1 + (2\theta_2)(2\theta_3)(2\theta_4)N_1 + \cdots \qquad (4.U)$$

It follows from Eq.(3) that $\gamma_n/\gamma_{n,e}$ is of common value to n(>1)'s at a given p for $\Delta\varepsilon_{n(>1)} = 0$. Function $\gamma_n = \gamma_n(\theta_n,\xi)$ as well as $\gamma_{n,e}$-value being universal among all n(>1)'s, we have a common $\theta_{n(>1)}$-value at a given p, hence $N_U = N_1 + (2\theta_2)N_1 + (2\theta_2)^2N_1 + (2\theta_2)^3N_1 + \cdots =$ $= N_1/(1 - 2\theta_2)$.

These conclusions are tested by isotherms observed for krypton, which appears favourable for the postulate, $\Delta\varepsilon_{n(>1)} = 0$, to hold on account of its rather big size. The θ_2 for $\gamma_2 \leq \gamma_{2,e}$ is $\theta_{2,e} = 0.0026$ at most for $\xi = 8.9$ of krypton at 77°K * by function $\gamma_n = \gamma_n(\theta_n,\xi)$ shown in fig.2. The N_U is now $N_1/(1 - 0.0052)$ at most; hence N remains almost on the same level with increase of p until the bulk condensation comes about all at once. This requirement is by no means met by observation either of stepwise[6] or smoothly[7] ascending isotherm of krypton at 77°K respectively reproduced in fig.3 or 4. Rejecting the postulate leading to the contradiction, account is now taken of the rise of $\Delta\varepsilon_n$ with separation of δ_a from the adsorbent along with the increase of n.

3.1.2 $\Delta\varepsilon_n$ of van der Waals sttraction. $\Delta\varepsilon_n$ varies as $\Delta\varepsilon_n \propto 1/r_n^3$ [8], where distance r_n of the n-th layer from the surface of adsorbent is given for a pile of uniform potential layers, in terms of radius r_o of δ, i.e. the distance of the first layer from the adsorbent surface and the ratio, 1.6330, of the distance between adjacent layers to r_o, as $r_n = r_o +$ $+ 1.6330(n - 1)r_o$, so that

* cf. footnote * on p.5.

$$\Delta\varepsilon_n = \alpha[1 + 1.6330(n - 1)]^{-3}, \tag{5}$$

where α is the proportional factor depending on the constant of van der Waals attraction and r_o. The $\Delta\varepsilon_n$ thus rises stepwise with increase of n, hence $p_{n,e}$ as well similarly by Eq.(2), so that $p_{n,e}$'s are surpassed by increasing p one after another in order of increasing n. As p attains just a certain $p_{n,e}$, γ_n reaches $\gamma_{n,e}$ by Eq.(1), where θ_n increases from $\theta_{n,I}$ to $\theta_{n,II}$ at constant $p_{n,e}$ or such γ_n with addition of δ_a's until the n-th layer is completed and so on, resulting in a stepwise ascending isotherm. Breadth of the tread is reduced with increase of n, as the rise of $\Delta\varepsilon_n$ to $\Delta\varepsilon_{n+1}$ diminishes with increase of n by Eqs.(2) and (5).

Attributing the steep ascent of isotherm around $p/p_o = 0.41$ in fig.3 to that at $p = p_{2,e}$, we have according to Eq.(2) $\Delta\varepsilon_2 = -136.2$ cal/mole, hence by Eq.(5) $\Delta\varepsilon_3 = -32.1$ cal/mole, $\Delta\varepsilon_4 = -12.1$ cal/mole, $\Delta\varepsilon_5 = -5.8$ cal/mole etc. It follows from $\Delta\varepsilon_3 = -32.1$ cal/mole that $p_{3,e}/p_o = 0.81$ by Eq.(2), the vertical through which point just hits the next steep ascent in the center.

The stepwise ascending isotherm is now theoretically reproduced fitting N_1 to experiment, generalizing Eqs.(4.L) and (4.U) as follows. Lower Limit N_L to N on completion of n - 1 layers is given, similarly to Eq.(4.L), as

$$N_L = (n - 1)N_1 + 2N_1\theta_n. \tag{6.L}$$

Upper limit to N on completion of n - 1 layers may be given generalizing Eq.(4.U), as $(n - 1)N_1 + 2\theta_n \cdot N_1 + 2\theta_n \cdot 2\theta_{n+1} \cdot N_1 + 2\theta_n \cdot 2\theta_{n+1} \cdot 2\theta_{n+2} \cdot N_1 + \cdots$ We see, however, from Eq.(3) that γ_n decreases with increase of n at a given p, on account of the universality of $\gamma_{n,e}$-value among n * and of the associated rise of $\Delta\varepsilon_n$. But since $\theta_n (<\theta_{n,e})$ decreases monotonously with γ_n, we have $\theta_n > \theta_{n+1} > \theta_{n+2} \cdots$, hence substituting $\theta_{m(>n)}$ with θ_n the upper limit N_U may be formulated as

$$N_U = (n - 1)N_1 + N_1 \cdot 2\theta_n/(1 - 2\theta_n). \tag{6.U}$$

The lower and upper limits are calculated at n = 2, 3, 4 and 5 with $\xi = 8.9$ for krypton at 77°K. The θ_n being not more than $\theta_{n,e} = 2.6 \times 10^{-3}$ as referred to in 3.1.1, N_L and N_U are, according to Eqs.(6.L) and (6.U), practically coincident, so that N/N_1 may be expressed, ignoring second order difference with respect to θ_n, as

$$N/N_1 = n - 1 + 2\theta_n, \tag{6.R}$$

where each integer n keeps between $p_{n-1,e}$ and $p_{n,e}$. Stepped broken line in fig.3 shows the isotherm of Eq.(6.R), with N_1 adjusted to the best fit of the latter to observed points. The monolayer capacity, N_1, thus determined practically reproduces N marked with $\theta = 1$ by the authors[6)].

Clark[9)] confirmed the stepped isotherm of the same adsorptive on the same adsorbent at 70.2°K by checking the adsorption equilibrium, often questioned, by desorption. He observed "vertical discontinuity" or infinite slope at the first and second steep ascents[9)]. The $\Delta\varepsilon_2$-value is calculated by Eq.(2), identifying p/p_o at the second "vertical discontinuity" with $p_{2,e}/p_o$, at -134.8 cal/mole which practically agrees with

* cf. 4.1, ref.3).

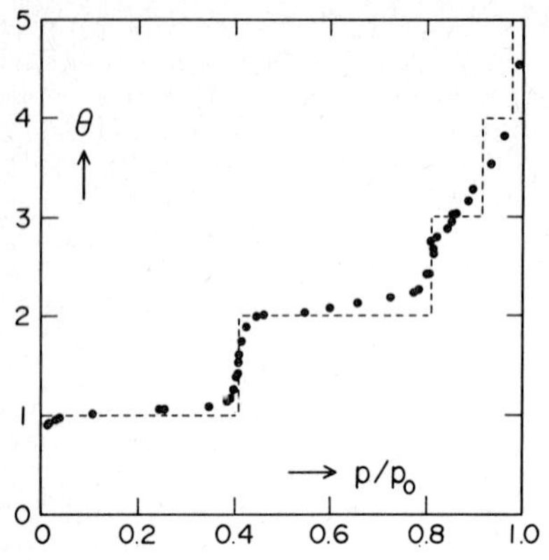

Fig.3 N vs. p/p_o of Kr on graphitized carbon black at 77°K. Dots represent the observation by Singleton and Halsey[6], who graduated the ordinate by setting $\theta = 1$ at the tread of the first step. The stepped broken line shows the theoretical isotherm constructed according to Eq.(6.R) with N_1 adjusted to its best fit to observed points.

$\Delta\varepsilon_2 = -136.2$ cal/mole derived above from the observation of Singleton and Halsey[6]. The $p_{3,e}/p_o = 0.80$ is obtained from the value of $\Delta\varepsilon_2$ by Eqs.(2) and (5), the vertical through which point hits the third steep ascent just in the center as in the former case.

The $p_{1,e}/p_o$ is derived, similarly from the latter value of $\Delta\varepsilon_2$ that $p_{1,e}/p_o = 2.1 \times 10^{-8}$, which is far wide of the observation, $p/p_o = 5.6 \times$ $\times 10^{-4}$, at the first "vertical discontinuity"[9], its quantitative validity being however questioned by the author[9] on account of thermal transpiration. A similar discrepancy is found as regards the observation of Singleton and Halsey[6], who found the first steep ascent at $p/p_o = 2 \times$ $\times 10^{-3}$. Whereas $p_{1,e}/p_o$ is thus quantitatively open to question, the $p_{1,e}/p_o$-values, anyway so minute, are in conformity with postulate (iii).

3.2 Smoothly ascending isotherm

We deal now with the smoothly ascending isotherm observed in most of cases. Since the postulate of uniform potential layers led to the stepwise ascending isotherm in accord with experiment with "exceptionally uniform surface"[9], but in contradiction to bulk of observations, it is now postulated with the latter case that $\Delta\varepsilon_n$ varies continuously with v according to the equation,

$$\Delta\varepsilon_n = \alpha[1 + 1.6330\, v/v_m]^{-3}, \qquad (7)$$

which follows by eliminating n from Eqs.(5) and (6.R), ignoring $2\theta_n$ with reference to 3.1.2 and replacing N/N_1 with v/v_m. The $p_{n,e}$, now varying continuously with $\Delta\varepsilon_n$ by Eq.(2), coincides with prevailing p in equilibrium, hence $\Delta\varepsilon_n = RT\ln p/p_o$. Equating the last two alternative expressions of $\Delta\varepsilon_n$, we have

$$v = (v_m\alpha^{1/3}/1.6330)(RT\ln p/p_o)^{-1/3} - v_m/1.6330. \qquad (8)$$

This procedure prescribes $\Delta\varepsilon_n$ to rise continuously from $\Delta\varepsilon_1$ at $v = 0$* sharply up to $\Delta\varepsilon_2 = 0.0548\ \Delta\varepsilon_1$ at $v = v_m$* by Eq.(7) rendering the compact first layer unstable or postulate (iii) untenable in comtrast with in the case of the uniform potential layers, where $\Delta\varepsilon_n$ is kept constant at $\Delta\varepsilon_1$ until the first layer is completed, where v approximates v_m.

Fitting the linear dependence of v upon $(RT\ln p/p_o)^{-1/3}$ of Eq.(8) to observation, v_m as well as α is determined. Figs.4 and 5 show the theoretical curves of N or v vs.p/p_o respectively drawn with constants determined as above, in comparison with observation represented by dots.

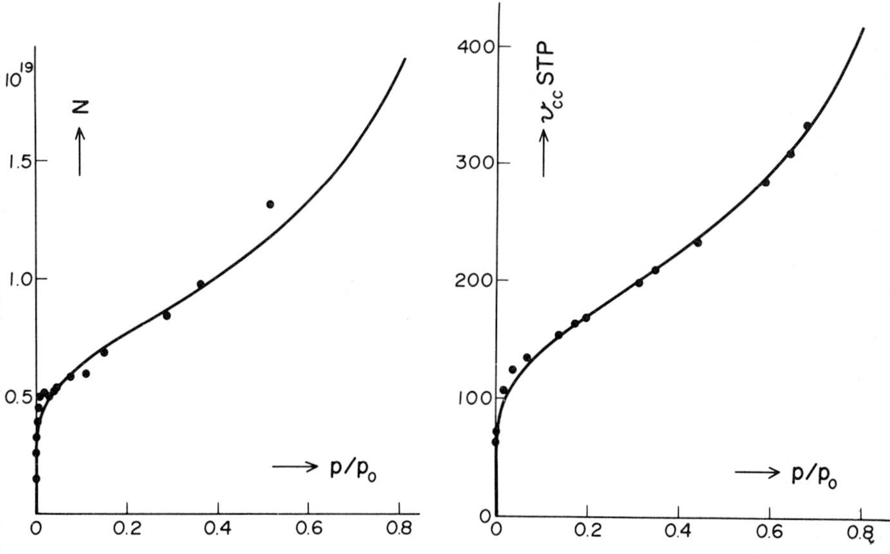

Fig.4 N vs.p/p_o of Kr on Mo-film at 77°K. Dots reproduce the observation[7], while the curve figures Eq.(8) with constants fitted to the observation.

Fig.5 v vs.p/p_o of N_2 on Fe-Al_2O_3 catalyst 954[10] at -195.8°C. Dots reproduce the observation[10], while the curve figures Eq.(8) with constants fitted to the observation.

The v_m-values thus obtained are tabulated below together with those from other sources in comparison with respective BET-values.

* In accordance with the uniform potential layers, $v = 0$ or $v = v_m$ corresponds respectively to $n = 1$ or $n = 2$ approximately by Eq.(6.R), where $v/v_m = N/N_1$.

Adsorptive, Adsorbent	Temp.	v_m		Sources of Obs.
		by Eq.(8)	by BET-method	
Kr, Mo-film	77°K	6.88×10^{18}molec.	6.55×10^{18}molec.[7)]	ref.7)
N_2, Fe-Al_2O_3 Catal.954	-195.8°C	157 cc STP	139 cc STP	ref.10)
N_2, Glaucosil	-195.8°C	160 cc STP	141 cc STP	ref.11)
A, Fe-Al_2O_3 Catal.954	-183°C	209 cc STP*	123.5 cc STP[12)]	ref.12)

It may be noted that the fit of Eq.(8) to observations is not partial to a particular region of p/p_0 and that BET-values are smaller than the present v_m-values by 5∼69 % so far as found, as expected from the over-estimation, in the former case, of adsorbed amount per site, as mentioned in the Introduction.

* p_0 = 1011.5 at -183°C of argon is derived from p_0 = 760 mmHg and ΔH = 1.558 kcal/mole at -185.87°C as given in "Selected Values of Chemical Thermodynamic Properties, Circular of the National Bureau of Standards, USA, 500 (1952)".

REFERENCES

1) J. Horiuti and T. Nakamura, Advances in Catalysis 17 (1967) 1.
2) J. Horiuti and C. F. Lin, J. Research Inst. Catalysis, Hokkaido Univ. 16 (1968) 395.
3) J. Horiuti and C. F. Lin, ibid. 16 (1968) 411.
4) J. Horiuti, ibid. 1 (1948) 8.
5) H. A. Bethe, Proc. Roy. Soc. London, A150 (1935) 552. R. E. Peierls, Proc. Cambridge Phil. Soc. 32 (1936) 471.
6) J. H. Singleton and G. D. Halsey Jr., J. Phys. Chem. 58 (1954) 1011.
7) Y. Delaunois, A. Frennet and G. Lienard, J. Chim. Phys. 63 (1966) 906.
8) T. H. Hill, "An Introduction to Statistical Thermodynamics", Addison-Wesley Pub. Co. Reading Mass. and London, 1960, Chapter 7.
9) H. Clark, J. Phys. Chem. 59 (1955) 1068.
10) P. H. Emmett, S. Brunauer and K. S. Love, Soil Science 45 (1938) 57.
11) S. Brunauer and P. H. Emmett, J. Am. Chem. Soc. 59 (1937) 2682.
12) S. Brunauer, P. H. Emmett and E. Teller, J. Am. Chem. Soc. 60 (1938) 309.

DISCUSSION

D. D. ELEY

This is a most important contribution. I wonder if Professor Horiuti would care to comment on the assumption of pair-wise interactions of molecules. If I remember rightly, J. W. Rowlinson has pointed out that many-body interactions must be considered if one is to get the right crystal structures of the rare gases. If so, in the next approximation it might be worth seeing if they could be introduced into this treatment and multilayers in the future.

As a practical point, is your equation simple to apply to determine v_m?

J. HORIUTI

The many body interactions, I_1, mentioned by Professor Eley emerges from the third order perturbation term of interaction, while that we have dealt with comes from the second order one. From the latter we have left out the interaction potential due to the second nearest neighbours, I_2, in the present calculation. With these neglects, however, the theory follows quantitatively $P_{n,e}/P_0$ of stepwise ascending isotherm as far as perceptible, except $P_{1,e}/P_0$, its reliability being questioned by the observer on account of the thermal transpiration as mentioned in the text. It follows, premising the absence of other significant neglects, that the effects of neglects of I_1 and I_2 are either insignificant individually or else compensatory. We might hence calculate I_2 as the extension of the present calculation. If its effect is negligible, the same must be the case with I_1, but if not, we might work out I_1 as well to investigate the presumed compensation.

To the last question the answer is yes; we have just to plot, according to Equation (8), observed v against $(RT \ln p/p_0)^{-1/3}$ and to produce the straight line thus obtained down to $(RT \ln p/p_0)^{-1/3} = 0$ to find $-v_m/1.6330$ on the ordinate.

J. W. HIGHTOWER

Can you explain the rather large discrepancy between the v_m value obtained by your method and that obtained from the BET equation for argon catalyst 954 in the last line in the last figure?

J. HORIUTI

The BET-method attributes the whole amount of vaporization heat of adsorptive to the adsorption potential of an adsorbate molecule even when attached on the top of a single, isolated kindred molecule; this leads to the overestimation of adsorbed amount per unit area, hence to the underestimation of the monolayer capacity from the observed amount of adsorption.

The question of Professor Hightower appears, however, to refer to the reason why v_m by BET-method is smaller than that by our method particularly in the case of adsorption of argon on catalyst 954. It is difficult to answer the latter question explicitly, but comparison might be made here especially with that of nitrogen on the same catalyst. The appreciable excess of ΔH of A(1.558 kcal mole^{-1}) as compared with that (1.33 kcal mole^{-1}) of N_2 might amplify the underestimation of monolayer capacity by BET-method particularly in case of A. This difference could, however, hardly be dealt with quantitatively, since the extent of misestimate depends on how far one follows the observation with p/p_0 according to the BET-method.

P. H. EMMETT

I am always happy to see attempts made to improve on B.E.T. or any other methods for measuring surface areas of finely divided or porous solids. I am especially glad to see Professor Horiuti turn his attention to this matter because he has a reputation of giving constructive and well conceived ideas to any subject with which he becomes concerned. I shall not attempt to deal here with the nature of his theoretical ideas mentioned in this paper except to remark that they bear some resemblance to some of the ideas advanced by Halsey and also some of the ideas of the potential theory proposed by Sydney Ross and his coworkers. I would, however, like to make some comments on the application of his theory to calculating the volume of gas in a monolayer on an iron synthetic ammonia catalyst as given in Table 1 of this paper.

To begin with I would like to suggest that isotherms be taken for some non-porous solids such as glass spheres, quartz spheres, or even the diamond dust mentioned by Ross in his book. We can obtain area estimates independently for these solids and thus be able to compare the surface area obtained by his porposed theory with those obtained by the B.E.T. method on substances of known particle sizes.

As for the results that he gives for iron catalysts, I am unconvinced that his values for v_m are any improvement over those obtained by the B.E.T. method. It will be noted for example that the v_m for argon is about 30% larger than for nitrogen by his calculations; it is about ten percent smaller than the nitrogen value by our B.E.T. interpretation. Which is to be preferred depends obviously on the molecular cross sectional area one uses for nitrogen as compared to argon. There is no good way of telling whether the area of the iron catalyst more nearly corresponds to the large value that he finds or the smaller one obtained by the B.E.T. calculations. Measurements of isotherms on non-porous particles of known particle size will be welcome as a better comparison.

J. HORIUTI

I quite agree with Professor Emmett that the result of BET-method should be compared with ours with non-porous solids of known surface area. Based on the results of comparison thus made, we might consider remarks of Professor Emmett in the latter half of his comment.

V. PONEC

I have a simple question for clarification. Am I right in saying that the equation (7) is essentially the equation suggested by Halsey and coworkers, in several of their papers for the adsorption potential? Therefore, is the isotherm (8) in principle the Halsey isotherm which was used by Halsey to explain the experimental data such as in Figure 3 of this paper?

J. HORIUTI

Halsey's isotherm reads $\ln p/p_0 = a/RT\ \theta^r$ [J. Chem. Phys. 16 (1948) 931], as derived by him semiquantitatively assuming that the potential energy of adsorptive molecules varies as inverse r-th power of its distance from the surface, where a is a constant. The latter equation approximates Equation (7) or (8) but not identical. Equation (8) might be alternatively derived below for reference.

The partition function of adsorbates was constructed by summing up relevant Boltzmann factors in accordance with postulate (i) and (ii) for 2^9 different cases of occupation of nine sites shown in Fig. 1 circumscribed by the hexagon of broken line, dealing with the potential of adsorbates outside the hexagon by Bethe-Peierls approximation. If then the potential of adsorbate molecules due to adsorbent is neglected in the second and higher layers, no appreciable adsorption occurs until p exceeds p_0, when suddenly infinite condensation of adsorptive takes place. This contradiction to experience obliged us to take into account the van der Waals potential of adsorbate due to adsorbent by the inverse third power law. The result was the stepped isotherms shown in Fig. 3, experimental results showing the tendency of higher steps to collapse into a smooth curve. This being interpreted as the perturbation by irregularity of the adsorbent's surface, usually observed smooth isotherms were approximated by replacing the $\Delta\varepsilon_n$ stepwise increasing with integral number n of layers according to Equation (5), shown in the appended figure, with that along the continuous straight line, -----, through tips of the steps; Equation (8) is obtained by substituting n-1 with v/v_m, hence $\Delta\varepsilon$ with RT ln p/p_0.

V. MAYAGOITIA

You have presented an important correction to the method of estimating v_m, the monolayer capacity. I wish to present now, a correction that must be achieved in order to estimate the S_0 constant in: $S_s = S_0 v_m$, where S_s is the specific surface area (M^2/gr), S_0 is the factor of proportionality (m^2/cm^3), v_m = monolayer capacity (cm^3/gr).

Normally, the factor S_0 is considered as: $S_0 = A_m N_m$, where A = transverse area of the molecule of adsorbate (m^2), N_m = number of molecules per cm^3 (SPT) of adsorbate (cm^{-3}).

Nobody has considered the influence of surface curvature for mesopores (200 Å > pore radius > 16 Å), where the correction may attain a value of 10% (at 20 Å as pore radius), for N_2 determinations at 76°K. A factor equal to: $2\bar{r}/2\bar{r}-\sigma$, where $\bar{r}$ is the mean pore radius, and σ is the corrected diameter of adsorbate molecule (for hexagonal packing), accounts for the influence of pore wall curvature on S_0. Then: $S_s = A_m N_m \cdot 2\bar{r}/2\bar{r}-\sigma \cdot v_m$.

A detailed discussion will be published in the Journal of Catalysis.

J. HORIUTI

I wish to thank Dr. Mayagoitia for his suggestion for a curved surface. When his detailed discussion will have appeared and we will be so far to deal with curved surfaces, his suggestion would be very helpful.

J. D. CARRUTHERS

With reference to the comment made by Professor Horiuti and Professor Paul Emmett for the need to assess Horiuti's method for monolayer capacity on <u>non-porous</u> solids of well-characterized surface area, may I make the following suggestion:

Professor Ken Sing of Brunel University, London, has published the results of an extensive study of the adsorption of nitrogen and argon on a non-porous standard, flame-hydrolyzed silica sample which had been examined by several coworkers using different techniques (including electron microscopy). This data: "Carruthers, J. D., Cutting, P. A., Day, R. E., Harris, M. R., Mitchell, S. A., and Sing, K. S. W., Chemistry and Industry (London) 1968, 1772;" is perhaps the most reliable data obtained on spherical, non-porous silica and would be ideally suited to examination using Professor Horiuti's method. Also see: "Sing, K. S. W., <u>et al</u>., Journal of Coll. & Int. Sci., <u>38</u>, 109 (1972).

J. HORIUTI

I am very grateful to Dr. Carruthers for his kind suggestion of the most reliable and suited data to examination of our theory. Both the papers in which the above data are published being available, we can readily practice the suggested examination of our theory.

Paper Number 38

ELECTRON TRANSFER FROM METAL TO ADSORBED MOLECULES*

JOHN TURKEVICH and TOSHIO SATO**
Frick Chemical Laboratory, Princeton University

*Supported by the U.S. Atomic Energy Commission
**Present address: Government Industrial Development Laboratory, Hokkaido, Japan

SUMMARY

Transfer of electrons from metals to adsorbed molecules is often postulated as the first step in heterogeneous catalysis. It seemed of interest to determine under what conditions a thin film of sodium containing many "free electrons" would transfer these electrons to adsorbed methyl iodide to produce methyl radicals. Further it was of interest to determine whether palladium, a transition metal, would catalyze this transformation and what effect the electrons present in organic complexes would have on this transfer, both in the absence and presence of light. For this purpose a sodium film was produced on porous Vycor glass by evaporation of sodium in liquid ammonia. A quantitative measure of the number of "free electrons" was furnished by the ESR signal of the free electrons. The methyl radicals were detected by their characteristic four line ESR spectrum. The film was stable up to 400°C and reacted at room temperature with oxygen and chlorine. It did not react with hydrogen, propylene, cis butene-2 or benzene up to 400°C. It transferred electrons to naphthalene, anthracene and phthalocyanine. While it did not react with methyl iodide at room temperature, the reaction proceeded at 40°C. However no methyl radicals were produced. Attempts to catalyze the reaction using phthalocyanine or palladium film were unsuccessful. However irradiation with 2537Å or with visible light when phthalocyanine was present produced a strong methyl radical signal accompanied by a marked decrease in the sodium signal.

As part of a program of studying the elementary processes taking part in heterogeneous catalysis, we have studied the preparation, properties and reactions of a sodium film produced in pores of Vycor glass (Corning No. 7930). The purity of this silica material, its high surface area and uniform porosity of 40Å make this material highly suitable for carrying out reactions, since the pores are so small that light can be transmitted through the body of the support and can be used to carry out photochemical reactions wtihin the pores. This property has been utilized effectively by Turkevich and Fujita[1] in the preparation and stabilization at room temperature of methyl radicals. The purpose of this investigation was to prepare a film of sodium of at most three monolayer thick in order to carry out reactions within the pores of the Vycor glass--decreasing thereby the diffusion rate of the primary reaction product and thus facilitating the study of active species on surface. It was hoped that methyl radicals could be prepared by direct reaction of sodium with methyl iodide rather than by photolysis. It turned out that this reaction does not take place at room temperature and thus led to the investigation of the basic problem: under what conditions do free electrons from metal transfer to an adsorbed reactive species.

MATERIALS AND PROCEDURE

Ammonia and deutero-ammonia were purified by distillation of the liquid kept over sodium. Sodium metal was purified by triple distillation in a high vacuum. Its purity was attested by its ESR line (14 gauss wide) observed at room temperature. Porous Vycor glass (Corning No.7930) was purified by washing with conc. nitric acid, water and then heating in oxygen gas at 600°C for twelve hours. It showed no ESR signal at 90°K after ten minute irradiation with 2537Å light. The gases and organic compounds were purified by conventional vacuum techniques.

A Varian ESR spectrometer (Model V4500) operating at 9.5 monocycles and with the cavity in the TE 102 mode was used. Care was exercised in tuning and in placing of the sample in the cavity to insure correct shape of the signal. A Varian nuclear resonance spectrometer (Model 4300) was used at 11.212 megacylces to study the sodium nuclear resonance at 293° and 90°K. A conventional high vacuum apparatus was used for sample preparation and reaction studies.

EXPERIMENTAL RESULTS

Sodium-Liquid Ammonia System

The sodium in liquid ammonia solution with ammonia to sodium ratio of 4500 had a g value of 2.0023 and a line width which varied from 75 milligrams at room temperature to 125 at 103°K. On freezing at 78°K its width became 1.9 gauss.[2] ESR measurements were carried out at 103°K on the sodium-liquid ammonia solution (Fig.1) in pores of Vycor glass as the ammonia is removed. The original solution of NH_3/Na ratio of 72 gave a g value and a ΔH as indicated above. At ratios between 63 and 54 the line broadens to 1.7 gauss, retaining its g value. When the ratio is 5.2 this line at g=2.0023 still has a width of 1.6 gauss, and another line appears at 2.0048 with a width of 8 gauss. At a ratio of 1.8 the signal at 2.0048 still has a width of 6 gauss, and a new signal 60 gauss wide appears at 1.97 and remains unchanged as the ammonia is removed. The signal at 1.97 is ascribed to the metallic sodium film. When this sample is examined at 78°K it breaks up into three peaks. A sharp one at g 2.00 (ΔH=4.4) is the same as that of frozen sodium-ammonia solution and also that of sodium metal dispersed in paraffin. However it is also the signal obtained with oxygen (see below). The other broader peaks at 1.99 (ΔH=32) and 1.97 (ΔH=80) seem to be characteristic of sodium in pores of Vycor glass. The g values do not depend on the ratio of silicon on surface to sodium for the range 16 to 2. Only the peak at 1.97 is observed at room temperature and at that temperature its width is 48 gauss. To check the metallic origin of this peak a nuclear resonance was measured at room temperature and at 78°K. The frequency shift of this signal from that of a saturated solution of NaCl $K = (\nu-\nu_0)/\nu_0$ was 1.24×10^{-3}, slightly higher than the most reliable value in the literature, 1.13×10^{-3} for metallic sodium.[3)] The metallic sodium nuclear resonance signal could not be detected after exposure of the sample to air, undoubtedly because the width of the resultant sodium ion signal is too broad.

Reactions of Sodium on Vycor Glass

The sodium film on Vycor glass characterized by the ESR signal is a convenient material for study of electron transfer from sodium to various molecules adsorbed on the surface. The sodium on Vycor glass (θ=0.07) was heated to successively higher and higher temperatures in a sealed sample tube, and ESR measurements were made at 78°K. The signal disappeared by heat treatment at 400°C for two hours, presumably due to the reaction of the sodium with the silica. Hydrogen (100mm pressure) produced no change in the thermal behavior of sodium which seems to react faster with silica than with hydrogen to form NaH. Oxygen (100mm) was introduced at 78°K to a sample (θ=0.25). After three hours at this temperature a sharp signal at g=2.016 (ΔH=2.5) and a broad signal at the same g (ΔH=28) has been observed in addition to the typical sodium-Vycor glass signal. These signals remained unchanged for twenty-five hours at 78°K. One

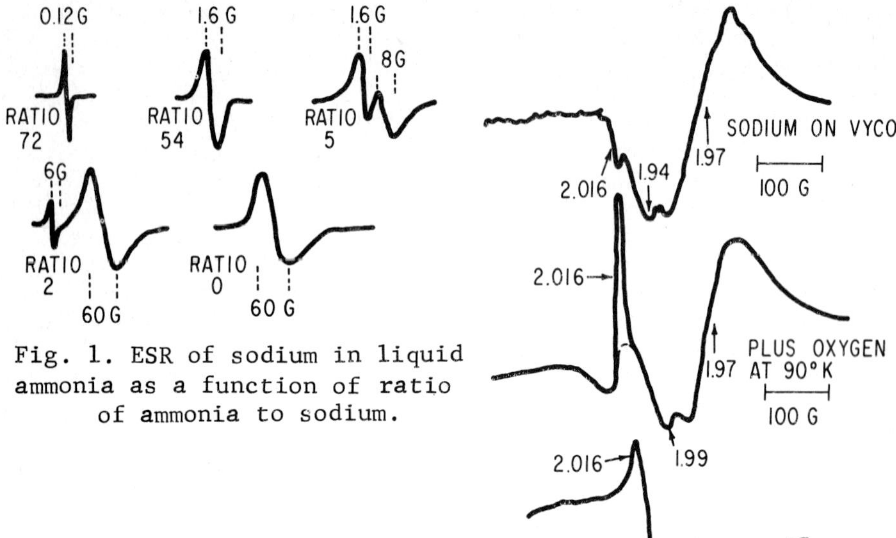

Fig. 1. ESR of sodium in liquid ammonia as a function of ratio of ammonia to sodium.

Fig. 2. Reaction of oxygen with sodium film

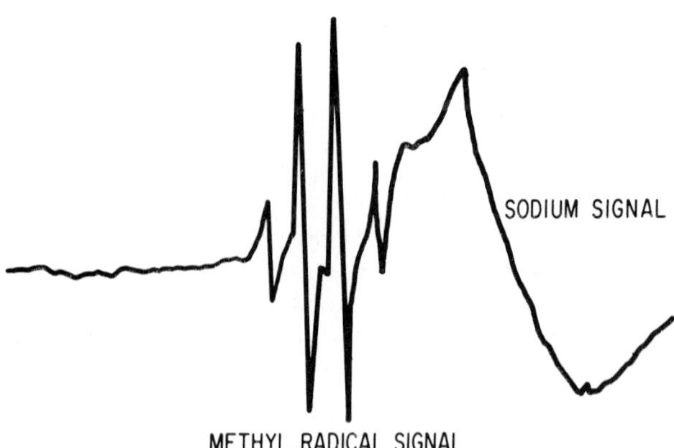

Fig. 3. Reaction of methyl iodide with sodium film in the presence of visible light.

and one-half hours at room temperature produced a change in the ESR signal with the appearance of a signal similar to O_2^- with g_x=2.016, g_y=1.999 and g_z=1.97. The sodium metal signal was still observed at g=1.97. The signal was stable at room temperature for one day and at 100°C for one hour. At 200°C the sodium signal at g=1.97 disappeared and the O_2^- signal decreased to one-tenth its value. Heat treatment at 300°C for two hours completely destroyed the signal.

Purified chlorine corresponding to θ=0.6 was contacted with sodium Vycor (θ=0.7) at 190°K for one minute and the ESR spectrum consisted of a doublet of four lines each which we ascribe to chlorine atoms (I for Cl being 3/2). The signal disappeared on raising the system to room temperature.

No change in the ESR signal was observed on contacting propylene, cis butene-2 or benzene (θ=0.2) for one week at room temperature with the sodium film. In the case of benzene no signal was produced by warming, by introduction of tetrahydrofuran, or by irradiation with 2537Å light. Introduction of 8×10^{-6} moles of naphthalene vapor to a system containing 1.4×10^{-5} moles of sodium in Vycor gave a strong unsymmetric signal with g_x=2.003 and g_y=1.999, and the intensity of the sodium signal decreased by one-half. This was taken to indicate a transfer of electron from sodium to naphthalene but that the species produced was immobile on the surface. Introduction of 0.03 ml of tetrahydrofuran developed a green color in the system and a spectrum of 21 sets of quartets of equal intensity and a separation of 1.04 gauss in the quartet.[4)] This spectrum is undoubtedly due to sodium ion--naphthalene ion pair system which is now mobile in the pore. Similar results were obtained with anthracene. A tetrahydrofuran solution of phthalocyanine 10^{-6} moles introduced to a system containing 1.4×10^{-6} moles of sodium produced a strong ESR signal with g=2.00 (ΔH=4 gauss). A weak signal was observed in the supernatant liquid. The system was extracted with tetrahydrofuran until there was no signal in the latter and then evacuated. This exhaustive extraction did not change the ESR spectrum. The system was heated with purified hydrogen at pressures of 1.5×10^{-2}, 1.0×10^{-1}, 1×10^{0}, and 1×10^{2} Torr. No change in the ESR spectrum was observed. This experiment was carried out to determine whether one could detect any interaction between hydrogen and the complex, since Tamaru[5)] found that a complex of this type catalyzed the hydrogen-deuterium equilibration.

Turkevich and Fujita[1)] showed that photolysis of methyl iodide in pores of Vycor glass produced a stable methyl radical. It was therefore of interest to see whether methyl radicals could be produced by transfer of electrons from sodium films to the methyl iodide. Contacting liquid CH_3I with sodium-Vycor systems for 45 minutes produced no change in sodium ESR signal nor any methyl radical signal. Warming the sample to 40°C for a minute produced disappearance of sodium signal but no methyl radical signal. This was taken as indication of the weak reactivity of the sodium with methyl radicals at room temperature and the strong reactivity of methyl radicals with sodium once the methyl radicals are formed. Attempts to catalyze the transfer were made in successive experiments in which tetrahydrofuran, oxygen, phthalocyanine and palladium films were used as catalysts for the transfer of electrons from sodium film to methyl radical. It was found that these materials had no catalytic activity for such transfer at room temperature.

Photochemical Transfer Reactions

An investigation was made of the effect of irradiation on the transfer. Turkevich and Fujita[1)] had shown that photolysis of methyl iodide could be carried out using 2537Å light. Comparison of the yield of methyl radicals produced by 20 minute irradiation using low pressure 500 watts mercury lamp using both pure Vycor, sodium-Vycor (θ=0.6), and methyl iodide (θ=0.2) gave at liquid nitrogen a yield of 10 to 100 arbitrary units, indicating that in the photolysis the efficiency of production of methyl radical is increased tenfold by the presence of sodium. Undoubtedly the sodium reacts with the iodine atoms produced in the photolysis to produce iodide ions,thereby lessening the back reactions. After irradiation,increasing the temperature to 190°K decreased the intensity of the methyl radical signal from 100 to 1 in the case of the sodium Vycor system, and 10 to 2.5 in the case of the Vycor alone emphasizing the reactivity of sodium with methyl radicals even at temperatures as low as 190°K. While irradiation with 3650Å G.E. type B-H6 16.9 watt/steread (incapable of dissociating methyl iodide in the vapor state) did not produce methyl radicals in pure Vycor, it did so in the presence of sodium at 90°K. The signal intensity of the methyl radical increased with irradiation time and was comparable in size in one hour with that of 2537Å irradiation in 20 minutes. A simultaneous decrease of the sodium signal was observed. Focused visible light of wavelength longer than 4200Å obtained from a 300 watt zirconium arc lamp produced a weak methyl ESR signal in two hours of irradiation. Addition of small amounts of water vapor or oxygen did not enhance the signal obtained by irradiation of the system at 90°K though it is stated that oxygen and water vapor enhance the photoemission of sodium films. We conclude from this that the photo reaction that takes place is not determined by the photo emission of electrons so much as by the change in the absorption characteristics of the methyl iodide on adsorption. When methyl iodide is adsorbed(θ=0.4) on a phthalocyanine-sodium-Vycor glass sample and irradiated with visible light at 90°K, a marked increase in intensity of the methyl radical is observed reaching constant value in fifteen minutes. The signal intensity of the phthalocyanine complex decreases while that of the sodium remains unchanged. These results show that the phthalocyanine complex is capable of transferring on irradiation its electron to the methyl iodide molecules to produce methyl radical.

DISCUSSION

The sodium in the film differs from the sodium in the bulk by having a smaller g value of 1.97 instead of 2.00 and by a larger Knight shift in nuclear resonance. Based on the argument that calcium has a low g value because of its low cohesive energy and that the Knight shift of sodium is inversely proportional to density, we conclude that the sodium films we produced are of low cohesive energy and that the sodium lattice is expanded.

The results obtained in many of the reactions are not surprising. However the non-reactivity of hydrogen with the sodium film was unexpected since sodium hydride is made by reacting hydrogen with sodium in an iron tube. The iron may play the role of a catalyst.

The non-reactivity of sodium with methyl iodide to produce methyl radicals can be rationalized on the following basis. The dissociation energy of carbon-iodine band is 54 kilocalories and the work function of sodium is 46 kilocalories. The electron affinity of iodine is 71 kilocalories. The production of methyl radical requires 29 kilocalories, which is furnished by visible light in the photo processes studied or by the electron affinity of the methyl radical to produce methyl ion in the subsequent reaction leading to the destruction of the methyl radical.[6)]

REFERENCES

1. J. Turkevich and Y. Fujita, Science 152, 1619-1621 (1966).

2. R. A. Levy, Phys. Rev. 102, 31 (1956); C.A. Hutchinson Jr. and R.C. Pastor, J. Chem. Phys. 21, 1959 (1953).

3. W.D. Knight, Phys. Rev. 76, 1259 (1949).

5. M. Ichikawa, M. Some, T. Onishi, and K. Tamaru, J. Catalysis 6, 336 (1966).

6. J. D. Roberts and M. C. Caserio, "Basic Principles of Organic Chemistry," W.A. Benjamin, Inc. N.Y. 1964, p.86.

DISCUSSION

J. R. ANDERSON

It is worth recalling that there is inferential chemical evidence for the formation of the adosrbed methyl ion as an intermediate during the reaction of methyl chloride at a sodium surface.[1]

If one can assume that the behavior of methyl iodide at a sodium surface under illumination is similar to that elucidated for methyl chloride,[2] I doubt whether the comments made in the present paper concerning the methyl iodide reaction are adequate. The methyl chloride/sodium reaction is photo-accelerated by illumination in what amounts to the F-center absorption band of the product layer (not a region of absorption of methyl chloride itself), and the action of the light is to promote the generation of F-center dimers which function as specific centers for the reaction of methyl chloride to give ethane and two incorporated chloride ions. There is no doubt that the reaction between methyl chloride and metallic sodium occurs.

1) Anderson, J. R. and McConkey, B. H., J. Catal. 9, 263 (1967).
2) Anderson, J. R. and McConkey, B. H., In "Reactivity of Solids" (J. W. Mitchell, R. C. Devries, R. W. Roberts and P. Cannon, eds.), Wiley-Interscience, p. 533, 1969.

J. TURKEVICH

We feel that the commentator has missed several significant details of our paper.

In the first place our system is quite different from the one he used. We studied the reaction of methyl iodide and not methyl chloride with sodium. Furthermore, our sodium film was less than a monolayer deposited in pores of uniform diameter of 40 Å, while his was about 2000 Å thick. In addition, we measured directly the transport of electrons from the sodium film with a sensitivity of about 10^{12} electrons, whereas the commentator inferred the transfer from analysis of gas product with a sensitivity of 10^{15} to 10^{17} molecules. Under special conditions we do observe transfer of electrons to the methyl iodide without production of methyl radicals.

No evidence is presented in reference (2) that the behavior of methyl iodide with a sodium surface is similar to that elucidated for methyl chloride. Methyl iodide is mentioned only in connection with its reaction with copper and silver.

There is no experimental evidence in (1) that an adsorbed methyl ion is formed. It is just an assumption made on general considerations of bond character and greater exothermicity of the reactions producing the ion. No evidence is presented that the radical may not be the intermediate. Postulating an uncharged radical as an intermediate would avoid the difficulty of two negative ions reacting with each other to form ethane as is postulated in (1).

J. MANASSEN

You state that no electron transfer takes place in the system sodium phthalocyanine/methyl iodide; with illumination electron transfer occurs. On the other hand, it might be possible that a charge-transfer complex is formed. Therefore, did you notice a change in color when the methyl iodide was brought in?

J. TURKEVICH

No observation was made of color change. If there was a charge transfer between the sodium phtyalocyanine and methyl iodide, we would expect the ESR of the sodium phthalocyanine to change.

K. TAMARU

May I ask what kind of phthalocyanine you have used in your experiments? We have studied hydrogen exchange reaction between propylene and hydrogen on the stoichiometric electron donor-acceptor (EDA) complexes of various phthalocyanines with alkali metals. The adsorbed state of propylene was examined by analyzing the reaction products, d_1- and d_2- species by means of microwave spectroscopy. Over metal-free phthalocyanine no exchange reaction took place between propylene and deuterium, although the H_2-D_2 exchange reaction proceeded at a considerable rate. Over the Ni-$Pc^{4-}4Na^+$n-propyl is the main reaction intermediate, while it is mainly isopropyl intermediate over the $CoPc^{5-}5Na^+$.

J. TURKEVICH

The phthalocyanine used was ultra pure, metal free phthalocyanine, obtained from Dr. Sol Harrison of the RCA Laboratories.

D. E. W. VAUGHAN

Acid treatment of silica glass followed by heat annealing (oven at 600°C) will result in some devitrification of the glass surface. Deposition of a Na film followed by heat treatment would also probably lead to devitrification with the formation of stuffed silica phases (cridymite or cristobalite), in which case the signal would disappear if the stuffed phase contained Na^+ and would remain if the trapped species was Na°. Did you observe either crystalline silica polymorph before or after the experiments (e.g. by X-ray photography)?

J. TURKEVICH

We did not carry out any X-ray examination of the silica after sodium treatment. The sodium film was made by crystallization from liquid ammonia solution. The effect you mention took place at 400°C when the sodium signal disappeared. Undoubtedly devitrification with stuffed silica phases took place.

Paper Number 39

CHEMICAL CHARACTERIZATION OF SUPPORTED EUROPIUM BY MÖSSBAUER SPECTROSCOPY

P. N. ROSS, JR. and W. N. DELGASS
Department of Engineering and Applied Science, Yale University
New Haven, Connecticut U.S.A.

ABSTRACT: Significant reduction of Eu^{3+} to Eu^{2+} on η-Al_2O_3 and SiO_2 supports has been observed after exposure to H_2 at 400-500°C. On η-Al_2O_3 the degree of reduction depended on Eu loading and was enhanced by the presence of Cl^-. Mössbauer and x-ray diffraction data indicated high dispersion of Eu on both supports, but Eu was more strongly bound to η-Al_2O_3 than SiO_2. Reduction of Eu^{3+} to Eu^{2+} in CO and reoxidation in H_2O were also observed at 400-500°C. Implications of the results to catalytic oxidation are discussed.

1. INTRODUCTION

A variety of oxidation-reduction reactions over oxide catalysts have been found [1] to proceed via a regenerative sequence in which the substance to be oxidized reacts with lattice oxygen of the catalyst surface and the surface oxygen content is restored by reaction of the catalyst with the oxidizing agent. The rate controlling step in this sequence is often the reduction of the surface, which is controlled by the strength of the surface cation-oxygen bond. The rate of heterogeneous oxygen exchange ($^{16}O_{lat}$ + 1/2 $^{18}O_2 \rightarrow {}^{16}O^{18}O$) has been found to be a direct measure of the surface cation-oxygen bond in transition metal oxides [2] and correlates with the activation energy for the oxidation of hydrogen, methane, carbon monoxide[2] and propylene[3]. The rate of heterogeneous oxygen exchange over rare earth oxides has been found [4] to be at least equal to the rate for the transition metal oxides, suggesting that rare earth oxides may have significant activity for oxidation reactions which proceed via a regenerative sequence.

Europium was chosen for a study of this interesting possibility because of its multiple valence and convenient Mössbauer effect parameters. Europium occurs most frequently in the 3+ oxidation state. The 2+ state is generally unstable in solids exposed to air, but the mixed oxide, Eu_3O_4,[5] has been found to be stable. The reduction of bulk Eu_2O_3 to either EuO or Eu_3O_4 is thermodynamically favored only at elevated temperatures (1200°C). This work, however, reports the reduction of supported, highly dispersed europium by hydrogen and carbon monoxide at moderate temperatures (400-500°C) as observed by Mössbauer spectroscopy. The Mössbauer effect and its use in the characterization of catalysts has been reviewed previously[6]. In the europium Mössbauer effect, the 2+ oxidation state ($4f^7 5s^2 5p^6$, $^8S_{7/2}$) is separated from the 3+ state ($4f^6 5s^2 5p^6$, $^7F_0$) by an isomer shift of 10-15 mm/sec and an unambiguous assignment of oxidation state is possible. The quadrupole splitting, produced by the interaction of the nuclear quadrupole moment with the gradient of the electric field at the nucleus, is less readily identified owing to the high nuclear spin (I_e=7/2, I_g=5/2) and the large number of allowed transitions. Although there is some disagreement on the interpretation of Eu^{3+} quadrupole splitting[7], Blok and Shirley[8] have shown that even considering the $^7F_2$ excited state the major contribution to the splitting should arise from asymmetry in the charge distribution external to the ion. Lattice sum calculations [9] have shown that electric field gradients in rare earth salts are 3-10 times smaller than those for transition metal salts. Together with the large number of resonant lines, the gradients in europic salts generally produce only a mild broadening of the spectral line. Magnetic dipole splitting resulting from the interaction of the nuclear spin with a magnetic field at the nucleus completely removes the degeneracy of the nuclear levels and results in a characteristic, magnetically-split spectrum. Since the ground state of Eu^{2+} is paramagnetic (J=7/2), paramagnetic relaxation phenomena can also occur. High dilution of Eu^{2+} ions can reduce the rate of spin-spin exchange[10] and produce an effective magnetic field at the nucleus resulting in magnetic broadening even at 300°K. The magnetic broadening in the Eu^{2+} line may thus give a qualitative indication of the extent of dilution of reduced ions on the catalyst surface.

This work reports the use of Mössbauer effect spectroscopy to observe the state of dispersion of europium on high surface area supports, the $Eu^{3+} \rightarrow Eu^{2+}$ reduction by hydrogen and carbon monoxide and reoxidation by air and water vapor, and to give some insight into the chemical nature of the Eu^{3+} - support interaction.

2. EXPERIMENTAL

Harshaw η-Al_2O_3, 325 mesh and having a measured BET surface area of 130 m^2/g , and Cab-O-Sil silica M-5, having a measured BET surface area of 170 m^2/g , were used as supports. The supported europium was obtained by impregnation at pH 7-7.5 with aqueous solution of the nitrate and chloride salts (Lindsay Rare Earths 99.9% purity) followed by air drying at 140°C for 12 hours. The europium content was determined by x-ray fluorescence. The sample powder was pressed into a wafer 5/8" in diameter, approximately 2mm. thick and weighing 0.25 g . The wafer was mounted in a stainless steel holder which was held in a tubular glass cell by stainless steel leaf springs. Thinned glass windows allowed good transmission of gamma rays through the cell, which had a path length of about 1". The cell was heated by an external heater, while the temperature was measured by a chromel-alumel thermocouple in contact with the wafer. The cell was connected to a system of gas cylinders at one end and a standard vacuum and gas-handling system at the other. Thus evacuation, or flow or batch exposure to gases at temperatures from 25°C to 500°C could be achieved. Matheson Ultra High Purity grade hydrogen (99.999% dew point -85°F) and Matheson C.P. grade carbon monoxide (99.5%) were used without further purification.

The Mössbauer spectra were obtained at room temperature from a constant acceleration electromechanical device with a scintillation detector. The source, ^{151}Sm in Sm_2O_3, was mounted on the drive, the absorber being the wafer mounted inside the glass cell. The isomer shift of ^{151}Eu in Eu_2O_3 was taken as zero velocity and the velocity scale of the spectrometer was calibrated using the ^{57}Fe Mössbauer spectrum of iron foil. Computer fitting of the data was done using the Variable Metric Minimization algorithm of Davidon[11].

3. RESULTS

3.1 Deposition of europium

The europium content by weight as determined by x-ray fluorescence and the Mössbauer parameters IS, the isomer shift relative to Eu_2O_3, and FWHM, full-width at half maximum, are given in table 1 for several freshly prepared samples. The Mössbauer parameters of some salts are included

TABLE 1

Mössbauer effect parameters for freshly prepared supported europium

Code	Support	Salt	Eu content Weight %	IS (mm/sec)	FWHM (mm/sec)
2.3 Eu(N)/A	η-Al_2O_3	nitrate	2.3	-0.62	2.85
5.5 Eu(N)/A	η-Al_2O_3	nitrate	5.5	-0.62	2.85
12.5 Eu(N)/A	η-Al_2O_3	nitrate	12.5	-0.58	3.05
27.0 Eu(N)/A	η-Al_2O_3	nitrate	27.0	-0.58	3.05
10.5 Eu(N)/S	SiO_2	nitrate	10.5	no effect	
5.5 Eu(Cl)/A	η-Al_2O_3	chloride	5.5	-0.84	2.75
		Eu_2O_3		0	3.40
		$Eu(OH)_3$		-0.55	3.62
		$Eu(NO_3)_3 \cdot 6H_2O$		-0.92	3.29
		$EuCl_3 \cdot nH_2O$		-0.89	2.93

for reference. For alumina, essentially all the europium placed in solution was deposited on the support, but with silica 20% of the europium in solution was not deposited on the support. The europium on silica as prepared showed negligible Mössbauer effect. After evacuation for twelve hours at room temperature, a resonant line was observed at -0.45 mm/sec with a width of 3.00 mm/sec and percent effect comparable to that for freshly prepared 5.5 Eu(N)/A. The narrow line widths, close to that observed for dilute europic ions in ice[12)], indicate that the europium in the freshly prepared samples is in a state with near-cubic symmetry. The isomer shifts indicate that after drying at 140°C the deposition from the nitrate salt probably does not produce a surface nitrate but the deposition from the chloride salt may result in an europium surface complex containing some Cl^- nearest neighbors.

3.2 Reduction of supported europium

The reduction of supported europium was accomplished by heating the wafer in flowing hydrogen (or, in one case, CO) for a fixed time and temperature, cooling to room temperature, and sealing the wafer in the hydrogen (CO)atmosphere while collecting the Mössbauer spectrum. No reduction was observed for any supported europium at a temperature below 400°C, while, in general, some Eu^{2+} was observed for all the preparations reduced at 400-500°C. For any given preparation and temperature, increasing the time of reduction from 2 hours to 6 hours increased the amount of 2+ observed, but after 6 hours no additional 2+ was formed. In order to compare the effect of support and preparation on the extent of reduction, 500°C and 6 hours were used as the standard conditions. Previous work [13)] on Eu_3O_4 has shown that the f-values for Eu^{2+} and Eu^{3+} are equal at 25°C and that the ratio of areas for the two resonant lines is the relative proportion of each ion in the crystal. In this work the percent reduction has been defined as Area (2+) / [Area(2+) + Area(3+)]. The extent of reduction at the standard conditions as a function of preparation is summarized in table 2.

The Eu^{3+} isomer shifts and line widths in table 2 were measured with a -6 to +6 mm/sec scan. Computer fits gave the 95% confidence limits as ± 0.05 mm/sec. The Eu^{2+} isomer shifts and line widths were measured with a -32 to +32 mm/sec scan and have computed 95% confidence limits of ± 0.25 mm/sec. The zero isomer shift of the unreduced Eu^{3+} in any preparation heated in H_2 above 400°C was observed regardless of the extent of reduction. The reduction of the silica supported europium exhibited a strong temperature dependence, the % reduction doubling from 425°C to

TABLE 2

Reduction of supported europium in hydrogen (CO where noted) at 500°C

Preparation	Mössbauer Parameters Eu^{2+} IS (mm/sec)	FWHM (mm/sec)	Area (mm/sec)	Eu^{3+} IS (mm/sec)	FWHM (mm/sec)	Area (mm/sec)	% Reduced
2.3 Eu(N)/A	-13.3	9.7	0.050	0	4.2	0.133	27
5.5 Eu(N)/A	-13.2	9.2	0.203	0	4.3	0.172	54
12.5 Eu(N)/A	-12.8	7.8	0.301	0	4.3	0.687	30
27.0 Eu(N)/A	-13.3	8.3	0.437	0	4.2	1.160	24
10.5 Eu(N)/S	-14.0	9.0	0.205	0	3.8	0.278	43
5.5 Eu(Cl)/A	-13.6	7.2	0.300	0	4.3	0.074	80
5.5 Eu(Cl)/A by CO	-14.0	8.6	0.282	0	4.3	0.114	71
Eu_3O_4	-12.6	3.9	0.584	+0.67	5.0	1.030	-

475°C, while the extent of reduction on alumina varied only slightly in 450-500°C range. Deposition of europium from the chloride salt produced a material which could be reduced to a considerably greater extent than that from the nitrate, as illustrated more dramatically by fig. 1. No reduction of a wafer of pure Eu_2O_3 mixed with η-Al_2O_3 was observed at conditions used for table 2.

The Eu^{2+} ion was observed to be oxidized essentially instantaneously on exposure to air at room temperature. Exposure of a prereduced 5.5Eu(N)/A sample to 22 torr water vapor for six hours at temperatures from 25-200°C did not oxidize the Eu^{2+}, but it was oxidized when the temperature was increased to 400°C.

With one exception, x-ray diffraction of all the wafers following reduction did not reveal the presence of any Eu containing phases. For sample 27Eu(N)/A, a broadening of the (222) line of cubic (C-Type) Eu_2O_3 was observed which, according to the simplest line broadening theory[14], corresponded to 400 Å particles.

3.3 Hydration and dehydration

In order to further examine the chemical state of supported Eu after various treatments, the effect of adsorption of water vapor on the isomer shift and line width was investigated. The result for a series of sequential treatments is shown in fig. 2. Dehydration in vacuo and reduction in hydrogen at high temperature both markedly increased the line width and isomer shift of the 3+ line, the effect being larger for the alumina support than for silica. The adsorption of water nearly reversed these effects particularly following reduction and reoxidation by air, as shown in fig. 3 for 5.5% europium on alumina. After treatment by hydrogen at 500°C and air at room temperature, the 3+ line width increased to 4.2 mm/sec. On adsorption of water the line width fell to 3.4 mm/sec. Qualitatively these effects suggest that many of the europium ions on the support are affected by the water and therefore that the dispersion must be high.

4. DISCUSSION

4.1 Reducibility of supported europium

The extent of reduction of supported europium is strongly influenced by the total amount of europium present on the support. This is shown by

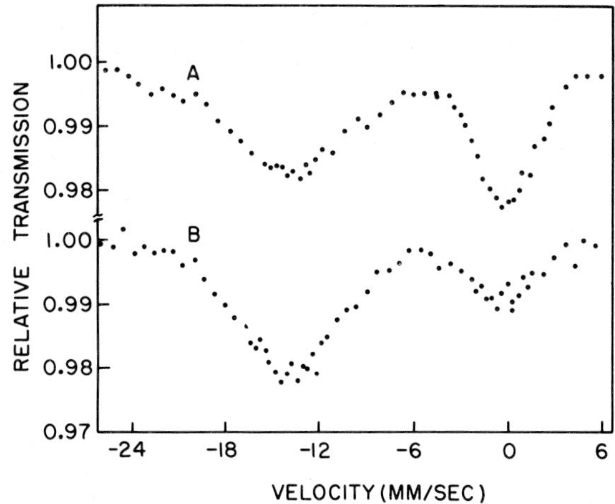

Fig. 1 Room temperature Mössbauer spectra of A.) 5.5Eu(N)/A, B.) 5.5Eu(Cl)/A reduced in hydrogen at 500°C

the first four entries of table 2 where the europium content increases from 2.3 to 27%. The total number of europium atoms reduced on treatment by hydrogen is proportional to the product of % reduction and the percent europium. Thus, the total number of atoms reduced increased monotonically with europium content. Increasing the europium content from 2.3 to 5.5% produced a five-fold increase in the number of europium atoms reduced, but further increasing the Eu content was not nearly so productive. In terms of an equivalent europium monolayer, an upper limit computed from the alumina surface area and the diameter of the europium and oxide ions, the total coverage varied from 7 to 114% of a monolayer, and the coverage by reduced Eu^{2+} varied from 2 to 27% of a monolayer. The fact that the highest extent of reduction was not observed for the lowest loading suggests that isolated Eu^{3+} ions are not the ones most easily reduced.

The europium content also appeared to affect the line width of Eu^{2+}, increasing concentration of europium leading to a narrower 2+ line. Assignment of the unusually large width of the Eu^{2+} line to paramagnetic relaxation is well founded in that the line is symmetric and too wide to be accounted for by either a distribution of isomer shifts or an unresolved quadrupole interaction. Furthermore, in Eu-Y zeolite, where the Eu^{2+} ions are expected to be more isolated, the Eu^{2+} width is about 12mm/sec at 25°C[15]. If the Eu^{2+} line width broadening is associated qualitatively with increasing mean Eu^{2+}-Eu^{2+} distance, then the dependence of the width on Eu^{2+} concentration follows directly. Using this correlation, 5.5Eu(Cl)/A was found to have a smaller mean Eu^{2+}-Eu^{2+} distance than 27 Eu(N)/A even though the latter had 50% more Eu^{2+} ion. More detailed analysis in terms of the state of aggregation of Eu on the surface must await a more complete theoretical development.

Although there exists a temperature range where reduction of europium on alumina can be observed while none is observed for europium on silica, at 500°C 10.5Eu(N)/S and 5.5 Eu(N)/A have similar extents of reduction. On η-Al_2O_3, however, loading and the presence of Cl^- have a marked effect on Eu reducibility. The reduction by carbon monoxide and re-oxidation by

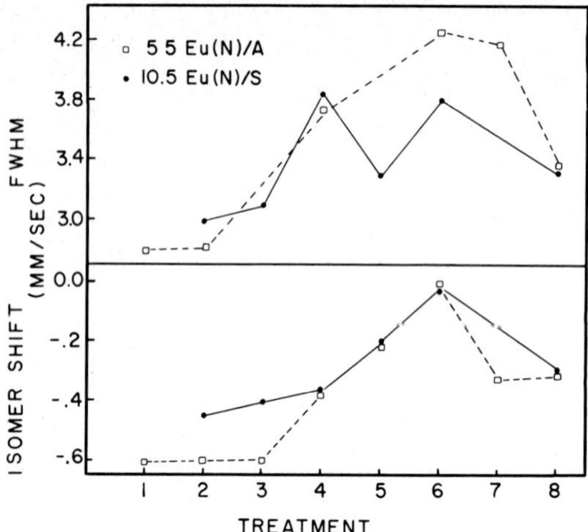

Fig. 2 Changes in europium 3+ line width and isomer shift with treatment: 1.) freshly prepared; 2.) dehydration in vacuo at 25°C; 3.) rehydration following 2.; 4.) dehydration in vacuo at 400°C; 5.) rehydration following 4.; 6.) reduction in H_2 at 500°C; 7.) reoxidation in air at 25°C; 8.) rehydration following 7.

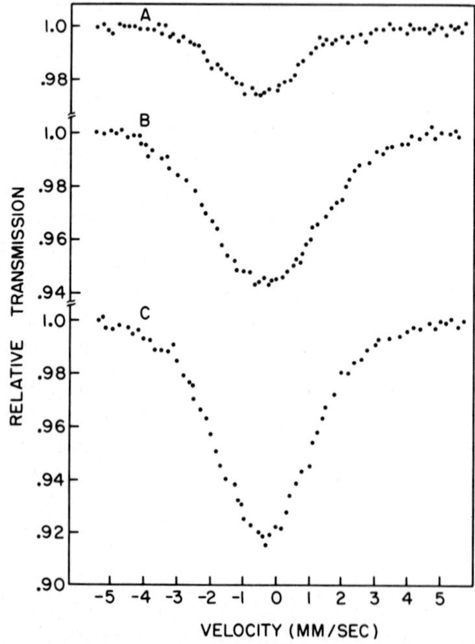

Fig. 3 Room temperature Mössbauer spectra of 5.5Eu(N)/A A.) as prepared, B.) following reduction at 500°C and reoxidation in air, C.) rehydrated after B).

water of supported europium at high temperature indicate that this material may be a good high temperature shift catalyst and may have promising activity for other redox reactions which proceed via the regenerative sequence.

4.2 Dispersion of europium on the support

At 27% europium, or more than three equivalent monolayers of 3+ ion, the result of heating in hydrogen at 500°C was the growth of large (400 Å) particles of Eu_2O_3, easily seen by x-ray diffraction. The x-ray diffraction for the other preparations following this treatment indicated there were no large particles, >100 Å, on the surface at <12.5% europium.

Since some europium salts with many waters of hydration[7] and europic ions in ice[12] have the lowest Mössbauer line widths, the narrow line width of the freshly prepared materials is consistent with a high degree of hydration of the europium surface complex. Dehydrating in vacuo at 400°C resulted in a broadened 3+ line (from 3.1 to 3.8 mm/sec.) and an isomer shift towards zero, as shown in treatment 3-4 of fig. 2. This loss of water was partially reversible since rehydration restored the line width to 3.3, but the isomer shift was different, indicating some chemical change in the Eu^{3+} environment had taken place. Again the magnitude of the rehydration effect indicates high dispersion since many of the Eu^{3+} ions were affected. It is tempting to assign the line broadening to quadrupole splitting, but nearly all trivalent europium compounds have an isomer shift in the range +0.5 to -1.0 mm/sec.[7] and this mild broadening could be caused by multiple sites of slightly different chemistry. Treatment of the alumina supported europium in hydrogen at 400-500°C gave a residual 3+ line that was shifted to zero velocity (± .05mm/sec) and broadened to 4.3 mm/sec for all degrees of reduction from 0-80%. In this case the isomer shift and the magnitude of the line width point to quadrupole broadening of the 3+ line, corresponding to an $|e^2qQ|$ of 8 mm/sec[16]. An e^2qQ of this magnitude for europium represents high asymmetry of the environment, such as would be expected of a surface ion. The magnitude of the Eu^{3+} line width changes were not as large for the silica support as for alumina support. This may represent less asymmetry of the europium environment on the silica surface.

There also appear to be some differences in the strengths of the interactions of the europium with the alumina and silica supports. The deposition of europium onto silica produced very weak bonding with the support since there was negligible Mössbauer effect for this state. Removal of weakly bound water strengthened the interaction, but rehydration reversed the effect. Only high temperature treatment in vacuo or in hydrogen produced strong interaction with the support. Even then the bonding was not as strong as with alumina, as shown by the difference in average recoil-free fraction. In table 2, 10.5Eu(N)/S has a total area of 0.483 and 12.5 Eu(N)/A a total area of 0.988, and the average recoil-free fraction, proportional to the factor area/Eu concentration, is clearly greater for europium on alumina than for europium on silica.

5. CONCLUSIONS

Mössbauer spectroscopy has been used to show that roughly half of the Eu^{3+} ions supported on SiO_2 or η-Al_2O_3 could be reduced to Eu^{2+} in hydrogen at 400-500°C. On η-Al_2O_3 the extent of reduction depended on Eu loading and was enhanced by the presence of Cl^-. Reduction of Eu was more strongly temperature dependent on the SiO_2 support than on η-Al_2O_3. Eu_2O_3 is thermodynamically stable under the conditions used and no reduction of Eu_2O_3 mixed with η-Al_2O_3 was observed. Eu^{2+} was quickly oxidized in air at 25°C and in H_2O at 400°C and could be produced by reduction in CO at 500°C. An Eu phase was observed by x-ray diffraction only for the highest loading. Changes in Mössbauer parameters as a function of sample treatment in H_2, air, and H_2O suggest that much of the Eu was affected by adsorption and that dispersion was high. The bonding of Eu^{3+} to the support was found to be considerably stronger for η-Al_2O_3 than SiO_2.

In view of the importance of the regenerative sequence in catalytic oxidation, the reducibility of a surface is a parameter of great interest. The fact that oxidation-reduction of Eu was observed under catalytically interesting conditions and that the reducibility was a function of treatment, loading, and sample chemistry suggests that supported europium will be an interesting oxidation catalyst.

ACKNOWLEDGEMENTS

This work has been supported by NSF institutional grant GU-3282 to Yale University and grants from the Chevron Research Company. Acknowledgement is also made to the donors of the Petroleum Research Fund, administered by the American Chemical Society for partial support of this research. One of us (PNR) thanks Texaco Inc. for a graduate fellowship during part of this work. The generous assistance of M. Kaplan and J.C. Love was most important in the successful construction of the Mössbauer spectrometer. We are indebted to R. C. Rau for supplying the Eu_3O_4 sample and to W. P. Wolf and S. Mroczkowski for $Eu(OH)_3$.

REFERENCES

1) G. K. Boreskov, Kinetika i Kataliz 11 (1970) 374.
2) V. A. Sazonov, V. V. Popovskii and G. K. Boreskov, Kinetika i Kataliz 9 (1968) 307.
3) M. Ya. Rubanik, K. M. Kholyavenko, A. V. Gershinghorina, and V. I. Lazukin, Kinetika i Kataliz 5 (1964) 666.
4) G. V. Antoshin, K. M. Minachev, and R. V. Dmitriev, Kinetika i Kataliz 9 (1968) 816.
5) R. C. Rau, Acta Cryst. 20 (1966) 716.
6) W. N. Delgass and M. Boudart, Catalysis Reviews 2(1) (1968) 129.
M. C. Hobson, Jr., Adv. Coll. Interface Sci. (in press).
V. I. Goldanskii and I. P. Suzdalev, Russ. Chem. Rev. 39 (1970) 609.
7) G. W. Dulaney and A. F. Clifford, in Mössbauer effect methodology (ed. I. J. Gruverman, Plenum Press, 1970) 65.
8) J. Blok and D. A. Shirley, Phys. Rev. 143 (1966) 278.
9) G. Burns, Phys. Rev. 128 (1962) 2121.
10) H. H. Wickman Phys. Letters 31A (1970) 29.
11) W. C. Davidon, Argonne National Laboratory Rept. ANL-5990 (1966).
12) A. Nozik, Unpublished PhD dissertation (Yale University 1967).
13) H. H. Wickman and E. Catalano, J. Appl. Phys. 39 (1968) 1248.
14) B. E. Warren, X-Ray diffraction (Addison Wesley 1969) 253.
15) W. N. Delgass and E. Samuel, ACS National Meeting (Boston, Mass. 1972).
16) E. Samuel, P. N. Ross, Jr. and W. N. Delgass, to be published.

DISCUSSION

B. DELMON

The results by P. N. Ross and W. N. Delgass are very interesting in many respects. I particularly noticed the possibility of an europium catalyst becoming a good high temperature shift conversion catalyst.

The solid state transformation approach in Professor W. N. Delgass's report is of particular interest for me. My question is in relation to the progressive increase of the proportion of reduced Eu^{2+} with increasing Eu_2O_3 content, followed by a region of much slower increase. For example, the proportion of Eu^{3+} ions which can be reduced is much higher with a sample containing 5.5% than with a sample with only 2.3% europium. This effect is very similar to what is observed in the reduction of many other supported oxides, with the additional implication of the final flattening of the curve.

There might be two explanations for this effect. One may assume that the first portions of the deposited oxide form new compounds with the carrier, and that this compound is more difficult to reduce. One can also emphasize the nucleation step in the reduction to lower oxides and assume that too high a dispersion prevents nucleation, with the higher oxide clusters being too small to allow the formation of a stable nucleus. We had to resort to this sort of explanation in our studies of the reduction of supported MoO_3 and NiO.[1-3]

What is the opinion of Professor Delgass concerning the possible explanation in the case of Eu_2O_3? The action of reduction catalysts, like palladium, gave us much valuable information in the reduction of supported MoO_3 or NiO. Does Professor Delgass think that catalytic effects similar to those which Professor Boudart observed in the reduction of tungsten oxide[4] or to our own results with MoO_3 or NiO could occur in his case and shed some additional light on the mechanism of the reduction of supported Eu_2O_3?

1) J. Masson, B. Delmon, J. Nechtschein, Compt. Rend. Acot. Sci., 266 Ser. C (1968), 428.
2) J. Masson, J. Nechtschein, Bull. Soc. Chim. Fr., (1968), 3933.
3) A. Roman, B. Delmon, Compt. Rend. Acot. Sci., 273, Ser. C (1971), 1310.
4) J. E. Benson, H. W. Kohn, M. Boudart, J. Catal., 5 (1966), 307.

W. N. DELGASS

These are both interesting points. It seems likely that reduction would be easier for Eu^{3+} doublets than singlets and that any favorable range of cluster size would be fairly narrow since thermodynamics do not permit bulk reduction of Eu_2O_3. Whether nucleation could control reduction in our samples depends on how large the europium clusters are. Though we have strong evidence that the dispersion of the europium is high we do not know the average cluster size as a function of loading and therefore cannot accurately assess the nucleation effects you mention. We have observed, however, that Eu^{3+} in single-exchanged Eu^{3+} Y zeolite is nearly 100% reducible to Eu^{2+} in H_2. This indicates that chemical interactions with the support can alter reducibility and that small europium oxide crystallites are not required for reduction.

We have also been intrigured by the effect of noble metal reducing catalysts on the reduction of oxides and in the phenomenon of hydrogen spillover. Early in this program we observed that co-impregnation of europium nitrate and chloroiridic acid on alumina led to higher reducibility of the Eu^{3+} than impregnation of europium nitrate alone. Interpretation of this experiment is now confused by the chloride ion enhancement of reducibility mentioned in this paper. We have not yet returned to the study of this question but plan to do so.

J. HAPPEL

Perhaps the success of alumina as a support compared with silica can be explained by the formation of europium aluminate, an ionic crystal of the perovskite type. Such a material might be reduced more readily than Eu_2O_3 to form an anion defect structure. This hypothesis could be tested by reduction at a temperature of perhaps 800-900°C with hydrogen, or perhaps at lower temperatures if the catalyst were prepared by co-precipitation.

W. N. DELGASS

We have not tried any high temperature reductions or catalyst preparations by co-impregnation. While europium aluminate surface compound formation is a definite possibility and could explain some of the chemical differences between supports, we feel that other considerations are likely to be more important in explaining Eu^{3+} reducibility since Eu(N)/A and Eu(N)/S have about the same reducibility at 500°C.

W. H. MANOGUE

It is well known that transition metals can be ion exchanged onto silica and alumina. For example on exhaustive washing of silica gel soaked in ferric nitrate solution, some iron cannot be removed. In view of this, I suggest that the behavior of your material on reduction implies the presence of some of the europium in the ion exchanged form. Did you wash your material after impregnation and then check for unremoved europium?

W. N. DELGASS

While direct cation exchange is possible on these materials, it is not favored at the low pH used in our experiments.[1)] Variation of the pH of the starting impregnation solution from pH = 4 to pH = 7.5 (at which $Eu(OH)_3$ is insoluble) produced no observable differences in alumina supported samples. Furthermore, the lack of a Mössbauer peak for freshly prepared 10.5 Eu(N)/S indicates that any ion-exchanged Eu is either low in concentration or weakly bound to the SiO_2 surface in the presence of water. Ion exchange on alumina can be checked nicely by your washing experiment, which we will try. A negative result of that experiment, however, would not rule out the possibility that Eu^{3+} moves into ion exchange positions after high temperature treatment.

1) S. C. Churms, J. South African Chem. Inst., 19, 98 (1966).

M. BOUDART

It is worth stressing that Mössbauer spectroscopy, in spite of its limitations, presents the great advantage to permit study of the catalyst during the catalytic reaction. Do you plan to carry out such an investigation?

W. N. DELGASS

We agree that the possibility of in situ studies of the catalyst at reaction conditions is one of the great strengths of Mössbauer Spectroscopy. We have been delayed in such experiments by some difficulties in fabrication of a suitable high temperature cell, but plan to begin such an investigation in the near future. It is well known from the work of Professor Tamaru and others that in addition to steady state experiments, spectroscopic studies in which the transient response of the system can be followed are particularly advantageous. In the case of europium, where the time required to collect a spectrum as long, dynamic experiments can be carried out only if a cyclic perturbation is applied to the system and several spectra are collected in phase with the perturbation. In the case ^{57}Fe Mössbauer spectroscopy with a high count rate spectrometer and a narrow scan, the direct transient response may be accessible.

J. TURKEVICH

What is the sensitivity of detection of various valence states of europium and what is the lower limit in total amount that can be used in the experiment?

W. N. DELGASS

The sensitivity of detection is, of course, somewhat a question of how long one is willing to wait. In this case, the Eu^{2+} line is considerably broadened by spin-spin relaxation. For the samples reported here, a practical lower limit of Eu^{2+} is of the order of 5% of the total Eu. In general, the lower limit for resonant absorption is about 10^{18} Eu atoms/cm^2 in the γ ray beam. To balance resonant and non-resonant absorption, however, the sample should be > 0.1 wt. % Eu.

Paper Number 40

HYDROGEN SPILLOVER ONTO INORGANIC OXIDES AND ZEOLITES

W. C. NEIKAM and M. A. VANNICE*
Sun Research and Development Company
Marcus Hook, Pennsylvania 19061

ABSTRACT: Much discussion in recent years has centered on the phenomena of hydrogen spillover (the diffusion of hydrogen from a surface capable of dissociating H_2 onto the surface of an adjoining solid), however unequivocal examples of spillover are rather sparse. In this paper we show clearly the presence of spillover from platinum black onto a number of inorganic oxides and zeolites at room temperature. It was found that a bridging compound between the platinum and the hydrogen accepting solid is necessary in order for spillover to occur. Various bridging compounds were studied and it was concluded that large organic ring compounds were particularly effective (i.e., perylene and perhydrophenanthrene). The magnitude of the hydrogen uptake suggests that hundreds and perhaps thousands of hydrogen atoms migrate through a single bridge. The nature of the ultimate sink for hydrogen is discussed with reference to the zeolites.

1. INTRODUCTION

There has been considerable discussion during the past few years on the subject of hydrogen spillover; however, few unambiguous examples of spillover have appeared in the literature. The clearest demonstrations of spillover are the Pt/WO_3 system[1)], the Pt/C system [2,3)], the Pt/CuO system[9)], and our recent results on the Pt/CeY zeolite system[4,5)]. In each of these systems, a bridging compound is required to promote spillover from the platinum particles to the hydrogen-accepting solid. Water serves as a bridge in the Pt/WO_3 and Pt/CuO systems, carbon dendrites in the Pt/C system, and perylene in the Pt/Ce-Y zeolite system. The nature of the ultimate sink for hydrogen was not considered during the study of the Pt/C system. Bulk reduction occurs for both WO_3 and CuO. For the Pt/CeY sieve mixtures, we have offered evidence which shows that the sink is associated with adsorbed oxygen, either as Ce^{4+} or as an adsorbed oxygen species.

In this paper, we report on hydrogen spillover onto a number of inorganic oxides and zeolites which suggests that spillover may be a more common phenomenon than expected. In addition, we show that a number of organic molecules can serve as bridges although aromatic hydrocarbons appear to be more efficient than other compounds tested.

*Present address: Esso Research & Engineering Company
Corporate Research Laboratories
Linden, New Jersey 07036

2. EXPERIMENTAL

Details of the sample preparation have been described previously[4]. All samples were pretreated at 500°C in flowing air for 2 hr. prior to adsorption of perylene and measurement of hydrogen uptakes. The amount of hydrogen taken up by the samples at room temperature was monitored via measurement of pressure change in a typical high-vacuum adsorption system incorporating a Texas Instrument Pressure Gage. The hydrogen was purified by passage through a palladium alloy thimble manufactured by the Milton Roy Company.

The lanthanum and cerium exchanged zeolites were prepared by ion exchange from NaY sieve and are 13.0 and 13.3 wt. % rare earth, respectively.

3. RESULTS AND DISCUSSION

3.1 Hydrogen spillover onto zeolites

For the Pt black/Ce-Y sieve/perylene system we have proposed that a reasonable sink for spilled-over hydrogen is Ce^{4+} according to the equation:

$$Ce^{4+} + H\cdot \longrightarrow Ce^{3+} + H^{+} \quad (1)$$

This sink is suggested by the amount of hydrogen taken up by the sample (9.4×10^{20} H·/g vs. 4.6×10^{20} Ce^{4+}/g), the necessity of activation in oxygen to achieve appreciable spillover, and the high oxygen uptake of the sample. It is clear that in this system the ultimate sink for hydrogen is an adsorbed oxygen species or a site closely associated with it, such as Ce^{4+}, or both. On the basis of these comments, one might expect CeY to be unique among the zeolites in its ability to accept large quantities of hydrogen. However, this is not observed experimentally; NaY and LaY also possess very large hydrogen uptakes. These uptake values are shown in Table I.

TABLE I
Hydrogen uptake on various zeolites after 18 hours and at room temperature

Catalyst	Uptake μ Moles H_2/g	Remarks
5% Pt Bl'k/CeY	785	8 μ mole perylene/g
4% Pt Bl'k/CeY	23	no perylene
CeY	-	8 μ mole perylene/g
5% Pt Bl'k/LaY	604	8 μ mole perylene/g
5% Pt Bl'k/NaY	754	9.24 μ mole perylene/g
4% Pt Bl'k/NaY	326	0.8 μ mole perylene/g
5% Pt Bl'k/NaY	14	no perylene
0.3% Pt on CeY	169	supported catalyst 8 μ mole perylene/g

The authenticity of the NaY sample was verified by x-ray diffraction and elemental analysis. The oxygen uptake for NaY at 500°C is quite low compared to CeY (37 vs. 320 μ mole/g) so that reaction with an adsorbed oxygen species does not account for the hydrogen uptake. We do not have a completely satisfactory explanation at this time to account for these large hydrogen uptakes although we might speculate on the possibility that hydrogen reacts with the aluminosilicate structure of the zeolite to give tricoordinate alumina according to the following scheme:

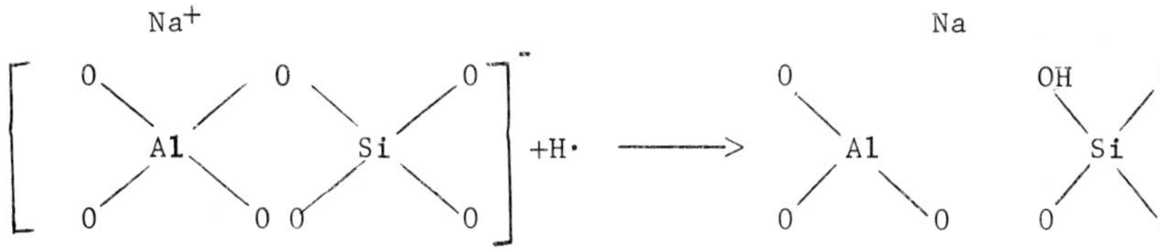

$$\left[\begin{matrix} & Na^{+} & & & \\ O & & O & & O \\ & Al & & Si & \\ O & & O\;O & & O \end{matrix} \right]^{-} + H\cdot \longrightarrow \begin{matrix} & Na & & & & \\ O & & & OH & & O \\ & Al & & & Si & \\ O & & O & O & & O \end{matrix} \quad (2)$$

Further research is clearly needed to characterize the sink on the NaY and LaY zeolites.

The initial rate of spillover appears to be highest on LaY and lowest on NaY. This is shown in Figure 1 where hydrogen uptakes per μ mole perylene per gram of catalyst are plotted versus time to the one-half power. All samples have a Pt loading of 5% and a perylene loading of 9 μ mole/g. The slopes of the three lines, which reflect relative rates, are 5.0, 3.3 and 2.2 for LaY, CeY and NaY, respectively. The error in the determination of the slope is estimated to be no greater than $\pm$ 0.4 and is determined from triplicate experiments on CeY.

3.2 Hydrogen spillover onto inorganic oxides

Five inorganic oxides were tested for the ability to accept hydrogen from platinum at room temperature. They are listed in Table II. Only Al_2O_3 and Ce_2O_3 do not show evidence of significant spillover. RuO_2 has a higher uptake in the absence of perylene. This is not surprising if the Pt/RuO_2 system behaves similarly to the Pt/CuO system where water formed during the bulk reduction of CuO acts as an efficient bridge between the Pt and CuO particles. If RuO_2 reduces to give H_2O, and this water is an effective bridge, the perylene present may not play a major role in the transport process. The result with ZnO is especially interesting because it is well known that O_2^- forms on the surface during activation in oxygen[6]. It may be possible to monitor *via* ESR the destruction of O_2^- radicals during reaction with spilled-over hydrogen.

3.3 The nature of the bridge

Up to this point we have not considered the nature of the bridging compound which facilitates hydrogen spillover. With perylene on CeY there are only two possibilities: positive ion radicals or hydrogenated perylene since all the perylene molecules adsorbed are oxidized to ion radicals[7]. When perylene is adsorbed on NaY, however, the ratio of neutral molecules to ion radicals

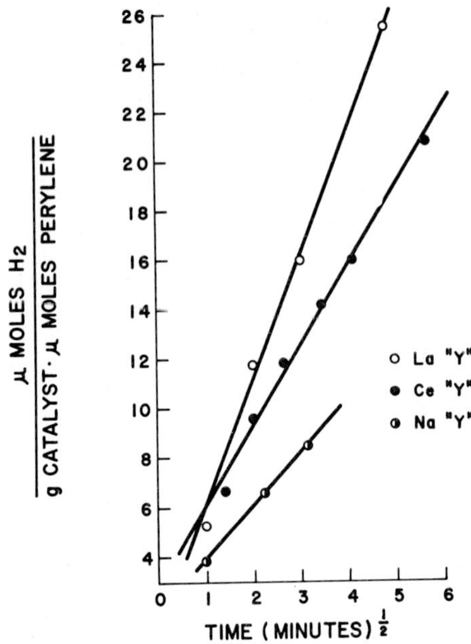

Figure 1
The dependence of the initial rate of hydrogen uptake at 24° on the exchangeable counter ion for the zeolites LaY, CeY and NaY

is near 1000 and yet the initial rate of hydrogen uptake on Ce-Y is only 1.5 times greater than that on NaY (Figure 1). Apparently little difference exists between ion radicals and neutral molecules in their ability to act as a bridge over the platinum-inorganic oxide interface. The possibility remains that the bridging species is perhydroperylene. To examine this possibility more thoroughly and to provide additional information about the nature of the bridge, hydrogen uptakes were measured in the presence of various adsorbed species. These results are given in Table III and plotted in Figure 2.

The initial rate of hydrogen uptake is greater with perylene preadsorbed on the surface than with anthracene preadsorbed. This is seen in Figure 2 by comparing the average slope for perylene (2.9) with the slope for anthracene (1.2). For perylene all the adsorbed molecules are on the exterior zeolite surface since perylene is too large to enter the zeolite supercages[7)]. However, anthracene readily adsorbs into the zeolite interior so that it is expected that fewer potential bridges exist on the exterior surface. Real differences between anthracene and perylene may be masked by differences in surface concentrations of the adsorbed species. A similar difficulty exists when the uptakes for perhydrophenanthrene and the n-C_{18} alcohol are compared with

TABLE II

Hydrogen uptake on various inorganic oxides, with 8 μ mole perylene/g after 18 hours and at room temperature

Catalyst	Uptake μ Moles H_2/g	Remarks
5% Pt Bl'k/Al_2O_3	15	Alcoa F-10
0.5% Pt on Al_2O_3	18	supported Engelhard
5% Pt Bl'k/SiO_2	75	---
5% Pt Bl'k/SiO_2	20	no perylene
5% Pt Bl'k/Ce_2O_3	7	---
5% Pt Bl'k/RuO_2	125	---
5% Pt Bl'k/RuO_2	240	no perylene
5% Pt Bl'k/ZnO	80	---
5% Pt Bl'k/ZnO	37	no perylene
ZnO	--	---

perylene; that is, the exterior surface concentrations of these molecules are not known since adsorption in the supercages can occur. It is clear, though, that both can function as a bridge with perhydrophenanthrene comparing favorablbly with perylene (slope of 1.9 vs. 2.9).

TABLE III

Effect of adsorbate on hydrogen uptake after 18 hours at room temperature on 4% Pt Bl'k/CeY catalyst

Adsorbate	Uptake μ Moles H_2/g	Remarks
Perylene	609	8 μ moles/g
Anthracene	472	11.3 μ moles/g
n-C_{18}-ol	257	40 μ moles/g
Perhydrophenanthrene	263	7.78 μ moles/g
CCl_4	435	sample pumped for 1 hour at 25°C
CCl_4	-	sample pumped for 1 hour at 60°C
Water	-	sample exposed to H_2O vapor for 5 min. at 25°C
Apiezon N	301	5.5 hours

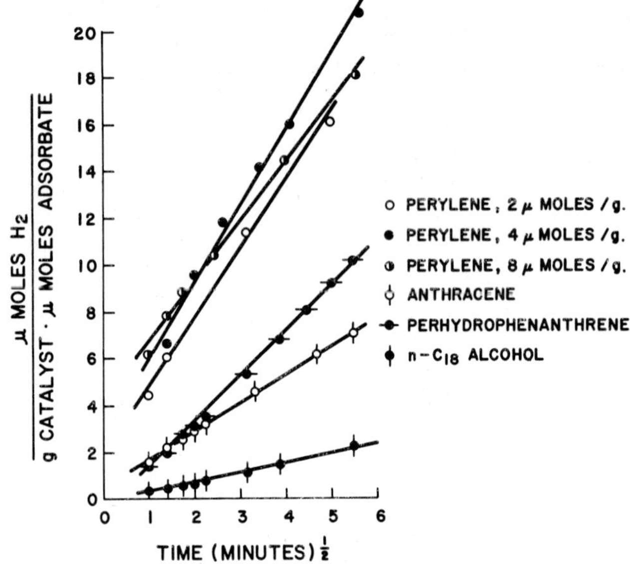

Figure 2
The variation of the initial rate of hydrogen uptake at 24° on the nature of the adsorbate

The results in Table III show that evacuation of the sample for 1 hr. at 60°C is sufficient to remove the CCl_4 solvent. This procedure was routinely followed for all hydrocarbon experiments where CCl_4 was used. Further evidence is supplied by Figure 2 where it is seen that normalized H_2 uptakes for 3 perylene loadings are coincident. The large hydrogen uptake that occurs when CCl_4 is not removed cannot unambiguously be attributed to spillover since it has recently been shown that CCl_4 can react with hydrogen over Pt at low temperatures[8)].

In contrast to the Pt/WO_3 system no uptake is observed after the sample is exposed to water vapor. Finally it is clear that Apiezon N stopcock grease is a very effective bridging compound. This result coupled with the variety of other organic materials which can act as bridges makes it clear that great care should be taken to protect samples from hydrocarbon impurities prior to hydrogen chemisorption measurements commonly used to measure metal surface areas. Particularly in zeolite systems, small amounts of such contaminants might result in artificially high uptake measurements.

4. CONCLUSIONS

Hydrogen spillover occurs readily onto a number of zeolites and inorganic oxides when suitable bridges for hydrogen migration are provided. Both aromatic and saturated hydrocarbons are effective bridging compounds although hydrogenation of the aromatic hydrocarbon may occur before the onset of spillover. The ultimate sink for hydrogen with CeY sieve is thought to be an adsorbed oxygen species or a site closely associated with it. The uptakes on RuO_2 may be explained by bulk reduction of this oxide which is thermodynamically feasible at room temperature. For the other inorganic oxides and the LaY and NaY sieves, the situation is less clear. The sink may be a defect site in the oxide surface or in the silica-alumina framework.

5. ACKNOWLEDGEMENT

We acknowledge the assistance of Mr. R. E. Ledley, III, who carried out much of the experimental work reported here.

REFERENCES

1) Benson, J. E., Kohn, H. W., and Boudart, M., J. Catal. 5, 307(1966).
2) Robell, A. J., Ballou, E. V., and Boudart, M., J. Phys. Chem. 68, 2748 (1964).
3) Boudart, M., Aldag, A. W., and Vannice, M. A., J. Catal. 18 46 (1970).
4) Vannice, M. A. and Neikam, W. C., J. Catal. 20, 260 (1971).
5) Neikam, W. C. and Vannice, M. A., J. Catal., Submitted.
6) Kohn, R. J., J. Phys. Chem. 66, 99 (1962).
7) Neikam, W. C., J. Catal. 20, 102 (1971).
8) Weiss, A. V., Gambhir, B. S., and Leon, R. B., J. Catal. 22, 245 (1971).
9) Boudart, M., Vannice, M. A., and Benson, J. E., Z. Phys. Chem. N.F., 64, 171, (1969).

DISCUSSION

B. DELMON

The authors state rightly that few unambiguous examples of spillover have been reported in the literature. I would like to mention another recent example, in addition to those cited in the communication.

It concerns the strong acceleratory action of supported platinum and palladium on the hydrogen reduction of nickel oxide and tungsten trioxide.[1)] Platinum (1% wt.) was supported on alumina; palladium (1% wt.) was on silica. Both carriers had high surface areas (300 $m^2 \cdot g^{-1}$). The supported metals promoted the nucleation stage in the reduction of nickel oxide, and the formation of the blue compound in the reaction with WO_3, even when the catalyst and the reactant were simply admixed (e.g. when the catalyst and reactant were separately introduced in a moving bed reactor and subsequently agitated) and when the catalyst was present in a very small amount (e.g. 5% wt. of total catalyst, corresponding to 0.05% wt. Pd respective to NiO).

These observations are interesting, because they show that the spillover (a) is effective even when the contact between the different solids is very loose and (b) can proceed through a succession of 3 solids (instead of 2 in other examples), namely Pd-SiO_2-NiO.

1) A. Roman, B. Delmon, Compt. Rend. Acad. Sci., 273 Ser. C (1971), 94.

W. C. NEIKAM

The comment is quite interesting. However, it should be noted that when Khoobiar first reported the accelerative effect of Pt on the low-temperature reduction of WO_3, it involved a mixture of WO_3 and a Pt/Al_2O_3 catalyst.[1)] Also a Pd/SiO_2 catalyst mixed with WO_3 has been shown to readily reduce WO_3 at room temperature by means of hydrogen spillover.[2)] In both cases, three solids were involved.

1) Khoobiar, S., J. Phys. Chem. 68, 411 (1964).
2) Boudart, M., Vannice, M. A., and Benson, J. E., Z. Physik, Chem. N. F. 64, 171 (1969).

R. LEVY

The work of Neikam and Vannice has clearly demonstrated that a bridging compound has to be added to the metal-nonmetal system in order to observe hydrogen spillover at low temperatures. I would like to present one example in which the chemical role of this bridging compound is understood. Recent work on the Pt catalyzed reduction of WO_3 to H_xWO_3 in the presence of water or alcohols has shown that these bridging compounds react with the dissociated hydrogen to form solvated proton intermediates. The protons are transported in an adsorbed layer from the Pt surface to the WO_3 sites where the proton-transport agent bond is broken and the proton penetrates into the oxide bulk. In the kinetic regime of the reduction this last process is rate determining and the rate of reduction can be correlated with the proton affinity of the transport agent. The bridging compound therefore not only provided a mobile intermediate for the hydrogen transport but also supplied a chemical bond that satisfied the energy requirements of low temperature hydrogen spillover.

J. A. RABO

You propose in your paper that your H_2 spillover gives rise to reduction of the zeolite cation and the formation of protons. I expect the protons to attack lattice oxides forming lattice OH groups which are also formed when noble metal loaded zeolites are exposed to H_2 at beyond 250° (Rabo, et al. Faraday Dis. 1966). Did you find increased amounts of OH groups following your spillover experiments? What was the quality of your zeolite relative to impurities?

W. C. NEIKAM

We did not attempt to study the OH groups by means of IR spectroscopy. This would be a good way to determine if OH groups are created during the spillover process. We have proposed that this process does occur within our zeolite system.[1)]

In regard to the second question, the NaY zeolite used for exchange was specially prepared to be iron-free by the Linde Company. Activation of the iron-free NaY results in a sample which is diamagnetic so that no appreciable paramagnetic impurities are present. Neither the NaY nor the final cerium exchanged material showed the typical ESR spectrom of Fe^{3+}.

1) Neikam, W. C., and Vannice, M. A., J. Catal., Accepted for publication.

M. SHELEF

The rate of uptake of H_2 is in all cases proportional to the square root of time. Are there any mechanistic implications in this fact? If so, the differences in the rates warrant a more detailed explanation.

W. C. NEIKAM

The proportionality of H_2 uptake to the square root of time is very strong evidence that one is observing a diffusion process. We have shown[1)] that increasing either the number of perylene molecules adsorbed on the zeolite or the number of Pt particles in the sample does enhance the rate of spillover. This is consistent with a model proposing the surface diffusion of hydrogen from the Pt surface, through the perylene "bridge," and hence across the zeolite surface.

1) Neikam, W. C., and Vannice, M. A., J. Catal., Accepted for publication.

W. K. HALL

Our published work with dehydroxylated hydrogen Y zeolites showed that the electron acceptor for radical ion formation was a form of oxygen which when adsorbed could not be pumped off at 10^{-6} torr and 500°C in 48 hours. This oxygen could be removed, however, by reaction with H_2 at high temperature, and there was a stoichiometric relation between the H_2O removed, calculated as O_2, and the decrease in radical ion concentration. I wonder if the "spillover" does not accomplish the same thing at room temperature?

Would you please summarize your evidence for the statement that perylene does not enter into the interior of the zeolite? If this is true, our results would necessarily pertain to the external surface of the zeolite crystals, and this seems unlikely.

W. C. NEIKAM

We feel that spilled-over hydrogen is quite likely to react with preadsorbed oxygen on the Ce exchanged zeolite. The Ce exchanged material had an oxygen uptake[1)] of 360 μ mole/g which is sufficient to account for the hydrogen uptakes observed assuming the final surface species to be hydroxyl groups or adsorbed water.

The evidence that perylene does not enter the supercage was published in an eralier paper.[2] Equal number of spins were generated when perylene or anthracene were adsorbed. There was a 1:1 correspondence between perylene adsorbed and perylene radicals formed, and yet there was twenty-one times as much anthracene adsorbed as free radicals formed. In addition, the spin generation versus time curve for perylene was coincident with the adsorption versus time curve. For anthracene the spin generation curve was coincident with the same curve for perylene but not with the anthracene adsorption curve.

This may not be in conflict with your data since additional radigenic sites exist in the interior of the zeolite. Molecules smaller than perylene can react at those.

1) Neikam, W. C. and Vannice, M. A., J. Catal., accepted for publication.
2) Neikam, W. C., J. Catal. 21, 102 (1971).

H. WISE

In view of the "bridging" properties of a number of compounds, as demonstrated in your work, one could consider a reactant, such as an olefinic hydrocarbon, to have similar "bridging" properties during a hydrogenation reaction. Also, do you have an explanation for the apparent non-zero intercept of your curves relating the mean uptake of hydrogen to the square root of time?

W. C. NEIKAM

The results of Sinfelt and Lucchesi[1] show that indeed olefinic hydrocarbons can act as an effective bridge.

The non-zero intercept occurs for two reasons. First of all, there is an immediate uptake as the hydrogen adsorbs on the Pt black and titrates the oxygen on the surface. Secondly, hydrogenation of the perylene bridge may occur prior to the onset of spillover. These processes may occur on a time scale very short compared to our diffusion measurements.

1) Sinfelt, J. H. and Lucchesi, P. J., JACS 85, 3365 (1963).

W. H. FLANK

I would like to note that the high dynamic pressures generated in the grinding process can cause extensive damage to zeolite crystallites. In view of the preparative technique described in your discussion, I would think that this might have a significant effect on your results, especially in the quantitative sense, although I would agree that the quantitative conclusions appear to be valid. Perhaps the defects that are created contribute to the mechanistic pathway.

W. C. NEIKAM

The point you raise is an interesting one. We prepared our mixtures both by grinding and by gently rolling the Pt black and zeolite together in a vial for short periods of time. The method of mixing had no effect on either the total H_2 uptake or the rate of spillover. We conclude from this that defects due to grinding, if present, do not appear to play a major role in the spillover process.

J. H. BLOCK

We have evidence for spillover of hydrocarbons in the Ca Y-zeolite/Pt system. Using field ion mass spectrometry with a Pt emitter which is partially coated by a zeolite surface, compounds present on the Pt surface differ appreciably from those of an uncoated Pt surface. With cyclohexane it could be shown by energy analysis of field ions, that C_6H_{11} surface radicals are formed on the zeolite. These radicals are mobile and show spillover to the Pt surface where they can be field desorbed. A detailed description will be given at the symposium on "Mobility of Surface Molecules," Louvain, Belgium, September, 1972.

Paper Number 41

ELECTRONIC STRUCTURE AND ENSEMBLES IN CHEMISORPTION AND CATALYSIS BY BINARY ALLOYS

D. A. DOWDEN
Imperial Chemical Industries Limited, Agricultural Division
Billingham, Teesside, England

Abstract: The electronic structure of alloys Pd X (X = Cu, Ag, Au) together with the concept of 'ensembles' and simple models are used to outline the variation in absorptive properties with composition. The effects of short and long range order are also discussed. The random alloys of Ni-Cu, where there may not be 3d-band filling, behave differently from the alloys Pd X and the differences are most marked in the alloys richest in group 1B metal.

1 INTRODUCTION

The influence of electronic factors on the surface chemistry of metals and alloys has been the subject of serious study for the last 25 years but the models employed by chemists have hardly changed during that period. Recent results have modified the phase diagrams of some important binary solid solutions (eg Ni-Cu[1]), new techniques have revealed unexpected ordering or clustering[2] and advances in the understanding of the electronic structure of metals have greatly improved upon the early, simple band theories[3,4]. There has been a marked trend away from the concept of rigid bands of electron levels according to which the characteristic properties of the transitional metals and alloys arose from two overlapping bands (d and s) which maintain their shape and relative positions independent of the electron to atom ratio. The Fermi surface (E_M) of a group 8 metal lies in the d-band where the density of states $n(F_M)$ is large but in the group 1B metals $n(E_M)$ is small.

In AB alloys (A = Ni, Pd, Pt; B = Cu, Ag, Au) it is now certain that there is no common d-band[5] and that d-band holes do not occur in B atoms. Palladium has recently been shown[6] to possess only 0.36 4d-band holes per atom, yet the band is just filled in its alloys PdX (X = Cu, Ag, Au) near the composition $Pd_{0.4}X_{0.6}$. On the other hand when A and B are neighbouring transitional elements the rigid band model still possesses useful features[7].

Following Friedel's work[8] it is evident that altervalent solutes in substitutional alloys, solute atoms in interstitial alloys and chemisorbates have their extra potentials shielded more or less locally by the electrons or holes of the solvent or absorbent depending upon the magnitude of $n(E_M)$ and $dn(E_M)/dE$. Bonding does not occur by charge transfer or exchange involving many atoms but is confined to near or next-near neighbours and the shielding radius may not much exceed that of the foreign atom itself.

These advances leave the elementary principles of the electronic theories of chemisorption and catalysis unaffected 9) but they greatly modify the application to real systems. For instance in a random alloy A_xB_{1-x} there will be groups of atoms of varying composition ranging from pure A to pure B, with volume fractions dependent upon x and the group size. If in addition there is shielding of the solute potential upon neighbouring atoms of the solvent then there will be present also atoms of electronic structure different from A and B. Chemically speaking, in the alloy Pd_1Ag_{1-x} there will be found on the Pd-rich side, atoms approximating to $Pd4d^95s$, $Pd4d^{10}$, $Pd^-4d^{10}5s$, $Ag4d^{10}5s$ and Ag^+4d^{10} although from Pauling's electroneutrality principle the excess charges should be shared by near neighbours.

Corresponding to the clusters in the bulk there will be 'ensembles'[10)] in the surface and the purpose of this paper is to show that the statistics of these ensembles together with the introduction of localised shielding lead from simple models to results for binary alloys which closely pattern experiment. The alloys noted above between the metals of groups 8 and 1B display many of the characteristics likely to have profound effects on catalytic activity - clustering, phase separation, order-disorder transformations, etc. Moreover, it is increasingly doubtful whether the Ni-Cu system shows evidence of 3d-band filling whereas the d-shells of the palladium atoms in the Pd-Ag and Pd-Au alloys are certainly full at $Pd_{0.4}X_{0.6}$, the critical composition x*.

2 THE STATISTICS OF THE ENSEMBLES

In a completely random binary alloy A_xB_y ($y = 1-x$) of face-centred cubic structure (cell edge = a_w) the (001) face is a square net (unit edge = $a_w/\sqrt{2}$) which together with the atoms of the penultimate layer, at a distance $a_w/2$ below the square centres, forms an array of square pyramidal interstices with a metal atom at each of the five apices. Such an ensemble contains five atoms and a small adatom situated near the centre of the square is said to be interstitially chemisorbed. Each surface atom possesses four nearest neighbours in the surface making another 5-ensemble and four below, in the penultimate layer, making a 9-ensemble. The number (n) of atoms in the ensemble can take any value but the most suitable is that corresponding to the pyramidal interstice or aggregates thereof or to the central surface atom with its shells of nearest, next-nearest, etc neighbours, ie n = 5, 9, 14...., taken so that the continuous square net of the (001) surface is preserved.

For any value of n the probability of finding m atoms of A in the ensemble is given by the binomial probabilities,

$$p_{n,m} = \binom{n}{m} x^m y^{n-m} \qquad (1)$$

where $\binom{n}{m}$ are the binomial coefficients. The probability of finding m or more atoms of A in the ensemble is equal to the sum of $p_{n,m}$ over m from m to m=n-

$$P_{n,m} = \sum_{m}^{m=n} p_{n,m} \qquad (2)$$

For ease of calculation at large values of n the binomial distribution has, in places, been approximated by the normal distribution[11] to obtain values of $P_{n,m}$; the error so introduced is not important for the purposes of this paper.

Similarly, the probability that a given atom will be type A and that it will have m nearest neighbours of the same kind in an n-ensemble is

$$p'_{n,m} = x\binom{n-1}{m} x^m \; y^{n-m-1} \qquad (3)$$

whereas the probability that the number will be $\geq$m is

$$P'_{n,m} = \sum_{m}^{m=n-1} p'_{n,m} \qquad (4)$$

If short-range ordering occurs [12] with a Cowley parameter of α_1, then

$$\alpha_1 = 1 - (1-\beta)/(1-x)$$

where β is the average atomic fraction of A in the nearest-neighbour shells around all A atoms, whence β must equal $(1-x)\alpha_1+x$. Therefore under these conditions x^m must be replaced by β^m and y by $(1-\beta)$ in equation (3). Use will also be made of some results of percolation theory[13] which give the probabilities $P_c(s)$ at different values of x that an atom A in a random fcc alloy is a member of an infinite chain of such atoms. Fig 1 indicates the several regimes; (i) P_c (s) $\sim$1 and any atom A is almost certainly a member of the one and only infinite cluster of A atoms which extends throughout the bulk and the surface; (ii) $P_c(s)\sim 1$ for the bulk and (111) but $\geq$ 0.4 on (001); (iii) $P_c(s)\sim 1$ for the bulk, $\geq$ 0.5 on (111) but zero on (001) and the A-atom cluster is continuous throughout the bulk but finite on the surfaces; (iv) $P_c(s)\sim 1$ for the bulk but zero for the surfaces; (v) $P_c(s)$ decreases rapidly to zero for the bulk; (vi) $P_c(s)\sim 0$, an atom A almost certainly belongs to a finite cluster or ensemble.

3 SURFACE COVERAGE BY VARIOUS ENSEMBLES

The fraction of surface covered by various ensembles depends upon n and m and is taken to be equal to the corresponding probabilities: $p_{n,m} = p'_{n,m} = 0$ at x = 0 and 1 and pass through maxima when $x \simeq n/m$. It is difficult to attribute the maximum found near x = 0.7 in the chemisorption of hydrogen on Pd-Au alloys[14] to preferences for ensembles with x = m/n = 0.7 and it will be supposed, as for the solubility of hydrogen[9], that this feature has more fundamental origins. On the other hand $P_{n,m} = P'_{n,m} = 1$ at x = 1 and zero at x = 0 falling off more or less sharply at intermediate values. Figs 1 and 2 show how the probabilities vary with x for different n and m; also indicated are the several percolation regions and the critical composition x* = 0.4 for the palladium alloys.

Fig 1 (see below) reveals that for ensembles of reasonable size ($5 \leq n \leq 22$) and $m/n \geq x^*$ the values of $P_{n,m}$ increase with n but lie fairly close together so that for absorbates of different size similar activity patterns may be observed. The values of

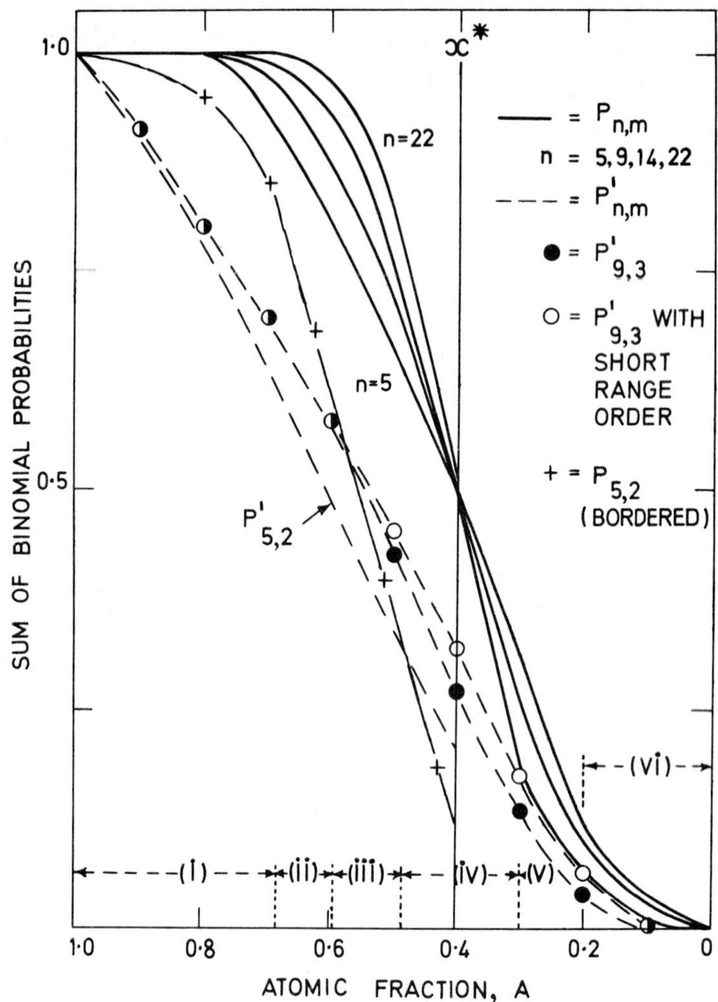

Fig 1 - Values of $P_{n,m}$ and $P'_{n,m}$ at various compositions x. For $P_{n,m}$ $m \geqslant 0.4n$

$P'_{n,m}$ fall much more linearly than $P_{n,m}$, with decreasing x. If heats of chemisorption fall rapidly for $x \leqslant x^*$ (because the d-band is full, see later) so that absorption on the corresponding ensembles falls to small values, then the $P_{n,m}$ curves cut off by the vertical line at $x=x^*$ best resemble that found for the uptake of hydrogen by the Pd-Au system. When there is no d-band filling the marked tail of $P_{n,m}$ in the B-rich regions would ensure that some strong absorption occurs well beyond what was once thought to be a critical composition, as perhaps in a truly random Ni-Cu alloy. Just such a situation must exist in the iron alloys where it is

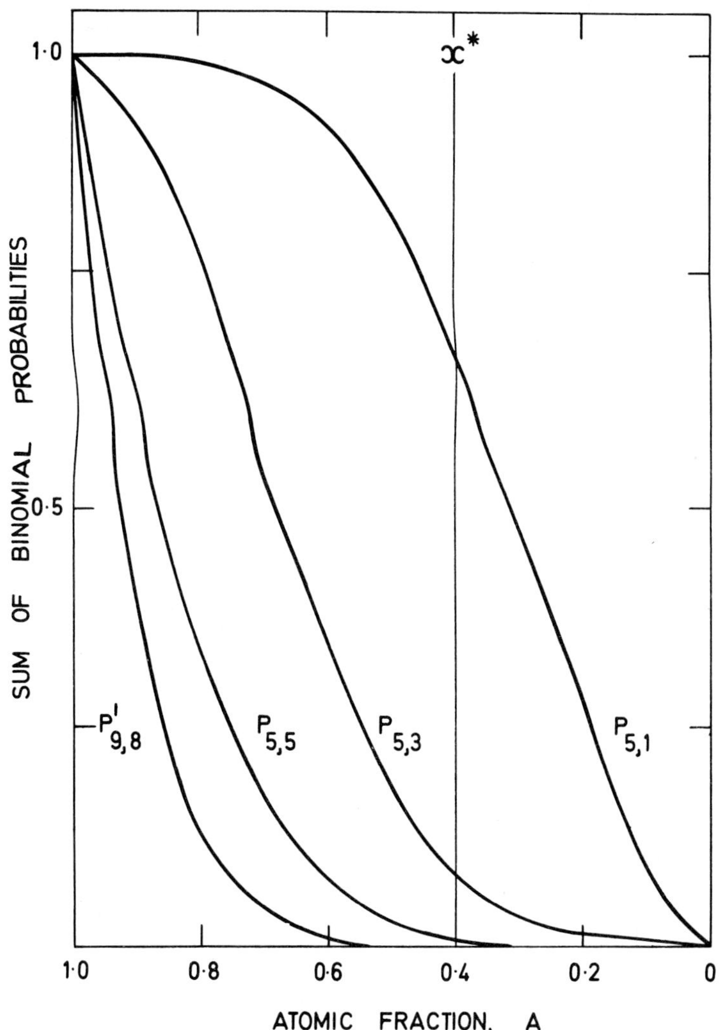

Fig 2 - Values of $P_{5,m}$ at various values of x and m; also $P'_{9,8}$

known that solutes like aluminium take substitutional positions without greatly affecting the occupancy of the iron d-orbitals.

Fig 2 shows the variation of $P_{5,m}$ with m (=1,3,5) and x, and the values of $P'_{9,8}$. The simple notion that adsorption takes place on A-like ensembles (n = m) does not fit the properties of the Pd-Au alloys because the curves for n=m=5 and n=m=9 fall off much too sharply with decreasing x; but it cannot be excluded that a different behaviour may be found for other adsorbates for which the heats decline much more rapidly with decreasing m/n in the ensembles.

Again, the shape of the $P'_{n,m}$ curves for the A-centred ensembles does not accord well with the data but may be relevant to a different kind of adsorption, not in interstices but atop metal atoms.

The value of n which defines the size of the ensemble required by a given adsorbate cannot be inferred with certainty although the usual simple devices, based on the geometry of the molecule and fixed bond angles, can be used. Whereas a 5-ensemble may be adequate for hydrogen, and indeed is related to the octahedral sites occupied by hydrogen in solid solution in palladium, larger molecules like benzene and the polynuclear hydrocarbons will certainly require larger values of n, especially when they lie flat on the surface. When n is large then Fig 1 shows that the drop in $P_{n,m}$, for a given value of m, becomes sharper and it is not possible to decide from the variation in adsorbate coverage with x alone whether there is an effect due to band filling.

The effect of short range ordering on $P_{n,m}$ and $P'_{n,m}$ is to shift the curves a short distance towards smaller values of x, as expected, without much affecting their shape. Values of α_1 for Ni-Cu were used as given by the proportionality $\alpha_1 \propto x(1-x)$ adjusted to the experimental values 0(Ni), 0.047($Ni_{0.8}Cu_{0.2}$), 0.121(NiCu) and 0(Cu)[15]. These changes must further increase the effects due to the large tails of the P's in the B-rich alloys.

In the absence of specific surface phenomena or adventitious segregations, the activity pattern can be predicted for the phase diagram which results from the miscibility gap produced below the critical temperature by the subsequent spinodal decomposition of the Ni-Cu solid solutions but the long-range ordering found in the Pd-Cu system gives activity changes which are not so easily understood unless the special composition of localised ensembles is recognised. Although $n(E_M)$ falls [22] as the ordered α-phase (Pd Cu, L20 type) forms and hardly changes with the appearance of the α-phase (Pd Cu_3, L12), both the activity and the activation energy for hydrogen exchange decrease[16]: electronic factors by themselves suggest only the first result because clustering should diminish on ordering. The β-phase however, is a superstructure of the A2 (bcc) type with each atom A having eight B nearest neighbours (like CsCl) so that (100) planes are alternately all A or all B. Then a (100) surface of palladium atoms contains pyramidal 5-ensembles-4Pd plus 1 Cu below - and the ensemble has an activity and an activation energy more like those of Pd_4Cu than PdCu. But the (100) plane is not the most densely packed, the surface energy of Cu(100) is probably less than that of Pd(100) and the excess copper in $Pd_{1-\delta}Cu_{1+\delta}$ may occupy sites in the palladium surface; consequently although the activation energy falls so does the activity because the area of exposed Pd(100) is small. The corresponding changes occurring on appearance of the α-phase, a superstructure of the A1 type, cannot be explained so directly because each Pd atom (at the cube corners) has 12 nearest neighbour Cu atoms and does not form Pd-rich regions. It must be assumed that misplaced Pd atoms exist in the (100) planes giving Pd-rich 5-ensembles or that for compositions $Pd_{1+\delta}Cu_{3-\delta}$ the extra Pd atoms form two-dimensional platelets[17]. Using computer methods[17] it should be possible to pursue these problems still further.

The percolation probabilities suggest that for 1>x>0.6 there is a continuous network of A-atoms in the surface and the bulk. Because $P_c(s) = 1$, on the average each A-atom has $\nless$ 2 A-neighbours in the surface and $\nless$ 1.5 in the bulk[13]; thus, the 5-ensembles of the surface and the clusters of the bulk have x>x* and d-band holes exist at Pd atoms in PdX alloys. These d-band holes and strongly chemisorbed species migrate most easily across and through the surface by way of the paths provided by the A-atoms. From the values of $P_c(s)$ given above corresponding conclusions can be drawn for other compositions. For alloys like Ni-Cu in which unfilled d-orbitals persist in the Cu-rich alloys, holes can migrate throughout the bulk as long as x>0.2. Even when x<x* adsorbed hydrogen atoms are expected to associate more strongly with the transitional atoms A[9].

Clearly the ensemble theory can also be applied to the B atoms of the alloy A_xB_y.

4 HEATS OF CHEMISORPTION

Given the values of $p_{5,m}$ with decreasing x, the properties of the alloys depend upon the heats of chemisorption (Δh) of the adsorbate and of the activated complexes on the various ensembles Pd_5, Pd_4Ag, ... Ag_5. No attempt will be made to calculate these heats, instead their values will be estimated empirically using two simple models ($\Delta h'$, $\Delta h''$) and combined with $P_{5,m}$ to show the likely trend of the weighted heats of chemisorption with alloy composition.

The first method takes no account of 4d-band filling. The initial heat* ($\Delta H_{\Theta} = 0$) of chemisorption at small coverage $\Theta = 0$, of hydrogen on palladium[18] is 26 corresponding to a bond energy ($D_{\Theta=0}$) of 65 but $D_{\Theta=0}$ (Ag-H) is put equal to 54.5 (see later). Then $\Delta h'$ for each ensemble is found by proportion from its composition, eg

$$\Delta h'(Pd_3Ag_2) = \Delta h'_3 = 2\,[0.6D(Pd\text{-}H) + 0.4D(Ag\text{-}H)] - D(H\text{-}H)$$

The heat for immobile adatoms on the alloy surface of composition x is then -

$$\Delta h'_x = P_{5,5}(x).\Delta h'_5 + p_{5,4}(x).\Delta h'_4 + \ldots .p_{5,0}(x).\Delta h'_0$$

and Fig 3 shows, as expected, that $\Delta H'_x$ falls linearly with decreasing x. If the adatoms are mobile they migrate to the ensembles richest in Pd so that for sufficiently small values of $\Theta, \Delta H'_x$ remains at 26 independent of x down to pure Ag where it drops stepwise to 5. More realistically Fig 3 shows $\Delta H'_x$ for $\Theta = 0.05$; the values remain at 26 as long as $p_{5,5} \geq 0.05$ but when $p_{5,5} < 0.05$ adatoms migrate to ensembles with $p_{5,4}$ and

* all heats and energies in kcal, $mole^{-1}$ (1kcal = 4.187 kJ).

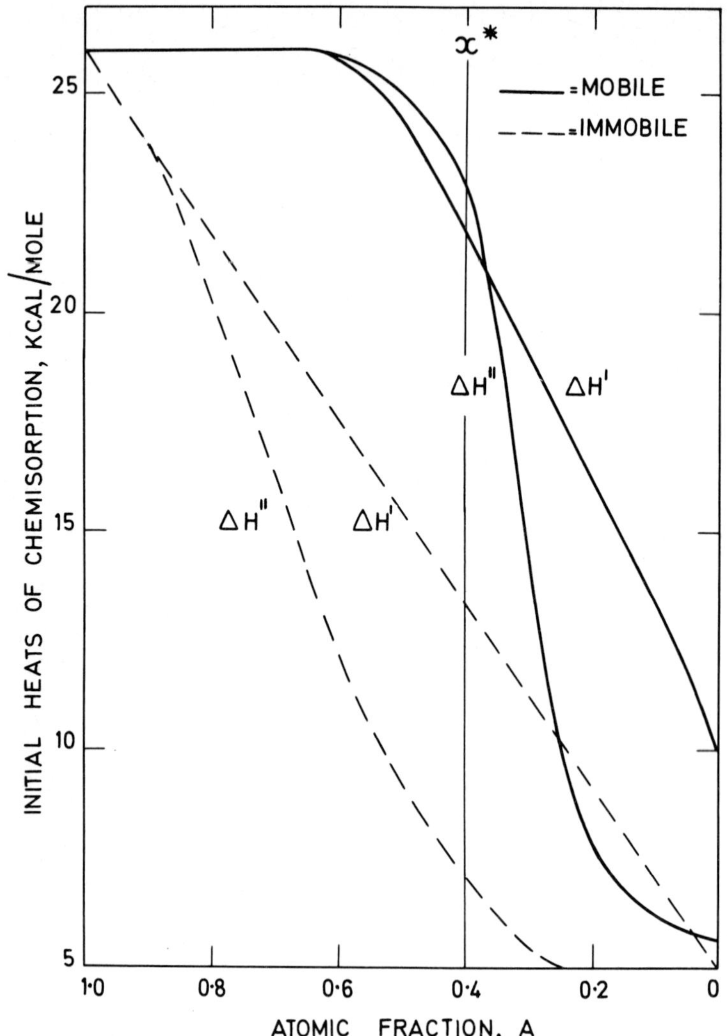

Fig 3 - Initial heats of chemisorption, ΔH' and ΔH"

the heat falls proportionately. As x decreases the fraction 0.05 of adatom coverage is accommodated on ensembles less and less rich in Pd.

The second method attempts to take account of d-band filling. It is assumed that the decrease in ΔH with x or with Θ (for pure Pd) is due entirely to the increase (ΔE) in E_M per H adatom as calculated from curves of $n(E_M)$ against E which are themselves derived from curves of $N(E_M)$ versus x for the Pd_xAg_{1-x} alloys[19], allowing for the fact that Ag atoms do not carry d-band holes[20]. The graph of $n(E_M)$ against E is not shown because of its similarity to that given by Burch[21]: almost the same graph can be given for the random Pd-Cu alloys in which, however, there is clustering of Pd atoms on the Pd-rich side[22].

Interstitial adsorption of hydrogen places the proton within the pyramidal 5-ensemble with local sharing of the electron on the Pd atoms only until the d-band is full, thereafter all 5 atoms share the electron (e). Thus the ensembles Pd_5, Pd_4Ag and Pd_3Ag_2 acquire 0.2e, 0.25e and 0.33e per Pd atom, respectively, whereas Pd_3Ag_3, $PdAg_4$ and Ag_5 have 0.2e per atom. The corresponding ΔE's are found, as described, and the value for pure Pd subtracted from each because D(Pd-H) = 65 for Pd_5 already includes this term; the D's for the other ensembles are found by subtracting the appropriate, adjusted ΔE from 65 and then $\Delta h''_5$, $\Delta h''_4$ .... $\Delta h''_0$, and $\Delta H''_x$ (mobile or immobile) as before. No allowance is made for variation in the energy of the proton which is assumed to remain constant at that implicit in D(Pd-H); this somewhat gross simplification prevents the curves showing a maximum at x = ca.0.7[9]. Because $\Delta H''$ = ca.5 at x = 0, D(Ag-H) is put equal to 54.5 in the calculations of $\Delta H'_x$

For mobile layers Fig 3 shows that $\Delta H''$ falls off sharply at x = 0.5, much more so than $\Delta H'$, so justifying the cut-off at x = x* for Pd alloys in Figs 1 and 2. $\Delta H''$ for both mobile and immobile layers falls off more rapidly than $\Delta H'$ with decreasing x and this may provide a method for checking the validity of the two models; unfortunately it is not always possible to experiment under one or the other of the two limiting conditions. Calorimetry may show the rise in ΔH which should occur as a more or less immobile layer slowly migrates to the more stable states.

Similar methods can be used to sketch curves representing the variation of $\Delta H''$ with Θ for mobile adsorption on the series of alloys. A net of nine squares (16 atoms) is drawn on graph paper and bordered with eight identical nets, then a hydrogen atom is added to the same square in each net in a systematic way. The first atom is added to the central 5-ensemble in each net, which for pure palladium distributes one electron among Pd_5 (ie 0.2 e/Pd) and ΔE is found as before. A second atom is placed in an ensemble in the diagonal of the square so that 2e are now shared by nine Pd atoms, ie 0.22e/Pd and the corresponding ΔE evaluated. The process is repeated until all four diagonally placed ensembles and then the remainder of the ensembles have been occupied; when the adsorption is complete, Θ = 9/9, and the electron concentration is 0.5e/Pd. By simultaneous additions to the bordering ensembles edge effects are eliminated and values of $\Delta H''$ can be estimated for increments of Θ by 1/9 (Fig 4). Two points are given for pure Pd at Θ = 1, the higher is obtained by the method outlined. The second rests on the assumption that migration is so slow that the last empty ensembles are occupied by hydrogen molecules whereby the electron of one atom is shared by 5 Pd atoms and the electron of the other by the four surface palladium atoms only; then the final electron ratio is 0.75e/Pd instead of 0.5e/Pd and $\Delta H''$ correspondingly smaller. The variation of the mobile heat $\Delta h''$ with Θ is found similarly for the ensembles Pd_4Ag, Pd_3Ag_2 and Pd_2Ag_3 for which, of course, ΔE increases and $\Delta h''$ decreases rapidly with x and Θ (Fig 4).

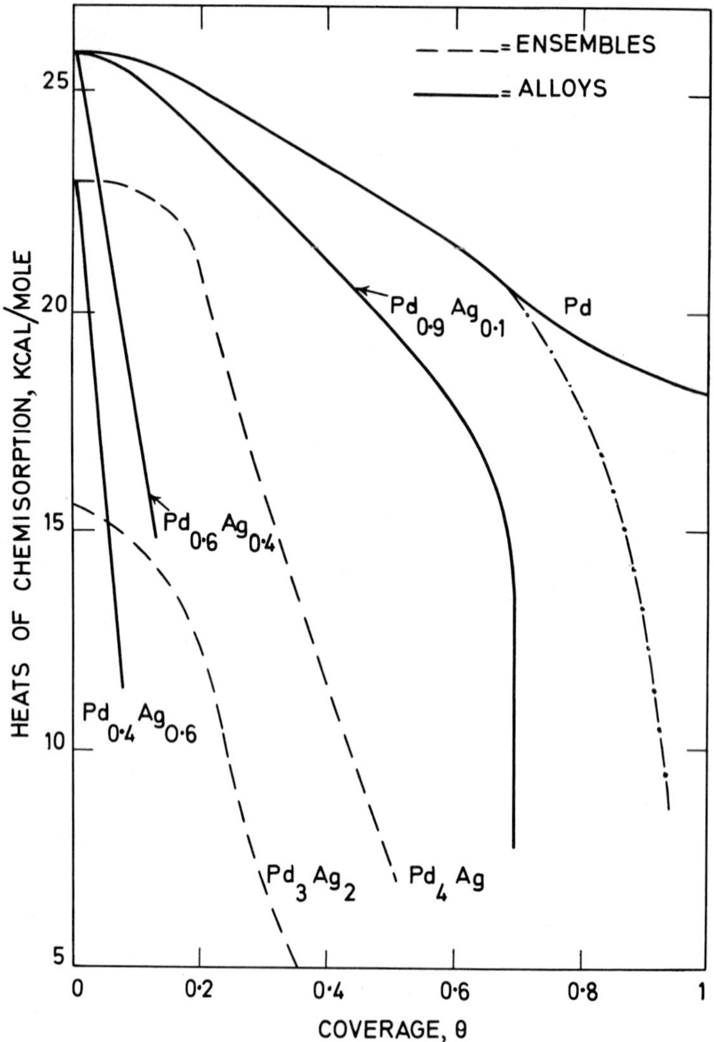

Fig 4 - Variation of heats of chemisorption with coverage and composition: ΔH" and Δh", mobile

It is supposed that ΔH" falls off as pure palladium (high heat at $\Theta = 1$) until the area ($p_{5,5}$) of Pd_5 ensembles is full when adsorption begins on Pd_4Ag ensembles and ΔH" falls off as $\Delta h''_4$ until that area ($p_{5,4}$) is also occupied, and so on. Taking into account the approximations used the only noteworthy feature is the rapid decrease of ΔH" with Θ as x decreases.

Heats of activation for adsorption and catalysis (H_2- D_2) can be estimated in the same way and yield similar patterns.

5 CONCLUSIONS

Recent results on the electronic structure of alloys and its modification by additional solutes and adsorbates suggest a review of the electronic factor in adsorption and catalysis. A suitable compromise combines Kobozev's ensembles with electron-band filling (where appropriate) and indicates that the surface chemistry of completely random binary alloys requires very detailed investigation. A careful comparison of Ni-Cu alloys with PdX alloys (X = Cu, Ag and Au), especially in the regions rich in Cu, Ag and Au, should help to elucidate the problem.

REFERENCES

1) W.M.H. Sachtler and M.E. Jongepier, J Catal., 4 (1965) 665.
2) T.J. Hicks et al, Phys. Rev. Letters, 22 (1969) 531.
3) N.F. Mott and H. Jones, The Theory of the Properties of Metals and Alloys (Clarendon Press, Oxford, 1936).
4) F. Seitz, The Modern Theory of Solids (McGraw-Hill, New York, 1940).
5) G. Longworth, J. Physics C., Solid State Physics, 3, Metal Physics, 1 (1970) S81.
6) J.J. Vuillemin and M.G. Priestley, Phys. Rev. Letters 14 (1965) 307.
7) E. Von Meerwall and D.S. Schreiber, Phys. Rev. B3 (1971) 1.
8) J. Friedel, Phil. Mag. 43 (1952) 339; Advan. Phys. 3, (1954) 446.
9) D.A. Dowden, J. Inst. Petrol (in the press).
10) N.I. Kobozev, Acta Physiocochim. URSS 9 (1938) 805.
11) A.J. Duncan, Quality Control and Industrial Statistics (R.D. Irwin, Inc Homewood, Illinois, 1959).
12) J.M. Cowley, Acta Cryst. A24 (1968) 557.
13) V.K.S. Shante and S. Kirkpatrick, Advan. Phys. 20 (1971) 325.
14) B. Tardy and S.J. Teichner, J. Chim. Phys. 67 (1970) 1968.
15) J.W. Cable, E.O. Wollam and H.R. Child, Phys. Rev. Letters, 22 (1969) 1256.
16) G. Rienäcker and G. Vorum, Z. Anorg. Allgem. Chem. 283 (1956) 287.
17) J.B. Cohen, Phase Transformation (Am. Soc. Metals, Ohio, 1970).
18) G.C. Bond, Catalysis by Metals (Academic Press Inc., New York, 1962).
19) H. Montgomery, G.P. Pells and E.M. Wray, Proc. Roy. Soc A 301 (1967) 261.
20) J.S. Dugdale and A.M. Guénault, Phil. Mag. 13 (1966) 503.
21) R. Burch, Trans. Faraday Soc. 66 (1970) 736.
22) Y. Sato, J.M. Sivertsen and L.E. Toth, Phys. Rev. B1 (1970) 1402.

DISCUSSION

V. PONEC

Professor Dowden pointed out in his paper the substantial difference which exists between Ni/Cu and Pd/Ag alloys. The literature on the electronic sturcture of these alloys brought, indeed, many pieces of evidence that d-holes persisted also in highly diluted Ni alloys, while the population of Pd energy levels varied strongly on alloying. It was, therefore, challenging to investigate whether the mentioned difference is also reflected in the catalytic reactions performed with both alloy systems under comparable conditions. R. de Jonge in our Laboratory has recently found that, indeed, a remarkable difference exists in cyclopentane-deuterium exchange reaction catalysis. On Ni/Cu alloys the (multiple-)/(single-) exchange selectivity remains constant, and high, for even very strongly diluted Ni-alloys, while on Pd/Ag the multiple exchange is slowed down and the D_5 and D_{10} peaks of initial product distribution, typical for Pd, disappear, already at low concentrations of Ag.

Further, I have the following question. It is known that hydrogen changes the Pd-electronic structure in a similar way as silver does--e.g. the paramagnetic susceptibility varies similarly, etc. I did not understand well from your paper if this effect has been taken into account in calculations of the $-\Delta H(\Theta)$ curve for hydrogen adsorption.

D. A. DOWDEN

Yes, this effect has been taken into account by using the density of states curve for the metallic alloy and supposing that one added electron from a hydrogen atom has the same effect as one added from an atom of a group 1B metal. The results you quote are most interesting.

W. M. H. SACHTLER

I certainly sympathize with your attempt to correlate catalytic activities of alloys with the abundance of various ensembles on the surface. There is, however, a practical difficulty in applying the statistical formulas, as the surface composition for annealed alloys differs from their bulk composition. For Pd-Ag we have found that the surface contains more Ag than the bulk because of the Gibbs adsorption phenomenon. Similarly, Bourdman found an agglomeration of Sn in the surface of P + Sn alloys. The surface composition changes again in the presence of a gas which forms stronger bonds with one metal than with the other. CO, for example, causes an agglomeration of Pt or Pd, respectively on the surfaces of Pt/Au or Pd/Ag.

D. A. DOWDEN

I am of course aware of your work on the composition of alloy surfaces. Twenty-five years ago I expected to find such an effect because even then it was well-known that the surface energy of group 1B metals is much less than that of transitional metals; early qualitative research did not reveal it. Adventitious segregation cannot be taken into account but equilibrium concentrations might be treated by a broken-bond model using methods similar to that employed to allow for clustering in nickel-copper alloys.

In the same way we had the vague idea that changes could be induced by strongly adsorbed gases but this phenomenon waited a long time before you found it; presumably it could be treated by similar methods, given the bond energies.

G. L. ERTL

Recent experiments on the adsorption of CO on Pd/Ag alloys (K. Christmann and G. Ertl, Surface Sci., in press) support the basic ideas of Dr. Dowden's paper. Clean single crystal planes with (111) orientation were prepared by expitaxial growth on mica substrates under UHV conditions. The surface compositions of the alloys were determined by a quantitative analysis of Auger electron spectra, and isosteric heats of CO adsorption were obtained by recording adsorption isotherms by means of measuring work function changes at different temperatures and pressures. The results are shown in Figure 1 where the heat of adsorption for different alloys is plotted vs.

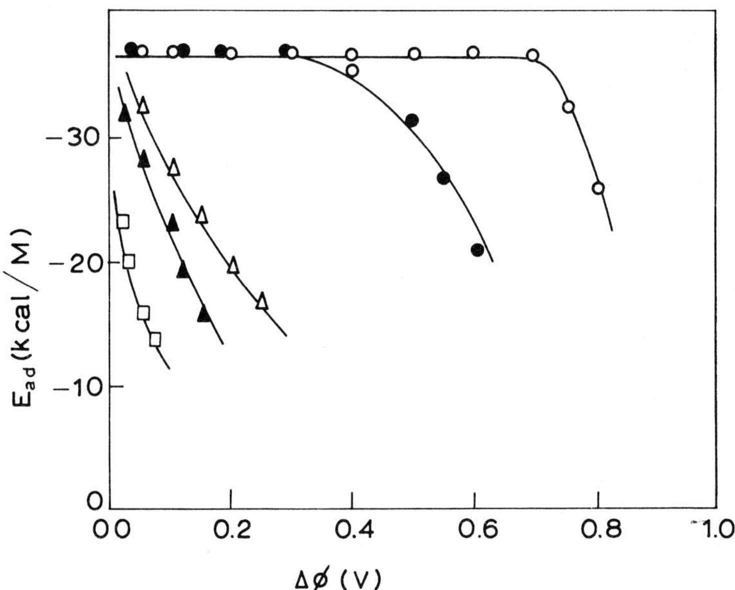

Fig. 1 Work function change during CO adsorption on Pd/Ag
o 0% Ag
• 5% Ag
Δ 22% Ag
30% Ag
45% Ag

the work function change which is a measure of the coverage. For pure Pd (111), E_{ad} is constant up to high coverages and then decreases strongly due to the onset of intermolecular repulsion. A surface containing 5% Ag exhibits at lower coverage the same constant value for E_{ad} as for Pd (111). Then the decrease in E_{ad} begins with increasing coverage, but it is not so sharp as for pure Pd. With the other alloys containing 22,30 and 45% Ag, this trend is further advanced. This example also shows that it may be questionable to "titrate" the surface composition of alloys by measuring the adsorbed amount of gas, since in this case almost no CO uptake at room temperature with alloys containing less than 50% Pd in the surface could be observed. Our conclusion was quite similar to that of Dr. Dowden, namely that chemisorption has to be considered with "ensembles" consisting of several surface atoms.

D. A. DOWDEN

Dr. Ertl has already informed me of his results--it is gratifying to see that my roughly calculated curves have the right kind of shape.

R. J. KOKES

If adsorbed hydrogen donates electrons to the d-band and this contributes to the fall in the heat as holes in the d-band are used up, one would expect a more rapid fall for small metal particles than for large ones. The existence of a large number of facile reactions requires that either this factor does not play a role or the change in energetics has no effect on these reactions. Could you please comment?

D. A. DOWDEN

I had intended to make a similar observation in the discussion of papers on crystal size effects. If (a) activity at a surface palladium atom is associated with one hole in its 4d-orbitals, (b) the electronic structure of very small crystals is not very different from that of larger crystals, and (c) the d-band holes are freely mobile throughout the solid and drift to the surface because of the chemical potential of chemisorption, then because the ratio of surface to bulk atoms is large in small crystals, there is a crystal size below which there are not enough d-band holes to provide surface atoms with one hole each. At that point the activity per unit area should begin to change. The effect is still more marked in alloys. I have a slide showing the trend but did not refer to it in my paper for lack of space.

It is interesting to calculate the relative change in sticking coefficient by considering the act of chemisorption to be a collision at the surface between an adsorbate molecule in a precursor state and a gas of d-band holes having the mobility found by experiment.

Paper Number 42

POISONING OF Ni-Cu ALLOY CATALYST FILMS BY ABSORBED HYDROGEN

A. FRACKIEWICZ, Z. KARPIŃSKI, A. LESZCZYŃSKI and W. PALCZEWSKA
Institute of Physical Chemistry, Polish Academy of Sciences,
Warszawa, Poland

ABSTRACT: The rates of heterogeneous recombination of hydrogen atoms, para-ortho hydrogen conversion and ethylene hydrogenation on Ni-Cu alloy (3% wt. Cu) evaporated films have been studied. The catalytic efficiency of the investigated films has been compared with that of the same films after their saturation in situ with hydrogen and transformation into the β-phase of the respective Ni-Cu-H system. For all reactions studied a substantial decrease in activity was found in passing from the hydrogen free Ni-Cu film catalyst to the β-Ni-Cu-H phase. The recombination coefficient of atomic hydrogen has been determined at temperatures from 473°K to 213°K: the spontaneous transformation of the Ni-Cu film catalyst into its less active form with a higher energy of activation occurred at lower temperatures. For the para hydrogen conversion (at 195°K) the absolute reaction rate decreased by a factor of $\sim 10^2$. The similar but smaller (by a factor of ~ 10) poisoning effect of the incorporated hydrogen has been observed for ethylene hydrogenation on Ni-Cu catalyst films. The observed phenomenon could be interpreted in terms of the change in the transition metal character of the investigated metal samples.

1. INTRODUCTION

The numerous attempts to relate catalytic activity to the electronic properties of metals were stimulated by a series of papers by Dowden [1]. They concentrated the main interest of research workers on the binary alloys of transition and IB metals.

Pd, Pd-Au, Pd-Ag systems and Ni, Ni-Cu systems became the conventional systems in these investigations.

The above mentioned metals saturated with hydrogen and transformed into their β-hydride phases give the possibility of an enlarged approach to the problem. The Pd-H system has been extensively investigated, the monograph by Lewis [2] presents an interesting review of the subject. Baranowski, Smialowski, Janko, Majchrzak et al. [3] investigated the Ni-H and Ni-Cu-H systems and proved their analogy to the Pd-H systems.

For different catalytic reactions involving hydrogen, Pd and Ni (and their alloys) to some extent lose their initial catalytic activity in consequence of absorption of hydrogen. This effect was observed in the case of Pd-H and Pd-Au-H systems for para-ortho hydrogen conversion [4] and hydrogen

atom recombination [5]; in the case of Ni-H and Ni-Cu-H systems for hydrogen-deuterium exchange [6] and hydrogen atom recombination [7].
The former study on the catalytic effect of the hydride formation in Ni and Ni-Cu alloys concerned metal foils saturated electrolytically with hydrogen for obtaining Ni-H or Ni-Cu-H β-phases. It seemed necessary, however, to examine the same systems under the conditions preventing contamination and assuring larger possibilities for catalytic research. It is only recently that this aim could be achieved, when a method has been developed for the transformation of evaporated Ni and Ni-Cu films into their β-hydride phase by action of pure hydrogen [8]. In the present study Ni-Cu system was chosen owing to its greater (than in case of Ni itself) ability of hydride formation. These films have been used for studying the catalytic effect of absorbed hydrogen mainly for hydrogen atom recombination, but the preliminary results concerning p-o hydrogen conversion and ethylene hydrogenation are also included.

2. EXPERIMENTAL

2.1. *Apparatus and Procedure*

2.1.1. Preparation of Ni-Cu and β-Ni-Cu-H-phase films. The Ni-Cu alloy catalyst films were deposited under high vacuum conditions ($p \leq 10^{-5}$ mmHg) by evaporation of electrically heated, previously outgassed, Ni-Cu alloy filament (1,37% wt. Cu, content of impurities < 0.15%, d=0.5mm) and by condensation of the respective metal vapour on the glass walls of the reaction vessel. The films obtained (marked further as Ni97Cu3) were at least 2000 Å thick, their copper content amounted to 3.0% wt. For the transformation into the respective hydride phase the film cooled at 193°K was exposed to the atomic hydrogen produced in the radio-frequency discharge in hydrogen of 0.1 mmHg pressure. The method was described in detail in the separate paper [8], where it has been stated that under these conditions Ni97Cu3 films (as well as pure Ni) transform partially or completely into the β-phase, identified by X-ray diffraction. The respective details concerning the condition of film sample deposition will be included in the description of the experimental procedure for every catalytic reaction investigated.

2.1.2. Heterogeneous recombination of hydrogen atoms. The recombination of H atoms on the Ni97Cu3 alloy film has been studied using the Smith-Linnett side-arm method [9]. The apparatus and the experimental procedure have been described in the earlier papers [5, 7] and do not need to be treated in detail here. The catalyst film deposited on the inner wall of a glass tube (120 mm long) was placed in the sidearm of the Smith apparatus surrounded by a glass jacket filled with a liquid thermostating the reaction zone. The H atoms produced by means of the radio-frequency discharge recombined on the alloy film and, though in smaller extent, on the glass walls of the side-arm itself. The decay of atoms was registered by the Cu-constantan thermocouple moveable along the side-arm axis. The ratio of the two thermocouple readings: the moveable one, E_x, and the fixed one, E_0, monitori the H atom concentration at the entrance to the side-arm, allow the coefficient of recombination, γ , to be determined. Under the appropriate experimental conditions discussed by Dickens, Linnett et al.[10]

$$\gamma = \beta^2 (2RD) / \bar{c}$$

where: β = the slope of the ln E_x/E_o versus x plot; x = the position of the moveable thermocouple; R = the side-arm tube radius; D = is the diffusion coefficient of H atoms [11], $\bar{c}$ = the mean atomic speed. The measurements were carried out at temperatures ranging from 473°K to 213°K, at a hydrogen pressure of 0.1 mmHg.
The alloy films were deposited in a separate high vacuum apparatus on the walls of a Pyrex glass tube. In order to obtain the homogeneous alloy film the substrate temperature was maintained at 723°K and the film, when deposited, was still annealed at the same temperature during 16 hr. After cooling to room temperature the section of the tube covered with alloy film was cut out and then slid into the side-arm of the atom recombination apparatus.

2.1.3. Para-ortho hydrogen conversion.
The kinetics of the p-oH_2 conversion was studied in a conventional reaction system essentially based on that of Eley and Rideal [12]. The Ni97Cu3 alloy film about 1000A thick was deposited in vacuo of 10^{-6} mmHg directly on the inner wall of a reaction vessel maintained at 195°K. The conversion rate was measured at the same temperature within a pressure range of 2 - 20 mmHg for the Ni97Cu3 film catalyst before and after its exposure in situ to atomic hydrogen produced in r.f. discharge during 75min.

2.1.4. Hydrogenation of ethylene.
The activity of Ni97Cu3 film for ethylene hydrogenation was determined in a constant volume system. The initial total pressure of 1 : 1 mixture of ethylene and hydrogen was 100 mmHg. For every run a fresh film catalyst sample was deposited in situ on the wall of a new cylindric glass reaction vessel at 233°K. The β-Ni-Cu-H phase was prepared also in situ by the exposure of the newly evaporated Ni97Cu3 alloy film to the H atoms generated by means of r.f. discharge. Separate fresh alloy and hydride films were used for each experimental run in order to avoid the film being poisoned by reaction products. The course of reaction at 233°K was followed during the first 15 min by reading the change of pressure as indicated by a mercury manometer.

3. RESULTS

3.1. Rate of hydrogen atom recombination.
The plots of lg E_x/E_o against distance x gave straight lines which confirmed the first order of the kinetics of the heterogeneous recombination and its Rideal-Eley mechanism Me-H + H = H_2. The coefficient of recombination, γ, was determined at each temperature in the given sequence of temperatures during everyone experimental run: 473°K, 363°K, 293°K, 247°K, 213°K and finally 473°K anew. Fig. 1 shows the variation of the values of γ obtained in

function of temperature. The relationship between $\lg(\gamma T)$ and $1/T$ is represented by two straight lines: the first – for the temperatures above 273°K and the second – for the lower range of T. The linear form of the $\lg(\gamma T)$ versus $1/T$ dependence results also from the theory of absolute reaction rates on account of the small activation energy of H atom recombination. The quantitatively different catalytic behaviour of the alloy film studied within the two ranges of temperature is expressed by two values of the energy of activation of the reaction investigated, i.e. $E_A^{I} = 0.51$ kcal/mol at higher temperatures and $E_A^{II} = 4.1$ kcal/mol for low temperatures. After returning to 473°K in every experimental run (crosses in fig. 1) the distinct increase of activity of the film investigated was observed in comparison with its initial value under the same conditions.

3.2. Rate of para-ortho hydrogen conversion.
The reaction rate studied at 2 - 20 mmHg and at 195°K followed the 0.7 order kinetic equation. The determined value of the absolute reaction rate, k_m^{I} in the case of the Ni97Cu3 catalyst film was 6.0×10^{15} molecules $cm^{-2}s^{-1}$ and decreased only slightly under the influence of interaction with molecular hydrogen admitted. After a 75 min. exposure of the sample to atomic hydrogen the r.f. discharge was switched out and the catalytic activity of the Ni97Cu3 film transformed in some extent into its hydride was investigated under the same conditions of p-T as formerly. The resulting value of k_m^{II} was 3.7×10^{13} molecules $cm^{-2}s^{-1}$. The high poisoning effect of the incorporated hydrogen may be expressed as $k_m^{I}/k_m^{II} = 160$.

3.3. Rate of ethylene hydrogenation.
In the course of each 15 min run the reaction of ethylene hydrogenation on Ni97Cu3 film as well as on the respective β -Ni-Cu-H phase was of the first order with respect to hydrogen and zero order with respect to ethylene at 233°K. Fig. 2 shows the experimental results for 3 specimens of each system investigated. In table 1 the values of the rate constants of the reaction catalysed by different film samples are collected. Although both groups of films under similar experimental conditions exhibit some quantitative differences in their activity, the decrease of the reaction rate after exposure to hydrogen atoms is beyond the order of irreproducibility of the systems investigated. The poisoning effect of the hydride formation is expressed in this case as a factor of 3 to 10 decrease in the value of the rate constant.

4. DISCUSSION

The catalytic effect of the Ni97Cu3 alloy film when "pure" or saturated with hydrogen till its transformation into the respective hydride has been studied in three reactions with different mechanisms but all involving hydrogen as a species interacting with the catalyst surface in a heterogeneous stage of the overall reaction. At low temperatures the prolonged exposure of the alloy film to the action of atomic hydrogen resulted for all reactions studied in a distinct decrease of the activity of film catalysts.
In the separate experiments carried out under the identical conditions as

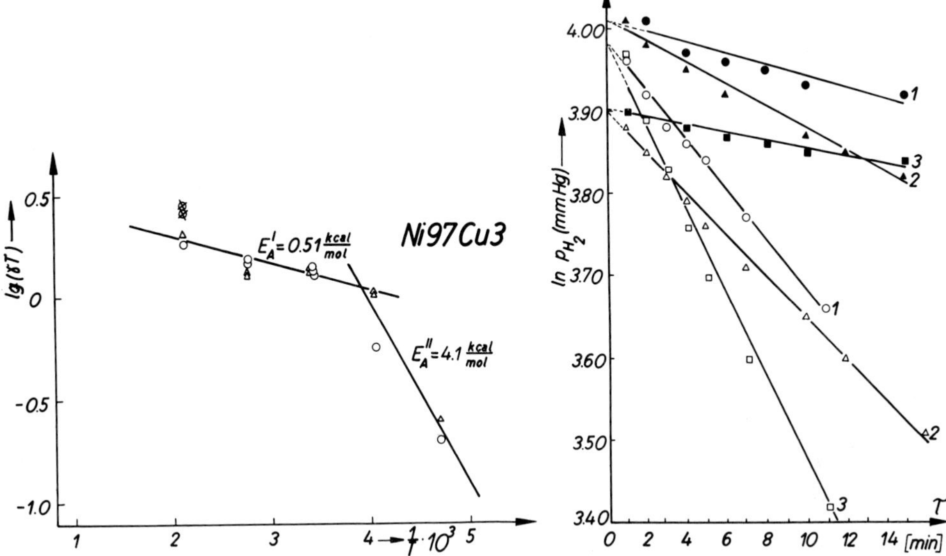

Fig. 1. Coefiicient of hydrogen atom recombination on the Ni97Cu3 film in function of 1/T.

Fig. 2. Kinetics of ethylene hydrogenation on Ni97Cu3 alloy films and on the respective hydride phase (black points), T=233°K

those related above for the reactions studied the formation of the β-Ni-Cu-H phase of the alloy was confirmed by means of the X-ray diffraction method [8]. Thus it seems to be justified to attribute the observed loss of catalytic activity of metal to its saturation with hydrogen and transformation into the respective hydride. These two sets of experimental facts give an evidence of a strong negative catalytic consequence of the "hydride" form of hydrogen incorporated in metal.

The increase of γ value at the end of a H atom recombination run at 473°K still supports the assumption of the former existence of the hydride phase in the sample involved at low temperatures. The decomposition of the hydride at high temperature should result in a marked desintegration of the film crystallites [13] and in consequence in an enhanced activity. This fact was confirmed by the observed increase of BET area and the decrease of crystallite sies as proved by the X-ray technique [14].

Under the experimental conditions of the present investigations a long exposure of films to an intense source of atomic hydrogen resulted in formation of thick hydride layers which could be checked by the X-ray diffraction. But the similar poisoning effect may be expected (and was observed) [15], though in a feebler extent, in different reactions involving hydrogen, when a metal catalyst is exposed to some "active" hydrogen form (particularly in presence of species enhancing its penetration into metals [16]).

Table 1
Rate constants of ethylene hydrogenation on Ni97Cu3 film catalysts and their hydride phase at 233°K.

Film		Mass of film, mg	$k \times 10^2$ min^{-1}	$k \times 10^2$ K values for 10mg of film
	1	8.9	3.0 ± 0.1	3.4 ± 0.1
Ni97Cu3	2	9.0	2.5 ± 0.1	2.8 ± 0.1
	3	21.5	5.1 ± 0.1	2.4 ± 0.1
	1	8.6	0.69 ± 0.05	0.80 ± 0.05
β-Ni-Cu-H	2	12.1	1.3 ± 0.1 *)	1.1 ± 0.1 *)
	3	16.0	0.50 ± 0.05	0.31 ± 0.05

*) after a r.f. discharge of low intensity

The problem of the catalytic inactivity of the hydride phase of at least some transition metals such as Pd or Ni (and their alloys) may be based until the present on the band theory of metals which associates catalytic activity with the partly empty d-bands in the electronic structure of transition metals. The 1s electron of hydrogen built into the Pd or Ni lattice would contribute to filling the d-band of the metal and in decreasing its initial transition metal character. Though the recent experimental and theoretical achievements both emphasise the necessity of the surface state approach to heterogeneous catalysis, in the case of alloys and even more of the β-hydride phases of Pd or Ni no data are available which could elucidate the electronic structure of their surface or forms and states of the reactant species adsorbed on it. Assuming that the modification of the electronic structure of a bulk metal brought about by the formation of a respective hydride influences directly the electronic states at the surface it would be possible to give the general, above suggested, explanation of the poisoning effect of the "hydride" form of hydrogen when present in the Ni, Ni-Cu or Pd lattice. At any rate the metal-hydrogen system seems to be in the case of nickel (and palladium) much more suitable than the alloy systems (particularly the NiCu alloy system with its tendency towards the phase segregation [17]) for the investigations of the influence of the change of a transition character of metal on its catalytic activity.

REFERENCES

[1] D. A. Dowden, J. Chem. Soc. (1950) 242;
D. A. Dowden, P. W. Reynolds, Disc. Faraday Soc. 8 (1950) 172;
D. A. Dowden, J. Res. Inst. Catalysis, Hokkaido Univ. 14 (1966) 1.

[2] F. A. Lewis, The Palladium Hydrogen System, Acad. Press, 1967.

[3] B. Baranowski, M. Śmialowski, Bull. Acad. Polon. Sci., Ser. sci. chim. 7 (1959) 633; J. Phys. Chem. Solids 112 (1959) 206;
A. Janko, Naturwiss. 47 (1960) 225;
B. Baranowski, S. Majchrzak, Roczniki Chem. 42 (1960) 225.

[4] A. Couper, D. D. Eley, Disc. Faraday Soc. 8 (1950) 172;
J. J. F. Scholten, A. A. Konvalinka, J. Catalysis 5 (1960) 1.

[5] P. G. Dickens, J. W. Linnett, W. Palczewska, J. Catalysis 4 (1965) 104.

[6] K. M. Zhavoronkowa, G. K. Boreskov, V. N. Nekipelov, Dokl. Akad. Nauk SSSR 171 (1967) 1124.

[7] W. Palczewska, A. Frackiewicz, Bull. Acad. Polon. Sci., Ser. sci. chim. 14 (1966) 67; W. Palczewska, A. Frackiewicz, Z. Karpinski IV Intern. Congress on Catalysis, Moscow, 1968; Bull. Acad. Polon. Sci. Ser. sci. chim. 17 (1969) 687.

[8] W. Palczewska, A. Janko, Bull. Acad. Polon. Sci. Ser. sci. chim. (in press).

[9] W. Smith, J. Chem. Phys., 11 (1943) 1110;
J. W. Linnett, D. G. H. Marsden, Proc. Roy. Soc. 234A (1956) 489;
J. W. Linnett, Trans. Faraday Soc., 55 (1959) 1355.

[10] P. D. Dickens, D. Schofield, J. Walsh, Trans. Faraday Soc. 56 (1960) 225.

[11] S. Weissman, E. A. Mason, J. Chem. Phys. 37 (1962) 1289.

[12] D. D. Eley, F. K. Rideal, Proc. Roy. Soc. 178A (1941) 429.

[13] A. Janko, Bull. Acad. Polon. Sci. Ser. sci. chim. 10 (1962) 613.

[14] Z. Karpinski, A. Janko, W. Palczewska, Bull. Acad. Polon. Sci. Ser. sci. chim (in press).

[15] J. H. Singleton, J. Phys. Chem. 60 (1956) 1606;
P. B. Schallcross, W. W. Russell, J. Am. Chem. Soc. 81 (1959) 4132;
W. K. Hall, P. H. Emmett, J. Phys. Chem. 63 (1959) 1102;
W. K. Hall, J. A. Hassell, ibid. 67 (1963) 636.

[16] M. Śmialowski, Hydrogen in Steel, Pergamon Press, 1961.

[17] W. M. H. Sachtler, G. J. Dorgelo, J. Catalysis 4 (1965) 654;
W. M. H. Sachtler, R. Jongepier, ibid. 4 (1965) 665;
P. Van der Planck, W. M. H. Sachtler, ibid. 12 (1968) 35;
L. Elford, F. Muller, O. Kubaschewski, Ber. Bunsen ges, physik. Chem. 73 (1969) 601.

DISCUSSION

J. J. F. SCHOLTEN

1) On page 5 of your paper you state that an increase in BET surface area was found due to disintegration of the film crystallites. What was the increase of surface area? Did you find a change of the activation energy due to this disintegration?

2) Changes in activity due to the $\alpha \rightarrow \beta$ transition can be explained also by a decrease in the number of free Pd sites due to this transition at the surface which process goes parallel with a change in the distribution of the various crystallographic planes. Do you agree with such an explanation?

W. PALCZEWSKA

1) The increase of surface area (caused by the desintegration of the initially large alloy crystallites) amounted to about 70% of the initial value (confirmed by isotherms and X-ray line broadening). This final film with smaller crystallites was not further investigated at different temperatures. The whole set of results represented on Fig. 1 was collected during about 48 hours of a continuous experiment, which could not be prolonged further.

2) When we are lowering the temperature (Fig. 1), we pass in a continuous way from the recombination on the large crystalline film of NiCu (or rather on the α-Ni-Cu-H solution) at higher temperatures, to the recombination on the β-Ni-Cu-H phase system at lower temperatures. We have obtained no evidence that the formation of the β-phase hydride is accompanied by the change of the crystallographic orientation of the Ni-Cu crystallites; however this change appears later in consequence of the crystallite cracking during the hydride decomposition. As the lattice parameter of the β-hydride phase is of about 6% higher the surface concentration of metal ion sites decreases, but the "structural," interstitial hydrogen forms eventually itself new sites for adsorption. In our opinion the change in electronic structure in β-hydride phase, when compared with that of the Pd metal, influences strongly the ability of surface atoms to bind H atoms and to activate them for the subsequent reaction.

D. D. ELEY

I should like to compliment Professor Palczewska on her paper and to draw attention to the very recent theoretical paper by Switendick, of Sandia Laboratories. Switendick[1)] has done electronic energy band calculations (APW) for Pd and PdH, and NiH etc. He shows the rigid band theory does not hold for Pd and PdH, but still it is clear that this only makes a quantitative change in earlier ideas, i.e., the percentage H for loss of paramagnetism no longer equates to the concentration of holes in the d-band of Pd. This does not change the idea advanced by Couper and myself in 1950, and supported today by Professor Palczewska, that the "poisoning" effect of dissolved H atoms in Pd may be associated with the loss of "holes" in the d-band necessary for a low energy activated complex in the catalysis.

1) A. C. Switendick, Bev. Bunschges. 76 (1972) 535.

W. PALCZEWSKA

Switendick's conclusions concerning the electronic structure of the NiH and PdH supply really the affirmation of the much higher ability of these metals to absorb and bind hydrogen than it could be anticipated from the electronic structure of the Pd or Ni themselves. This paper really gives a new support to the link between the d-band structure and the catalytic activity of these transition metals.

C. G. HARKINS

In paper 36, Takeuchi introduced H^3 into Ni and Ni-Cu alloy, showed its pressure by autoradiography, and found an enhancement of the activity for ethylene hydrogenation.

You report introduction of hydrogen into Ni-Cu alloy by H atom exposure, confirm its presence in the solid by X-ray evidence to the hydride phase, and show a decrease in activity for ethylene hydrogenation. Please comment on these apparently paradoxical results on the effect of metal phase hydrogen on the surface reactivity.

W. PALCZEWSKA

In our investigations we could prove hydrogen to be built into the Ni-Cu host lattice as the β-hydride bulk phase. Professor T. Takeuchi's experiments showed the accumulation of H^3 in the grain boundaries. For the 1:1 Cu-Ni alloy, used by T. Takeuchi and coworkers, it is difficult to state whether the hydride is forming (Roczniki Chem., 42, (1960) 225) because of the small lattice parameter change induced by the eventual hydride formation. But even if this Ni50 Cu50 alloy hydride exists, it will probably decompose under T. Takeuchi experimental conditions and hydrogen will diffuse through dislocations to the grain boundaries (Bull. Acad. Pol. Sci. sér. sci. chim., 20 (6) 611 (1972)). The hydride formation and decomposition sequence may in consequence increase the lattice imperfections and the number of microcracks (ibid. 20, (5) 487 (1972)).

Desorbing hydrogen accumulated at the crystallite surfaces may be catalytically active as e.g. hydrogen permeating through the Pd membrane.

D. J. C. YATES

I would like to add my congratulations to those of Professor Eley on your very interesting paper. I have three questions: 1) Have you made any measurements of the surface composition of your alloy films (before hydriding)? As Cu has a lower melting point than Ni, it might be expected that the Cu would evaporate first, thus giving a Cu-rich film. This would seem to be shown by the statement on Page 2 of the paper that films evaporated from a filament with 1.37% Cu gave films containing 3.0% Cu. It seems to me that there is a good chance of a very homogeneous alloy being formed under these conditions. 2) There is also the possibility of segregation of either Cu or Ni as the metals condensed to form the film--what was the temperature of the glass wall while the film was being condensed? 3) Do you think that surface segregation could have occurred during your stress relieving (723°K for 16 hours)? Could you perhaps mention whether the annealing was done in vacuo or in H_2?

W. PALCZEWSKA

Being conscious of the possibility of the Ni-Cu alloy segregation and of the formation, in consequence, of two Sachtler's equilibrated alloys, we evaporated our films at the temperature in which the system Ni-Cu is homogeneous (Ber. Buns. Ges. Phys. Chem., 73 (1969) 601). The glass walls which were a support for our film were kept at the same temperature during the evaporation of the films and their subsequent annealing (723°K).

The annealing was done under high vacuum conditions ($p<10^{-5}$Tr). During the cooling of the film to the room temperature in vacuo the mobility of metal ions was probably not sufficient under these conditions for the formation of an equilibrium two phase system of the alloy investigated.

In the X-ray diffraction pattern of the films investigated only the series of lines corresponding to the homogeneous Ni97Cu3 alloy were registered.

T. TAKEUCHI

We had studied also the effects of the preadsorbed or dissolved hydrogen on the catalytic activity for the hydrogenation reaction of ethylene using a nickel film. Our results were quite different from yours. In our case, the film was prepared by means of deposition of nickel in hydrogen (various pressures) on a glass wall cooled by liquid nitrogen. The film was sintered at 30°C and evacuated to 10^{-6} torr. The catalytic activity of the film prepared in 5 torr hydrogen was Ca. 140 times greater than the film prepared in vacuum.

Such discrepancy seems to be caused by the copper which coexisted in nickel. In your case, the copper-nickel alloy was evaporated on a wall whose temperature was higher than ours. In this case copper would evaporate at the initial stage and a non-uniform alloy may be formed. It is possible that the nickel crystal may be covered with copper.

W. PALCZEWSKA

In our experiments we did not observe the formation of nickel hydride during a simple deposition of nickel in a molecular hydrogen atmosphere, though the amount of adsorbed hydrogen could be quite high under these conditions. The high activity of samples prepared by this way may be caused by the desorption of this, previously dissolved, hydrogen under the actual reaction experimental conditions (c.f. the preceding comment and answer). We had no evidence in favor of the inhomogeneity of our alloy films; X-ray diffraction pattern was that of one phase alloy system only. The evident transformation of the alloy into its β-hydride phase, as stated directly in a separate experimental setup, confirms that the alloy film grains could not be covered by Cu or rich in Cu alloys. In any case however, the same alloy film (built from the same crystallites) lost to some extent its catalytic activity after its transformation into β-hydride phase.

M. SIMANSKA

In the interesting paper by Professor Palczewska a significant change in catalytic activity of Ni-Cu alloy after its saturation with hydrogen and transformation into the respective hydride has been reported. Since in the hydrogenation of reactant mixtures we have also organic molecules, I would like to ask what could be--if any--the influence of these organic reactants on the process of hydride formation and the subsequent metal catalyst poisoning? Could organic molecules inhibit or catalyze the hydride formation?

W. PALCZEWSKA

To my knowledge there are no informations about the influence of organic compounds on the kinetics of the β-hydride phase formation in Ni, Pd or alloys.

H_2S, H_2Se, H_2Te as well as AsH_3, SbH_3, BiH_3 are the well-known positive catalysts which enhance the rate of penetration of hydrogen into these metals and of the respective hydride formation (M. Śmialowski, Hydrogen in Steel, Pergamon Press, 1962).

Paper Number 43

REACTIONS OF HEXANE ISOMERS ON NICKEL/COPPER ALLOYS

V. PONEC and W. M. H. SACHTLER
Gorlaeus Laboratoria, Rijksuniversiteit, Leiden, Netherlands

ABSTRACT: Reactions of hexane isomers have been studied in an open flow reactor on nickel-copper powder alloys prepared from coprecipitated carbonates. The main results are: (a) between 0-23% Cu the activity of the catalysts decreases with the copper content and the selectivity for C_6-products increases to the value previously found for extremely thin Ni-films; (b) between 20-60% Cu the activity remains almost constant while the selectivity increases and reaches the values common for platinum. An interpretation is suggested for the region (a) in terms of changing number of "ensembles". For the alloys sub (b) an explanation essentially based on the electronic structure of alloys is suggested.

1. INTRODUCTION

An important hitherto unsolved problem of heterogeneous catalysis is the marked difference in selectivity between similar metals in interconversion of hydrocarbons[1]. Even such closely related metals as Ni, Pd and Pt exhibit very strong difference with respect to the hydrogenolysis, isomerisation and dehydrocyclisation. Such fundamental problems in selectivity patterns form a challenge in heterogeneous catalysis research.

An inspection of the hypotheses under consideration shows that a systematic study of the catalytic selectivity of certain alloys is likely to provide the relevant information required to decide between various existing theories[2-10].

In a recent paper[1] we have shown that whereas e.g. the activity for single and multiple exchange of cyclopentane with deuterium was relatively similar for pure Ni and Ni-Cu alloys, the hydrogenolytic reactions were strongly affected by alloying. On the basis of these results and the above considerations it appeared desirable to study the reactions of hexane isomers over Ni and Ni-Cu alloys.

In the present work the temperature of catalyst preparation and of subsequent reaction were chosen above the temperatures closing the miscibility gap of Ni-Cu alloys[11]. By consequence alloys of different bulk compositions should present surfaces of different compositions, although the bulk and surface Ni-Cu ratios need not to be equal.

2. EXPERIMENTAL

2.1 Materials used

The hydrocarbons used for reactions and calibrations were of "GLC puriss" grade, ex. Fluka, Switz. When necessary, corrections were applied for the small content of impurities (<0.5%). As reactants methylcyclopentane (MCP), n-hexane (n-HEX) and 3-methylpentane (3MP) were used. Catalysts were prepared by the method of Best and Russell[12]. It has been shown[13] that this method produces alloy phases near the equilibrium. Coprecipitated carbonates were decomposed in air at 400°C, the resulting oxides were reduced at 350-400°C for several hours. Subsequently, the powders were passivated by CO_2 at room temperature, finely ground and brought to the reactor where they were reduced in situ by a flow of hydrogen for cca 24 hours. Powders with less than 60% Cu were reduced at 400°C, the other alloys at 360°C. Composition of alloys was determined by complexometric titrations; in this paper all Cu concentrations are given in atomic percents.

2.2 Apparatus and procedure

Experiments were performed in an open flow system. The stream of hydrogen purified by Pd-asbestos and cold traps was saturated at 0^oC with the saturation pressure of the hydrocarbons under study (40-60 Torr), the total pressure being 1 atm. The all-glass apparatus was equipped with metallic (HOKE) or Teflon valves, essential parts of the apparatus could be heated by heating tapes. Samples were taken close to the exit of the reactor through the septum and then injected into the GLC apparatus (Becker, Delft). Results shown below represent steady state with respect to the rate and products distribution.

Separation of all C_1-C_5 hydrocarbons as well as of MCP, HEX, benzene (Be) and cyclohexane (Cy-HEX) is easily achieved following Ref. 14), at 80^oC using a o.6 mm diam., 7m long column filled with 10% wt squalane on Chromosorb S, 60-80 mesh. However, 2MP, 2,3-dimethylbutane (2,3DMB) and cyclopentane (CP) could not be separated. Therefore, the ratio (2MP + 2,3DMB) /(MCP) was measured using a capillary column (50 m long, 0.25 mm diam., with Apiezon) so that the CP concentration could be calculated. No attempts were made to separate 2,3DMB from 2MP. Concentrations of individual products could be determined from peak heights and empiric sensitivity factors. For necessary calibrations C_1-C_4 hydrocarbons were injected at 1 atm. C_5-C_6 hydrocarbons at their saturation pressures (at 0^oC) in a mixture with air. Calibration graphs were always linear.

The GLC data converted in molar concentrations of species (C_i denotes concentration of a hydrocarbon with i-carbon atoms) were used to calculate the total conversion α in %:

$$\alpha = 100\Sigma_{i=1}^{6}\Sigma_{j,j\neq k}iC_i^{(j)}/(\Sigma_{i=1}^{6}\Sigma_{j,j\neq k}iC_i^{(j)} + 6C_6^{(k)}) \tag{1}$$

Summation over j is performed over all detected compounds for each C_i hydrocarbon, the subscript (k) denotes the hydrocarbon in the feed.

The rate of the overall reaction was determined from the measured conversion α and the flow rate (F, molecules/sec). For conversions <20% the function $\alpha = f(F^{-1})$ was found to be linear and, therefore, the rate per cm^2 of the total surface r_S, or the rate per g catalyst r_W are:

$$r_S = 10^{-2}d\alpha/d(SW/F) \doteq 10^{-2}(SW)^{-1}\alpha F \tag{2}$$

$$r_W = 10^{-2}d\alpha/d(W/F) \doteq 10^{-2}W^{-1}\alpha F \tag{3}$$

where the surface area of 1 g of catalyst is denoted by S and the weight of the catalyst by W. The surface areas of the catalysts were measured by the Perkin Elmer-Shell Sorptometer 2120.

The selectivity in producing other C_6-hydrocarbons is characterised by the selectivity parameter S (in %):

$$S = 600\Sigma_{j,j\neq k}C_6^{(j)}/(\Sigma_{j,j\neq k}\Sigma_{i=1}^{6}iC_i^{(j)}) \tag{4}$$

In the chain splitting the following possibilities may be discerned.

a) A single terminal fission running as

$$C_6 \rightarrow C_5 + C_1 \rightarrow C_4 + 2C_1 \rightarrow C_3 + 3C_1 \rightarrow C_2 + 4C_1 \rightarrow 6C_1.$$

The fission parameter M defined as:

$$M = (\Sigma_j\Sigma_{i=2}^{5}(6 - i)C_i^{(j)})/(C_1)_{measured} \tag{5}$$

which can be evaluated from the experimental data, should be near unity for this reaction if step 5 can be neglected.

b) If the multiple fission $C_6 \rightarrow 6C_1$ is important, a value of M lower than unity will result.

c) If a disproportionation such as $C_6 \rightarrow 2C_3$ and $C_6 \rightarrow C_4 + C_2$ is important,

M will be much higher than unity.

3. RESULTS

After an initial period of changing activity which lasted several hours the performance of the catalysts with less than 70% Cu was quite satisfactory in stability and reproducibility provided that $T < 350^{o}C$. Measurements performed at higher temperature were in most cases accompanied by a decay of activity which could partially or even fully be restored by hydrogen reduction at $T < 300\text{-}350^{o}C$. The fluctuation of activity for a given catalyst are mostly due to differences in the preexponential factor.

TABLE 1
Activity and selectivity of NiCu alloys in hexane reactions

Cu %	[1] S%	[1] M	[1] 2MP/3MP	[2] A_1	[3] A_2	[4] E_{act}
[7] 0	-	0.7	-	18.3	14.1	44
[8] 0	4.5	0.8	1.7-3.2	17.2-17.5	-	44
5	8-9	1.05	2.5-3.5	16.2-16.6	11.2-11.6	44
23	12-9	1.15	1.8-2.0	16.3	11.5	47
[5] 40	43	1.5	3.7	-	-	-
47	68	3.0-4.5	4.7-3.5	15.8-16.1	10.9-11.2	55
55	60	2.5-3.5	3.7-3.3	15.6-15.7	10.9-11	57
73	64	3.0-4.0	3.5	16.3	11.4	57
[6] 85	∼ 40	∼ 1.0	∼ 0.5	∼ 14.5	-	-

1) Selectivity S and fission parameter M, see text. Where two values are given, the parameter increases, approximately linearly, or decreases within the limits (see Fig. 2).

2) $A_1 = \log r_W$, where r_W is the rate (in molecules/sec) of the overall reaction per 1 g catalyst, at $330^{o}C$.

3) $A_2 = \log r_S$, where r_S is the rate (in molecules/sec) of the overall reaction per 1 cm^2 of the total surface area of the catalyst, at $330^{o}C$.

4) Activation energy of the overall reaction, in Kcal/mol.

5) Measured at one temperature, $344^{o}C$.

6) Measured at $337^{o}C$. An unstable catalyst where only approximate values could be determined.

7) Measured with pure nickel, 6 g, at 240-280°C; value for A determined by extrapolation to $330^{o}C$.

8) Measured with 0.1 g nickel diluted by 1 g SiO_2, directly in the region 300-330°C.

The conversion of hexane was studied most extensively and the data are summarized in Table 1 and Fig. 1. In Fig. 2 the parameter M is plotted as a function of temperature for n-hexane on 23% Cu alloy. At higher temperatures ($T > 380^{o}C$) M decreases to zero.

The C_6-isomers studied did not differ very much in their reactivities, for pure Ni and 5% Cu alloy. The values of the parameter M on pure Ni were 0.9 for 3MP and 1.1 for MCP. On 5% Cu alloy M was 1.05 for MCP and 1.1 for 3MP. These values were constant in the range of the total conversion 1-50%.

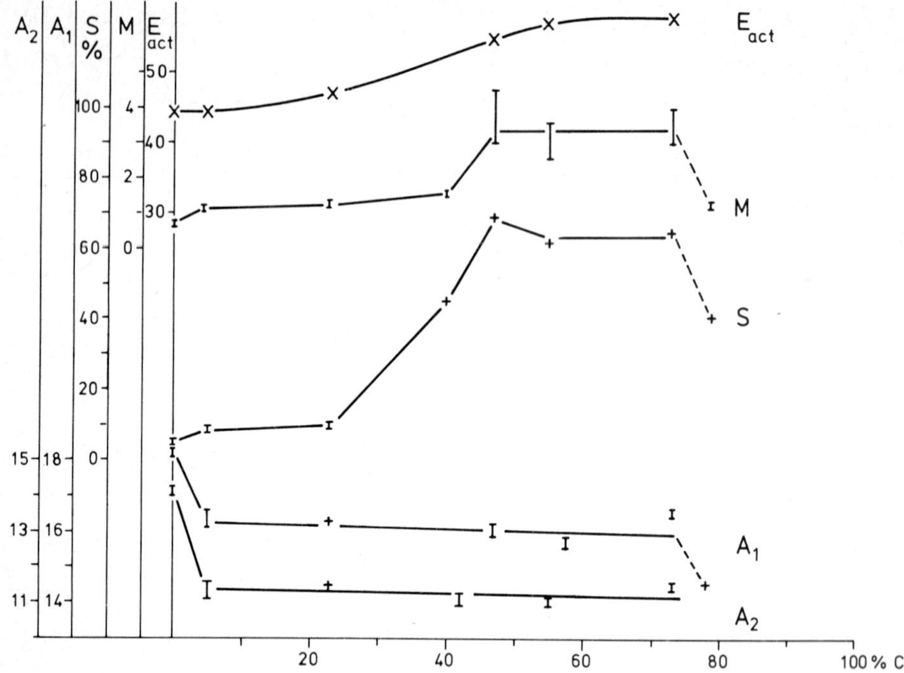

Fig. 1 Reaction parameters of n-hexane conversion by Ni and Ni-Cu alloys. A_1 = log r_W at 330°C, A_2 = log r_S at 330°C, activation energy of the overall reaction E_{act}, fission parameter M, selectivity parameter S; all as a function of alloy composition (in atomic % Cu).

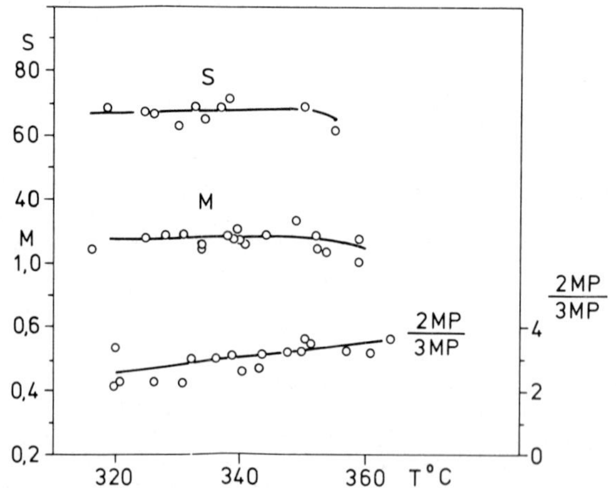

Fig. 2 Reaction parameters as a function of temperature. Selectivity in C_6-hydrocarbon production (S).- for 47 at% Cu alloy, fission parameter (M)- for 23 at% Cu; 2MP/3MP ratio for 5 at% Cu alloy; all as a function of the reaction temperature.

TABLE 2

Products distribution (%) in conversion of n-hexane on Ni and NiCu alloys

At% Cu	°C T	% α	Concentrations in %											
			C_1	C_2	C_3	iso C_4	n-C_4	iso C_5	n-C_5	2MP	3MP	MCP	Be	Cy-Hex
0[1]	251	5.5	69.4	1.2	2.7	-	8.5	-	18.2	-	-	-	-	-
0[2]	316	11.6	74.6	6.3	6.0	0.4	5.4	0.4	5.5	0.7	0.3	0.1	0.3	-
5	320	7.7	57.3	3.0	2.7	0.4	8.5	0.3	24.1	2.1	1.0	0.4	0.2	(0.05)
23	333	8.9	52.1	1.8	1.0	-	6.9	0.1	32.5	2.6	1.5	1.1	0.4	*)
47	337	5.5	17.2	6.0	8.6	0.2	6.1	0.5	10.6	35.1	9.7	4.6	1.3	-
55	349	5.6	23.0	6.5	8.3	-	7.3	0.7	13.1	27.9	8.4	3.6	1.1	-

*) traces 1) pure Ni, 6 g
2) 0.1 Ni diluted by 1 g SiO_2

The ratio isopentane/pentane in the hydrogenolytic products of 3MP which should be 2 for random splitting is found to be 1.8-3.5 (depending on temperature) for clean Ni, but 2.8-4.7 for a 5% Cu alloy. A similar trend characterises also the difference between Ni and Pt[3].

For hydrogenolysis of n-HEX the ratio 2MP/3MP is found higher than the expected thermodynamic value (cca 1.6) and the value (2) predicted for random splitting of a postulated MCP intermediate[7]. With MCP in the feed this ratio is 0.7 for pure Ni, but 1.4-1.7 for a 5% Cu alloy.

The selectivity parameter S, the fission parameter M and the value for the ratio 2MP/3MP were derived from graphs as Fig. 2 where three representative plots for three different alloys are shown. The results for other alloys are essentially of the same character and the plots are similar also for 3MP and MCP reactions.

In Table 2 examples are listed of initial product distributions for hexane reactions on several alloys. The feed was the same in all these experiments.

4. DISCUSSION

The striking results of the present work as illustrated by Figs.1 and 2 are:

a) Between 0-23% Cu the activity parameters A_1 and A_2 decrease sharply and this is accompanied by an increase of S to the values previously found for extremely thin Ni films[4].

b) A much more pronounced increase in S and M is observed for 40-73% Cu, where S reaches values common for platinum. In this region the A-parameters change only little.

c) Selectivity and fission parameter show only a small temperature dependence.

These results confirm the conclusions of our previous work[1] where it was reported that for reactions involving C-H bond rupture (i.e. single and multiple exchange of cyclopentane with deuterium) the reaction rate per surface Ni atom remains essentially the same on alloying with Cu (assuming that the surface Ni atoms are correctly counted by hydrogen chemisorption[15]), but the hydrogenolytic reactions are influenced much more severely by alloying. Likewise, the rates of benzene[15] and cyclopropane[16] hydrogenations are only little affected when Ni is compared to Ni-Cu alloys, but the hydrogenolytic reaction of cyclopropane yielding methane and ethane is altered substantially[16]. On the basis of these results it appears to be a general phenomenon that presence of Cu influences the reactions involving C-H bonds to a much lesser extent than the reactions where C-C bonds are broken or formed (metal cracking, isomerisation, possibly dehydrocyclisation).

Let us first focus our attention to alloys with 0-23% Cu (see (a) above). Here the activation energy of the total conversion of n-HEX conversion is only marginally influenced and the observed effects are consequently connected with the preexponential factors. This justifies an interpretation in terms of number of sites.

When considering the concentration of various sites e.g. isolated Ni atoms, pairs, triplets and other "ensembles" of Ni atoms in the alloy surfaces, it should be further kept in mind that for homogeneous equilibrated alloys the surface composition differs in general from the bulk composition in the sense that the element lowering the surface energy accumulates in the surface (Gibbs adsorption). Thus, it has been found in this laboratory[17] in agreement with that the surface of Pd-Ag alloy films is enriched in Ag. As one must expect that in general the alloy with a lower heat of sublimation will be agglomerated in the surface, the implication is that for the equilibrated Cu-Ni alloys at temperatures above the miscibility gap the surface should be enriched with Cu. This, indeed, was recently confirmed[18] by means of Auger spectroscopy. We may, therefore, expect that also in our Ni-rich alloys the surface Ni atoms are diluted.

With this in mind, it is tempting to introduce a simple model for explaining the selectivity effects. One may assume that for certain hexane reactions bonding to one or to a small number of Ni atoms is sufficient, while for other reactions ensembles of more adjacent Ni atoms are required. More specific the ensembles required for those catalytic reactions which affect C-H bonds, or can lead to isomerisation contain a smaller number of adjacent Ni atoms than those ensembles which are required for hydrogenolytic splitting. Since alloying with an inert element reduces more strongly the concentration of those ensembles which contain more Ni atoms, an increased selectivity of non-hydrogenolytic reactions is the necessary consequence. Recent data on homogeneous catalysts[19] indeed show that the C-H bond reactions are possibly performed on an isolated atom whereas it is a general consensus that hydrogenolytic splitting implies several adjacent adsorption sites.

The model suggested is very similar to that used by Anderson et al.[4]. These authors observed an increased selectivity for C_6 products on highly dispersed Pt and Ni. They explain their results by postulating that the very small crystals contain a relatively large fraction of corner atoms in their surface. Upon further assuming that these "isolated" corner atoms favour the formation of carbocyclic intermediates of isomerisation, whereas hydrogenolysis is postulated to require two or three adjacent platinum atoms in a crystal plane, the authors show that their results can be rationalized by this simple geometrical principle. It is obvious that alloys of transition metal with non-transition metals are ideal objects to check Anderson's hypothesis, provided we assume that transition metal atoms diluted by non-chemisorbing atoms fulfill the same role as the corner atoms of Anderson. The selectivity of Ni diluted by Cu is indeed near to the value found by Anderson et al.[4] for highly dispersed Ni films.

The alloys with 40-70% Cu show another effect. While the overall activity changes marginally, the selectivity and fission parameters change dramatically. The increase in activation energy for alloys with 47-85% Cu shows that in addition to the "dilution effect" also energetic effects are involved. The bond strength of adsorbed fragments with e.g. a Ni atom depends on its neighbours in and below the surface. This in turn affects the activation energy and possibly also the performance of alloys in other respects as well. In the region of 40-70% Cu the selectivity S, the parameter M, the ratio 2MP/3MP or the isopentane/pentane ratio (in experiments with 3MP) show that the catalytic behaviour of these alloys is somewhere between Ni and Pt, actually rather nearer to platinum.

Several theories have been proposed to rationalize the marked difference in selectivity of transition metals with respect to hydrocarbons interconversion. Recent papers[2-10, 20, 21] show that in this respect the transition metals can be classified into two groups: one involving Pt and Pd and the other encompassing all the rest of transition metals with Ni as a typical representative. One attempt to explain the difference is based on the Balandin's "multiplet theory". Applying it to the hydrogenolytic and isomerisation reactions some authors[9, 10] postulate that Pt favours a sextet-doublet mechanism, Ni a simple doublet mechanism. The former mechanism should lead as a rule to isomerisation reactions in general and to the hydrogenolytic reactions of strained adsorbed rings, whereas the doublet mechanism leads to hydrogenolysis of alkanes. This theory, evidently, is unable to rationalize the present results. The crystallographic structure (f.c.c.) remains the same for all alloys studied and the lattice constant changes only marginally, so there is no obvious reason for the transition from one mechanism to the other. Moreover, low Cu content alloys reveal both isomerisation and hydrogenolysis and it would be difficult to understand why dilution of an active by an essentially non-active metal would lead to the transition from doublet to the sextet mechanism.

An alternative postulate proposed by Matsumoto et al.[3] suggets a carbonium ion mechanism for Pt catalysts, the ions being favoured by the high work function of Pt and its affinity towards hydrogen. For nickel, there the authors postulate a mechanism involving a homolytic bond rupture. However, this idea is not confirmed either. The work function of Cu-rich alloys is lower than the work function of pure Ni[22] (and of Pt as well). The heat of hydrogen chemisorption reveals the same trend among alloys[23]. Although different polarities of chemisorption bonds on different metals and alloys may play a role, in general, the rules suggested for a carbonium ion mechanism fail to rationalize the experimental data on alloys.

In connection with the two forms of chain rupture observed, terminal fission for Ni and random fission for Pt, it is interesting to note that quantum mechanical calculations have shown that terminal bonds are thermodynamically more stable than internal bonds[24]. However, the reactivity indexes, such as polarity of bonds, are highest for the terminal bonds[25]. Thus the splitting on Ni and Cu lean alloys seem to be rather "kinetically" controlled while on platinum (and 40-70% Cu alloys) which adsorbs hydrocarbons less firmly, the adsorbed molecules can react in a manner less determined by the metal, the reaction is then more "thermodynamically" controlled.

Besides this essentially energetic effect another factor may also play a role. Namely, when nearing 60% Cu composition, ferromagnetism disappears and this may lead to a greater availability of the bonding orbitals of surface Ni atoms so that more complicated reactions can proceed even on isolated Ni atoms.

It is interesting to note that Reman, Ali and Schuit[14], when investigating the conversion of n-hexane on alloys formed on the surface of zeolites, found essentially the same change in product distribution with alloying as in the present work. From the results presented in Fig. 6 of their paper it appears that the alloy with low Cu content supported by a zeolite behaves as an unsupported alloy with a higher Cu content. It appears that in the work of Reman et al.[14] the acidity of the support was not the reason for the high isomerisation activity as essentially the same results are found for alloys without support.

Acknowledgements. The investigations were supported by the Netherlands Foundation for Chemical Research (S.O.N.) with financial aid from the Netherlands Organisation for the Advancement of Pure Research (Z.W.O.). For the technical assistance in preparation and analysis of catalysts our thanks are due to Messrs A. Roberti and J.A. Smit and Miss J. van Beelen.

REFERENCES

1) V. Ponec and W.M.H. Sachtler, J. Catalysis, in press.
2) H. Matsumoto, Y. Saito and Y. Yoneda, J. Catalysis 19 (1970) 101.
3) H. Matsumoto, Y. Saito and Y. Yoneda, J. Catalysis 22 (1971) 182.
4) J.R. Anderson, R.J. McDonald and Y. Shimoyama, J. Catalysis 20 (1971)147.
5) J.L. Carter, J.A. Cusumaro and J.H. Sinfelt, J. Catalysis 20 (1971) 223.
6) E. Kikuchi, M. Tsurumi and Y. Morita, J. Catalysis 22 (1971) 226.
7) G. Maire, G. Plouidy, J.C. Proudhome and F.G. Gault, J. Catalysis 4 (1965) 556.
8) G. Maire, C. Collolleur, D. Juttard and F.G. Gault, J. Catalysis 21 (1971) 250.
9) A.L. Liberman, D.V. Bragin and B.A. Kazanskii, Dokl. Akad. Nauk USSR 156 (1964) 1114.
10) O.V. Bragin and A.L. Liberman, IV Internat. Congress Catalysis, Moscow 1968, preprint No. 27.
11) L. Elford, F. Müller and O. Kubaschewski, Ber. Bunsenges. Phys. Chem. 73 (1969) 601.
12) J.R. Best and W.W. Russell, J. Amer. Chem. Soc. 76 (1954) 838.
13) D.A. Cadenhead and N.J. Wagner, J. Phys. Chem. 72 (1968) 2775.
14) W.G. Reman, A.H. Ali and G.C.A. Schuit, J. Catalysis 20 (1971) 374.
15) P. van der Plank and W.M.H. Sachtler, J. Catalysis 12 (1968) 35.
16) J. van Beelen, Leiden University, Private Communication (will be published soon).
17) R. Bouwman, Thesis, Leiden University, 1970.
18) K. Nakayama, M. Ono and H. Shimizu, Internat. Conf. Solid Surfaces, Boston, Mass., USA, Oct. 11-15, 1971.
19) R. Ugo, La Chim. e L'Industria, 51 (1969) 1319.
R. Ugo, Engelhard Ind. Techn. Bull. 11 (1970) 45.
20) R. Merta and V. Ponec, IV Internat. Congress Catalysis, Moscow 1968, preprint No. 50.
21) R, Maurel and G. Leclerc, Bull. Soc. Chim. France (1971), 1234.
22) W.M.H. Sachtler and G.J.H. Dorgelo, J. Catalysis 4 (1965) 654.
23) H. Matsuyama, M. Yoshinaka and M. Tanigaku, Dashisha Daigaku Kenkyu Hokoku 10 (1969) 287; Chem. Abstr. Japan 73 (1970) 18756 t.
24) K. Hirota and K. Fueki, Nippon Nagaku Zashi 81 (1960) 209; quoted in Ref. 3).
25. A. Julg, J. Phys. Chim. (1956) 548.

DISCUSSION

J. H. SINFELT

In agreement with the work of Ponec and Sachtler, we have observed that copper suppresses markedly the hydrogenolysis activity of nickel (Sinfelt, Carter, and Yates, J. Catalysis, 24, 283 (1972)). In the same investigation, we found that copper had very little effect on the cyclohexane dehydrogenation activity of nickel for alloys containing as much as 95 atom % copper. This demonstrates a high degree of specificity in the behavior of copper-nickel alloys with regard to the type of reaction.

In general, we have observed that Group IB metals inhibit markedly the hydrogenolysis activities of Group VIII metals. This is true even for metal combinations exhibiting very low miscibility in the bulk, e.g., ruthenium-copper, ruthenium-silver, and osmium-copper. In such bimetallic combinations, we believe that the Group IB metal concentrates in the surface. Of particular interest, we have observed the marked inhibition of hydrogenolysis activity with Group VIII-Group IB metal pairs in a state of very high dispersion on a carrier, where the total metal concentration is of the order of only 1-2%. We have referred to these systems as supported "bimetallic clusters" rather than alloys, since they are not limited to metal combinations corresponding to known bulk alloys.

V. PONEC

Thank you very much for this remark. We have also observed the high degree of specificity you mentioned, caused by alloying with Cu. For example, in the cyclopentane-deuterium exchange the activity and multiplicity is only marginally changed by alloying whereas hydrogenolysis, at higher temperatures, is strongly suppressed (Ponec, Sachtler, J. Catalysis 24, 250 (1971)). Another example is the conversion of cyclopropane which allows to follow the two mentioned functions of the catalyst, i.e. the activity in the C-H bond reactions and the fission of C-C bonds, with one molecule under identical reaction conditions. Also here the same picture emerges: Cu enhances the activity in hydrogenation only slightly, but it depresses the hydrogenolysis strongly (Beelen, Ponec, Sachtler).[1] This phenomenon seems thus to be quite a general one.

We appreciate very much the close agreement of the results by Sinfelt and Yates and our own, as it was only four years ago in the presidential closing speech of the IVth International Congress on Catalysis in Moscow that the work on alloys was quoted as a notorious example for great discrepancies in the results found by different investigations. The agreement achieved now is certainly encouraging.

1) J. Catalysis, in print.

V. HAENSEL

As the Cu content is increased, the relative amount of benzene is reduced and the relative amount of 2-MP is increased. Is this due to the concurrent change in absolute amounts of products (benzene and 2-MP) formed?

Is there a chance that the larger amount of benzene formed is covering the surface and thus preventing the isomerization reaction?

V. PONEC

Your question refers to data which I mentioned in my oral presentation and which I have compiled in the first 3 columns of the following Table for the benefit of the reader.

%Cu	T°C	Benzene, %*	R-benzene**
0	335	33.1	16.5
23	333	7.7	0.92
63	334	1.2	1.04

*Benzene among C_6 products

**Absolute rate of benzene formation in arbitrary units

As can be seen from the Table the formation of benzene decreases not only relatively (to other C_6 hydrocarbons) but also in absolute values. This does not prove, however, that an intermediate of benzene structure blocks the sites for isomerization. We rather believe that the further fate of an absorbed n-C_6 structure will depend on the strength of the chemisorption bond with the surface. On pure nickel this bond is very strong and the molecule will leave the surface predominantly as a dehydrocyclized or cracked product. On Cu-Ni alloys, however, chemisorption is weaker and accordingly the chance increases that the C_6 adsorbate will leave the surface with the same number of carbon and hydrogen atoms as before the adsorption as an isomerized product.

F. G. GAULT

To support the interpretation given by Dr. Ponec and Sachtler for the "dilution effect" on selectivity in copper-nickel alloys, I wish to mention the fact that similar changes of selectivity are obtained for the same reactions on nickel-alumina catalysts prepared by impregnation with various nickel contents.

We noticed a long time ago a large change of selectivity in the isomerization of hexanes[1] and hydrogenolysis of methylcyclopentane[2] on supported platinum-alumina catalysts of various platinum contents (0.2 to 10%), and we attributed this dispersion effect to a change of the metal particle size.[3] Later on, Dr. Corolleur observed the same affect with other metal catalysts supported on alumina.

The results for nickel are as follows: using the same terminology as yours, when the content of nickel in nickel-alumina catalysts was increased from 1% to 10%, the selectivity S for the isomerization of 2-methylpentane decreased from 2.6 to 1.2%, and, for the isomerization of 3-methylpentane from 6.3 to 2.3%. Simultaneously the fission parameter M does not change much. We observe then the same trends as the ones you obtain using nickel-copper alloys with 0 to 20% copper.

It is interesting to note that similar results showing a close connection between the change of product distributions in isomerization or hydrogenolysis and the dispersion of the metal may be obtained by three independent methods: changing the size of metal particles in supported catalysts, changing the size of metal particles in platinum films, as Dr. Anderson does,[4] and changing the composition of binary alloys.

The three methods indicate a very big effect of the dispersion of the metal and of the metal particle size on the reaction mechanisms.

1) Barron, et al., J. of Catal., 5, 928 (1966).
2) Gault, Comptes Rendus Acad. Sc. 245, 1620 (1957).
3) Gault, Annales Chimie, 645 (1960).
4) Anderson and Shimoyama, Paper No. 47.

V. PONEC

I agree with your remark to which I like to add only a short comment. As can be seen from our paper we believe that there are "geometrical" effects present in alloying. However, one should also remember that e.g. the fundamental difference between Pt and Ni (similar to the difference between Ni and Cu rich Ni alloys) cannot be satisfactorily explained by a different density or geometry of sites. Similarly, alloying of Ni with Cu will cause several parallel changes the combination of which results in the observed changes of catalytic activity and selectivity. At approximately constant surface concentration of Ni-sites (see hydrogen adsorption results[1,2,3]), the size of Ni clusters decreases[4] and so does the strength of the metal adsorbate bond.[5] As the Table in our answer to Dr. Haensel shows, Cu alloying leads also to a decrease in the dehydrogenation activity of the catalysts, most likely 1) because there is less room on smaller clusters to accommodate hydrogen split off from the molecules and 2) because at lower heats of adsorption the degree of dissociation will be lower also for energetic reasons. Less dehydrogenation and lower bond-strength lead to particles which are more reactive and which have greater chance to react in a more random way (see increase in M, increase in 2MP/3MP ratio, etc.) and with less cracking (see increase in S). Beside these effects, it is also possible that the skeletal rearrangements are better catalyzed by metal atoms which are less involved in metal-metal interactions, as in the case e.g., with non-ferromagnetic, 63% Cu-alloys.

1) P. van der Plank, W. M. H. Sachtler, J. Catalysis 12 35 (1968).
2) V. Ponec, W. M. H. Sachtler, J. Catalysis 24, 250 (1972).
3) J. H. Sinfelt, J. L. Carter, D. J. C. Yates, J. Catalysis 24, 283 (1972).
4) E. Vogt, Phys. Stat. Sol. (b) 50, 653 (1972).
5) L. S. Shield, W. W. Russell, J. Am. Chem. Soc. 64, 1592 (1960).

J. R. ANDERSON

The results in this paper are extremely interesting, and the work well illustrates the strength of the technique of the use of alloy catalysts for the study of alternative reaction paths, particularly when these can be understood on a geometric basis.

In connection with the authors' interpretation (with which I have no quarrel), one needs to remember that, although it happens to be so for nickel, it is not generally true that a quasi-isolated metal atom necessarily disfavors hydrocarbon hydrogenolysis. As pointed out in another paper at this meeting,[1] with a metal such as platinum there is a well-identifiable single atom hydrogenolysis mechanism.

It would be very interesting and mechanistically quite important to examine (say) the reactions of n-hexane over a suitable range of platinum alloy catalysts.

Secondly, the selectivity of a metal catalyst for hydrogenolysis can be expected to depend on the surface concentration of adsorbed hydrogen (as well as other factors), and in the case of alloys this will certainly be dependent on the alloy composition.

1) Anderson, J. R. and Shimoyama, Y., Paper 47, this meeting.

V. PONEC

The mechanism of hydrogenolysis suggested in your very interesting paper (No. 47, this meeting) and based on the reactions of single Pt atoms seems to be a very attractive alternative to other possibilities. However, I am afraid it is extremely difficult to prepare catalysts with only isolated atoms and when there are pairs and triplets etc., it is, I think, impossible to estimate the extent of the real single atom hydrogenolysis.

As far as your suggestion is concerned, R. P. Dessing and C. J. Boogerd from our laboratory are already studying the behavior of Pt-Au alloys in butane, pentane and n-hexane reactions. Very briefly said, the preliminary results are: at low temperature there are, indeed, only rather small shifts in selectivity for destructive and nondestructive reactions. However, at higher temperatures (above 300°C) when the surface of Pt is either poisoned by C-residues, or to an even higher extent when Pt is alloyed with Au, some unexpected reactions were found to occur, e.g. disproportionation of butane to propane and pentane.[1)] In analogy to results on oxidic catalysts, we would ascribe this disproportionation reaction to single Pt atoms.

We did our experiments at a constant hydrocarbon/hydrogen ratio; it would be of course desirable to see if the variation in this ratio also brings about different changes for Ni and, say, Cu rich alloys.

1) Dessing, R. P., Ponec, V. and Sachtler, W. M. H., J. Chem. Soc. (Chem. Commun.)(1972) 880.

J. K. A. CLARKE

Ponec and Sachtler have reemphasized here a factor which may have the effect of modifying an otherwise understandable activity/composition relationship in an alloy series, namely that reactions such as hydrocracking (e.g., T. J. Plunkett and J. K. A. Clarke, J. Chem. Soc., Faraday I, 68, 600 (1972)) require more than one site. By extension, the paper prompts us to bear in mind that texture changes through a given set of polycrystalline alloys--specifically, variations in the relative predominance of exposed crystal faces--may affect measured patterns for demanding reactions. We have demonstrated such an example in the case of formic acid dehydrogenation on Pd-Au alloy films (J. K. A. Clarke and E. A. Rafter, Zeits für Physikal Chem., N.F., 67, 169 (1969)).

In view of this, it is gratifying that the type of relationship found by Ponec between cracking activity and % Ni agrees with that found in our laboratory and by Sinfelt's group.

V. PONEC

I completely agree with you. In our work the distribution of the crystallographic planes in the surface was not controlled. Although it would, undoubtedly, be ideal to do all measurements with clean and well-defined monocrystal, this is experimentally extremely difficult. For Ni-Cu alloys we have reason to assume that the participation of crystal planes in the surface remains fairly constant as both metals have the same crystallographic structure and similar lattice constants. Qualitatively, very similar results are also obtained when working with alloy films and alloy powders, although the way of preparation is very different in both cases.

L. GUCZI

I would like to make a comment on the importance of hydrogen on the structure of ensembles. If the hydrogenolysis of n-hexane were carried out in the presence of relatively low amount of hydrogen on Ni catalysts, the rate of the formation of cyclic compounds (benzene and toluene) was relatively high to that of the product of hydrogenolysis (see Figure 1). However, if the hydrogen pressure was increased, Figure 2), the distribution of products completely changes and, which is more important, the ratio of these products decreases. It shows that a higher hydrogen concentration on the surface diminishes the concentration of the sites which are available for cyclization. In addition, another experiment on platinum carried by Dr. Paal showed that the dehydrocyclization of n-hexane yields only benzene in the presence of helium; however, a large amount of methylcyclopentane is produced in the presence of hydrogen.

Both of these experiments show that the structure of ensembles responsible for different routes can be changed similarly to alloying nickel by copper.

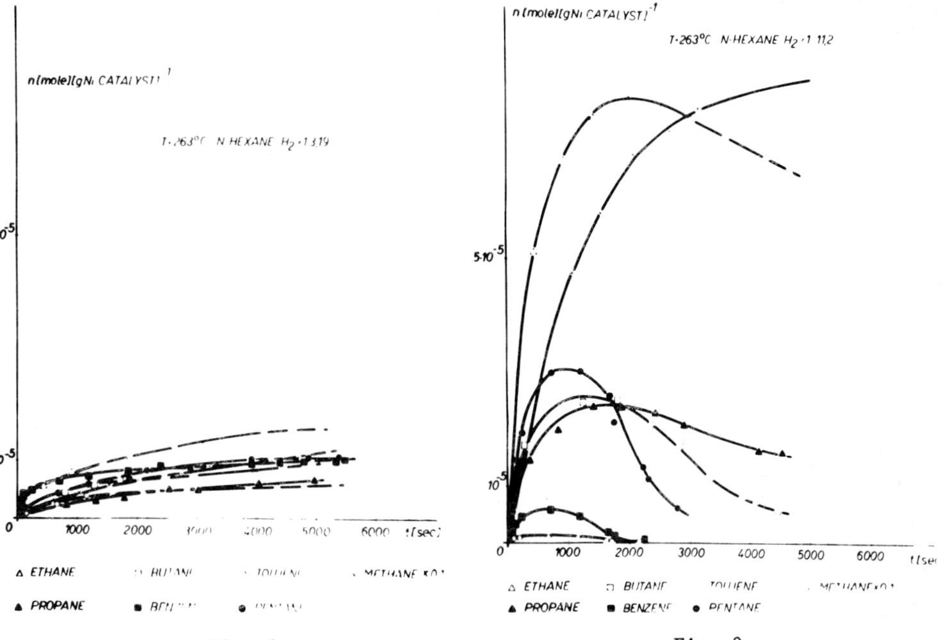

Fig. 1

Fig. 2

V. PONEC

You touch a problem similar to that mentioned in the last paragraph of Dr. Anderson's question. Your results clearly show that more work with different hydrocarbon/hydrogen ratios is certainly desirable.

D. V. MUSHENKO

It is stated that the catalysts reach the values characteristic of co-union for platinum. But platinum isomerization leads to formation of 2,2- and 2,3-dimetylbutane. In Table 2 these isomers are not mentioned. Are they formed? In other cases, CuNi catalysts are inferior for this isomerization reaction.

V. PONEC

For 2,3-dimethyl-butane (see the "Experimental" part) Ni-Cu alloys are by no means equivalent to Pt catalysts, but we have never claimed that.

The comparison between Pt and Cu rich alloys is used in our paper only to illustrate the general trend of changes which Ni undergoes by alloying. I stressed this point also in my oral presentation.

Paper Number 44

STRUCTURE-SENSITIVE HYDROGENATION: TWO FORMS OF NICKEL CATALYST WHICH SHOW DISTINCTIVE SELECTIVITIES FOR THE HYDROGENATION OF 1,2- AND 1,3-BUTADIENE[1]

R. G. OLIVER and P. B. WELLS
Department of Chemistry, The University, Hull, HU6 7RX, England
and (in part) J. GRANT
Department of Chemistry, College of Technology,
Hull, HU1 3DG, England

ABSTRACT: Further investigations have been made of two forms of nickel which differ in their selectivity for 1,3-butadiene hydrogenation. Type A nickel provides 1-butene as the major product, whereas trans-2-butene is the major product over Type B nickel.

The dependence of catalyst type upon (a) conditions of catalyst preparation, especially temperature and duration of reduction, (b) metal particle size, and (c) the nature of the support, has been investigated. Nickel formed by the reduction of nickel(II) oxide is of Type A, whereas that formed by the reduction of mixed nickel(II,III) oxide is of Type B.

Products of 1,2-butadiene hydrogenation over Types A and B nickel are reported and are compared with predictions based on the known mechanisms of 1,3-butadiene hydrogenation. The formation of pi-allylic intermediates at the Type B nickel surface is confirmed.

It is argued that nickel-catalyzed butadiene hydrogenation should be classed as a structure-sensitive reaction.

1. INTRODUCTION

It has become evident in recent years that the selectivity obtained in a metal-catalyzed reaction may depend critically upon the physical form of the metal particles that make up the catalyst; particularly is this so when the metal crystallite size is in the so-called mitoedrical region[2] i.e. 10 to 50Å. Evidence that the specific activity for a given reaction (and hence the selectivity in a system of competitive reactions) may depend upon crystallite size has been reviewed by Boudart[3] and by one of the present authors[4]. Where such an effect is observed the reaction is described as being "demanding" or "structure-sensitive"; where no such effect is found the reaction is classed as "facile" or structure-insensitive"[3]. Structure-sensitive reactions reported so far include the exchange of hydrogen for deuterium in benzene catalyzed by nickel-silica[5], the isomerisation of neopentane to isopentane catalyzed by platinum[6], and perhaps the hydrogenolysis of toluene to benzene and methane catalyzed by nickel-alumina[7]. No instance of a structure-sensitive hydrogenation has yet been reported. This prompted the present authors to extend their examination of the curious behavior of nickel as a catalyst for diolefin hydrogenation.

The composition of butene formed by 1,3-butadiene hydrogenation at $100^{o}C$ over cobalt or nickel powder depends upon the temperature at which the metal is formed from the oxide by reduction in hydrogen. Wilson first demonstrated[8)] that nickel powders formed by reduction at $310^{o}C$ or below catalyzed 1,3-butadiene hydrogenation at $100^{o}C$ to give a product distribution in the region: 1-butene, 60%; trans-2-butene, 25%; cis-2-butene 15% (Type A behavior), whereas reduction at $400^{o}C$ or above gave catalysts for which the corresponding product distribution was approximately: 1-butene, 30%; trans-2-butene, 60%; cis-2-butene, 10% (Type B behavior). Alumina-supported nickel and cobalt were also observed to give Types A and B behavior. Deuterium tracer studies[8)] established that, for reaction over each type of catalyst, 1-butene and 2-butene were each formed entirely by direct 1,2- and 1,4-addition of hydrogen to the diolefin. No isomerisation of butene occurred before its desorption, and butane yields were small (usually less than 0.2%). Furthermore, the product composition was independent of reactant pressures and conversion, and only weakly dependent upon temperature. Thus, simply to describe a catalyst as Type A or Type B is a significant description of the character of the catalyst.

The deuterium tracer studies also revealed[8)] that a Type B distribution of products such as that shown above was probably a combination of Type A behavior (contributing about one third of the product) and of a product composition in the region 1-butene, 7%; trans-2-butene, 86%; cis-2-butene, 7% (contributing two thirds of the product). Consequently, it was hoped that a more fundamental understanding of the nature of the Type B surface would enable catalysts having very high selectivities for trans-2-butene formation to be prepared.

Mechanisms for Type A and Type B behavior are proposed in reference 8. The basic difference appears to be that chemisorbed 1,3-butadiene and certain of the half-hydrogenated states are able to attain pi-allylic configurations at the Type B surface, whereas this is not the case at the Type A surface.

No firm interpretation was offered concerning the differences in nature between the Type A and Type B nickel or cobalt surface. Type A cobalt powders were of close-packed hexagonal structure whereas the Type B cobalt powders were face-centered cubic; however, the bulk structure appeared to be irrelevant, because all of the nickel powders, whether of Type A or of Type B, were face-centered cubic[9)]. Electron microscopy revealed that Type A powders contained very large numbers of particles in the range 20 to 100 Å, whereas particles in this range were almost absent in the Type B powders[9)]. There was thus the possibility that the product composition was dependent upon metal particle size. It was also shown that the behavioral type could be changed by suitable oxidation and further reduction[8)], thus suggesting that oxygen contamination might influence the reaction mechanism.

The purpose of the present work was to confirm and extend the observations in this system (i) by a more detailed examination of Types A and B nickel using 1,3-butadiene hydrogenation as the test reaction, and (ii) by examining 1,2-butadiene hydrogenation with a view to determining whether or not the products of this reaction were predictable on the basis of the published mechanisms for 1,3-butadiene hydrogenation.

2. EXPERIMENTAL

2.1 Catalyst designations

Twenty seven nickel catalysts have been examined. Each is described by a number of the type Ni-X-Y. X = A denotes alumina-supported nickel, X = S denotes silica-supported nickel, and X = P describes nickel powder. Y is the sample number.

2.2 Nickel-alumina

Nickel-alumina was prepared as previously described[8]. Peter Spence Type A alumina of nominal formula $Al_2O_3.H_2O$ was used as support. Impregnation with nickel nitrate in neutral aqueous solution was followed by calcination at 650°C for 24h. The particles of alumina-supported nickel oxide so formed were blue-green or light grey. This material was reduced in a hydrogen stream at temperatures and for periods of time described in Section 3. The total surface area of the catalyst (B.E.T. method) was 193 $\pm$ 2 m^2 g^{-1}. Completely reduced catalysts contained 10% by weight of nickel.

2.3 Nickel-silica

Nickel-silica catalysts containing 10%, 20%, and 50% by weight of metal were prepared by impregnation of Aerosil with nickel nitrate in neutral aqueous solution, followed by calcination at 650°C for 7h (10% and 20% nickel-silica) or 20h (50% nickel-silica). The calcined material was medium grey in color. This material was reduced in a hydrogen stream which was maintained for 6h. Reduction temperatures are given in Section 3.

Particle size distributions of the nickel crystallites were measured by electron microscopy. The 50% nickel-silicas reduced at 250°C and at 475°C exhibited similar distributions, the former having a maximum in the region 30 to 70Å, and the latter in the region 60 to 130Å. The distributions were broad, and each catalyst contained one or two percent of particles in the range 450 to 600Å. The 20% nickel-silica contained particles in the narrow range 15Å (the lower limit of detection) to 40Å; a large proportion of particles were probably not detected. No metal particles were detected in micrographs of 10% nickel-silica, from which it was concluded that all particles were below about 15Å in size.

2.4 Nickel powders

Nickel powders were prepared by the reduction of an appropriate sample of nickel oxide in a hydrogen steam at an appropriate temperature for 24h. Further details are given in Section 3.

2.5 Apparatus, methods and materials

The static high-vacuum system, the chromatograph, and the general methods employed have been described[8]. 1,3-butadiene was purified by distillation; as used it contained 0.2% 1-butene. 1,2-butadiene as received contained many impurities all of which were removed by g.l.c. before use. Hydrogen used for the reduction of catalysts was used without purification. Hydrogen used as a reactant was purified by diffusion through a heated palladium-silver thimble before use.

2.6 The test reaction

A standard test was used to determine whether nickel catalysts were of Type A or of Type B. The test involved examination of the product composition of 1,3-butadiene hydrogenation carried out at or near 100°C. Reactant pressures were typically: 1,3-butadiene, 50 torr; hydrogen,

100 torr; conversion was normally 10%. However, the product composition in this reaction is independent of reactant pressures over a wide range, is independent of conversion until the near removal of reactant, and is only weakly temperature dependent[8]. Product compositions typical of Types A and B behavior are given in the Introduction. The terms "Type A" and "Type B" are used as adjectives to describe both catalysts and product compositions.

3. RESULTS AND INTERPRETATION

When the present work commenced, the dependence of behavioral type upon catalyst reduction temperature had been established for a series of eight nickel powders. Only two alumina-supported nickel catalysts had been prepared, one of Type A and one of Type B[8]. In consequence, the nickel-alumina system was the first to be studied more extensively. Five nickel-aluminas were prepared, each being reduced at a different temperature for six hours, and were submitted to the standard test. Figure 1 (circular points, firm curves) shows that Type A behavior was obtained when reduction temperatures were below 310°C, and Type B behavior when reduction temperatures were 448°C or above. These results were almost identical to those reported previsouly for reactions catalyzed by nickel powders; thus previous work appeared to have been confirmed.

Next, a series of eight nickel-silica catalysts of known metal particle size characteristics (see Section 2.3) was investigated, in order to determine whether the transition from Type A to Type B behavior was associated with changes in particle size distribution of the metal crystallites. Figure 2 shows that the behavioral type was independent of reduction temperature, of nickel crystallite size, and of the concentration of nickel on the support. That all catalysts were of Type A was an unexpected result.

These observations prompted a reconsideration of the processes occurring during the preparation of supported catalysts. As mentioned in Section 2, alumina-supported nickel oxide was blue-green or light grey, whereas silica-supported nickel oxide was medium to dark grey. Reduced catalysts were darker grey or black, except that some granules of nickel-alumina retained their blue-green appearance due to incomplete reduction. The formation of blue-green nickel oxide at the surface of alumina has been reported by Hill and Selwood[10] who showed that a proportion of the nickel ions had undergone valency inductivity and had achieved oxidation states of (III) and perhaps (IV). (The methods and materials used by Hill and Selwood in catalyst preparation were closely similar to those used by the present authors). Such valence inductivity does not occur at the silica surface. Thus, the differences in behavior of high-temperature -reduced nickel-silica (Type A) and of nickel-alumina (Type B) appears to be related to the oxidation state of nickel before reduction. More explicitly, it appears that the reduction of Ni(II) oxide in the range 430 to 730°C gave Type A catalysts, and that reduction of mixed Ni(II,III) oxide within the same temperature range gave Type B catalysts. This postulates was amenable to further test.

First, the manufacturers (British Drug Houses) of the nickel oxide used for the previous preparations of eight nickel powders[8] confirmed that the material was a mixed Ni(II,III) oxide, and this concurred with our own (belated) analysis. Thus, the observed formation of Type B nickel powders by reduction of this material above 400°C is in accord with the above postulate.

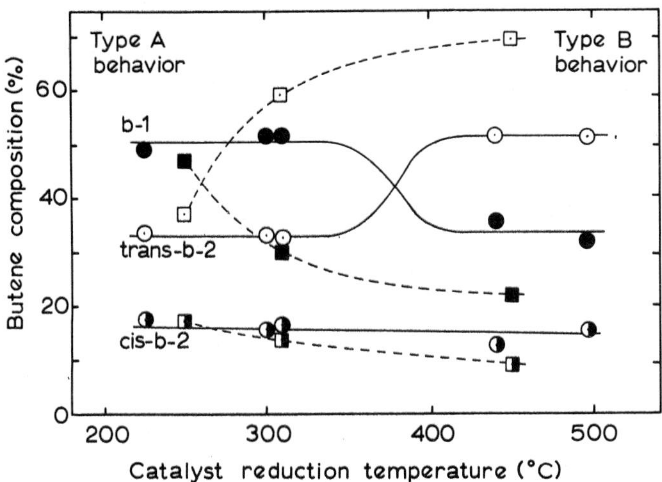

Fig. 1. Nickel-alumina catalysts. Dependence of behavioral type upon the reduction temperature employed during catalyst preparation. Test reaction: 1,3-butadiene hydrogenation at 100°C (for further details see Section 2). Circular points represent Ni-A-1 to 5 for which the reduction period was 6h. Square points: Ni-A-6 reduced at 250°C for 44h; Ni-A-7 reduced at 313°C for 25h; and Ni-A-8 reduced at 450°C for 18h.

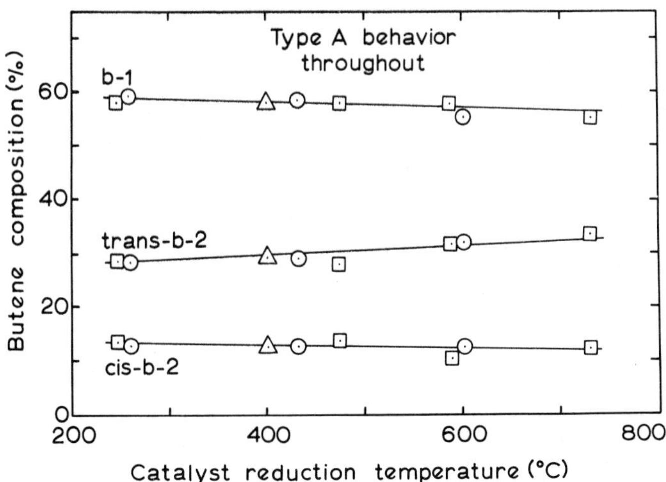

Fig. 2. Nickel-silica catalysts. The independence of behavioral type upon the reduction temperature employed during catalyst preparation. Test reaction as for Figure 1. Weightings of nickel on the support: 10%, circles; 20%, triangles; 50%, squares. All catalysts were reduced for 6h before use.

Secondly, six nickel powders were prepared from various samples of nickel oxide. Ni-P-1 and Ni-P-2 were prepared by reduction at 250°C and 450°C respectively of the B.D.H. mixed nickel oxide as received; this constituted a repetition of previous work. As expected, these catalysts were respectively Type A and Type B in character when subjected to the standard test (see Table 1). Ni-P-3 was prepared by heating the B.D.H.

TABLE 1
Catalytic behavior of various nickel powders

Catalyst	Reduction temp. (°C)	Butene Composition (%)			Behavior
		1-b	trans-2-b	cis-2-b	
Ni-P-1	250	53	31	16	Type A
Ni-P-2	450	35	50	15	Type B
Ni-P-3	450	43	40	17	Intermediate
Ni-P-4	450	56	31	13	Type A
Ni-P-5	450	60	29	11	Type A
Ni-P-6	450	31	56	13	Type B

mixed nickel oxide in air at 650°C for sixteen hours followed by reduction at 450°C for 16h. It was intermediate in behavior. Thus the heating in air, which reduced the Ni(III)-content of the mixture, caused a movement of behavior from Type B towards Type A. Ni-P-4 and Ni-P-5 were prepared from samples of nickel(II) oxide made by the calcination of nickel nitrate hexahydrate at 650°C and 1000°C respectively. Their reductions at 450°C for six or sixteen hours gave Type A nickel catalysts, again in accord with the above postulate.

Thirdly, a black hydrated nickel oxide of nominal formula $Ni_2O_3.2H_2O$ was prepared by the method of Cairns and Ott[11]. The colloidal material was ultracentrifuged, dried, and reduced to metal at 450° (reduction time, 1.5h). Table 1 shows that the resulting catalyst, Ni-P-6, was of Type B again in accord with the postulate.

These three tests demonstrate the basic soundness of the postulate that Type A nickel is formed by the reduction of nickel(II) oxide, and that Type B nickel is formed by the reduction of nickel(II,III) oxide.

Attempts to reduce the black hydrated nickel(III) oxide at 250°C in hydrogen did not give an active catalyst. This may interpret why all catalysts formed by reduction at 300°C or below are of Type A. This interpretation, too, was amenable to further test. The nickel-aluminas were greenish, which indicated that they were only partially reduced. Consequently, extended treatment of such catalysts in hydrogen at elevated temperatures should increase the proportion of nickel(III) oxide that is reduced to metal. This in turn should cause a movement towards Type B behavior. Three catalysts were prepared, Ni-A-6 by reduction at 250°C for 44h, Ni-A-7 at 313°C for 25h, and Ni-A-8 at 450°C for 18h. (Ni-A-1 to 5 had been reduced for 6h.) Their behavior in the test reaction is recorded in Figure 1 (square points, dashed curves). Reduction for an extended period at 250°C achieved only a marginal movement towards Type B behavior, which is consistent with very slow reduction of nickel(III) oxide at this temperature. Extended reduction at 313°C provided a Type B catalyst, whereas Type A behavior had been observed after only 6h reduction at this temperature. This dramatic change is in the sense predicted above. Long reduction at 450°C gave the best Type B catalyst yet obtained; indeed the product composition approaches the best attainable theoretically, which is quoted in the Introduction. Thus, slow reduction of the Ni(III)-component in mixed Ni(II,III) oxide at the alumina surface is confirmed.

The experiments described thus far have established a pattern of behavior which has been interpreted in terms of a simple model. However, as yet, the model is not informative concerning the basic difference in surface structure of Types A and B nickel. Information concerning surface structure can be derived indirectly from a consideration of the mechanisms of reactions at these surfaces. Studies of 1,3-butadiene hydrogenation have shown that pi-allylic intermediates are formed at the Type B surface but not at that of Type A. Before utilising this information a further test of its validity was sought; the test applied consisted of a comparative examination of 1,2-butadiene hydrogenation catalysed by Types A and B nickel.

1,2-butadiene hydrogenation has been little studied. Reaction over palladium in a flow system at about 35°C has been reported by Meyer and Burwell[12)] who observed high yields of 1-butene and of cis-2-butene, and a low yield of the trans-isomer (see Table 2). We repeated this experiment in our static system with catalyst Pd-A-1 and obtained a similar result. The mechanism advanced by Meyer and Burwell is similar in principle, in regard to the natures of the species formed, to that proposed by ourselves for 1,3-butadiene hydrogenation over Type A nickel. Thus, we expected that 1,2-butadiene hydrogenation over Type A nickel would yield similar products. Table 2 shows that this expectation was correct. (The products of 1,2-butadiene hydrogenation over nickel were independent of reactant pressures and of conversion until the near-removal of reactant, and were only weakly dependent upon temperature.) Finally, Table 2 shows the product composition obtained by 1,2-butadiene hydrogenation over Type B nickel. The yield of trans-2-butene is increased nine-fold over that afforded by the Type A catalyst. The interpretation is that a new route to the formation of butene is available at the Type B surface due to the ability of intermediates to attain pi-allylic configurations when chemisorbed at this surface. Chemisorbed syn- and anti-1-methyl-pi-allyl may be formed from chemisorbed 1,2-butadiene by hydrogen atom addition to the sp-hybridised carbon atom, and are converted to trans- and cis-2-butene respectively by the addition of a further chemisorbed hydrogen atom. Thus, the comapratively substantial yield of trans-2-butene over Type B nickel is considered to constitute further proof of the facile formation of pi-allylic intermediates at this surface.

TABLE 2

Products of 1,2-butadiene hydrogenation†

Catalyst	Reaction temp. (°C)	Product Composition (%)			
		1-b	trans-2-b	cis-2-b	butane
$Pd-Al_2O_3$[12)]	35	40	7	52	1
Pd-A-1	21	37	3	57	3
(Type A)Ni-P-7	75	37	3	58	2
(Type A)Ni-S-9	115	36	2	61	6
(Type B)Ni-A-9	46	29	26	44	1

The sum of the information concerning mechanism provides a slender clue as to the difference between the surfaces of Types A and B nickel. The co-ordination of the ligand 1-methyl-pi-allyl to a metal atom is generally regarded as requiring two vacant co-ordination positions. Nickel atoms at the Type B surface apparently fulfil this requirement whereas those at the Type A surface do not. A lower mean co-ordination number for atoms at the Type B surface might imply that it is a metastable structure, perhaps retaining some disorder initially attributable to the presence of the Ni(III)-component of the mixed oxide. This situation may arise for two reasons. First, f.c.c. nickel formed from nickel(II) oxide should be comparatively well-ordered because this oxide has a sodium chloride structure, and thus during reduction simple relaxation of the metal ions of the nickel oxide lattice provides equivalently ordered metal atoms in the metal lattice. However, the structure of the Ni(III)-phase is unlikely to provide as easy a transition to f.c.c. nickel. Its sturcture is not known, although an attractive possibility is the orthorhombic layer structure of boehmite, AlO(OH), which is the only crystallographic phase present in the support, and in which certain hydrated oxides of iron and copper are known to crystallize[13]. Reduction of metal ions in such a lattice might well provide relatively disordered nickel crystallites, containing surface atoms of particularly low co-ordination number.

Supposing this to be so, this metastability should be removable by heat treatment at suitably elevated temperatures, thus converting Type B catalysts to Type A. Now the Tamman temperature for nickel is about 300°C. Consequently, catalyst Ni-A-8 was heated in hydrogen for a total of 200h at 313°C, and its behavior was monitored periodically using the standard test; it remained Type B. Further heating under hydrogen at 480°C for 10h brought about a change to Type A behavior which was accompanied by a five-fold diminution in the activity. This experiment was repeated twice, and the behavioral change and the loss of activity were reproducible. We therefore conclude that Type B nickel is metastable with respect to Type A nickel.

4. IN CONCLUSION

Boudart has provided an operational definition of a structure-sensitive reaction as "one for which the specific activity of the catalyst is dependent upon its mode of preparation". Such a definition was conceived as applying to catalysts which contained surface metal atoms of low co-ordination number by virtue of very high dispersion of metal. For a reaction giving a single product, this definition is excellent. It is but a small step to redefine the term for a reaction giving a number of products by parallel routes. We may then say that "a structure-sensitive reaction is one for which the specific activities of the various product-forming steps, or alternatively the selectivity, is dependent upon the mode of preparation of the catalyst". Nickel-catalyzed butadiene hydrogenation appears to be structure-sensitive according to this definition. Furthermore, it has been shown that the stabilisation of surface metal atoms of low co-ordination number can be achieved by means other than increasing dispersion to the limit.

ACKNOWLEDGEMENTS

We wish to acknowledge the award of a Science Research Council studentship to R.G.O.

REFERENCES

1) This paper is to be considered as Part VI of the series "Hydrogenation of Alkadienes" by P.B. Wells and co-workers, previous Parts of which have appeared in J. Chem. Soc. See particularly reference 8 which is Part III of the series.
2) O.M. Poltorak and V.S. Boronin, J. Phys. Chem. (U.S.S.R.), 39, (1965), 1329.
3) M. Boudart, Advan. Catalysis, 20, (1969), 153.
4) P.B. Wells, Specialist Periodical Reports on the Surface and Defect Properties of Solids, volume 1, to be published by The Chemical Society, London, 1972.
5) R. van Hardeveld and F. Hartog, Paper No. 70 presented to the Fourth International Congress on Catalysis, Moscow, 1968.
6) M. Boudart, A.W. Aldag, L.D. Ptak, and J.E. Benson, J. Catalysis, 11, (1968), 35.
7) K. Morikawa, T. Shirasaki, and M. Okada, Advan. Catalysis, 20, (1969), 98.
8) J.J. Phillipson, P.B. Wells, and G.R. Wilson, J. Chem. Soc., (1969), 1351.
9) B.J. Joice, J.J. Rooney, P.B. Wells, and G.R. Wilson, Discuss. Faraday Soc., No.41, (1966), 223.
10) F.N. Hill and P.W. Selwood, J. Amer. Chem. Soc., 71, (1949), 2522.
11) R.W. Cairns and E. Ott, J. Amer. Chem. Soc., 55, (1933), 534.
12) E.F. Meyer and R.L. Burwell, J. Amer. Chem. Soc., 85, (1963), 2881.
13) R.W.G. Wyckoff, in "Crystal Structures" volume 1, pp.294, 295 (Interscience, London and New York, 1963).

DISCUSSION

R. van NORDSTRAND

First my question, then the preamble. Have you attempted to interpret your results on the basis of the Woodward-Hoffman rules for concerted reactions?

As I recall, these rules predict 1-4 addition of hydrogen in the thermal reaction with 1,3 butadiene in its ground state, predict 1-2 addition if the 1,3 butadiene is in its first excited state. From Professor Ugo's paper yesterday, I gather that molecules adsorbed on metals often resemble the first excited state.

P. B. WELLS

The published mechanisms for butadiene hydrogenation catalyzed by Types A and B nickel (see reference 8 of the paper) are not concerted in the sense required by the Woodward-Hoffman rules. Nevertheless, one may indeed speculate that the forms of butadiene chemisorbed at these surfaces correspond to the ground state and the first excited state. We have considered this possibility in detail in the context of cyclopentadiene hydrogenation; this hydrocarbon is planar in the ground state and angular in the first excited state.

M. M. BHASIN

Dr. Wells, as you mentioned in your presentation, nickel powders and nickel supported on aluminas or silicas are not fully reduced. Do you feel that partially reduced nickel or nickel in an oxide environment gives type "A" behavior while fully reduced nickel--no matter whatever the source--gives type "B" behavior where butene-1 is isomerized to t-butene-2?

P. B. WELLS

First let me correct your statement. I claimed that my nickel-alumina samples were not fully reduced, but that my nickel powders and nickel-silica samples were fully reduced.

The answer to your question is "No." Recent work in our laboratory has confirmed that specimens of pure nickel powder may be either Type A or Type B. We are convinced that the structure of the precursor oxide in some way influences the behavioral type of the metal formed by its reduction.

Finally, in a previous publication (J. Chem. Soc., 1969, p. 1351), we have shown that Type B behavior involves genuine 1:4-addition of hydrogen to the diolefin to give 2-butene as the primary product; the mechanism does not involve any significant formation of 2-butene as a secondary product by the isomerization of 1-butene.

M. SCHIAVELLO

Recently we have shown[1,2] that when NiO is supported on γ- or η- Al_2O_3 a "surface spinel" is formed. The conditions of preparation were 600°C, 24 hr in air or dry atmosphere. The Ni^{2+} distribution in tetrahedral and octahedral sites is affected by the atmosphere of firing, by the Ni content and by the type of the support. Ni^{3+} was not present. The reducibility of the Ni^{2+} is strongly dependent on the cation distribution. Specimens with a high content of tetrahedral Ni^{2+} are hardly reducible even at 520°C for 24 hr in H_2. It is tempting to suggest that also the formation of your Type A or Type B catalysts in the Ni^{2+}/Al_2O_3 specimens is dependent on the amount of Ni^{2+} in octahedral and tetrahedral sites in the spinel phase. Normally type A and B can be formed either from an imperfect reduction of tetrahedral and octahedral nickel or from spinels with different initial contents of tetrahedral and octahedral nickel ions. I would appreciate a comment on this remark.

1) M. L. Jacone, M. Schiavello and A. Cimino, J. Phys. Chem. 75, 1044 (1971).
2) M. Schiavello, M. L. Jacone and A. Cimino, J. Phys. Chem. 75, 1051 (1971).

P. B. WELLS

Because our Types A and B behaviors have been observed using (unsupported) nickel powders, the participation of alumina as a support in the manner described by Dr. Schiavello cannot be a critical feature of the system.

However, following the publication of the above-cited papers we have made a very brief study of the effect of cation distribution in the supported nickel oxide upon the behavioral type of the resultant nickel catalysts. Our preliminary conclusion is that the manner of the distribution of Ni^{2+}, whether it be mostly in the octahedral sites or mostly in the tetrahedral sites, does not influence the Type of nickel catalyst subsequently obtained.

M. L. JACONO

1) On page 6, after Table 1, you write "Thus the heating in air, which reduced the Ni^{+++}-content of the mixture, caused a movement of behavior from Type B toward Type A." I would like to know what proof you have to say that the heating in air at 650°C reduced the Ni^{+++} content of the B.D.H. mixed nickel oxide. In other words, I wish to know if you have done any analytical test in order to measure the Ni^{+++}-content of your different catalysts.

2) Do you think that the two NiO catalysts made by calcination of nickel nitrate hexahydrate at 650°C and 1000°C were quite similar? Surely the oxygen vapour pressure on the two specimens (650°C, 1000°C) during the preparation was different, giving consequently two different stoichiometries for NiO. How do you obtain the same type A catalysts? What is your comment on these two points? Thank you.

P. B. WELLS

1) When the paper was written, we considered the Ni(III)-content of the commercial oxide to be in the region of 30%, both from information passed on by the supplier and from our own wet analysis. Very recently, results obtained using thermogravimetric analysis have indicated a rather lower value. This matter is now being further investigated.

2) Yes, we consider that the samples of nickel oxide prepared by calcination of nickel nitrate hexahydrate at 650°C and at 1000°C each correspond closely to NiO.

J. A. BETT

If, as the authors suggest, the formation of Type A or B Nickel is the result of the prior structures of the nickel oxide precursor, then there ought to be a relationship between the size of the nickel metal particles and the size of the nickel oxide crystals from which they were formed. Was this in fact seen?

P. B. WELLS

I regret that we have not compared particle size distribution of nickel oxide and nickel.

T. KWAN

The hydrogenation of 1,3-butadiene by $Co(CN)^{3-}$ in aqueous solution is well known to yield excess 1-butene over trans or cis-2-butene when CN/Co is greater than 5. The product distribution changes drastically, however, when the CN/Co is slightly less than 5; 2-butenes turn out to be the major products.

Since I understand that the authors have found a structure-sensitive reaction similar to those reported here also with cobalt-catalyzed butadiene hydrogenation, it must be interesting to see whether a common interpretation can be made on the structure-sensitive hydrogenation of butadiene in terms of co-ordination numbers.

P. B. WELLS

The interesting feature of our work, and of that quoted in the question is that, in each system, the 1:4-addition reaction is favoured when the transition metal attains an unusually low co-ordination number. In my view, the situations of low co-ordination number are, by their nature, the ones that permit the formation of π-allylic intermediates; these intermediates, in turn, are necessary for the achievement of 1:4-addition.

Paper Number 45

DEPENDENCE OF ACTIVITY FOR BENZENE HYDROGENATION OF NICKEL ON CRYSTALLITE SIZE

J. W. E. COENEN, R. Z. C. VAN MEERTEN and H. Th. RIJNTEN
Catholic University, Nijmegen, The Netherlands

ABSTRACT: The effect of crystallite size on activity per unit area in benzene hydrogenation was investigated for a range of silica supported nickel catalysts. From 50 Å down to 12 Å specific activity increases with decreasing crystallite size, while for still smaller crystal size activity appears to go down.

From a kinetic study in the liquid phase a dual site stepwise hydrogenation mechanism was derived with the second hydrogen atom addition rate determining. Activity differences between catalysts appear to be reduced by internal compensation effects.

Hydrogen kinetic isotope effects for H/D-mixtures show a parallel trend to specific activity. The information derived from the isotope effect is consistent with the kinetic considerations, showing that the entropy of adsorbed hydrogen plays an important role.

1. INTRODUCTION

Since H.S. Taylor in 1925 highlighted the importance of surface heterogeneity for catalytic activity of solids (1, 2) the reality of geometric and energetic heterogeneity of solid surfaces has become well established, but the effect of this heterogeneity on catalytic activity remains controversial.

Heterogeneity of metal surfaces may arise from different crystallographic planes in the surface and from intersection lines and points of these planes. The effect of various types of lattice defects may be superimposed. For supported metals the interaction of metal and support may stabilise an abnormal state of the metal. The contact line between metal crystallite and support where metal ions are bonded to foreign atoms will provide further abnormal surface sites.

With decreasing crystallite size the extent of homogeneous crystal planes must decrease and the proportion of surface atoms of low coordination will increase. The probability of lattice defects will be less in small crystals and they will anneal out with greater ease. The lattice in small crystallites may be compressed by surface tension effects (3), interaction with the support may induce lattice extension (4). It has been argued that for small metal crystallites (< 50 Å) the proportion of high energy B5 surface sites may be expected to increase (5, 6).

There are thus abundant reasons for surface heterogeneity and for differences therein for different preparations of the same metal. It is then the more surprising that for many reactions differences in specific activity appear to be slight. Although some authors find indications that certain types of lattice disorder increase catalytic activity (7, 8) other sources (9, 10) do not confirm these findings. Different crystal faces have been shown to have different catalytic activity (9, 11, 12). Differences are generally not very large. We may expect surface heterogeneity to change with crystallite size. Nonetheless the effect of crystallite size on specific activity appears to be small for many reactions. In hydrogenolysis of C_2H_6 Yates (13) found no influence of crystallite size, Taylor (14) found large differences, whilst Carter (15) found smaller crystals to be more active. In cyclopropane hydrogenation Boudart (16) likewise found small crystals to be somewhat more active than larger ones. In hydrogenation of benzene Aben (17), Dixon (18), Nikolajenko (19) and Taylor (20) found no influence of crystallite size on specific activity, Hill (21) and Krivanek (22) found small crystals to be less active than larger ones, whereas Selwood (23) concluded the reverse. Poltorak (24, 25, 26) found for a number of reactions specific activities which did not vary by more than a factor of 2 with crystal size. In all of the cited investigations variation of activity with crystal size remained within a factor of 3. There are some exceptional reactions known where the influence of the surface geometry on the specific activity is much stronger.

This surprising situation led Boudart (27) to divide catalytic reactions into two groups, facile and demanding reactions, the latter being those which are sensitive to surface structure. The majority of catalytic reactions appears to be facile under this definition, an unexpected result, which poses a number of problems:

1. What is the exact shape of the slight dependence of specific activity on crystallite size.
2. It appears likely that the virtual absence of a crystallite size dependence is the result of compensation effects: although the specific rates for different crystallite sizes are almost the same it is improbable that the detailed kinetics are identical. A kinetic study on a range of crystallite sizes is therefore desirable.
3. Can the data obtained under 2 be correlated with data on adsorptive behaviour, derived from independent measurements on the same catalysts.

To study these problems we chose the hydrogenation of benzene on silica supported nickel catalysts.

Benzene hydrogenation, though a 'facile' reaction under Boudart's definition, is a slow hydrogenation, which facilitates avoiding mass transport limitation. To safeguard further that our conclusions will refer to true chemical kinetics we did rate studies both in liquid and in gas phase. Since with acidic supports like alumina hydrogenation activity may not be strictly confined to the metal surface, due to 'spillover' effects (28-31) we chose a silica support. To obtain a range of crystallite sizes we preferred varying nickel content over thermal sintering, so that all catalysts could receive the same heat treatment.

To study the effect of crystallite size on specific activity reaction rates at standard conditions of temperature and pressure were measured in gas and liquid phase on catalysts with a range of nickel contents, having crystallite sizes between 5 and 50 Å.

A more detailed kinetic study was done in the liquid phase on four catalysts with rate measurements at 16 T-p-combinations.

To obtain information on surface structure isotope effects for simultaneous equilibrium adsorption of hydrogen and deuterium on the same four catalysts were measured.

2. EXPERIMENTAL

2.1. Ingredients: Pure thiophene-free benzene ex Merck was further purified by 3 hrs refluxing over NaPb-alloy and subsequent distillation into a receptacle, directly connected to the hydrogenation apparatus. Cylinder hydrogen was purified by passage over a deoxo catalyst and a molecular sieve drier and finally brought to high purity by diffusion through a PdAg-thimble. A series of silica supported nickel catalysts was prepared by precipitation from p.a. nickel nitrate solution onto 'Aerosil' silica. Precipitates were filtered, washed, dried at 120°C, ground and stored in a desiccator. Samples as needed for surface area determination, hydrogenation or deuterium exchange were reduced in a glass vessel with sintered glass disk at 450°C for 4 hrs with 60 ℓ/h.g. cat. hydrogen flowing through the catalyst bed.

2.2. Apparatus and measurements

a. Catalyst characterisation: Total nickel contents were determined, to refer all data to unit weight of nickel. Nickel surface areas were determined by chemisorption of hydrogen in a mercury free volumetric apparatus. After reduction the sample was evacuated during 2 hrs at 450°C and equilibrated with 1 atm hydrogen at 25°C during 16 hrs. The degree of reduction of the sample was obtained by measuring hydrogen evolution on dissolving the reduced catalysts in acid in hydrogen atmosphere and correcting for adsorbed hydrogen.

b. Liquid phase hydrogenation was performed in a constant pressure stirred tank apparatus. The double walled glass reactor was equipped with baffles and a turbine stirrer driven by means of a magnetic coupling. Thermostat liquid circulated through the jacket. All connections in contact with benzene were greaseless with teflon sleeves and viton O-ring seals. The glass reactor gas space was connected via a reflux condenser to a series of jacketed hydrogen burettes and to an oil manometer. The other leg of the manometer was connected to a thermostated gas space in which a selected reference pressure between 0.1 and 1.5 ata could be adjusted, making working pressures independent of barometric changes. A photoelectric sensor on one of the manometer legs controlled via a relay system the fluid supply to the burettes, thus maintaining a chosen pressure setting within ± 0.1 torr. Burette readings at timed intervals, corrected for temperature and pressure were used to compute rates in micromoles hydrogen per minute. Hydrogen pressures were derived from the set reference pressure, corrected for oil manometer differential and vapor pressure. Reaction temperature was read from a thermometer immersed in the liquid. On all catalysts standard rates were measured at 25°C and 600 Torr hydrogen pressure. On four catalysts rates were measured at 16 T-p-combinations (25, 45, 65, 85°C; 75, 150, 300, 600 Torr).

c. Gas phase hydrogenation was performed in a differential flow reactor. The reaction system, consisting of series connected benzene saturator, condensor, fixed bed differential reactor, GLC sampling valve and cold trap, was enclosed between two precision needle valves. The entry valve was connected to a constant pressure high purity hydrogen source, the exit valve to a vacuum line. Using the same reference pressure control system as under b, constant flow and constant reactor pressure were maintained. Incoming hydrogen flow was measured with a ball float and with a soap film flow meter. Saturator and condensor were thermostated, the latter at 10-15 degrees lower temperature than the former, ensuring saturation vapour pressure for benzene. The reactor, equipped with a preheater spiral was also thermostated. Reaction temperature was measured with an uncovered thermocouple in the catalyst bed. Degree of conversion was obtained from GLC analysis (Carbowax column, catharometer detection) of reactor effluent. Standard rates were measured at 25°C and 600 Torr hydrogen pressure.

d. Hydrogen isotope effect: Isotope effects in simultaneous adsorption of hydrogen and deuterium were measured on the four catalysts used in the kinetic study in a thermostated mercury free system consisting of a sample vessel and a number of calibrated volumes connected to H_2, D_2 and vacuum lines, Bourdon and membrane pressure gauges and an AEI MS 10 mass spectrometer. A total coverage with hydrogen isotopes could be selected and values of Θ_H, Θ_D, p_{H_2}, p_{D_2} and p_{HD} were determined.

3. RESULTS

3.1. Crystallite size and specific activity: From V_m values (ml S.T.P.) for hydrogen adsorption nickel surface areas (m^2) were obtained, assuming one H-atom and an average area of 6.33 $Å^2$ per surface nickel atom: $S_{Ni} = 3.41\ V_m$. Nickel areas were used to standardise rate data on unit area. Areas per gram nickel metal were also used to obtain a measure of crystallite size D from D = 4310/S_{Ni}, based on hemispherical crystallites, attached with the aequatorial plane to the support. Excellent correlation with X-ray line broadening crystallite sizes was obtained earlier on 22 Ni/SiO_2 catalysts (4), with crystallite sizes between 30 and 250 Å. Below 30 Å the model used gradually loses meaning and crystallite size should be regarded as a parameter to indicate nickel dispersion. A more direct experimental parameter N_H/N_{Ni} giving number of H-atoms adsorbed per nickel metal atom is also given.

Data on the investigated catalysts is given in table 1. The crystallite sizes cover the range from below 50 Å, by several authors (5, 6, 25) said to be most interesting from the surface structure point of view.

Table 1 - Analytical and standard activity data

Cat.	% Ni	S_{Ni} m^2/g cat	Reduction degree %	S_{Ni} m^2/g Ni_{met}	Cryst. size Å	N_H/N_{Ni}	Standard activity	
							L	G
NZ 1	0.82	4.53	68.2	810	5.3	1.25	1.27	
NZ 2	2.03	6.71	84.8	390	11.0	0.60	4.60	
NZ 4	4.26	12.86	84.0	359	12.0	0.55	6.14	6.2
NZ 5	4.88	14.4	83.8	352	12.2	0.34	4.68	
NZ 10	10.20	26.2	89.9	286	15.1	0.44	3.41	4.2
NZ 29	28.9	56.6	87.6	224	19.2	0.34	2.80	3.6
NZ 54	52.5	43.1	96.3	85	50.6	0.13	2.67	3.8

In the same table standard activities A_s in liquid and gas phase are given in $\mu mol\ H_2.m^{-2}.min^{-1}$. It is noted that the gas phase activities are somewhat higher than those obtained in liquid phase. They show, however, the same trend, as shown in fig. 1.

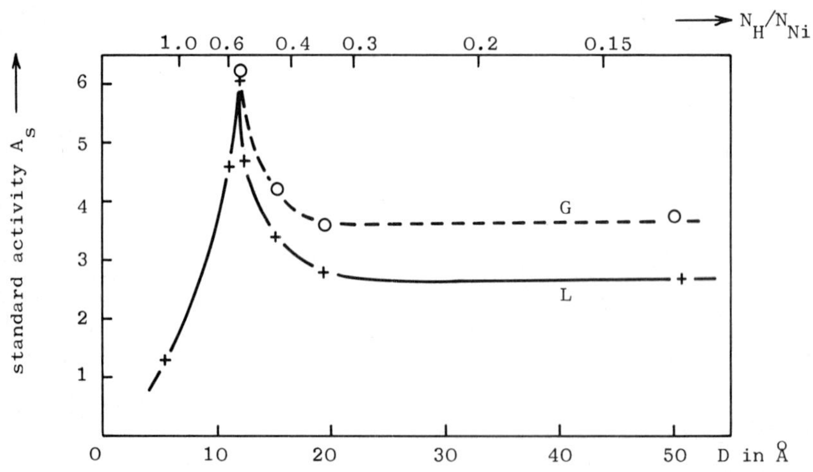

Fig. 1 - Relation between activity per unit nickel area and crystallite size

3.2. Kinetics of liquid phase hydrogenation for four catalysts: In gas phase hardly any activity decline with time was observed. In liquid phase apparent activity declined slowly with time. Therefore kinetic measurements were interspersed with runs at standard (25°C, 600 Torr) conditions, which were used to correct data for activity fall off.

Rates in liquid phase were proportional to the amount of catalyst added. The reaction order in benzene was found to be strictly zero at least down to 10% benzene. For a given catalyst at any one temperature the rate per gram catalyst or per unit nickel surface area is thus solely a function of hydrogen pressure: $R_T = k_o f(p)$. Although a slight curvature was evident in logarithmic plots of the exponential rate expression $R_T = k.p^n$, reaction orders n in hydrogen were obtained as given in table 2.

Table 2 - Reaction orders n in hydrogen and apparent activation energy E_a

Cat.	Reaction order n				Apparent activation energy E_a			
	25°C	45°C	65°C	85°C	75	150	300	600 Torr
NZ 4	0.52	0.67	--	0.81	9.5	10.0	10.7	11.6
NZ 10	0.55	0.65	0.80	0.80	9.3	10.4	10.8	11.5
NZ 29	0.56	0.60	0.68	0.78	10.5	10.8	11.2	11.7
NZ 54	0.61	0.67	0.78	0.73	11.1	11.5	11.8	11.8

The temperature dependence of the rates at a particular hydrogen pressure was analysed in Arrhenius plots, which yielded reasonably straight lines. The resultant apparent activation energies E_a are also shown in the table.

Differences between catalysts for these two empirical kinetic constants n and E_a are small and give little evidence for the expected differences in kinetic behaviour. For all catalysts there is a trend that n increases with temperature and that E_a increases with hydrogen pressure. Both trends as well as the fractional order in hydrogen are suggestive of a Langmuir type of behaviour. Therefore the rate data for the catalysts were fitted to three Langmuir type rate equations, which for the moment should be regarded purely as empirical correlation functions:

a. $R = k_o \left[\frac{(Kp)^{\frac{1}{2}}}{1 + (Kp)^{\frac{1}{2}}} \right]$ b. $R = k_o \frac{Kp}{1 + Kp}$ c. $R = k_o \frac{Kp}{[(1 + (Kp)^{\frac{1}{2}}]^2}$

with linear plots:

R^{-1} versus $p^{-\frac{1}{2}}$ R^{-1} versus p^{-1} $R^{-\frac{1}{2}}$ versus $p^{-\frac{1}{2}}$

Function b was successfully used by Aben et al (17) at higher p and T in gas phase. We obtained sixteen good fits (4 catalysts, 4 temperatures) with positive intercept only with function c, and this gave also in all cases the best fit, which was generally excellent. The logarithms of the resulting k_o and K values were then plotted against reciprocal temperature, yielding per catalyst four empirical parameters, A_o, E_o, ΔS^o and ΔH^o, which correlate the 16 rates. These are shown in table 3.

From the four parameters for each of the catalysts the standard rates at 25°C and 600 Torr were recalculated (A_s calc) to be compared with the directly observed data (A_s exp). Their agreement demonstrates both the reliability of individual rate measurements and the quality of the correlation.

3.3. Isotope effects: From the data for four catalysts an isotope effect parameter K_I was calculated as $K_I = (\Theta_D/\Theta_H)^2 . (p_{H_2}/p_{D_2})$, given in table 3.

Table 3 - Kinetic parameters for liquid phase hydrogenation

Cat.	D_{cryst}	A_s exp	ln A_o	E_o	ΔS^o	ΔH^o	A_s calc	K_I
NZ 4	12.0	6.14	31.23	16.77	-29.61	-9.45	6.38	3.05
NZ 10	15.2	3.41	30.18	16.53	-27.92	-8.94	3.34	1.55
NZ 29	19.2	2.80	29.92	16.64	-26.85	-8.90	2.53	1.14
NZ 54	50.6	2.67	26.82	14.61	-14.24	-4.54	2.35	0.91
Units:	Å	$\mu mol.min^{-1}.m^{-2}$		kcal/mol	e.u. (1 atm)	kcal/mol	$\mu mol.min^{-1}.m^{-2}$	

4. DISCUSSION

Reproducibility of measurement of rates and nickel areas was generally better than ± 4%. We therefore believe the differences in specific activity to be significant.

Before we attach undue significance to our data we should enquire whether mass transport effects may play a role. The fact that the rates in liquid phase hydrogenation are strictly proportional to catalyst amount definitely precludes an influence of gas/liquid transport of hydrogen. With respect to intraparticle transport reasonable values for particle size (2×10^{-4} cm), pore radius (20 Å), diffusion coefficient (7×10^{-5} cm^2 sec^{-1}) and solubility of hydrogen (2.5 $\mu mol.cm^{-3}$) yield for the fastest standard hydrogenation (NZ 4) a Thiele modulus h = 0.15, equivalent to 99.3% effectiveness. For the higher rate at 85°C h = 0.375 corresponding to 95.5% effectiveness. For the gas phase hydrogenations the situation is even more favourable. We may thus conclude that in the rate data in this paper mass transport limitation is insignificant.

As a further check on reliability a comparison with literature data is indicated. The ratio of average specific rate observed by other workers to the specific rate observed by us is 0.33 (17), 0.4 (18), 1.6 (32), 0.3 (22), 0.2 (34), 0.7 (33) and 0.4 (35). These ratios are very close to unity, considering that in most comparisons an extrapolation was involved, all workers use slightly different definitions for nickel surface area and also the ranges of nickel crystallite size differ. We may thus conclude that with present day techniques specific rates can be compared between laboratories.

The parallel behaviour of gas phase and liquid phase data is a further indication that the trend of activity with crystallite size must be real. Why then are they not identical? Three effects may explain the differences:
i. Selfpoisoning may be more serious in liquid phase, where the catalyst may be subjected to low hydrogen supply over lengthy periods. Also mechanical catalyst loss through spattering may operate. These effects may make liquid phase data too low.
ii. The reaction is strongly exothermic. It is imaginable that the temperature measured in the gas phase catalyst bed is still lower than the actual temperature of the catalysing nickel surface. This effect may be less in the liquid phase where the individual catalyst particles are immersed in the liquid in which the temperature is measured. A temperature error of 3°C would explain the average deviation. This effect would make the gas phase data slightly too high.

iii. From gravimetric measurements we know that under the conditions of gas phase hydrogenation at 45°C the catalyst is covered with about 0.7 statistical layers of physisorbed benzene, at 25°C about 1.5 layers. This situation is still not identical with immersion in liquid benzene. Also the kinetic behaviour is different in detail: in gas phase the apparent activation energy is somewhat higher than in liquid phase, the reaction order in hydrogen is slightly higher in liquid phase than in gas phase, the more so the higher the temperature; also the order in benzene differs: zero in liquid phase, 0.1-0.2 in gas phase. All these differences may be expected to disappear at a lower temperature, where the catalyst pore system gets entirely filled with liquid benzene by capillary condensation, making the situations identical.

Returning now to the dependence of specific activity on crystallite size we may conclude that the differences in activity may certainly be considered significant. The upward trend in activity, going from 50 to 12 Å crystallite size, we consider as established. The fall in activity below 12 Å, if real, is of great interest. Ideas that a minimum crystallite size is required appear to find confirmation.

Whether the effect shown is purely due to differences in crystallite size, resulting in differences in surface character, is less certain. Going from right to left in Fig. 1 not only crystallite size but also nickel content goes down. This means that unit nickel surface area is accompanied by an increasing silica area. This silica surface may help by scavenging catalyst poisons or possibly by a spill-over effect. Of course the silica surface could never explain a maximum. We should also note that the average crystallite size may be an inadequate parameter to describe the nickel surface characteristics, especially with crystallite sizes as close as 11.0, 12.0 and 12.2 Å. Differences in crystallite size distribution or crystal habit may well contribute to the curious shape of the curve, which may thus not be universal.

We have seen that the liquid phase kinetic data is well described by

$$R = k_o\left[\frac{(Kp)^{\frac{1}{2}}}{1 + (Kp)^{\frac{1}{2}}}\right]^2 \quad \text{(i)}; \qquad K = e^{\Delta S^o/R}\, e^{-\Delta H^o/RT} \quad \text{(ii)}; \qquad k_o = A_o\, e^{-E_o/RT} \quad \text{(iii)}$$

We introduced k_o, K; A_o, E_o; ΔH^o, ΔS^o as empirical parameters. To give them mechanistic meaning we will postulate a stepwise hydrogenation mechanism:

$B + xS_1 \underset{}{\overset{1}{\rightleftharpoons}} B_a$	where B denotes benzene
$H_2 + 2S_2 \overset{2}{\rightleftharpoons} 2H_a$	index a indicates adsorbed state
$B_a + H_a \overset{3}{\rightleftharpoons} BH_a$	BH_n denotes C_6H_{6+n}
$BH_a + H_a \overset{4}{\rightleftharpoons} BH_{2a}$	S_1, S_2 are two types of surface site
etc	

We will assume that elimination of hydrogen from the surface by desorption is much faster than by benzene hydrogenation, which will be verified later. Then equilibrium 2 will be undisturbed.

It is then attractive to rewrite eqn. i as $R = k_o \vartheta_H^2$ (iv), which gives ΔH^o and ΔS^o the usual meaning of adsorption enthalphy and entropy for hydrogen. The observed zero order in benzene precludes a Rideal mechanism with reaction of H_a with benzene from the liquid. Chemisorption of benzene has been established by magnetic measurements (36). Further assuming that the degree of occupation with organic entities attains a fractional coverage Θ_o, independent of hydrogen pressure, this leaves $1-\Theta_o$ free for hydrogen adsorption. ϑ_H in iv is then the fraction of the remaining surface covered with adsorbed H.

To refer Θ_H to the entire surface k_o must then contain the factor $(1-\Theta_o)^2$. The proportionality to ϑ_H^2 strongly suggests addition of the second H to adsorbed benzene to be rate determining. We found experimentally that cyclohexene hydrogenation is 3500 x faster than benzene hydrogenation at 25°C in agreement with ref. (37), cyclohexadiene may be expected to be even more reactive, making our choice of reaction 4 for the slow step reasonable.

If equilibrium 3 is established $\Theta_{BH} = K_3\Theta_B\Theta_H$ and $\Theta_B = \Theta_o/(1 + K_3\Theta_H)$. The resonance stability of benzene may be partly conserved in the adsorbed state causing equil. 3 to lie to the left at moderate hydrogen pressure, so that $K_3\Theta_H \ll 1$. We note that $\Theta_o(1 - \Theta_o)^2$ varies by only ± 9% around its mean value of 0.136 for $0.19 < \Theta_o < 0.51$. From magnetic measurements we found that benzene adsorption at the considered temperatures tends to stop at $\Theta = 0.2 - 0.3$. The rate equation then becomes

$$R = k_4\Theta_{BH}\Theta_H = k_4K_3\Theta_B\Theta_H^2 = k_4K_3\Theta_o(1-\Theta_o)^2\vartheta_H^2/(1+K_3\Theta_H) = 0.136\ k_4K_3\vartheta_H^2 = k_o\vartheta_H^2 \quad (v)$$

which is of the experimentally observed form. By assuming equilibrium established for reactions 2 and 3 the kinetics become identical with concerted addition of two H_a to B_a rate determining.

Returning to the data A_o, E_o, ΔS^o and ΔH^o of table 3 we note that they show a qualitatively consistent trend: From NZ 4 to NZ 54 the heat of adsorption goes down and the weaker adsorption shows a smaller entropy loss. We also find that the less tightly bound hydrogen requires a smaller activation energy E_o to combine with adsorbed benzene, as is to be expected. We further recall that $\ln A_o$ must contain an activation entropy term for the surface reaction. A larger entropy of the adsorbed hydrogen clearly works unfavourably in the formation of the activated complex and we find A_o accordingly smaller.

We may make this reasoning more quantitative by using transition state theory

$$R = \frac{kT}{h}\, e^{\Delta S^{\ddagger}/R}\, e^{-\Delta H^{\ddagger}/RT}\, n_s\Theta_B\Theta_H^2 = \frac{ekT}{h}\, e^{\Delta S^{\ddagger}/R}\, e^{-E_o/RT}\, n_s\Theta_B\Theta_H^2 \quad (vi)$$

which on introduction of the experimental ϑ_H transforms to

$$R = 0.136\ \frac{ekT}{h}\, n_s\, e^{\Delta S^{\ddagger}/R}\, e^{-E_o/RT}\, \vartheta_H^2, \text{ so that } A_o = 0.136\ \frac{ekT}{h}\frac{A}{\sigma}\, e^{\Delta S^{\ddagger}/R} \quad (vii)$$

where A is the nickel surface area on which the reaction is taking place, $1\ m^2 = 10^{20}\ Å^2$ and σ the area taken up by one activated complex. For the latter we will choose a value of 45 $Å^2$, which is the cross-sectional area for benzene in adsorption on metals (38, 39). We thus find $\ln A_o = 70.8 + \Delta S^{\ddagger}/R$ for a mean temperature of 328°K, so that we can calculate the activation entropy for the surface reaction $\Delta S^{\ddagger} = S^{\ddagger} - S_{B_a} - 2S_{H_a}$ from the experimental A_o values, after converting them to molecules benzene per m^2.sec. Probably $S^{\ddagger} \approx S_{B_a}$; $\Delta S^{\ddagger} \approx -2H_a$.

Table 4 - Entropy per g at. adsorbed hydrogen during hydrogenation

Catalyst	$\ln A_o$	$\Delta S^{\ddagger}/R$	$\Delta S^{\ddagger}$	S_H from A_o	S_H from $\Delta S^o = S^o_{H_2} - 2S_H$
NZ 4	67.0	- 3.8	- 7.5	3.75	1.2
NZ 10	65.9	- 4.9	- 9.7	4.85	2.0
NZ 29	65.6	- 5.2	-10.3	5.15	2.6
NZ 54	62.6	- 8.2	-16.3	8.15	8.9

There is a gratifying parallelism in the two sets of entropy data and for catalyst NZ 54 there is even quantitative agreement. For the other three catalysts there is a difference of about 2.5 entropy units. We should of course realise that in this confrontation all experimental errors, imperfections in empirical correlations as well as effects introduced by the theoretical assumptions come together so that possibly the agreement is as good as we may expect. The discrepancy is equivalent to a combined error in E_o and ΔH^o of a little over 1 kcal, which is not excessive. In sofar as the observed discrepancy is not trivial one may note that the observed A_o values are too low. One might imagine that on the catalysts with the smaller crystallite sizes part of the surface is so active that it becomes blocked with carbonaceous residues, by which it is poisoned irreversibly. This effect might then also help to explain the drop in activity below 12 Å. It is then difficult to explain why this does not occur at all on NZ 54.

Also for the activation energies there is the possibility of an internal consistency check, although less fundamental in nature. One may easily derive by logarithmic differentiation of eqn. (i) that the Arrhenius activation energy $E_a = RT^2$ (d lnR/dT) also equals $E_a = E_o + (1 - \vartheta_H)\Delta H^o$ (viii). The values for ϑ_H for 600 Torr and 328 °K, calculated from the ΔS^o and ΔH^o values of table 3, together with the E_a values from eqn. (viii) and the directly obtained E_a values of table 2, are shown in table 5.

Table 5 - Comparison of directly observed and calculated activation energies

Catalyst	NZ 4	NZ 10	NZ 29	NZ 54
ϑ_H	0.415	0.427	0.490	0.444
E_a from viii	11.2	11.4	12.1	12.1
E_a direct	11.6	11.5	11.7	11.8

The agreement is in all cases within 0.5 kcal/mol, again showing the quality of the empirical correlation functions and the consistency of the data.

The assumption of equilibrium (3) being undisturbed is fundamental to the present kinetic description. With absolute rate theory one may estimate the rate of desorption of hydrogen and compare it with the rate of hydrogenation at the low temperatures considered. We found the assumption amply justified.

Considering now the isotope effect data (table 3) we note first of all the striking parallelism with standard activity figures. The fact that the K_I values, which were measured in a clean system where the catalysts were only exposed to hydrogen isotopes, correlate with catalytic activity in benzene hydrogenation indicates that the differences in activity really stem from differences in surface properties.

The detailed interpretation of the isotope effect on a heterogeneous surface is quite as complicated as the adsorptive behaviour of these surfaces. However, the fact that in isotope effects at least part of imperfect description tends to cancel out led us originally to choose this technique to obtain information on surface properties. For a homogeneous surface the K_I as defined here is equal to the ratio of Langmuir constants for deuterium and hydrogen for fixed site adsorption and to the ratio of Volmer constants for two dimensional gas adsorption. It can further be shown that K_I may be expected to decrease with increased mobility or entropy of the adsorbed hydrogen. This is in agreement with the entropy values obtained from the kinetic data (compare tables 3 and 4). More detailed discussion of the isotope effect will be published elsewhere (40)

The problems posed in the introduction under points 2 and 3 have now been partially answered. Differences in detailed kinetic behaviour do show compensation: NZ 54 and 29 show significant differences in A_o, E_o, ΔS^o and ΔH^o but they cancel out almost entirely. The large difference between NZ 4 and NZ 10 is entirely due to a different preexponential factor A_o, due to different mobility of adsorbed hydrogen. For the catalysts with very small crystallites a faint indication was obtained that increased selfpoisoning may diminish activity.

ACKNOWLEDGEMENT: Thanks are due to research students J.A.B.C. Wolf and P.W.A. van Poppel whose careful measurements were used in this paper.

REFERENCES

1. H.S. Taylor, Proc. Roy. Soc. A108 (1925) 105
2. H.S. Taylor, J. Phys. Chem. 30 (1926) 145
3. F.W.C. Boswell, Proc. Phys. Soc. (London) 64A (1951) 465
4. J.W.E. Coenen and B.G. Linsen, Phys. and chem. aspects of adsorbents and catalysts, Acad. Press (1970)
5. R. van Hardeveld, F. Hartog, Surface Science 15 (1969) 189
6. E.G. Schlosser, B. Bunsenges. 73 (1969) 358
7. M.J. Duell, A.J.B. Robertson, Trans. Farad. Soc. 57 (1961) 1416
8. I. Uhara et al, J. Phys. Chem. 66 (1962) 1374
9. J. Bagg, H. Jaeger, J.V. Sanders, J. Catalysis 2 (1963) 449
10. D. Graham, J. Phys. Chem. 66 (1962) 510
11. J. Völter, H. Jungnickel, G. Rienäcker, Z. anorg. allg. Chem. 360 (1968) 300
12. A.T. Gwathmey, R.E. Cunningham, Adv. Catalysis 10 (1958) 57
13. D.J.C. Yates, W.F. Taylor, J.H. Sinfelt, J. Am. Chem. Soc 86 (1964) 2996
14. W.F. Taylor, J.H. Sinfelt, D.J.C. Yates, J. Phys. Chem. 69 (1965) 3857
15. J.L. Carter, J.A. Cusumano, H.H. Sinfelt, J. Phys. Chem. 70 (1966) 2257
16. M. Boudart, A. Aldag, J.E. Benson, N.A. Dougharty, C.G. Harkins, J. Cat. 6 (1966) 92
17. P.C. Aben, J.C. Platteeuw, B. Stouthamer, Rec. Trav. Chim. 89 (1970) 449
18. G.M. Dixon, K. Singh, Trans. Farad. Soc. 65 (1969) 1128
19. V. Nikolajenko, V. Bosacek, V. Danes, J. Catalysis 2 (1963) 127
20. W.F. Taylor, H.K. Staffin, Trans. Farad. Soc. 63 (1967) 2309
21. F.N. Hill, P.W. Selwood, J. Am. Chem. Soc. 71 (1949) 2522
22. M. Krivanek, V. Danes, V. Nikolajenko, Coll. Czech. Chem. Comm. 29 (1964) 2726
23. P.W. Selwood, S. Adler, T.R. Phillips, J. Am. Chem. Soc. 77 (1955) 1462
24. O.M. Poltorak, V.S. Boronin, J. Phys. Chem. USSR 39 (1965) 1329
25. O.M. Poltorak, V.S. Boronin, J. Phys. Chem. USSR 40 (1966) 1436
26. O.M. Poltorak, V.S. Boronin, A.N. Mitrofanova, 4th Int. Congr. Cat. Moscow (1968) preprint 68
27. M. Boudart, Adv. Catalysis 20 (1969) 153
28. J.H. Sinfelt, P.J. Lucchesi, J. Am. Chem. Soc. 85 (1963) 3365
29. J.L. Carter, P.J. Lucchesi, J.H. Sinfelt, D.J.C. Yates, Proc. 3rd Int. Congr. Cat. 664 (1965)
30. D.V. Sokolskii, B.O. Zhusupbekov, Kinetics & Catalysis 9 (1968) 164
31. K.M. Sancier, J. Catalysis 20 (1971) 106
32. R. van Hardeveld, F. Hartog, 4th Int. Congr. Cat. Moscow (1968), preprint 70
33. G. Lyubarskii et al, Kinetics & Catalysis 1 (1960) 235; 5 (1964) 277
34. W.F. Taylor, J. Catalysis 9 (1967) 99
35. P. van der Plank, thesis, Leiden (1968)
36. P.W. Selwood, Adsorption and collective paramagnetism, Acad. Press (1962)
37. W.F. Madden, C. Kemball, J. Chem. Soc. 54 (1961) 302
38. B.L. Harris, P.H. Emmett, J. Phys. Chem. 53 (1949) 811
39. A.I. Sarakov, Proc. Acad. Sci USSR 112 (1957) 464
40. J.W.E. Coenen et al, to be published

DISCUSSION

N. B. BHATTACHARYYA

1) We have done some work on crystallite size and activity in connection with nickel reforming catalysts and found that catalytic activity decreases with increasing crystallite size. The crystallite size involved in our study ranges from 180 Å to 960 Å.

2) In your paper you have made a correlation between specific activity and crystallite size. While variation was made in crystallite size (Table 1), nickel concentration also varied simultaneously. It is well known that activity of a catalyst depends not only on its physical and structural parameters, it also depends greatly on concentration of the active component. Don't you think that inferences drawn on the basis of crystallite size above may be misleading when concentration of nickel is also changing? Curves similar to Figure 1 could be obtained also when activity is plotted against nickel concentration. In my opinion the correlation of crystallite size with activity could be more significant and meaningful if the concentration of nickel in all the samples were kept fixed. This could be done through progressive sintering of the parent preparation.

3) Secondly, degrees of reduction are different for different samples. Don't you think with incomplete reduction the measured activity may be different from its real activity?

J. W. E. COENEN

1) It is difficult to comment on your statement without knowing further details. I would not expect significant differences in specific activity per unit area in crystals as large as 180 Å and 960 Å respectively. Is it quite certain that no extraneous effects such as mass transport limitation can play a role?

2) The fact that with crystallite size other parameters vary in the series of crystals, notably the nickel content is explicitly stated in the paper. To achieve a variation in crystallite size some variation in composition or history will have to be varied. We chose to keep the heat treatment constant and chose to vary nickel content. This is very much a matter of arbitrary choice and the choice of different heat treatments would be no less arbitrary and if it were chosen it could be the basis of a discussion remark, very similar to the one you made.

3) Indeed, the degrees of reduction vary a little. To attain complete reduction much more severe heat treatments would have to be chosen, which would have made the smallest crystallite sizes inaccessable. We feel that in that way the loss would be greater than the gain. It is not quite clear what is meant by "real activity." If reduction were complete the support would be exclusively silica, instead of a combination of silica and nickel silicate in our catalysts.

J. J. F. SCHOLTEN

1) In your thesis (Delft University) you introduced the concept of the

Ni crystallites being bound to the carrier via a small layer of Ni-silicate.

The extent and strength of this "chemisorption" of Ni on silica may influence the crystallographic plane distribution of the free nickel as it influences the total surface energy, and hence it influences the catalytic behavior. This effect (though resulting from the binding of Ni) may be a function of crystallite diameter (your thesis). It may be that the relation found in Figure 1 of your paper can be explained in this way.

J. W. E. COENEN

As further explained in reference (4) of the paper we may indeed assume that the nickel crystallites have a preferred orientation with (111) parallel to the silica surface. In our earlier work on Ni/SiO_2-catalysts, we found evidence for a f.c.c. nickel lattice with slightly enlarged lattice parameter, the effect being more pronounced for the smallest crystallite sizes. It is indeed an interesting suggestion to bring this observation to bear on the present activity dependence. Qualitatively the maximum in the activity/crystallite size correlation would thus be transformed into a maximum in an activity/lattice constant relation and this would be less difficult to rationalize.

G. C. BOND

I should like to ask Professor Coenen about the rather unusual dependence of crystallite size upon nickel content shown in Table 1. It does not conform to any of the expected or usually observed relations. The smallest particle size seen (5.3 Å) deserves comment: five nickel atoms in a trigonal bipyramid would be of about this size.

J. W. E. COENEN

That lower nickel contents produce smaller average crystallite sizes in this series is certainly not unexpected behavior. I am not sure in how far the more detailed behavior is unexpected. For five out of seven catalysts it is a linear relation, though not through the origin. This behavior is certainly no cause for surprise. Remains consideration of the two extremes. To understand the large crystallite size for the 52% Ni we should recall the function of the support: provided the intermediate nickel compound is closely associated with the support and evenly distributed the support restricts sintering of the metal formed in reduction. It becomes progressively more difficult to achieve homogeneous distribution and close association between intermediate nickel hydroxide and support as the ratio of the two increases. Remains the very small crystallite size for the other end. Here we should point out that due to excessive nickel silicate formation reduction is more difficult and incomplete sintering inhibition is exceptionally effective. Here moreover crystallite size losses meaning end is more an arbitrary yardstick. For our definition of crystallite size see the reference quoted for question 2.

T. KWAN

You showed a striking parallelism between the benzene-hydrogenation activity of nickel catalysts of different crystallite size and the hydrogen isotope effect in the chemisorption of H_2 or D_2 on these catalysts at equilibrium.

I cannot see immediately how these two quantities are correlated. Could you explain more about the chemisorptive character of your nickels for hydrogen?

J.W. E. COENEN

So far we have no detailed explanation for the parallelism between benzene hydrogenation activity and hydrogen isotope effect. We can only say the following:

With rather drastic simplifying assumption one can calculate K_J by statistical thermodynamics. For catalyst NZ 54 the calculated value can be made to equal the measured K_J by suitable choice of vibration frequencies. Then we find that also the temperature dependence of K_J is the same as calculated between -80 and +100°C. At higher temperatures the experimental K_J becomes much too high. This we interpret by assuming that exchange of adsorbed D with protons on silica occurs, which simulates high K_J. The high value of K_J on catalyst NZ 4 at room temperature cannot be understood by statistical thermodynamics: here we assume that already at room temperature some exchange occurs with the large SiO_2-surface.

Why then parallelism with benzene activity? Here we then have to call in a similar effect: spill-over to the large support area. This explanation is still very qualitative and unsatisfactory. We are working on a more quantitative picture.

Paper Number 46

SEARCH FOR SURFACE-STRUCTURE SENSITIVITY IN METAL CATALYSIS. PART I. THE NORBORNANE/DEUTERIUM EXCHANGE AND THE CIS-1,4-DIMETHYLCYCLOHEXANE EPIMERISATION REACTIONS ON PALLADIUM

J. K. A. CLARKE, E. McMAHON and A. D. O'CINNEIDE
Department of Chemistry, University College, Belfield,
Dublin 4, Ireland

ABSTRACT: The norbornadiene deuterogenation reaction and the exchange reaction between norbornane and deuterium were examined on palladium catalysts of differing surface structure. In the former reaction only products resulting from exo addition are found. In the latter both exo and endo hydrogens are exchanged, the exo at a faster rate; no evidence is found within the limits of catalyst reproducibility for structure sensitivity of the relative rates of exchange of exo and of all other hydrogens.

In addition, the epimerisation of cis-1,4-dimethylcyclohexane was studied with several supported palladium catalysts of differing metal dispersion. Specific activities, measured at 60-110°C and at atmospheric pressure of hydrogen, are not appreciably dependent on dispersion, supporting the view that partial or complete detachment of the reaction intermediate from the surface occurs during 'roll-over'.

1. INTRODUCTION

As is well known, the chemistry of the norbornane system is dominated by exo attack by reagents on account of steric protection of the endo positions. This is also the case in metal catalysed reactions, both homogeneous and heterogeneous[1)]. Endo attachment of a metal atom is, however, possible as shown by the preparation of many endo-bidentate norbornadiene complexes[2)]. It was of interest to find whether surface metal atoms of low coordination were more effective than others in causing endo reaction. The approach adopted was to study the norbornadiene deuterogenation reaction and the exchange reaction between norbornane and deuterium on several palladium catalysts of differing surface characteristics.

Recent studies[3)] have established that epimerisation of dimethylcycloalkanes on palladium catalysts occurs by 'roll-over' of an intermediate having a planar structure at one of the methyl-substituted ring carbons. It is still unclear whether the intermediate retains some measure of bonding to the surface during the roll-over process. Because residual bonding probably implies surface-structure sensitivity[4)] it was decided to measure the activities for epimerisation of cis-1,4-dimethylcyclohexane of a series of supported palladium catalysts of differing metal dispersion.

2. EXPERIMENTAL

2.1. Catalysts

Three forms of palladium were used: (i) evaporated films, (ii) alumina-supported and (iii) carbon-supported. (i) and (ii) were employed in the norbornane work and (iii) for epimerisation runs. Details of these catalysts follow.

2.1.1. Films (designated A) were formed by evaporation *in vacuo* at a wall temperature of 0°C and areas were taken to be approximately those measured by Kemball[5)]. In addition, films (B) were formed at a wall temperature of 250°C and sintered for 1 hr in 3-4 torr deuterium also at 250°C. For these films the area was assumed to be that of the substrate, i.e. 400 cm^2.

2.1.2. Following reduction at 160°C (see 2.2.1) two γ-alumina-supported samples C (Degussa) and D (Johnson, Matthey) were characterised as follows. Electron microscopy gave estimates of average diameter, and spread in diameter, of the metal crystallites. Carbon monoxide chemisorption measurements yielded values for metal area, by assuming a 1:1 Pd/CO stoichiometry, and permitted independent estimates of mean crystallite size. Results are given in Table 1.

TABLE 1

Characteristics of supported catalysts

Catalyst type	Pd/alumina		Pd/charcoal				
Designation	C	D	E	F	G	H	H'
w/w %	0.5	5.0	4.8	4.8	5.3	19.2	19.2
Metal area (CO)[a], M^2/g catalyst	1.09	1.02	5.4	6.4	5.6	17.9	-
Mean crystallite size (CO), A	23	240	37	31	40	45	-
Mean crystallite size (e.m.)[a], A	40	200	-	37[c]	-	-	250[c]
Metal area (e.m. estimate[b]), M^2/g catalyst	-	-	-	-	-	-	3.8

[a] CO and e.m. denote values based on carbon monoxide chemisorption and electron microscopy, respectively;

[b] assuming spherical crystallites;

[c] following epimerisation reaction.

2.1.3. Four charcoal-supported catalysts, E-H (Johnson, Matthey), were subjected to carbon monoxide chemisorption measurements following room temperature reduction. The procedure followed was that described in a similar study[6)]. Values of metal area and of average metal dispersion were derived as in 2.1.2 (Table 1). Electron microscopic estimates of dispersion were obtained (Table 1) for a catalyst F sample following reaction (see 2.2.2) and for a sample of H which had been given severe heat treatment (H'). The value of 37A for F is considered to be in good agreement with that of 31A derived from carbon monoxide chemisorption, indicating that the dispersion of metal did not change during reaction. The estimate of average particle size of 250 A for H' clearly depends to some extent on the number of very large particles which are selected for counting, and is therefore probably not as accurate as the value obtained for catalyst F. However, it is clear that H' is composed of large crystallites. Its metal area was estimated from the measured average particle size.

2.2. Reaction systems and procedures

2.2.1. Studies with norbornadiene and norbornane

Reactions were carried out in a static system[7)] with capillary bleed to a mass-spectrometer. In all experiments the reaction mixture contained 2.2 torr of hydrocarbon and an eight-fold excess of deuterium (hydrogen). Supported catalysts were activated by evacuation (170°C, 2 h) followed by reduction in 5 torr deuterium (170°C, 2.5 h) in conjunction with a liquid nitrogen trap. In the exchange experiments corrections were made to peak heights for naturally occurring isotopes and fragmentary ions. At 13.5 V fragmentation of norbornane with loss of one hydrogen atom was 17%. Because there are both secondary and tertiary hydrogens the fragmentation corrections could not without question be made on the usual statistical basis. As a test, fragmentation patterns of norbornane and exo-d_4-norbornane were compared. By assuming that all secondary hydrogens were equivalent for this purpose, it was calculated that the chance of a norbornane molecule losing a tertiary hydrogen was 1.3 times the chance of losing a secondary. The additional correction is, therefore, small and it was ignored for the exchange runs with norbornane; it was, however, included for all runs with exo-d_4-norbornane (see 3.1). The reproducibility of rates with the supported catalysts was about ±25% and with films about ±10%.

Norbornadiene (Koch-Light) and norbornane (Aldrich) were subjected to freeze-thaw cycles and to multiple sublimations, respectively, *in vacuo* before use. Deuterium and hydrogen were palladium-diffused.

2.2.2. cis-1,4-Dimethylcyclohexane epimerisation.

Use was made of a single-pass fixed bed reactor in the form of a vertical pyrex tube (1 cm dia.) which contained 0.1-0.3 g of catalyst resting on a porosity 3 sintered-glass disk. This arrangement satisfied the geometric criteria for 'piston-type' flow[8)].

cis-1,4-Dimethylcyclohexane (Aldrich, >99.5%), contained in a saturator, was degassed by bubbling through it successively nitrogen and hydrogen which had been pretreated in each case in a deoxo unit. The saturator was then isolated and thermostatted at 0°C. Prior to use each catalyst was reduced in a flow of hydrogen (100 ml/min, 2 h, 25°C). The hydrogen flow was decreased to 5-10 ml/min, still at 1 atm, and diverted through the saturator (hydrocarbon pressure, 4.1 torr) before its passage through the catalyst bed. The reaction was followed by g.l.c. tail-gas analysis. A period of 3-6 h was allowed for the attainment of steady-state conditions at each temperature before the conversion was measured. Conversions were always kept below 10%.

3: RESULTS AND DISCUSSION

3.1. Studies with norbornadiene and norbornane

The reaction of norbornadiene with deuterium at <0°C on catalysts A, B, C, D and F gave d_4-norbornane (>99%) as product; no exchange of the olefinic hydrogens accompanied deuterogenation. We note that norbornane does not undergo exchange with deuterium at these temperatures. Continuous analysis of a norbornadiene/deuterium reaction mixture at -23°C showed that the alkane was produced by successive reaction: diene→monoene→alkane. Endo-bidentate attachment of the diene to a surface metal atom could, in principle, lead to alkane as initial product: initial alkane production was not observed within the admittedly low (5%) sensitivity of the method. The NMR spectrum of norbornane[9] shows three distinguishable proton signals attributable to bridgehead, exo- and bridging + endo-hydrogens. No exo-hydrogens (<0.5%) could be detected from NMR analysis of the d_4-norbornane product from deuterogenation on any of the catalysts noted, showing that deuterium addition was completely cis, exo in all cases. The same finding resulted from similar tests with norbornene. Thus, endo addition is not found on any of the catalysts irrespective of their different surface characteristics.

In the exchange reaction between norbornane and deuterium on palladium two hydrogen atoms at most exchange initially[10]. We carried out this reaction on catalyst A and allowed it to proceed at *ca*. 200°C until only 20% of the hydrocarbon remained unreacted. NMR analysis of

TABLE 2

Initial product distributions in exchange reactions

Reaction	Catalyst	Temp., °C	Initial products,%	
			d_5	d_6
exo-d_4-norbornane/ deuterium exchange	A	43	84	16
	B	100	81	19
	C	78	86	14
	D	114	81	19
			d_1	d_2
Norbornane/deuterium exchange	A	33	28	72
	B	57	29	71
	C	82	77	23
	D	82	35	65
			d_3	d_2
exo-d_4-norbornane/ hydrogen exchange	A	63	10	90
	C	115	72	28

the product showed[1] the exchange to have taken place predominantly at the exo positions. Since 4 is the maximum deuterium number that can arise from exchange of exo-hydrogens after several sojourns at the surface, the presence of some d_5-norbornane showed that other hydrogens had also exchanged. To identify at least partially the positions involved the exchange reaction between exo-d_4-norbornane and deuterium was examined. The presence of d_6-norbornane as an initial product (see Table 2 and remarks to follow) signified that endo-hydrogens had reacted and NMR product analysis showed that bridgehead hydrogens had also undergone exchange. Because neither NMR nor mass-spectrometry distinguish bridge hydrogens we have no information on their rate of exchange. The participation of endo-hydrogens in the exo-d_4-norbornane/deuterium reaction suggested that the rate of this exchange reaction might be usable as a measure of the ease of approach of the endo side of the norbornane molecule to surface sites; the predominance of exchange at exo relative to other positions in the norbornane/deuterium reaction suggested that the rate of exchange in this reaction could be used as a measure of the ease of approach of the exo side to site atoms.

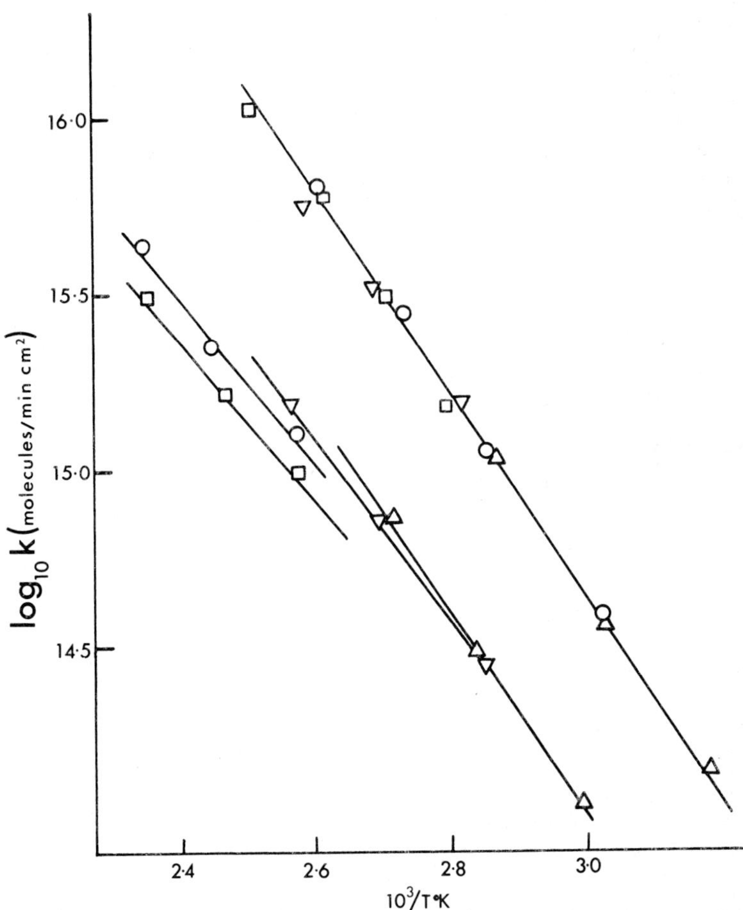

Fig. 1. Arrhenius plots for exchange reactions. Identifying symbols for catalysts: △, A; ○, B; ▽, C; □, D. Upper line is a common plot (see text) for norbornane/deuterium exchange. Other lines refer to the exo-d_4-norbornane/deuterium exchange. Activity units (legend) are appropriate to catalyst A.

Both rates were measured for catalysts A, B, C and D. The kinetics obeyed the usual first-order rate laws[11)] and Arhenius plots were linear. Fig. 1 shows the relation between the rate constants of the two reactions for catalysts tested. The representation was obtained by (i) adjusting empirically through the frequency factor the measured rate constants (k_n) for each catalyst for the norbornane exchange to bring all four sets of results to a common Arrhenius plot, and (ii) applying the same adjustment factors to the rate constants (k_r) for exo-d_4-norbornane exchange. No significant difference in the ratio of the two rates is revealed and therefore, contrary to prior speculation, no clear difference in surface-structure sensitivity between exo and non-exo exchange can be discerned. The possibility that each reaction is 'demanding'[12)] and to much the same extent (so that the ratio k_r/k_n showed little variation) may be tested by reference to estimated specific activities for both reactions. The overall spread is six-fold for k_n and ten-fold for k_r (Table 3). However, the area estimates for C and D are the most reliable and, since the specific activities of these catalysts differ by only a 2 factor, it appears that neither exchange reaction is notably crystallite-size dependent. A qualification of detail is discussed later.

The exo-d_4-norbornane/deuterium exchange gives mainly d_5 product compatible with a large contribution from bridgehead and perhaps bridge hydrogens. The relative intensity of bridgehead and endo + bridge NMR signals for a product sample (catalyst C) showed the rate of exchange of a tertiary hydrogen to be 4 times the rate of exchange of a secondary (assuming all secondary hydrogens to be equivalent). As there are 6 secondary and only 2 tertiary hydrogens, the relative chance of a molecule exchanging a tertiary to a (bridging or endo) secondary is 1.3/1. To the extent that bridgehead reactivity is marked the exchange of exo-d_4-norbornane with deuterium does not provide a fully sensitive measure of endo-face attack. A more clear-cut test might be based on measurement of protium exchange with the difficulty synthesised endo-d_4-norbornane. Efforts to prepare this hydrocarbon have so far been unsuccessful.

TABLE 3

Arrhenius parameters and specific activities for exchange reactions.[a]

Reaction	Catalyst	E_a, kcal/mole	$\log_{10}A$, A in mol/cm^2 min	[b] k, 112°C mol/min cm^2 x 10^{-14}
Norbornane/deuterium exchange	A	13	23.2	65
	B	14	23.5	38
	C	12	21.9	12
	D	14	22.9	11
exo-d_4-norbornane/deuterium exchange	A	13	22.2	15
	B	11	21.2	6.4
	C	12	21.2	2.4
	D	11	20.6	1.4

[a] average values from a number of determinations in each case.

[b] k(i.e. k_n or k_r) is the rate constant defined in Ref. 11.

In the exchange of norbornane d_2 is the main product (except with catalyst C), consistent with the predominantly exo exchange. The large d_1 found with catalyst C - which contains highly dispersed metal - could be due to either an extra large contribution from bridgehead exchange or a decreasing tendency of exo-monoadsorbed norbornane (say at C-2) to become diadsorbed (C-2, C-3). That the latter is the cause is shown by the distribution pattern obtained from the exchange of exo-d_4-norbornane with hydrogen (protium) (Table 2). Only exo positions can undergo exchange in this reaction, yet a large h_1 (that is, d_3-norbornane) is similarly detected with catalyst C. The failure to find such a pronounced change in the d_1'/d_2' (that is, d_5/d_6) ratio in the exo-d_4--norbornane/deuterium exchange may be due to the limited contribution which endo exchange makes in this reaction. Dependence on surface structure of simple to multiple exchange has been noted previously for simple alkanes[13,14].

3.2. Epimerisation experiments

Results are presented in Table 4. In the case of catalyst G linearity of the Arrhenius plot was relatively poor and derived parameters must be regarded as approximate. Reproducibility was satisfactory (±10%) with the remaining catalysts.

The present experiments do not support a crystallite-size dependence of activity, a mere four-fold spread in the specific activity k_e at 73° being found for all catalysts. In particular, F and H', which have average crystallite diameters of 31 and 250A, respectively, differ in specific activity by no more than a factor of 3. While it is true that the magnitude of the specific activity variations among the catalysts depends on the temperature chosen for the comparison (Table 4), it is nevertheless clear that at the several working temperatures k_e is not dependent on dispersion to any pronounced extent and that most of the activity variations are scarcely outside the limits of accuracy of catalyst characterisation and activity measurement.

TABLE 4

Arrhenius parameters, and specific activities at 73°C, for epimerisation of cis-1,4-dimethylcyclohexane

Catalyst	E_a, kcal/mole	$\log_{10} A$, A in mol/min cm^2 Pd	$k_e^{73°C}$, mol/min cm^2 Pd x 10^{-12}
E	18.1	24.2	4.7
F	15.3	22.3	3.8
G	15.3	21.8	1.3
H	26.8	29.1	1.2$_5$
H'	20.4	25.0	1.2

This conclusion, particularly if confirmed by results at lower hydrogen pressures, would tend to argue against the suggestion[4] that intermediates 'roll over' while still bonded to the surface. Such a process should be facilitated on low-coordination sites and, therefore, on catalysts of high dispersion. Siegel proposed some years ago[15] that intermediate olefins may desorb transiently from the surface during hydrogenation of aromatics and cycloalkenes. Rooney and his co-workers have recently suggested[3] that similar intermediates may be involved in epimerisation and related reactions of cycloalkanes, that is, that roll-over may occur with partial or complete detachment from the surface. If this is so, then surface geometry should not be as critical a factor as might be the case if the roll-over were occurring on the surface. The present results are in harmony with this description.

ACKNOWLEDGEMENTS

We thank Dr. T. Baird (Glasgow) and Dr. H. S. Inglis (I.C.I., Agricultural Division) for help with, respectively, electron microscopic measurements and carbon monoxide chemisorption determinations.

REFERENCES

1) E. McMahon and J.K.A. Clarke, Tetrahedron Letters (1971) 1413 and references therein.
2) E. O. Fischer and H. Werner, Metal π-Complexes (Elsevier) Amsterdam 1966 p. 81.
3) H.A. Quinn, J.H. Graham, M. A. McKervey and J. J. Rooney, J. Catalysis 22 (1971) 35.
4) R.L. Burwell,Jr. and K. Schrage, Discuss. Faraday Soc. 41 (1966) 215.
5) C. Kemball, Proc. Roy. Soc. (London) A 214 (1952) 413.
6) D. Pope, W. L. Smith, M.J. Eastlake and R.L. Moss, J. Catalysis 22 (1971) 72.
7) E. McMahon, P.F. Carr and J.K.A. Clarke, J. Chem. Soc. A (1971) 2012.
8) R. B. Anderson, Experimental Methods in Catalytic Research (Academic Press) New York 1968,pp 8, 19.
9) e.g. A.P. Marchand and N.W. Marchand, Tetrahedron Letters (1971) 1365.
10) R.L. Burwell, Jr., B.K.C. Shim and H.C. Rowlinson, J. Amer. Chem. Soc. 79 (1957) 5142.
11) J.R. Anderson and C. Kemball, Proc. Roy. Soc. (London) A 223 (1954) 361.
12) M. Boudart, A. W. Aldag, L.D. Ptak and J. E. Benson, J. Catalysis 11 (1968) 135.
13) R.L. Burwell, Jr. and R. N. Tuxworth, J. Phys. Chem. 60 (1956) 1043.
14) J.R.Anderson and R.J. MacDonald, J. Catalysis 13 (1969) 345.
15) S. Siegel, Adv. Catalysis 16 (1966) 123.

DISCUSSION

P. B. WELLS

What was the basis of the normalization procedure used in Figure 1? I fear that, by presenting the information in this way, you may be obscuring genuine structure sensitivity. Could you give us original (i.e. unadjusted) data for catalysts C and D for the exchange of Norbornane with deuterium? Any need "to adjust" these data is equivalent to saying that the acvitities, as measured, were dependent upon crystallite size, and hence this reaction should be considered structure-sensitive.

There is the possibility that, by concentrating on relative structure sensitivity, evidence for structure sensitivity in your individual reactions has been obscured. The data in column 4 of Table 4 may constitute another example.

J. K. A. CLARKE

If the exo-d_4-norbornane/deuterium exchange reaction gives an adequate measure of endo-hydrogen reaction rates (limitations are discussed in the text), any special dependence of the latter on edge site availability--the question of interest to us--should reflect in a variation in the specific activity. The unadjusted data for C and D, presented in Table 3, represent evidence against such special dependence (your second question). We do not know at all precisely the areas of films A and B, which are the catalysts giving the more reproducible activities. Results for A and B can, however, be included in the discussion if we look for some significant variation in the ratio of the rate of the exo-d_4-norbornane/deuterium exchange to that of the norbornane/deuterium exchange which is ~ 80% exo-hydrogen and only ~ 6% endo-hydrogen reaction. This is essentially what has been done in plotting Figure 1 (your first question). That there is no such variation through catalysts A-D confirms the conclusion reached from individual rates on C and D only. We argue, incidentally, in the text on the basis of "unadjusted" specific activities of C and D that there is no surface structure sensitivity for the greater part of the norbornane exchange (exo and bridgehead hydrogen atoms) on palladium.

The issue of relative structure sensitivity does not arise in the epimerization study. Metal areas being available in all cases specific activities and not relative reaction rates were measured for five catalysts (Table 4) and, in addition, for samples of C (not included in the paper). The four-fold--at most--spread in specific activity would appear to justify the limited conclusion we have given. Further work on this system has been carried out by Mr. T. J. Plunkett using more extended reaction conditions and this will be reported in a forthcoming publication.

Paper Number 47

EFFECT OF PLATINUM PARTICLE SIZE ON HYDROCARBON HYDROGENOLYSIS

J. R. ANDERSON* and Y. SHIMOYAMA**
*CSIRO Division of Tribophysics, University of Melbourne;
**School of Physical Science, Flinders University, Adelaide

ABSTRACT: Hydrogenolysis of n-hexane, 2- and 3-methylpentane and methylcyclopentane has been studied in the presence of excess hydrogen over ultra-thin and thick platinum film catalysts with a range of crystallite sizes. The rates of hydrogenolysis per unit platinum area decrease with increasing crystallite size. The proportion of reaction leading to hydrogenolysis as opposed to isomerization and dehydrocyclization decreases with increasing crystallite size in the range <15Å-40Å. The results are discussed in terms of the likely contributing reaction pathways.

1. INTRODUCTION

The evidence concerning the effect of metal particle size on the course of hydrocarbon hydrogenolysis reactions tends to be equivocal[1]. Thus, for instance, while Carter, Cusumano and Sinfelt[2] noted that the specific activity of nickel/silica-alumina catalysts for ethane hydrogenolysis appeared to decrease by a factor of 20 as the relative particle size increased by a factor of 4, Yates, Taylor and Sinfelt[3] reported that there was no dependence of the specific activity of nickel/silica catalysts for ethane hydrogenolysis on particle size. However, with conventional supported catalysts there is the possibility that the metal surface can vary both in structure and in degree of purity as the result of varying degrees of thermal sintering which may be used for adjustment of average particle size. It thus seemed worthwhile to study the influence of particle size by a method which eliminates surface impurity as a significant experimental variable, and in which metal particles of varying sizes could be prepared at the same temperature.

We have studied hydrocarbon reactions on thick and on ultra-thin evaporated platinum films; the latter consist of arrays of discrete metal crystals of which the average particle diameter may be readily controlled down to 15Å or so. Ultra-thin films are, in effect, models for highly dispersed conventional supported catalysts, but are prepared and used under conditions where adventitious contamination is negligible. They also have the advantage that control of the average metal particle size is easy (without the need for drastic thermal treatment), while direct study of the metal particles in the electron microscope is also easy. The thick highly sintered films were used as models to represent catalysts of very large metal crystal size. Reactions have been studied with n-hexane, 2-, and 3-methylpentane, and methylcyclopentane.

2. EXPERIMENTAL

The apparatus and general experimental details have been

described previously[1,4] for the preparation and use both of ultra-thin and of thick, highly sintered, "fringe-free" metal films.

The reaction mixtures contained hydrogen and hydrocarbon in the molar ratio 10/1 and were introduced to the catalyst at room temperature and to a total pressure of 100 torr. The total reaction volume was 1.5ℓ.

Total metal surface areas for ultra-thin films were estimated by a combination of hydrogen chemisorption and electron microscopy. Areas estimated from the chemisorbed hydrogen uptake were generally some 20% in excess of the value obtained electron microscopically. The possible reasons for this include the presence of some very small platinum particles not resolved in the electron microscope, and the transfer of chemisorbed hydrogen from the platinum to the substrate: both factors may well operate. We have adopted the hydrogen chemisorption areas as the correct values, although the conclusions we shall draw from the data remain unaffected if the electron microscope areas are used instead. The dependence of platinum particle size and total platinum area on the weight of platinum per unit substrate area, was of good reproducibility and hydrogen chemisorption measurements were only made on a fraction of the total number of film catalysts actually used. Surface area values quoted for thick film catalysts were measured by hydrogen chemisorption. In both cases, a monolayer of chemisorbed hydrogen was achieved at 10^{-2} torr at 0°C and it was assumed that 1.25×10^{15}H atoms were chemisorbed per cm^2.

3. RESULTS

Ultra-thin films were deposited variously on Pyrex glass, silica glass and mica substrates. These substrates had no effect on catalyst structure or behaviour, and we therefore feel confident that the substrate was, in all cases, inert. Moreover, no reaction could be detected in blank experiments using only bare substrate.

In all cases, measurement of reactant composition versus time showed that there was an initial fast reaction, the rate of which decreased rapidly with time until the conversion reached about 0.1-0.2%, after which a steady rate was achieved. This behaviour is no doubt due to self-poisoning of the catalyst with strongly adsorbed hydrocarbon residues, and in the subsequent parts of this paper, these are distinguished as "initial" and "steady" rates respectively. In fact, this self-poisoning to a "steady" rate is, in large measure, reversible if a mixture which is reacting under "steady" conditions is cooled to about room temperature and the reaction vessel then thoroughly pumped. One must conclude that, under these circumstances, the poisoning residues are desorbable.

The results are collected in tables 1-4. Duplicate experiments were in good agreement, and the results have been averaged in constructing the individual entries in the tables. In the main, reactions were confined to 273°C: however, some reactions with 2-methylpentane were carried out over the range 275-350°C. Using thick film catalysts it was observed that the proportion reacting by hydrogenolysis increased with increasing temperature, being greater by a factor of 5.5 at 350°C compared with 275°C. On the other hand, using ultra-thin film catalysts of 1μg cm^{-2}, the proportion of hydrogenolysis was independent of temperature over this range.

The distributions of reaction products which are also recorded in tables 1-4 were essentially constant, irrespective of whether the reaction was under "initial" or "steady" conditions. For comparison, some results obtained by Matsumoto et al.[5] for reactions over an 0.8% platinum/silica supported catalyst are contained in tables 1-3.

4. DISCUSSION

A comparison of the results obtained by Matsumoto et al.[5] clearly show that, as expected, the behaviour of a 0.8% platinum/silica catalyst closely resembles that of ultra-thin film catalysts, but is substantially different from that of massive metallic platinum (thick films).

There are important general trends to be discerned from tables 1-3 as a result of variation of platinum particle size. With n-hexane and to a lesser extent with 3-methylpentane, the proportion of hydrocarbon reacting by hydrogenolysis as opposed to isomerization and dehydrocyclization, decreases with increasing average particle size up to about 40Å: however, increasing the particle size beyond this towards massive platinum (thick films) leads to an increase in the proportion of hydrogenolysis. In other words, the proportion of hydrogenolysis is a minimum at about 40Å. With 2-methyl-pentane the increase at larger particles is not apparent.

The most striking feature in the behaviour of methylcyclo-pentane (table 4) is the very small extent to which, irrespective of catalyst structure, hydrogenolysis leads to products in the range C_1-C_5. C_6 products never make up less than 97% of the reaction. There is a small tendency for the C_1-C_5 products to be a minimum for ultra-thin films with average platinum particle sizes in the region of 40Å (reaction to C_6 products >99%). Over most of the ultra-thin film catalysts, the only C_1-C_5 products are methane and cyclo-pentane. However, over thick film catalysts, very small amounts of products with intermediate carbon numbers appear, and there is a tendency in this direction with the thickest of the ultra-thin films (2.5 μg cm^{-2}).

The fact that the product distributions were much the same irrespective of whether the reaction was occurring in "initial" or "steady" conditions, leads to the conclusion that self-poisoning occurs generally over the catalysts surface, and is not confined to specific types of sites.

The product distributions are consistent with the assumption that, in the main, only one C-C bond is broken in any given molecule. This assumption has been verified previously[7] for hydrogenolysis of aliphatic hydrocarbons on platinum.

The dependence of hydrogenolysis rate on particle size is also evident from tables 1-4. In all cases, there is a trend for the rate per unit platinum surface area to decrease with increasing particle size. This trend holds both for the

TABLE 1. HYDROGENOLYSIS OF n-HEXANE AT 273°C

	1	2	3	4	5	6	hydrogenolysis reaction products (mole %)				
							CH_4	C_2H_6	C_3H_8	$n-C_4H_{10}$	$n-C_5H_{12}$
UTF*, 0.020	<15	-	1.2	16	-	-	11.7	14.5	45.1	15.6	13.1
" 0.060	15	17.5	1.7	14	4.0	0.48	12.1	15.7	41.6	16.1	14.1
" 0.080	20	29	1.0	11	2.2	0.22	13.7	16.3	41.1	14.2	14.7
" 0.125	20	46	3.4	9.4	1.45	0.18	13.4	16.9	38.3	18.0	13.4
" 0.25	36	108	3.6	6.0	0.55	0.073	16.0	19.2	33.6	18.4	12.8
" 0.50	38	130	0.9	5.4	0.53	0.060	12.7	15.7	41.2	19.6	10.8
" 1.0	43	163	4.0	6.0	0.35	0.037	13.4	14.7	40.0	17.2	14.7
" 2.5	58	264	1.6	7.9	0.35	0.045	15.2	15.2	40.6	15.9	13.1
thick film	-	300	0.6	22**	0.37	0.039	24.9	11.4	28.5	11.4	23.8
0.8% Pt/SiO_2 †	-	-	1.8	7.8	-	-	12.6	18.2	42.8	15.1	11.3

TABLE 2. HYDROGENOLYSIS OF 3-METHYLPENTANE AT 273°C

	1	2	3	4	5	6	hydrogenolysis reaction products (mole %)				
							CH_4	C_2H_6	$n-C_4H_{10}$	$n-C_5H_{12}$	$i-C_5H_{12}$
UTF*, 0.125	20	46	1.0	17	15	2.2	37.3	9.7	11.8	9.0	32.2
" 0.60	40	145	1.4	4.1	13	1.7	35.9	12.8	15.4	7.7	28.2
" 2.5	58	264	1.1	4.1	8.8	1.3	27.6	18.4	18.4	6.6	29.0
thick film	-	300	2.4	7.8	6.1	0.84	28.1	23.7	22.6	4.4	21.2
0.8% Pt/SiO_2 †	-	-	2.9	9.4	-	-	28.7	22.2	20.1	6.4	22.6

1. average platinum crystallite diameter (Å)
2. total platinum surface area (cm^2)
3. maximum conversion (%)
4. proportion (%) reacting by hydrogenolysis
5. initial hydrogenolysis rate (10^{12} molec $sec^{-1} cm^{-2}$)
6. steady hydrogenolysis rate (10^{12} molec $sec^{-1} cm^{-2}$)

* ultra-thin film, density in μg cm^{-2}
** values of >20% were found previously on sintered platinum films[1)]
† Matsumoto et al.[5)]

TABLE 3. HYDROGENOLYSIS OF 2-METHYLPENTANE AT 273°C

	1	2	3	4	5	6	hydrogenolysis reaction products (mole %) CH_4	C_2H_6	C_3H_8	i-C_4H_{10}	n-C_5H_{12}	i-C_5H_{12}
UTF*, 0.02	<15	-	0.7	36	-	-	15.7	17.6	27.4	15.7	7.9	15.7
" 0.04	<15	-	0.6	19	-	-	16.1	16.1	29.0	16.1	6.6	16.1
" 0.08	20	29	0.8	5.8	29	5.2	19.0	16.2	24.8	16.2	8.6	15.2
" 0.5	38	130	1.4	3.5	14	1.9	17.0	17.0	29.7	15.6	7.4	13.3
" 1.0	43	163	2.1	9.0	13	1.7	16.7	16.7	33.2	16.7	5.6	11.1
" 2.5	58	264	3.9	5.2	8.5	1.3	20.6	18.6	29.4	15.7	6.9	8.8
thick film	-	300	2.2	7.5	5.0	0.5	11.8	16.5	41.9	18.6	6.5	4.7
0.8% Pt/SiO_2†	-	-	2.6	5.4	-	-	22.9	11.9	32.1	13.8	10.1	9.2

TABLE 4. REACTION OF METHYLCYCLOPENTANE AT 273°C

						reaction products (mole %)												
						C_1-C_5 products								C_6 products**				
	1	2	3	5	6	CH_4	C_2H_6	C_3H_8	n-C_4H_{10}	i-C_4H_{10}	n-C_5H_{12}	i-C_5H_{12}	c-C_5H_{10}	2-C_6H_{14}	3-C_6H_{14}	n-C_6H_{14}	c-C_6H_{12} + C_6H_6	7
UTF*, 0.02	<15	-	1.4	-	-	2.8	-	-	-	-	-	-	2.2	45.8	18.8	23.5	6.9	2.4
" 0.08	20	29	1.7	26	6.1	3.9	-	-	-	-	-	-	2.6	47.2	17.4	28.9	-	3.0
" 0.60	40	145	2.9	14	1.7	<1†	-	-	-	-	-	-	0.3	57.6	21.2	18.0	2.9	~0.3
" 1.0	43	163	1.0	14	2.4	<1†	-	-	-	-	-	-	0.6	57.6	23.0	18.4	0.4	~0.6
" 2.5	58	264	1.9	-	-	1.4	0.8	1.1	-	-	-	0.5	0.7	51.6	20.4	20.2	3.4	2.3
thick film	-	300	1.6	8.4	2.2	~1†	1.1	1.1	0.7	0.4	0.2	0.2	0.3	61.3	32.2	1.8	-	2.5

1. average platinum crystallite diameter (Å)
2. total platinum surface area (cm^2)
3. maximum conversion (%)
4. proportion (%) reacting by hydrogenolysis
5. initial hydrogenolysis rate (10^{12} molec $sec^{-1} cm^{-2}$)
6. steady hydrogenolysis rate (10^{12} molec $sec^{-1} cm^{-2}$)
7. proportion (%) reacting to C_1-C_5 products

** also about 0.5% 2,3-dimethyl butane
† methane analysis inaccurate at <1%
* ultra-thin film, density in μg cm^{-2}

"initial" and "steady" rates to about the same relative extents.

We turn now to consider hydrogenolysis pathways, and possible effects of catalyst structure upon them. In a mechanistic sense, hydrogenolysis is a carpet-bag term, and it is already recognised that more than one mechanism can occur[6)]. On platinum, three mechanisms can be distinguished[6)] involving (a)a 1-2 adsorbed intermediate; (b) a 1-3 adsorbed intermediate; (c)a π-adsorbed intermediate. The evidence for (a) and (b) is the observed reaction from ethane and neo-pentane, since if these are adsorbed at two carbon atoms, this is only possible in the 1-2 and 1-3 modes respectively. On thick platinum films the activation energy for hydrogenolysis is very much larger for ethane (57 kcal $mole^{-1}$) than for larger hydrocarbons ($\ngtr$ about 20 kcal $mole^{-1}$), and ethane reacts in a considerably higher temperature range (cf. ref.6). Inasmuch as the C-C bond energy in ethane is not more than a few kcal $mole^{-1}$ different from that in other aliphatic hydrocarbons, it is reasonable to conclude that mechanism (a) will not be of much importance in hydrogenolysis reactions over platinum for hydrocarbons other than ethane. On the other hand, mechanism (b) certainly operates with other hydrocarbons in addition to neopentane, as judged from the equality of the kinetic parameters[7)] and the latter also shows that the 1-3 adsorbed species is the same as that involved in skeletal isomerization by bond shift. The slow step is the formation of a bridged intermediate (B) from the 1-3 adsorbed precursor (A), followed by hydrogen attack on (B)[6,7)].

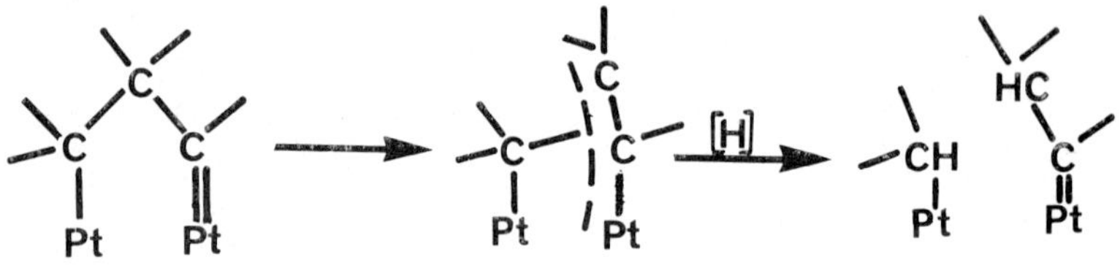

A B C

Further hydrogen attack then results in the desorption of the species in C. We shall refer to this as 1-3 hydrogenolysis The process requires two nearest-neighbour surface platinum atoms and the formation of the bridged intermediate is energetically favoured by partial electron transfer from the hydrocarbon to the metal. Thus, one expects it to occur preferentially on low index surface planes[7)].

There is good evidence, based on steric considerations, that there must be at least one other important mechanism possible. It has already been noted[6)] that ring opening in molecules such as cyclopentane and cyclohexane cannot proceed by 1-3 hydrogenolysis. Furthermore, in the reaction of 3-methylpentane in the present work, the formation of isopentane (a major product) cannot be accounted for by 1-3 hydrogenolysis. Simple ring opening from cyclopentane or cyclohexane can be formulated as the reverse of ring closure, and mechanisms for this have been suggested by Shephard and Rooney[8] based on analogous processes in organometallic chemistry[9,10] and rather similar processes have been suggested by Barron et al.[11] for ring closure and opening during hydrocarbon isomerization reactions. Although these various possibilities differ somewhat in detail, they all involve the use of π-allyl and/or π-olefin adsorbed species with the catalytically active site consisting of a single platinum atom. This latter concept is also supported by a recent study[1] of the effect of catalyst structure on n-hexane dehydrocyclization and isomerization.

However, there is no reason to believe that a π-allyl or π-olefin mechanism for C-C bond rupture on platinum is limited to ring opening. Rather it should be a general process occurring in straight and branch-chain aliphatics as well. For the purpose of illustration we use the essentials of Shephard and Rooney's original proposals, which in general form may be represented as in reactions (2) or (3).

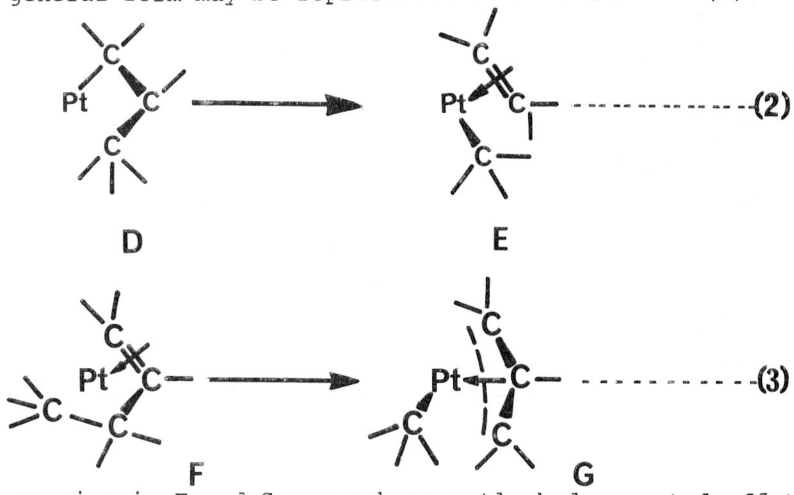

D → E ------------ (2)

F → G ------------ (3)

The species in E and G are subsequently hydrogenated off the surface. We shall refer to mechanisms such as (2) or (3) as π-olefin/allyl hydrogenolysis.

We consider first the behaviour of methylcyclopentane. To a good approximation only π-olefin/allyl hydrogenolysis occurs since the reaction is very largely confined to ring opening for which, for steric reasons, 1-3 hydrogenolysis cannot operate. Thus, the falling reaction rate with increasing platinum crystallite size must refer to this reaction path alone. With n-hexane, and 2- and 3-methylpentane, we consider that both reaction paths operate. The overall hydrogenolysis rate decreases with increasing crystallite size and it is reasonable to infer that, as with methylcyclopentane, the rate of the π-olefin/allyl pathway decreases in the same way.

Our data are insufficient, relative to the obvious

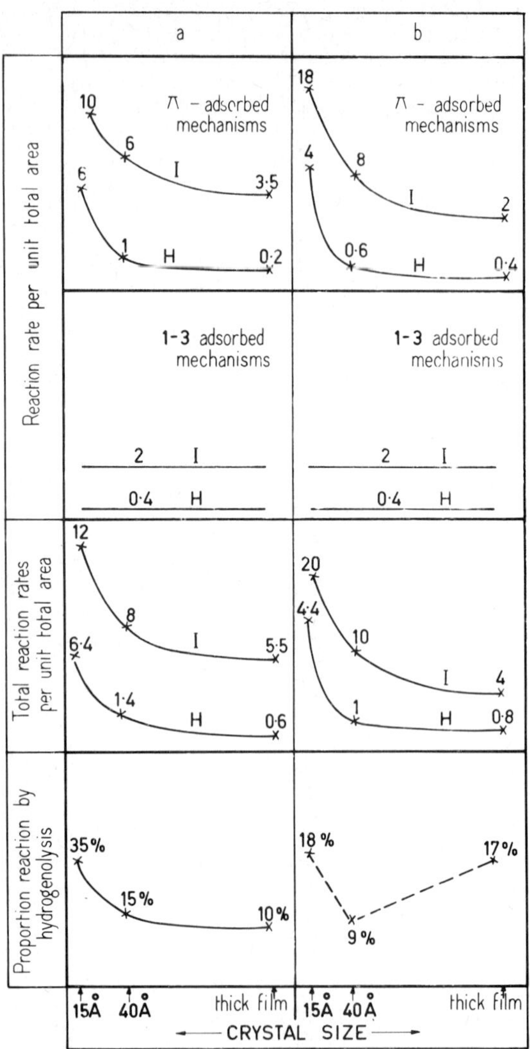

Fig.1 Schematic representation of the variation of individual reaction rates with platinum crystallite size. a, (left) corresponds to the type of behaviour of 2-methylpentane; b, (right) to n-hexane.

complexity of the situation, to permit the construction of a model in quantitative detail.However, we show in fig. 1 a scheme which reproduces the main features of the observed results in a semi-quantitative way. Fig.1a (left) corresponds to the type of behaviour observed with 2-methylpentane, while that in 1b (right) corresponds to n-hexane.

The numbers given against various points on the curves represent assumed rate values on an arbitrary scale. There obviously are variations possible while remaining consistent with the known data. In particular, we have chosen to assume that the rate of the 1-3 hydrogenolysis pathway is independent of particle size. There are, in fact, no relevant data on this point. Such data could be obtained from a study of neopentane hydrogenolysis, but this has not yet been done. It is worthwhile noting that, provided the crystallites are geometrically ideal, the proportion of face atoms in the surface of crystallites of size >15Å is high (>~80%) and so the proportion of surface available for 1-3 hydrogenolysis would not change very much as the size increases above 15Å. On the other hand, electron removal is expected to be easier with very small crystallites than with larger ones, and one would expect partial electron transfer from a 1-3 adsorbed species to the platinum to occur to a lesser extent with very small crystallites, so that reaction by a 1-3 mechanism may well have a lower intrinsic rate over very small crystallites.

There are some ancillory factors with which the representation such as fig.1, must remain consistent. These include: (a) judged from the behaviour of neopentane[7)], the proportion of reaction by a 1-3 adsorbed mechanism to hydrogenolysis is not high (~20%) compared with isomerization; (b) isomerization via an adsorbed cyclic intermediate (i.e. involving π-adsorbed processes) with these hexanes is never trivial compared with isomerization by a 1-3 adsorbed mechanism (bond shift) and on very highly dispersed platinum reaction via an adsorbed cyclic intermediate is dominant[13)]; one would expect a similar relative importance for hydrogenolysis by the two paths. The relative importance of reaction by an adsorbed cyclic intermediate is also emphasized by the high proportion of methylcyclopentane formed from hexanes, particularly over ultra-thin platinum films[1)].

Despite the fact that the representations in fig.1 are essentially schematic it is impossible to avoid the conclusion that the rates per unit total area of both hydrogenolysis and isomerization via π-adsorbed species decrease rapidly with increasing platinum particle size. It was previously suggested [1)] that, in these reactions, for a single platinum atom to act as a catalyst site it should preferably (but not necessarily) be of low co-ordination to other platinum atoms, for instance as a corner atom in a crystallite. The proportion of corner atoms decreases rapidly as the crystallite size increases: the exact behaviour depends on the assumed crystallite geometry, but assuming geometrically ideally shaped crystallites, the proportion of surface atoms existing in corner positions falls from the region of 20% to about 1% or so, for an increase in crystallite size from roughly 15 to 40Å. It is obviously tempting to associate the fall in reaction rate via π-adsorbed mechanisms with the fall in the proportion of corner atoms. In fact the above decrease in the proportion of corner atoms with increasing size is much more severe than the decrease in rate (per unit area). One would expect real crystals not to be of ideal geometry so that, having a much higher proportion of atoms in low co-ordination positions, their proportion would decrease much less rapidly with increasing size than would be the case for ideal crystallites. In addition, one cannot discount the possibility that the intrinsic rate (i.e. rate per site) changes with particle size.

REFERENCES

1. Anderson, J.R., Macdonald, R.J. and Shimoyama, Y. J. Catal. 20 (1971), 147.
2. Carter, J.L., Cusumano, J.A. and Sinfelt, J.H. J. Phys. Chem. 70 (1966), 2257.
3. Yates, D.J.C., Taylor, W.F. and Sinfelt, J.H. J. Amer. Chem. Soc. 86 (1964), 2296.
4. Anderson, J.R. and Macdonald, R.J. J. Catal. 19 (1970), 227.
5. Matsumoto, H., Saito, Y. and Yoneda, Y. J. Catal. 19 (1970), 101.
6. Anderson, J.R. and Baker, B.G. In "Chemisorption and Reactions on Metallic Films: (J.R. Anderson ed.) Academic Press, London, 1971, Vol.2, p.63.
7. Anderson, J.R. and Avery, N.R. J. Catal. 5, (1966), 446.
8. Shephard, F.E. and Rooney, J.J. J. Catal. 3 (1964), 129.
9. Wilke, G. Augew. Chem. Intern. Ed. 2 (1963), 105.
10. Fisher, E.O. and Werner, H. Augew. Chem. Intern. Ed. 2 (1963), 80.
11. Barron, Y., Maire, G., Muller, J.M. and Gault, F.G. J. Catal. 5 (1966),428.
12. van Hardeveld,R. and Hartog, F. Surface Sci. 15 (1969), 189.
13. Gault, F.G. private communication.
    Corolleur, C. Ph.D. thesis, University of Caen, 1969.

DISCUSSION

YU. I. YERMAKOV

As this session was devoted mainly to the problem of the structure sensitivity in hydrogenation, I should like to use this chance to communicate a new method of preparing catalysts, containing very high dispersed metal on support. This method is based on interaction of metallorganic compounds of transition metals with the surface hydroxyls of supports. In the case of bis-π-allyl nickel this interaction proceeds according to the following main equation:

$$\underset{/|\backslash}{\overset{OH}{\overset{|}{Si}}} + Ni(C_3H_5)_2 \rightarrow \underset{/|\backslash}{\overset{ONiC_3H_5}{\overset{|}{Si}}} + C_3H_6$$

During reduction of such catalysts by hydrogen, the surface metallorganic compound decomposes with the formation of zero-valent nickel:

$$\underset{/|\backslash}{\overset{ONiC_3H_5}{\overset{|}{Si}}} + H_2 \rightarrow \underset{/|\backslash}{\overset{OH}{\overset{|}{Si}}} + Ni^\circ + C_3H_6$$

The catalysts obtained by this procedure have the following peculiarities:

1) All nickel atoms in these catalysts are accessible for oxygen chemisorption, and during chemisorption the mole ratio O:Ni is unity; this ratio was conserved in spite of the broad change of nickel concentration in catalysts (from 0.2 to 20%) and temperature of chemisorption (from 20 to 400°C).

2) These catalysts have a very low degree of magnetization in comparison with the samples obtained by conventional techniques (by decomposition of $Ni(NO_3)_2$ on SiO_2). When this nickel content was 3.5% the degree of magnetization (after subtraction of magnetization which was caused by impurities of iron oxide in the carrier) was ~ 0.3 ($gs \cdot cm^3/g$ Ni) at a magnetic field strength ~ 8000 Oersted. This corresponds to the case where there are nickel species in the catalyst with average magnetic moment ~ 2.2 m.B.

The data on activity in benzene hydrogenation are given in Table 1.

TABLE 1

Benzene hydrogenation at 100° by catalysts

Exp. N°	Content of Ni, wt.%	Specific surface per 1 g of Ni	Average particle size, Å	Reaction Rate	
				$g\ C_6H_6$ / g Ni·hour	$g\ C_6H_6$ / 1 m^2 Ni·hour
1	3,4	650	< 10 Å	370	0,55
2	7,2	650		350	0,50
3	9,0	650		350	0,50
4	9,2	110	~ 50 Å	25	0,22

$SiO_2 - Ni(C_3H_5)_2$ (N°1,2,3); $SiO_2(NO_3)_2$ (N° 4)

The catalytic activity does not depend on nickel concentration. The total catalytic activity of 1 g of Ni (see lines 1-3 in table) was more than 1 order higher than in conventional catalysts (see line 4), but there was little difference in the specific activity (per 1 m^2 of Ni) of two catalysts.

J. R. ANDERSON

In my view, the chance of obtaining anything other than a very low concentration of zero valency, single nickel atoms (Ni°) in a supported nickel catalyst, is very small. Atoms Ni° will be thermodynamically unstable with respect to the formation of larger aggregates. For instance, the equilibrium energy of attraction between the atoms in a pair cluster of transition metal atoms is roughly one-half the heat of sublimation,[1)] and this is ample to ensure that if one had equilibrium between pairs and dissociated pairs, the proportion of the latter at (say) 700°K will be quite negligible. The question then arises if equilibrium is likely in practice. The removal of Ni° by cluster formation requires the surface diffusion of Ni° atoms through a sufficient distance. A simple example is indicative of what might be expected, but is not conclusive. If nickel atoms really were present as Ni°, their interaction with the substrate would only involve van der Waals forces, and an interaction energy of no more than 8-10 kcal g. atom^{-1} seems likely by comparison with other known data.[1)] From this one would expect an activation energy for surface diffusion on Ni° on a silica substrate of perhaps 3 kcal g. atom^{-1}. If one then assumes a normal value for the pre-exponential factor in the equation relating the diffusion coefficient to temperature, one obtains an estimate for the diffusion coefficient of Ni° on silica at (say) 700°K of the order of 10^{-3} cm^2 sec^{-1}. From this it would follow from the Einstein relation that the RMS migration distance in (say) 10 minutes at 700°K would be many times the value required for complete aggregation.

It is, of course, always possible that some Ni° atoms remain trapped in some way at special surface site, but it would also seem reasonable to believe that the concentration of such special sites would not be large.

So little is known about the properties of very small clusters of transition metal atoms with respect to either reaction with oxygen or to magnetic properties, that I would hesitate to accept the data cited by Dr. Yermakov as being necessarily reasonable evidence for the existence of Ni° as the major species.

1) Geus, J. W., in "Chemisorption and Reactions on Metallic Films" (J. R. Anderson, ed.) Vol. 1, p. 129, Academic Press, London, 1971.

P. TETENYI

I want to refer to our earlier investigation[1)] concerning the exchange of methane with tritium and the tritiation of ethylene and propylene on platinum. Two different platinum catalysts with mean particle size 80 and 700 Å were used. A significant difference in the behavior of these to catalysts was observed. The rate of exchange as well as the hydrogenation rate was substantially higher in the case of the catalyst with smaller particle size. This is in accordance with the data presented in the paper. We have stated, however, that the quantity of adsorbed hydrogen and methane was substantially higher in the case of the catalyst with the larger particle size. The initial hydrogen adsorption heats were 49 kcal mole^{-1} for large particles and 34 kcal mole^{-1} in the case of small particles.

I think that the effect can be explained by inhibition of the reaction with hydrogen, as was found in the case of hydrogenolysis of ethane, propane and some of the hydrocarbons on nickel.[2),3)]

1) L. Guczi, P. Tetenyi: Acta Chim. Hung. 71, 341 (1972).
2) L. Guczi, A. Sarkany, B. S. Gudkov, P. Tetenyi, J. Catalysis 24, 187 (1972).
3) L. Guczi, A. Sarkany, P. Tetenyi, Proc. this Congr., paper No. 78.

J. R. ANDERSON

I wonder if the differences in the initial heats of hydrogen adsorption quoted by Dr. Tetenyi for platinum blacks with mean particle sizes of 80 and 700 Å, might have originated, at least in part, from differences in surface cleanliness, since they appear to have been subjected to different temperatures for hydrogen reduction.

F. G. GAULT

In this important paper, where very difficult techniques are used, Professor Anderson gives a further proof of the existing relationship between the metal particle size and the relative importance of the various reaction mechanisms.

However I disagree with several points presented in his conclusions. First I do not believe that the ring opening of methylcyclopentane is so simple as presented in the paper. Very good evidences have been provided that two at least and probably three different mechanisms contribute to this reaction.[1)]

One of these mechanisms corresponds to an entirely selective hydrogenolysis of the CH_2-CH_2 bonds and occurs only on the catalysts with very large metal crystallites (higher than 150 Å).

Another mechanism, taking place on small crystallites, corresponds to an equal break of any cyclic bond and can be explained by the reaction scheme 2 of the paper. However the direct rupture of an $\alpha\beta$-diadsorbed species, demonstrated in the case of the ring opening of cis and trans disubstituted cyclobutanes, could as well explain the nonselective hydrogenolysis of cyclopentane molecules.[2)]

I believe then that the situation in the case of methylcyclopentane hydrogenolysis and ring opening is very complex and should not be over simplified.

The second point of disagreement concerns the statement that the presence of a high percentage of methylcyclopentane in the product distribution means automatically an important contribution of the cyclic mechanism of isomerization. That might be and that might not be; only the tracer techniques allow one to give the correct answer: In the case of the "thick" films, the results obtained in the isomerization of n-hexane-2-^{13}C show that the experiments where large amounts of methylcyclopentane are formed are connected with a predominance of the bond shift mechanism, while the experiments giving a typical cyclic mechanism distribution correspond to a very small yield of methylcyclopentane. This result may be easily explained if one considers the relative rates of desorption of the cyclic intermediates and of the dehydrocyclization/ring opening process. When the rate of methylcyclopentane is large enough, the bond shift isomerization becomes predominant, and that is the case indeed of the heavy films.[3)]

In the case of ultrathin films, however, preliminary work done in collaboration between the Australian and the French teams seems to indicate that the cyclic mechanism is the main reaction path, and methylcyclopentane is also large.

I wish thirdly to comment about the nature of the sites responsible for the different mechanisms. We were able to prepare a platinum-alumina catalyst with a high content of platinum (10%) and with a high dispersion (mean size of the crystallites:40 Å). On such a catalyst, the unselective type of ring opening of the methylcyclopentane takes place--Now if this catalyst is heated during several hours in a flow of hydrogen at 500°C, the size of the crystallites grows up to 200 Å, the overall rate decreases, but the selectivity remains constant.[4)]

This experiment shows that the regular crystal faces are not directly involved in the catalytic process and does not fit either with the picture of the active sites being corners and edges of cubo-octahedral crystallites.

It seems that two types of sites (or more) do exist on the catalyst, associated with the various reaction mechanisms, and that the relative importance of these sites, whatever they are, is determined by (or related to) the size of the crystallites appearing virtually on the surface during the early stage of the preparation.

1) G. Maire, G. Plouidy, J. C. Prudhomme and F. G. Gault, J. Catalysis 4, 556 (1965).
2) G. Maire, F. G. Gault, Bull. Soc. Chim. (France), 894 (1967).
3) J. M. Muller, Ph.D. Thesis, Caen 1969.
4) D. Juttard, C. Corolleur, J. M. Muler, G. Maire, F. G. Gault, J. of Catalysis (in press).
D. Juttard, Ph.D. Thesis, Caen 1971.

J. R. ANDERSON

I agree with Professor Gault that, in principle, more than one ring-opening mechanism is possible with methylcyclopentane. To reiterate; I believe the following processes for C-C bond rupture have to be considered: (a) 1-3 hydrogenolysis; (b) 1-2 hydrogenolysis; (c) π-olefin/allyl hydrogenolysis. As pointed out in the paper, we ruled out 1-3 hydrogenolysis on steric grounds, and this is beyond question for all ring C-C bonds not adjacent to the methyl substituent. In fact, it is possible to put methylcyclopentane into something approaching the geometry for 1-3 hydrogenolysis of the ring C-C bond adjacent to the methyl substituent, although the structure is rather misalined compared to the structure possible from (say) isobutane where a much greater degree of free rotation is possible: this would involve forming a double bond between the methyl substituent and a surface platinum and on bond energy grounds might be expected to be more difficult than double bond formation at a CH_2 group. A 1-3 adsorbed structure in the other sense, that is with a double bond from ring carbon number two to a surface platinum is sterically disallowed, so demethylation of methylcyclopentane by this means should be impossible. The results over thick platinum film catalysts where a 1-3 mechanism should operate if at all, do show that demethylation is virtually negligible (0.3% cyclopentane), and the extent of ring opening adjacent to the substituent is somewhat greater (1.8% n-hexane) however, even the latter is a quite minor process over this catalyst.

In judging the likelihood of 1-2 hydrogenolysis, the following points need to be considered.

In the case of ethane hydrogenolysis where this reaction mode is mandatory, hydrogenolysis is much more difficult over platinum than with larger hydrocarbons where alternative hydrogenolysis pathways are possible. Furthermore, in the hydrogenolysis of neohexane over platinum catalysts, carbon-carbon bond rupture within the ethyl group is very difficult compared to the other bonds in the molecule.

It would be very illuminating to study hydrogenolysis with 1,1,3,3-tetramethylcyclopentane, since the bond C_4-C_5 in this molecule could only be ruptured by some sort of 1-2 hydrogenolysis mechanism. The reverse of this reaction, that is ring closure by dehydrocyclization from 2,2,4,4-tetramethylpentane has recently been studied by Gault and Muller[1] and processes involving either one or two platinum atoms are possible, but no π-olefin/allyl mechanism can operate because of blockage by the two quarternary carbon atoms. However, the fact that this closure is roughly as facile as others in which these steric limitations do not occur, does not necessarily mean that ring opening will also be easy. In fact, I believe the chances are that ring opening in this molecule will prove to be relatively difficult.

I do not believe that the earlier work on methylcyclopentane hydrogenolysis to which Professor Gault refers, of itself <u>proves</u> that more than one mechanism operates. In fact, what was done in this previous work was to assume that more than one mechanism operated, and then to show that the observed product distributions were generally consistent with this assumption. However, one can just as well assume only a π-olefin/allyl mechanism to operate, but at the same time to propose that the position of ring opening is dependent on the interplay between the influence of a ring substituent and the nature of the catalyst site, and in this interplay, both steric and electronic factors would doubtless operate. This is the position adopted in the present paper, and it does seem to offer some economy of hypothesis compared to the alternative of assuming more than one basic mechanism. I should add, however, that on other metals than platinum, other mechanisms may well be important.

I agree with the substance of Gault's second point, to the effect that the presence of a high proportion of methylcyclopentane in the products from the reactions of the hexanes over platinum may not necessarily mean that the dominant pathway for isomerization lies via an adsorbed carbocyclic intermediate. The important factor is whether desorption as methylcyclopentane is extensive enough grossly to disturb the surface concentration of adsorbed carbocyclic intermediate. As Gault points out the results obtained to date to illuminate this point are contradictory and more work is needed. However, the main theme of the present paper will remain essentially unaltered by whatever finally emerges.

Gault's third point emphasizes that one has to be very careful in attempting to correlate platinum crystallite size with the details of metal surface topography. Provided one is operating above the size limit where variations in purely electronic factors become important (and this limit is certainly < 40 Å), it is the surface topography and not the particle size as such which is the important factor. In other words, there is no guarantee at all that the metal crystallites in a dispersed catalyst will be geometrically ideal, and this may be particularly the case when particles have been generated by aggregation of smaller ones. One can readily envisage a situation where one has two catalysts of substantially different average particle size but where, for instance, the porportions of surface metal atoms of low co-ordination are much the same.

1) Muller, J. M. and Gault, F. G., J. Catal. 24, 361 (1972).

L. GUCZI

In your recent paper you compared the behavior of two films of different concentration. Here one can find a decrease of selectivity (characterized by isomerization to hydrogenolysis) if you compare the reaction of n-hexane on ultrathin film to that on thick film. In the present paper, however, as the concentration of Pt is decreasing the importance of hydrogenolysis is also diminished. Is there any explanation for this?

J. R. ANDERSON

First it should be pointed out that somewhat different definitions for the selectivity factor were used by Anderson, Macdonald and Shimoyama[1)] and in the present paper by Anderson and Shimoyama. In the earlier paper[1)] the selectivity was defined as the ratio of the number of hexane molecules reacting to C_6 products (I) to the number reacting to give C_1-C_5 products (H); that is, the ratio I/H. In the present paper, the selectivity has been defined as H/(I+H). Provided the same definition is used for the selectivity, and provided comparison is made for catalysts of similar density or structure, the data in the two papers are in agreement. Thus, using the definition of the selectivity of H/(I+H) (that is, the definition used in the present paper), we find that for an ultra-thin film of 0.25 $\mu g\ cm^{-2}$ the data in the previous paper[1)] for the reaction of n-hexane gives a selectivity of 5.5%, compared with the value of 6.0% given in the present paper for an ultra-thin film catalyst of similar density. Again, from the previous paper the selectivity for n-hexane reaction over thick films is about 25-30%, compared with the figure of 22% given in the present paper.

It should be noted that in the present paper the reactions were studied over ultra-thin films down to much lower film densities than had been quoted in the previous work, and it is with these extremely low density films ($<0.25\ \mu g\ cm^{-2}$) where increasing particle size results in decreasing values for the selectivity (cf. Table 1) with n-hexane reaction.

1) Anderson, J. R., Macdonald, R. J. and Shimoyama, Y., J. Catal. 20 (1971), 147.

W. M. H. SACHTLER

From the hydrogenolysis product distribution in Table 2 the 0.8% Pt/SiO_2 catalysts appears intermediate between the thick film and the 58 Å catalyst, but drastically different from the 20 Å catalyst. As the Pt particle size in 0.8% Pt/SiO_2 is presumably < 20 Å, the question arises whether particle size is the only cause of the different catalytic performance of the samples studied. Supported catalysts and evaporated films are not simple mechanic mixtures of a metal powder and a support, but the metal-support interaction is strong. For metal-organic complexes it is well known that the electron energy levels of the metal atom are markedly shifted by the ligands. Similarly, chemisorption studies on alloys have revealed that the metal-adsorbate bond strength can depend significantly on the nature of the atoms adjacent to the adsrobing metal atom. It therefore seems not unreasonable to expect that the adsorptive and catalytic behavior of a Pt atom in the surface of a Pt crystal differs from that of a Pt atom on a SiO_2 surface. While negligible for thick films this support effect might be significant for very thin films and SiO_2 supported metals. Does the fact that you prefer to analyze your results in geometric terms only imply that you have reasons to exclude the possibility that metal-support interaction be another cause for the observed differences in catalytic behavior?

J. R. ANDERSON

It is impossible to eliminate with complete certainty the factor raised by Dr. Sachtler, that is that catalytic sites of special reactivity exist in the region of the metal particle near its junction with the support. Perhaps two points may be made. First, the catalytic behavior of our ultra-thin films appears to be largely independent of the substrate, at least for substrates of pyrex glass, silica glass and mica. Second, there is a good deal of evidence which has recently been summarized by Geus[1)] to the effect that, provided one is operating under conditions where the formation of compounds such as metal oxide can be neglected, the interaction between a metal particle and a support such as silica is essentially physical in nature, that is, it involves van der Waals forces only. In this circumstance, the degree of perturbation of the chemical properties of the metal atoms in contact with the support would be quite small. We believe this to be the situation with ultra-thin films, but it is also certainly possible that a conventional supported platinum catalyst has a somewhat oxidized metal-silica interface, despite hydrogen reduction.

The other point raised by Dr. Sachtler concerns the properties of Matsumoto's 0.8% Pt/silica catalyst. In the course of preparation, this catalyst had been heated to 550°C for 4 hours, and I would certainly not feel confident to predict that it would have had an average platinum particle diameter < 20 Å. By way of comparison for instance, a 2.5% Pt/silica catalyst prepared (like Matsumoto) by impregnation, and reduced at 210°C is known by direct electron microscopic examination to have an average platinum particle diameter of about 45 Å. However, the result will certainly be much dependent on the support structure, platinum loading and on the preparative details, and all that can be said with any reliability is that Matsumoto's catalyst functioned as expected, in roughly the same was as our ultra-thin film catalysts, but markedly differently from thick film catalysts.

1) Geus, J. W., in "Chemisorption and Reactions on Metallic Films" (J. R. Anderson, ed.) Vol. 1, p. 129, Academic Press, London, 1971.

D. J. C. YATES

With regard to the comment by Dr. Anderson that he knew of no data on existence of isolated single metal atoms on supported catalysts, there are some data that show dispersions approaching this value. Dr. Prestridge and I have been able to photograph individual atoms of rhodium in a 1% Rh on silica catalyst (Nature 234, 345, 1971). In most cases the atoms were clustered together in a ring of 5 or 6 atoms, with spacings between the atoms of 3.5 to 4 Å. The size of the atoms were measured to be 2.7 ± 0.2 Å; the accepted metallurgical value is 2.69 Å. We have found pairs of atoms also, but the contrast problem becomes more severe as the number of atoms in a cluster decreases. Similar observations have been made with other metals of Group VIII.

J. R. ANDERSON

I am grateful to Dr. Yates for reminding me of this extremely elegant electron microscopic work which I omitted to mention in my reply to Dr. Yermakov. As I understand it, and from the context in which it was made, the implication of Dr. Yates' comment is that if clusters down to a size of only two atoms are known to be present, then surely some single atoms can also be expected. I would not attempt to claim that the density of single atoms is absolutely zero. My contention is only that their concentration is likely to be small, and if atom mobility over the substrate is high enough for the cluster distribution to approach equilibrium, their concentration will be extremely small indeed.

N. G. SAMMAN

According to Professors Anderson and Gault the essential intermediate for the bond-shift rearrangement of alkanes and cycloalkanes, e.g. neopentane to isopentane, is $\alpha\alpha\gamma$-triadsorbed species attached to two adjacent metal atoms, as in the 111 face of Pt.

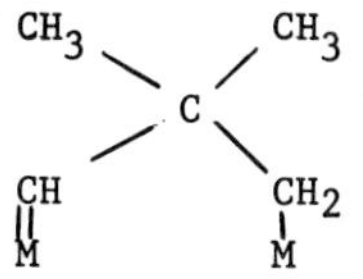

CH_3 CH_3
C
CH CH_2
‖ |
M M

We have investigated the isomerizations of compounds such as bicyclo (2,2,1)heptane, bicyclo(3,2,2)octane, and protoadamantane on Pd and Pt catalysts and find that the ease of the bond-shift rearrangement increases in that order, analogous to Carbonium ion rearrangements. Thus protoadamantane is selectively converted to admantane on Pd (E_a=24·1 kcal·mole^{-1}, 200-300°C) and on Pt (E_a=10·1 kcal·mole^{-1}, 160-300°C). (See Figure 1).

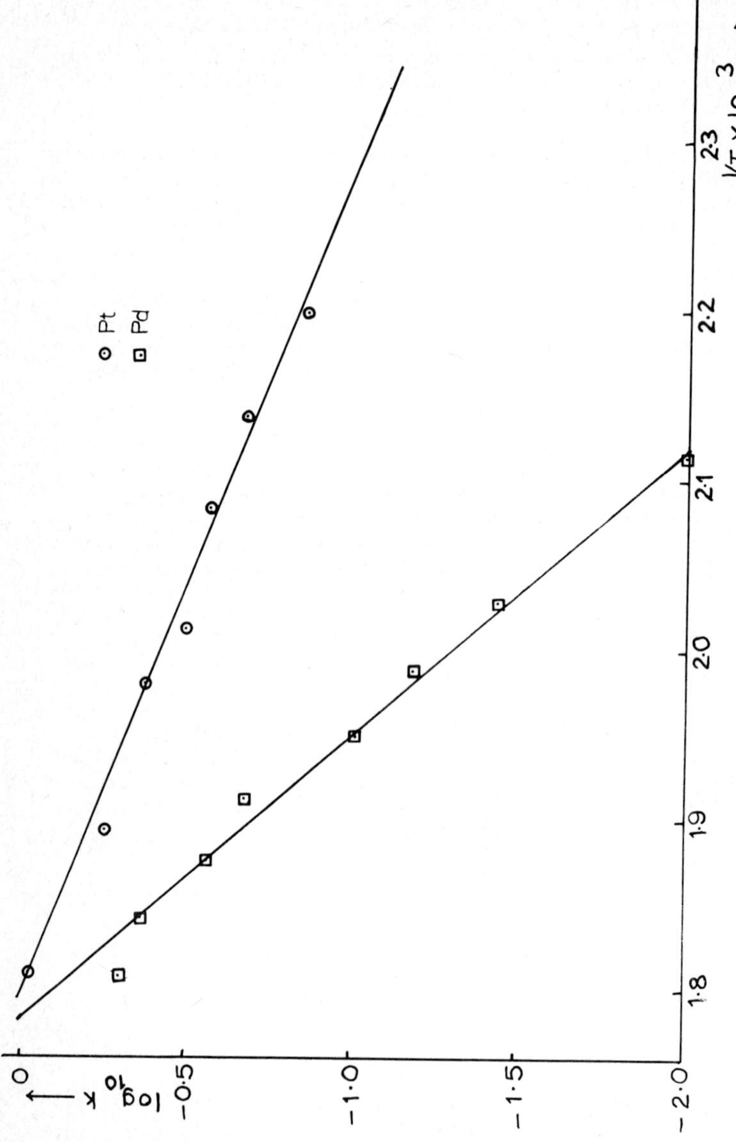

Fig. 1 Arrhenius plots for protoadamantane conversion to adamantane over Pt and Pd

Models show that ααγ-triadsorbed species as described above are geometrically impossible for this series of compounds.

One possible mechanism is simply the formation and hydrogenolysis of cyclopropane via α,γ-diadsorbed species. The required α,γ,σ-bonds to the surface would almost be parallel for bicyclo(2,2,1)heptane but inclined towards each other to such an extent for protoadamantane that bonding to one metal atom only seems most probable.

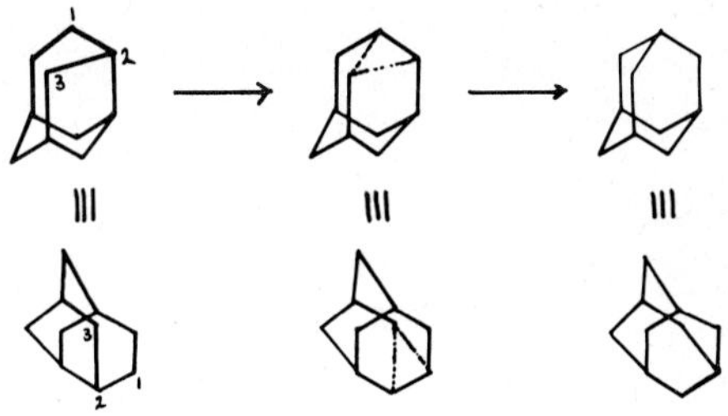

However, we are not too satisfied with this mechanism either, since one highly strained model compound rearranges during exchange with deuterium on Pd and Pt, but both products and reactant have only a simple exchange pattern. This clearly shows that a monoadsorbed species is rearranging. Since Pt-alkyl is undoubtedly a covalent bond, how do we explain carbonium ion behavior?

We suggest the following novel idea, as exemplified by the isomerization of neopentane.

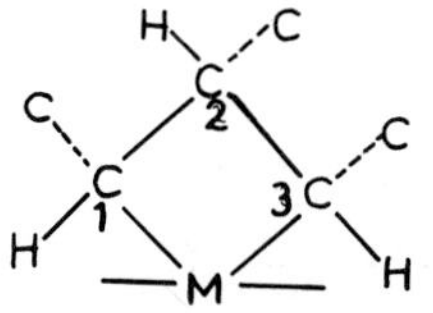

The nature of the transient π-bonded olefin complex is the key to the driving force for carbonium ion character and rearrangement. The components of the olefin metal bond, are donation from the filled pπ-orbital to a σ-orbital on the metal and back-donation from a filled dπ-orbital to the empty pπ*-antibonding orbital. The CH_3 group is bonding to the pπ-orbital but is nonbonding to the pπ*-orbital.

Fig. 2

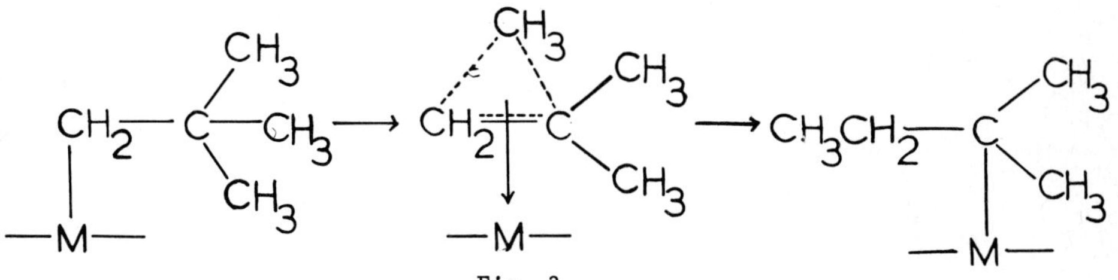

Fig. 3

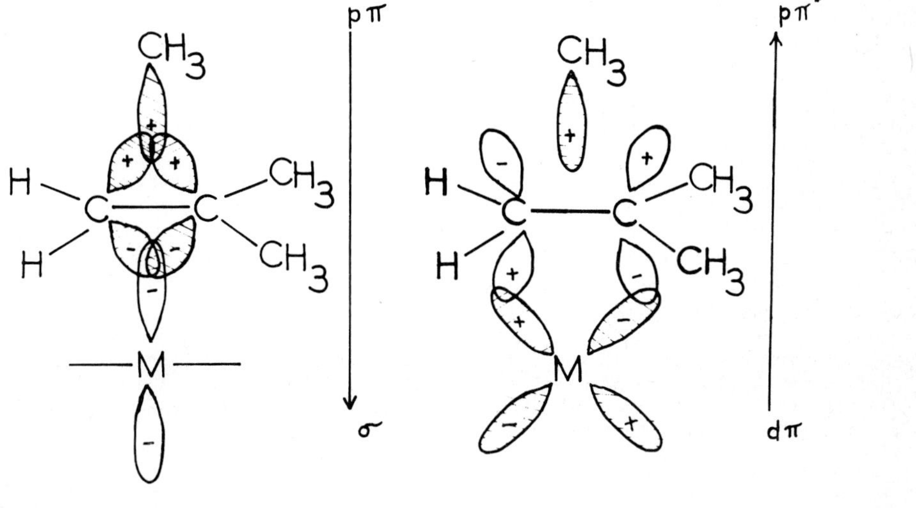

Fig. 4

Since the bonding pπ-orbital is electron deficient as a result of donation to the metal the carbonium ion behavior is understood (See Olah, G., A., Chem. in Britain, 8, 281, 1972), even though the hydrocarbon part of the complex may have no net positive charge. Pt is also a better catalyst than Pd because stronger backdonation enhances donation.

J. R. ANDERSON

I am grateful to Dr. Samman for his interesting report of platinum catalyzed bond shift reactions in condensed-ring hydrocarbons. I shall confine my remarks to the conversion of protodamantane to adamantane which Samman illustrates in detail in his Figs. 1 and 2.

I have examined molecular models and I do not believe that the Anderson-Avery mechanism is necessarily ruled out on steric grounds, although it is true that the bonds to the catalyst surface in the $\alpha\alpha\gamma$ adsorbed species would be somewhat more skewed away from normal to the surface than in the analogous structure from simpler aliphatic molecules. On the other hand however, protoadamantane is itself significantly strained but adamantane is strain-free, so one would reasonably expect higher intrinsic reactivity for this conversion than for conversions between unstrained but smaller aliphatic molecules.

The mechanism proposed by Samman certainly appears quite plausible at first glance. However, it does seem to have some inadequacies when compared with all the experimental facts. Using this mechanism one would expect the conversion of a tertiary to a quaternary carbon atom to be relatively easy. In fact it is quite difficult. For instance, the amount of neopentane formed from isopentane is quite small, much smaller than the amount allowed thermodynamically. It is, however, often not absolutely zero. An analogous situation exists with the reaction of some C_6 aliphatics with respect to the formation of neohexane.[1] In fact, it seems probable that the extent to which the (small) amounts of quaternary hydrocarbons are formed, is dependent in some way on catalyst structure, but the details are yet unknown.[2] The Anderson-Avery mechanism would not allow the formation of any quaternary hydrocarbon at all so, even if the Anderson-Avery mechanism were correct for the bulk of the reaction, one still has to admit a parallel ancillary mechanism, and the one proposed by Samman ought at least to be seriously considered for this. Secondly, it would be quite difficult on Samman's mechanism to account for high activity of the platinum (111) plane for the bond shift reaction. Thirdly, it should be pointed out that Samman's mechanism would be fairly hard to reconcile with Boudart's[3] results on the effect of catalyst sintering on neopentane isomerization.

Finally, one last point: if one admits that more than one basic mechanism exists for "bond shift" reactions, it is by no means inconceivable that in special circumstances, such as with reactant molecules which are strained or which have other special steric requirements, a particular mechanism may be forced into prominance, although it may be relatively unimportant in less unusual circumstances.

1) Barron, Y., Maire, G., Muller, J. M. and Gault, F. G., J. Catal. 5, 428 (1966).
2) Anderson, J. R., Advances in Catalysis, in press.
3) Boudart, M., Aldag, A. W., Ptak, L. D. and Benson, J. E., J. Catal. 11, 35 (1968).

J. R. ANDERSON

The paper by Anderson and Shimoyama has prompted a number of comments on the influence of metal particle size on the behavior of a catalyst. One obvious question is how the electron availability varies with particle size. This may well be a matter of some importance in as much as, depending on the type of reaction, the rate controlling step may be assisted or impeded by the degree of electron availability. It seems likely that the effect to be expected will depend on the range of particle size considered, since there are two distinct factors operating. If the particles are sufficiently large that their electronic properties are not intrinsically dependent on size as such, and the lower limit to this size range is probably in the vicinity of 15-20 Å, one then expects electron removal to become easier with decreasing size, basing the argument on the expected decrease in work-function with decreasing radius of curvature of the surface. This is the size range considered in the paper by Anderson and Shimoyama. However, for extremely small particles, say below 15-20 Å, one must envisage the opposite trend to become established, because the lower limit to the particle size is (in principle, if not in reality) a single metal atom for which the ionization potential is always a good deal higher than the work-function of the corresponding metal.

Paper Number 48

THE CHARACTERIZATION OF THE METAL SURFACE OF ALUMINA-SUPPORTED PLATINUM CATALYSTS BY TEMPERATURE-PROGRAMMED DESORPTION OF CHEMISORBED HYDROGEN

P. C. ABEN, H. VAN DER EIJK and J. M. OELDERIK
Koninklijke/Shell-Laboratorium, Amsterdam, the Netherlands

ABSTRACT: The metal surface area present in alumina-supported platinum catalysts appears to be very heterogeneous with respect to the adsorption of hydrogen. This is manifest from a very complex desorption picture which indicates at least three different adsorption sites. Different hydrogen pretreatments not only change the "overall" H/Pt value but can also drastically change the ratio in which the various sites contribute to the "overall" H/Pt value. A linear correlation of the latter parameter with the rate of a catalytic reaction is not precluded but should not be considered a hard and fast rule. This is illustrated by means of benzene hydrogenation activity tests. An interesting correlation of hydrogenation activity for benzene at 50 °C with a specific low-temperature desorption peak in the TPD chromatograms is indicated.

1. INTRODUCTION

An important parameter for the catalytic activity of supported platinum (and catalytic metals in general) is the metal surface area per gram of metal (specific metal surface area). In recent years a discussion has developed among many investigators[1 to 7, 9 to 11] about the question whether and, if so, to what extent heterogeneity of a catalytic metal surface area plays a role in its catalytic behaviour.

Hydrogen chemisorption is a technique widely used for the determination of specific metal surface areas, and in many studies catalytic activity is related either to micromoles of H_2 per g (or gat) of Pt or to the atomic H/Pt ratio calculated from it. Such parameters do not account for surface heterogeneity. Recently Cvetanović et al.[8] have shown by means of temperature-programmed desorption that different adsorption sites can be discerned on the metal surface area of platinum*. This evidence induced us to extend earlier work of our laboratory[9] on the relationship between H/Pt and the rate of benzene hydrogenation by a study of temperature-programmed desorption of hydrogen from supported platinum catalysts.

2. EXPERIMENTAL PART

2.1. Temperature-Programmed Desorption (TPD)

A. Apparatus

The apparatus used for the TPD measurements is schematically shown in fig. 1.

B. Materials

Electrolytic hydrogen was purified by diffusion through a palladium membrane. Argon was purchased from AGA Nederland N.V., Amsterdam. Maximum impurity levels in ppmv: H_2O 5, N_2 30, O_2 3, H_2 0.1, hydrocarbons 0.2. Traces of oxygen present in the argon/hydrogen mixture (carrier gas in TPD) were converted over a platinum catalyst at 400 °C to water, which was trapped at -78 °C in molecular sieves (Linde 5A). The catalysts tested are described under 2.2.A and details are shown in table 1.

----

* unsupported metal.

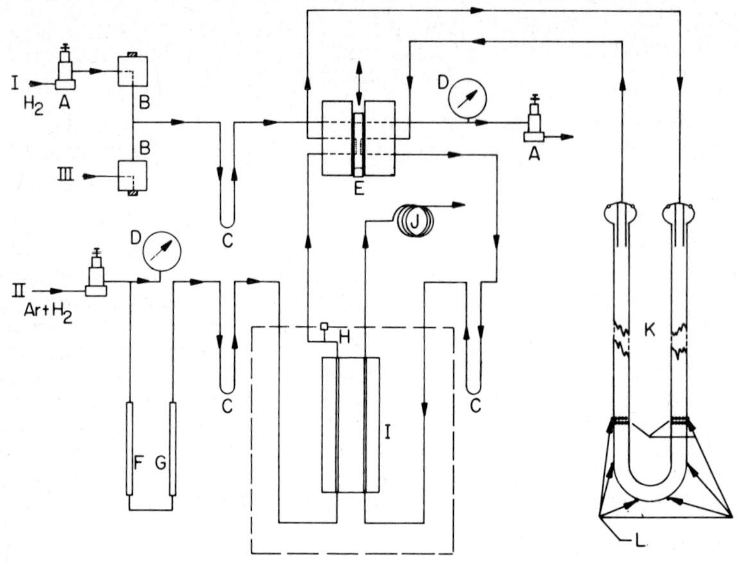

Fig. 1 Flow scheme of the adsorption-desorption apparatus

A. Reducing valve B. Three-way valve
C. Cold trap (-78 ^{o}C), filled with molecular sieves 5A (30/80 mesh)
D. Precision manometer E. Gas-flow switch
F. Tube filled with molecular sieves 5A (30/80 mesh)
G. Oxygen scavenger, filled with a platinum/alumina catalyst, operating at 400 ^{o}C
H. Injection point, for calibration purposes
I. Katharometer J. Brake capillary
K. Quartz U-tube reactor L. Iron/constantan thermocouples

TABLE 1
Survey of catalyst heat treatments applied to obtain various atomic H/Pt ratios
(Hydrogen at 1 to 1.5 bar)

Catalyst	Temperature, ^{o}C	Time, h	H/Pt Atomic Ratio
0.76 %w Pt on UOP Alumina	400/500	$\frac{1}{2}$/5 min	1.10
	550	16	0.71
	650	16	0.49
	850	16	0.21
0.46 %w Pt on Péchiney Alumina	400	2	1.0
	600	2	0.6

C. Procedure

The procedure followed in the hydrogen adsorption-desorption measurements is described in detail in table 2. The total amounts of hydrogen adsorbed and desorbed were calculated from the integrated peak areas measured by means of the heat conductivity cell, which was calibrated by injection of known volumes of H_2 into the carrier gas (Ar + 0.1 %v H_2) stream.

The volumes of hydrogen desorbed agreed within 5% with the volumes adsorbed.

TABLE 2
Temperature-programmed hydrogen desorption procedure
Survey of conditions applied during
pretreatment of samples, adsorption and desorption

Gas rate in all steps: 600 ml (STP)/h
Samples : 2 ml, 0.2-1.5 mm particle size

Stage of the Procedure	Gas comp., %v	Total Pressure, bar	Hydrogen Pressure, bar	Temperature, °C	Duration of Treatment, min
Sample Pretreatment	↕ 100 H_2	↕ 1.5	↕ 1.5	25⟶400 400 400⟶500 500	25 30 5 5
Adsorption**	0.1 H_2 + 99.9 Ar	↕ 1.5	↕ 0.0015*	500⟶-78 -78	30 30****
Desorption (TPD)	0.1 H_2 + 99.9 Ar	↕ 1.5	↕ 0.0015	-78⟶500*** 500	33 20****

* Results are identical when a much higher hydrogen pressure is applied in the final adsorption step at -78 °C.

** Hydrogen uptake is monitored by a calibrated heat conductivity cell; adsorption conditions are maintained until no more hydrogen pick up is registered (see under ****).

*** Linear heating rate of 17.5 °C/min. Desorption of the final amount of hydrogen can be speeded up by continuing heating to e.g. 650 °C, but this does measurable harm to the sample. To be on the safe side we chose not to heat beyond 500 °C and to continue desorption. The amount of hydrogen left adsorbed at 500 °C is negligible.

**** This period is sufficient to completely return to the baseline of the heat conductivity cell.

2.2. Benzene hydrogenation

A. Materials

Benzene (ar grade) was obtained from Union Chimique Belge. Its analysis indicated: residue on evaporation < 0.0005 %w, sulfur content < 1 ppmw, water content < 0.02 %w.

Before the benzene feed was introduced into the reactor, it was passed over a mixture of reduced copper catalyst (BTS catalyst from BASF) and Linde 5A molecular sieves to remove possible trace amounts of oxygen and water. Electrolytic hydrogen was dried prior to use over Linde 5A molecular sieves.

As catalyst supports we used UOP and Péchiney alumina. Their physical properties are as follows.

Support	Density, g/ml	Surface area, m^2/g	Pore volume, ml/g
UOP alumina	0.43	185	0.75
Péchiney alumina	0.64	135	0.41

Platinum catalysts were prepared by impregnation of alumina with an aqueous solution of chloroplatinic acid. Catalyst was dried at 120 °C for three hours and subsequently calcined at 500 °C in air for three hours. By a reduction treatment in hydrogen at various severities (400-850 °C) samples were obtained with various platinum surface areas. Details are shown in table 1.

B. Procedure

Hydrogenation experiments with benzene were conducted in a flow system using 1-2 ml of catalyst of 0.2-1.5 mm particle size. Prior to the hydrogenation test the catalysts were treated in hydrogen in situ at 400 °C for one hour. The reaction conditions were as follows:

temperature : 50 °C
benzene partial pressure : 0.07 bar
hydrogen partial pressure: 1.43 bar
LHSV : 4 ml benzene $(ml\ cat.\ h)^{-1}$

3. RESULTS, INTERPRETATION AND DISCUSSION

3.1. Temperature-programmed hydrogen desorption

Fig. 2(a) represents typical TPD chromatograms of hydrogen adsorbed on the series of 0.76 %w Pt on UOP alumina catalysts of different H/Pt values. The results in this figure are based on equal amounts of catalysts i.e. equal amounts by weight of platinum. Fig. 2(b) shows the same data based on equal amounts of hydrogen desorbed, so as to facilitate comparison of the relative contributions of the various "peaks" to the total amount of H_2 adsorbed on platinum of a given overall H/Pt value.

With the bare support of these catalysts essentially no adsorption of hydrogen is observed.

Figs. 2(a) and 2(b) in the first place show the complex nature of the chemisorption phenomena. Apart from a distinct peak at about -20 to -30 °C, little differentiation can be made in terms of peaks in the range of 25 to 400 °C. At least two (largely overlapping) major peaks occur in the temperature ranges of 100 to 200 °C and 200 to 300 °C, respectively.

Figs. 2(c) and 2(d) represent TPD results obtained with a 0.46 %w Pt catalyst based on Péchiney alumina. Comparison of these results with those of the first series shows similar peaks at corresponding temperatures.

It is interesting to note in figs. 2(a) and 2(c) that a heat treatment in hydrogen which reduces the overall H/Pt from 1.1 to 0.7 (UOP alumina-supported catalyst) and from 1.0 to 0.6 (Péchiney alumina-supported catalyst) in both cases gives rise to a significant loss of adsorption capacity for the strongly adsorbed hydrogen - the peaks between about 100 and 400 °C show considerable losses in surface area - whereas the peak below 0 °C remains virtually constant.

For the UOP alumina-supported catalyst we see in fig. 2(a) that a further reduction of overall H/Pt below 0.7 is attended with a more or less random loss in surface area for all "peaks". This reflects itself in fig. 2(b) in a fairly uniform distribution of the total peak area over the various "peaks" for the three catalysts of H/Pt 0.7, 0.5 and 0.2.

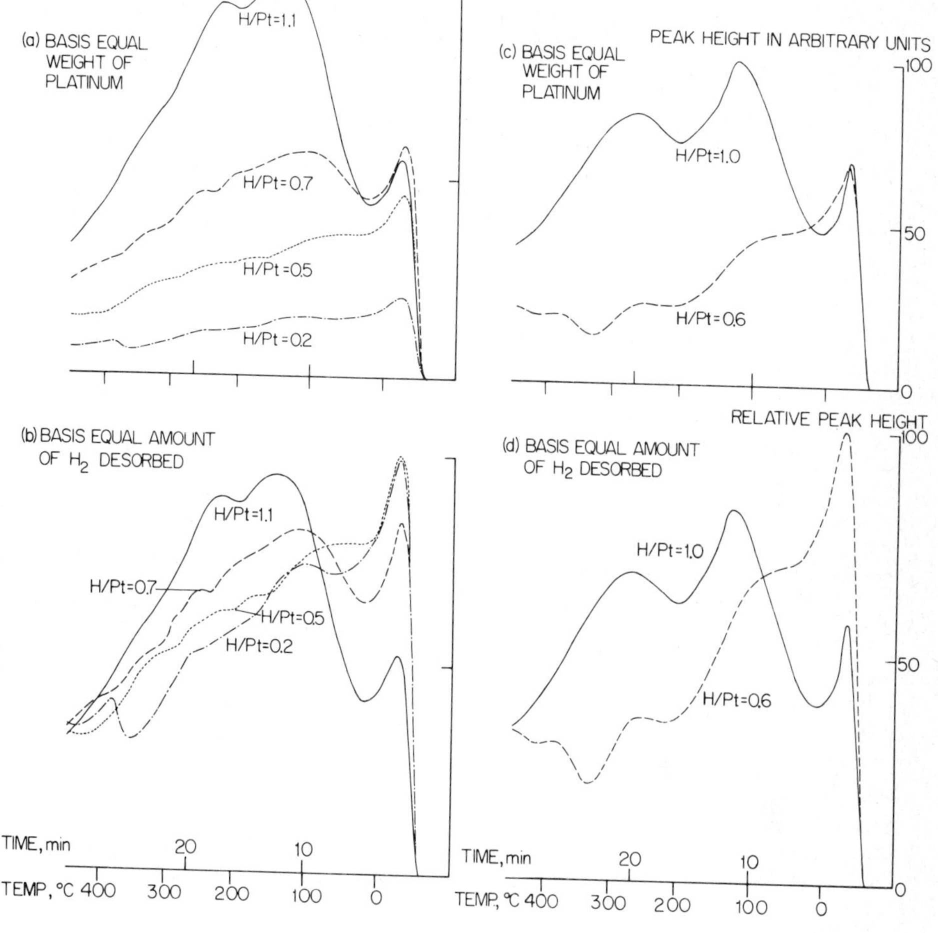

Fig. 2 Temperature-programmed desorption of hydrogen from Pt/alumina catalysts. (a) and (b) 0.76 %w Pt/UOP alumina; (c) and (d) 0.46 %w Pt/Péchiney alumina

The main conclusions from figs. 2(a)-(d) are:

(a) The platinum surface area obtained after a given heat treatment in hydrogen is very heterogeneous with regard to the adsorption of hydrogen. This heterogeneity manifests itself in the occurrence of a complex composition of desorption peaks representing at least three different species of adsorbed hydrogen. We assume therefore that Pt is exposed in at least three different fashions.

(b) The heat treatment in hydrogen by which the overall H/Pt is changed, at the same time changes the ratio in which the various sites contribute to this overall H/Pt.

Cvetanović et al.[8] have tentatively assigned configurations to the modes of adsorption which in their experiments with Pt black reflect themselves in peaks at about -20(β), 90(γ) and 300 °C(δ) viz.:

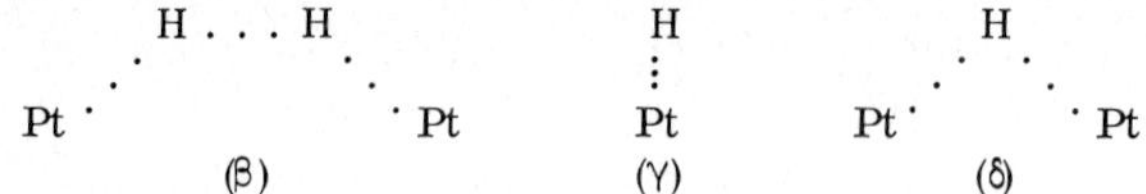

If one visualizes the various sites in a fashion as suggested by Cvetanović et al., it becomes at once clear that one can hardly assume a direct relation between a reaction rate and the sum total of these various sites as measured by overall H/Pt ratio where the ratio of the sites varies together with their total number. In cases where a pretreatment only changes the total number of sites (overall H/Pt) but leaves the ratio of the various species reasonably constant, a linear relation between reaction rate and overall H/Pt would seem less improbable. Still even then one must bear in mind that for reactions other than H_2-D_2 exchange, such as e.g. hydrogenation of benzene or hydrogenolysis of neopentane, one is left with the uncertainty as to how the adsorption of hydrocarbon and that of hydrogen influence each other and if and how the number of "sites" counted by means of hydrogen also relates to the number of adsorbed species on whose concentration the reaction rate depends under actual reaction conditions.

Anyhow, it would seem that in the ranges of H/Pt values where the ratio of the different hydrogen chemisorption sites is not even constant their grand total expressed as "overall" H/Pt can hardly be a useful measure of the number of actual reaction sites that matter in the catalytic reaction.

3.2. Benzene hydrogenation

Under the conditions of the test benzene conversions were kept below 10 %m and the reaction rate calculated as the product of conversion and space velocity appeared to be independent of space velocity, i.e. differential conditions apply.

By variation of the partial pressures of benzene and hydrogen around the values adopted for the standard tests, it was shown that the reaction is zero order in benzene and follows a Langmuir adsorption isotherm with respect to hydrogen. For the latter no distinction can be made between the kinetic rate equations

$$v = k\frac{bP_{H_2}}{1 + bP_{H_2}} \quad \text{and} \quad v = k'\left(\frac{b'\sqrt{P_{H_2}}}{1 + b'\sqrt{P_{H_2}}}\right)^2$$

(k, b and b' being constants), which refer to molecularly and atomically adsorbed hydrogen, respectively: in other words, a plot of the reciprocal rate versus reciprocal P_{H_2} gives an equally good fit of the data as does a plot of the square roots of these parameters.

The results of the benzene hydrogenation tests are shown in table 3 and fig. 3. From fig. 3 one sees that it is possible to represent the results obtained with the three catalysts having H/Pt values of 0.2, 0.5 and 0.7, respectively, in terms of a straight-line relationship between hydrogenation rate and overall H/Pt value in similar fashion to earlier results obtained in our laboratory by Aben, Platteeuw and Stouthamer[9]. It would seem that in this range of H/Pt values the overall H/Pt reasonably reflects the extent to which the platinum surface area is available and suitable for benzene hydrogenation. Hence one is inclined to conclude, regardless of the complex "composition" of these overall H/Pt values and despite the undoubtedly still greater complexity of the true situation on the platinum surface when benzene is present together with hydrogen, that the presence of more platinum surface (as expressed in a higher overall H/Pt value) goes hand in hand with a roughly proportional increase in reaction sites for the hydrogenation reaction.

TABLE 3
Benzene hydrogenation tests on 0.76 %w Pt/alumina (UOP)
Conditions: temp.: 50 °C; hydrogen/benzene partial pressures: 1.43 bar/0.07 bar

H/Pt Ratio ex TPD exp.	0.21	0.49	0.71	1.10
Hydrogenation Activity, mmole C_6H_6 . $(gat\ Pt.s)^{-1}$	16 15	53 58 54	74 78 78 76 75**	40 35* 34* 45

* Between these two experiments we performed the experiment marked ** on the sample with H/Pt = 0.71 to show once more that the results were not affected by any aberration of technique or chemicals.

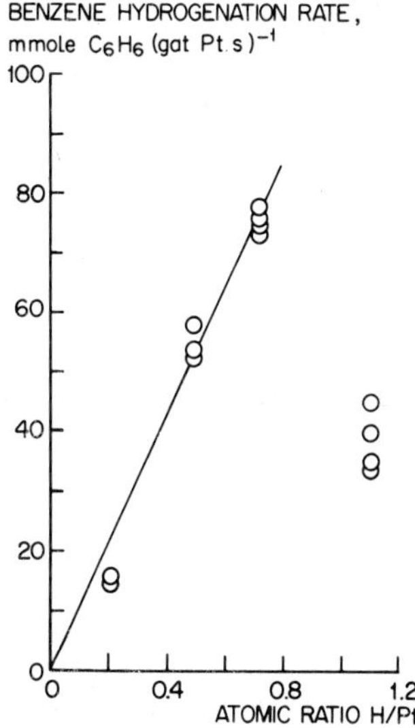

Fig. 3 Plot of benzene hydrogenation rate versus overall H/Pt (catalyst 0.76 %w Pt on UOP alumina)

The most striking observation in the present work is undoubtedly the low activity of the catalyst with the highest H/Pt ratio of the series (i.e. H/Pt = 1.1), which is the starting material from which the other samples were prepared by the various heat treatments specified in table 1. Here we must conclude that a large part of the platinum "surface area" measured by means of hydrogen chemisorption fails in one way or other to contribute to the number of sites essential for benzene hydrogenation under the present reaction conditions.

Although one can indeed visualize that a platinum particle could be too small to accommodate a "flatwise" adsorbed benzene molecule we were somewhat suspicious of our observation. However, the catalyst in question showed its high H/Pt ratio in the TPD equipment before and after determination of H/Pt = 0.7 on the more active sample and it also showed its lower activity before and after testing of the more active sample in the microflow reactor. Thus reproducibility and repeatability have been demonstrated and the result cannot be ascribed to an abnormality in equipment or materials. It must therefore be possible to find an explanation for the phenomenon.

Now there is an essential difference between the catalyst of overall H/Pt = 1.1 and the other three samples (H/Pt values 0.2, 0.5 and 0.7). This difference results from the procedure applied for the preconditioning of the platinum. The latter three catalysts had been reduced at 400 °C and then conditioned in hydrogen at (far) more severe conditions (see table 1) than applied during subsequent in-situ pretreatment of these samples for the TPD and benzene hydrogenation experiments. These subsequent pretreatments do not affect the conditions of the platinum attained in the previous severe heat treatments. In the case of the H/Pt = 1.1 sample, however, we charged unreduced catalyst (precalcined at 500 °C for 3 h) to the TPD and benzene hydrogenation reactors and consequently the condition of the platinum in the two tests was not necessarily the same (although the mild reduction conditions applied are known to give what is generally called "well-dispersed" platinum of H/Pt values of about 1). Therefore we ran a TPD experiment using the pretreatment conditions of the benzene hydrogenation test (400 °C, 1 h) and a benzene hydrogenation experiment using the TPD-pretreatment conditions (400 °C, $\frac{1}{2}$ h; 500 °C, 5 min.).

TABLE 4
Benzene hydrogenation tests on 0.76 %w Pt/alumina (UOP)
Effect of different catalyst reduction conditions
Reaction conditions of benzene test: see table 3

Pretreatment in H_2	Hydrogenation Rate, mmole C_6H_6 (gat Pt. s)$^{-1}$	Overall H/Pt
400 °C, 1 h	38 (average from table 3)	1.1
400 °C, $\frac{1}{2}$ h; 500 °C, 5 min	73	1.1

The results of fig. 4 and table 4 are very interesting indeed: the overall H/Pt value is the same as before, viz. 1.1; nevertheless the hydrogenation activity is doubled. At first sight this looks equally puzzling as the first set of results on the H/Pt = 1.1 catalyst reported in table 3. But closer inspection of the desorption curves in fig. 4 reveals that the two catalysts of equal overall H/Pt ratio show a distinct difference with respect to their low-temperature (-20 °C) desorption peak. Combining this with the activity data in table 4 we see that the hydrogenation activity of the two catalysts with H/Pt = 1.1 runs parallel with the extent of their low-temperature hydrogen desorption peak. The same parallel applies for the other peaks (fig. 2a) and activities (table 3) measured on this series of catalyst samples.

In fig. 5 we have plotted activity versus the height of the low-temperature peaks. (Peak height is used as a measure of peak surface area on the reasonable assumption of equal peak width at the base.) The fair straight-line relationship in fig. 5 is a strong indication that the low-temperature hydrogen desorption peak is a relevant yardstick of catalytically active sites for benzene hydrogenation at 50 °C. The overall H/Pt value appears to be a useful parameter for activity as long as the essential part of the chemisorbed hydrogen is a constant fraction of the total amount irrespective of H/Pt value.

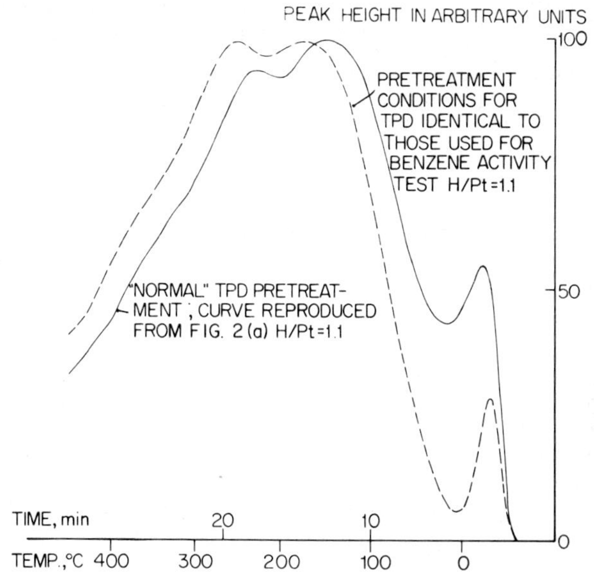

Fig. 4 Temperature-programmed desorption of H_2 from Pt on UOP alumina. Basis equal weight of platinum

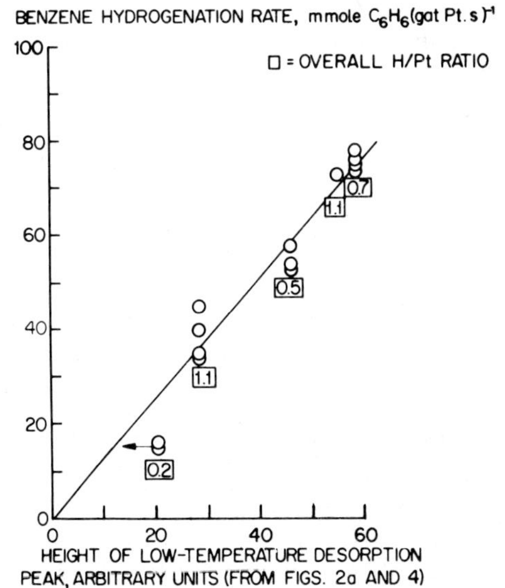

Fig. 5 Relationship between low-temperature desorption peak in TPD chromatogram and hydrogenation activity (catalyst 0.76 %w Pt on UOP alumina)

The lower activity of samples with H/Pt values in the range of 1.1 to 1.2 (measured in the way described in this article) has also manifested itself in a series of 0.46 %w Pt catalysts based on Péchiney alumina in tests with benzene in the presence of a controlled amount of H_2S (which greatly suppresses catalytic activity and necessitates a reaction temperature of 300 °C). Here the sample with the highest overall H/Pt (1.2) was less active than samples prepared from it by hydrogen treatment (H/Pt values of 1.1 and 0.7; relative activities being 10, 15 and 14, respectively). In our opinion it is worthwhile to further explore the merits of the correlation indicated, in the first place by trying to improve the separation of the desorption peaks in the TPD chromatograms.

REFERENCES

1. V. Nikolajenko, V. Bosaček and Vl. Daneš, J. Catalysis 2(1963)127.
2. H. E. Swift, F. E. Lutinski and H. H. Tobin, J. Catalysis 5(1966)285.
3. F. N. Hill and P. W. Selwood, J. Am. Chem. Soc. 71(1949)2522.
4. F. A. Dorling and R. L. Moss, J. Catalysis 5(1966)111.
5. D. J. C. Yates, W. F. Taylor and J. H. Sinfelt, J. Am. Chem. Soc. 86(1964)2996.
6. J. L. Carter, J. A. Cusumano and J. H. Sinfelt, J. Phys. Chem. 70(1966)2257.
7. M. Boudart, A. Aldag, J. E. Benson, N. A. Dougharty and C. Girvin Harkins, J. Catalysis 6(1966)92.
8. S. Tsuchiya, Y. Amenomiya and R. J. Cvetanović, J. Catalysis 19(1970)245.
9. P. C. Aben, J. C. Platteeuw and B. Stouthamer, Rec. Trav. Chim. 89(1970)449 (IVth Int. Congr. Catal. Moscow 1968, paper 31).
10. F. M. Dautzenberg and J. C. Platteeuw, J. Catalysis 19(1970)41.
11. J. R. Anderson, R. J. Macdonald and Y. Shimoyama, J. Catalysis 20(1971)147.

DISCUSSION

M. BOUDART

It is a measure of progress in concept and technique, as well as a tribute to the experimental skill of the groups in Amsterdam and in Nymegen, that we are today in a position to discuss the meaning of a factor of two or three in rates of hydrogenation of benzene per unit surface area. What still strikes me, though, is the amazing constancy of this quantity as the particle size of Ni or Pt is varied between about 10 and 50 Å. This fact still stands and for that reason, I still prefer to consider benzene hydrogenation as a structure insensitive reaction. Now, both groups of workers report a smaller rate per unit surface area when Ni or Pt particles are smaller, perhaps of the order of 5 Å. This may be due to a number of reasons, one of them being discussed in our paper at this Congress. Another one is the difficulty to measure surface area by chemisorption when particles approach 100% dispersion. There is a great need to develop alternative methods of counting surface sites in this domain, and T.P.D. may well be such a method as suggested by the findings of Aben *et al*.

G. C. BOND

Dr. P. A. Sermon and I have recently developed a new method for estimating the amount of hydrogen adsorbed on platinum/silica catalysts. By passing 1-pentene at 1 torr in nitrogen over a reduced catalyst, complete conversion to n-pentane is first observed. After a certain time, the conversion falls abruptly and is followed by a long "tail" of partial conversion. We believe the abrupt fall corresponds to the consumption of hydrogen atoms chemisorbed on the metal particles, and (except for very small particles) good agreement is obtained with particle size estimated by X-ray diffraction and electron microscopy, assuming a H/Pt_s ratio of unity. We further believe that the "tail" arises from slow-migration of "spilled-over" hydrogen back to the metal.

J. A. BETT

Electrochemical potentiodynamic sweep techniques have been used to observe hydrogen adsorption and desorption on platinum surfaces. These yield results markedly similar to data obtained by TPD (Fig. 1). Here the surface free energy of the catalyst is varied by controlling the potential rather than temperature as a linear function of time. Strongly and weakly chemisorbed hydrogen appear on adsorption with approximately the same relative peak heights as in TPD. Low temperature (physically) adsorbed hydrogen cannot be observed electrochemically. With the increased resolution possible for the electrochemical method, it is possible to show an interaction peak in the desorption sweep, as well as changes in the relative peak heights with increase in crystallite size (Fig. 2). Will[1] has associated these peak height changes with the different crystallite faces of platinum. In addition, analysis of the potentiodynamic sweep curves give equilibrium isotherms for each species, showing[2] departures from Langmuir conditions.

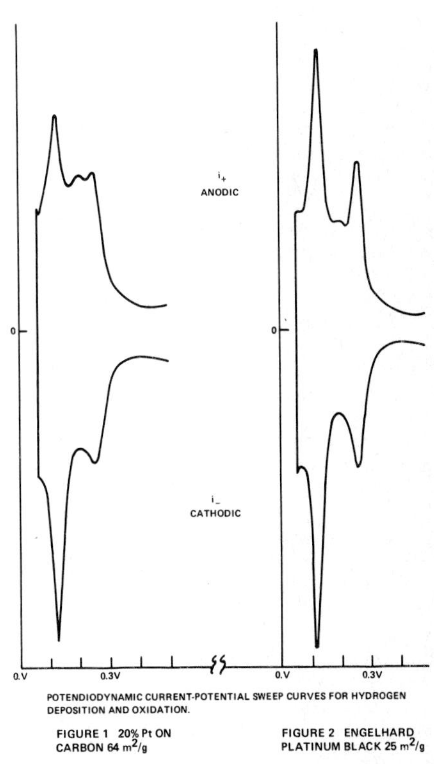

POTENDIODYNAMIC CURRENT-POTENTIAL SWEEP CURVES FOR HYDROGEN DEPOSITION AND OXIDATION.

FIGURE 1 20% Pt ON CARBON 64 m^2/g

FIGURE 2 ENGELHARD PLATINUM BLACK 25 m^2/g

Additional information on the kinetic parameters of the surface reactions is available[3] by examining the departure of the adsorption isotherms from equilibrium at high potentiodynamic perturbations, where the reverse reaction is negligible. For the oxidation of adsorbed hydrogen, a linear perturbation rate of $d\Delta G/dt = 50 k.cal.s^{-1}$ is required[3,4] before nonequilibrium kinetics can be obtained.

1) F. G. Will, J. Electrochem. Soc. 112, 451 (1965).
2) P. Stonehart, Electrochim. Acta 15, 1853 (1970).
3) P. Stonehart, H. A. Kozlowska, B. E. Conway, Proc. Roy. Soc. A 310, 541 (1969).
4) F. G. Will, C. A. Knoor, Z. Electrochem. 64, 258 (1960).

J. W. E. COENEN

1) By reworking the data of Table 3, dividing the average rate/g Pt by the H/Pt ratio, which we may assume to be proportional to specific Pt area, we obtain

H/Pt Ratio	0.21	0.49	0.71	1.10
Activity per unit area (arbitrary units)	74	112	107	35

In our work we found a maximum at H/Ni ratio of 0.55. It appears a distinct possibility that also in your work a still higher specific rate might be found near that ratio.

2) It is striking that in your work--like we did in ours--you found H/metal ratios above unity.

The crucial question is: Is really more than one H atom bonded to a single metal atom?

An alternative explanation might be spillover. In view of our isotope effect data, we feel inclined to the latter viewpoint. In that case the weakly bonded hydrogen species might be spillover hydrogen.

J. M. OELDERIK

1) In our paper we have stressed the observation that the rate of benzene hydrogenation per gram of platinum is lower at H/Pt=1.1 than at H/Pt = 0.7.

This we consider the most important point, as it indicates the sensitivity of the test reaction to surface topology or crystal structure. As to the case of H/Pt = 0.21, we have to realize that the specific activity obtained by dividing measured rate by measured H/Pt value is the quotient of two small numbers.

In line with the observations reported in the previous paper of our laboratory on this subject, which was presented at the World Catalysis Congress in Moscow, we have a slight preference for the plot presented in Fig. 3, which suggests that within the experimental accuracy and in the range of H/Pt ratios up to 0.7 the activity is proportional to the number of Pt atoms detectable by hydrogen chemisorption.

2) There are several strong arguments against your suggestions that our weakly bound hydrogen is to be attributed to spill-over.

In the first place, Cvétanovic *et al*. observed this weakly bound hydrogen (the desorption peak at -20 to -30°C) on platinum black where there is no support to accomodate the spill-over. The same observation was made with Pt-black in our own laboratory.

Secondly, it is highly improbable that the benzene hydrogenation rate would show such a clear correlation with "spill-over" hydrogen and that at H/Pt ratios of 0.2, 0.5 and 0.7 the spill-over would remain proportional to the Pt content while at H/Pt = 1.1 it does not show any proportionality at all.

H. WISE

In conjunction with Dr. Coenen's comment on your paper, I would like to add that ultra-high vac. sorption studies demonstrated a maximum surface coverage of hydrogen on platinum corresponding to H/Pt = 0.5. This condition is very near that found by you for maximum activity in benzene hydrogenation.

Also, may I ask how the weakly-bound hydrogen that readily desorbs between -30°C to -20°C can be involved in benzene hydrogenation at +50°C?

J. M. OELDERIK

In our opinion, it is hard to figure how strongly bound hydrogen (which needs a temperature above 200°C to desorb in a sweep-stream of low-pressure gas) could be involved in low-temperature benzene hydrogenation at relatively high hydrogen partial pressures (1.43 bar).

We consider it quite reasonable, though, to associate benzene hydrogenation activity with the weakly bound hydrogen since under the conditions prevailing during the benzene hydrogenation tests (1.5 bar) the low-temperature adsorption sites will be covered with adsorbed hydrogen to a reasonable extent. This can be illustrated as follows:

If we assume that during TPD of hydrogen the adsorption-desorption equilibrium is established, then for a symmetrical peak the hydrogen partial pressure corresponding with the peak height relates to a coverage $\Theta = 0.5$. We find the adsorption coefficient b for these sites from $bP_{H_2} = \Theta/1-\Theta$. In our TPD experiments a typical value for b(-30 °C) is 300 bar^{-1}.

In the benzene hydrogenation experiments at 50°C b values were of the order of 0.5 bar^{-1}. The temperature dependence of b corresponds to a heat of adsorption of about 10-11 kcal/mole and extrapolation to -30°C indicates a b value of about 100 bar^{-1}. This is the same order of magnitude as calculated from the TPD experiments. A value of b at +50°C of 0.5 bar^{-1} still allows a fair coverage of the weak-adsorption sites at the P_{H_2} of the benzene experiment, which is several orders of magnitude larger than those prevailing in the TPD experiemtns.

This reasoning is not rigid in that it does not take into account any effect of benzene on the adsorption of hydrogen. Still, these "broad-brush" calculations show nothing to contradict our hypothesis that weakly bound hydrogen is involved in benzene hydrogenation at 50°C.

D. D. ELEY

A minor point in the paper, is that I think it is not possible from TPD to distinguish unequivocally the different fractional-bonded model specified by β, γ and δ on page 6 of your paper. The reason is that even if we consider only the single-bonded mode, γ, i.e. Pt-H, the strength of this bond

$$E(Pt\text{-}H) \simeq \frac{1}{2}\{E(H\text{-}H)+(Pt\text{-}Pt)\}+23(X_{Pt}-X_H)^2$$

will depend on the lattice plane, 100, 111, etc., since the electronegativity X_{Pt} as given by the electron work function will depend on the lattice plane concerned, as will also the possibilities for forming multiple bonds as in modes β and δ. By combining TPD with other techniques, the distinction may, however, be possible.

J. M. OELDERICK

We fully agree with Professor Eley that other techniques than TPD alone are required if one tries to assign configurations to the various TPD peaks. The assignments given in the paper are merely illustrations cited from the work on Pt-black by Cvétanovic et al. These authors also stress the speculative nature of such assignments and indeed base their choice on additional data such as those derived from infrared analysis.

Paper Number 49

THE HYDROGEN EXCHANGE REACTION BETWEEN BENZENE AND PERDEUTERIOBENZENE CATALYZED BY EVAPORATED METAL FILMS. PART II

R. B. MOYES, K. BARON and R. C. SQUIRE
Department of Chemistry, University of Hull, Hull, U.K.

ABSTRACT: The exchange reaction between benzene and perdeuteriobenzene at 133.3 N m^{-2} (1 torr) and 0°C has been studied using evaporated metal films of the first transition series.

The reaction occurs over the metals titanium to nickel, and the activity reaches a maximum at cobalt in Group VIII_2. The rate constants for the reaction show a linear relationship with percentage d-character of the metals and the variation in activity of these metals with metallic radius has been examined. The relative surface area of the metals were compared by the extent to which films of each metal adsorbed benzene.

The reaction is explained by a dissociative mechanism.

The results for the first transition series considered along with the results for the second and third transition series, given in a previous publication[1)] have been examined together to give a more complete understanding of the exchange reaction.

1. INTRODUCTION

Previous examination of the exchange of deuterium for hydrogen in aromatic compounds includes the work of Crawford and Kemball[2)], who favoured an associative mechanism and Garnett and Sollich-Baumgartner[3)] who preferred a dissociative mechanism. The dissociative mechanism was also favoured by Fraser and Renaud[4)] who studied the exchange of deuterium oxide with monosubstituted benzenes, and Hirota and co-workers[5)], who examined the self exchange reaction in monodeuteriotoluenes.

The adsorption of benzene on metal catalysts has been studied by D'or[6)] who suggested two types of adsorbed species to explain the hydrogenation of benzene, and Tetenyi[7,8)] who reported experiments on ^{14}C-labelled benzene. The latter workers suggested the dissociative chemisorption of benzene on nickel. Erkelens and his co-workers[9)] studied the infra-red spectra of chemisorbed benzene on silica-supported nickel and copper and interpreted their results in terms of a loss of aromatic character and so dismissed the possibility of a π-bonded species.

These present studies were designed to provide more information on the reactivity of metal films, in a state of high purity, for the exchange reaction and hence elucidate which of the two suggested mechanisms occurred.

2. EXPERIMENTAL

2.1 Apparatus

The main features of the apparatus have been outlined before[10)]. It consisted of a high-vacuum gas-handling system with a reaction vessel of volumes 250, 390, or 880 cm^3. The change of the percentages of the products and reactants with time were analysed mass spectrometrically.

2.2 Materials

Puriss grade perdeuteriobenzene (Koch-Light Laboratories Ltd.) and Analar grade benzene (dried over calcium chloride) were used. The hydrocarbons were mixed in equal molar proportions by weight and purified by repeated freezing, pumping and thawing. Metals of high purity were obtained from four sources: Johnson-Matthey Ltd., Engelhard Industries Ltd., Koch-Light Laboratories Ltd., and New Metals Ltd.

2.3 Technique

Preliminary treatment of the reaction vessels included degassing by heating to 450°C for about three hours and electrical heating of the metals to just below the evaporation temperature during the final hour of this treatment.

The metals titanium, iron, cobalt and nickel were in a wire form. Vanadium and chromium films were produced by electrically heating pellets of these metals on tantalum wire supports. Manganese films were produced by electrically heating flakes of this metal on tantalum wire. Copper wire was also supported by a tantalum wire.

The pressure just before evaporation of the film was 10^{-4} N m^{-2} (10^{-6} torr) or better, measured with an ionization gauge near to the reaction vessel. Benzene and perdeuteriobenzene were allowed to react on the catalyst at 0°C and a total hydrocarbon pressure of 133.3 N m^{-2}. The mass spectrometer (A.E.I., M.S.3) was operated at 10 ev. Under this condition no material at mass 77 was observed. Corrections were made for ^{13}C and the amount of C_6HD_5 in C_6D_6 at the start of the reaction. The adsorption of benzene on the metal films was measured in separate experiments using an LKB thermal conductivity gauge.

3. RESULTS

The reaction should proceed by a randomisation process similar to that described in a previous paper 1), in which the percentage of species at equilibrium was calculated theoretically and compared with experimental values obtained using rhodium and iridium films. This process was repeated for a cobalt film examined for this work and a similar agreement was found.

The nomenclature of Anderson and Kemball 11) has been adopted, with modifications, to derive rate constants for this reaction. The variables $\emptyset_F$ and $\emptyset_R$ are given by

$$\emptyset_F = d_1 + 2d_2 + 3/2d_3 \qquad (1)$$

and $$\emptyset_R = d_5 + 2d_4 + 3/2d_3 \qquad (2)$$

Where d_1 is the percentage of C_6H_5D in the mixture at any time d_2 is the percentage of $C_6H_4D_2$ in the mixture etc. This sub-division of the reaction into two halves yields two rate constants k_F and k_R given by the following:

$$-\ln(\emptyset_\infty - \emptyset_F) = (k_F\, t/\emptyset_\infty) - \ln(\emptyset_\infty) \qquad (3)$$

$$-\ln(\emptyset_\infty - \emptyset_R) = (k_R\, t/\emptyset_\infty) - \ln(\emptyset_\infty) \qquad (4)$$

k_F and k_R are the rate constants for the initial entry of deuterium or hydrogen into 100 molecules of hydrocarbon per minute per milligram of catalyst at the beginning of the reaction and $\emptyset_\infty$ is the equilibrium value of $\emptyset_F$ and $\emptyset_R$.

Two more rate constants can be defined viz: k_o and k_6 which represent the initial disappearance of benzene and perdeuteriobenzene respectively, and are obtained from the following equations:

$$-\ln(d_o - d_{o_\infty}) = k_o t/(100 - d_{o_\infty}) - \ln(100 - d_{o_\infty}) \qquad (5)$$

$$-\ln (d_6 - d_{6_\infty}) = k_6 t/(100 - d_{6_\infty}) - \ln (100 - d_{6_\infty}) \quad (6)$$

where d_{0_∞} and d_{5_∞} are the equilibrium values of d_0 and d_6.

The multiplicity of exchange factors M_F and M_R are given by

$$M_F = k_F/k_0 \quad (7)$$

$$M_R = k_R/k_6 \quad (8)$$

Plots in ln $(\emptyset_\infty - \emptyset_F)$, ln $(\emptyset_\infty - \emptyset_R)$, ln $(d_0 - d_{0_\infty})$ and ln $(d_6 - d_{6_\infty})$ against time, were used to evaluate the rate constants.

Least squares procedures, by computational analysis, were used to find the best straight lines, correlation coefficients were typically 0.995. The results for metals of the first transition series are shown in Table 1 in descending order of activity.

TABLE 1
Randomisation rates for benzene and perdeuteriobenzene at 0^oC

Metal	k_F^*	k_R^*	k_0^*	k_6^*	M_F	M_R	No. of films examined
	*units, % µM min^{-1} mg^{-1}						
Cobalt	13.4	11.0	7.9	6.2	1.8	1.9	4
Nickel	11.6	11.5	7.3	7.0	1.6	1.6	6
Iron	11.2	11.5	7.0	6.6	1.6	1.7	5
Manganese	4.9	4.8	2.7	2.5	1.7	1.6	7
Chromium	3.5	3.6	2.3	2.7	1.6	1.5	7
Vanadium	1.0	3.1	0.6	0.5	1.7	1.7	7
Titanium	0.6	0.7	0.2	0.2	3.0	3.5	5
Copper	I N A C T I V E						4

The mean monolayer coverages of benzene on films of the first transition series metals are given in Table 2. These are compared with molybdenum which, for convenience, is given the relative value 100.

TABLE 2
Monolayer benzene coverages on films of the first transition series metals and molybdenum

Metal	Monolayer (µM mg^{-1})	Relative surface area (Mo = 100)
Ti	0.60	92
Cr	0.30	46
Mn	0.30	46
Fe	0.25	38
Co	0.08	12
Ni	0.08	12
Mo	0.65	100

The variation in these relative surface areas, together with the variation in $\log_{10} k_F$ from metal to metal in the first transition series is shown in Figure 1.

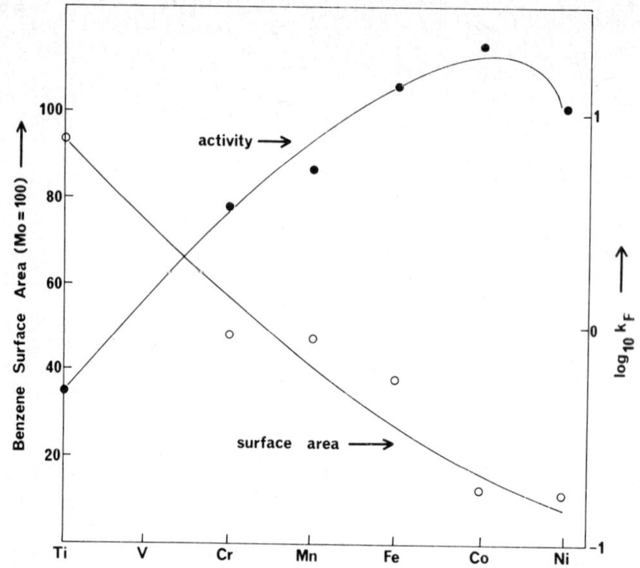

Fig. 1. The variation, from metal to metal in the first transition series of a) $\log_{10} k_F$ (●), b) relative surface area (○),

4. DISCUSSION

Of the eight metals outlined for this paper, seven displayed activity for the exchange reaction at 0°C. These metals, along with the nine metals described in a previous paper [1] show that a wide range of metals is active for this reaction. The results of the exchange reaction on the first transition series will be considered in relation to the other transition metals examined.

The activity of all the metals examined can be arranged in order of decreasing activity.

Rh > Ir > Mo > Re > W > Co > Ni > Fe > Pt > Mn > Cr > Pd > Ta > V > Ti > Ag

Cu, Au and Hf are inactive.

Few workers have covered such an extensive selection of metals but the order of activity for benzene hydrogenation (W > Pt > Ni > Fe > Pd) by Anderson and Kemball [11] agrees well with the above sequence, except for platinum which we consider to be anomalous. The activity series found by Schuit and Van Reijen [12] for the hydrogenation of benzene on silica-supported metals also agrees if platinum and palladium are excluded (Rh > Co > Ni > Fe).

The plot of the logarithm of the rate constant, k_F, against periodic group number for all the metals examined is shown in Figure 2. The activity was found to increase from group VA, to reach a maximum at group VIII, and then decrease again. This pattern is the same as that found for the hydrogenation of ethylene on metal films and on supported metals [13].

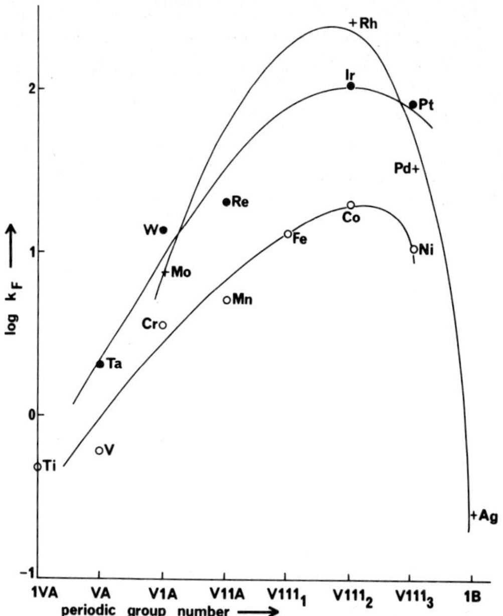

Fig. 2. Variation of log k_F from metal to metal in the three transition series (O) 1st Transition series, (+) 2nd Transition series, (●) 3rd Transition series.

The valence-bond theory has been used by many authors to relate catalytic activity to some electronic parameter. This is usually the percentage d-character of the intermetallic bonds, and a plot of $\log_{10} k_F$ against this parameter for all the metals, except platinum and palladium, is shown in Figure 3.

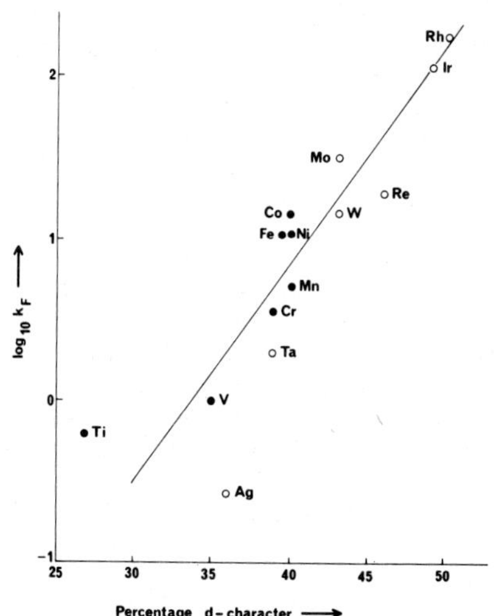

Fig. 3. Log_{10} k_F as a function of percentage d-character of the intermetallic bonds.

Thus the rate constant k_F can be predicted from the equation of the straight line shown in Figure 5. This line is

$$\log_{10} k_F = 0.13\,(\%d) - 4.38 \qquad (9)$$

Percentage d-character is calculated from a knowledge of the single bond radius of the metal and a plot of $\log_{10} k_F$ against metallic radius for most metals of the three transition series is given in Figure 4.

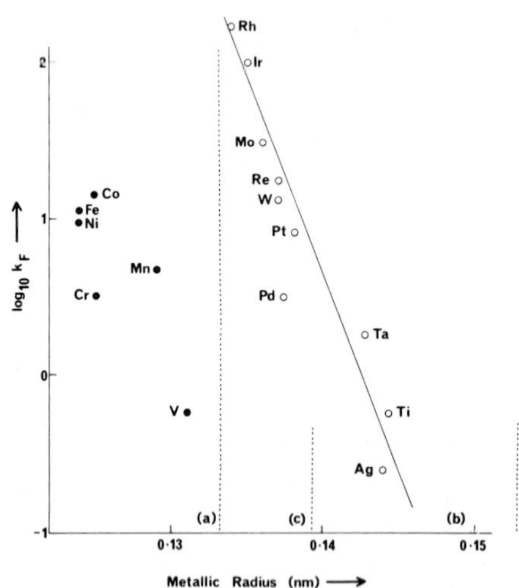

Fig. 4. $\text{Log}_{10}\ k_F$ as a function of the intermetallic radius.

The results depicted in this figure fall into two distinct sections; metals with radii less than 0.132 nm and metals with radii above this value. If platinum is removed from the latter series (because of its unusual kinetic behaviour), a straight line of negative slope is obtained. Metals of the former type (all of which are in the first transition series) do not show the same linear relationship. To explain the linear portion of the graph, the values of the carbon-carbon double and single bond distances and the carbon-carbon bond distance in benzene are required. They are: 0.133, 0.154 and 0.139 nm respectively, and are marked by the dotted lines a, b, and c in Figure 6.

Thus a striking feature of the linear portion of the graph is that it lies between the two lines a and b representing the C-C bond distance in ethylene and ethane respectively. The chemisorption of benzene on the surface of most metals would seem to involve the accommodation of the carbon-carbon distance to fit the lattice of the metal. This could suggest the presence of a di-adsorbed benzene species. Metals with metallic radii less than the carbon-carbon bond length in an isolated double bond system may form this di-adsorbed species less readily, and this could explain the rate effect on the metals iron, cobalt, nickel, chromium and vanadium

5. MECHANISM

A pictorial scheme of suggested mechanisms for this exchange reaction is shown in Figure 5.

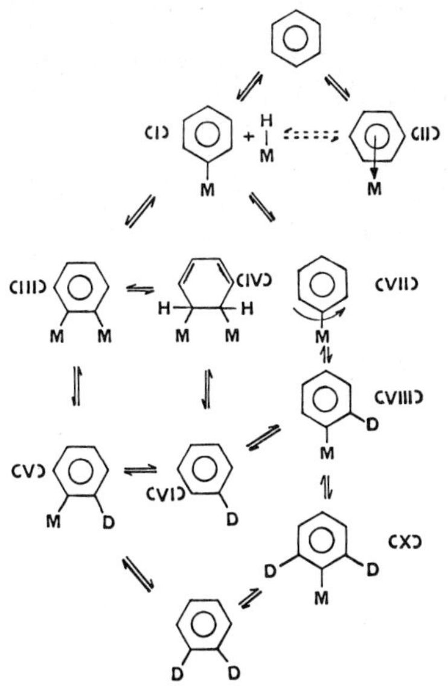

Fig. 5. Possible interconversions between adsorbed forms of benzene.

The reaction is initiated by the chemisorption of benzene at the surface of the metal to form a σ-bonded species (I) or a π-bonded species (II). Interconversion between species (I) and (II) may occur according to the views of Garnett and his co-workers[3]. The σ-bonded species is formed by a dissociative process liberating the hydrogen atom essential to permit exchange of benzene with perdeuteriobenzene. This active component of the reaction (I) can turn to form an α-β di-adsorbed species (III). The exchange could be completed by the addition of a deuterium atom to give intermediate (V) which can form a singly exchanged benzene (VI) by the addition of hydrogen, or double exchanged benzene (IX) by the addition of deuterium. Evidence for this step is provided in the value of the multiplicity factors M_F and M_R of about two on most metals examined.

Platinum and palladium could form species (IV) by the opening of a double bond and subsequent exchange could occur by a stepwise mechanism, as values of M_F and M_R for these metals are about one.

A third type of interaction between benzene and deuterium is shown in species (VII). Moyes and Wells[14], studying the exchange of pyridine with deuterium, propose a rotation of the mono-σ-adsorbed pyridine about an axis perpendicular to the metal surface to explain exchange at the two and six positions in pyridine. A similar mechanism could occur for the reaction of benzene with the metals vanadium, chromium, manganese, iron, cobalt and nickel. The sigma-adsorbed benzene could rotate to interact with an adsorbed deuterium atom which is positioned more favourably than an immediate neighbour. This mechanism would be expected to yield M values of three, and the values observed for the above metals are closer to two but this could be a contributing factor in explaining the exchange reaction on these metals. However, the conditions chosen, and the analysis of the results make it difficult to observe large M values and such an effect may be cloaked.

Investigations of the relative amounts of benzene adsorbed on the metals of the first transition series show that the reaction rate constant is approximately inversely proportional to the relative amount of benzene adsorbed. One explanation of this finding is that these measurements are of associatively adsorbed benzene (II in Fig. 6). In common with measurements of the adsorption of other gases by metal films made by other workers,[15] we find that the amount of benzene adsorbed decreases from titanium to nickel, while the catalytic activity increases. This suggests that such adsorption is not a measure of the active form of the adsorbed benzene, which we ascribe to form I in Figure 6, and points to the difficulties in comparing one metal with another on the basis of rates calculated per unit surface area of metal. In fact, correction of the results in Figures 4 and 5 by applying relative surface areas obtainable from the literature for the adsorption of krypton, xenon, hydrogen or oxygen, make little difference to these relationships which continue to show the same trends.

The hydrogenation of benzene on metal films is now being investigated to find if the same trends exist in this system.

ACKNOWLEDGEMENTS

Two of us, K.B. and R.C.S., wish to thank the Science Research Council for awards. The mass spectrometer was purchased with the aid of a grant from the former Department of Scientific and Industrial Research.

REFERENCES

1) R.B. Moyes, K. Baron and R.C. Squire, J. Catal. 22 (1971), 333.
2) E. Crawford and C. Kemball, Trans. Faraday Soc. 58 (1962), 2452.
3) J.L. Garnett and W.A. Sollich-Baumgartner, J. Phys. Chem. 68 (1964), 3177.
4) R.R. Fraser and R.N. Renaud, J. Amer. Chem. Soc. 88 (1966), 4365.
5) K. Hirota, T. Ueda, T. Kitayama and M. Itoh, J. Phys. Chem 72 (1968), 1976.
6) L. D'Or, Bull. Classc. Sci. Acad. Roy. Belg. 45 (1959), 387.
7) P. Tetenyi, Proc. Int. Congr. Catal. 3rd, A (1965), 1223.
8) P. Tetenyi and L. Babernics, J. Catal. 8 (1967), 215.
9) J. Erkelens and S.H. Eggink-du Burck, J. Catal. 15 (1969), 62.
10) C. Horrex, R.B. Moyes and R.C. Squire, Preprint Proc. Int. Congr. Catal. 4th Moscow (1968).
11) J.R. Anderson and C. Kemball, Advan. Catal. Relat. Subj. 9 (1957), 51.
12) G.C.A. Schuit and L.L. van Reijen, Advan. Catal. Relat. Subj. 10 (1958), 242.
13) G.C. Bond, R.I.C. Reviews 3 (1970), 1.
14) R.B. Moyes and P.B. Wells, J. Catal. 21 (1971), 86.
15) D. Brennan and R. Graham, Phil. Trans. A258 (1965), 325, 347 and 358.

DISCUSSION

J. L. GARNETT

In this exceedingly interesting paper, the authors have presented further evidence to show that exchange in benzene is a dissociative process on a wide variety of metal films. My question refers to the reaction scheme depicted on page 7 of the paper. This shows possible interconversion between the adsorbed forms of benzene, in particular σ-bonded phenyl radicals interconverting with 1,2-diadsorbed benzene. Do you feel that metal lattice distances on the surface would render difficult the possibility of forming the diadsorbed species with benzene? In this respect, I again draw attention to the value of the homogeneous platinum catalyzed exchange of benzene where M values greater than unity are observed, but only one platinum moiety is involved (Garnett et al., this conference). Thus, with heterogeneous systems it is still possible to rationalize M values greater than unity using a single metal site, and one could propose a suitable relevant reaction scheme to explain the results in the present work using one metal site.

R. B. MOYES

I am grateful to Professor Garnett for this comment. It raises two matters. First the M values, i.e. the multiplicity of the reaction. I would point out that M values usually assume a high preponderance of deuterium on the surface. In our case the ratio H:D on the surface was 1:1, which suggests that M values should, perhaps, be doubled to give their true meaning, the number of deuterium atoms exchanged for hydrogen atoms at one residence on the surface. With regard to the second point, the 1-2 diadsorbed benzene and metallic radius, we feel we have found a relationship. It is clear that the benzene molecule must "roll" on one edge, adsorbed at one or two points on the surface, or be adsorbed parallel to the surface. We favour the "roll" mechanism which involves successive single- and double-point adsorption. It is, of course, possible that the adsorption to a single atom could lead to exchange in more than one position on the ring as suggested by Professor Garnett in his paper. We have tried to suggest this in our reaction scheme.

G. C. BOND

Metallic radius cannot of itself be the determining factor in the catalytic activity of metals. The best evidence for the truth of this statement is found in the activities of alloys, where abrupt rate changes can occur without a corresponding change in lattice parameter. Metallic radius, like percentage d-bond character and many other physical properties, reflects the number of valence electrons and the manner of their use. The chemical and catalytic similarities between the second and third row metals are directly attributable to the Lanthanide constraction: This suggests that the polarizability of electrons in surface orbitals may be of importance. We should not fool ourselves that, because we can find reasonable correlations between catalytic activity and some physical parameter, we really understand catalysis.

R. B. MOYES

Professor Bond is entitled to his view. We feel that we have found an interesting relationship between metallic radius and the logarithm of the reaction rate constant. Our explanations may or may not be invalid.

In order to compare reactions on different metals it is necessary to be sure that the effects are not due to changes in mechanism, or relative surface area. We are happy that, except for platinum and palladium, the mechanism is likely to be the same for all metals. The relative constancy of the M values is one aspect of this. The surface areas of films were measured by benzene chemisorption and other means. Correction of the values in Figs. 2 and 3 and 4 do not alter the essential features of these relationships. The energy of activation for the reaction is virtually the same for all the metals.

K. TANAKA

Looking at Fig. 4, I would like to ask you if you have observed any significant differences in the isotopic distribution patterns of the exchanged benzenes between the open-circle metals and the solid-circle metals? I am asking this because I suspect that the exchange mechanism might be different between the two metal groups, and thereby you might get different isotopic distribution patterns.

R. B. MOYES

No difference on the isotopic exchange patterns have been noted between the first transition series metals and other transition metals when used as catalysts for this reaction. However, it should be noted that, because equal concentrations of C_6H_6 and C_6D_6 were used, the production of highly exchanged species in the reaction cannot be observed. We looked for, but found no differences between the rates at which H was exchanged for D or D for H. We are re-examining the exchange of benzene with gaseous deuterium catalyzed by these metals and, in this case, we observe both simple and multiple exchange.

P. TETENYI

Some data obtained in our recent works with L. Babernics lend support to the presented paper.

We have investigated the reversible and irreversible adsorption of benzene, and the dissociative adsorption as well. The temperature range was 50-200°C. The catalysts were Fe, Co, Ni, Cu and Pt.

i) It was stated that the mean quantity of benzene adsorbed irreversibly was one order of magnitude smaller than the total adsorption.

ii) There was not observed any correlation between the catalytic activity for benzene hydrogenation and the total quantity of adsorbed benzene. However definite parallelism was found between the catalytic activity and quantity of benzene adsorbed dissociatively by rupture of C-H bonds.

R. B. MOYES

I am very grateful for this information. It supports our view that, as shown for the first transition series in this paper, a separate measure of what might be called "actively adsorbed" benzene is required. By this we mean benzene able to take part in reaction, as compared with benzene adsorbed, but not apparently reacting. Radioactive- and deuterium-labeled benzene are likely to give rise to this useful information, but not measurements of benzene chemisorption alone.

S. SIEGEL

In general, correlations between reaction rates and a particular parameter may or may not be found. If not found, this may mean that no relationship exists, or it may imply different rate controlling steps applying to different groups of reactions (or a single group) which might correlate with the selected parameter. We simply emphasize Dr. Tanaka's comment that perhaps you are not dealing with single mechanism. His paper at this meeting is an interesting illustration.

R. B. MOYES

See reply to comment No. 2. Clearly, more work will have to be done before we can state absolutely that the mechanism is the same for all the metals, other than platinum and palladium. We can find no evidence to suggest that the mechanism is not the same.

Paper Number 50

CONSECUTIVE SKELETAL REARRANGEMENTS OF HYDROCARBONS IN THE ADSORBED PHASE. ISOMERIZATION OF 2-3-DIMETHYLBUTANES 1-^{13}C AND 2-^{13}C ON PLATINUM FILMS

J. M. MULLER and F. G. GAULT
Institut de Chimie - Université de Strasbourg, France

ABSTRACT : The isomerization of unlabeled dimethylbutanes and of 2-3-dimethylbutane I-^{13}C and 2-^{13}C was studied over platinum films at 275-350°C. In the isomerization of 2-3-dimethylbutane, besides the 2-methylpentane and the neohexane expected by a bond shift mechanism, 3-methylpentane, n-hexane, methylcyclopentane and benzene are formed initially, which can only be obtained by the succession of at least two consecutive skeletal rearrangements in the adsorbed phase. In the tracer experiments, the labeling of 3-methylpentane is best explained by the succession of a bond shift and a cyclic mechanism, while n-hexane is partly formed by two consecutive bond-shift rearrangements. The paper shows then that, on platinum films as on platinum-alumina catalysts, the skeletal rearrangement of saturated hydrocarbons consists of a multistep process and not of a single elementary reaction.

1. INTRODUCTION

It was shown in a number of papers that platinum itself in supported platinum catalysts catalyses the rearrangement of saturated hydrocarbons. This reaction, effective even at a rather low temperature (250-300°C) is quite distinguishable from the classical carbonium ion rearrangement of intermediate olefins (1) ; the latter reaction, taking place at 350-450°C on the acidic carrier of industrial catalysts, probably competes with the metal catalysed reaction under the reforming conditions.

The skeletal rearrangement of hydrocarbons on platinum metal, first suggested by Liberman, Kazanskii and alii (2, 3) in the case of complex molecules such as trimethylcyclopentanes and trimethylpentanes, was extensively studied by several group of workers during the past ten years. From these studies, it appears that two mechanisms basically may explain the skeletal rearrangements on platinum :

- The bond shift mechanism was first proposed by Anderson and Baker to explain the isomerization of neohexane to isopentane on platinum films (4, 5, 6). This mechanism involves an $\alpha\alpha\gamma$ triadsorbed precursor, which leads, via a bridged intermediate (7) (or an adsorbed cyclopropane intermediate (8)) to the reaction products.
- The cyclic mechanism involves a cyclic intermediate of cyclopentane structure. It was first introduced by Maire, Gault et alii (9) to explain the identity of the initial product distributions in the isomerization of methylpentanes and n-hexane and in the hydrogenolysis of methylcyclopentane on a dispersed supported catalyst. The use of methylpentanes labeled with carbon 13 allowed to investigate further the relative importance of either mechanism and their interrelationship according to the size of the metal particles on a supported catalyst. While a simple cyclic mechanism takes place on a highly dispersed

catalyst (size of the crystallites below 50 Å) (10), both bond shift and cyclic mechanisms combine in the case of a concentrated catalyst (mean size of the particles between 150 and 200 Å). On the latter catalyst, the initial product distributions of the various isotopic species may be explained only if one assumes that several successive rearrangements, cyclic or non-cyclic, takes place in the adsorbed phase before desorption (11).

The scope of the present paper is to show the generality of such a "multistep" mechanism. Platinum films were chosen in order to avoid the classical objection of diffusion into the pores, which may be opposed to the results on platinum-alumina catalysts. Dimethylbutanes were selected as hydrocarbons, since these molecules, in a single step, may only be rearranged according to a bondshift mechanism. For example, 2-3-dimethylbutane is expected to yield only 2-methylpentane and 2-2-dimethylbutane according to this process. The appearance of 3-methylpentane and n-hexane in the initial distribution will then be considered as the clue of a "multistep" mechanism. Lastly, labeling with carbon 13 of these hydrocarbons was also used in order to determine as far as possible the various consecutive steps, cyclic or non-cyclic, which may occur during this reaction.

2. EXPERIMENTAL

2.1 Materials : The purity of the unlabeled dimethylbutanes (obtained from Fluka) was better than 99.9 %. Labeled 2-3-dimethylbutanols were prepared by Grignard synthesis : 2,3 dimethyl 2-butanol-I-^{13}C was obtained from labeled methylmagnesiumiodide and unlabeled isopropyl-methyl-ketone. 2-3-dimethyl 2-butanol-2-^{13}C was prepared from 2-propanone-2-^{13}C and isopropylmagnesiumbromide. Similar synthesis are described extensively in reference (12). The tertiary alcohols were then dehydrated on alumina at low temperature (240°C), to avoid any skeletal rearrangement, and the resulting olefins hydrogenated at 100°C on platinum-alumina. The overall yield was about 50 % (0.1 m Mole of hydrocarbon obtained from 0.2 m Mole of labeled compound). The isotopic purity of the labeled hydrocarbons was a little lower than the one of the starting materials (50 to 60 %).

2.2 Catalysts and procedure : Platinum films (about 10 mg) were evaporated in a conventional vacuum apparatus (reaction vessel of about 200 cc). A leak to a mass spectrometer allowed to follow the reaction and to stop it at a low conversion. The reaction mixture, 5 torr of hydrocarbon and 50 torr of hydrogen, was introduced on the film at 0°C. It is worthwhile to note that two purifications, the first one on a sodium film to eliminate any moisture, the second one on a first platinum film to eliminate some unknown contaminant, were necessary to obtain a reasonable reaction rate.

2.3 Analysis : a) Gas chromatography : The analysis of the reaction mixture was easily performed on a five meters SE 30 silicone column at 0°C. A similar column, but of ten meters, was used to separate the various isomers. 2-3-dimethyl and 2-methylpentane were not resolved enough to prepare the 2-methylpentane with a sufficient degree of purity. In the experiments with labeled compounds, the mass spectrometrical analysis of the 2-methylpentanes could not then be effected.

b) Mass spectrometry : The 3-methylpentanes and n-hexanes separated by gas chromatography were analysed mass-spectrometrically to determine the location of the carbon 13 in the molecule (12). For the analysis of the neohexanes, it was assumed that the ions ($M-CH_3$) and ($M-C_2H_5$) resulted from a fragmentation at the quaternary carbon atom. The mole fractions of the various isotopic species are represented as shown in Figure I.

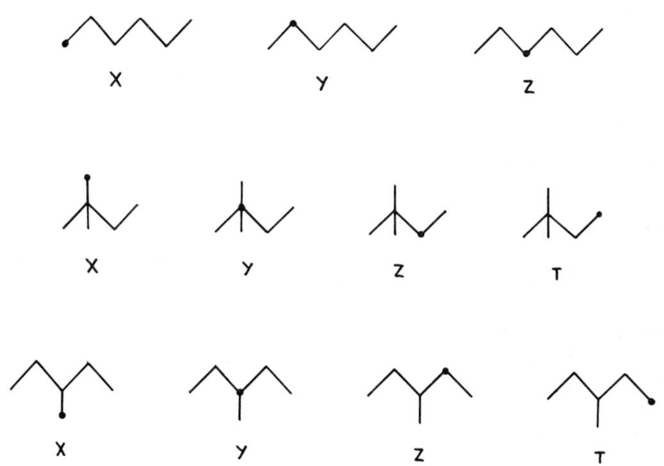

Figure I : Labeled hexanes

3. RESULTS

3.1 Isomerization of unlabeled dimethylbutanes :

The isomerization of 2-3-dimethylbutane (I) was studied over platinum films at various temperatures (275 to 350°C). The analysis of the reaction products at very low conversions are reported in table 1. All hexane isomers, methylcyclopentane and benzene are formed initially. A regular decrease of the amounts of 2-methyl (II) and 3-methylpentane (III) is observed when the temperature is raised from 275 to 350°C, corresponding to a regular increase of benzene. Simultaneously 2-2-dimethylbutane (VI) seems to go through a maximum, while n-hexane (IV) and methylcyclopentane (V) vary on a rather erratic way . At any temperature the hydrocracking to C_5 hydrocarbons is about as important as the skeletal rearrangement.

The isomerization of 2-2-dimethylbutane (VI), studied at two temperatures, shows similar trends : decrease of both methylpentanes and increase of benzene when the temperature is increased. However methylcyclopentane and n-hexane at any temperature are formed from 2-2-dimethylbutane in much smaller amount than from 2-3-dimethylbutane ; cracking to pentane isomers, mostly neopentane, is much more important.

TABLE 1

Product Distributions in the reaction of dimethylbutanes on platinum films

Reacting hydrocarbon	2-3-dimethylbutane				2-2-dimethyl-	
Temperature °C	275	310	325	350	275	350
Conversion : % of C_6	1.34	3.17	4.14	2.52	1.04	1.12
2-2-dimethylbutane	7.8	14.0	10.0	10.65	-	-
2-3-dimethylbutane	-	-	-	-	33.2	27.3
2-methylpentane	28.1	16.0	14.1	7.7	37.4	9.3
3-methylpentane	18.9	18.9	14.8	6.5	12.6	4.0
n-hexane	16.2	23.4	13.8	27.1	5.9	6.9
methylcyclopentane	29.0	21.4	28.1	13.8	3.7	3.2
benzene	0.0	6.3	19.2	34.1	7.2	49.3
C_6/C_5	0.87	0.90	1.40	1.07	0.69	0.59

3.2 Isomerization of 2-3-dimethylbutanes 1-^{13}C and 2-^{13}C at 300°C

The distribution of the isotopic varieties in 3-methylpentane, n-hexane and neohexane are given in Table 2, together with the relative concentrations of these various hydrocarbons plus methylcyclopentane. The relative abundances a_5 and a_4 for ions $C_5H_9^+$ and $C_4H_8^+$ in methylcyclopentane are given in this table.

On account of the very small conversion, 3-methylpentane and n-hexane contained some amount of 2-3-dimethylbutane as impurities. The $C_5H_{11}^+$ fragment of (III) and (IV) being very small in comparison with the corresponding fragment of (I), the corrections due to this impurity were then very often high. For this reason only the deethylated fragment was used for n-hexane and for 3 - methylpentane in the second experiment ; x+y and z+t could be determined in the case of 3-methylpentane, x'+y' and z' in the case of n-hexane.

TABLE 2

Labeling of the isomerization products

Reacting Hydrocarbon		2-3-dimethylbutanes $1-^{13}C$	2-3-dimethylbutanes $2-^{13}C$
3-methylpentane	x	0.245±0.005	0.49±0.01
	y	0.002±0.005	
	z+t	0.752±0.005	0.51±0.01
	%[1]	33.4	32.3
2-2-dimethylbutane	x	0.72	0.155
	y	0.045	0.465
	z+t	0.235	0.380
	%	22.8	27.2
n-hexane	x+y	0.76 ±0.005	0.67±0.02
	z	0.24±0.005	0.33±0.02
	%	29.8	15.6
methylcyclopentane	a_5	0.817	0.920
methylcyclopentane	a_4	0.673	0.715
	%	14.0	24.9

[1]Percent of the C_6 reaction products.

4. DISCUSSION

If a single step is assumed for the isomerization of dimethylbutanes, only the following bond shift rearrangements may be envisaged :

For 2-3-dimethylbutane (Scheme 1) and 2-2-dimethylbutane (Scheme2)

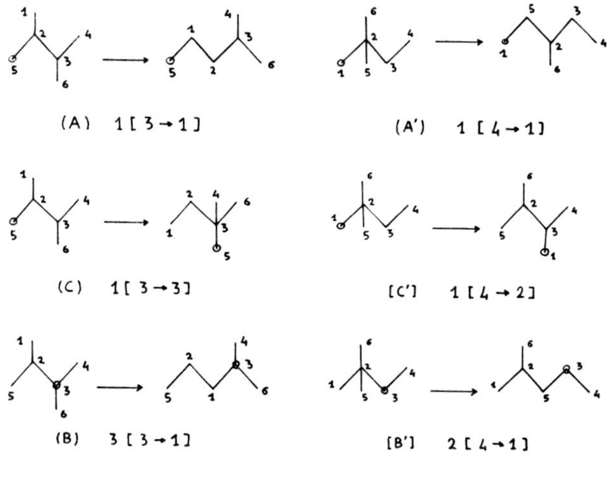

(A) 1[3→1]

(A') 1[4→1]

(C) 1[3→3]

(C') 1[4→2]

(B) 3[3→1]

(B') 2[4→1]

In the schemes, circled carbon atom represents the pivot carbon atom γ in the $\alpha\alpha\gamma$ triadsorbed precursor. The sigla 1[3 → 1] indicates that the pivot carbon atom is a primary one and that the bond shifts from a tertiary to a primary carbon atom.

From the schemes it appears that 2-methylpentane and neohexane only should be formed from 2-3-dimethylbutane and that n-hexane and benzene should not be obtained from neohexane. When one compares these expectations with the results shown in table 1, it is obvious that, except in the isomerization of 2-2-dimethylbutane at 275°C, where 85% of the reaction products may be accounted for by a simple bond shift rearrangement, in any other case, most of the products (60 to 80 %) can only be explained by the succession of several rearrangements in the adsorbed phase. The presence of large amounts of methylcyclopentane in the products of 2-3-dimethylbutane suggests that, in the case of this molecule, the cyclic mechanism of rearrangement or at least its first step, the dehydrocyclization, is very important as an elementary reaction consecutive to the bond shift rearrangement. It would be therefore interesting to examine whether the succession of a bond shift and of a cyclic mechanism could explain the distributions of the various isotopic species.

4.1 3-methylpentane. The succession of a bond shift mechanism A or B and of a cyclic rearrangement of the resulting hydrocarbon, (A+Cyc.) or (B+Cyc.) does explain indeed the distributions of the 3-methylpentane varieties, as shown in table 3. But it should be noted that a methyl shift T' following reactions A or C would also account for the same product distributions.

TABLE 3

Labeled 3-methylpentanes
Comparison between the observed and the calculated values

Reacting Hydrocarbon	3-methyl Pentane-		A+Cyc	B+Cyc	A+T'	B+T'	C+T'	obs
2-3-dimethyl-butane 1-^{13}C	Me-^{13}C	x	0,25	0,25	0,25	0,25	0,25	0,25
	3-^{13}C	y	0	0	0	0,25	0	0
	2-^{13}C	z	0,25	0,50	0,25	0	0,25	0,75 (z+t)
	1-^{13}C	t	0,50	0,25	0,50	0,50	0,50	
2-3-Dimethyl butane 2-^{13}C	Me-^{13}C	x	0	0	0	0	0	0,49 (x+y)
	3-^{13}C	y	0,50	0,50	0,50	0	0,50	
	2-^{13}C	z	0,50	0	0,50	1	0,50	0,51 (z+t)
	1-^{13}C	t	0	0,50	0	0	0	

We believe that most of the 3-methylpentanes formed by isomerization of 2-3-dimethylbutane result from the succession of two rearrangements according to a bond shift and a cyclic mechanism. In the isomerization of 2-2-dimethylbutane at 275°C indeed, where there is no evidende for an important multistep mechanism, the shift B' (2[4 → 1]) producing 2-methylpentane and invoving a secondary carbon atom as pivot, seems to be more favoured than the shift A' (1[4→1]) producing 3-methylpentane and involving a primary atom. One does not see therefore why reaction B (3[3→1]) should be less plausible than reaction A (1[3→1]). Mechanisms (A+Cyc.) and (B+Cyc.) seem therefore the most correct mechanisms to explain the 3-methylpentane distributions.

4.2 n-hexane. The distributions of the n-hexanes varieties on the contrary cannot be explained on a simple way by one of the previous mechanisms ((A+Cyc.) or (B+Cyc.) (table 4). It is necessary in this case to introduce also the succession of two bond shift rearrangements.

TABLE 4

Labeled n-hexanes-Comparison
between the observed and the calculated values.

Reacting Hydrocarbon	n-hexane	A+Cyc	B+Cyc	A+A"	A+B"	B+A"	B+B"	Obs
	1-^{13}C x	0,375	0,625	0,5	0,5	0,5	0,5	
2-3-dimethyl butane-1-^{13}C	2-^{13}C y	0,25	0	0,5	0,25	0,25	0	0,76
	3-^{13}C z	0,375	0,375	0	0,25	0,25	0,5	0,24
	1-^{13}C x	0,25	0	0	0	0	0	
2-3-dimethyl butane-2-^{13}C	2-^{13}C y	0,50	0,75	0	0,5	0,50	1	0,67
	3-^{13}C z	0,25	0,25	1	0,5	0,50	0	0,33

In table 4, A" and B" represent the two possible bond shift rearrangements of 2-methylpentane into n-hexane (Scheme 3) :

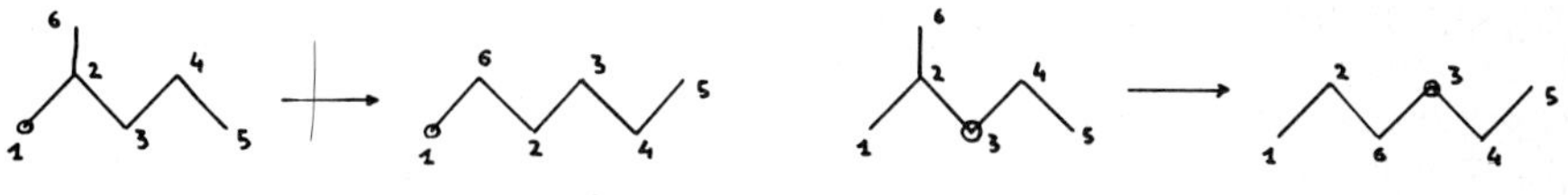

(A") 1 [3→ 1] (B") 2 [3→ 1]

Scheme 3

The importance of the cyclic mechanism for forming n-hexane, relatively small, is entirely consistent with the fact that the hydrogenolysis of methylcyclopentane yields methylpentanes rather than n-hexane (13).

4.3 Neohexanes. As shown in table 5 the distribution of the neohexanes formed from 2-3-dimethylbutane-1-^{13}C is consistent with a simple bond shift mechanism. In the second experiment however, it appears that some of the molecules result from a multistep mechanism.

TABLE 5

Labeled neohexanes-Comparison between the observed and the calculed values.

Reacting Hydrocarbon	2-2-dimethyl butane		B	A+T'	B+T'	Obs
2-3-dimethyl butane 1-^{13}C	1-^{13}C	x	0,75	0,50	0,75	0,72
	2-^{13}C	y	0	0	0	0,04
	3-^{13}C	z	0	0,25	0	0,24
	4-^{13}C	t	0,25	0,25	0,25	
2-3-dimethyl butane 2-^{13}C	1-^{13}C	x	0	0,5	0	0,15
	2-^{13}C	y	0,5	0,5	0,5	0,47
	3-^{13}C	z	0,5	0	0,5	0,38
	4-^{13}C	t	0	0	0	

T' represents the reverse step of B'.

4.4 Methylcyclopentane. The mass spectra of the methylcyclopentanes labeled in the various positions were recently obtained and the relative abundances of the heavy fragments $^{13}C\ C_3H_8^+$ and $^{13}C\ C_4H_9^+$ a_4 and a_5 determined for these molecules with an accuracy of ±10 % (14). It is therefore possible to calculate the a_4 and a_5 values expected if methylcyclopentane results, as we believe, from the dehydrocyclization of the 2-methylpentanes obtained by a first bond-shift reaction and to compare these values with the observed ones (Table 6). The agreement between the calculated and the observed values is quite good.

4.5 Benzene. The initial formation of benzene from 2-3-dimethylbutane and 2-2-dimethylbutane necessitates at least two skeleta rearrangements, which follows either a 1-6 ring closure of an adsorbed n-hexane or a ring enlargement of an adsorbed methylcyclopentane. At the reverse of what is observed on a highly dispersed catalyst, where the first reaction seems to occur (15 on platinum films, both reactions are equally possible (16).

IN CONCLUSION, the "multistep" mechanism, demonstrated in the case of a highly concentrated catalyst (10% platinum-on-alumina with a average particle size of 170 Å) is also taking place on platinum films. However, while on platinum-alumina the number of surface reactions before desorption is rather large, on film it does not seem to be larger than two or three. Moreover the formation of 3-methylpentane, where the distribution is best explained by the succession of a bond shift and a cyclic mechanism is quite distinct from the formation of n-hexane. In this case, two consecutive bond shifts occur. The difference between the mechanisms for forming either hydrocarbon could explain the variation with temperature of the product distributions.

TABLE 6

Relative abundances a_5 and a_4 for labeled methylcyclopentane-Calculated and observed values.

	2-3-D.M.B. 1-^{13}C			2-3-D.M.B. 2-^{13}C		
	Product distributions			Product distributions		
	A+Cyc	B+Cyc	Obs	A+Cyc	B+Cyc	Obs
Me-^{13}C	0.25	0.25		0	0	
I-^{13}C	0	0		0.50	0.50	
2-^{13}C	0.25	0.50		0.50	0	
3-^{13}C	0.50	0.25		0	0.50	
	Relative abundances			Relative abundances		
a_5	0.76	0.75	0.82	0.88	0.90	0.92
a_4	0.67	0.66	0.67	0.72	0.71	0.71

The calculated values of a_5 and a_4 are obtained from the following values of the pure methylcyclopentanes.

	Me-^{13}C	1-^{13}C	2-^{13}C	3-^{13}C
a_5	0.45	0.94	0.83	0.87
a_4	0.96	0.86	0.58	0.56

REFERENCES

1) G.A. MILLS, H. HEINEMANN, T.H. MILLIKEN and A.G. OBLAD, Ind. Eng. chem. 45 (1953), 134.
2) B.A. KAZANSKII, A.L. LIBERMAN, T.F. BULANOVA, V.T. ALEKSANYAN and Kh.E. STERIN, Dokl. Akad. Nauk. SSSR. 95-1 (1954), 77 ; 95-2 (1954), 281.
3) A.L. LIBERMAN, T.V. LAPSHINA and B.A. KAZANSKII, Dokl. Akad. Nauk. SSSR. 105 (1955), 727.
4) J.R. ANDERSON and B.G. BAKER, Nature 187 (1960), 937 ; Proc. Roy. Soc. (London) A 271 (1963), 402.
5) J.R. ANDERSON and N.R. AVERY, J. Catal. 2 (1963), 542 ; 5 (1966) 446.
6) M. BOUDART, A.W. ALDAG, L.D. PTAK and J.E. BENSON, J. Catal. 11 (1968),35.
7) J.R. ANDERSON and N.R. AVERY, J. Catal. 7 (1967), 315.
8) J.M. MULLER and F.G. GAULT, 4th Inter. Congress on Catalysis, Moscow, 1968. Symposium on "Mechanism and kinetics of Complex catalytic reactions". Paper 15.
9) Y. BARRON, G. MAIRE, J.M. MULLER and F.G. GAULT, J. Catal. 2 (1963), 152 ; 5 (1966), 428.
10) C. COROLLEUR, S. COROLLEUR and F.G. GAULT, J. Catal. in Press.
11) C. COROLLEUR, D. TOMANOVA and F.G. GAULT, J. Catal. in Press.
12) C. COROLLEUR, S. COROLLEUR and F.G. GAULT, Bull. Soc. chim. Fr. (1970),158.
13) G. MAIRE, G. PLOUIDY, J.C. PRUDHOMME and F.G. GAULT, J. Catal. 4 (1965), 556.
14) C. COROLLEUR, S. COROLLEUR and F.G. GAULT, unpublished work.
15) F.M. DAUTZENBERG and J.C. PLATTEUW, J. Catal. 19 (1970), 41.
16) J.M. MULLER and F.G. GAULT, unpublished work. Ph. Thesis Caen (1969).

DISCUSSION

J. R. ANDERSON

First, I should like to congratulate Professor Gault and his group on a very fine piece of work. The reality of multistep surface processes in these sorts of reactions is, I believe, beyond question. It remains now to consider what the implications are. One most important projection one can make is that the overall nature of this sort of reaction is likely to be highly sensitive to catalyst structure, both as regards to surface topography and surface poisons. There is now good evidence which is summarized elsewhere[1,2] that bond shift processes require catalytic sites consisting of two or three adjacent platinum atoms as on a low index crystal face, while reaction via an adsorbed carbocyclic intermediate uses a single platinum atom site which is for preference of low co-ordination as at a crystal corner.

On this basis, the favoured route for a sequence involving (say) a bond shift followed by a carbocyclic intermediate would require the transport of the reacting entity from a site on a crystal face to a corner site. While this would obviously be possible on a single platinum crystal, there is also the possibility of transfer between crystals. Since inter-site transport may involve surface diffusion and also possibly gas-phase transport in a porous supported catalyst, one would expect the reaction details to be dependent on general catalyst morphology, as well as on the structure of the metal particles themselves. Such evidence as is currently available[1,2] does indicate considerable sensitivity of this sort of reaction to the various aspects of catalyst structures, but a lot more exploration in this area remains to be done. One might add that inter-site transfer of adsorbed hydrocarbon can presumably occur via chemisorbed species (as well as via the more obvious physically adsorbed species), by making use of repeated transitions between non-adsorbed and diadsorbed structures; the adsorbed residue then "walks" across the surface.

1) Anderson, J. R., to appear in Advances in Catalysis.
2) Anderson, J. R., Macdonald, R. J. and Shimoyama, Y., J. Catal. 20, 147 (1971).

F. G. GAULT

1) I agree entirely with the idea that the cyclic mechanism of isomerization, at least on the highly dispersed platinum catalysts, is connected with one-atom sites, while the bond shift rearrangement involves several metal atoms.

Species adsorbed on single metal atom sites were already proposed to explain the non-selective hydrogenolysis of methylcyclopentane and the cyclic type of rearrangement on highly dispersed catalyst[1,2] and further evidences were then given to support this idea.[3,4] Recently an intermediate species for dehydrocyclization, with two carbon atoms bonded to a single metal atom, was described; such an intermediate was strongly supported by the identity of the cyclization rates of several trimethyl and tetramethylpentanes.[5]

However it was shown that a second type of hydrogenolysis of methylcyclopentane corresponding to a selective break of CH_2-CH_2 cyclic bond does occur on catalysts with large crystallites.[1] It is quite possible, as stated,[1] that the species associated with this selective ring opening and with the corresponding dehydrocyclization process involve several metal atoms, as in the bond shift rearrangement.

2) Concerning the multistep process and the nature of the sites, it is interesting here to relate two experiments.[6]

a) The multistep process occurs only on platinum-Al_2O_3 catalysts with sufficiently large crystallites, beyond 120 Å.

b) The second experiment was already quoted during the discussion of Professor Anderson's paper.

When a highly dispersed 10% platinum-alumina catalyst is sintered at 500°C in hydrogen, the size of the crystallites increases, the activity decreases, but the selectivity in methylcyclopentane hydrogenolysis does not change at all. That shows that the number of sites have decreased, but the nature of the sites did not change.

The best way, in our opinion, to interpret the second experiment is to admit that the low-index regular faces do not play a catalytic role and that two types of defects (or more) involving "ensembles" of one-atom or several atoms are present on the surface. The relative abundances of these defects are determined by the size of the crystallites formed in the early stage of the reduction of the catalysts.

One may speculate about the nature of the defects: corners and edges, superficial atoms in various oxidation states, etc. It remains that they are only incidentally related to the regular crystal lattice itself and, in the case of large crystallite catalysts, most probably not contiguous.

This last consideration could explain experiment (a), showing that a critical size is necessary for a multiple process. Several deffects may be present on a single crystallite, provided it is large enough. The multistep process could be therefore pictured as an intersite transfer on a single crystal. The experimental results do not allow to say more about this intersite transfer: participation of mono and diadsorbed species or surface diffusion via a van der Waals phase seems equally likely.

1) G. Maire, G. Plouidy, J. C. Prudhomme, and F. G. Gault, J. Catal. 4, 556 (1965).
2) Y. Barron, G. Maire, J. M. Muller, and F. G. Gault, J. Catal. 5, 428 (1966).
3) C. Corroleur, S. Corroleur, D. Tomanova, F. G. Gault, J. Catal. 24, 385, 401 (1972).
4) G. Maire, C. Corroleur, D. Juttard, F. G. Gault, J. Catal. 21, 250 (1971).
5) J. M. Muller, F. G. Gault, J. Catal. 24, 371 (1972).
6) Unpublished results from this laboratory.

J. W. HIGHTOWER

This report reflects a tremendous amount of very excellent work, and the authors are to be congratulated for their results. I notice that in all cases there is an increase in selectivity for benzene as the temperature is increased. How do equilibrium effects enter into the picture? Is it possible that you are near equilibrium between benzene and methylcyclopentane and that the increase in benzene selectivity with temperature simply is a consequence of thermodynamics and not necessarily a result of intrinsic selectivity changes?

F. G. GAULT

The question is a very interesting one.

Since we are dealing with initial distributions (1 to 4%) an equilibrium between methylcyclopentane and benzene reached by a desorption-readsorption process should be ruled out. Moreover cyclohexane was never obtained, even at the lowest temperature 275°C, as it would be.

Nevertheless, the ratio methylcyclopentane over benzene is not very far from the equilibrium one, if one assumes that the concentration of hydrogen is much higher on the surface than in the gas phase, which is quite possible. On the other hand the variation of enthalpy for the reaction:

$$\rightleftharpoons \quad + 3H_2$$

obtained from our results if one assumes equilibrium, 8.6 kcal/mole, is not very different from the one obtained from the table (ca 11 kcal/mole). An equilibrium between adsorbed methylcyclopentane and benzene is therefore formally possible.

However, if such an equilibrium is achieved, one should have observed, in experiments previously reported on the contact reactions of methylcyclopentane on platinum film (1), that the aromatization of this compound is very fast in comparison with its ring opening: just the reverse is true.

H. PINES

The data reported in this paper could also be explained by assuming the intermediate formation of 1,2-dimethylcyclobutane from 2,3-dimethylbutane, and 1,1-dimethylcyclobutane from 2,2-dimethylbutane. From the 1,2-dimethylcyclobutane, n-hexane, 3-methylpentane and benzene could be obtained. 1,1-Dimethylcyclobutane could give 2-methylpentane. The intersection between 2- and 3-methylpentane could also be explained by assuming cyclobutane intermediates.

Did you consider cyclobutane intermediates in evaluating the mechanism of the dimethylbutanes isomerization?

F. G. GAULT

The isomerization via a cyclobutane intermediate has not been considered in this paper, mainly because desorbed cyclobutanes could never be observed, even under what was supposed to be the most favorable conditions (during the contact reactions of 2,2,3,3-tetramethylbutane in example).

However this mechanism cannot be ruled out really; the isomerization of 2,3-dimethylbutane via a cyclobutane intermediate would predict a distribution of the 3-methylpentanes identical to the one corresponding to the succession of a double bond shift A and a cyclic mechanism, which is very close indeed to the observed one.

On the contrary, the distribution of the n-hexanes formed via a cyclobutane intermediate would be very different from the observed distribution. But since it is not possible to interpret the n-hexanes distribution by a single mechanism, the "cyclobutane mechanism" is not excluded even in this case.

In the isomerization of the 2,2-dimethylbutane, however, the ring hydrogenolysis of a supposed gemdimethylcyclobutane intermediate would only reproduce neohexane, since, in cyclobutanes as in cyclopentanes, it is very difficult on platinum to open a cyclic quaternary-secondary C-C bond. Even if that was possible, the cyclobutane mechanism would not account for the 3-methylpentane and the n-hexane formed in appreciable amounts.

C. KEMBALL

We (Davie, Whan and Kemball, J. Chem. Soc. Faraday I, in press) have examined reactions of n-butane with deuterium on platinum films at temperatures above 300°C. Exchange is very rapid compared with either deuterolysis or isomerization and are very extensive. Therefore, reactions involving the reversible dissociation of C-H bonds and the interconversion between various adsorbed n-butane species occur much more readily than reactions involving a change of structure of the molecule. Under these circumstances, migration of the hydrocarbon between sites, suggested by Dr. J. R. Anderson as a factor in the results of Muller and Gault, would be expected.

S. SIEGEL

As you know, certain transition metal complexes break open strained cyclics, such as three and four membered rings, under rather mild conditions. Presumably the reverse change would occur easily at higher temperatures, as in your experiments. Might not such intermediates (cyclopropanes), be involved in the mechanism of isomerization?

F. G. GAULT

An adsorbed cyclopropane intermediate has been previously proposed[1] to replace in the bond shift mechanism the bridged intermediate suggested by Anderson and Avery.[2]

In favor of this cyclopropane intermediate is the fact that the dueterolysis of gemdimethylcyclopropane on platinum and irridium* yields large amounts of neopentane 1,1,3d_3[3] which shows that the ring opening of gemdimethylcyclopropane yields an $\alpha\alpha\gamma$ triadsorbed neopentane, i.e. the precursor species in neopentane isomerization.

The reverse reaction of this dissociative hydrogenolysis of cyclopropane was then proposed as a step in the bond shift rearrangement.

The calculations made by Anderson and Avery[4] on the isomerization rates of various hydrocarbons by using the simplified Hückel m.o. theory and the very nice agreement obtained between the observed and calculated relative rates convinced us that their model was correct and not ours.

However, since then, the study of the isomerization of labeled pentanes seemed to indicate that the hypothesis of a cyclopropane intermediate for the bond shift rearrangement could be correct.

F. Garin in Strasbourg has shown that 2-methylbutane 2-^{13}C yields mostly n-pentane 2-^{13}C and only very few n-pentane 3-^{13}C on platinum films and 10% Pt-Al_2O_3 catalyst, which suggests that the bond shift rearrangement involves carbon atom 3 and not 1 as a "pivot atom."

This result is well explained if one assumes the formation of an adsorbed cyclopropane.

I

ads.

ads.

II

Intermediate II in the figure is expected to be much more stabilized than intermediate I, on account of the number of substituents to the cyclopropane ring, and that would explain the observed distribution of the n-pentanes.

1) J. M. Muller, and F. G. Gault, 4th International Congress on Catalysis, Moscow, 1968. Symposium on "Mechanism and kinetics of complex catalytic reactions," Paper 15.
2) J. R. Anderson, and N. R. Avery, J. Catal. 5, 446 (1966).
3) J. C. Prudhomme et F. G. Gault, Bull, Soc. Chim. 832 (1966).
4) J. R. Anderson, and N. R. Avery, J. Catal. 7, 315 (1967)

---

*Among eleven investigated metals, only two, Pt and Ir yielded neopentane d_3, and these two metals are precisely the ones on which the skeletal rearrangement of neopentane takes place.

G. R. LESTER

The results in Table 4 for the 2-C^{13} compound do not distinguish between sequences involving cyclopentane as opposed to those involving only bond shifts, while the results for the 1-C^{12} reactant support the latter sequence. The authors conclude that the bond shift mechanism is the important one for both compounds, and support the conclusion with the statement that little n-hexane is formed by hydrogenolysis of methylcyclopentane on these films. The results of Anderson and Shimoyama[1)] suggest that such thick films differ donsiderably from "ultra thin films" in the hexane made by hydrogenolysis of methylcyclopentane. Does this suggest that the multiple bond shift sequence might be less important than the one involving a cyclopentane on such thick films, and possibly also on commercial supported catalysts?

1) J. R. Anderson and Y. Shimoyama, Preprint No. 47, Vth International Congress on Catalysis, Palm Beach, Florida, 1972.

F. G. GAULT

The results presented by Anderson and Shimoyama concerning thick platinum films are in very good agreement with the one we obtained in similar conditions. Very little n-hexane is formed in the hydrogenolysis of methylcyclopentane. The results obtained on ultrathin films differ considerably and are much closer to the one which can be obtained on highly dispersed supported catalysts, such as the one used in reforming.

The study of the isomerization of 2,3-dimethylbutane on supported catalysts with small particle size would be interesting indeed. One might expect a larger contribution of the cyclic mechanism in the multistep sequence for the formation of n-hexane.

Paper Number 51

CATALYTIC DEUTERATION OF 4-tert-BUTYLCYCLOHEXANONE AND ALLIED REACTIONS OVER PLATINUM METALS

YUZURU TAKAGI, SHOUSUKE TETRATANI and KAZUNORI TANAKA
The Institute of Physical and Chemical Research
Wako-shi, Saitama, Japan

ABSTRACT: With intent to clarify the mechanism of alicyclic ketone hydrogenation to the corresponding alcohol, the title ketone was deuterated at 80°C in cyclohexane under pressure using platinum metal catalysts. The resulting alcohol was separated into isomers, and analyzed by mass spectrometry and nmr spectroscopy. The use of shift reagents in nmr spectroscopy enabled the positions of the incorporated deuterium to be determined.

Simple addition of two deuterium atoms to the carbonyl linkage predominated over Ru, Os, Ir, and Pt, while considerable deuterium incorporation into C_2 (or C_6) position accompanied the double bond saturation over Rh and Pd. The isotopic distribution pattern of the cis-alcohol was very similar to that of the trans-alcohol whichever catalyst used. The striking result with Rh and Pd was in the equality between the axial- and equatorial-deuterium contents at C_2 or C_6. A mechanism is presented which correlates these findings.

1. INTRODUCTION

The factors controlling the cis/trans ratio in the alcohols obtained on catalytic hydrogenation of substituted alicyclic ketones has been the subject of many investigations[1,2]. Whenever an alicyclic ketone in the chair conformation is adsorbed with its C_1-oxygen bond roughly parallel to the catalyst surface, one could envisage the possibility of two distinct modes of adsorption, with either the ring placed toward or away from the catalyst surface. These two modes would be different in hindrance to adsorption. The simple assumptions of the less hindered adsorption and the cis-addition of two hydrogen atoms to the carbony linkage from the catalyst side lead to the exclusive formation of the axial alcohol. In order to account for the actual formation of both of the axial and equatorial alcohols, a variety of mechanistic interpretations have been presented. These include the accompanying cis-trans isomerization of alcohols[3], the mechanisms involving direct trans-addition of hydrogen and the like modes[4-6], the hydrogenation from two distinct conformations of ketone[3], and the interaction of ketone with the produced alcohol[7].

In an attempt to gain some insight into the mechanism of alicyclic ketone hydrogenation using a tracer technique, cyclopentanone was deuterated by Kemball and Stoddart[8], 2-methylcyclopentanone by Cornet and Gault[6], and more recently cyclohexanone in our laboratory[9]. These research groups agreed on an important finding that the deuterium addition to the carbonyl linkage were accompanied more or less by deuterium incorporation into the alicyclic ring, especially into the positions β to the hydroxyl group. However, their mechanistic interpretations were different. Kemball and Stoddart emphasized that ketone and alcohol compete for the surface with the formation of a common adsorbed species. Cornet and Gault proposed a mechanism involving the trans addition of deuterium, while we accounted for our own results in terms of the cis addition.

In these tracer studies, the measurement of the contents of the axial and equatorial deuteriums at each carbon position would provide valuable information concerning the reaction mechanism. A lack of appropriate experimental techniques has not allowed any attempt at such a measurement. However, the recent advent of para-magnetic shift reagents for use in nmr spectroscopy promises to enable this type of measurement to be made. We therefore undertook a re-study of alicyclic ketone deuteration. 4-tert-Butylcyclohexanone was selected for this study because it is conformationally rigid. Some additional reactions were also conducted to decide whether or not the process of deuterium incorporation into the β positions is involved in the sequence of steps of ketone deuteration.

2. EXPERIMENTAL

All reactions were conducted using cyclohexane as a solvent while in our earlier work for cyclohexanone deuteration[9)] no solvent was employed. Most of the experimental procedures were similar to those in this earlier work, and are only briefly described with special reference to the modifications employed to allow the use of solvent, the separation of isomeric products, and the measurements of the nmr shifted spectra.

2.1. Materials

The metal blacks used as catalyst were the same preparations as used previously[9)] for cyclohexanone deuteration. In brief, the blacks were prepared by reduction of the appropriate oxide or hydroxides with hydrogen.

4-tert-Butylcyclohexanone was prepared by reduction of p-tert-butylphenol with hydrogen to give tert-butylcyclohexanol, and by the subsequent partial oxidation with sodium dichromate. The procedure for partial oxidation was similar to that reported for 2-methylcyclohexanone preparation[10)]. The crude ketone obtained was distilled under reduced pressure, and the fraction of b.p. 128-130°C/30 mmHg was collected. Commercial deuterium was stated to be 99.9% in isotopic purity. Other chemicals, including nmr shift reagents, $Eu(dpm)_3$ [tris(dipivalomethanato)europium(III)] and $Pr(dpm)_3$, were also commercially obtained, and used without further purification.

2.2. Deuteration

A sample of 5 g (0.032 mol) of 4-tert-butylcyclohexanone dissolved in 20 ml of cyclohexane, was stirred in a 100-ml autoclave at 80°C with 20 mg of a catalyst and deuterium at an initial pressure of 19-22.5 kg/cm^2. The reaction times employed, i.e., the stirring times at the reaction temperature, are listed in Table 1 together with the initial pressures. After reaction the residual deuterium was analyzed directly in a mass spectrometer for isotopic dilution. The produced alcohol was separated into the cis- and trans-isomers by gas chromatography prior to analyses by mass spectrometry and nmr spectroscopy.

2.3. 4-tert-Butylcyclohexanol-Deuterium Exchange

About 2.5 g of 4-tert-butylcyclohexanol, 10 ml of cyclohexane, and 10 mg of a catalyst were stirred in a 50-ml autoclave at 80°C and a pressure of 23-25 kg/cm^2 of deuterium. For each of the catalysts used, the reaction time was set equal to that of deuteration.

2.4. Cyclohexanone-Deuteriocyclohexanone Exchange

Immediately before conducting the exchange reaction, 10 mg of a catalyst sample was suspended in 10 ml of cyclohexane, and re-reduced with hydrogen in the smaller 50 ml autoclave. After reduction the hydrogen was purged, and a mixture of 0.5 g of cyclohexanone-2,2,6,6-d_4 and 2 g of normal cyclohexanone was added to the catalyst-cyclohexane suspension. The exchange reaction was conducted at 80°C and 24 kg/cm^2 of helium employing the same reaction time as in the case of deuteration.

2.5. Isomerization

About 2.5 g (0.015 mol) of a mixture of cis- and trans-4-tert-butylcyclohexanol not isomerically equilibrated was placed in the 50-ml autoclave together with 1.5 g (0.015 mol) of cyclohexanone, 10 ml of cyclohexane and 10 mg of a catalyst. This mixture was stirred under hydrogen at a pressure of 24 kg/cm^2 employing the same reaction temperature and time as in the case of deuteration.

2.6. Analyses

Nmr spectra were obtained on a Varian HA-100D spectrometer with a 10-mg sample of the cis- or trans-alcohol dissolved in 0.5 ml of CCl_4 containing a shift reagent; 25 mg of $Eu(dpm)_3$ for cis-alcohol and 15 mg of $Pr(dpm)_3$ for trans-alcohol. Proton assignments for the former system have been reported by Demarco et al.[11)], and those for the latter system are given elsewhere[12)]. The axial and equatorial deuterium contents in each carbon position were calculated from the integrals obtained from the shifted nmr spectra.

Mass spectra were taken on a Hitachi RMS-4 type spectrometer at 8 V for alcohol, and at 80 V for gas phase deuterium. Isotopic compositions were calculated from the molecular ion M^+ for cis-alcohol, and from the fragment ion $(M\text{-Water})^+$ for trans-alcohol. When trans-alcohol was highly deuterated, the relative intensity of its M^+ was too low to allow the calculation of isotopic compositions. Satisfactory agreement between the isotopic composition calculated from M^+ and that from $(M\text{-Water})^+$ was preliminarily confirmed using trans-alcohol samples moderately deuterated.

The separation of isomeric alcohols by gas chromato-

Table 1

Isotopic and Isomeric Analyses of Reaction Products in 4-tert-Butylcyclohexarone Deuteration.

Cat.	t (min)	P	X (%)	R	Isomer	d_0	d_1	d_2	d_3	d_4	d_5	d_6	D_m	D_g (%)
					Isotopic alcohol composition (%)*									
Ru	90	21.8	100	1.5	cis	2.0	93.3	4.2	0.5	-	-	-	1.03	96.3
					trans	2.8	91.9	4.5	0.8	-	-	-	1.03	
Rh	60	19.0	92	2.1	cis	20.4	40.5	20.7	11.2	5.0	1.8	0.4	1.47	70.6
					trans	18.3	41.7	20.7	11.5	5.4	1.9	0.5	1.52	
Pd	120	21.5	42	1.6	cis	20.6	32.4	25.1	15.0	5.9	1.0	-	1.56	61.8
					trans	17.5	30.7	26.1	16.8	7.3	1.6	-	1.70	
Os	10	21.7	100	1.3	cis	1.7	98.0	0.3	-	-	-	-	0.99	96.6
					trans	1.5	97.9	0.6	-	-	-	-	0.99	
Ir	65	22.5	100	0.91	cis	2.6	96.2	1.2	-	-	-	-	0.99	97.1
					trans	2.0	96.8	1.2	-	-	-	-	0.99	
Pt	120	21.0	92	0.38	cis	4.0	90.3	5.1	0.6	-	-	-	1.02	95.2
					trans	3.2	91.3	5.1	0.4	-	-	-	1.03	

t : Reaction time.
P : Initial pressure of reaction mixture in kg/cm^2.
X : Conversion based on reacted ketone.
R : cis/trans Ratio.
* : Hydroxyl deuterium was replaced by hydrogen before mass spectrometric analysis.
D_m: Mean number of deuterium atoms introduced into an alcohol molecule.
D_g: Isotopic purity of the residual gas-phase deuterium.

graphy caused complete replacement of hydroxyl deuterium by hydroxyl hydrogen. Accordingly, the isotopic distributions of alcohols obtained by mass spectrometry refer to the cyclohexyl ring and tert-butyl group rather than to the whole molecule inclusive of the hydroxyl group.

3. RESULTS

3.1. Deuteration

The mass spectrometric data are listed in Table 1. In the inspection of these data, it should be remembered that the hydroxyl deuterium is not involved. In agreement with our earlier work[9] on cyclohexanone deuteration, d_1 species was predominant over the third series metals (Os, Ir, Pt). This suggests that deuteration proceeds almost exclusively by simple addition of deuterium to the carbonyl linkage. Further evidence for this mechanism is provided by the nmr data of Table 2 where the alcohol samples obtained over these metals are deuterated only at the C_1 position. Although the formation of small amounts of d_0 and d_2 suggests the presence of a little hydrogen in C_1 and a little deuterium in positions other than C_1, the nmr data of Table 2 does not indicate these isotopic substitutions. The disagreement is probably due to lower sensitivity for deuterium of nmr spectroscopy in comparison with mass spectrometry.

Table 2
Stereochemical Isotopic Distributions within molecule for deuterated 4-tert-butylcyclohexanols

Cat.	Isomer	Number of axial- and equatorial-deuterium atoms							D_m
		C_1	Eq C_2+C_6	Ax C_2+C_6	Eq C_3+C_5	Ax C_3+C_5	Ax C_4	t-Bu	
Ru	cis	1.0	0.0	0.0	0.0	0.0	0.0	0	1.0
	trans	1.0	0.0	0.0	0.0	0.0	0.0	0	1.0
Rh	cis	0.57	0.44	0.44	0.0	0.0	0.0	0	1.45
	trans	0.59	0.50	0.50	0.0	0.0	0.0	0	1.59
Pd	cis	0.40	0.64	0.64	0.0	0.0	0.0	0	1.68
	trans	0.34	0.68	0.68	0.0	0.0	0.0	0	1.70
Os	cis	1.0	0.0	0.0	0.0	0.0	0.0	0	1.0
	trans	1.0	0.0	0.0	0.0	0.0	0.0	0	1.0
Ir	cis	1.0	0.0	0.0	0.0	0.0	0.0	0	1.0
	trans	1.0	0.0	0.0	0.0	0.0	0.0	0	1.0
Pt	cis	1.0	0.0	0.0	0.0	0.0	0.0	0	1.0
	trans	1.0	0.0	0.0	0.0	0.0	0.0	0	1.0

Eq, equatorial; Ax, axial; D_m, defined in Table 1.

The results obtained with Ru were very similar to those of the above three metals, although in our earlier work on cyclohexanone there was a much more significant exchange during the deuteration. The exchange over Rh and Pd was quite extensive, forming considerable amounts of d_0 to d_5 alcohols, as in the case of cyclohexanone deuteration. The nmr data of Table 2 indicate that the incorporated deuterium atoms are located in the three positions C_1, C_2, and C_6.

A characteristic feature of the mass spectrometric data was the close similarity in isotopic distribution pattern for the cis- and trans-alcohols. This was true with all the catalysts used. A striking result with Rh and Pd was in the equality between the axial- and equatorial-deuterium contents at the C_2 (or C_6) position.

3.2. Exchange and Isomerization

Rh and Pd were chosen for the exchange studies described in Sections 2.3 and 2.4 because the exchange during deuteration was remarkable over these metals. In the reaction of 4-tert-butylcyclohexanol with deuterium, no significant deuterium incorporation into the cyclohexyl ring occurred, although the hydroxyl group was found to be in exchange equilibrium with the gas phase. This indicates that the exchange during deuteration, i.e., the deuterium incorporation into the cyclohexyl ring, cannot be explained in terms of dissociative adsorption of the alcohol. Table 3 shows that neither metal catalyzed significantly the exchange between ordinary and deuterated cyclohexanones. Therefore, the exchange during deuteration is

Table 3
Exchange between Ordinary and Tetradeuterated Cyclohexanones

Isotopic species	Composition over Rh (%)		Composition over Pd (%)	
	Initial	Final	Initial	Final
d_0	79.9	80.0	80.5	80.2
d_1	0.3	0.2	0.1	0.4
d_2	0.2	0.2	0.2	0.2
d_3	1.2	1.7	1.2	2.0
d_4	18.4	17.9	18.0	17.2

neither attributable to the dissociative adsorption of ketone nor to the tautomerization. The results of these two exchange experiments can be summarized in the conclusion that the exchange during deuteration is not an independent parallel path, but involved in the sequence of steps of deuteration.

The study of isomerization in Section 2.5 was made using Rh, Pd, and Pt. None of these metals catalyzed the cis/trans isomerization. A selection of the data is shown in Table 4.

Table 4
cis-trans Isomerization of 4-tert-butylcyclohexanol in the presence of cyclohexanone and hydrogen.

Species	Composition over Pd (%)*		Composition over Pt (%)*	
	Initial	Final	Initial	Final
cis-ol	60	62	61	61
trans-ol	40	38	39	39
C_6-one	100	96	100	6
C_6-ol	0	4	0	94

* cis-ol plus trans-ol = 100, C_6-one plus C_6-ol = 100

4. DISCUSSION

4.1. Assumed Mechanism

Any proposed mechanism of the hydrogenation of 4-tert-butylcyclohexanone has to accommodate the following results: (i) the considerable isotopic exchange during deuteration over Pd and Rh, and a much less exchange over the rest of the metals, (ii) the equality between the axial- and equatorial-deuterium contents at C_2 or C_6, (iii) the formation of both of the stereoisomers, and (iv) the close similarity in isotopic distribution pattern for the cis- and trans-alcohols. Although a mechanism involving trans-addition of hydrogen may play a significant role, we demonstrate that these results can be satisfactorily explained in terms of cis-addition. Another basic assumption involved is that 4-tert-butylcyclohexanone is locked in the chair conformation with an equatorial tert-butyl group.

Fig. 1 illustrates the mechanism presented in our earlier work for cyclohexanone hydrogenation. The results

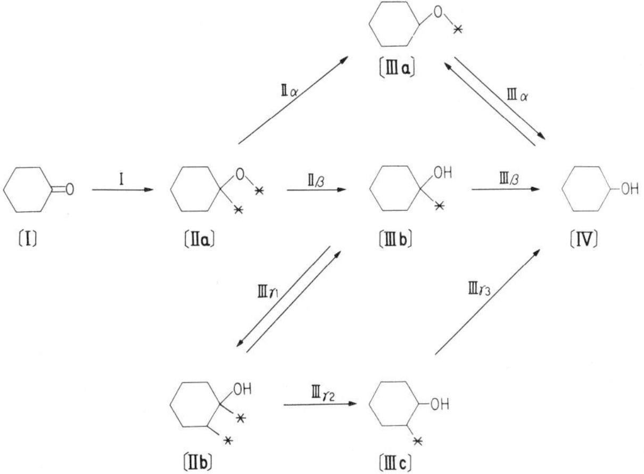

Fig. 1. Reaction scheme for alicyclic ketone hydrogenation.

obtained with 4-tert-butylcyclohexanone are also explained on the basis of this reaction scheme, which apparently consists of three hydrogenation paths:

path 1; sequence I, II_α, III_α
path 2; sequence I, II_β, III_β
path 3; sequence I, II_β, $III_{\gamma 1}$, $III_{\gamma 2}$, $III_{\gamma 3}$.

4.2. Interpretation of Kinetic Data

Ignoring hydroxyl deuterium, let us consider what isotopic alcohol is formed through each reaction path when an alicyclic ketone is deuterated. Apparently both paths 1 and 2 lead to the exclusive formation of the d_1 alcohol deuterated at the C_1 position. Path 3, on the other hand, gives only the d_2 species with one deuterium atom in C_1 and the other in C_2 or C_6. The appearance of the d_3-d_5 species must be attributed to the alternation between intermediates II_b and III_b followed by the reduction through step III_β or sequence $III_{\gamma 2}$, $III_{\gamma 3}$. Thus, an explanation for result (i) is that over Os, Ir, Pt, and Ru, deuteration proceeds almost exclusively by the straight-through paths 1 and 2, while over Rh and Pd, the path 2 or 3 each intervened by the alternation plays an important part.

In order to explain the result (ii), it is necessary to consider the orientation of the adsorbed intermediate III_b. If its C_1-catalyst bond is equatorial and perpendicular to the catalyst plane, the axial and equatorial hydrogens at C_2 and C_6 will be at exactly the same distance from the surface. Consequently, these four hydrogen atoms could be replaced by deuterium with the same probability, in conformity with result (ii).

The result (iii) can be explained by assuming the existence of two distinct modes of adsorption for the diadsorbed intermediate II_a, as was mentioned in the INTRODUCTION. These are represented in Fig. 2 as II_a(E), and II_a(A), where symbols (A) and (E) refer to the axial and equatorial C_1-oxygen bonds,

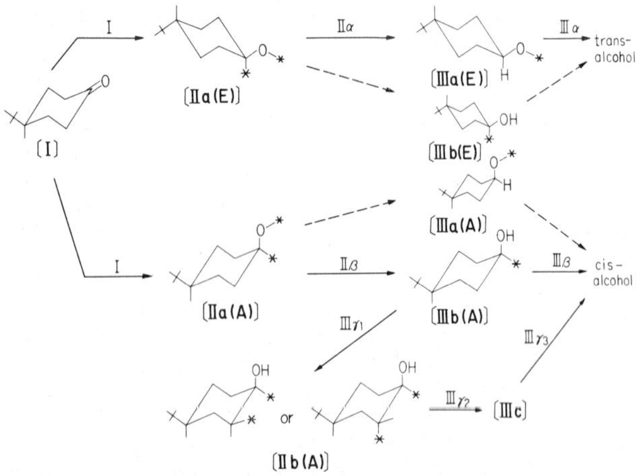

Fig. 2. Stereochemical representation of reaction scheme.

respectively. By cis-addition of hydrogen, the intermediate II_a(A) is invariably converted to the cis-alcohol, and II_a(E) to the trans-alcohol. As a consequence, the cis- and trans-isomers would be formed in the presence of the two adsorption modes.

To account for the result (iv), we have to take a closer look at the reaction scheme of Fig. 2. By addition of one deuterium atom, the diadsorbed species II_a(E) is converted into the half-hydrogenated III_a(E) or III_b(E). Similarly, II_a(A) gives III_a(A) or III_b(A). Among these four conceivable half-hydrogenated species, we exclude III_b(E) on the ground that it is subject to a far greater hindrance to adsorption, even as compared with the hindered diadsorption species II_a(E). The larger hindrance of III_b(E) is readily seen on inspection of suitable molecular models which show that the bulky tert-butyl group is inclined more toward the catalyst plane on III_b(E) than on II_a(E). We also neglect the formation of III_a(A) from II_a(A) by assuming that the most difficult step in the whole hydrogenation process is the rupture of the C_1-catalyst bond. This assumption leads to the view that the steps II_α, III_β, and $III_{\gamma 2}$ are rate-determining for paths 1,2, and 3, respectively. In this situation, all of the steps preceding these rate-determing ones, i.e., steps I, II_β, and $III_{\gamma 1}$, should be in quasi-equilibrium. Since the intermediates II_a(E) and III_b(A) are interconverted through two of these reversible steps, both intermediates must be nearly the same in deuterium distribution in the cyclohexyl ring. This similarity, in turn, should lead to the similarity between the cis- and trans-alcohols as stated above as result (iv). In support of the above quasi-equilibrium, our recent experiments show that even in the initial stage of deuteration 4-tert-butylcyclohexanone exchanges with deuterium.

Table 1 shows that Ru and Os, out of the four metals over which the straight-through paths 1 and 2 operate almost exclusively, have cis/trans ratios greater than unity, while the other two, i.e., Ir and Pt, give values less than unity. This suggests that path 2 is preferred to path 1 for Ru and Os, and vice versa for Ir and Pt.

4.3. Comments on Other Mechanisms

In their study on deuteration of 2-methylcyclopentanone, Cornet and Gault[6)] found that the resulting cis- and trans-2-methylcyclopentanols are nearly the same in isotopic distribution. They explained this finding by assuming the formation of the O,C_1,C_2- and O,C_1,C_5-triadsorbed species followed by both cis- and trans-additions of hydrogen to these species. This mechanism requires the significant deuterium incorporation into the positions β to the hydroxyl of the produced alcohol. Exchange between ordinary and deuterated ketones even in the absence of hydrogen would also be required by this mechanism. In our studies, Ru, Os, Ir, and Pt did not meet the former requirement, and Rh and Pd did not fulfil the latter requirement. Therefore, the mechanism of Cornet and Gault would not be true of any of these catalysts under our reaction conditions.

As has been cited in the INTRODUCTION, cis-trans isomerization via ketone intermediate has been pointed out by Wicker[3]. On the other hand, Newham and Burwell[7] have suggested the interconversion of ketone and alcohol. That neither of these reactions is operating in our case is clearly indicated by lack of isomerization in Section 2.5.

4.4. Concluding Remark

Although our mechanism explains our own kinetic data satisfactorily in terms of cis-addition, it is probably over-simplified, and further work is required to elaborate and verify it.

ACKNOWLEDGMENT

The authors wish to express their appreciation to Professor Emeritus K. Hirota (Osaka Univ.) and Dr. K. Taya (Tokyo Gakugei Univ.) for their valuable discussion. Thanks are also extended to Rev. C. D. McMahan for his linguistic comments on the original manuscript.

REFERENCES

1) R.L.Augustine, Catalytic Hydrogenation (Marcel Dekker, New York, N.Y., 1965) p.86.
2) P.N.Rylander, Catalytic Hydrogenation over Platinum Metals (Academic Press, New York, N.Y., 1967) p.284.
3) R.J.Wicker, J. Chem. Soc., (1956), 2165.
4) J.H.Brewster, J. Am. Chem. Soc., 76 (1954), 6361.
5) R.L.Augustine et al., J. Org. Chem., 34 (1969), 1075.
6) D.Cornet and F.G.Gault, J. Catalysis, 7 (1967), 140.
7) J.Newham and R.L.Burwell,Jr., J. Am. Chem. Soc., 86 (1964), 1179.
8) C.Kemball and C.T.H.Stoddart, Proc. Roy. Soc., A246 (1958), 521.
9) Y.Takagi, S.Teratani, and K.Tanaka, to be submitted to J. Catalysis.
10) Org. Syntheses Coll. Vol.4 (1963), 164.
11) P.V.Demarco et al., J. Am. Chem. Soc., 92 (1970), 5734.
12) Y.Takagi, S.Teratani, and J.Uzawa, submitted to Chem. Communications.

DISCUSSION

S. SIEGEL

The distribution of deuterium in the cyclohexanols obtained over rhodium and palladium poses a question of mechanism similar to the one met in the exchange of deuterium and cyclohexene in which an intermediate such as cyclohexanes or the roll-over mechanism of Burwell, is apparently required to account for exchange of both sides of the cycle.

In your experiments, the corresponding intermediate is the enol. You showed that cyclohexnone and 2,2,6,6-tetrodeuterocyclohexanone do not exchange in the presence of the rhodium catalyst (but in the absence of hydrogen) and concluded that enolization was not promoted by the catalyst. However in the presence of hydrogen, a reaction path is provided via the desorption of the diadsorbed intermediate IIb (Figure 1). Although desorption might be slow, it cannot be limited by the thermodynamics of the change because about one percent of the enol would be present at equilibrium.

K. TANAKA

Paraphrasing Professor Siegel's comment, we could say that in the presence of metal catalyst hydrogen may act as a cocatalyst for keto-enol tautomerization. In this connection, let me point out the generally accepted view that in enolization the axial hydrogen is usually abstracted preferentially and in ketonization hydrogen addition to the enol also occurs preferentially from the axial position.[1] In fact we found that when 4-t-butylcylcohexanone is shaken with CH_3OD using Na_2CO_3 as catalyst, the axial D content at C_2 (or C_6) is more than twice as much as the equatorial D content at the same carbon position.[2] In the light of this axial preference, the observed identity of the axial and equatorial D contents during our ketone deuteration suggests that the isotopic exchange at C_2 and C_6 is brought about by some process other than keto-enol tautomerization. In other words, even if hydrogen acts as a cocatalyst for tautomerization, its contribution to the isotopic exchange should be negligible under our experimental conditions.

1) E. L. Eliel, Stereochemistry of Carbon Compounds, p. 241, McGraw-Hill, New York, 1962.
2) S. Teratani and Y. Takagi, unpublished data.

H. van BEKKUM

Regarding your elegant work, I would like to remark that the suggestion on page 8 of your paper stating that the chair conformation of the ring is retained in an α, β- diadsorbed species such as IIb seems wrong to me. Exchange and other data have shown that eclipsed or near-eclipsed geometry is required. Now this conflicts with your idea of equal probability of dissociation of axial and equatorial C-H bonds at C_2. In my opinion you have to include boat-like conformations in your mechanism and also roll-over or desorption-adsorption to account for deuterium of both sides of the ring at C_2 and C_6.

In view of the equality of axial/equatorial deuterium contents of the cis- and trans-alcohol, one could think about separate surface sites for exchange and for hydrogenation on Rh and Pd.

K. TANAKA

First of all I am not convinced of your argument against an α,β-diadsorbed species such as (II_b) in which the adsorption bonds are staggered to each other. On the principle of maximum overlapping for bonding, the staggered adsorption may be less stable than the corresponding eclipsed adsorption. But even so the species (II_b) could be stable enough as at least a transient intermediate.

As for your suggestion concerning mechanism that includes boat-like conformations and a roll-over process, let me point out that such a mechanism would be difficult to explain the observed equality of axial and equatorial deuterium contents unless you assume a rather dubious roll-over process. Fig. a illustrates such a dubious process which is characterized

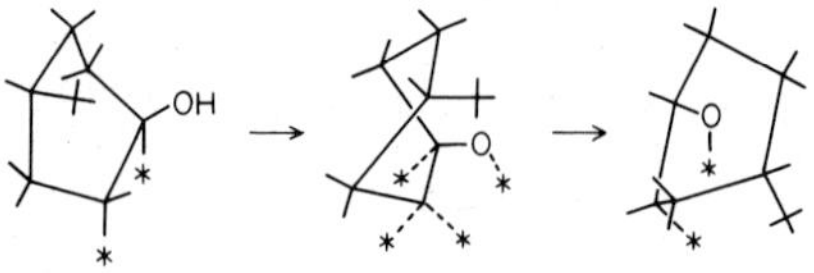

Fig. a Roll-over mechanism

by release of the hydroxyl hydrogen and participation of the O-M adsorption bond in roll-over. Aside from the roll-over, however, we suspect that alternation between the monoadsorbed boat form (I_s) and the diadsorbed boat form (II_s) plays some role in the deuterium incorporation into the trans alcohol (see Fig. b). Apparently the intermediate (I_s) is led to the trans

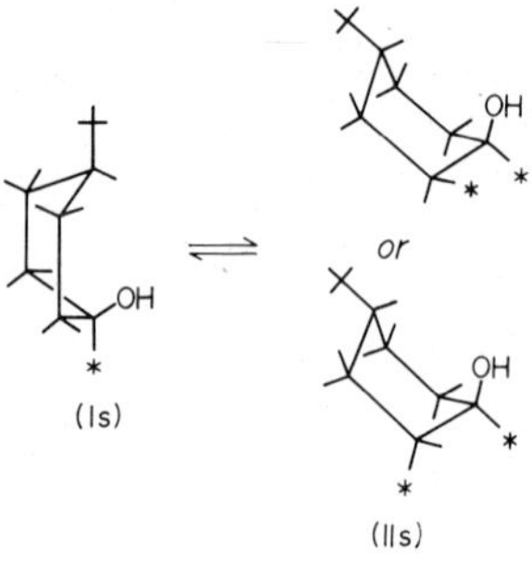

Fig. b Boat form mechanism

alcohol by the rupture of the adsorption bond and the subsequent transformation into the chair form.

Coming to Professor Bekkum's remarks on two distinct surface sites for exchange and for hydrogenation, I think it is very difficult to say whether one has to assume only one type of surface sites or more, and only further research can solve this problem.

In conclusion we believe that there is no conclusive evidence against the mechanism presented by us.

R. L. BURWELL, JR.

The diadsorbed species in this paper are all assumed to be in the staggered conformation rather than in the eclipsed conformation. That the staggered conformation is unconventional does not prove that it is incorrect, but it does suggest that further evidence is desirable. It would be interesting to know what would happen with cyclopentanones where staggered conformations are excluded.

K. TANAKA

We entirely concur with Professor Burwell's remarks. Incidentally, it occurs to me that deuteration of bicyclic ketones such as norcamphor would also be worth studying; for norcamphor can be regarded as invilving, as part of its structural framework, a staggered conformation or boat form of cyclohexanone, and its transformation into the chair forms is completely prohibited.

J. L. GARNETT

My comment is relevant to the preceding point raised by Professor Siegel. Several years ago we exchanged cyclohexanone and a number of derivatives on metal surfaces with D_2O (Calf and Garnett, unpublished work). The results could be interpreted in terms of a contribution to the adsorption of the cyclohexane in the enol form, i.e., as a π-olefin complex (species (I)).

OH
M
(I)

My question then is do you consider that cyclohexane in your present work is the enol complex (I) a possible intermediate in your deuteration?

K. TANAKA

Before answering your question, let me talk about the equilibrium enol content for pure liquid cyclohexanone. According to Gero,[1] the content amounts to only 1.18% at room temperature. This value would not vary too much even at the reaction temperature of 80°C nor by substituting tert-butyl for hydrogen at the C_4 position. On the other hand, keto-enol tautomerization is practically frozen under our reaction conditions as we wrote in the Preprint. Therefore, even if there exists the enol complex (I) that you proposed, its content must be negligible as compared with that of the keto counterpart (II) unless the former has a greater adsorption coefficient. This in turn suggests that the enol complex (I) could play only a negligible role as an intermediate in our deuteration.

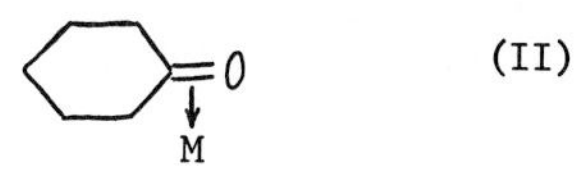

(II)

1) A. Gero, J. Org. Chem., 26, 3156 (1961).

C. KEMBALL

My first comment on this very interesting paper concerns the product distributions reported for palladium in Table 1. These agree satisfactorily with distributions calculated for 5 exchangeable positions equilibrated with 1.67 D atoms and 3.33 H atoms (i.e. the observed deuterium content). This agreement supports the interpretation of the results given by the authors, but it draws attention to the fact that the exchange of these 5 positions with the gas phase deuterium is not complete; the gas phase is richer in D than H.

My second comment is to support one of Professor Siegel's remarks. I too am surprised that no exchange (apart from its trivial reaction of the hydroxyl group) was observed between the alcohol and deuterium over rhodium or palladium. It would be advisable to look for such exchange (which ought to show 5 exchangeable H atoms) at slightly higher temperatures. The metals ought to speed up this back reaction just as they catalyze the forward process of ketone hydrogenation.

K. TANAKA

We thank Professor Kemball for his interesting interpretation and suggestion.

F. G. GAULT

One of the many results presented by Professor Tanaka in his interesting paper is the identity, observed in every case, of the deuterium distributions of the cis and trans alcohols. This result confirms the observation made by Cornet and Gault (Reference 6) in their study of the deuteration of methylcyclopentanone on nickel.

In the reaction mechanism he proposed, and represented in Fig. 2, Professor Tanaka explains this important feature by the existence of three slow reactions II_{α}, III_{β} and $III_{\gamma 2}$ and of a quasi-equilibrium achieved for all the other elementary steps.

According to this mechanism, one would expect a fast desorption and readsorption of the ketone, and therefore an exchange between light and heavy molecules. In other words, the very same objection presented against the mechanism proposed by Cornet and Gault holds also for the mechanism proposed by Professor Tanaka.

We believe that a feasible explanation for lack of the isotopic exchange between light and heavy ketones could be in the fact that the dissociatively adsorbed ketone species, whatever they are, are certainly much less numerous on the surface than are the associatively adsorbed ketones.

Now, in the results presented in this paper, there is not really any major objection to the mechanism proposed by Cornet and Gault. Only the description of the intermediate responsible for the multiple exchange is doubtful and should probably be revised. This intermediate indeed was presented as a "pseudo-π-allylic species," and the recent works of Professor Burwell and Dr. Rooney in the field of exchange has extensively shown that the concept of π-allylic species should be handled very carefully in the case of metals.

K. TANAKA

The argument raised by you against our mechanism is not valid. Professor Gault insists that according to our mechanism one would expect a fast desorption and readsorption of the ketone, and therefore an exchange between the light and heavy molecules. We admit that our mechanism does really imply the rapid desorption and readsorption of the ketone during its reduction. However, this does not mean that the light ketone-heavy exchange occurs even in the absence of hydrogen. Looking at Fig. 1, you will see that mere ketone adsorption and desorption, namely, alternation between species (I) and (II_a), does not cause any isotopic exchange. In order to have the exchange you have to go all the way to species (II_b), which is brought about through the half-hydrogenated species (III_b). Our mechanism for ketone reduction, therefore, suggests that in the absence of hydrogen no isotopic exchange occurs between the light and heavy ketones.

Incidentally, let me remind you of the boat-form mechanism (Fig. b) that we proposed in answer to Professor Bekkum's comments. If you attribute the deuterium incorporation into the trans alcohol entirely to this mechanism, then you can explain the similarity in isotopic distribution between the cis- and trans-alcohol without assuming the quasi-equilibrium of steps I, II_{β}, and $III_{\gamma 1}$.

Paper Number 52

THE EFFECT OF ADSORBED ACETYLENE ON THE ETHYLENE HYDROGENATION BY PALLADIUM CATALYST

I. YASUMORI, H. SHINOHARA and Y. INOUE
Department of Chemistry, Tokyo Institute of Technology
Ookayama, Meguro-ku, Tokyo, Japan

Abstract: In the catalytic hydrogenation of ethylene on palladium foil at 0°C it was observed that adsorbed acetylene promotes and stabilizes the activity of ethylene-covered surface. The role of the adsorbed acetylene was investigated by following the behavior of adsorbed ethylene and acetylene labelled with C^{14} during the course of the reaction. By repeating reaction, the activity diminished to less than one fiftieth of the original value of the clean surface, 2.5 x 10^{17} (±0.3) molecules/cm^2min. This reduced activity, however, increases linearly with the amount of adsorbed acetylene, approaching a maximum value, 4.6 x 10^{16} (±0.5) molecules/cm^2min, at the monolayer coverage of acetylene.

The study revealed that acetylene replaces a small fraction of preadsorbed ethylene, inducing the stabilized active sites whose area corresponds to about 2% of the total surface.

1. Introduction

Numerous investigations have been devoted to the study of the mechanism of the hydrogenation of ethylene on transition metals such as nickel, platinum and palladium. There still remain, however, some ambiguities on the nature of the reaction, since a large fraction of the surface is covered with acetylenic compounds which form from ethylene during the course of hydrogenation. This self-contamination is very distinct on nickel as observed by Beeck [2] and Jenkins and Rideal [3] and on palladium by Thomson et al. [4] and makes the quantitative analysis of the reaction difficult.

The hydrogenation of acetylene on cold-worked palladium surface has recently been investigated in our laboratory. By the use of C^{14} tracer, pressure-jump and hydrogen-deuterium replacement techniques, the role of adsorbed acetylene in the reaction and the dynamic character of the hydrogenation were revealed. The existence of two kinds of active sites was verified; one is stable below 250°C and the other of lower activity appears on annealing at temperatures above 600°C [5]. Adsorbed and dissolved hydrogen were found to be responsible for the reaction on these sites [6].

By extending the research to the ethylene hydrogenation on the palladium surface, it was found that adsorbed acetylene restores and stabilizes the activity diminished by ethylene contamination. The present study aimed at verifying the role of adsorbed acetylene in this phenomenon by studying the behavior of acetylene and ethylene during the hydrogenation and the state of adsorbed acetylene on the basis of IR spectra.

2. Experimental

For the kinetic and adsorption measurements, we used a reaction system of 420 cc composed of a circulating pump, a glass Bourdon gauge and a G.M. counter.

The catalyst foil was set below the window of the counter so that the change in the adsorbed amount of ethylene or acetylene labelled with C^{14} could be followed during the course of the hydrogenation. By the comparison of surface count rate with the adsorbed amount of carbon monoxide labelled with C^{14}, the efficiency of the counter was evaluated. The surface densities of acetylene and ethylene per 1 c.p.m. were estimated as 2.68×10^{11} and 1.07×10^{11} molecules/cm^2 respectively.

The adsorption of acetylene was performed at 0°C under pressures of acetylene between 5×10^{-3} and 20 Torr. Gaseous acetylene was evacuated before the hydrogenation of ethylene. The reaction was studied at 0°C using an equimolar mixture of ethylene and hydrogen at a total pressure of 20 Torr. The formation of ethane was followed by means of a gas-chromatograph connected directly to the system.

Palladium of 99.99% pure was obtained from Johnson Matthey and Co., Ltd. and used in the form of foil, 30 mm in length, 20 mm in width and 0.2 mm in thickness. Before each run, the foil was oxidized with oxygen of 15 Torr at 340°C, subsequently reduced with hydrogen of 30 Torr at 150°C and then annealed under vacuum at 200°C and finally reduced again at 150°C. Each treatment was carried out for 1 hr. The surface prepared this way is called "clean" surface hereafter.

The non-radioactive gases ethylene (99.8%), acetylene (99.6%), hydrogen (99.99%) and deuterium (HD less than 0.5%) were purchased from Takachiho Co.. Radioactive acetylene and ethylene of 19.5 and 54.0 mC/mM respectively were obtained from Radiochemical Center, Amersham, England.

3. Results

3.1. *Activities of various surfaces on ethylene hydrogenation*

The hydrogenation reaction of ethylene proceeded quantitatively to give ethane under the present conditions. The activity was then estimated from the initial rate of the increase in the ethane pressure. The change in activity after various treatments is shown in fig. 1.

The activity of the clean surface, 2.5×10^{17} ($\pm$ 0.3) molecules/cm^{2}min is extremely high but not stable and fluctuates within ten percent. (point A) By repeating the kinetic run without any treatment, the activity diminished drastically to less than one fiftieth of the original value for the clean surface. (Points B and C) However, when the surface is exposed to acetylene of 15 Torr at 0°C for 20 min, the activity recovers to 4.2×10^{16} ($\pm$.05) molecules/cm^{2}min. (point D). On the other hand, the direct adsorption of acetylene on the clean surface attenuated the activity to the definite value denoted by point D. When the reaction on this stabilized surface is repeated after evacuation, the

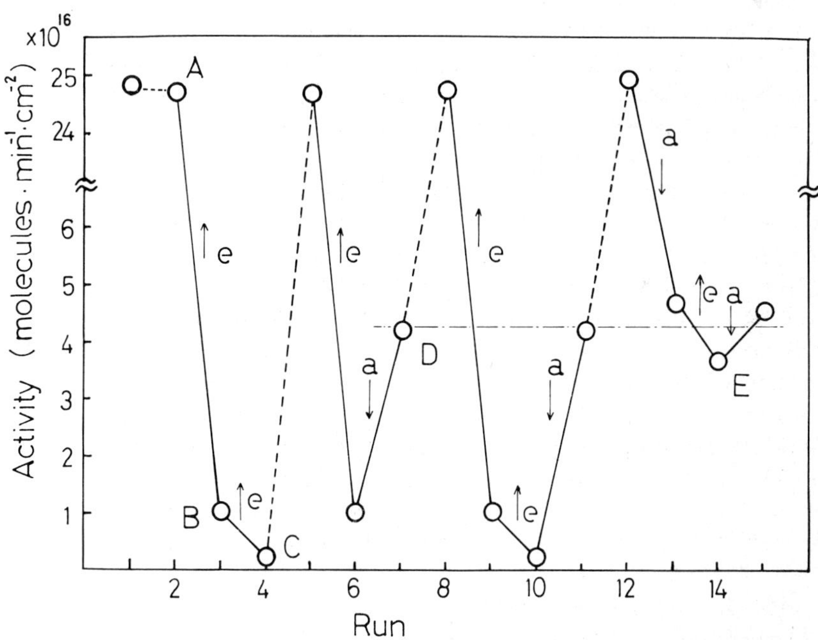

Fig. 1. Change in activity with various treatments. P (total) = 20 Torr. P (hydrogen)/P (ethylene) = 1. Reaction temperature = 0°C. Treatments a and e mean the introduction of acetylene and the evacuation of reactant respectively.

activity decreased by 20% (point E), but the reduced activity comes back again to the value of point D by exposing the surface to acetylene. These findings evidently show that the acetylene is adsorbed strongly on the surface and stabilizes the active sites on which the hydrogenation proceeds steadily.

3.2. *Kinetics of the reaction on the acetylene-covered surface*

Because of high but less stable activity of the clean surface, it was rather difficult to determine the precise rate expression on the surface. Accordingly, the kinetics on the acetylene-covered surface was examined at pressures of hydrogen and ethylene up to 20 Torr. At 0°C, the reaction order was found to be one half with respect to hydrogen and slightly negative with respect to ethylene.

The reaction of ethylene with deuterium was performed at 0°C by using the equimolar mixture of ethylene and deuterium at a total pressure of 20 Torr, and the distribution of deuterium was determined at the 9% conversion. It was recognized that the hydrogen in ethylene does not appear in the reactant deuterium, since the fraction of HD is smaller than 1% of the total deuterium. The deuterium distribution in ethylene (d_0 = 94%, d_1 = 4.4%, d_2 = 0.3%) shows that the degree of exchange with deuterium is considerably smaller than that reported previously by Bond et al. [7]. In the case of ethane, the fraction of ethane-d_0 is not negligible. This is quite different from the previous results.

3.3. *Activity change with the adsorbed amount of acetylene*

The behavior of adsorbed ethylene was followed during the course in hydrogenation as shown in fig. 2a. The adsorbed amount decreased gradually as the hydrogenation proceeded. But some fraction of this species still remained on the surface even after the completion of the run and evacuation. The introduced acetylene replaced ethylene to some extent but no substitution occured by the subsequent addition of either the reaction mixture or of ethylene. The behavior of preadsorbed acetylene during the course of ethylene hydrogenation is also shown in fig. 2b. After the decrease on evacuation, the adsorbed acetylene was replaced to a considerable extent by the reaction mixture but most of acetylene remained when only ethylene was introduced. These data indicate that the acetylene is adsorbed more strongly than ethylene and determines the topological configuration of the surface.

The correlation of the rate of hydrogenation with the amounts of adsorbed ethylene and acetylene is shown in fig. 3 which was derived from separate measurements of acetylene adsorption, replacement of ethylene by acetylene and of activity change with acetylene adsorption.

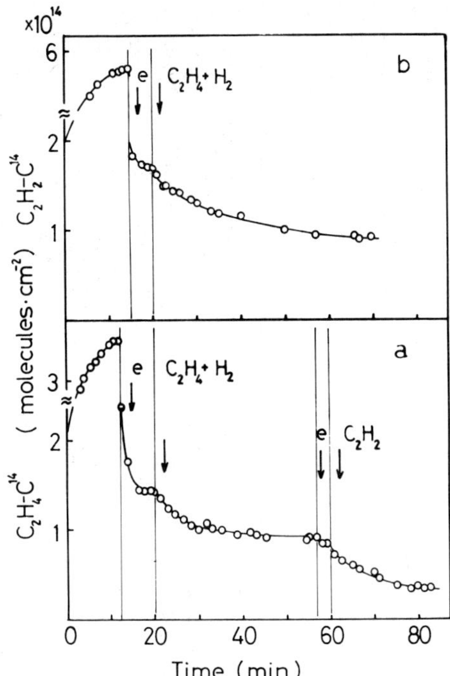

Fig. 2. Replacement of preadsorbed ethylene and acetylene by reaction mixture at 0°C. Total pressure of reaction mixture = 20 Torr. P (hydrogen)/P (ethylene) = 1. (a) Pressure of introduced ethylene = 1×10^{-1} Torr, Pressure of introduced acetylene = 20 Torr, (b) Pressure of introduced acetylene = 6×10^{-1} Torr.

The amount of adsorbed ethylene decreases by 20% on the adsorption of 3×10^{14} molecules/cm^2 of acetylene and then decrease only slightly when additional C_2H_2 is adsorbed. On the other hand, the activity changes little when initially a small amount acetylene is adsorbed but increases linearly on further adsorption of acetylene. This correlation holds up to monolayer coverage of acetylene at which point the activity attains a maximum value, 4.2 $\times 10^{16}$ (± 0.5) molecules/cm^{2}min. (point D in fig. 1) The total amount of ethylene removed in the region where the activity increases linearly was estimated to be 4.5×10^{13} molecules/cm^2 or about 2% of the surface area, assuming that each molecule corresponds to a palladium atom on surface.

The results of measurement carried out at the beginning of the reaction are shown in fig. 4.

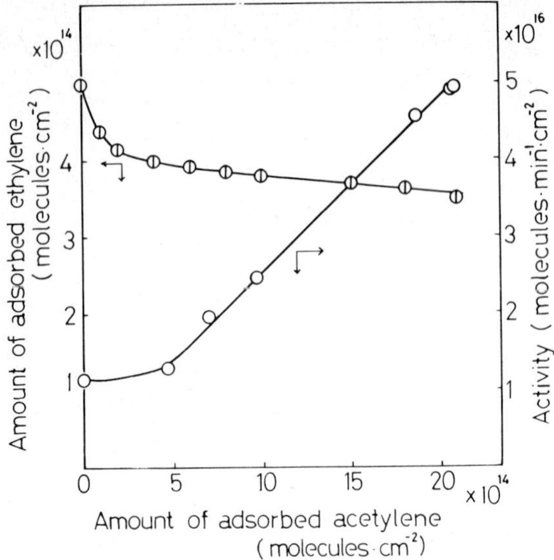

Fig. 3. Correlation of rate with the adsorbed amounts of ethylene and acetylene.

A definite fraction of preadsorbed ethylene is replaced by adsorbed acetylene. Very high count rate of acetylene includes the contribution from the acetylene in gas phase of high pressure. After evacuation and the introduction of the reactant mixture, adsorbed acetylene is partly desorbed from the surface. The amount of acetylene desorbed while a steady reaction is reached is nearly equal to that of replaced ethylene.

3.4. *The adsorbed state of acetylene*

In order to examine the state of adsorbed acetylene, a separate IR experiment was carried out using 15% of palladium dispersed on Aerosil. This dispersed palladium was prepared by the reduction of impregnated and decomposed palladium chloride and was treated in the same way as the foil catalyst. The introduction of acetylene to the clean surface did not show any spectrum of adsorbed species, but the subsequent addition of hydrogen or of the reactant mixture generated three absorption bands in regions 2961, 2927 and 2860 cm^{-1} as observed by Little et al. [8]. These bands are identified as CH_3 asymmetric, CH_2 asymmetric and CH_3 or CH_2 symmetric stretching vibrations respectively. The intensities of these spectra were not altered by evacuation after hydrogenation at 30°C or by the reduction with hydrog n at 100°C for

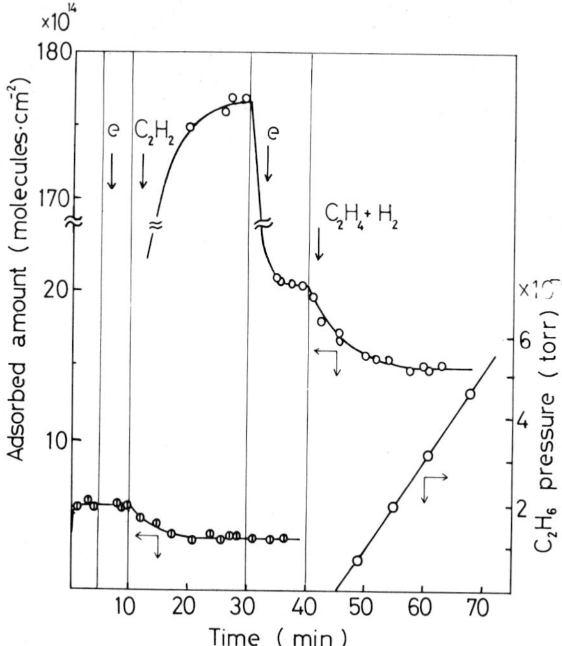

Fig. 4. Behavior of adsorbed acetylene and ethylene at the beginning of reaction. θ; Initial pressure of ethylene labelled with C^{14} = 5.5 Torr, O; Initial pressure of acetylene labelled with C^{14} = 4.8 Torr. Total pressure of reaction mixture = 20 Torr.

1 hr. The spectra disappeared only by the oxidation treatment described in the experimental section. From the intensity ratio of stretching vibration of CH_3 to CH_2, it was concluded that the surface acetylene treated with hydrogen was in the form of a polymer having more than six carbon atoms. By introducing deuterium into gas phase, CD stretching vibration spectra appeared and, the intensities of CH vibration spectra decreased. These observations on the adsorbed species were also confirmed by the results on the system of deutero-acetylene and deuterium which shows characteristic bands, 2210, 2180 and 2090 cm^{-1}. However, the exchange rate estimated from the intensity change of CD band was about ten times as slow as the hydrogenation rate at 30°C.

The X-ray photoelectron spectrum of acetylene adsorbed surface was examined, and the adsorption of acetylene was found to increase the peak intensity due to 4d electron levels located closely to the Fermi surface.

4. Discussion

4.1. *Function of adsorbed acetylene*

The fact that adsorbed acetylene on the clean surface does not show any CH vibration spectrum in the present catalyst suggests that acetylene is in a highly dissociated state, i.e. as carbonaceous compound. In the case of ethylene-covered surface, it is expected that the introduced acetylene reacts with the adsorbed ethylene to form a polymerizate. The amount of removed ethylene in the linear part in fig. 3 is able to cover 2% of the surface area. Fig. 4 shows that the amount of removed ethylene is nearly equal to that of acetylene desorbed in the period in which the reaction becomes stationary. The previous study on acetylene hydrogenation by palladium foil indicated that the active sites occupied about 2% of the surface area. It is remarkable that these values of effective surface area coincide with each other.

These considerations lead us to the following model of surface; by repeating ethylene hydrogenation, active sites are almost occupied by the dissociated ethylene,

$$C_2H_4 + S \rightarrow C_2H_{4-x}S + XH(a) \tag{1}$$

where S denotes the active site and H(a) the hydrogen atom adsorbed on surface metal atom or in the interstitial positions of lattice. As the adsorbed acetylene covers the surface uniformly, the dissociatively adsorbed ethylene desorbs and the acetylene adsorbed near the site forms polymerizate. This species can stabilize the active site by fixing surrounding metal atoms.

$$nC_2H_2(a) + C_2H_{4-x}S \rightarrow (\text{polymerizate}) + S + C_2H_4 \tag{2}$$

On this vacant site acetylene can be adsorbed weakly. Thus the introduced reaction mixture of ethylene and hydrogen removes this weakly bound acetylene and the hydrogenation proceeds with its steady rate on this vacant site. It is a complicated problem to determine the real structure of the active site at present. The drastic decrease in the hydrogenation activity of acetylene by annealing, however, strongly suggests that the structure of the active sites is closely correlated to such surface irregularities as lattice defects [5].

In addition to the geometrical effect of adsorbed acetylene, the preliminary result of X-ray photoelectron spectroscopy suggests that acetylene contributes to some electron transfer processes which participate in hydrogenation. The excess electron in the valence electron band of metal may bring hydrogen into a reactive hydride state as is assumed in the case of homogeneous catalysis by platinum-tin complex catalyst [9].

4.2. *Reaction mechanism on the acetylene-covered surface*

The reaction order of one half with respect to hydrogen is close to the value of 0.66, previously obtained by Schuit and von Reijen [10]. The Deuterium distribution in the products shows that hydrogen (or deuterium) does not revert from the half-hydrogenated state of ethylene into gaseous ethylene and deuterium and that, on the other hand, the exchange of hydrogen with ethyl radicals proceeds considerably. These results are not explained either by the associative mechanism or by a mechanism involving collisions of ethylene with adsorbed hydrogen. The plausible mechanism, therefore, must be one in which the formation of the half-hydrogenated state is a rate controlling step. The observed order in hydrogen is readily explained by assuming a quasi equilibrium for hydrogen adsorption. However it is necessary to assume an intermediate state between ethyl radical and ethane in order to explain the wide distribution of deuteroethanes.

4.3. *The role of surface residuals*

The present results show that the adsorbed acetylene is not a simple poison but stabilizes the active site for ethylene hydrogenation. Such an effect is found in some cases. Thermal desorption studies reveal a similar behavior of nickel catalyst for methanol decomposition [11]; Methanol adsorbed on the clean surface gives hydrogen and carbon dioxide without forming carbon monoxide up to 350°C. However, when the methanol-covered surface is heated to 150°C to remove only hydrogen, methanol introduced on this modified surface gives hydrogen and carbon monoxide as the temperature is increased up to 300°C. This fact shows that the surface residual, a precursor of carbon dioxide stabilizes the nickel surface because the products of the decomposition are hydrogen and carbon monoxide. These findings indicate that the surface residual takes a role of producing and stabilizing active sites rather than acting merely as a poison and, requires additional investigation for complete understanding the catalytic action by solid surface.

References

[1] G.C. Bond and P.B. Wells, Advances in Catalysis 15 (1964) 91.
[2] O. Beeck, Discussions Faraday Soc. 8 (1950) 118.
[3] G.I. Jenkins and E.K. Rideal, J. Chem. Soc. (London) (1953) 2490, 2496.
[4] D. Cormack, S.J. Thomson and G. Webb, J. Catalysis 5 (1966) 224.
[5] Y. Inoue and I. Yasumori, J. Phys. Chem. 73 (1969) 1618.
[6] Y. Inoue and I. Yasumori, J. Phys. Chem. 75 (1971) 880.
[7] G.C. Bond, J.J. Philipson, P.B. Wells and J.M. Winterbottom, Trans. Faraday Soc. 162 (1966) 443.
[8] L.H. Little, N. Sheppard and D.J.C. Yates, Proc. Roy. Soc. A259 (1960) 242.
[9] I. Yasumori and K. Hirabayashi, Trans. Faraday Soc. 67 (1971) 3283.
[10] G.C.A. Schuit and L.L. von Reijen, Advances in Catalysis 10 (1958) 242.
[11] I. Yasumori and E. Miyazaki, J. Chem. Soc. Japan 92 (1971) 659.

DISCUSSION

A. D. O'CINNEIDE

We have investigated the influence of acetylene on the ethylene hydrogenation activity of a supported platinum catalyst. The study was performed in a slug-type flow reactor with H_2 as carrier gas (35 c.c/min.). The catalyst sample was 4.0 grm. of 4.6% Pt./Filtros FS-140-L, prepared by a standard impregnation technique. Filtros FS-140-L is a low area, ceramic silica support. The catalyst was oxidized in situ at 450°C and reduced in flowing H_2 at 400°C. The reaction temperature was -45°C. The CO adsorption capacity of the sample at this temperature (measured in a separate experiment with identical pre-treatment conditions) was found to be 6.73 x 10^{-2} c.c. N.T.P. The sizes of the ethylene and acetylene pulses were 15.13 x 10^{-2} and 4.84 x 10^{-2} c.c--corresponding to about 10 and 2 surface monolayer equivalents, respectively. The results are shown in Table 1. The surface

TABLE 1

Enhancement effect of acetylene on ethylene hydrogenation

Pulse No.	Type	Product Composition (%)			Percentage of Pulse eluted
		C_2H_6	C_2H_4	C_2H_4	
1	C_2H_4	53	47		
2	C_2H_2	68	28	4	91.3
3	C_2H_4	72	28		
4	C_2H_2	58	30	11	98.8
5	C_2H_4	76	24		
6	C_2H_2	56	30	14	100.0
7	C_2H_4	78	22		
8*	C_2H_4	79	21		
9*	C_2H_4	78	22		
10**	C_2H_4	61	39		
11***	C_2H_4	56	44		

*After a 30 minute interval.
**After heating to 200°C in flowing H_2.
***After heating to 400°C in flowing H_2.

is about 50% more active after it has been exposed to a pulse of acetylene. The effect is independent of time since pulses 8 and 9 were made after 30 min intervals. Also, it cannot be accounted for by the removal (as C_2H_6) of surface residues during the passage of the C_2H_4 pulse. Prior to pulses 10 and 11 the catalyst was heated in H_2 to 200°C and 400°C respectively. This caused the activity to decline back towards the original value. CH_4 0.13 x 10^{-2} c.c.), C_2H_6 (0.21 x 10^{-2} c.c) and C_3H_8 (0.08 x 10^{-2} c.c.) desorbed from the surface during these heat treatments. Therefore enhanced activity seems due to adsorbed hydrocarbons.

I. YASUMORI

The present study shows that the diminished activity of Pd annealed at 200°C recovers partly, 17% of the original value, by the adsorption of acetylene. Dr. O'Cinneide found similar increase of activity due to added acetylene on Pt reduced at 400°C. In his case, however, the increased activity is 1.5 times as high as that of the original surface. These results suggest that the effect of adsorbed acetylene depends strongly on the way of pretreatment. In a recent preliminary study on Pd annealed at higher temperature such as 800°C, we have observed that the activity of acetylene-covered surface becomes higher than that of clean surface by a factor of about 3. Kinetics on this surface is quite different from that on Pd which is stabilized with acetylene after annealing at 200°C; the reaction orders in hydrogen and ethylene are 1.0 and 0.8 respectively. This finding means that the rate-determining step shifts to the reaction between adsorbed hydrogen and ethyl radical. If the increase in activity is due to only the removal of some contaminants by acetylene, it will be hard to understand this shift. Therefore, the structure of exposed surface should be more responsible than contaminants for the effect of adsorbed acetylene.

W. PALCZEWSKA

Under the experimental conditions (e.g. p_{H_2} = 10 torr, T = 0°C) the β hydride phase of the Pd-H system should form. In consequence, the Pd foil could--or even should--be poisoned (R. S. Mann, T. R. Lien, J. Catalysis 15 (1969) 1; our report--this Congress). The authors did not register this effect. It seems to me that the well known low rate of penetration of the H_2 into massive Pd samples, as e.g. foils, under the experimental conditions favored the investigations. The formation of the β-Pd-hydride did not really occur and the reactions were occuring at the Pd surface. Did you, however, notice perhaps the difference, when studying dispersed Pd?

I. YASUMORI

We cannot completely eliminate the possibility that β-hydride phase is produced to some extent during the course of reaction under the present conditions. However, if absorbed hydrogen is in equilibrium with gaseous hydrogen and contributes to the reaction, simple half order in hydrogen will not be expected. In the previous study on the acetylene hydrogenation using the same Pd foil, we found that the rate of the reaction is almost unchanged over a wide range of the concentration of preabsorbed hydrogen and the products scarcely involve absorbed hydrogen as shown in the H_2-D_2 replacement (J. Phys. Chem., 75, 880 (1971)). On the other hand, for Pd foil annealed at 800°C, the rate increased with the amount of preabsorbed hydrogen and the hydrogenation proceeded via reaction with absorbed hydroben. Professor Kokes also observed similar effect of dissolved hydrogen in the hydrogenation of ethylene and propylene (J. Phys. Chem., 70, 2543 (1966)). Pd foil catalyst prepared at 200°C or lower temperature has many surface imperfections which are correlated with the active sites for reaction. Such structures may exist largely on small particles of Pd and, then, the similar behavior is expected in dispersed Pd catalysts.

H. C. EGGHART

Dr. Yasumori described a rate increasing effect of an adsorbed species which could have been expected to be a poison. I like to point out very briefly that a rate increasing effect of well known poisons like sulfur, selenium, tellurium or lead was observed by Binder, Kohling and Saniltstede[1] in the electrocatalytic oxidation of carbon monoxide and of formic acid on platinum electrodes. In the electrocatalytic oxidation of olefins, little effect of the adsorbed poisons was found. In the electrocatalytic oxidation of saturated hydrocarbons, where dissociative adsorption appears to be particularly important, the usual poison effect of sulfur, etc. was observed.

1) H. Binder, A. Köhling and G. Sandstede, "From Electrocatalysis to Fuel Cells," edited by G. Sandstede, The University of Washington Press, Seattle and London, p. 59; Advances in Chemistry Series 90, p. 128, American Chemical Society, Washington, D.C. (1969).

I. YASUMORI

Dr. Egghart pointed out that added sulfur in the electrocatalytic oxidation of formic acid alters the nature of electrode and increases the rate of the electron transfer process, $H^+ + e \rightarrow 1/2\ H_2$. This conclusion and the present result of preliminary ESCA study predict that the action of the adsorbate as a promoter is to modify not only the geometric factor but also the electronic factor in catalytic activity.

Paper Number 53

INTERMOLECULAR HYDROGEN TRANSFER IN LINEAR OLEFINS AND DIOLEFINS CATALYZED BY GROUP VIII METAL CATALYSTS

M. M. BHASIN
Union Carbide Corporation, South Charleston, W. Va., USA

ABSTRACT: Linear olefins (α and internal) have been found to undergo an intermolecular hydrogen transfer at 100-200°C over palladium yielding the corresponding paraffin and diolefins in roughly equivalent amounts. The yield of C_6-diolefins from 1-hexene at 200°C was 2.0%, from 2-hexene, 1.0%. The yield of C_{12}-diolefins from 1-dodecene was 8%. Only part of the diolefins were conjugated - about 90% of the hexadienes, and 20-30% of the dodecadienes. In an equivalent type of reaction, 1,11-dodecadiene was partially converted to dodecenes and dodecatrienes. Palladium catalyzed these reactions at temperatures well below dehydrogenation temperatures for these compounds. Ruthenium and rhodium were much less active, and platinum was inactive. It is proposed that such a hydrogen transfer occurs via a concerted multi-center mechanism.

1. INTRODUCTION

Numerous studies have been made of intramolecular hydrogen transfer (abbreviated as intramol.-HT) in various olefins (also called double bond isomerization). However, the only mention of intermolecular hydrogen transfer (abbreviated as intermol.-HT) is in the case of cyclohexene and its derivatives[1,2]. Intermol.-HT processes are also known to occur in the catalytic cracking of hydrocarbons. However, in this case olefins are acceptors of hydrogen atoms from other olefins that have been converted to coke or of hydrogen released from the conversion of naphthenes to aromatics[3]. This study reports on the intermol.-HT observed in linear olefins and diolefins over various noble metal catalysts. The role of the support and the behavior of various noble metals is also discussed.

2. EXPERIMENTAL

2.1 Catalysts

All catalysts were obtained from Engelhard Industries, Inc. In all cases the metal was deposited on the shell of the alumina pellets. The physical properties of these catalysts are given in Table I. The γ-alumina was obtained by calcining Conoco's (Catapal-N) alpha alumina monohydrate (99.9% pure) at 550°C overnight. All catalysts were dried at 200°C for 30-60 minutes in a N_2 purge unless otherwise stated.

2.2 Apparatus and Materials

All reactions at atmospheric pressure were carried out in a 250 ml flask equipped with an electric stirrer, a condenser, and a thermowell. In intermol.-HT experiments with

TABLE I

PALLADIUM CATALYSTS

Catalyst Description	Catalyst Code
γ-Alumina, Conoco-N, 250 m^2/g	Al
0.5% Pd on γ-alumina, 100 m^2/g, 0.12" pellets	Pd-1
0.5% Pd on γ-alumina, 400 m^2/g, 0.06" pellets	Pd-2
0.5% Pd on α-alumina, 5 m^2/g, 0.12" pellets	Pd-3
Pd Black, powder, ~ 1 m^2/g	Pd-4

olefins and other hydrocarbons, the low-boiling olefins were separated by distillation in the same vessel using the condenser as a rough distillation column.

Reactions with hexenes were performed at higher pressures in a 250 ml. capacity Magna-drive autoclave (Autoclave Engineers, Inc.). A constant stirring speed of ~1200 r.p.m. was used in all experiments.

All α-olefins and paraffins were obtained from Phillips Petroleum Co. They were all found to be better than 98-99% pure by vapor phase chromatography (VPC). All reference compounds and 1,11-dodecadiene were obtained from Chemical Samples Co. (99% purity). All other olefins and chemicals were reagent grade. Random linear tetradecenes were prepared by isomerization of 1-tetradecene over a Na/Al_2O_3 catalyst.

2.3 Procedure and Analysis

The catalyst and the reactor (glass or autoclave) were purged with N_2, heated to 200°C for 30-60 minutes and then cooled to room temperature. The liquid reactants were then charged to the reactor using a hypodermic syringe. A N_2 purge was maintained at all times until the reactor was sealed. The reactor and its contents were heated to the desired operating temperature, and the samples withdrawn as needed.

The samples were analyzed by VPC and by wet-methods. Some samples were also analyzed by mass spectrometry (MS) and ultraviolet spectrometry (UV). The VPC analysis of C_{12}-C_{14} products was done on 15% FFAP columns. These columns were operated at 100°C and programmed at 20°C/min. after 2 minutes. Analysis of C_6 paraffin, olefin, and conjugated dienes was accomplished by the use of a column containing 1% $AgNO_3$ + 18% CARBOWAX 20M on Chromosorb W. The analysis of C_6-olefin isomers, 1,5-hexadiene, and 1,4-hexadiene was done on a dual column system consisting of a UCON LB-550X and Dimethylsulfolane columns[4)].

Bromine numbers and diene values for conjugated dienes were determined by standard ASTM procedures[5)]. Analysis by UV was done using the conjugated hexadienes and heptadienes as standards (abs. max. at 227 mμ).

3. RESULTS AND INTERPRETATION

3.1 Intermolecular Hydrogen Transfer in 1-Hexene

Intermol.-HT over a Pd-1 catalyst gave n-hexane and all the conjugated dienes. No 1,4-hexadiene was observed. 1,5-Hexadiene was formed in some runs at 200°C. 2,4-Hexadienes (trans, trans and cis, trans-2,4) were the predominating dienes in the product (see Tables II and III). Small amounts of cis, cis-2,4-, and 1,3-hexadienes were also formed. The reaction was slow and became slower after 2-3% conversion of 1-hexene.

TABLE II

INTERMOLECULAR HYDROGEN TRANSFER IN 1-HEXENE

OVER Pd-1 CATALYST

REACTANT VOLUME, 100 ml; REACTION TIME, 1 HR.

Run No.		1			2	3	4*
		Pd-1 (25g)			Pd-1 (12g)	Pd-1 (12g) Powder+	Pd-1 (12g) Powder
Components	Feed	100°C	150°	200°	200°	200°	200°
<C_6	0.13	0.11	0.07	0.05	0.09	0.10	----
n-Hexane	1.69	2.75	2.83	3.55	3.70	3.60	3.97
Hexenes	98.03	96.30	95.77	94.61	93.70	94.45	93.63
Hexadienes							
1,3-(c and t)	0.03	0.06	0.14	0.18	0.23	0.20	0.38
t,t-2,4-	0.00	0.32	0.60	0.77	0.92	0.80	0.95
c,t-2,4-	0.00	0.25	0.49	0.67	0.75	0.65	0.76
c,c-2,4-	0.13	0.17	0.09	0.14	0.17	0.17	0.15
1,5-	0.00	n.d.	n.d.	n.d.	0.43	0.05	n.d.
Conversion	----	1.73	2.26	3.42	4.33	3.58	4.40

* In this run the reactant consisted of a 20 vol % solution of 1-hexene in benzene.

\+ Powdered catalyst in all cases was 100-300 mesh size.

3.1.1 Thermodynamic Considerations

Intermol.-HT of 1-pentene to n-pentane and trans-1,3-pentadiene was taken as an example of such intermol.-HT since thermodynamic data on these C_5-hydrocarbons were readily available[6]. The values of the equilibrium constant (K_f) for such an intermol.-HT at 100°C and 350°C are 5 x 10^2 and 16 respectively. However, the value of K_f for intermol.-HT to yield non-conjugated 1,4-pentadiene and n-pentane is about 0.4 over this temperature range. Thus the formation of conjugated dienes is highly favored over the non-conjugated dienes.

3.1.2 Mass Transfer Effects

The catalyst concentration was changed from 25 gm to 12 gm (in 100 ml of reactant) without any appreciable effect on the conversion of 1-hexene at 200°C (see Runs 1 and 2, Table II). The Pd-1 catalyst was crushed to 100-325 mesh, and the results of Run No. 3 (Table II) shows very little effect on the yield of dienes and paraffins. Dilution of 1-hexene to

TABLE III

INTERMOLECULAR HYDROGEN TRANSFER IN 1-HEXENE OVER VARIOUS PALLADIUM CATALYSTS

REACTION CONDITIONS:

200°C; 1.0 HR; 1-HEXENE, 100 ml; CATALYST[(a)], 12g

Run No.		1	2	3	4	5
Components	Feed	Al	Pd-1	Pd-2	Pd-3	Pd-4
$<C_6$	0.13	0.09	0.09	0.09	0.11	0.11
n-Hexane	1.69	1.61	3.70	3.54	3.10	5.84
1-+t-3, $C_6^=$ (b)	96.33	21.35	39.62	40.29	73.47	56.51
t-2+ c-3, $C_6^=$	1.70	36.44	32.22	32.40	13.08	19.57
c-2, $C_6^=$	0.00	39.86	21.86	21.73	8.80	14.56
Hexadienes						
1,3-(c and t)	0.03	0.04	0.23	0.25	0.27	0.71
t,t-2,4-	0.00	0.04	0.92	0.84	0.49	1.17
c, t-2,4-	0.00	t	0.75	0.73	0.49	1.28
c,c-2,4-	0.13	0.03	0.17	0.12	0.19	0.24
1,5-	0.00	0.54	0.43	<.05	0.00	0.00
Conversion	----	0.38	4.33	3.61	2.68	7.38

(a) 12g of catalyst used in all runs except (1) where 25g was used.

(b) $C_6^=$ is an abbreviation for Hexene.

a 20% solution in benzene also did not alter the yields. No hydrogen transfer to benzene was observed at 100-200°C (Run No. 4, Table II). These results show that the reaction was not limited by catalyst concentration nor by mass transfer resistance in the bulk or inside the pores of the γ-alumina.

3.1.3 Activity of Group VIII Metals and High Purity γ-Alumina

The activity of alumina-supported Pd, Rh, Ru, and Pt was investigated at 100°, 150° and 200°C (not tabulated). The activity of all metals other than Pd was very low. The 1-hexene feed contained 0.13% cis, cis-2,4-hexadiene, and 0.03% of 1,3-hexadiene (called impurity dienes). Some 1,5-hexadiene (0.2-0.6%) was formed over the Rh and Ru catalysts at 200°C but none over Pt. If 1,5-hexadiene is formed by hydrogen transfer to the alumina (or spillover of hydrogen from Pd to alumina), then these metals are inactive for intermol.-HT. However, if the 1,5-hexadiene is formed by transfer on metal surfaces, Rh and Ru are at best weakly active catalyst. Both of these catalysts did form small amounts of conjugated dienes which could have been formed by intramol.-HT of the cis, cis-2,4-hexadiene impurity in the feed. The total amount of conjugated dienes in the product was approximately equal to the impurity dienes within experimental error. Also, unlike the results with Pd catalysts, formation of dienes was not accompanied by the formation of an equivalent amount of n-hexane. The Pt catalyst did not significantly isomerize the impurity dienes. All the metal catalysts for this activity comparison were supposedly made on the same γ-alumina.

A high purity γ-alumina (not the same as that used in the Engelhard catalyst) behaved similar to the Rh and Ru catalysts in isomerizing the impurity dienes (Run No. 1, Table III) and forming 1,5-hexadiene ($\sim$0.5%).

3.1.4 Effect of Catalyst Support

Palladium supported on two γ-aluminas, an α-alumina and in the unsupported form, was tested for intermol.-HT of 1-hexene (see Run Nos. 2, 3, 4 and 5, Table III). The Pd on 400 m^2/g alumina showed somewhat less activity than that supported on 100 m^2/g alumina. Palladium on α-alumina was the least active, while Pd black was the most active catalyst. In the absence of the knowledge of exposed Pd, it is difficult to ascertain whether these differences in activity are due to exposed Pd atoms or due to the size of the Pd crystallites. It is interesting to note that no 1,5-hexadiene was formed on Pd/α-alumina and Pd black.

3.2 Intermolecular Hydrogen Transfer in 2-Hexene Over Palladium

Intermol.-HT also occurred in 2-hexene (cis and trans) over the Pd-1 catalyst (see Table IV). The product spectrum was similar to that obtained with 1-hexene. The conversion of 2-hexene was about one-half of that observed with 1-hexene.

TABLE IV

INTERMOLECULAR HYDROGEN TRANSFER IN 2-HEXENE

CATALYST: Pd-1 POWDERED; CONDITIONS: 200°C, 5.5 HRS.

Component	Hexane	Hexenes	ΣHexadienes	Conversion
Feed	0.2	99.80	0.06	----
Product	1.31	97.45	0.80	2.35

3.3 Intermolecular Hydrogen Transfer in 1-Dodecene Over Palladium

Hydrogen transfer of 1-dodecene resulted in the formation of dienes and n-dodecane (see Table V). The dienes were of the conjugated and non-conjugated types. Analysis of these products was quite difficult. However, an FFAP column did separate many of the hydrocarbon types. Lacking standard compounds, tentative identification of dienes, trienes, etc., was based on the behavior of this column towards double bonds. This column, however, tended to deteriorate with age and occasionally had to be replaced. The presence of dienes and paraffins (also trienes and tetraenes) was also confirmed by MS, even though the results were not very quantitative in the absence of standards. The presence of conjugated dienes was confirmed by UV spectrophotometry and diene value determinations. These results agree very well with each other and the VPC analysis. The only disagreement was in the feed analysis; UV indicated only trace dienes (<0.1) while diene value indicated 2.0% dienes. This disagreement can be resolved if one considers the fact that the terminal olefins are known to condense (at a slower rate) with the maleic anhydride used in diene value determinations[7]. UV analysis and diene value of

TABLE V

INTERMOLECULAR HYDROGEN TRANSFER IN 1-DODECENE AND 1,11-DODECADIENE OVER A PALLADIUM CATALYST (Pd-1) AT 210°C

Reactant	1-Dodecene				1,11-Dodecadiene		
Time, hrs.	0 (Feed)	6			0 (Feed)	2	
Analysis	VPC	VPC	VPC	MS	VPC	VPC	MS
n-C_{12}-ane	0.46	8.64	8.14	~5	0.00	0.74	<1
1-C_{12}-ene	99.28						
C_{12}-enes	0.00	87.72	87.51	~85	0.41	21.20	10-15
C_{12}-dienes	0.00				99.57		
Conj. C_{12}-dienes	0.01	3.48	3.53	6-8	----	62.76	60-70
C_{12}-trienes	0.00	0.18	0.41	<1	0.02	13.19	10-15
C_{12}-tetraenes	0.00	----	0.32	1-3	0.00	1.97	2
Other Analysis							
Conj. Dienes[a)]	2.0	3.3	----	---	----	----	---
Conj. Dienes[b)]	<0.01	3.0	----	---	----	9.8	---
Bromine No.	94.2	87.8	----	---	96.1	84.4	---
Conversion	----	~16	~16	~15	----	30-40	20-30

(a) Conjugated dienes (wt %) by diene value.
(b) Conjugated dienes (wt %) by UV.

the product agree well because the terminal olefin had been isomerized to internal isomers. The reduction in the bromine number is also an indication of the formation of conjugated dienes since such dienes pick up only one-half of the theoretical amount of bromine.

3.3.1 Effect of Particle Size and Residence Time

A bried study of the effect of these variables was made using the amount of conjugated dienes as a measure of 1-dodecene conversion. There was very little or no effect of the particle size on the performance of the Pd-1 catalyst. Changing the residence time from 1.5 to ~10 hours increased the amount of conjugated dienes by only 0.5% (data not in Tables). Reduction of the Pd-1 catalyst prior to the N_2 purge did not affect the conversion either.

3.4 Intermolecular Hydrogen Transfer in 1,11-Dodecadiene

Intermol.-HT of 1,11-dodecadiene over Pd-1 at 210° yielded 10-15% each of the corresponding olefins and trienes (see Table V). However, UV showed only 9.8% as conjugated dienes. Thus intramol.-HT was not complete even after 2 hours of reaction time. There was also an indication of the formation of ~ 2% tetraenes. The results by VPC were confirmed by MS. Thus relatively large proportions of dienes and trienes must have been of the non-conjugated type.

3.5 Intermolecular Hydrogen Transfer in Random Tetradecenes

Random tetradecenes also underwent an intermol.-HT over Pd-1 giving the corresponding dienes and the paraffin (see Table VI). The conversion of random tetradecenes was about one-half that observed in 1-dodecene reaction.

TABLE VI

INTERMOLECULAR HYDROGEN TRANSFER IN RANDOM TETRADECENES

CONDITIONS: 237°C; 4 hrs; Pd-1 Catalyst

	Feed	Product
Lights	2.0	2.0
C_{14}-Paraffin	4.8	8.6
C_{14}-Olefins	92.9	85.2
C_{14}-Diolefins	0.3	4.0
Conversion	---	8.0

3.6 Intermolecular Hydrogen Transfer Between Olefins, Diolefins, and Other Hydrocarbons

No hydrogen transfer took place over the Pd-1 catalyst between α-olefins (1-hexene and 1-dodecene) and other paraffins (C_{14} and C_{17}); between diolefins and paraffins; between α-olefins and stearic acid; between α-olefins and methyl stearate; and between diolefins and methyl stearate. However, intermol.-HT in olefins and diolefins did occur at the same rate as if other compounds were absent.

3.8 Kinetics

The extent of intermol.-HT of 1-hexene over Pd-1 catalyst was followed with time at 200°C. The data are shown graphically in fig. 1. After the first one-half hour, the rate of hydrogen transfer slows down appreciably. The initial one-half hour is an approximate warm-up time at 200°C. Hydrogen transfer was also studied as a function of temperature (data not in Tables). Assuming pseudo first order behavior, the rate constants at 100°, 150°, and 200° were 1.5×10^{-3}, 8.5×10^{-3} and 1.5×10^{-2} hr^{-1}, respectively. An Arrhenius plot of the rate constant (fig. 2) gives an activation energy of 9 kcal/mole. However, this estimate of activation energy is only an approximate one. The true activation energy may vary by as much as 5 kcal/mole.

4. DISCUSSION

It has been shown in the preceding section that linear olefins and diolefins undergo an intermol.-HT over Pd catalysts yielding the corresponding hydrogen-rich and hydrogen-poor hydrocarbons. The yield of intermol.-HT products was 2% from 1-hexene as compared to 8% with 1-dodecene. 2-Hexene gave one-half the yield of hydrogen transfer product as 1-hexene, while random tetradecenes gave about one-half the yield of hydrogen transfer product as 1-dodecene. The reaction rates were very slow, and the rate constant increased only slowly with temperature. The apparent activation energy for the intermolecular hydrogen transfer process was found to be ~ 9 kcal/mole. It appears that the hydrogen transfer was inhibited by the desorption of conjugated dienes, since the conjugated dienes are known to adsorb more strongly than the olefins. However, surface diffusion (or migration) of H atoms may also

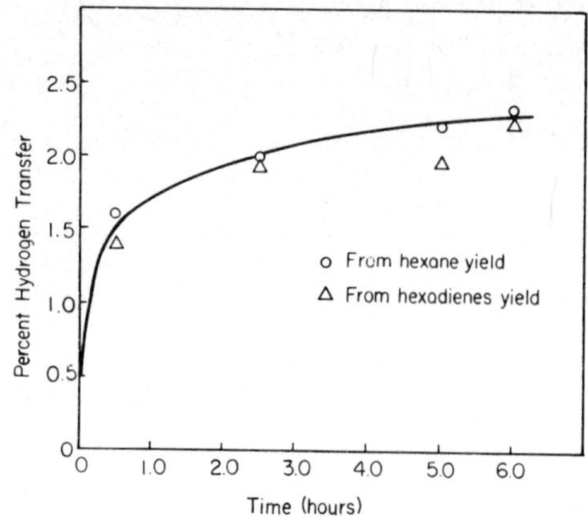

Fig. 1. INTERMOLECULAR HYDROGEN TRANSFER IN HEXENE-1 AT 200°C OVER Pd-1 CATALYST - YIELD VERSUS TIME CURVE

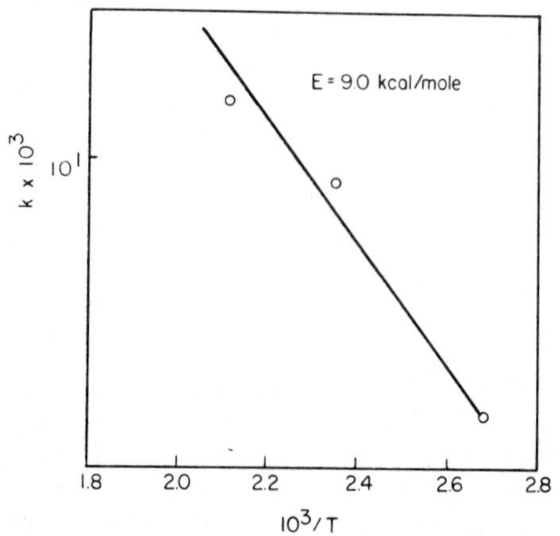

Fig. 2. ARRHENIUS PLOT FOR INTERMOLECULAR HYDROGEN TRANSFER REACTION OF HEXENE-1 OVER Pd-1 CATALYST

play a significant role. The activation energy for surface and bulk diffusion of H atoms is reported to be 5-7 kcal/mole[6]. It should also be pointed out that substantial amounts of 1-hexene were left in the product even though Pd is known to be an efficient double-bond migration catalyst. Furthermore, although the dienes in the hexene runs were >90% conjugated, the dienes in the dodecene runs were only 20-30% conjugated. Similarly, in the dodecadiene run, only ~15% of the total dienes were conjugated. Thus a substantial amount of the olefins and diolefins did not undergo an intramol.-HT to form thermodynamically favored internal olefins and conjugated dienes, respectively. The absence of hydrogen transfer between olefins/paraffins, olefins/stearic acid, etc., but the unperturbed intermol.-HT in olefins itself, suggest that olefins are so strongly adsorbed that they prevent other non-olefinic hydrocarbons from adsorbing and undergoing a cross intermol.-HT. The following mechanism is proposed for the intermol.-HT in linear olefins and diolefins:

1) Adsorption of olefins via the double bond (π or σ bonding).

2) Activation of some paraffinic hydrogens of the adsorbed olefin.

3) Formation Pd-C bond followed by surface migration of H atoms. The surface intermediates are represented below. Surface palladium atoms are represented by *.

```
        C       C
R-C   /   \   /   \   C = C
  |         C
  *      *      *      *
            H
  *      *      *      *
  *      *      *      *
R-C       C       C = C
    \   /   \   /
      C       C
```

4) Hydrogen transfer from the Pd-C bonded olefin to Pd to form a diene followed by hydrogenation of the second adsorbed olefin to form a paraffin.

5) Desorption of the paraffin and the diene.

Such a concerted multi-center mechanism can explain the formation of relatively greater amounts of non-conjugated dienes as the carbon number of the olefin increases. With long carbon chains there is a greater statistical probability of activating paraffinic hydrogen at the non-conjugated position than at the conjugated position. The extent of metal dispersion and the surface topography of Pd would influence such a hydrogen transfer. However, in the absence of such information on the surface it is difficult to say whether such effects influence the rate on the various Pd catalysts.

4.1 Behavior of Other Group VIII Metals and the Alumina Support

Rhodium and Ru were much less active for such a transfer, and Pt was inactive. The only activity observed with Rh and Ru was in the formation of 1,5-hexadiene at 200°C (0.2-0.6%). Similar activity was also exhibited by high purity γ-alumina dried at 200°C. It appears that the formation of 1,5-hexadiene took place on γ-Al_2O_3 by a H-transfer to the alumina surface. Although the γ-alumina of the Pd-1 catalyst is not the same, 1,5-hexadiene could form as a result of a spillover phenomenon. The lack of activity of Rh, Ru and Pt may be related either to their inability to activate the paraffinic hydrogens or the inability to migrate H atoms. The high initial rate of hydrogen transfer may be due to the higher concentration gradient of H atoms and thus a higher diffusion rate.

An important aspect of the proposed mechanism is that multicenter sites are required for intermol.-HT to take place in olefins and diolefins. Thus it can be predicted that such intermol.-HT will not take place with homogenous palladium catalysts which do not have multiple Pd atoms. Similarly, atomically dispersed Pd will not catalyze intermol.-HT.

5. ACKNOWLEDGMENTS

The author wishes to thank Drs. H. G. Davis, G. E. Keller and K. D. Williamson for the many helpful discussions he has had with them. The author also expresses his appreciation to Union Carbide Corporation for approval to publish this work.

6. REFERENCES

(1) L. M. Jackman, Advan. in Org. Chem., Vol. 2, 329 (1960).

(2) "Palladium, Recovery, Properties, and Uses", by E. M. Wise, p.p. 171 (1968).

(3) H. H. Voge, Catalysis, Vol. VI, 407 (1958).

(4) M. M. Bhasin, to be published.

(5) ASTM Standards, Petroleum Products, Vol I, D-1158 (1961) and D-1961 (1961).

(6) F. D. Rossini, et al, "Selected Values of Physical and Thermodynamic Properties", Carnegie Press (1953).

(7) A. S. Onishchenko, "Diene Synthesis", English Translation of "Dienovyi Sintez", Izdatel'stvo Akad. Nauk. SSR, Moskva, p.p. 6-7 (1963).

(8) F. A. Lewis, "Palladium-Hydrogen System", Academic Press, p.p. 106-108 (1967).

DISCUSSION

P. B. WELLS

I am intrigued that (i) the reaction virtually stops after about 2% conversion (Fig. 1), and (ii) that palladium is the only active Group VIII element. Is it likely that the disproportionation reaction is facilitated by a small amount of atomic hydrogen diffusing from the bulk, this having remained in the catalyst from the preparation? This might well occur for the hydrogen-poor phase of palladium at 200°C.

Is this 2% reaction reproducible; would a succession of identical experiments each give 2% conversion?

M. M. BHASIN

I do not feel that the intermolecular hydrogen transfer is facilitated by a small amount of hydrogen left in the bulk of palladium crystallites. I have done experiments in which the catalyst was used as such without reduction and the conversion to hydrogen transfer product was found to remain unchanged.

In answer to your next question, the 2% reaction (with hexene-1-only; with dodecene the hydrogen transfer was ~ 8%) with hexene-1 is very much reproducible. However, I have not done any succession of experiments with the same catalyst. In connection with this question and the question as to why the reaction stops after 2% hydrogen transfer, I feel that either the product diolefins or their further dehydrogenation products poison the catalyst because they are more strongly adsorbed than the olefins.

V. PONEC

If ethylene is admitted without H_2 to the metal catalyst in a closed volume, self hydrogenation occurs. Hydrogen is split off from the first adsorbed molecules, and other molecules are hydrogenated by it to ethane which appears in the has phase. However, the surface is successively covered by dehydrogenated species, and the process of self-hydrogenation does not proceed further; it is stopped when the surface is completely covered. There is no steady-state hydrogen redistribution which would lead in a flow system to a stationary production of acetylene and ethane from ethylene.

When looking at the results of this paper, I have a feeling that the situation is much the same as described above: only part of the dehydrogenated species, namely the dienes, desorbs in this case. I am afraid there is no steady-state hydrogen transfer among the olefins. Could you comment on this?

M. M. BHASIN

I am quite aware of the ethylene adsorption work you described in detail in your question. However, I feel that although the situation in the reaction I described may be the same to some extent, it is quite different in degree to which side reactions take place. By that, I mean that highly unsaturated hydrocarbons formed may account for the limiting low conversion

I observed, but the extent of reaction accounts for much more than an essentially stoichiometric reaction observed in the case of the reaction of ethylene with metals. Therefore, I feel that under proper conditions of temperature, pressure and flow rate, one would observe a steady state formation of diolefins and paraffins, even though one may need to regenerate the active catalytic surface when it gets poisoned by the hydrogen-poor hydrocarbons or carbon itself.

C. E. FRANK

Since the conversion is substantially higher over Pd black than over Pd-on-Al_2O_3, it suggests the Al_2O_3 may actually inhibit the transfer. Have you tried other supports such as carbon or silica?

M. M. BHASIN

Although alumina may be partly responsible for the inhibition of the hydrogen transfer reaction, I feel that the diolefins or the highly unsaturated hydrocarbons formed on the palladium crystallites may be more responsible for the inhibition of the reaction than the alumina support. This point needs to be clarified in further work. In answer to your next question, I have not made any experiments with palladium supported on carbon or silica gel.

J. W. HIGHTOWER

I am impressed by the apparent sensitivity of your analytical equipment as indicated by the number of significant figures in your tables. With this sensitivity, did you observe any cracking products which might be expected at these temperatures?

M. M. BHASIN

The sensitivity indicated in my tabulated data is not a measure of accuracy, but it does indicate the ability to detect trace components. The level of detection is in the range of 0.01 weight per cent. I have not observed any significant amount of cracking products in the liquid product recovered.

H. S. BLOCH

I wonder whether, under your reaction conditions, there might not be some polymerization or condensation of your polyenic products to heavier materials which accumulate on the surface and stop the reaction. Your material balances show a deficiency, in general, of dienes relative to paraffin indicating some disappearance of the dienes.

M. M. BHASIN

I do not think that there is a significant difference between the analyses of diolefins and the amount of paraffins within experimental error (~5% of each component), however, there may be a real difference. Thus, polymerization or the condensation of polyenic materials or the formation of hydrogen deficient materials may be responsible for the inhibition of the reaction.

Paper Number 54

SOME ASPECTS OF CATALYTIC RING EXPANSION AND CONTRACTION REACTIONS IN CYCLOHEXANE

G. PARRAVANO
Department of Chemical Engineering, University of Michigan,
Ann Arbor, Michigan, U.S.A.

Abstract: The rate of the catalytic redistribution of isotopic carbon between mixtures of cyclohexane-cyclopentane, and cyclohexane-cycloheptane was measured in the range of temperature 164° to 383°C and ratios of hydrocarbon partial pressures 3×10^{-2} to 30. Conversions up to 60% were obtained without interference from side reactions. The catalysts used were: $Pt\text{-}Al_2O_3$ with and without acid additions, $Rh\text{-}SiO_2$, $Ru\text{-}SiO_2 \cdot Al_2O_3$, $Re\text{-}Cr_2O_3$, Re-C and pyrolytic carbon. From the experimental results information on the chemisorption of the hydrocarbon mixtures and on the reactions of ring expansion and contraction of cyclohexane, was derived, and correlated with properties of the catalysts employed.

Previous work from this laboratory on the equilibrium hydrogenolysis of cyclohexane (CHA), indicated the possibility that reactions of ring expansion and contraction in cycloalkanes may occur under moderate conditions at the surface of typical hydrogenolysis catalysts.[1)] This conclusion is consistent with results on the hydrogenolysis of methylcyclopentane, substituted cyclobutanes, and the isomerization of n-hexane[2].

To observe directly the individual steps of ring expansion and contraction of CHA it is obvious that another cycloalkane must be used to provide to or accept from CHA CH_2 groups. Specifically, it is necessary to find conditions under which it is possible to study the rate of contraction of CHA into cyclopentane (CPA) by employing CPA itself as a methylene acceptor, and the rate of CHA expansion by employing cycloheptane (CTA) as methylene donor. Stoichiometrically during the transfer steps there is no change in the total number of CHA, CPA and CTA molecules, since for every CHA molecule transformed into CPA or CTA, one of the latter is converted to CHA. To follow the fate of the individual molecules it is convenient to label with isotopic carbon one of the two species, namely for the contraction of the CHA ring the reaction is:

$$*CHA(g) + CPA(g) \rightarrow CHA(g) + *CPA(g) \qquad (1)$$

and, for the expansion:

$$*CHA(g) + CTA(g) \rightarrow CHA(g) + *CTA(g) \qquad (2)$$

where * refers to an isotopic carbon atom, and g indicates the gas phase. Formally, reactions (1) and (2) may be written in terms of CH_2 transfer steps, namely for reaction (1):

$$*CHA(g) \rightarrow *CPA(g) + CH_2(s) \qquad (1a)$$

$$CPA(g) + CH_2(s) \rightarrow CHA(g) \qquad (1b)$$

Since during reactions (1) and (2) a surface-gas phase equilibrium is established, the rates of reactions (1a), (1b) and (1) are equal. Reaction (1) may be conveniently used instead of directly studying the individual steps (1a), (1b) which cannot be easily followed.

1. EXPERIMENTAL

Reagent grade CHA, CPA and CTA were used without further purification and high purity He gas was employed as a carrier. A stock solution of ^{14}C labeled CHA was made from milligram portion of a radioactively concentrated sample (1 mCurie). The catalysts employed include: Pt(0.4w%)-Al_2O_3,Pt(0.4w%)-Al_2O_3 fluorided,Rh(1w%)-SiO_2,Rh(10w%)-SiO_2,Ru(0.5w%)-$SiO_2 \cdot Al_2O_3$,Re(4.5w%)-Cr_2O_3,Re(5w%)-C,and pyrolytic carbon.The Pt catalysts were commercial samples in the form of 1/16" spheres.The Rh catalysts were supporte on Cabosil HS5, 300m^2/g. The metal surface area of these catalysts, measured by H_2 and CO chemisorption, was 40.8 and 19.6 m^2/g respectively.[3] The Ru catalyst was prepared by impregnation of $SiO_2 \cdot Al_2O_3$ with a solution of Ru(NO)$(NO)_3$ and subsequent thermal decomposition. Re-Cr_2O_3 was prepared by impregnation of chromia xerogel with a slightly acidified solution of $ReCl_3$, followed by drying under an IR lamp. Re-C was prepared by impregnation of pyrolytic carbon with the proper amount of a solution of NH_4ReO_4 and, subsequently, dried under an IR lamp. Pyrolytic carbon was obtained by treatment of polyfurfural alcohol at 2000°C for 16 hours in flowing N_2. The product had an ash content of 0.023%, a density of 1.54 g/cc, and showed a broad X-ray pattern. Pt-$Al_2O_3$was heated at 450°C for two hours at 0.1 Torr and, in situ, in a stream of H_2 at 400°C for an additional two hours. Pt-Al_2O_3 fluorided was pretreated in a similar manner except that the final treatment was carried out at 450°C. Re-Cr_2O_3 was heated in a stream of purified H_2 at a rate of 50°C/hour up to 300°C. Above this temperature H_2 was replaced by N_2 up to 400°C. Ru and Rh catalysts and the pyrolytic carbon preparations were pretreated in a manner similar to that of Pt-Al_2O_3. In a second sample of pyrolytic carbon the final heat treatment was carried out in air. The rate of reactions (1) and (2) was studied in a flow system at atmospheric pressure. Samples from the exit stream were fractionated by gas chromatography, and radioactive analysis carried out on the fractions by liquid scintillation techniques.[4] The experimental conditions were carefully chosen so that in the exit stream no products from isomerization, dehydrogenation, aromatization, reactions were present. The establishment of the equilibrium during the course of reactions (1) and (2) was inferred from the invariability of the hydrocarbon feed as it passed through the catalyst bed and from the reversibility and reproducibility of the results. In blank runs, performed with an empty reactor, the reaction conversion was found to be <0.05%. In the catalytic experiments, conversions up to 60% were measured. The error in the value of the rate constant was estimated between 15 and 25%, depending on the conversion level.

2. EXPERIMENTAL RESULTS

As discussed previously,[4] the rate of reaction step (1a), or that of reaction (1) is given by:

$$-\frac{1}{w}\frac{dn_{*CHA}}{dt} = k_c p_{*CHA} - k'_c\, p_{*CHA} \qquad (3)$$

where w, n_{*CHA} are the catalyst weight and moles of *CHA, respectively, and k_c and k'_c are the rate coefficients of forward and reverse step (1a). Integration of equation (3) for a flow reactor gives:

$$k_c = \frac{\dot{V}}{wRT} \; \frac{1}{1+\beta} \ln \frac{1}{1-\alpha} \qquad (4)$$

where V is the volumetric flow rate at temperature T, $\beta = \frac{p_{CHA}}{p_{CPA}}$, and the

reaction conversion $\alpha = \frac{p_{*CPA}}{(p_{*CPA})_e} = \frac{1+\beta}{\beta} \delta$ where $\delta = \frac{p_{*CPA}}{p_{*CPA} + p_{*CHA}} =$

$\frac{p_{*CPA}}{(p_{*CHA})_o}$. The suffixes o and e refer to initial and equilibrium conditions, respectively. For reaction (a), a similar derivation gives:

$$k_c = \frac{\dot{V}}{wRT} \; \frac{1}{1+1/\beta} \ln \frac{1}{1-\alpha} \qquad (5)$$

where $\beta = \frac{p_{CTA}}{p_{CHA}}$ and $\alpha \frac{p_{*CTA}}{(p_{*CTA})_e}$.

At constant temperature the rate coefficient is generally dependent upon β. This is expressed by $k_c = k\beta^{\pm m}$ (6) where k is the reaction rate constant and m is a constant.

Values of k_c on $Pt\text{-}Al_2O_3$ for the hydrocarbon mixture CHA-CTA at 233°C, 323°C and 385°C are reported in Figure 1, while in Figure 2 we have summarized the observations on the same catalysts for the mixture CHA-CPA. Plots of k_c versus β obtained from measurements on $Pt\text{-}Al_2O_3$ fluorided, 380°C, and on $Rh\text{-}SiO_2$ for the mixture CHA-CTA, 233°C and 325°C, are collected in Figures 3 and 4, respectively. Figure 5 presents the results on $Ru\text{-}SiO_2\cdot Al_2O_3$ for the same hydrocarbon combination at 325°C. The rate measurements on $Re\text{-}Cr_2O_3$ are summarized in Figure 6, while Figure 7 reports the results on pyrolytic carbon in the temperature range 233°C to 325°C for the mixture CHA-CPA. From these plots the values for the rate constant, k, were calculated by means of equation (6). In those instances in which the logarithmic plots were curvilinear, the linear parts of the plots were used to compute average values of m. The computed values of m are reported in Tables 1 to 4.

Table 1

Average Values of m (equation (6))and of the Rate Constant, k, for Reactions (1) and (2) Catalyzed by $Pt(0.4w\%)\text{-}Al_2O_3$.

Hydrocarbon Mixture	Temperature °C	m	$k \times 10^8$ $\left[\frac{\text{mole}}{\text{g(cat)} \times \text{sec} \times \text{atm}}\right]$
CHA-CTA	233	-0.26,0.73	3.8
CHA-CTA	327	-0.36,0.28	8.5
CHA-CTA	383	-0.16,0.60	16.0
CHA-CPA	233	-1.0	0.19
CHA-CPA	327	-1.0	0.54
CHA-CPA	383	-0.30	2.3
CHA-CTA[a]	380	-0.20,0.36	8.0
CHA-CPA[a]	380	-1.45,0.50	0.18

* a ; $Pt\text{-}Al_2O_3$ fluorided

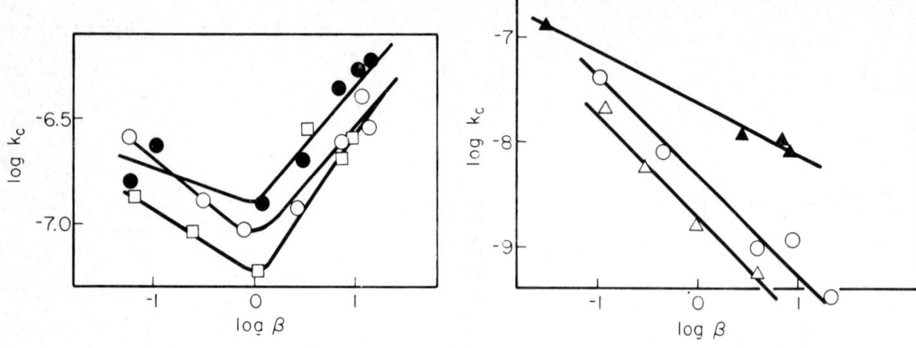

Fig. 1. Values of k_c versus β for reaction (1) between CHA and CTA on $Pt-Al_2O_3$; □ 233°C, ○ 323°C, ● 385°C.

Fig. 2. Values of k_c versus β for reaction (1) between CHA and CPA on $Pt-Al_2O_3$; △ 233°C, ○ 327°C, ▲ 380°C.

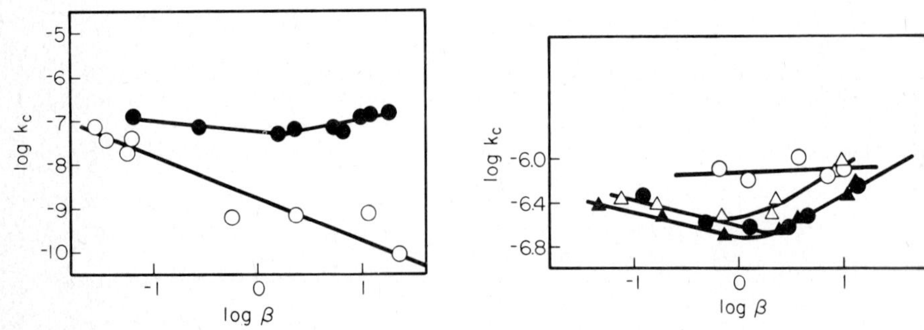

Fig. 3. Values of k_c versus β for reaction (1) catalyzed by $Pt-Al_2O_3$ fluorided O CHA-CPA, ● CHA-CTA, 380°C.

Fig. 4. Values of k_c versus β for reaction (1) between CHA and CPA; ● Rh(1%)-SiO_2 233°C, ⊙ Rh(1%)-SiO_2 325°C, ▲ Rh(10%)-SiO_2 233°C, △ Rh(10%)-SiO_2 325°C.

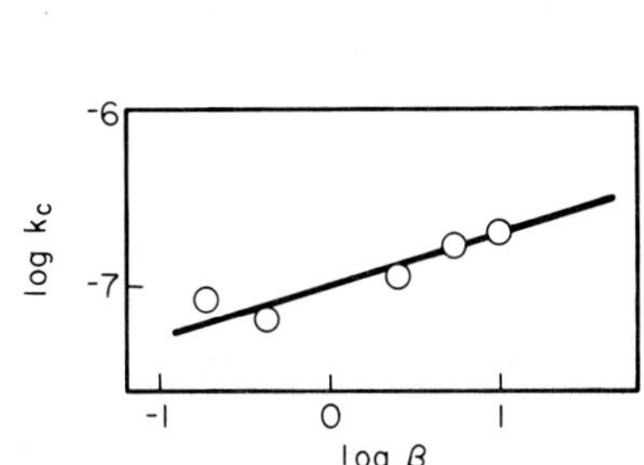

Fig. 5. Values of k_c versus β for reaction (1) on Ru(0.5%)-$SiO_2 \cdot Al_2O_3$ for the combination CHA-CTA, 325°C.

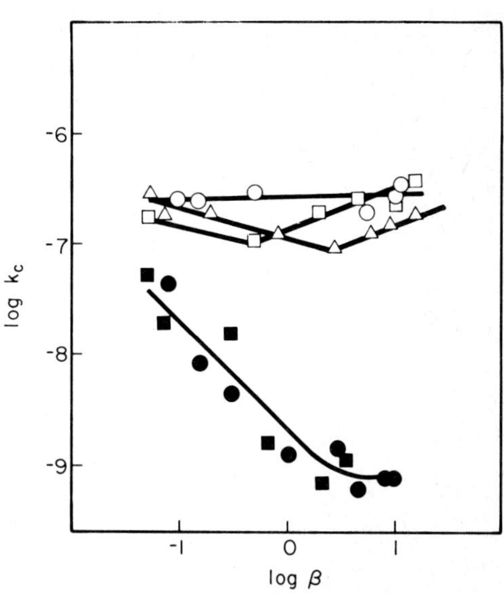

Fig. 6. Values of k_c versus β for reaction (1) on Re(5%)-Cr_2O_3, ■ CHA-CPA 243°C, ● CHA-CPA 355°C, □ CHA-CTA 164°C, △ CHA-CTA 243°C, ○ CHA-CTA 353°C.

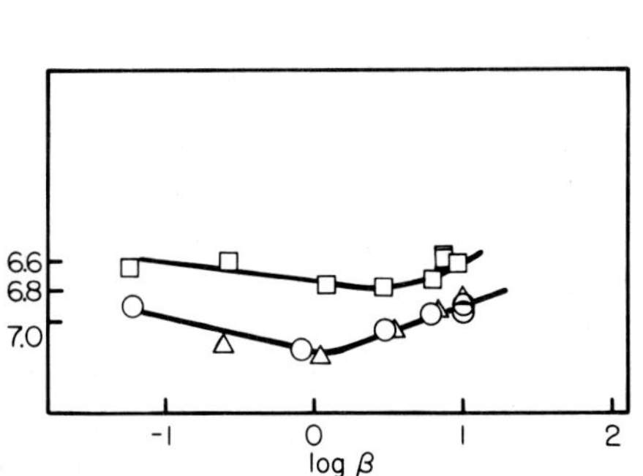

Fig. 7. Values of k_c versus β for reaction (1) for the combination CHA-CTA; ■ pyrolytic carbon 233°C (H_2 treated); pyrolvtic carbon 325°C H_2 treated); pyrolytic carbon, pyrolytic carbon with 5% Re, 325°C.

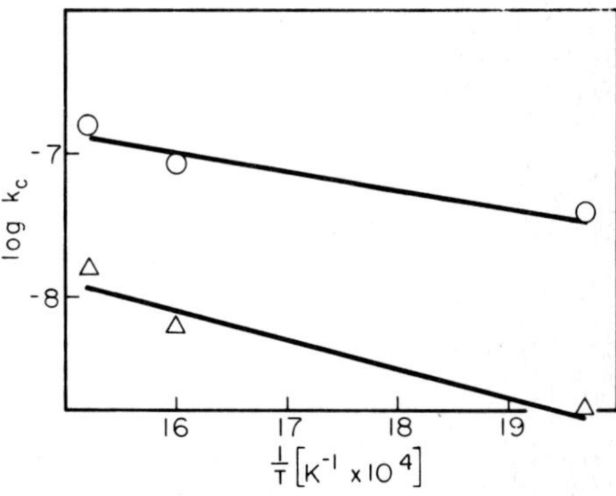

Fig. 8. Arrhenius plot for reaction (1) A and reaction (2) ⊙ catalyzed by Pt-Al_2O_3.

Table 2
Average Values of m (equation (6)) and of the Rate Constant, k, for Reaction (2) Catalyzed by Rh-SiO_2 and Ru-$SiO_2Al_2O_3$.

Metal	w%	Temperature °C	m	k x 10^8 [$\frac{\text{mole}}{\text{g(cat) x sec x atm}}$]
Rh	1	230	-0.24,0.55	19.0
Rh	1	327	∿ 0	30.0
Rh	10	230	-0.25,0.55	20.0
Rh	10	327	∿ 0.03	76.0
Rh	0.5	325	-0.38,0.36	1

Table 3
Average Values of m (equation (6)) and of the Rate Constant, k, for Reactions (1) and (2) Catalyzed by Re-Cr_2O_3.

Hydrocabon Mixture	Temperature °C	m	k x 10^8 [$\frac{\text{mole}}{\text{g(cat) x sec x atm}}$]
CHA-CTA	164	-0.25,0.30	14.0
CHA-CTA	243	-0.25,0.30	12.0
CHA-CTA	353	∿ 0	25.0
CHA-CPA	243	-0.96,0.14	0.25
CHA-CPA	353	-0.96	0.21

Table 4
Average Values of m (equation (6)) and of the Rate Constant, k, for Reaction (2) Catalyzed by Pyrolytic Carbon

Temperature °C	m	k x 10^8 [$\frac{\text{mole}}{\text{g(cat) x sec x atm}}$]
233	-0.11,0.41	5.8
325	-0.11,0.35	5.8
325[a]	-0.32,0.20	18.0
325[b]	-0.25,0.25	5.8

a, pretreated in air
b, containing 5% Re

3. DISCUSSION

The significant results presented in the previous section are the following:

1. Relatively high conversions for reactions (1) and (2) were obtained in the presence of different solids acting as catalysts and in the absence of a methylene donor or acceptor different from a cycloalkane.

2. In the majority of the cases investigated, the isothermal reaction coefficient, k_c, was found dependent upon the ratio of the partial pressures of the hydrocarbon feed. The analytical relation as described by the value and the sign of the sign of the exponent m, varied with the nature of the ring modification (expansion or contraction), catalyst, temperature and acidity. With CHA-CPA feed the exponent m had a negative value. Generally m (CHA-CPA)>m(CHA-CTA) with m(CHA-CPA) ≅ 1. With CHA-CTA feed two average values of m with opposite signs were observed.

3. The value of the reaction rate constant showed some significant differences according to catalyst employed and pretreatment.

The reported experiments are concerned with the simultaneous chemisorption of two cycloalkanes, followed by ring size modification. The nature of the reactive chemisorption is empirically characterized by the exponent m. A complete and unambiguous theoretical interpretation of the sign and value of m cannot be given for the lack of an independently justifiable model of the catalyst surface layer during the course of reactions (1) and (2). It is then, necessary to restrict the discussion to general trends only.

The main factors which influence the sign and value of m are: the structure of the adsorbed layer and its stoichiometry, the competition for surface between the hydrocarbons, and the relative proportions of active and nonactive adsorbate configurations for the reaction under consideration. Differentiation among these configurations may be made in terms of the number of adsorbate species per surface site(localized adsorption or by the fraction of surface area occupied by each adsorbate species(nonlocalized adsorption). Furthermore, since C-C bond breaking is necessary for reactions (1) and (2), adsorbate configurations with adsorbate/surface site ratios $\leq 1/2$ and with on carbon atom doubly bonded to the surface are important.[6] The correct configuration or the adsorbed cycloalkanes must allow the rapid formation and decomposition of the reaction intermediate, which should be reversibly formed and destroyed without undergoing fragmentation to more strongly adsorbed products.

Let us consider two adsorption configurations, R' and R'', of the species R, formed by the interaction of the hydrocarbon mixture with the catalyst surface and assume that [R']<<[R''] and [R''] ≅ [R'']max. These conditions mean that R'' is strongly and R' is weakly adsorbed. Using a model of nonlocalized adsorption the concentration of R' is given by.[7]

$[R'] = Cp_R^{1-\frac{\omega'}{\omega''}}$ or, introducing the step reaction equilibrium: CHA(g) + R'(s) $\rightleftharpoons$ CTA(g) : $[R'] = C''\left(\frac{p_{CTA}}{p_{CHA}}\right)^{1-\frac{\omega'}{\omega''}}$ (7)

where C, C' are constants and ω', ω'' the area of the surface occupied by one R' and R'' species, respectively. An essentially similar relation (except for the numerical value of the exponent in equation (7)) is obtained assuming a model of localized adsorption.[7] Equation (7) shows that for $\omega' > \omega''$, R' varies inversely with the ratio $\frac{p_{CTA}}{p_{CHA}}$. Since the rate coefficient of reaction (1) is dependent upon [R'], or

$$k_c = k[R'] = k \left(\frac{p_{CTA}}{p_{CHA}}\right)^{1-\frac{\omega'}{\omega''}} \qquad (8)$$

Comparison with equation (8) with equation (6) shows that $m = 1 - \frac{\omega'}{\omega''}$.

Despite its qualitative nature, the reasoning is helpful to suggest the meaning of high (>1) or low (<0) values of m. Whenever there is large difference between ω' and ω", m 1 and conversely for low values of m. A change in the algebraic sign of m indicates a basic shift in the controlling adsorption (surface stoichiometry, fractional area of the adsorbate).

The experimental results show that larger difference between ω' and ω" were present for CHA-CPA than for CHA-CTA (item 2), and that the presence of extrinsic acidity on the catalyst did not have a detectable effect in modifying this condition. For Rh there was no measurable influence of the metal concentration on ω' and ω". Defferences between ω' and ω" tended to decrease as the temperature was raised.

The values of the reaction rate constants are reported in Tables 1 to 5 in the last column. A comparison among these values reveal some interesting points. On the same catalyst, the reaction was considerably faster between CHA-CTA rather than between CHA-CPA (Table 1). On $Pt-Al_2O_3$ at 233°C, the reaction of CHA with CTA was about 20 times faster while on $Re-Cr_2O_3$ at 353°C it was about two orders of magnitude fater. Most likely these differences are the result of the higher stability of the five as compared with that of the seven membered ring compound. This effect is consistent with values of the activation energy for reactions (1) and (2) computed in the temperature range 233 to 383°C (on $Pt-Al_2O_3$). For reaction (1), under otherwise similar conditions, it was found equal to 10 kcal/mole, or larger by a factor of about two (Figure 8).

At 325°C, the activity sequence among the metals tested for reaction (2) was Rh>Pt>Ru. The sequence is similar to that found in a study of the alkylation step in aromatics.[7] Rh is also the most active metal for hydrogenolysis.[8] In the case of this metal there was no substantial difference in activity with the metal particle size(Table 3). It is rather puzzling that molecules whose dimensions are in the same range of the metal particles are not influenced by the size of the latter. A similar lack of dependence was previously found in the case of the hydrogen transfer between bezene and cyclohexane.[4]

On pyrolytic carbon, the rate constant for reaction (2) was increased by a factor of 5 by substituting air to H_2 during the activation procedure. The air activated carbon samples compared vary favorably in catalytic activit with Rh, the most active of the metals tested.

In previous study on the equilibrium hydrogenolysis of CHA rate constant between (0.20 to 2.6) x 10^{-8} $[\frac{mole}{g(cat)\cdot sec\cdot atm}]$ were measured at 376°C on $Pt-Al_2O_3$, depending upon the type of hexane isomer formed. On the same catalyst at 383°C the rate constant for CHA-CPA reaction was 2.3 x $10^{-8}[\frac{mole}{g(cat\cdot sec\cdot atm}]$ and 16.0 x $10^{-8}[\frac{mole}{g(cat)\cdot sec\cdot atm}]$ for the CHA-CTA reaction (Table 1). Thus the rates of the two reaction steps involving cyclohexane, hydrogeneolysis and ring size modification are comparable, seemingly indicating a common primary step(ring opening).

The existence of independent,adsorbed CH_2 groups,as formally represented in reaction steps(1a) and (1b), under reaction conditions is unlikely. One might conceive surface CH_2 groups originating from, but still attached to, adsorbed cycloalkanes, viz. for the case of CHA-CPA:

$$*C_6H_{12}(g) + C_7H_{14}(g) \xrightarrow{-2H} *C_6H_{12}{-}{-}{-}{-}\underset{H}{\overset{H}{\underset{|}{\overset{|}{C}}}}{-}{-}{-}{-}C_6H_{12}(s) \;(I) \xrightarrow{+2H}$$

$$C_6H_{12}(g) + *C_7H_{14}(g)$$

For the formation of the intermediate species (I), corresponding to adsorbed cycloalkymethanes, there is no direct evidence. However, it should be recalled that arylmethanes have been repeatedly considered as intermediate species in various catalytic reactions.[9)]

4. CONCLUSION

The study has demonstrated that the rate of expansion and contraction of the CHA ring may be easily followed at surfaces of supported Rh, Pt, Ru, Cr_2O_3 and pyrolytic carbon in the temperature range 164 to 383°C. This result was obtained by employing CPA and CTA as methylene acceptor and donor, respectively. The study of this system has made possible to observe thermodynamic (reactive adsorption isotherm) and kinetic (surface reactive efficiency) aspects of adsorption-desorption processes and given the possibility of measuring directly the rate of ring size modification. To translate the conclusions from this work into predictions for the net rate of catalytic ring expansion and contraction reactions of cyclohexane, it is necessary to complement the conclusions from this study with information on the activation of a suitable methylene donor. If the rate of the latter in relation to the reaction step (1a) at a given surface is rapid, the predictions gathered in this work will be closely followed in the net reaction of ring size modification.

ACKNOWLEDGEMENT

We wish to express our thanks to Professor E. E. Hucke for a gift of the sample of pyrolytic carbon. We gratefully acknowldege the support of this work by Grant GK-2013 of the National Science Foundation.

REFERENCES

1) G. Parravano, Journal of Catalysis, 22, 96, (1971).
2) G. Maire, G. Ploudy, J. C. Prudhomme, F. G. Gault, ibid. 4, 556, (1965); Y. Barro, G. Maire, J. M.Muller, F. G. Gault, ibid. 5, 428, (1966).
3) D. J. Yates, J. H. Sinfelt, ibid. 8, 348, (1967).
4) G. Parravano, ibid. 16, 1,(1970).
5) C. Kemball, Catalysis Reviews, 5, 33, (1971).
6) C. Wagner, Berichte der Bunsen Gesellschaft fur Physik. Chemie 74, 398, (1970)
7) G. Parravano, J. of Catalysis in press.
8) J. R. Anderson, B. G. Baker, Proc. Roy, Soc. A271, 402 (1962).
9) M. A. Lanawala, A. P. Bolton, paper presented at the First North American Meeting of the Catalysis Society Atlantic City, February 1969.

DISCUSSION

C. DIMITROV

Reported data by Professor G. Parravano are very interesting and useful for explanation of such reactions as ring expansion and ring contraction of hydrocarbons.

I have 2 questions:

1) Under investigating conditions ($Ru-SiO_2-Al_2O_3$, 380°C) didn't you detect any products from side reactions?

2) Do you think that any type of carbon redistribution takes place in the catalytic ring contraction and ring expansion of tetraline and indane?

G. PARRAVANO

In the ranges of temperature and cycloalkane partial pressure reported in the paper for each catalyst studied there was practically no formation of side products, as detected by gas chromatography. Extensive formation of them, however, was detected outside these ranges. This includes the silica-alumina supported Ru catalyst.

It is difficult to extrapolate the conditions employed for the three cycloalkanes studied to tetraline and indane. Our past experience with carbon redistribution reactions in alkanes, cycloalkanes and aromatics on a variety of supported metal catalysts shows that these reactions occur surprisingly easy under mild catalytic conditions.

F. G. GAULT

The reaction discovered by Professor Parravano is extremely interesting and throws a new light upon the mechanism of the cracking reactions of hydrocarbons.

As I understood, the exchange of ^{14}C between labeled and unlabeled hydrocarbon has been performed in absence of hydrogen.

It would be very interesting to know whether the same reaction would happen in presence of hydrogen. In this case, by using molecules labeled with ^{13}C, one should observe doubly labeled molecules. Although the accuracy for determining the parent peaks is not always good on account of their small intensity in the mass spectra, we never observed such doubly labeled molecules in our experiments. An upper limit of 2 to 4% could then be given for the participation of the carbon exchange reaction in the isomerization and ring enlargement of hydrocarbons in presence of hydrogen.

However, it seems likely that a similar intermediate which, in absence of hydrogen, leads to Professor Parravano's exchange reaction, probably determines rearrangement or cracking in presence of hydrogen.

Would you agree with this suggestion?

G. PARRAVANO

We did not perform experiments in the presence of hydrogen. The suggestion of Professor Gault is very interesting and we agree with his line of thinking that the presence of molecular hydrogen may drastically change the results obtained in its absence. In fact, many new and easily occurring reaction paths are opened up by the ready availability of adsorbed hydrogen.

G. R. LESTER

Are any methyl cycloparaffins observed in your reaction products?

G. PARRAVANO

No amounts of methyl cycloalkanes of practical significance were detected (see reply to C. Dimitrov's comment).

J. R. ANDERSON

Professor Parravano's interesting paper illuminates an important but little discussed sort of reaction which is known to occur over (inter alia) non-acidic platinum catalysts: this is the formation of hydrocarbon reaction products with a higher carbon number than the parent. Examples include the reports by Shuikin[1)]--e.g., the occurrence of methylcyclopentane, toluene and the xylenes in the reaction of cyclopentane parent, and by Csicsery and Burnett[2)]--e.g., the occurrence of diethylbenzene, 1-methyl-2-isopropylbenzene, 1-methyl-2-n-propylbenzene and 1,4-dimethyl-2-ethylbenzene in the reactions of 1-methyl-2-ethylbenzene parent.

Although it has previously been suggested (c.f. Refs. 1,3) that these reactions may occur by surface reaction with adsorbed C_1 residues such as CH_2, I agree with Professor Parravano that this is a most unlikely pathway, and this point is made in some detail in a recent review[4)] which also suggests a mechanism not dissimilar from that offered in the present paper. In this connection, it may be noted that all attempts to obtain exchange between C-labeled methane and other hydrocarbons over platinum catalysts, have been completely fruitless,[5)] and this is in clear agreement with the comments made above.

1) Shuiken, N. I., Advances in Catalysis, 9, 783 (1957).
2) Csicsery, S. M. and Burnett, R. L., J. Catal. 8, 75 (1967).
3) Anderson, J. R. and Baker, B. G., Proc. Roy. Soc. A271, 402 (1962).
4) Anderson, J. R., to appear in Advances in Catalysis.
5) Anderson, J. R. and Avery, N. R., J. Catal. 5, 446 (1966).

G. PARRAVANO

I am glad to hear the supporting evidence cited by Dr. Anderson, and his agreement on the possible pathway for the exchange reactions studied.

Paper Number 55

REACTION OF HCl WITH AN ALUMINA CATALYST AT ELEVATED TEMPERATURES

F. E. MASSOTH and F. E. KIVIAT
Gulf Research & Development Company
Pittsburgh, Pennsylvania, U.S.A.

ABSTRACT: The course of the chlorination of a Pt/Al_2O_3 catalyst with HCl/H_2 mixtures was studied in the temperature range 450° to 750°C with determination of weight changes, infrared spectra and catalyst property data. Reaction was rapid, showing a maximum weight pickup in five minutes. With continuing reaction time, a slow weight loss ensued, which was accompanied by loss in Cl content and surface area. Changes in concentration of various surface species were calculated from the available data. Infrared spectra of the chlorided catalyst, taken at reaction temperature, revealed the loss of a hydroxyl band and the formation of a new band. A mechanism of surface reaction is advanced to explain the results.

1. INTRODUCTION

Chlorine-promoted aluminas containing platinum are effective isomerization catalysts. Many methods of adding chlorine to the catalyst are available, among which may be mentioned: (1) impregnation with an aqueous solution, usually NH_4Cl or HCl; (2) direct reaction with gaseous HCl at elevated temperatures[1)]; (3) vapor-phase reaction with various chlorine-containing agents[2)]; and (4) combination of (2) and (3)[3)]. The work reported herein is concerned with the second mode of addition.

Myers[1)] reported decreases in chloride content and surface area with increasing treatment temperatures for a commercial eta-alumina using an $HCl-H_2$ mixture. Giannetti and Sebulsky[3)] found parallel decreases with reaction treatment time at constant temperature using a commercial (eta + gamma)-alumina. Both catalysts contained about 0.5% platinum. We undertook a study of this reaction using a flow microbalance technique in order to gain a better understanding of the surface reactions involved.

2. EXPERIMENTAL

2.1 Catalyst and Reagents

The catalyst was a commercial Sinclair-Baker, RD-150 catalyst, which contained 0.6% platinum on alumina. It was in the form of 1/8" pellets, but was ground to pass a 100 mesh sieve. Its surface area was 350 m^2/g; it contained 0.55% Cl and 0.25% S. Pure HCl gas was mixed with H_2 in a separate cylinder to make up a 20 vol. % HCl blend. The H_2 was purified by passing over a deoxo purifier to remove traces of O_2 and subsequently through 5A molecular sieves to remove H_2O. House N_2 was passed through a trap containing heated copper turnings to remove O_2. Residual H_2O in the purified N_2, as well as instrument air, was removed by molecular sieves.

2.2 Equipment and Procedures

A flow microbalance assembly, employing a Cahn RG electrobalance, was used to monitor catalyst weight changes during reaction. Catalyst samples of about 1 g were contained in a quartz bucket, which was suspended from one arm of the balance by quartz fibers. The bucket was positioned in the center of a split-shell furnace, being enclosed in a quartz tube of 1" diameter. A thermocouple well was located directly under the bucket, serving as temperature reading and thermoregulator sensor ($\pm$ 3°C). A special constriction tube, consisting of a 6" length of 1/4" ID glass tubing, was inserted between the balance case and the reactor which served to prevent contact of HCl with the metallic parts of the balance mechanism. This was achieved by using a nitrogen purge flow through the balance case, which exited below the constriction tube. Flow through the reactor was upflow, with a bed of quartz chips located below the bucket to provide preheating of the incoming gases. Gas flows were metered using calibrated rotameters. Flow rates were 200 cc (STP)/min through the reactor and 400 cc/min through the balance top.

Prior to reaction, the catalyst was heated in a flow of N_2 for 16-20 hours at reaction temperature to attain a constant weight. This was followed by a one-hour treatment in H_2. After reaction with the HCl mixture for the desired time, flows were switched to H_2 for 1/2 hour and then N_2 for 1/2 hour. Finally, the catalyst was cooled in N_2 to room temperature before removal for analyses. Catalyst analyses included: surface area by a single point BET method; % $AlCl_3$; and % S.

Infrared measurements were made on a sample reacted in situ in a high temperature cell similar to that described by Cant and Hall[4]. Approximately 80 mg of freshly powdered catalyst were pressed into a wafer under a pressure of 15 tons/sq. in. Spectra were obtained with a Beckman IR-12 instrument using a spectral slit width of 4 cm^{-1} and with the chopper between the high temperature cell and monochromator disconnected. Wafer temperature was maintained at 475 $\pm$ 25°C. The wafer was heated in vacuo for 2 hours followed by an H_2 reduction for 1 hour at a flow rate of approximately 20 cc/min. The wafer was then subjected to a stream of 20% HCl in H_2 for one-half hour. Spectra were taken at temperature separately after the H_2 treatment and after the HCl/H_2 treatment. Prior to recording the spectra, the cell was evacuated for about one-half minute in order to remove the gas phase.

3. RESULTS

3.1 Course of The Reaction

Reaction of the Pt/Al_2O_3 catalyst with the HCl/H_2 mixture resulted in a very rapid and large weight gain, followed by a slower subsequent loss. Typical weight change-time profiles are displayed in fig 1. The maximum displayed in the weight curve signifies that more than one process is occurring - a rapid increase due to adsorption and/or surface reaction, and a slower decrease due to secondary reactions. Use of 4% HCl in the reacting gas gave a less pronounced profile; a lower maximum weight gain and slower weight loss occurred. Increasing the reaction temperature also diminished the maximum weight pickup, but the subsequent weight loss became faster. It is significant that at the highest temperature, the catalyst actually suffered a net weight loss after about one-half hour.

Results of catalyst inspections after reaction are presented in table 1. Time of Reaction, HCl concentration, and temperature all had negative effects on both chloride retention and surface area. Comparison of the data with those of fig. 1 suggests that the initial weight gain can be identified with addition of Cl to the catalyst and the subsequent decrease with H_2O loss accompanying a loss in surface area. Although the chloride analyses varied with treatment temperature, time and gas concentration, the chloride surface concentration was almost constant for all runs at ca. 0.1 mg Cl/m^2. The only exception was when the catalyst was cooled down in the HCl/H_2 reaction mixture (last entry in table 1); now more Cl was retained by the catalyst, in agreement with the finding of Meyers[1].

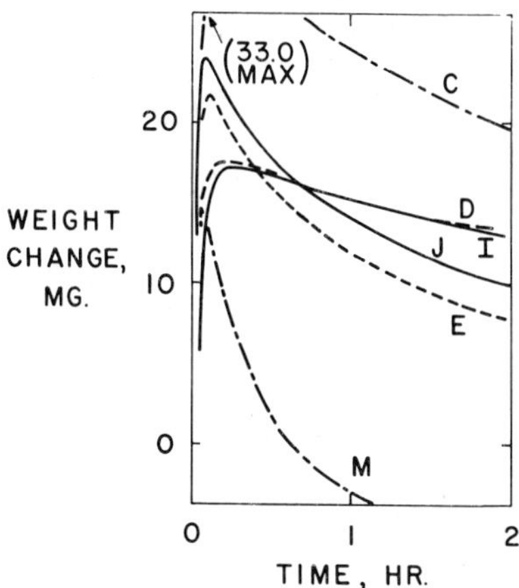

Fig. 1 Effect of Temperature and HCl Concentration on Weight Change Profiles. Catalyst Weight, 1.10 g. Letters Refer to Run Numbers in Table 1

A repeat of one of the runs was made with the same base alumina without platinum present. Essentially the same chloride addition was obtained as with the platinum-containing catalyst. Furthermore, the weight-change profiles were identical in the two runs, showing that the rate of the chlorination reaction was unaffected by platinum.

3.2 Adsorbed HCl

After completion of a specified reaction time period, a significant weight loss was always obtained upon replacing the HCl mixture with H_2 (or N_2). This effect is shown in fig. 2, where a series of identical runs for different time periods is displayed. The post-reaction decay period lasted about one-half hour, after which the catalyst weight remained constant. The difference in weight after post-treatment from that with the HCl present at the same total time period is ascribed to HCl adsorbed (complexed) on the catalyst.

Further insight into the nature of the adsorbed HCl was obtained in a separate adsorption experiment. The catalyst, after pretreating in H_2 at 565°C and cooling to room temperature, was exposed to the 20% HCl/H_2 mixture for 2 hours; a 15% weight gain was experienced. Following an overnight purge in N_2, the net weight gain was reduced to only 11%, showing that the adsorbed HCl was strongly held. Upon raising the temperature in N_2 to 565°C, about 60% of the weight gain was lost. Independent analysis of HCl removed in the heat-up step revealed that the bulk of the weight loss was from HCl (80%), the remainder from water. These results suggest formation of an HCl-complex on the catalyst surface at low temperature. This complex is not stable at high temperature in the absence of gaseous HCl, and may well account for the weight loss when HCl was removed from the gas phase (fig. 2).

TABLE 1

Results of Reaction of HCl/H_2 with Pt/Al_2O_3 Catalyst

Run	Temp °C	Time(1) Hr	HCl Vol %	Surf Area m^2/g	Cl Wt %	Cl mg/m^2
A	475	0	--	372	0.52	--
B	475	1/2	20	279	2.80	0.100
C	475	4	20	193	1.96	0.102
D	510	2	4	224	2.43	0.108
E	510	2	20	157	2.09	0.133
F	565	0	--	298	0.49	--
G	565	1/2	20	195	2.04	0.105
H	565	1	20	171	1.96	0.115
I	565	2	4	199	2.16	0.109
J	565	2	20	147	1.74	0.118
K	565(2)	2	20	154	1.83	0.119
L	565	4	20	135	1.77	0.131
M	870	2(3)	20	38	0.46	0.12
N	870	2(3)	20(4)	40	0.69	0.17

(1) Pretreatment: air, 20 hr + H_2, 1 hr at temperature.
Post-treatment: H_2, 1/2 hr + N_2, 1/2 hr at temperature.
(2) Al_2O_3 base without Pt.
(3) Pretreatment: H_2, 2 hr.
(4) Post-treatment: cooled in HCl/H_2.

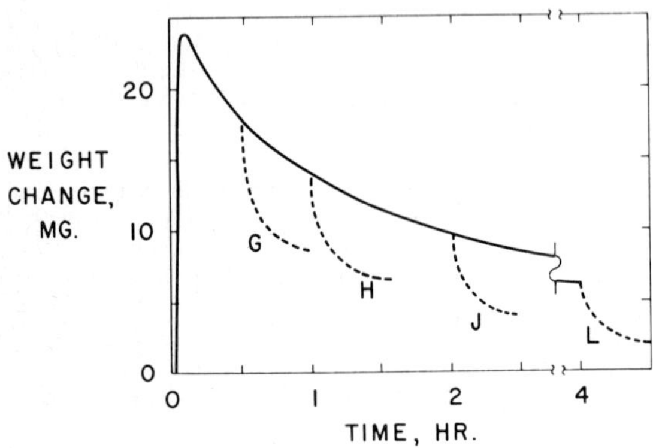

Fig. 2 Weight Change Profiles for HCl Runs at 565°C. Catalyst Weight, 1.10 g. Letters Refer to Run Numbers in Table 1. Dotted Lines Signify H_2 Post-treatment Time.

3.3 Infrared Spectra

The spectra were recorded from samples which were subjected to reduction only and to reduction followed by reaction with HCl. Fig. 3A presents the spectrum of the H_2-reduced sample. Distinct absorption bands occurred at 3670 and 3765 cm^{-1}, and diffuse shoulders were observed at ~3700 and ~3600 cm^{-1}. These latter features are not distinct, although

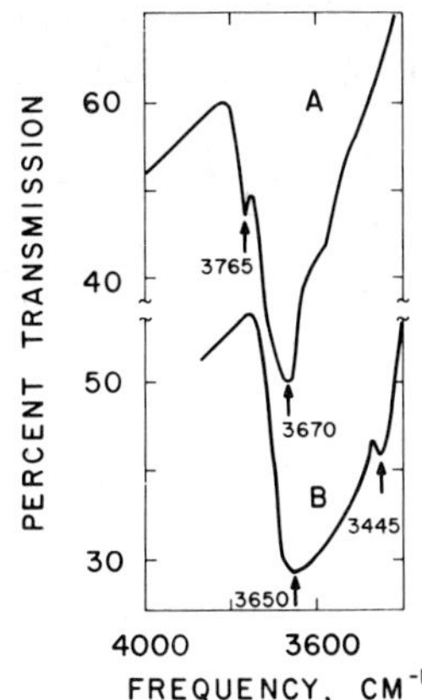

Fig. 3 Infrared Spectra of Catalyst in Hydroxyl-stretching Region. Spectra Taken at 475°C. a) after H_2 Reduction b) After HCl/H_2 Exposure for 1/2 Hr.

they appear to be real. These values are in agreement with those reported by Cant and Hall[4)] for a commercial eta-alumina. The absence of absorption bands in the 1600 to 1650 cm^{-1} region indicates negligible adsorbed water.

Fig. 3B presents the spectrum of the H_2/HCl-treated sample. Distinct absorption bands occurred at 3650 and 3445 cm^{-1}. The low wave-number side of the 3650 cm^{-1} band showed a broad absorption. Notably, the 3765 cm^{-1} band present in the H_2-reduced catalyst disappeared and the 3670 band was shifted down to 3650 cm^{-1}. These results are qualitatively in agreement with those of Peri[5)]. Again, no evidence for adsorbed water was found.

3.4 Analysis of Surface Concentrations

The total weight change of the catalyst at any time during reaction must be due to changes in adsorbed HCl, chemically bound Cl and H_2O loss only. Since the adsorbed HCl is reversibly removed, the net weight loss corrected for desorbed HCl is due to bound Cl and H_2O loss. Also, since adsorbed H_2O was not found in the infrared spectra at reaction temperature, we need only consider the changes occurring in the surface groups, Cl^-, OH^- and O^{2-}. An independent knowledge of net weight change and Cl retained then allows calculation of all three surface groups by use of the following weight and ion balances:

$$\Delta w_{net} = \Delta w_{Cl} + \Delta w_{OH} + \Delta w_{ox}$$

$$0 = \frac{\Delta w_{Cl}}{35.5} + \frac{\Delta w_{OH}}{17} + \frac{2\Delta w_{OX}}{16}$$

The results of these calculations are presented in table 2. In making the calculations, Run L was used for weight changes. These were corrected for small losses due to sulfate impurity (determined in a separate run with H_2 only), and for adsorbed HCl (determined from curves of fig. 2). Catalyst chloride and surface area values were taken from runs F, G, H, J and L; the values were considered to be applicable at the time at which the catalyst weights remained constant in the post-treatment period (reaction time + 1/2 hr.).

The weight changes of the various groups with reaction time are shown in fig. 4, which also includes the net weight change and that for adsorbed HCl. It can be seen from this figure that the initial surface reaction involves reaction of surface -O- atoms to form -OH and -Cl groups, viz.

```
                  H
O + HCl ---> O Cl
/\           / /
```

Subsequent secondary reaction then involves mainly dehydration of the catalyst surface by reaction between -OH groups, viz.

```
H   H
O + O ---> O + H2O
/   /      /\
```

as well as a smaller loss of Cl. It is noteworthy that, although initially -OH groups are formed, in a short time there is a net loss of -OH groups from the catalyst. This is consistent with the supposition that the chlorination reaction is terminated after initial rapid reaction and with the general correlation found between overall loss in surface area with subsequent weight loss.

In calculating surface concentrations of the various groups, a knowledge of the OH concentration prior to reaction is required. The value of 1.8×10^{14} cm^{-2} was obtained from separate measurement of weight loss in heating the fresh catalyst from 565° to 900°F in N_2, with suitable correction made for losses due to sulfur and chloride impurity. The total surface site concentration was taken to be 12.5×10^{14} $sites/cm^2$, which corresponds to a close packed oxide structure. Following Gerberlich, Lutinski and Hall[6], it was assumed that each O^{2-} was associated with a surface Al^{3+} ion. The resultant calculated surface concentrations are given in the last three columns of table 2. Despite a weight loss of chloride with treatment time, the chloride surface concentration actually increased slightly owing to the attendant loss in surface area. However, the surface hydroxyl concentration, after the initial reaction period, continually declined, indicating that the main reaction taking place during the weight decay period was simple catalyst sintering. In view of this, it seems likely that catalyst activity would suffer from long treatment.

TABLE 2

Variation of Surface Concentrations with Time at 565°C

Time, Min	Wt Change(1), mg/g net	Cl^-	OH^-	O^{2-}	Total Wt, mg/g Cl^-	OH^-	Surf Conc x 10^{-14}, cm^{-2} Cl^-	OH^-	O^{2-}, Al^{3+}
0	0	0	0	0	4.9	23.2	0.3	1.8	5.2
5(2)	17.1	16.5(3)	8.3	-7.6	21.4	31.5	1.2	2.8	4.2
60	8.1	15.5	-5.6	-1.8	20.4	17.5	1.7	1.7	4.5
90	7.2	14.8	-8.1	0.5	19.7	15.1	1.9	1.5	4.5
150	5.3	12.5	-8.5	1.2	17.4	14.7	2.0	1.6	4.4
270	3.9	12.9	-11.5	2.5	17.8	11.7	2.2	1.0	4.6

(1) Exclusive of adsorbed HCl, corrected for S (impurity) loss during reaction.
(2) At maximum weight gain.
(3) Calculated from maximum weight gain exclusive of adsorbed HCl.

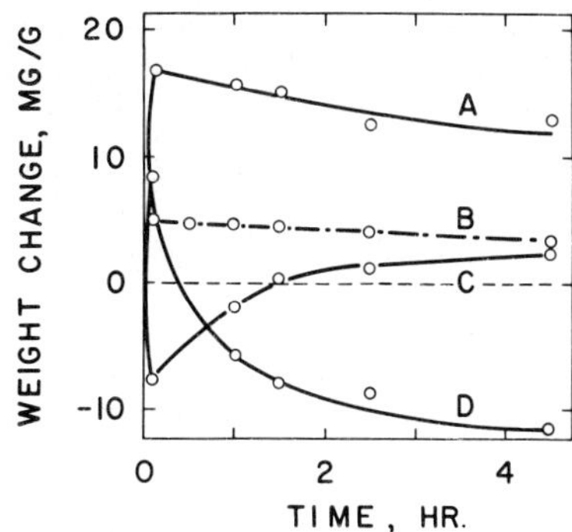

Fig. 4 Weight Change Profiles of Surface Species. Temperature, 565°C. a) Cl^-, b) Adsorbed HCl, c) O^{2-}, d) OH^-.

4. DISCUSSION

Peri[7] has shown that there are at least five types of hydroxyl groups on a pure gamma-alumina surface, each type corresponding to a distinct absorption maximum. Cant and Hall[8] have also obtained five distinct hydroxyl bonds in the spectrum of a pure eta-alumina, although the frequencies were shifted to lower wave numbers relative Peri's data. The differences in the Sinclair-Baker, RD-150 spectrum relative to that of a pure alumina are probably due, in part, to the contamination of the former. The catalyst used in this study contained Cl, S, and Pt. Also, different catalyst treatments undoubtedly contribute to spectral differences. Consequently, the diffuse shoulders at 3700 and 3600 cm^{-1} are due to at least two, and possibly more, hydroxyl groups. The distinct bands at 3765 and 3760 cm^{-1} are clearly due to two distinct hydroxyl groups.

Previous investigators[5,9] have shown that when gamma-alumina is exposed to HCl at low temperature, the following processes occur: water is produced; HCl reacts with the oxide surface to form new hydroxyl groups; and hydrogen exchange occurs between existing hydroxyl groups and HCl. Our spectral data taken at high temperature showed the disappearance of one hydroxyl band, and the appearance of a distinct new, lower frequency band. Deo et al[10] observed the disappearance of the hydroxyl group characterized by the highest wave number absorption based on gamma-alumina when treated with H_2S. These authors suggested that this was due to hydrogen-bonded chemisorption of H_2S. However, it cannot be conclusively ascertained in our case whether the hydroxyl group in question is hydrogen bonded to the HCl, or if the environment of the hydroxyl group is changed by modification of the surface. Nonetheless, the disappearance of the 3765 cm^{-1} band in the H_2-treated sample upon treatment with H_2/HCl does suggest that HCl has adsorbed on the catalyst. The appearance of the 3445 cm^{-1} band in the H_2/HCl treated catalyst indicates that a new species, not present on the untreated catalyst resulted from the treatment. This frequency is low for a surface hydroxyl group. The possibility that sintering produced internal hydroxyl groups, such as in boehmite, cannot be excluded. However, we favor the assignment of this band to a surface group because there is no evidence for hydroxyl deformations in the spectrum in the region 1100 to 1200 cm^{-1}. Deformational absorptions in this region accompany the OH stretching absorptions for internal hydroxyl groups in boehmite[11]. The broadness of the low wave number side of the 3650 cm^{-1} band suggests hydrogen bonding and/or an increase in the number of surface hydroxyl group(s) giving rise to the 3600 cm^{-1} shoulder in the untreated catalyst. The above can also be responsible for the 20 cm^{-1} shift of the band maximum before to that of 3650 cm^{-1} after HCl treatment.

Our microbalance results indicate that reaction of HCl with the catalyst was almost instantaneous, chloriding being complete within five minutes. Thereafter, at the higher temperatures, slow loss of water and some chloride occurred with continued reaction time. It seemed surprising at first that in the presence of HCl, chloride is first added than gradually lost. The reason probably is related to the decline in surface area with continued reaction time. Collapse of the surface structure entails loss of hydroxyl groups and probably some chloride would be lost in the ensuing changes. Also, the data suggest a maximum surface density of Cl^- to which the surface may be raised. Actually, the surface chloride concentration increased very slightly with reaction time despite an overall weight loss. This arose because the surface area decline was greater than the chloride loss.

From the foregoing analysis, reaction of HCl with the Pt/Al_2O_3 catalyst is proposed to occur according to the series of reactions given in fig. 5. The primary reaction is a direct addition of HCl to a surface oxygen atom via equation (1), creating new hydroxyl groups. The oxide ions active in this reaction may be surface "strained" sites, since we estimate some 20% of the total surface oxide ions participate (table 2). Possibly, a vacancy adjoining the oxide is necessary to provide a space for the chloride ion to reside.

Reaction of HCl with an hydroxyl group may result in formation of a complex, according to equation (2). The spectra of HCl/H_2-treated catalysts offers evidence for the adsorption of HCl on either a hydroxyl group or in its immediate environment. This reaction occurs to lesser extent than the direct addition reaction and is postulated mainly to account for reversibly adsorbed HCl. A similar type of complex, but involving two Cl ions was proposed by Goble and Lawrence in the chlorination of alumina with CCl_4[2]. In the absence of gaseous HCl, the complex is stable at low temperatures but dissociates at high temperatures. The complex may also decompose according to reaction (4) with liberation of water. It should be noted that combination of reactions (2) and (4) yield the following simple replacement reaction,

$$\overset{\text{OH}}{\diagup}\!\!\!\text{Al} + \text{HCl} \longrightarrow \overset{\text{Cl}}{\diagup}\!\!\!\text{Al} + H_2O$$

However, the present evidence does not support this direct exchange reaction.

The major secondary reaction is dehydroxylation via reaction (3) forming water. This reaction is strongly temperature dependent in contrast to reaction (1). The hydroxyl groups must be adjacent for reaction to occur. Water has an accelerating effect on sintering of catalyst surfaces, and may operate through a hydrogen-bonding complex sequence such as the one suggested by reaction (6). The net effect of reaction (6) is

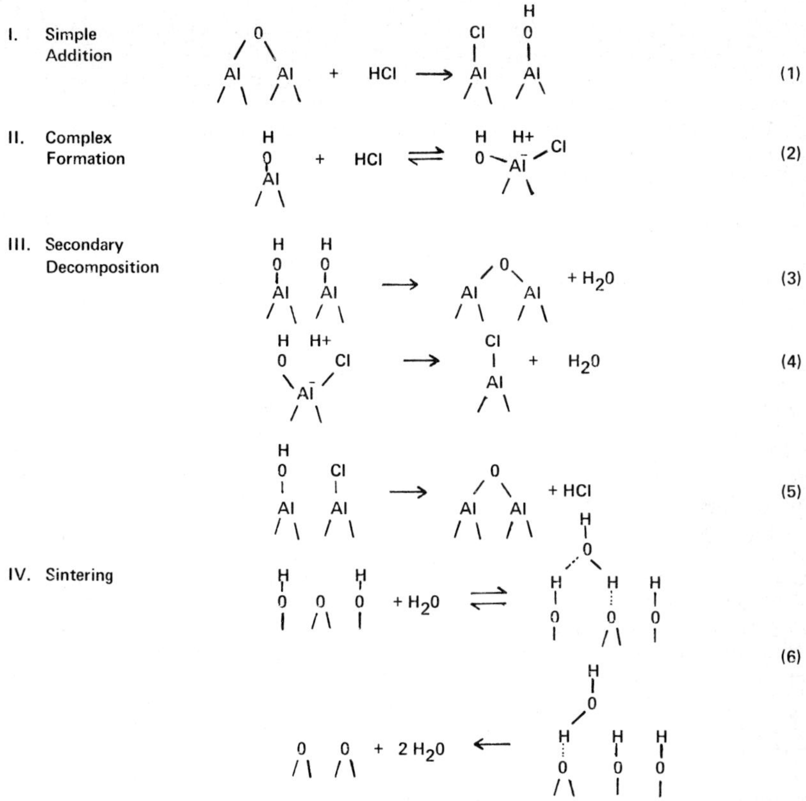

I. Simple Addition (1)

II. Complex Formation (2)

III. Secondary Decomposition (3) (4) (5)

IV. Sintering (6)

Fig. 5 Proposed Mechanism for HCl Reaction

a surface migration of hydroxyl groups, which may then dehydroylate via equation (3) when favorably located. The process of surface consolidation represented by reaction (3), regenerates oxide sites. However, the surface is annealed in the process reducing the number of accessible sites; thus, continued addition of HCl via reactions (1) and (3) sequence is precluded. Gradual loss of Cl may occur through reaction (5).

5. ACKNOWLEDGEMENT

The authors wish to thank Dr. W. Keith Hall for helpful discussions and Mr. W. E. Faust for the microbalance experiments.

REFERENCES

1) J. W. Meyers, Ind. Eng. Chem. Prod. Res. Develop. 10, 200 (1971).
2) A. G. Goble and P. A. Lawrance, Proc. Intern. Congr. Catalysis, 3rd, Amsterdam, 1964, 1, 320 (1965).
3) J. P. Giannetti and R. T. Sebulsky, Ind. Eng. Chem. Prod. Res. Develop., 8, 356 (1969).

4) N. W. Cant and W. K. Hall, Trans. Faraday Soc. 64, 1093 (1968).
5) J. B. Peri, J. Phys. Chem. 70, 1482 (1966).
6) H. R. Gerberlich, F. E. Lutinski and W. K. Hall, J. Catalysis 6, 209 (1966).
7) J. B. Peri, J. Phys. Chem. 69, 211 (1965).
8) W. K. Hall, private communication.
9) M. Tanaka and I. Ogasawara, J. Catalysis 16, 157 (1970).
10) A. V. Deo, I. G. Dalla Lana and H. W. Habgood, J. Catalysis 21, 270 (1971).
11) J. J. Fripiat, H. Bosmans and P. Rouxhet, J. Phys. Chem. 71, 1097 (1967).

DISCUSSION

R. J. BERTOLACINI

At the Paris International Congress we presented some evidence for the formation of a platinum-alumina-chloride complex on fresh catalysts. Did you, in your work, attempt to follow any changes taking place with the platinum? Was there any visible color change in the treated catalysts as the treating conditions changed, specifically, did the catalyst become lighter or yellow in color?

F. E. MASSOTH

We did not specifically monitor any property of the platinum during this study because a control run on the alumina support gave identical weight change and chloride concentration results to that of the catalyst. Hence, the chlorination reaction per se was independent of the presence of platinum. The support contained about 0.3% Cl whereas the catalyst had 0.55% Cl, so some additional chloride had been retained after platinum had been added. The color of the catalyst and treated samples was white.

T. R. HUGHES

Your reaction (2) is the formation of a surface complex which may be a Brönsted acid. Several years ago, we found that vapor phase treatment of γ-alumina with HF produced Brönsted acid sites, as shown by the infrared spectrum of adsorbed pyridine.[1] These protonic sites could be converted to Lewis acid sites by dehydration and reconverted to Brönsted sites by rehydration. It seems possible that your reaction (4) and its reverse may represent a similar interconversion between Brönsted and Lewis sites. Did you examine the infrared spectrum of your adsorbed ammonia in order to see whether the presence of Brönsted sites was indicated by the formation of ammonium ion?

1) T. R. Hughes, H. M. White and R. J. White, J. Catal. (1969).

H. KNOZINGER

I briefly comment on the question of Dr. Hughes in relation to the creation of Brönsted-sites through incorporation of chlorine into the surface. We carried out IR spectroscopic studies of pyridine on chlorine-containing alumina surfaces, recording the spectra at temperatures up to approximately 300°C. No Brönsted acidity was detected even at elevated temperatures; Lewis acidity was strongly enhanced.

F. E. MASSOTH

Ammonia adsorption at 175°C, which showed about 20% greater total acidity for the chlorided catalysts, was done volumetrically. We also thought the surface complex suggested may exhibit Brönsted acidity and carried out IR studies with pyridine at 150°C. Our results are in agreement with those of Dr. Knozinger, in that we did not observe Brönsted acidity on the chlorided catalysts. Either the Brönsted acidity was too weak or the complex sufficiently decomposed by the high-temperature evacuation prior to recording the spectra.

P. PICHAT

In an attempt to obtain more information on the sharp band at 3445 cm^{-1}, did you try to exchange the OH groups responsible for this band or to use DCl instead of HCl?

F. E. MASSOTH

No, neither of these was done.

C. H. AMBERG

I should like to report that in recent work on H_2S adsorption on alumina[1] we have obtained weight changes with time which are similar to those found by the present authors. Also, have the authors considered a hydrogen-bonded species as being responsible for the band at 3445 cm^{-1}? It seems to me that certain O-H...Cl structures could generate such a frequency.

1) T. L. Slager and C. H. Amberg, Canadian J. Chem., in press.

F. E. KIVIAT

Assignment of the 3445 cm^{-1} absorption bond to the structure O-H...Cl, as suggested by Dr. Amberg, is reasonable, but not definitive. An oxygen-chlorine distance of 3.3 Å, corresponding to the absorption at 3445 cm^{-1}, is obtained from the Lippincott-Schroeder (LS) semi-empirical curves[1] (which correlate, inter alia, OH with oxygen-chlorine distances). This value, which assumes a linear structure, is consistent with recently compiled values of hydrogen bonded oxygen-chlorine distances.[2]

However, there is no *a priori* reason to believe that the O-H...Cl structure should be either linear or approximately linear. This, as well as the lack of definitive knowledge of the surface structure of the catalyst, would introduce a degree of arbitrariness in results obtained from application of the LS potential curve method. Consequently, we do not believe application of the LS method to the catalyst would offer definite evidence of hydrogen bonding. This is contrary to a partially fluorided alumina hydrosyapatite, where use of the above methodology does offer definitive evidence for hydrogen bonding.[3]

1) R. Schroeder and E. Lippincott, J. Phys. Chem. 61, 921 (1957).
2) G. Pimertal and A. McClellan, Ann. Rev. Phys. Chem. 22, 347 (1971).
3) B. Menzel and C. Amberg, J. Colloid Interface Sci. 38, 256 (1972).

H. PINES

Do you have any experimental data to evaluate the catalytic activity of the chlorinated platinum aluminas with conversion of some specific hydrocarbon reactions? Without such data it is difficult to determine the various catalytic sites of your platinum aluminas.

F. E. MASSOTH

Catalytic activity of the HCl-treated catalysts was small for hexane isomerization under realistic processing conditions. However, subsequent treatment with other chlorinating agents, e.g. $SOCl_2$, S_2Cl_2, gave higher catalytic activities than the latter treatments alone[1]--thus, our interest in studying the effect of the HCl treatment.

1) Reference 3 of paper number 55.

Paper Number 56

ON THE REACTIVITY OF SUBSTITUTED OLEFINS AND THE MECHANISM OF THE DOUBLE BOND ISOMERIZATION ON ALUMINA

Gy. GATI+ and H. KNÖZINGER
Institute of Physical Chemistry, University of Munich, Munich (German Federal Republic)

ABSTRACT: The reactivity of 14 terminal olefins has been measured on an η-Al_2O_3. A mechanism is proposed for the isomerization of olefins on the grounds of the hundred fold differences in the reactivities and regarding results of other research groups. The reaction proceeds presumably on exposed aluminium ions with the assistance of neighboring oxygen ions by an intramolecular proton transfer. The cis preference is explained by the relative stability of cis- and trans-conformations of carbanions.

1. INTRODUCTION

A carboniumion mechanism is widely accepted for the isomerization of 1-butene over SiO_2 - Al_2O_3 [1)]. In spite of intensive work, the mechanism of olefin isomerizations on alumina is still not clear [1-22)]. It is only certain that the reaction proceeds by a different mechanism on Al_2O_3 than it does on SiO_2 - Al_2O_3 [15, 18-21)]. Various mechanisms have been proposed. Associative intermediates involving protonic surface sites are pictured as carboniumions [2-7)]. Proton addition and abstraction can occur in consecutive steps [2-6)] or simultaneously [8-10)]. For dissociative mechanisms π-allylic intermediates are suggested of either carbonium [13, 14)] or carbanion character [11, 14)]. The hydrogen shift can proceed either with the aid of certain surface sites [14)] or intramolecularly [13, 14)]. A model mechanism concerning geometric requirements of reactant and catalyst surface has recently been discussed [15, 16)].

An interesting feature of olefin isomerization is the

+) Permanent Adress: High Pressure Research Institute, Budapest XI, Gellert Ter 3, Hungary generally high cis-preference which has been explained by the assumption of an intermediate π-complex [5, 6] or of two simultaneous mechanisms for double bond and cis/trans isomerization involving different sites [10, 12, 17].

Even though reactivity studies of substituted olefins may assist in the interpretation of the isomerization mechanism there is only one systematic study with this concern [7]. In the present work, the reactivity of 14 terminal olefins of the general structure

$$H_2C^{(1)}{=}\underset{\displaystyle R_1}{\overset{}{C^{(2)}}}-\underset{\displaystyle R_2}{\overset{\displaystyle H}{C^{(3)}}}-R_3$$

has been studied on an η-Al_2O_3 and an isomerization mechanism proposed.

2. EXPERIMENTAL

2.1 Materials

The alumina was prepared by hydrolysis of aluminium isopropoxide (Fluka)and calcination at 700° C in air of the resulting hydroxide. The oxide showed the X-ray pattern of the η-phase, its BET-surface area was 178 m^2/g. The particle size fraction of 0.3 - 0.4 mm was used. 0.3 g portions were diluted with quartz particles for reactivity measurements and 3 g portions used for the adsorption studies. The oxide was treated 1 hour in dry helium (4.8 lt/h, dried at 77° K over molecular sieve Linde 3A) at 300° C before use.

All reactants and products (Fluka) were 98 mole % pure and were used without further purification.

2.2 Apparatus and Procedure

2.2.1 Adsorption studies: these were carried out by means of the pulse technique [23] using a tube of 4 mm inner diameter as adsorption column. With 0.4 μl liquid pulses or the corresponding amounts of vapor and with a flow rate F of 2.4 lt helium/h, the eluted peaks showed only insignificant tailing except for the olefins of highest molecular weight at low temperatures. For most of the substances a linear adsorption isotherm can therefore be adopted. Adsorption coefficients b are then calculated from the retention times Δt (methane was used as reference) from equ. (1) [24]

$$b = \Delta t_r \cdot F/P \cdot W \ . \qquad (1)$$

P is the average total pressure (atm) in the column, W the weight (g) of the adsorbent. Heats of adsorption are obtained from the temperature dependence of b [23].

The peaks were only slightly affected by low conversions in some cases of the most reactive olefins. The adsorption coefficients of isomeric olefins differed only insignificantly.

2.22 Kinetic studies: a microcatalytic pulse reactor [25] was used in connection with a 4 m dimethylsulfolane/chromosorb column (4.8 lt helium/h, 0° C). The conversion is independent of pulse size (0.1-6 μl) and unaffected by the addition of a second reactant. The isomerization is therefore a first order reaction as frequently shown [5, 7, 10, 12, 17, 43]. For a reversible first order surface reaction

$$A \underset{k_B}{\overset{k_A}{\rightleftharpoons}} B$$

conversion x and reciprocal specific flow rate W/F are related by equ. (2):

$$\ln \left[1 - (1 + \frac{b_B}{K\ b_A})\ x \right] = - k_A b_A (1 + \frac{b_B}{K\ b_A})\ W/F, \qquad (2)$$

where $K = k_A/k_B$ = equilibrium constant. Equation (2) can be simplified for the present experimental conditions ($x \leq 0.1$):

$$x \approx k_A b_A W/F, \qquad (3)$$

since $1 \leq b_B/b_A \leq 1.4$ and $K \geq 2$ [26] for all possible combinations. This approximation leads to errors of only 7 - 8% in the most unfavorable case. Equation (3) is therefore used throughout as a measure of the factor $k_A b_A$. Relative conversions x_{rel} are referred to 1-butene and relative reactivities

$$x_{rel}^{corr} = k_A/k_A^{stand}$$

are obtained by correction of x_{rel} for the ratio of adsorption coefficients.

2.3 Errors

Inhibition by diffusion is absent even for the most reactive olefins as shown by means of calculations of limiting conditions according to Horak [27] and of measurements with catalysts of different particle size. The mean error of the conversion data is about 5 - 6%. Adsorption coefficients show mean errors of about 10%. which may rise to about 20% for the most reactive olefins.

3. RESULTS AND INTERPRETATION

The experimental results are summarized in Table 1. The reactants are grouped into 6 classes as indicated in column 1 of Table 1. The reactions were always measured at low conversions, for which equ. (3) is valid. Measurements were carried out in different temperature ranges depending on the reactivity of the respective olefin. For comparison a standard temperature of 160° C is chosen and extrapolation in an Arrhenius plot has been made where necessary. Conversions x are total conversions with respect to the reactant. Skeletal isomerization was never observed under the present experimental conditions. Values given in brackets in Table 1 exhibit errors higher than usual.

3.1 Adsorption

Relative adsorption coefficients b_{rel} are given in column 4 of Table 1. Dispersion forces play an important role in the adsorption interaction; contributions of specific interactions are superimposed. Chromatographic heats of adsorption are equal for all compounds. The value of 9 (± 1) kcal/mole compares quite well with literature data [12, 24, 28, 29].

3.2 Reactivity of the olefins:

The value of x_{rel}^{corr} given in column 5 of Table 1 is a relative rate constant and thus a measure of the relative reactivity. The use of chromatographic adsorption coefficients to eliminate the influence of differing adsorptivities is justified for a first order reaction at low surface coverage Kochloefl et. al.[44] found identical b-values from chromatographic and kinetic measurements for the dehydration of alcohols. Corrections for the number of allyl hydrogens have also been made where necessary. Cis/trans ratios are given in column 6 of Table 1. Generally they show a cis-preference. With increasing carbon chain length and even more pronounced with

TABLE 1

Relative adsorption coefficients and reactivities for 6 classes of terminal olefins at 160° C.

Class	Substituent (R_1)	R_2	x_{rel}	b_{rel}	x_{rel}^{corr}	cis/trans ratio
I. $H_2C=CH-CH_2-R$	Me		1.0	1.0	1.0	2.1 - 2.3
	Et		1.13	1.43	0.79	
	Pr		1.34	2.60	0.52	1.2 - 1.4
	i-Pr		0.61	2.63	0.26	
	t-Bu		0.26	3.76	0.07	0.5 - 0.6
II. $H_2C=CH-CH(CH_3)-R$	H		1.0	1.0	1.0	2.1 - 2.3
	Me		1.06	1.31	0.81	
	Et		1.37	2.14	0.64	1.3 - 1.6
III. $H_2C=C(R)-CH_2-CH_3$	H		1.0	1.0	1.0	2.1 - 2.3
	Me		9.75	1.72	5.67	
	Et		18.70	(2.46)	(7.60)	1.7
IV. $H_2C=C(CH_3)-CH_2-R$	Me		9.75	1.72	5.67	
	Et		10.50	2.89	3.64	
	Pr		12.10	(8.32)	(1.45)	
	t-Bu		3.58	(10.38)	(0.35)	
V. $H_2C=C(R_1)-CH(R_2)-CH_3$	R_1	R_2				
	H	H	1.0	1.0	1.0	2.1 - 2.3
	Me	H	9.75	1.72	5.67	
	Et	H	18.70	(2.46)	(7.60)	1.7
	H	Me	1.06	1.31	0.81	
	Me	Me	11.60	(2.46)	(4.70)	
VI. $H_2C=C(R_1)-CH(R_2)-C_2H_5$	H	H	1.13	1.43	0.79	
	Me	H	10.50	2.89	3.64	
	H	Me	1.37	2.14	0.64	1.3 - 1.6
	Me	Me	13.40	4.05	3.30	

increasing chain branching the cis-preference is reduced. For 4,4-dimethyl-1-pentene a cis/trans ratio of 0.5-0.6 is obtained at 160° C. The same trends are reported by Maurel et al.[7].

It can easily be seen from the reactivities of class I and II olefins that alkyl substituents at $C^{(3)}$ diminish the reactivity. On the contrary, alkyl substituents at $C^{(2)}$ greatly increase the reactivity (class III), even a methyl group overcompensates the effect of the strongest $C^{(3)}$ substituents (class IV). 2,4,4-trimethyl-1-pentene is more reactive than 4,4-dimethyl-1-pentene by a factor of 5. Class V and VI give analogous examples. The apparent activation energies, which contain only a constant contribution from the structure independent heats of adsorption, increase from 10.5 (± 1) kcal/mole for the most reactive 2-ethyl-1-butene to 15.3 (± 1) kcal/mole for the least reactive 4,4-dimethyl-1-pentene. The difference of 4 - 5 kcal/mole may well account for the 110 fold reactivity of 2-ethyl-1-butene as compared to 4,4-dimethyl-1-pentene, if the preexponential factors are assumed to be constant.

The influence of steric effects on the reactivity can most probably be excluded as shown by linear free energy relationships which give a straight line fitted even by such

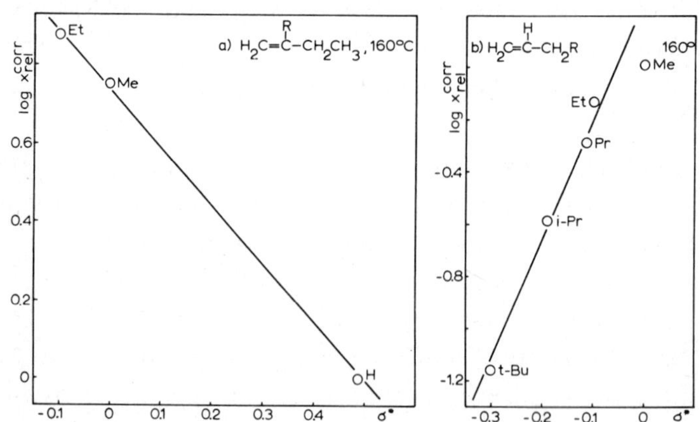

Fig. 1 log x_{rel}^{corr} versus the inductive constant σ* of Taft: a) vinyl substitution, b) allyl substitution.

bulky substituents as tertiary butyl and fitted also in the case of chain branching at $C^{(2)}$ (fig. 1a and b). Hence, inductive effects of the substituent groups predominantly influence the reactivity. Electron donation to the allyl $C^{(3)}$ atom diminishes the reactivity, whereas electron donation to the vinyl $C^{(2)}$ atom increases the reactivity.

4. DISCUSSION

The observed substituent effects are in contradiction to allyl carbanion [30, 31] as well as allyl carbonium ion [32] and allyl radical [33, 34] mechanisms. Particularly, the strong negative influence of allyl substituents can hardly be understood through an associative mechanism with carbonium ion intermediates, the more so as it has been shown that η-alumina does not develop Brönsted acidity with the much stronger base pyridine even at temperatures up to 300° C [37].

The most striking results regarding olefin isomerization on alumina reported in the literature are as follows:

a) surface hydroxyls do not participate in the reaction [14, 17, 36]; hydrogen transfer must therefore proceed intramolecularly;

b) active sites are aprotonic or Lewis acid sites [14, 22, 38];

c) a large kinetic isotope effect of $k_H/k_D \approx 4$ in the isomerization indicates, that C - H bond cleavage is involved in the rate determining step [1, 20];

d) the terminal vinyl H atoms undergo exchange with D_2 most easily [16, 35].

To explain these and the present results on olefin reactivity, a model mechanism will now be deduced. Among the pssible Lewis acid sites of the alumina surface those formed by two adjacent exposed aluminium ions [14] are supposed to be active sites since they may accommodate the terminal vinyl group of the 1-olefins, most probably by a co-ordination bond. The molecule must then lie "flat" on the surface because of the symmetry of its π-orbitals. In this position the vinyl hydrogens come close to neighboring oxygen ions and may interact with them, the vinyl $C^{(1)}$ - H bond becoming polarized.

Thus. the deuterium exchange of these vinyl H-atoms may easily be explained assuming the simultaneous adsorption of a stongly polarized D_2 molecule on the same acid site below the olefin molecule. This exchange reaction is still a process independent of the isomerization as stated by Hightower and Hall [16]. The polarization of the vinyl $C^{(1)}$-H bond, however, together with electron donation to $C^{(2)}$, facilitates the proton transfer to $C^{(1)}$ because of the resulting increase in electron density in the initial double bond. Analogous effects have been discussed by Cvetanović [(39)] for the attack of electrophilic reagents at the double bond. The transfer of a protonic hydrogen follows from the inhibiting effect of electron donation to the allyl $C^{(3)}$ atom. Neighboring oxygen ions on the surface may assist the intramolecular proton transfer [14, 16]. The rate determining step is suggested to be the proton transfer as indicated by the linear free energy relationships (fig. 1 a and b). A large kinetic isotope effect must then be expected as experimentally observed [1, 20].

Since the allyl hydrogen is transferred as a proton, the carbon skeleton must have some carbanion character during this process, i.e. the allylic grouping should be nearly planar and lie "parallel" to the alumina surface. The cis conformation in allylic carbanions seems to be more stable by far than the trans conformation [40]. Thus, the cis preference can be explained. An analogous interpretation was given by Dent and Kokes [41] for the cis preference in the isomerization of 1-butene over ZnO. The higher stability of the cis conformation of allylic carbanions is explained by charge dipole interactions [42] between the negative charge of the allylic grouping and the positive end of the dipole of an attached alkyl group. If steric interactions are also taken into consideration the stability of the cis conformation might decrease for more bulky substituents R as compared to the trans conformation. The decrease of cis/trans ratios with increasing complexity of the reactant (see Table 1) becomes thus understandable.

Further work, particularly measurements of kinetic isotope effects and deuterium distributions for the isomerization of allyl deuterated olefins will certainly be helpful in testing the proposed mechanistic model.

A scholarship of the Deutsche Akademische Austauschdienst (DAAD) is gratefully acknowledged. The authors also wish to express their gratitude to Dr.K.Kochloefl for valuable discussions.

REFERENCES

1) J.W.Hightower and W.K.Hall, in Kinetics and Catalysis, P.B.Weisz and W.K.Hall, Ed., Chem.Engng. Progr.Symposium Series, Vol. 63 (1967), 122
2) C.L.Thomas, Ind.Eng.Chem. 41 (1949), 2564
3) A.G.Oblad, J.U.Messenger and J.N.Brown, Ind.Eng.Chem. 39 (1947), 1462
4) F.C.Whitmore, Chem.Eng.News 26 (1948), 668
5) W.O.Haag and H.Pines, J.Am.Chem.Soc. 82 (1960), 2488
6) P.J.Lucchesi, D.L.Baeder and J.P.Longwell, J.Am.Chem.Soc. 81 (1959), 3235
7) R.Maurel, M.Guisnet and P.Perot, Bull.Chem.Soc.France 1970, 573
8) J.Turkevich and R.K.Smith, J.Chem.Phys.16 (1948), 67
9) J.Horiuti, Shokubai (Sapporo)1(1948), 67
10) D.M.Brouwer, J.Catalysis 1(1962), 22
11) N.F.Foster and R.J.Cvetanović,J.Am.Chem.Soc.82(1960) 4274
12) S.Ogasawara and R.J.Cvetanović, J.Catalysis 2(1963), 45
13) M.Tanaka and S.Ogasawara, J.Catalysis 16 (1970) 164
14) J.B.Peri, Actes II.Congr.Intern.Catalyse, Paris 196o Technip. 1961, p. 1333
15) H.R.Gerberich and W.K.Hall, J.Catalysis 5(1966), 99
16) J.W.Hightower and W.K.Hall, Trans.Faraday Soc.66(1970)477
17) J.W.Hightower and W.K.Hall, J.phys.Chem. 71(1967), 1014
18) A.Ozaki and K.Kimura, J.Catalysis 3(1964), 395
19) H.R.Gerberich, J.G.Larson and W.K.Hall, J.Catalysis 4 (1965), 523
20) J.W.Hightower and W.K.Hall, J.Am.Chem.Soc.89(1967), 778
21) A.Clark and J.N.Finch, IV.Intern.Congr.on Catalysis, Moscow 1968, Preprint No. 75

22) D.S.MacIver, W.H.Wilmot and J.M.Bridges, J.Catalysis 3 (1964) 502.

23) A.V.Kiselev and J.I.Yashin, Gas-Adsorption Chromatography Plenum Press 1969.

24) K.Kochloefl, P.Schneider, R.Komers and F.Jošt, Collect. Czechoslov.Chem.Commun. 32(1967), 2456.

25) P.Steingaszner, in Ancillary Techniques of Gaschromatography, Ed.Ettre-McFadden, J.Wiley &. Sons, Inc.1969,p.13.

26) J.E.Kilpatrick, E.J.Prosen, K.S.Pitzer and F.D.Rossini, J.Res.Natl.Bur. Std. 36(1946), 559.

27) J.Horak, Chemicky Prumysl 17(1957), 57.

28) R.D.Oldenkamp and G.Houghton, J.phys.Chem. 67(1963), 597.

29) H.Spannheimer and H.Knözinger. Ber.Bunsenges.phys.Chem. 70(1966) 575.

30) A.Schriesheim, C.A.Rowe and L.Naslund. J.Am.Chem.Soc. 85(1963) 2111.

31) Y.Schächter and H.Pines. J.Catalysis 11(1968), 147

32) P.B.D.de la Mare, in Molecular Rearrangements, P.de Mayo, Ed., Interscience, N.Y.1963, Vol.1, p.43.

33) E.W.R.Steacie, Atomic and free radical reactions, Reinhold, New York, 1954, Vol.2, p. 501.

34) C.R.Adams, Proc. III. Intern. Congr. on Catalysis, Amsterdam 1964. North Holland Publ. Comp., 1965, p. 240.

35) J.W.Hightower and W.K.Hall, J.Catalysis 13(1969), 161.

36) Y.Kaneda, Y.Sakurai, S.Kondo, E.Hiroto, T.Onishi, Y.Morino and K.Tamaru, IV.Intern.Congr.Catalysis Moscow, 1968, Symposium on Mechanism and Kinetics of Complex Catalytic Reactions, Preprint No. 12.

37) H.Knözinger and C.P.Kaerlein, J.Catalysis, in press.

38) H.Knözinger and B.Aounallah, unpublished.

39) R.J.Cvetanović, J.chem.Phys. 30(1959), 19.

40) S.Bank, A.Schriesheim and C.A.Rowe, J.Am.Chem.Soc. 87(1965), 3244.

41) A.L.Dent and R.J.Kokes, J.Phys.Chem. 75(1971), 487.

42) S.Bank, J.Am.Chem.Soc., 87(1965), 3245.

43) D.Kalló and G.Schay. Acta Chim.Hung. 39(1963), 183.

44) R.Komers and K.Kochloefl, Coll.Czech.Chem.Commun. 32(1967), 3679

DISCUSSION

H. P. LEFTIN

This is a very interesting paper and the data that you have provided on this series of olefins will, I am sure, be a valuable addition to the catalytic literature on olefin isomerization. Unfortunately your approach of using the linear free energy treatment of substituent effects is ambiguous in all cases except for one and in that one case the approach is completely invalid. I will take these points up separately below:

1) Those compounds in your classes I, II, and IV are all compounds in which the number of allyl hydrogens are constant within the individual classes. In these cases, the effects of the several substituents on the relative reaction rates are exactly parallel to that expected on strictly steric grounds. The data also appear to correlate reasonably well, for each of these groups taken separately, with the Taft substituent constants. However, the range of σ^* in these cases is very small and from such limited data you cannot determine if the observed effects are due to inductive effects of the groups or to steric hindrance.

2) I would like to also point out that Tafts σ^* measures the influence of a substituent in a molecule relative to hydrogen in the same molecule and at the same position in that molecule. Since this is the case, then you have selected the wrong reference compound for your class I olefin series. The proper compound would be propylene, in which the R- group is in fact hydrogen. Alternatively, you could select butene as your reference compound, viz.

$$CH_2 = \overset{\overset{\displaystyle H}{|}}{C}\text{-}CH_2CH_2R'$$

In this case, however, the first compound is 1-butene in which the substituent to be considered, R', is hydrogen. In the next compound R' is methyl, not ethyl. Then R' is ethyl rather than propyl. The fourth compound in the series has two substituents each of which is a methyl group and the last compound contains three methyl substituents. This would be the correct and proper treatment of the data in class I, and when treated in this way, the correlation shown in Fig. 1 does not exist.

H. KNOZINGER

As pointed out by Dr. Leftin, the relative rates of the reactants of the classes I, II and IV parallel qualitatively the sequence expected on steric grounds. Quantitatively, however, the correlation coefficient is only 0.9 for the correlation of the logarithms of the relative rates with the steric substituent constants E_s, whereas a correlation coefficient of 0.99 is obtained for the correlation with the inductive constants.

With respect to the second point, Dr. Leftin is certainly in error, since Taft's σ^*-constant measures the inductive effect of a substituent relative to the methyl group (see e.g., E. S. Gould "Mechanism and Structure in Organic Chemistry," 1959, p. 229; or P. R. Wells "Linear Free Energy Relationships," 1968, p. 37). Consequently, 1-butene has been chosen as a reference compound with the methyl group (and not hydrogen) as the reference substituent. The correlation shown in Fig. 1 is thus correct.

J. MANASSEN

Some twelve years ago we published for the first time the cis-preference on isomerization of butene-1 in 0.5 N perchloric acid.

You propose a carbanion type intermediate, which is difficult to envisage in a 0.5 N perchloric acid solution. Therefore, we think there is some danger in connecting the existence of a carbanion intermediate with the preferential cis-formation, which is not only occurring on catalytic surfaces but also in homogeneous acidic solutions.

H. KNOZINGER

I agree that a carbanion mechanism cannot occur in a 0.5 N perchloric acid. However, since our experimental evidence seems to favor a carbanion intermediate, we believe that the most plausible explanation of the preferential cis-formation is through the relative thermodynamic stabilities of cis- and trans-conformations of allylic carbanions. An analogous interpretation has been given by Kokes (Ref. 41) for the cis-preference in the isomerization on ZnO and by Kemball et al. (J. R. Shannon, C. Kemball, H. F. Leach, in P. Hepple, Ed. "Chemisorption and Catalysis," p. 46).

The latter authors used MgO as catalyst and observed cis/trans-ratios of up to 23. We do certainly not suggest this interpretation to be a general one for all systems and experimental conditions. Most probably steric interactions between reactant and catalyst surface or solvent molecules may be the predominating effect under certain conditions.

W. K. HALL

In our experience, isomerization and the exchange of surface OD groups with the hydrogen atoms of hydrocarbons are independent reactions. In fact, the facile nature of the latter reaction has prevented using these techniques in mechanistic considerations. Your pre-treatment conditions were somewhat different. Will you please summarize the observations which lead you to believe that you do not encounter these problems in your work?

H. KNOZINGER

After the submission of our contribution we have indeed obtained experimental evidence, that supports the view that on alumina surfaces with sufficiently high hydroxyl content the D_2-exchange proceeds through the isomerization steps via participation of the surface hydroxyl groups. We studied the reaction of a 3-D-2,3-dimethyl-1-butene on an η-alumina pretreated at 300°C. This olefin consisted mainly of a singly deuterated species besides some undeuterated and some multiply deuterated species. The average number of D-atoms per molecule was 1.45 as determined by mass spectroscopy and NMR. 60% of the reactant contained a D-atom in the $C^{(3)}$-allyl position. On isomerization the product 2-olefin lost 0.5 D-atoms per molecule on the average. This value compares quite well with the 0.6 D-atoms per molecule in allyl position. We may thus conclude, that the isomerization proceeds through an intermolecular proton transfer with participation of surface hydroxyl groups. This conclusion is confirmed by the fact, that one observes deuterium uptake in the product olefin on isomerization on a deuterated catalyst surface. D_2-exchange experiments also confirm the intermolecular mechanism, since exchange only occurs at temperatures above 200°C, when D_2-exchanges with surface hydroxyl groups to a certain degree. Furthermore tertiary butyl ethylene, which cannot undergo double bond shift, does not exchange with D_2 even at 250°C. For the interpretation of these results, we suggest an "active site" containing a Lewis acid base pair and a hydroxyl group in suitable configuration, on which a cyclic carbanion-like intermediate may be formed.

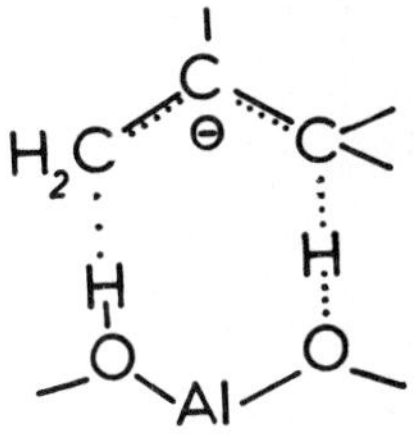

The discrepancy of this mechanistic concept with the observations mentioned by Dr. Hall can easily be explained by the different pretreatment of the respective alumina catalysts. The presently proposed mechanism requires a certain hydrogen content of the surface, in particular hydrogen in the suitable configuration in the "active sites." Since these contain hydroxyl groups and Lewis acid base pair sites, they will probably be formed during the initial stages of surface dehydroxylation at comparatively low temperatures. They will, however, exist in only very low number or not at all on surfaces of low hydroxyl content, which are obtained after pretreatment above 500°C. On such surfaces the presently proposed intermolecular mechanism will certainly contribute to only a small degree or not work at all. Instead an intramolecular isomerization mechanism may predominate under such conditions (see also response to comment No. 4 by Dr. Haag).

W. O. HAAG

It was interesting to hear that Dr. Knozinger's most recent results indicate that surface hydroxyls on alumina are involved in the double bond shift (d.b.s.) in 1-butene. We have found for η-alumina that dehydroxylation reduces the rate of d.b.s. by several orders of magnitude in agreement with the present conclusions. We also noted that cis - trans isomerization in butenes is not affected much by dehydroxylation and proceeds rapidly. As a consequence, in the isomerization of 1-butene on the dehydroxylated catalyst, the 2-butenes are formed in thermodynamic equilibrium. I would like to ask Dr. Knozinger if he has information regarding mechanistic aspects of cis-trans isomerization on alumina and in particular, the possible role of hydroxyl groups in this reaction.

H. KNOZINGER

We have also studied in some detali the influence of the degree of surface hydroxylation of η-alumina on the d.b.s. An activity pattern with two maxima at approximately 470°C and 600°C for the dehydroxylation temperature was observed. We therefore believe that the d.b.s. proceeds through the presently proposed carbanion mechanism with participation of surface hydorxyl groups on surfaces which have been treated at temperatures below approximately 500°C. A second mechanism--most probably that proposed by Hightower and Hall (Reference 1)--without participation of surface hydroxyls works on surfaces with lower hydroxyl content, which have been pretreated at temperatures above 500°C.

With respect to Dr. Haag's question regarding the mechanism of the cis-trans isomerization, I must say that we have still not carried out detailed experimental work.

Paper Number 57

SURFACE BASICITY AND ALCOHOL DEHYDRATION ACTIVITY. A COMPARATIVE STUDY OF η, θ and α ALUMINAS

Z. G. SZABÓ and B. JÓVÉR
Institute of Inorganic and Analytical Chemistry,
L. Eötvös University, Budapest, Hungary

Abstract: The aim of this paper was to prepare pure and from structural point of view uniform bayerite η, θ and α aluminas and to compare several of their properties supplementing the earlier works with some new aspects. These were: studying the morphology by electron scanning microscopy, determining surface basicity and measuring the catalytic activity of η and θ Al_2O_3 at the i-PrOH dehydration in the range of 240-310°C. The reflexion spectra of adsorbed indicators were also recorded.

Results show great similarity between η and θ Al_2O_3 the latter being a little more basic, and less active in i-PrOH dehydration, indicating that a two-center mechanism may be correct but the number of the acidic sites are decisive. α Al_2O_3 looks quite inert in all these investigations.

These properties are in accordance with morphology shown by the electron scanning pictures very rich in details.

1. INTRODUCTION

From theoretical point of view it is interesting to compare the polymorphic modifications of the same substance, thus to study the effect of structure on surface properties separately from other factors. The several modifications of alumina,including the η and θ Al_2O_3, too, has been studied by many authors. This great interest is understandable, if we consider that not only the alumina itself, but also as a supporter it plays a very important role in catalysis. Thus, the comparative study is important also from technological point of view. In most of the practical cases, as a consequence of the preparation procedure, Al_2O_3 catalysts are not unique phases, and it has considerable significance to know the influence of the contaminant phases on catalytic processes.

Now we draw our attention to the following questions: 1/ We studied the morphology of our samples by electron scanning microscopy obtaining more information, as by the usual electron microscopy, 2/ we determined the surface basicity of the samples. It has already been measured at η Al_2O_3,[1/] but comparative studies with θ Al_2O_3 have only been made concerning the surface acidity,[2/] 3/ we compared the catalytic activity of η and θ aluminas in the iso-propanol dehydration reaction. It also has been done by many authors[3-6/], but the activity tests were made only at a fixed temperature and in some cases samples contained other alumina modifications, too.

2. EXPERIMENTAL

2.1. Preparation of the samples

η, θ and α aluminas were prepared by heating bayerite, produced by amalgamation technique[7,8/] for 6 hours at 650°C, 1000°C and 1350°C, respectively.

2.2. Specific surface area determination

The usual BET method was applied using an apparatus proposed by Bliznakov et al.[8/] The surface 33 m^2/g at bayerite, 166 m^2/g at η, 61 m^2/g at θ and 8 m^2/g at α alumina was determined.

2.3. Water-content determination

Water-content of η and θ Al_2O_3 was determined by gravimetric method igniting the air dry sample at 1300°C for 5 hours and measuring the loss of weight: 3.80% at η and 1.50% at θ Al_2O_3 was found.

2.4. Determination of the sodium-contamination

Less than 5 ppm Na was found by emission spectroscopy.

2.5. Thermoanalysis

The process of calcination of bayerite was examined with a MOM Derivatograph[9/]. The rate of heating was 12°C/min, α Al_2O_3 being used as reference-material. The sample was placed in thin layer on platinum plates. During the measurement, dry N_2 flowed through the system.

2.6. X-ray analysis

X-ray diffraction diagrams were made by means of a Philips-Müller-Micro-111 Diffractometer and proportional counter, CuK_α radiation being used. The accelerating voltage used was 36 kV.

2.7. I.R. spectroscopy

I.R. spectra were made with a UR-10 Zeiss I.R. spectroscope in the range of 400-4000 cm^{-1}.

2.8. Electron-scanning microscopy

JSM-03 JEOL apparatus was used. The accelerating voltage applied was 25 kV and the sample-current was $0.4 - 1x10^{-11}$ A. The samples were coated with layers of carbon and gold subsequently, in vacuum at $5x10^{-5}$ torr. The thickness of the layers was a few hundreds Å. The carbon layer helped the adhering of gold, which increases the secondary emission and prevents the charge of the samples. As a maximum dissolution 200 Å was achieved.

2.9. Titration of surface basicity with acetic acid

The measurements were carried out in dry benzene. The procedure applied was analogous to that of Benesi[10/] for acidity determination. To prevent the adsorption of moisture and CO_2, experiments were carried out in dry box filled with purified N_2. Before titration the samples were activated for 3 hours at 500°C /bayerite at 150°C/. The used H_- indicators are listed in Table 1.

Table 1

Indicator	P_k	Colour change
Bromo-cresol-green	4.6	yellow→blue
Bromo-cresol-purple	6.0	yellow→purple
Bromo-thymol-blue	7.1	yellow→blue
Phenolphthalein	9.3	colourless→red
Nitramine	11.9	yellow→reddish-brown

2.10. Reflexion spectroscopy

Opton-type spectrophotometer with a RA-II reflection appurtenance and α Al_2O_3, as reference substance was used. The spectrum of phenolphthalein was recorded in a vacuum--sample-holder, too, because its colour faded in air. The tests were carried out as follows: indicator dyes were adsorbed from benzene solution to the catalyst activated in the same manner as previously for titration. After shaking for 5 hours, they were washed three-times with benzene and the solvent was expelled in vacuo at room temperature.

2.11. Surface basicity determination by CO_2 adsorption

The same BET apparatus was used as for surface area determination. The adsorption isotherms were evaluated using the Langmuir equation.

2.12. Catalytic activity measurements

The iso-propanol dehydration activity measurements were carried out in a modified Schwab reactor[11/] at an alcohol pressure of 1 atm, between 240 and 310°C. The conversion was always less than 10%.

3. RESULTS

3.1. Purity and phase identity of the samples

The negligible Na-content of the samples /less than 5 ppm/, is not sufficient to form notable quantities of β Al_2O_3. The surface acidity poisoning effect[12/] is also negligible. This Na-quantity on the surface covers at most only 10^{15} sites/m^2, while the dimension of the acidity is 10^{17} sites/m^2.

X-ray diffraction and I.R. spectroscopy data show that the substances are well crystallized and structurally uniform. Only the θ Al_2O_3 contains traces of α Al_2O_3, but because of the inertness of the latter, it does not exert any influence on the results.

Under the experimental conditions of thermoanalysis, the water could easily escape. Probably that is why the decomposition of bayerite took place at a little lower temperature than usual /305-310°C/. DTA shows neither the presence of boehmite in the original sample, nor its formation during the heating process. The formation of boehmite would be indicated by a small endothermic peak or shoulder between 250 and 300°C, and its decomposition by a small endothermic peak between 500 and 550°C.[13]

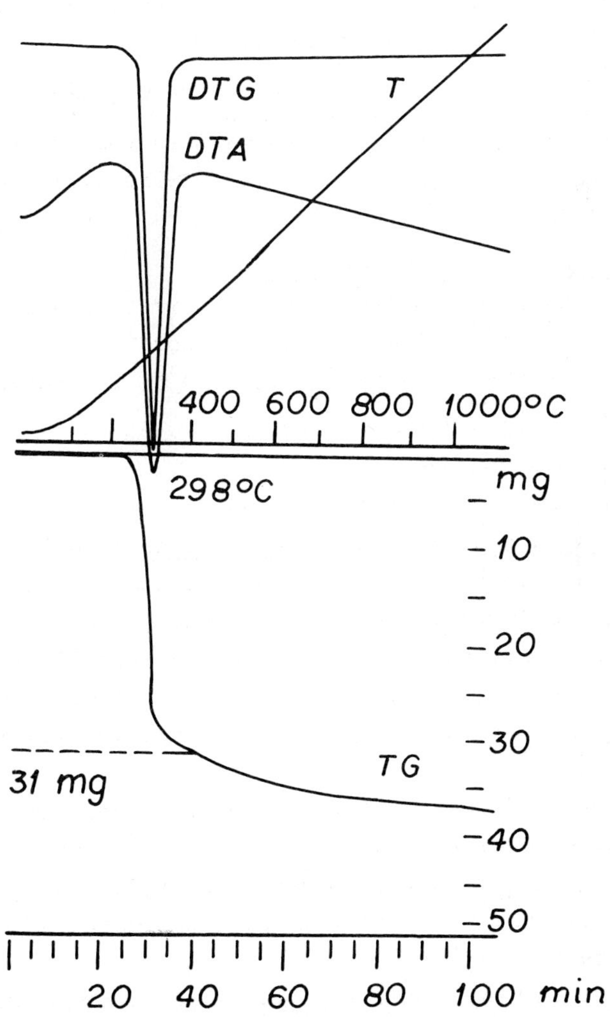

Fig.1. Thermoanalytical curves of bayerite.

The presence of boehmite would give way for formation of γ and δ Al_2O_3, contaminating the η and θ phases wanted. Such examples are frequently found in the literature[4] at catalytic activity measurements. It seems that the probability of boehmite formation is the greatest when bayerite is prepared through aluminium hydroxide gel. In our case, as a consequence of preparation procedure, bayerite was well crystallized and its particle size was very small, thus excluding the development of hydrothermal conditions inside the particles and thus the possible amorphous aluminium hydroxide → boehmite transformation.[13]

3.2. Electron-scanning microscopy

The texture and morphology of bayerite and different aluminas has been studied thoroughly by means of electron-microscopy. /See e.g.[14-16]/. By electron-scanning microscopy we got similar pictures but much richer in details. Fig.2 shows well the platelet structure of the "triangular" and "hour--glass" shaped bayerite somatoids. It is interesting to observe small particles at the surface of the large faces. As shown in Figs.3,4, η, and θ Al_2O_3 has a very similar morphology to bayerite, but here in some places new, cylindrical, stalactite-like forms also appear. The external similarity is curious if we take into consideration that the bayerite → η Al_2O_3 and the η Al_2O_3 → θ Al_2O_3 transformations are accompanied with radical inner rearrangements. In the first case half of the lattice oxygens is expelled and the hexagonal packing of O^{2-} ions is replaced by cubic symmetry. In the second case the formation of θ Al_2O_3 is accompanied with a complete recrystallization of the substance.[14]

α Al_2O_3 /Fig.5/, however, presents quite a different picture. Its perfectly rounded, dumb-bell shaped particles remind of the picture of a plastisized and suddenly congealed substance. This is understandable, considering the high ignition temperature.

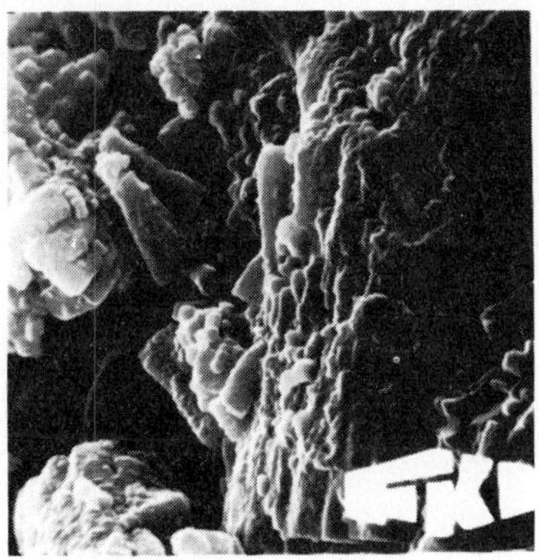

Fig.2. Bayerite x 10000

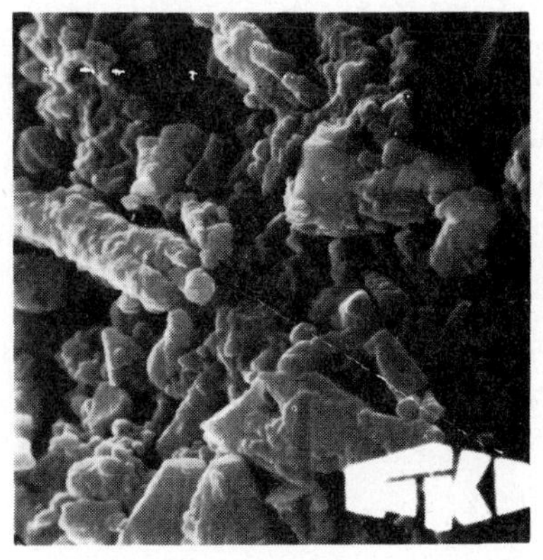

Fig.3. η Al_2O_3 x 1000

Fig.4. θ Al_2O_3 x 10000

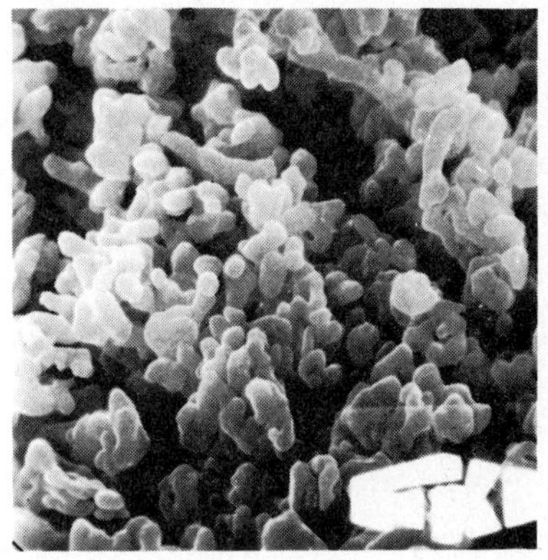

Fig.5. α Al_2O_3 x 1000

3.3. Determination of surface basicity

It can be seen from Table 2 that at room temperature CO_2 indicates centres with basicity of about $P_k \geq 6$, and not all of them, as it has hitherto been assumed.[17/] If we want to measure the weakest sites too, it is necessary to work at lower temperatures. This is in accordance with Boehm's[1/] results, who found $25.8x10^{17}$ sites/m^2 basicity on η Al_2O_3 by CO_2 adsorption at 0^oC and -14^oC. Comparing this result with the present data, we can state that sites with basicity of about $P_k \geq 4.6$ were determined by him.

Table 2
Surface basicity determination

Basicity Sites $m^2.10^{-17}$	CH_3COOH titration					CO_2 adsorption
	$P_k \geq 4.6$	$P_k \geq 6.0$	$P_k \geq 7.0$	$P_k \geq 9.3$	$P_k \geq 11.9$	
Bayerite	21.15	10.45	6.82	1.37	0.00	8.40
η Al_2O_3	20.95	8.82	5.44	0.68	0.00	10.03
θ Al_2O_3	22.20	9.32	6.20	0.95	0.00	10.50
α Al_2O_3	0.00	0.00	0.00	0.00	0.00	0.00

Studying the agreement between the titration and the gas adsorption method, we see that on η and θ Al_2O_3 the adsorption, while on bayerite, the titration method indicates a little more centers. This small fluctuation is probably due to the different nature of the surfaces. It may be assumed that on bayerite CO_2 is predominantly present as HCO_3^-, while on the oxides also as CO_3^{2-}. The different stabilities of these surface complexes may well cause such differences. Anyway, at the similar η and θ Al_2O_3 there is a good agreement.

Data show that bayerite, η and θ Al_2O_3 contain almost the same quantity of basic centers, the number of which is approximately 1/5 of all the surface anions, with slightly different strength distribution. All of these samples must be considered as moderately basic ones, since nitramine does not indicate any basicity on them, though it does on KOH, CaO or MgO.

The similar basicity of these substances is in accordance with the similarity of the external forms, shown by electron scanning microscopy, indicating that the type of crystal imperfectness and of its roughness plays a more important role than the symmetry of the elementary cell.

It is interesting to note that there exists a greater difference in acidity, η Al_2O_3 being 7-8-times more acidic than θ Al_2O_3.[2]

The inertness of α Al_2O_3 may be attributed to the lack of surface OH^- groups /its water content is practically zero/ and to the vigorous self-diffusion at the high temperature of its formation. The diffusion promotes the healing of the basic O^{2-} defects. This is in accordance with the levelling of all the unevennesses on its surface.

Table 3
Absorption and reflection spectra of the indicators

Indicator	Method	Characteristic maxima $1/\lambda.10^{-3}$ $/cm^{-1}/$		Ref.	Remark
Bromo-cresol-green	Absorption spectrum	16.13	25.00	/18/	pH = 8.2
	Reflection spectrum	η 16.00	25.10		
		θ 15.90	24.50		
Bromo-cresol-purple	Absorption spectrum	17.09		/19/	pH = 7.00
	Reflection spectrum	η 16.95			
		θ 16.85			
Bromo-thymol-blue	Absorption spectrum	16.12		/19/	pH = 7.8
	Reflection spectrum	η 16.10			
		θ 16.10			
Phenol-phtha-lein	Absorption spectrum	18.00		in 0.1N NaOH	
	Reflection spectrum	MgO 18.00			
		θ 18.2 and 19.4			

3.4. Reflection spectroscopy of the used H_- indicators

To control the mechanism of the colour change at the used H_- indicators the remission spectra of the adsorbed indicator dyes were taken and compared with those of absorption. This method has already been applied for the investigation of some H_o indicators[20-22].

Table 3 shows that bromo-cresol-green, bromo-cresol--purple and bromo-thymol-blue works in the same way on the surface of η and θ Al_2O_3 as in aqueous medium. The observed small differences in $1/\lambda$ amount only to some tenths of kcal/mole. Thus it can be assumed that P_k values of them, determined in aqueous solution will be much the same at the oxide surfaces, too.

With the phenolphthalein, however, we found a split and greater shift of the characteristic maximum. It has been put[2] that it adsorbs not only at the basic, but at the Lewis acid sites, too. If it is true, indeed, the nature of the spectrum is understandable but in order to eliminate the contradiction with titration data, we must assume that acetic acid adsorbs with similar affinity on the strongly basic and strongly Lewis acid centers. It means at the same time that the values in Table 2, belonging to $P_k \geq 9.3$ are overestimated.

3.5. The measurement of i-PrOh dehydration activity

At the applied alcohol pressures, the reaction is of zero-order[23] and the rate of the reaction can be written by the Arrhenius equation. The preexponential factor and activation energy were determined by the method of the least squares. /Table 4/. Also the reliability interval of the constants was estimated. /The level of significance was 0.1/.

The data of Table 4 suggest that the surface of η Al_2O_3 contains more and slightly stronger active sites than that of θ Al_2O_3. Comparing this fact with those discussed previously in connection with basicity, we may state that the proposed two-center mechanism[23,24], emphasizing the necessity of both the acidic and the basic centers for alcohol dehydration reactions may be correct, but in this case the number of the acidic sites is more decisive.

Table 4
Preexponential factors and activation energies in i-PrOH dehydratation reaction

	η Al_2O_3	θ Al_2O_3
log A	11.60 ± 0.82	9.42 ± 0.84
A $\left[\frac{ml\ C_3H_6}{m^2.min}\right]$	$6.0x10^{10} \leq A \leq 2.6x10^{12}$	$4.0x10^{8} \leq A \leq 1.8x10^{10}$
E /$\frac{Kcal}{mole}$ /	27.81 ± 1.96	24.15 ± 2.00
Residual scattering	4.55×10^{-2}	6.43×10^{-2}

4. CONCLUSION

As we have made all efforts to handle exactly defined substances the results found points to the fact that the catalytic activity is much less bound to the structure of the bulk but more to the surface properties. Pseudomorphism can play a great role in this respect. It means that the catalytic activity can remain between some limits even if the bulk structure is transformed. The variation of the surface seems to be less in extent than it might be assumed.

ACKNOWLEDGEMENT

The authors express their sincere thanks to the Research Institute of Non-Ferrous Metals for the electron-scanning pictures and for the X-ray data.

REFERENCES

1/ H.P.Boehm, Faraday Discussion, Brunel University, Sept. 1971.
2/ H.Pines, W.O.Haag, J.Am.Chem.Soc., 82 /1960/ 2471.
3/ A.M.Rubinstein, Izv.Akad.Nauk, SSSR, 30 /1960/ 31.
4/ Tadao Shiba, Kogyo Kagaku Zasshi, 69 /1966/ 2249.
5/ A.Sedzimir, Roczniki Chemii, 41 /1967/ 655.
6/ D.Treibmann, A.Simon, Z.anorg.allg.Chem., 350 /1967/ 281.
7/ H.Schmäh, Z.Naturforsch. 1 /1946/ 323.
8/ G.M.Bliznakov, I.V.Bakardjiev, E.M.Gocheva, J.Catalysis, 18 /1970/ 260.
9/ F.Paulik, J.Paulik, L.Erdey, Z.anal.Chem., 160 /1958/ 241.
10/ H.A.Benesi, J.Phys.Chem., 61 /1957/ 970.
11/ Z.G.Szabó, F.Solymosi, Actes du deuxième Congrées International de Catalyse, Paris, 1960, p.1631, /1961/.
12/ H.Bremer, Z.anorg.allg.Chem., 366 /1969/ 130.
13/ K.C.Mackenzie, "Differential Thermal Analysis, Academic Press, London, 1970.
14/ B.C.Lippens, J.H.de Boer, Acta Cryst., 17 /1964/ 1312.
15/ L.Moscou, G.S.Van der Vlies, Kolloid Zeitschr., 163 /1959/ 35.
16/ H.L.Watson, J.Parsons, A.Vallejo-Freire, P.Souza Santos, Kolloid Zeitschr., 140 /1955/ 102.
17/ S.I.Malinowski, S.Siczepanska, J.of Catalysis, 7 /1967/ 67.
18/ E.E.Sager, A.A.Maryott, M.R.Schovley, J.Am.Chem.Soc., 70 /1948/ 735.
19/ M.G.Mellon, G.W.Ferner, J.Phys.Chem., 35 /1931/ 1031.
20/ G.-M.Schwab, H.Kral, Proc.3rd Intern.Congr.on Catalysis, Amsterdam, 1964, p.433.
21/ W.C.Kocarenko, L.G.Koracsiev, V.A.Oliszko, Khim.i.Kataliz. IX.I /1968/ 158.
22/ Hitoshi Hattori, Shokubai, 7 /1965/ 486.
23/ J.R.Jain, C.N.Pillai, J.Catalysis, 9 /1967/ 322.
24/ M.Pines, J.Manassen, Adv.in Cat., 16 /1966/ 49.

DISCUSSION

D. BARTHOMEUF

As to the basicity results, you noted

1) a correlation between the basicity and the external form of the samples, and

2) the importance of the external form rather than the type of the alumina (η, θ...).

Could you comment on the fact that the large differences in the acidity results given in your paper for η- and θ- aluminas seem to indicate that there is no correlation between the acidity and the external form?

Z. G. SZABÓ

We think that basic and acidic sites are of different nature. The many, weak basic centers are elements of the "ideal surface," hence determined by the external form. The less, stronger acidic sites are imperfections of the "ideal surface" and their number is not depending on the external pseudo-symmetry.

It is likely, that the situation at the strong basic centers is different. We have mentioned in the introduction of this lecture that one can prepare alumina with stronger basic centers, as well. Here there may exist a greater difference between η-and θ-Al_2O_3 .

H. KNÖZINGER

I would like to ask Professor Szabó a question with respect to the correlation of surface basicity and catalytic activity. Surface basicities reported in his paper have been measured on surfaces which were dehydroxylated at 500°C. The catalytic activity for the dehydration of isopropyl alcohol, on the other hand, was determined between 240 and 310°C. Under these conditions the surface not only contains a higher hydroxyl content than after heating at 500°C in vacuo, but surface alkoxide groups are also formed by dissociative adsorption of the alcohol. Acidic and basic properties of the surface which is "seen" by a reacting alcohol molecule, are therefore certainly different from those of a vacuum surface after dehydroxylation at 500°C. I am interested whether the authors have any information concerning the basicity of their aluminas under catalytic conditions. An approximate determination might probably be possible on surfaces, which had been treated under catalytic conditions and then evacuated at the reaction temperature.

Z. G. SZABÓ

The determination of surface acidity and basicity has been carried out in two ways: i) by titration according to Benesi and ii) by gas chromatographical impulse method. The titration procedure gives good orientation, especially for comparison, but does not at all simulate donditions during the catalytic reaction. Just therefore, we have developed the gas chromatographic procedure, which allows us to work nearer the catalytic conditions. We are aware of this being an approximation only, but a much better one, than any others.

In addition, we note that the temperature of activation very likely does not influence the ratio of basicities at η and θ samples, since thermogravimetric studies[1] show that the temperature dependence of residual water content is approximately the same at these modifications.

1) H. Yanagida, G. Yamaguchi, J. Kubota, Bull. Chem. Soc., 38, 2194 (1965).

S. K. BHATTACHARYYA

Alumina can be prepared by different methods--1) by precipitation from aluminum nitrate by ammonia and 2) by decomposition of sodium aluminate by CO_2. The activity and the basicity of aluminas depend on the method of preparation and their thermal treatment. I should like to ask Professor Szabó if he has done any experiments on these effects. We have done considerable work on the dehydration and dehydration-dehydrogenation of ethyl alcohol and have found that the maximum activity of alumina corresponds to the temperature where most of the adsorbed and bound water molecules are removed and the γ-modification is complete (therefore, peak temperatures by DTA is 380°C). This temperature also corresponds to the maximum surface area (235 m^2/g at 380°C). Further increases in temperature leads to poor activity and decreases in surface area. Another important factor for dehydration activity of aluminas is that Al_2O_3 should contain a small quantity of water (for ether formation from ethanol it is 6%, below which the activity decreases). I would like to ask Professor Szabó whether he noted any observation on the limit of basicity for dehydration of alcohols.

S. G. SZABÓ

We have not made use of classical procedures mentioned by Professor Battacharyya, because the aluminum hydroxide formed in these ways cannot be dehydrated unitarily. We did not intend to study the effect of the way of preparation on surface properties. These effects are mostly connected with the presence of ionic contaminants and crystallographical imperfections. We, on the contrary, wanted to produce contaminant-free and well-crystallized samples to have the effect of structural differences between η, θ, and α aluminas appear distinctly. Our starting substance produced by an amalgamation procedure yields a product of very small and homodisperse particle size. If this is ignited, in turn, it gives pure phases, because due to the fine distribution, no hydrothermal effect can take place.

Ether formation was always negligible.

H. BREMER

Based on our own experience, we agree with Professor Szabó that the differences in the catalytic activity of Al_2O_3-catalysts in the isopropanol-dehydration are not primarily dependent on their modification but more on their surface chemical and textural properties. We have demonstrated this--as he does--with η- and θ-Al_2O_3 and in addition with γ-, δ- and χ-Al_2O_3. We found[1] the following sequence in specific catalytic activities of i-propanol dehydration:

$$\eta \approx \gamma > \theta \approx >> \chi$$

In our opinion[2] the centers of the Al_2O_3-surface responsible for the 2-center-mechanism are acidic Lewis-sites (incompletely coordinated Al-ions) and basic (or only very weakly acidic) surface hydroxyl groups. As the number of acidic sites is lower by 1 to 2 orders of magnitude than that of the hydroxyl groups, relations frequently are found between the number of acidic sites and the catalytic activities.

What is your opinion about the detailed mechanism of i-propanol dehydration?

1) H. Bremer, J. Glietsch, Z-anorg. allg. Chem. (in press), 1972.
2) H. Bremer, K. H. Steinberg, J. Glietsch, H. Lusky, U. Werner, K. H. Wendlandt, A. Chemie 10, 161 (1970).

Z. G. SZABÓ

Our attention was focussed on the catalyst itself and less to the mechanism of the chemical reaction, e.g. in the sense of Dr. Knozinger's paper. Your remarks are, however, welcome and useful to us.

I. V. KRYLOVA

In connection with the paper of Szabó and Jover I would like to report some new data which have been obtained in a study of dehydration on both single crystals and powdered solids. We discovered the emission of negative charges during the catalytic decomposition. The decomposition of metanol and of formic acid adsorbed on the powdered alumina and on a single crystal of sodium chloride catalysts has been investigated. After dehydration at 400°C formic acid or methanol vapor was allowed to be adsorbed on the catalyst at room temperature. Then the adsorbed vapors were pumped off and the catalyst was heated at a rate of 10°/min. The emission of negative charges was recorded with a secondary electron multiplier. In the course of the catalytic decomposition, the emission glow peaks appeared at temperatures near 140, 220, and 330°C. The deflection of emission current in a magnetic field shows that heating gives rise to electrons and negative ions. Analysis of the evolved gas showed the presence of dimethylether and water, that is, dehydration took place. We have proposed that methanol and formic acid were adsorbed with radical formation. The recombination of radicals leads to desorption of reaction product (water) and emission of electrons. We have established earlier that TSE centers are at the same time the centers of activated desorption at water. These data may be useful in a further discussion of the very interesting results of Szabó and Jover.

Z. G. SZABÓ

This is an interesting contribution but it is more pertinent to the next two papers.

A. G. OBLAD

I should like to suggest that the techniques of selective poisoning be used to determine the nature of the active sites for isopropanol dehydration. Since a mechanism is proposed involving both acidic and basic sites, it would be necessary to utilize both acidic and basic poisons. Have you considered this, and have you done work along this line?

Z. G. SZABÓ

Thanks for your remark. It is quite obvious for us that the well planned and selected poisoning could afford further information about the nature of the active sites. Some of them are already finished or are just in progress.

F. S. STONE

In agreement with Dr. Knozinger, I also feel that the effect on activity of the OH content of the alumina surface should be studied per se. A convenient way is to examine the adaptation of activity of a well-outgassed surface as successive pulses of reactant are passed; the whole transition region to steady-state flow can, if necessary, be investigated.

Although the scanning electron microscope is valuable for studying texture, as shown in the paper, it is not very useful in understanding the origin or nature of the basic (and acid) centers. I believe we need more attempts to describe the geometry of the centers at the atomic level. If one in five of the surface anions of η- and θ-Al_2O_3 is a basic center, then some acidic and basic sites are very close and are probably generated together in a single step. Incidentally, this proximity will be an obstacle to the study of the activity of these surfaces by selective poisoning using large molecules.

Z. G. SZABÓ

In order to have as much information and correlations as possible, we have also taken scanning electron microscopic photographs. But perhaps I have not emphasized enough that this was done to see something about the change in morphology and not to evaluate them as microscopic, i.e. atomic source of information. In connection with this, I should like to make a remark referring to paper 2 by Professor Stone and Pepe. The variability of properties of the different alumina samples which are not caused by contaminants can perhaps be elucidated by the concept of Professor Stone: the different phases of alumina can be considered as analogous substances which, in turn, form solid solutions. On the surface these solid solutions can display disorder in atomic dimensions between wide limits and thus the change of activity is quite understandable. I must admit that I like very much this idea!

H. PINES

The study of dehydration of isopropyl alcohol as a model compound for differentiation of active sites on aluminas has many limitations. This alcohol does not permit characterization of the sites which are responsible for the removal of hydrogen from carbon atoms residing in γ-position relative to the carbon containing the hydroxyl group. The use of stereoisomers of alkylcyclohexanols or even 3-methyl-2-pentanol could furnish valuable data regarding additional active catalytic sites present on aluminas.

Z. G. SZABÓ

Your remark is a very useful one. At first we would not have the model to be very complicated. Therefore, the more simple isopropanol-decomposition. To apply more complicated alcohols needs to apply more devices and measurements. Now we are on the way to do this. Thanks for your remark.

K. TANABE

I would like to ask about the validity of your basicity measurement. We use benzoic acid as a titrating acid which gives reproducible basicity values. However, when acetic acid is used instead of benzoic acid, no reliable data are obtained for CaO and MgO. It is likely that acetic acid is so strong an acid that it dissolves the oxides. To what extent are the basicity values of Al_2O_3 in Table 2 reproducible?

Z. G. SZABÓ

We do not think that the acetic acid is the stronger of the two acids. Its K_d is 1.8×10^{-5}, while this value is 6×10^{-5} at benzoic acid.

Our previous study has shown that MgO does not dissolve even in the much stronger CCl_3COOH in benzene.[1)] Al_2O_3 is much more resistant--you can hardly dissolve it even in HNO_3 or H_2SO_4. We have proved that this effect can be entirely neglected. We placed the samples into 1N CH_3COOH benzene overnight and after filtration extracted the benzoic phase with water. We could not detect any Al^{3+} by Alizarin-S in the water phase.

Our results were fairly well reproducible. The uncertainty of the data of Table 2 is about $\pm$ 0.06 x 10^{17} sites/m^2. (At 100 m^2 sample, 2 x 10^{-2}N CH_3COOH, 5 x 10^{-2} ml portions.)

1) Z. G. Szabó, Berichte der Bunsengesellschaft, 75, 1127 (1971).

J. M. PARERA

Regarding the relationship between basicity and catalytic activity, I should like to add some comments. The determination of the acidity or basicity by the Benesi method measures the capacity of the bare solid surface to react with a base or an acid at room temperature. These conditions are quite far from the ones of reaction. During the reaction the catalyst surface is covered with reactants, and they remain adsorbed even evacuating for many hours; we found 1) that after the alcohol dehydration reaction the alumina surface remains covered by the reactants, and it is unable to give acid color to acid indicators and are then non-acid according to the method. Many authors found a reasonable proportionality between catalytic activity and acidity or basicity on samples of alumina from the same origin. For different aluminas the proportionality factor is different, being influenced strongly by the sodium content.[2] We found[3] that the amount of η-butylamine necessary to poison methanol dehydration on silica-alumina at reaction conditions is quite smaller than the acidity value, and that adding the same amount of base at different temperatures causes a greater decrease of acidity at higher temperatures. Then the acidity at room temperature measures active and non-active sites, the number of the last being proportional to the active ones. On alumina also the amount of base is smaller than the acidity value, but the poisoning is reversible.[4]

1) Figoli, N. S., and Parera, J. M., J. Res. Inst. Catal. Hokk. Univ. 18, 142 (1970).
2) Parera, J. M. and Oberto, S. C., unpublished results.
3) Parera, J. M. Hillar, S. A., Vincenzini, J. C. and Figoli, N. S., J. Catal. 21, 70 (1971).
4) Figoli, N. S., Hillar, S. A. and Parera, J. M., J. Catal. 20, 230 (1971).

G. SZABÓ

My reply is the same which was given to Dr. Knozinger.

ERRATUM

The figure below was erroneously not printed in its proper place. It should appear on page 24 -418 above the legend to fig. 6.

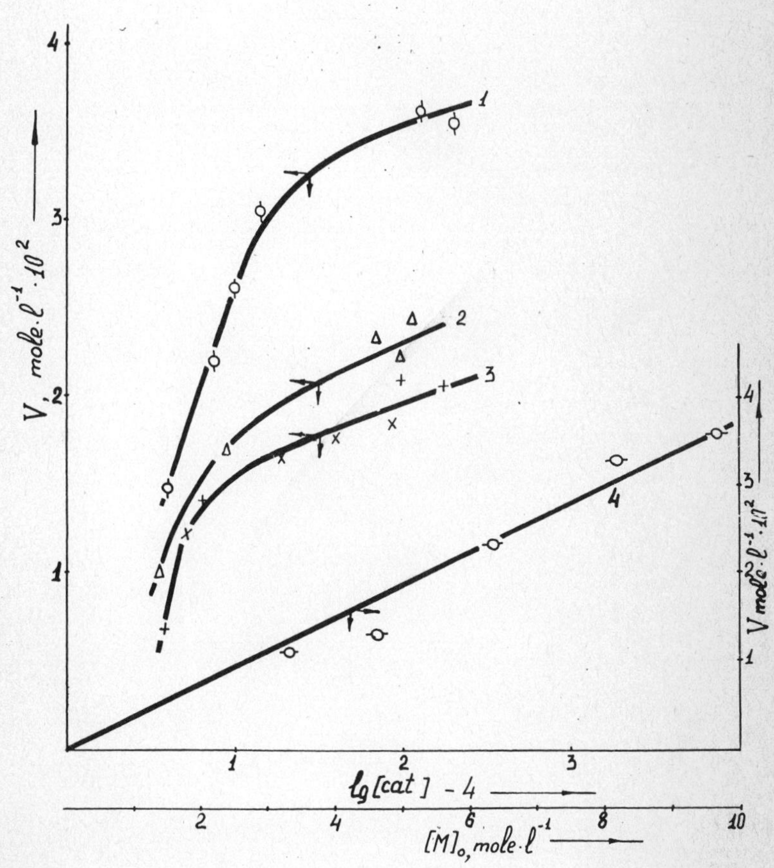